MATHEMATICS

FOR ELEMENTARY TEACHERS

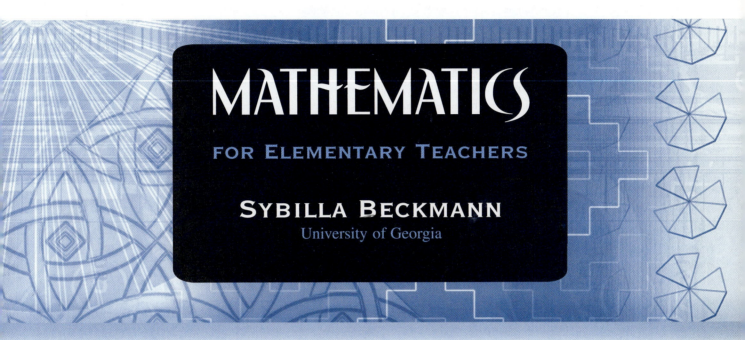

MATHEMATICS

FOR ELEMENTARY TEACHERS

SYBILLA BECKMANN
University of Georgia

PEARSON

Addison
Wesley

Boston San Francisco New York
London Toronto Sydney Tokyo Singapore Madrid
Mexico City Munich Paris Cape Town Hong Kong Montreal

Publisher: Greg Tobin
Acquisitions Editor: Carter Fenton
Associate Editor: RoseAnne Johnson
Managing Editor: Karen Wernholm
Production Supervisor: Kathleen A. Manley
Production Coordination: WestWords, Inc.
Senior Marketing Manager: Becky Anderson
Marketing Coordinator: Carolyn Buddeke
Senior Author Support/Technology Specialist: Joe Vetere
Senior Manufacturing Buyer: Evelyn Beaton
Rights and Permissions Advisor: Dana Weightman
Design Supervisor: Barbara T. Atkinson
Interior Design and Cover Design: Leslie Haimes
Composition and Illustrations: Techsetters, Inc.
Cover Photo: King Penguins/J. David Andrews/Masterfile

Photo Credits: p. 5 (top left, bottom left) Photo Disc Blue; **p. 5 (top right, bottom right), 172, 411 (top)** Photo Disc; **p. 326, 375, 410, 411 (bottom), 421, 477, 530, 534, 658** Courtesy of the author; **p. 376** Internet

Material on pages 598–602 is adapted with permission from *Primary Mathematics*, Vol. 4A, pp. 40, 41; Vol. 4B, pp. 73, 100; Vol. 5A, pp. 60, 79, 90; Vol. 6A, pp. 38, 67, 71; Vol. 6B, pp. 34, 58, 63, 64; and from *Primary Mathematics Workbook*, Vol. 3A part 1, pp. 20, 55; Vol. 4A part 1, p. 62; Vol. 4B part 1, p. 38 by Ministry of Education and Times Media Private Limited.

Many of the designations used by manufacturers and sellers to distinguish their products are claimed as trademarks. Where those designations appear in this book, and Addison-Wesley was aware of a trademark claim, the designations have been printed in initial caps or all caps.

Library of Congress Cataloging-in-Publication Data
Beckmann, Sybilla.
   Mathematics for elementary school teachers / Sybilla Beckmann.
       p. cm.
   Includes index.
   ISBN 0-201-72587-8
      1. Mathematics–Study and teaching (Elementary) I. Title.
QA135.6.B43 2005
372.7'044–dc22   2003063866

2 3 4 5 6 7 8 9 10—VH—07060504

Contents

3   FRACTIONS   57

8 GEOMETRY   317

11 MORE ABOUT AREA AND VOLUME   471

14 STATISTICS   647

15 PROBABILITY   679

To Will, Joey, and Arianna

Preface

INTRODUCTION

I wrote this book to help future elementary school teachers develop a deep understanding of the mathematics that they will teach. It is easy to think that the mathematics of elementary school is simple and that it shouldn't require college-level study in order to teach it well. But to teach mathematics well, teachers must know more than just *how* to carry out basic mathematical procedures; they must be able to explain *why* mathematics works the way it does. Knowing why requires a much deeper understanding than knowing how. For example, it is easy to multiply fractions—multiply the tops and multiply the bottoms—but why do we multiply fractions that way? After all, when we *add* fractions, we can't just add the tops and add the bottoms. The reasons we can't in the one case and can in the other are not obvious; they require study. By learning to explain why mathematics works the way it does, teachers will learn to make sense of mathematics. I hope they will carry this "sense of making sense" into their own future classrooms.

The book focuses on "explaining why." Prospective elementary school teachers will learn to explain why the standard procedures and formulas of elementary mathematics are valid, why nonstandard methods can also be valid, and why other seemingly plausible ways of reasoning are not correct. The book emphasizes key concepts and principles, and it guides prospective teachers in giving explanations that draw on these key concepts and principles. In this way, teachers will come to organize their knowledge around the key concepts and principles of mathematics, so that they will be able to help their students do likewise.

A number of problems and activities examine common misconceptions. I hope that, by having studied and analyzed these misconceptions, teachers will be able to explain to their students why an erroneous method is wrong, instead of just saying "You can't do it that way."

In addition to knowing how to explain mathematics, prospective teachers should also know how mathematics is used. Therefore, I have included various examples of how we use mathematics, such as using proportions to compare prices across years with the Consumer Public Index, using visualization skills to explain the phases of the Moon, using angles to explain why spoons reflect upside down, and using spheres to explain how the Global Positioning System works.

I believe that this book is an excellent fit for the recommendations of the Conference Board of the Mathematical Sciences regarding the mathematical preparation of teachers. I also believe that the book helps prepare teachers to teach to the principles and standards of the National Council of Teachers of Mathematics.

PEDAGOGICAL AND CONTENT FEATURES

- **Class Activities** are integral to the text and are designed to be done in class or outside of class. They promote critical thinking and discussion, as well as give students a depth of understanding and perspective on the concepts presented in the text. The activities are presented in the separate activities manual that is correlated with the main textbook.

- **Practice Problems** provide students with the opportunity to work through problems. Solutions appear in the text immediately after the Practice Problems. These solutions provide students with numerous examples of the kinds of good explanations they should learn to write. By attempting the Practice Problems themselves, and then checking their solutions against the solutions provided, students will be better prepared to provide good explanations in their homework.

- **Problems** are opportunities for students to work through the mathematics they have learned, without being given an answer at the end of the text. Problems are typically assigned as homework. Solutions appear in the Instructor's Solutions Manual and can be provided on-line for students at the discretion of the instructor.

- **Decimals and fractions** are covered earlier than in most texts. This early coverage allows students to become gradually used to explaining why the standard procedures we use with decimals and fractions are valid, so that the students won't be overwhelmed when they get to multiplying or dividing fractions. The importance of paying attention to the "whole" when working with fractions is repeatedly emphasized.

- Students will use **visualization** not only in traditional mathematical contexts, but also in order to understand basic astronomical phenomena, such as the phases of the Moon, the reason for the seasons, and the rotation of the Earth around its axis every day.

- There is ample coverage of **volume and surface area**. Both the text and the activities manual promote a deeper understanding of these concepts with the aid of hands-on exploration.

- **Unique content** in the chapter on functions and algebra (Chapter 13) introduces U.S. teachers to the impressive diagrammatic method presented in the math texts for grades 3–6 used in Singapore (whose children get the top math scores in the world). This method helps students make sense of and solve a variety of algebra and other word problems without using variables. The text helps students see the relationship between the Singaporean diagrammatic method and standard algebraic problem-solving methods.

SUPPLEMENTS

Class Activities

0-321-12378-6

This activities manual includes activities designed to be done in class or outside of class. These activities promote critical thinking and discussion and give students a depth of understanding and perspective on the concepts presented in the text.

Instructor's Resource and Testing Manual

0-321-12375-1

This manual includes teaching tips in each chapter, tips and solutions for the activities, sample test problems, and course organization and grading suggestions.

Instructor's Solutions Manual

0-321-12376-X

The Instructor's Solutions Manual contains worked-out solutions to all problems in the text.

Addison-Wesley Math Tutor Center

0-201-72170-8

The Addison-Wesley Math Tutor Center provides free tutoring through a registration number that can be packaged with a new textbook or purchased separately. The center is staffed by qualified college mathematics instructors and is accessible via toll-free telephone, toll-free fax, e-mail, and the Internet. For more information, visit *www.aw-bc.com/tutorcenter*.

Acknowledgments

In writing this book, I have benefited from much help and advice, for which I am deeply grateful. I would like to thank Malcolm Adams, Richard Askey, Ed Azoff, David Benson, Danny Breidenbach, Cal Burgoyne, Jason Cantarella, Roger Howe, Denise Mewborn, Ted Shifrin, Shubhangi Stalder, Ginger Warfield, Gale Watson, Paul Wenston, and Hung-Hsi Wu.

I would also like to express my appreciation to the following reviewers' class testers for many helpful comments on earlier drafts:

Gerald Beer, California State University, Los Angeles
Barbara Britton, Eastern Michigan University
Larry Feldman, Indiana University of Pennsylvania
Richard Francis, Southeast Missouri State University
Judy Kidd, James Madison University
Michelle Moravec, McLennan Community College
Shirley Pereira, Grossmont College
Kevin Peterson, Columbus State University
Ginger Warfield, University of Washington
Carol Yin, LaGrange College

I thank Marjorie Anderson for helping me express my ideas clearly. I also owe special thanks to Kevin Clancey, John Hollingsworth, and Tom Cooney for their advice and encouragement. I am very grateful for the support and the excellent work of everyone at Addison-Wesley who has contributed to this project. I especially want to thank my editor, Carter Fenton, for his enthusiastic support, as well as Joe Vetere, RoseAnne Johnson, Kathleen Manley, Becky Anderson, and Dana Weightman for invaluable assistance. It is only through their dedicated hard work that this book has become a reality. In addition, I want to thank Bill Poole, Anne Kelly, and Marjorie Anderson for their help and for their faith in the project. I thank my family—Will, Joey, and Arianna Kazez, and my parents Martin and Gloria Beckmann—for their patience, support, and encouragement. My children long ago gave up asking "Mom, are you still working on that book?" and are now convinced that books take forever to write.

Sybilla Beckmann
Athens, Georgia

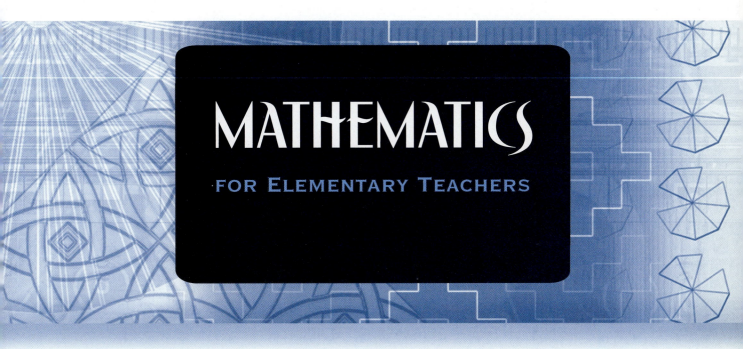

MATHEMATICS

FOR ELEMENTARY TEACHERS

Chapter 1

Problem Solving

I n this introductory chapter, you will learn why problem solving is central to mathematics. We begin by presenting a simple but sensible set of guidelines for solving problems. These guidelines will help you organize your thinking when developing a solution. In addition to providing a solution, you should be able to explain why your solution is valid. Thus, in the second part of the chapter, we present criteria for a good explanation.

A main theme of this book is explaining why: Why are the familiar procedures and formulas of elementary mathematics valid? Why is a student's response incorrect? Why is a different way of carrying out a calculation often perfectly correct? In Chapter 1, we set the groundwork to answer these questions and to develop good explanations for the mathematics you will teach.

1.1 Solving Problems

The main reason for learning mathematics is to be able to solve problems. Mathematics is a powerful tool that can be used to solve a vast variety of problems in technology, science, business and finance, medicine, and daily life. The potential uses of mathematics are limited only by human ingenuity. Solving problems is not only the most important *end* of mathematics, it is also a *means* for learning mathematics. Mathematicians have long known that good problems can deepen our thinking about mathematics, guide us to new ways of using mathematical techniques, help us recognize connections between topics in mathematics, and force us to confront mathematical misconceptions we may hold. By working on good problems, we learn mathematics better. This is why the National Council of Teachers of Mathematics advocates that instructional programs for prekindergarten through grade 12 focus on problem solving.

The National Council of Teachers of Mathematics (NCTM)—which is over 75 years old and comprises over 100,000 members, advocates the highest quality mathematics education for all students. (See www.aw-bc.com/beckmann for links to NCTM.) To this end, the NCTM has prepared *Principles and Standards for School Mathematics* [44], including the following standard on problem solving:

Instructional programs from prekindergarten through grade 12 should enable all students to—

NCTM Standard: Problem Solving

- build new mathematical knowledge through problem solving;
- solve problems that arise in mathematics and in other contexts;
- apply and adapt a variety of appropriate strategies to solve problems;
- monitor and reflect on the process of mathematical problem solving.

POLYA'S FOUR PROBLEM-SOLVING STEPS

In 1945, the mathematician George Polya presented a four-step guideline for solving problems in his book *How to Solve It* [53]. These four steps are simple and sensible, and they have helped many students improve their problem-solving abilities. Polya's four steps are as follows:

Polya's Steps

1. Understand the problem.
2. Devise a plan.
3. Carry out the plan.
4. Look back.

The first step, *understand the problem*, is the most important. It may seem obvious that if you don't understand a problem, you won't be able to solve it, but it is easy to rush headlong into a problem and try to do "something like we did in class" before you think about what the problem is asking. So, *slow down* and read problems carefully. In some cases, drawing a diagram or a picture can help you understand the problem.

Try to be creative and flexible in the second step, when you *devise a plan*. There are many different types of plans for solving problems. In devising a plan, think about what information you know, what information you are looking for, and how to relate these pieces of information. The following are a few common types of plans:

- Draw a diagram or a picture.
- Guess and test—make a guess and try it out. Be prepared to try again. Use the results from your first guess to guide you.
- Look for a pattern.
- Solve a simpler problem or problems first—this may help you see a pattern you can use.
- Use direct reasoning.
- Use a variable, such as x.

The third step, *carry out the plan*, is often the hardest part. If you get stuck, modify your plan or try a new plan. It often helps to cycle back through steps 1 and 2. Monitor your own progress: if you are stuck, is it because you haven't tried hard enough to make your plan work, or is it time to try a new plan? Don't give up too easily. Students sometimes think they can only solve a problem if they've seen one just like it before, but this is not true. Your common sense and natural thinking abilities are powerful tools that will serve you well if you use them.

The fourth step, *look back*, gives you an opportunity to catch mistakes. Check to see if your answer is plausible. For example, if the problem was to find the height of a telephone pole, then answers such as 2.3 feet or 513 yards are unlikely—it would be wise to look for a mistake somewhere. *Looking back* also gives you an opportunity to make connections: Have you seen this type of answer before? What did you learn from this problem? Could you use these ideas in some other way? Is there another way to solve the problem? When you *look back*, you have an opportunity to learn from your own work.

WHY DO I NEED TO SOLVE PROBLEMS MYSELF WHEN THE TEACHER COULD JUST TELL ME HOW?

Students sometimes wonder why they need to solve problems themselves. Why can't the teacher just *show* the students how to solve the problem? Of course, teachers do show solutions to many problems, but sometimes teachers should step back and *guide* their students, helping them to use fundamental concepts and principles to *figure out* how to solve a problem. Why? Because the process of grappling with a problem can help students understand the underlying concepts and principles. Therefore, teachers who are too quick to tell students how to solve problems may actually rob them of valuable learning experiences.

When you solve problems yourself, you will inevitably get stuck some of the time, and you will try things that don't work. This may seem unproductive, but it is not. Even if you are not able to solve a problem, the process of getting stuck and trying to get unstuck primes you to better understand a solution presented by someone else.

Stop for a moment and think about something that you know well. This could be an academic topic, or it could be from sports, music, religion, or any aspect of your life. Chances are that the topic you know well is something about which you have thought deeply, something with which you have grappled and maybe even struggled. You probably learned a lot about it from others, but it is likely that you have put effort into making sense of it for yourself. The process of making sense of things for yourself is the essence of education—use it in mathematics and do not underestimate its power.

CLASS ACTIVITY NOW TURN TO CLASS ACTIVITIES MANUAL

1A A *Clinking Glasses* Problem p. 1

1B Problems About Triangular Numbers p. 2

1C What Is a Fair Way to Split the Cost? p. 3

1.2 Explaining Solutions

After solving a problem, the natural next step is to explain why your solution is valid. Why does your method work? Why does it give the correct answer to the problem? As a teacher, you will need to explain why the mathematics you are teaching works the way it does. However, there is an even more compelling reason for providing explanations. When you try to explain something to someone else, you clarify your own thinking, and you learn more yourself. When you try to explain a solution, you may find that you don't understand it as well as you thought. This exercise is valuable because it is an opportunity to learn more, to uncover an error, or to clear up a misconception. Even if you understood the solution well, you will understand it better after explaining it. Therefore, the National Council of Teachers of Mathematics (NCTM) includes the following standard, communication, in [44]:

Instructional programs from prekindergarten through grade 12 should enable all students to—

NCTM Standard: Communication

- organize and consolidate their mathematical thinking through communication;
- communicate their mathematical thinking coherently and clearly to peers, teachers, and others;
- analyze and evaluate the mathematical thinking and strategies of others;
- use the language of mathematics to express mathematical ideas precisely.

Communicating about mathematics gives both children and adults an opportunity to make sense of mathematics. According to NCTM [44, p. 56],

From children's earliest experiences with mathematics, it is important to help them understand that assertions should always have reasons. Questions such as "Why do you think it is true?" and "Does anyone think the answer is different, and why do you think so?" help students see that statements need to be supported or refuted by evidence.

When we communicate about mathematics in order to explain and convince, we must use *reasoning*. Logical reasoning is the essence of mathematics. In mathematics, everything but the fundamental starting assumptions has a reason, and the whole structure of mathematics is built up by reasoning. Therefore, NCTM includes the following standard on reasoning and proof [44]:

Instructional programs from prekindergarten through grade 12 should enable all students to—

NCTM Standard: Reasoning and Proof

- recognize reasoning and proof as fundamental aspects of mathematics;
- make and investigate mathematical conjectures;
- develop and evaluate mathematical arguments and proofs;
- select and use various types of reasoning and methods of proof.

Explaining "why" is so fundamental to mathematics that it is the reason we emphasize it in every topic in this book.

FIGURE 1.1

Why are there
seasons?

Spring

Summer

Winter

Autumn

WHAT IS AN EXPLANATION IN MATHEMATICS?

What qualifies as an explanation? The answer depends on the context. In mathematics, we seek particular kinds of explanations: those using logical reasoning which are based on initial assumptions that are either explicitly stated or assumed to be understood by the reader or listener.

Explanations can vary according to different areas of knowledge. There are many different kinds of explanations—even of the same phenomenon. For example, consider the question: Why are there seasons? (See Figure 1.1)

The simplest answer is "because that's just the way it is." Every year, we observe the passing of the seasons, and we expect to see the cycle of spring, summer, fall, and winter continue indefinitely (or at least for a very long time). The cycle of seasons is an observed fact that has been documented since humans began to keep records. We could stop there, but when we ask why there are seasons, we are searching for a deeper explanation.

A poetic explanation for the seasons might refer to the cycles of birth, death, and rebirth around us. In our experiences, nothing remains unchanged forever, and many things are parts of a cycle. The cycle of the seasons is one of the many cycles that we observe.

Most cultures have stories that explain why we have seasons. The ancient Greeks, for example, explained the seasons with the story of Persephone and her mother, Demeter, who tends the earth. When Pluto, god of the underworld, stole Persephone to become his bride, Demeter was heartbroken. Pluto and Demeter arranged a compromise, and Persephone could stay with her mother for half a year and return to Pluto in the underworld for the remaining half of every year. When Persephone is in the underworld, Demeter is sad and does not tend the earth. Leaves fall from the trees, flowers die, and it is fall and winter. When Persephone returns, Demeter is happy again and tends the earth. Leaves grow on the trees, flowers bloom, and it is spring and summer. This is a beautiful story, but we can still ask for another kind of explanation.

Modern scientists explain the reason for the seasons by the tilt of the earth's axis relative to the plane in which the earth travels around the sun. When the northern hemisphere is tilted toward the sun, it is summer there; when it is tilted away from the sun, it is winter. Perhaps this settles the matter, but a seeker could still ask for more. Why are the earth and sun positioned the way they are? Why does the earth revolve around the sun and not fly off alone into space? These questions can lead again to poetry, or to the spiritual, or to further physical theories. Maybe they lead to an endless cycle of questions.

WRITING GOOD EXPLANATIONS

While an oral explanation helps you develop your solution to a problem, written explanations push you to polish, refine, and clarify your ideas. This is as true in mathematical writing as in any other kind of writing, and it is true at all levels. You should write explanations of your solutions to problems, and your students should write explanations of their solutions, too. Some elementary school teachers have successfully integrated mathematics and writing in their classrooms and used writing to help their students develop their understanding of mathematics.

Like any kind of writing, it takes work and practice to write good mathematical explanations. When you solve a problem, do not attempt to write the final draft of your solution right from the start. Use scratch paper to work on the problem and collect your ideas. Then, write your solution as part of the *looking back* stage of problem solving. Think of your explanation as an essay. As with any essay that aims to convince, what counts is not only factual correctness but also persuasiveness, explanatory power, and clarity of expression. In mathematics, we persuade by giving a thorough, logical argument, in which chains of logical deductions are strung together connecting the starting assumptions to the desired conclusion.

Good mathematical explanations are thorough. They should not have gaps that require leaps of faith. On the other hand, a good explanation should not belabor points that are well-known to the audience or not central to the explanation. For example, if your solution contains the calculation $356 \div 7$, a college-level explanation need not describe how the calculation is carried out, except when it is necessary for the solution. Unless your instructor tells you otherwise, assume that you are writing your explanations for your classmates.

On the next page, you will find characteristics of good mathematical explanations. When you write an explanation, check whether it has these characteristics. The more you work at writing explanations, and the more you ponder and analyze what makes good explanations, the better you will write explanations, and the better you will understand the mathematics involved.

SAMPLE PROBLEMS AND SOLUTIONS

The solutions to the problems in the next section demonstrate good written explanations. In addition, most of the answers to the practice problems at the end of the sections throughout this book provide additional examples of well-written explanations.

CHARACTERISTICS OF GOOD EXPLANATIONS IN MATHEMATICS

1. The explanation is factually correct, or nearly so, with only minor flaws (for example, a minor mistake in a calculation).

2. The explanation addresses the specific question or problem that was posed. It is focused, detailed, and precise. There are no irrelevant or distracting points.

3. The explanation is clear, convincing, and logical. A clear and convincing explanation is characterized by the following:

   a. The explanation could be used to teach another (college) student, possibly even one who is not in the class.

   b. The explanation could be used to convince a skeptic.

   c. The explanation does not require the reader to make a leap of faith.

   d. Key points are emphasized.

   e. If applicable, supporting pictures, diagrams, and equations are used appropriately and as needed.

   f. The explanation is coherent.

   g. Clear, complete sentences are used.

A PUZZLE PROBLEM AND THREE SOLUTIONS

Puzzle Problem: There are a total of 9 coins, some of them pennies, some nickels, and some dimes. If you collect all the pennies and nickels, there are 7 coins. If you collect all the nickels and dimes, there are 5 coins. How many of the 9 coins are pennies, how many are nickels, and how many are dimes?

This problem can be solved in many different ways.

Solution 1: If 7 of the 9 coins are pennies and nickels, then 2 must be dimes. Since the nickels and dimes together make 5 coins and we know there are 2 dimes, there must be 3 nickels. But once again, there are 7 pennies and nickels, so since 3 of them are nickels, 4 must be pennies. So, there are 2 dimes, 3 nickels, and 4 pennies. As a check, add the total number of coins: $2 + 3 + 4 = 9$, which agrees with our starting data. Also, there are $3 + 4 = 7$ pennies and nickels and $3 + 2 = 5$ nickels and dimes; so, these check, too.

Solution 2: There are 7 pennies and nickels and there are 5 nickels and dimes, so if we add 7 and 5, that will count all the nickels twice and all the pennies and dimes once. Therefore, $7 + 5 = 12$ represents the total number of coins plus the number of nickels. Since the total number of coins is 9, the number of nickels must be $12 - 9 = 3$. We know that there are 7 pennies and nickels together; so, since there are 3 nickels, there must be 4 pennies. We also know that there are 5 nickels and dimes together, so since there are 3 nickels, there must be 2 dimes. Thus, there are 3 nickels, 4 pennies, and 2 dimes. As a check, add the total number of coins: $3 + 4 + 2 = 9$, which agrees with our starting data. Also, there are $3 + 4 = 7$ pennies and nickels, and there are $3 + 2 = 5$ nickels and dimes, so these check, too.

Solution 3: Let P be the number of pennies, N the number of nickels, and D the number of dimes. The problem tells us the following:

$$P + N + D = 9$$
$$P + N = 7$$
$$N + D = 5$$

From the second and third equations,

$$P = 7 - N$$
$$D = 5 - N$$

Substituting for P and D in the first equation we get the following:

$$(7 - N) + N + (5 - N) = 9$$

Therefore,

$$-N + 12 = 9$$

so that

$$12 - 9 = N$$

Thus, $N = 3$. Since $P + N = 7$, it follows that $P + 3 = 7$, so $P = 4$. Since $N + D = 5$ and $N = 3$, we have $3 + D = 5$ and $D = 2$. So, there are 3 nickels, 4 pennies, and 2 dimes. Since $3 + 4 + 2 = 9$, this checks with the initial information that there are 9 coins. Also, there are $3 + 4 = 7$ pennies and nickels, and there are $3 + 2 = 5$ nickels and dimes, so these check, too.

CLASS ACTIVITY   NOW TURN TO CLASS ACTIVITIES MANUAL

1D Who Says You Can't Do Rocket Science? p. 4

PRACTICE PROBLEMS FOR SECTION 1.2

1. Joanne and Sahar make plans to attend ten concerts together. Sahar buys 7 pairs of tickets, which cost a total of $350. Joanne buys 3 pairs of tickets, which cost a total of $63. Who owes whom how much?

2. The shape in Figure 1.2 is called an "arithmogon." Find numbers to put in the circles. Choose these numbers so that when you add numbers in circles connected by a line segment, your total is the number in the box next to the line segment.

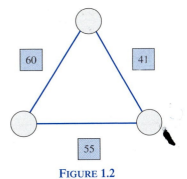

FIGURE 1.2

An Arithmogon

---

3. If there are 25 people at a party and everyone "clinks" glasses with everyone else, how many "clinks" will there be? Describe *two different ways* to solve this problem.

4. Explain how you can use what you learned from Class Activity 1A on *clinking glasses*, Class Activity 1B on triangular numbers, and the previous practice problem to help you determine the 500th triangular number.

ANSWERS TO PRACTICE PROBLEMS FOR SECTION 1.2

Most of the answers here, as well as in other sections, include an explanation. Keep in mind that each explanation is generally only one of many possible valid explanations.

1. Joanne owes Sahar $143.50. To figure this out, notice that in combination, Joanne and Sahar have spent $350 + $63 = $413. Therefore, in the end, each woman should spend half that amount, namely $206.50. Because Joanne already spent $63, she needs to spend an additional $206.50 − $63 = $143.50.

2. See Figure 1.3. One way to solve this problem is to make an initial guess for the three numbers, picking them so that two of the pairs add up correctly. Then, revise your guess, maintaining the two correct sums. You can also solve this using algebra.

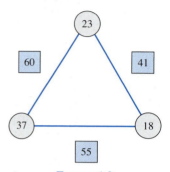

FIGURE 1.3

An Arithmogon

---

3. One way to solve this problem is to add

$$24 + 23 + 22 + \cdots + 4 + 3 + 2 + 1 = 300$$

to find the total number of "clinks." To see why, imagine the 25 people lined up in a row. The 25th person in the row could go down the row and clink glasses with all other 24 people (and then stand in the corner). The 24th person in the row could go down the row and clink glasses with the other 23 people left in the row (and then stand in the corner also). The 23rd person could clink with the remaining 22, and so on, until the 2nd person clinks with the first. Counting up all the clinks, you get the sum $24 + 23 + \cdots + 2 + 1$ above.

Another way to solve this problem is to calculate

$$(25 \times 24) \div 2 = 300$$

This method is valid because each of the 25 people "clinks" with 24 other people. This makes for 25×24 clinks—except that each clink has been counted *twice* this way. Why? Because when Anna clinks with Beatrice, it's the same as Beatrice clinking with Anna, but the 25×24 figure counts those as two separate clinks. So, to get the correct count, divide 25×24 by 2.

4. Imagine adding up the number of dots in the 500th triangle by starting with the bottom row and working your way up, one row at a time. According to this way of thinking, the 500th triangular number must be equal to

$$500 + 499 + 498 + 497 + \cdots + 3 + 2 + 1$$

It would be tedious to add up all those numbers! Fortunately, there's a quicker method. To see how it works, think about the explanation for Exercise 3. There are two different ways of counting the number of clinks when there are 25 people at a party. Since it's the same number of clinks, no matter which way you count them,

$$24 + 23 + 22 + \cdots + 4 + 3 + 2 + 1 = (25 \times 24) \div 2$$

Now think about doing the same analysis, but when there are 501 people at a party. By the same reasoning as before, one way of counting the number of clinks is

$$500 + 499 + 498 + 497 + \cdots + 3 + 2 + 1$$

while another way is

$$(501 \times 500) \div 2$$

However, there must be the same number of clinks either way you count them; therefore,

$$500 + 499 + 498 + 497 + \cdots + 3 + 2 + 1$$
$$= (501 \times 500) \div 2$$

Even though it appears to be entirely unlrelated, we can apply this to the triangular number problem and conclude that there are

$$(501 \times 500) \div 2 = 125,250$$

dots in the 500th triangle.

PROBLEMS FOR SECTION 1.2

1. Denise, Sharon, and Hillary go out to dinner. The total bill is $45 (including tax and tip). They decide to split the bill evenly, so each woman owes $15. Sharon wants to put the total on her credit card and have the other two give her cash or a check. Hillary only has one check, three $1 bills, and a $5 bill. Denise only has a $10 bill and a $20 bill. Explain how the women can settle among themselves so that the waiter only has to take Sharon's credit card for the total amount (and so that nobody owes anything). Explain your reasoning.

2. The Bradley Elementary School cafeteria has twelve different lunches that they can prepare for their students. Five of these lunches are "reduced fat." On any given day, the cafeteria offers a choice of two lunches. How many different pairs of lunches, where one choice is "regular" and the other is "reduced fat," is it possible for the cafeteria to serve? Explain your answer.

3. Mr. Roland, an elementary school teacher, has 24 students. Every day he picks two students to be his "special helpers" for the day. How many pairs of students are there in the class? If the school year is 190 days long, will there be enough days for every pair of children to be the "special helpers"? (Notice that the problem asks about every *pair* of children, not every individual child.)

4. a. Explain how to determine the sum

$$1 + 2 + 3 + \cdots + 100$$

using ideas from the *clinking glasses* activity (Class Activity 1A).

b. Another way to determine the sum

$$1 + 2 + 3 + \cdots + 100$$

is to add the following "vertically":

$$
\begin{array}{r}
1 + 2 + 3 + 4 + \cdots + 99 + 100 \\
+100 + 99 + 98 + 97 + \cdots + 2 + 1 \\
\hline
\end{array}
$$

Explain how to use this approach to determine the sum $1 + 2 + 3 + \cdots + 100$.

5. An auditorium has 50 rows of seats. The first row has 20 seats, the second row has 21 seats, the third row has 22 seats, and so on, each row having one more seat than the previous row. How many seats are there all together? Explain your reasoning.

6. A candy factory has a large vat into which workers pour chocolate and cream. Each ingredient flows into the vat from its own special hose, and each ingredient comes out of its hose at a constant rate. Workers at the factory know that it takes 20 minutes to fill the vat with chocolate from the chocolate hose, and it takes 15 minutes to fill the vat with cream from the cream hose. If workers pour both chocolate and cream into the vat at the same time (each coming full tilt out of its own hose), how long will it take to fill the vat? Before you find an exact answer to this problem, find an approximate answer, or find a range, such as "between . . . and . . . minutes." Explain your reasoning.

7. Jay and Mark run a lawn-mowing service. Mark's mower is twice as big as Jay's, so whenever they both mow, Mark mows twice as much as Jay in a given time period. When Jay and Mark are working together, it takes them 4 hours to cut the lawn of an estate. How long would it take Mark to mow the lawn by himself? How long would it take Jay to mow the lawn by himself? Explain your answers.

8. The Joneses and the Smiths take a trip together. There are four people in the Jones family and six in the Smith family. They board a ferry boat to get to their destination. The boat tickets cost $12 per person, and the Joneses pay for it. The Smiths pay for dinner at a lodge that costs $15 per person. If the Joneses and the Smiths want to divide the costs fairly, then who owes whom how much? Explain your answer.

9. Jane, Kayla, and Mandelite want to give their friend Nadine a potted plant for her birthday. Jane spends $34 on a pot, Kayla spends $3 on potting soil, and Mandelite buys a plant for $24. If Jane, Kayla, and Mandelite want to share their costs equally, who should give whom how much? Explain your answer.

10. The country of Taxalot has a different method of taxing income. If you make $1000 per year, your tax rate is 1%. For each additional $1000 you make per year, your tax rate goes up by 1%. For example, if you made $24,000 per year you would pay 24% in taxes. If you could set your salary anywhere between $0 and $100,000, what would you want it to be and why?

11. Suppose that a fireworks rocket is shot up into the air so that after 4 seconds, it reaches a height of 1600 feet. Two seconds after being shot into the air, is the fireworks rocket exactly 800 feet high, less than 800 feet high, or more than 800 feet high? Explain your reasoning.

Chapter 2

Numbers and the Decimal System

I n this chapter and the next, we discuss how to write and represent numbers, how to interpret the meaning of written numbers, and how to compare sizes of numbers. Writing, representing, interpreting, and comparing numbers are fundamental skills that are essential not only for the study of mathematics itself, but for effective functioning in today's society. For example, when we go to a store, our knowledge of how numbers are represented, interpreted, and compared affects our ability to make informed choices in the following decisions:

- Do I have enough money to buy this item?

- Is brand X a better buy than brand Y?

- Did the clerk give me the 35% discount that was advertised?

When we open a newspaper, we may read about a federal debt in trillions of dollars or a population increase of a certain percent. The more skilled we are at working with numbers, the better we can think critically about the information that comes our way.

A good understanding of numbers and of mathematics in general ensures a range of opportunities in the workplace. As a teacher, you will be entrusted with preparing your students for good citizenship and for productive employment. One hundred years ago, knowing how to add and subtract numbers rapidly was a marketable skill; today, any calculator can do as much and more. One hundred years ago, a large proportion of jobs were unskilled; today, unskilled work is a shrinking portion of our economy. Much of today's work requires an ability to think critically and to do more than just routine tasks. Some of the fastest growing segments of our economy, such as health care, high technology, and finance, require an increasingly sophisticated understanding of how to work with numbers in particular, and mathematics in general. Schooling that *only emphasizes the routine, mechanical aspects of working with numbers will leave students unprepared for the kind of critical thinking that will provide them with a wide range of opportunities for good employment. Schooling that asks students to interpret, analyze, and explain will equip them not only with skills for a wide range of choices for employment, but even more importantly, with the most empowering asset of all: that of deep knowledge and understanding. So that you can empower and guide your students, you must first empower yourself with a deeper understanding of numbers, arithmetic, and mathematics in general.*

The National Council of Teachers of Mathematics (NCTM) has set standards for the mathematics that children from grades prekindergarten through 12 should know and be able to do [44]. Concerning numbers and operations, the NCTM recommends the following:

Instructional programs from prekindergarten through grade 12 should enable all students to—

NCTM Standards

- *understand numbers, ways of representing numbers, relationships among numbers, and number systems;*
- *understand meanings of operations and how they relate to one another;*
- *compute fluently and make reasonable estimates*

For NCTM's specific number and operation recommendations for grades Pre-K–2, grades 3–5, grades 6–8, and grades 9–12, see

www.aw-bc.com/beckmann

This chapter and the next several chapters are devoted to deepening your own understanding of numbers and operations, so that you will be prepared to teach these topics for understanding. Even if you plan to teach younger children, keep in mind that you will be laying the foundation for your students' later learning; therefore, you must know how a topic develops beyond the grade level you teach.

2.1 Introduction to the Number Systems

A number system is a collection of numbers that fit together in a natural way. Our objectives in this section are to define various number systems in use today, to interpret and represent those numbers, and to provide an overview of the relationship between different number systems.

What humans consider to be numbers has evolved over the course of history, and the way children learn about numbers parallels this development. When did humans first become aware of numbers? The answer is uncertain, but it is at least many tens of thousands of years ago. Some scholars believe that numbers date back to the beginning of human existence, citing as a basis for their views the primitive understanding of numbers observed in some animals. (See [15] for a fascinating account of this as well as of the human mind's capacity to comprehend numbers.)

THE COUNTING NUMBERS

counting numbers · The simplest and most basic system of numbers is the **counting numbers**, also called the **natural numbers**.   The counting numbers are as follows:

$$1, 2, 3, 4, 5, \ldots$$

When young children first learn to count, they simply recite the familiar sequence 1, 2, 3, ... in order without associating quantities to these numbers. When the counting numbers are used in this way, as a sequence in order, or to denote order of something in a sequence, as in

$$1^{st}, 2^{nd}, 3^{rd}, 4^{th}, 5^{th}, \ldots$$

ordinals · then the counting numbers are referred to as **ordinals** or **ordinal numbers.** (See Figure 2.1.)

Later in their development, children learn that the counting numbers are not just a sequence to be said in order, but also a device to represent how many things are in a set. In other words, children learn that a number such as 3 is not just "what you say after 2," but it can represent how many cookies are on a plate or how many balls are in a box, and so on. When counting

cardinals · numbers are used in this way, to represent an amount, then they are called **cardinals** or **cardinal numbers.** (See Figure 2.1.)   To distinguish between ordinals and cardinals, remember that *ordinals* are related to putting things in *order*.

Ordinals and cardinals are closely related to each other because, in order to determine how many objects are in a set, we count them off one by one.

FIGURE 2.1

Ordinal Numbers and Cardinal Numbers

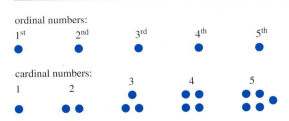

The Whole Numbers and the Integers

whole numbers   The **whole numbers** are the counting numbers together with zero as follows:

$$0, \ 1, \ 2, \ 3, \ 4, \ 5, \ldots$$

integers   The **integers** include the counting numbers, zero (0), and the negatives of the counting numbers as shown in the following:

$$\ldots, \ -5, \ -4, \ -3, \ -2, \ -1, \ 0, \ 1, \ 2, \ 3, \ 4, \ 5, \ldots$$

negative   The **negative** of a counting number is represented using a **minus sign** ($-$). For example, -4 is
minus sign   the negative of 4. The number -4 can be read *minus four* or *negative four*. When describing
positive integers   the integers, the counting numbers are often referred to as **positive integers**, and their negatives
negative integers   as **negative integers**.

The notions of zero and of negative integers may seem natural to us today, but our early ancestors struggled to discover and make sense of these concepts. Although humans have always been acquainted with the notion of "having none" as in having no sheep or having no food to eat, the concept of 0 as a number was introduced far later than the counting numbers—not until sometime before 800 A.D. (See [7].) Even today, the notion of 0 is difficult for many children to grasp. This difficulty is not too surprising: Although the counting numbers can be represented nicely by sets of objects, such as in Figure 2.2, you have to show *no* objects in order to represent the number 0 in a similar fashion. But how does one *show* no objects? We might use a picture as in Figure 2.3, for example.

Similarly, the introduction of negative numbers was relatively late in human development. Although the ancient Babylonians may have had the concept of negative numbers around 2000 B.C., negative numbers were not always accepted by mathematicians even as late as the 16th century A.D. ([7]). The difficulty lies in interpreting the meaning of negative numbers. Once again, starting with the representation of the counting numbers in Figure 2.2, how can negative numbers be represented in a similar fashion? This problem may seem perplexing at first, but in fact there is a nice interpretation of negative numbers, namely, as amounts owed. For example, -7 can represent *owing* 7 marbles. In comparison, 7 can represent *having* 7 marbles. (See Figure 2.4.)

A clear but more abstract way of representing the integers is on a number line, as in Figure 2.5. On a number line, the positive integers, 0, and the negative integers are all displayed on an equal footing. The positive integers are placed at equal spacings, heading infinitely to

FIGURE 2.2

Representing
the Counting
Numbers

FIGURE 2.3

Representing
the Whole
Numbers

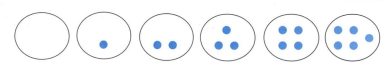

FIGURE 2.4

Representing
Negative
Integers

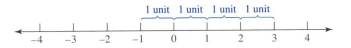

Having 7 marbles represents 7:

Owing 7 marbles represents –7:

FIGURE 2.5

Representing
Integers on a
Number Line

the right, the negative integers are placed at the same equal spacings, heading infinitely to the left, and 0 lies exactly between the positive integers and the negative integers.

THE RATIONAL NUMBERS

fraction   A **fraction** is an expression of the form

$$\frac{A}{B} \quad \text{or} \quad A/B$$

where B is not zero.

rational numbers   The **rational numbers** are all the numbers that can (at least theoretically) be written as a fraction

$$\frac{A}{B} \quad \text{or} \quad A/B$$

where A and B are integers and B is not zero. So, for example,

$$\frac{7}{8}, \quad \frac{43}{7}, \quad \text{and} \quad \frac{-2}{3}$$

are examples of rational numbers. You can remember the term *rational numbers* by remembering that *rational* has *ratio* as its root, and ratios and fractions are closely related.

A number that can be expressed as a fraction of integers is still considered to be a rational number, even when the number is written in a different form. For example, we can write

$$\frac{7}{8}$$

as the decimal number

$$0.875$$

by dividing 8 into 7. Therefore, the number 0.875 is a rational number, because it can also be written as a fraction of integers. Similarly, we can write

$$\frac{43}{7}$$

as the mixed number

$$6\frac{1}{7}$$

Again, $6\frac{1}{7}$ is a rational number because it can also be represented as a fraction of integers. We discuss in detail these and other facts about representing rational numbers and interpreting the meaning of rational numbers in Sections 3.1 and 4.3.

It's easy to imagine how rational numbers first came into use. A farmer filling bags with grain might have had only enough to fill the last bag half full, or perhaps a vintner filling barrels with grape juice for wine filled the last barrel only $\frac{3}{4}$ full. Even a relatively primitive barter economy would need fractions such as $\frac{1}{2}$ and $\frac{3}{4}$ to make trades. Today, even very young children can develop a basic understanding of very simple fractions such as $\frac{1}{2}$, $\frac{1}{3}$, and $\frac{1}{4}$.

THE REAL NUMBERS

real numbers

decimal notation

The **real numbers** are all those numbers that can, at least theoretically, be written in decimal notation. A number is in **decimal notation** if it is written in one of the following forms:

$$2001$$
$$-583$$
$$37.49$$
$$-.000792$$
$$33.33333333\ldots$$
$$-127.8524693019375\ldots.$$

Specifically, a number is in decimal notation if it is written in the form

$$\ast\ast\ast\ldots\ast\ast\ast.\ast\ast\ast\ast\ast\ast\ast\ldots$$

or

$$-\ast\ast\ast\ldots\ast\ast\ast.\ast\ast\ast\ast\ast\ast\ast\ldots$$

where each $\ast$ can be any one of the ten **digits**

$$0, 1, 2, 3, 4, 5, 6, 7, 8, 9$$

and where there are finitely many nonzero entries to the left of the **decimal point** (.) and either finitely or infinitely many nonzero entries to the right of the decimal point.

The word *decimal* comes from the Latin word *decem* for *ten*. It is appropriate because decimal notation makes use of the ten digits 0 through 9. When a number is written in decimal notation, we may simply call the number a **decimal**, or a **decimal number**, or we may say that the number has a **decimal representation** or **decimal expansion**.

decimal number

decimal representation

decimal expansion

We encounter decimal notation and real numbers every day when we see a price such as $1.99. So, when young children learn about prices, they are beginning to learn about decimal notation and real numbers.

CONVENTIONS OF DECIMAL NOTATION

There are several conventions to observe when writing numbers in decimal notation. Leading zeros are usually not written. So, for example, we do not write

$$0034.7$$

but instead we write

$$34.7$$

However, we may choose to write

$$0.35$$

instead of

$$.35$$

to emphasize that there is a decimal point present. Zeros at the end of a number in decimal notation are often not written. For example, instead of writing

$$589.32000000000000\ldots$$

or

$$589.32000$$

we may write

$$589.32$$

(However, it is sometimes not proper to drop zeros on the right. When a number records a physical measurement, such as 12.0 meters, zeros to the right indicate the precision of the measurement and should not be dropped.) When there are no nonzero entries to the right of the decimal point, the decimal point itself is usually dropped. For example, rather than writing

$$2005.0000000\ldots$$

we can simply write

$$2005$$

without any decimal point.

Another convention that is often used in writing decimal numbers is to place commas to make the decimal number more readable. This notation is especially helpful when there are many digits in the decimal number. Commas are only used to the left of the decimal point. If commas are used, they are placed in front of every third digit, starting from the decimal point, going to the left. For example, we may write the decimal number

$$1234567.89$$

as

$$1,234,567.89$$

in order to make it more readable.

The conventions that we use in the United States are not universal. In some countries the roles of the comma and the decimal point are interchanged.

EXAMPLES OF REAL NUMBERS

The real numbers consist of all the numbers that can, at least theoretically, be written in decimal notation, therefore

$$1998 \quad \text{and} \quad -123.456789$$

are examples of real numbers. The number

$$33.3333\ldots$$

where the threes continue forever, is another real number. The rational number

$$\frac{1}{11}$$

is also a real number, because $\frac{1}{11}$ can be written as the decimal number

$$.09090909\ldots$$

by dividing 1 by 11. (The 09 pattern repeats forever.) Similarly, *every* rational number is also a real number because every fraction of integers has a decimal expansion.

The square root of two,

$$\sqrt{2}$$

is another example of a real number because it can be written as a decimal number, namely,

$$1.414213562\ldots$$

It turns out that the digits in this decimal number go on forever, without repeating. The symbol

$$\sqrt{}$$

square root   stands for **square root**. If N is a number, then $\sqrt{N}$ stands for the positive number M such that $M \times M = N$. So, for example, $\sqrt{9} = 3$ because $3 \times 3 = 9$.

NEGATIVE REAL NUMBERS

negative   As with integers, the **negative** real numbers are those real numbers that have a minus sign in front when written in decimal notation. The **positive** real numbers are those that do not have a minus sign in front when written as a decimal number, with the exception of the number 0. It is understood that $-0 = 0$, and that 0 is neither positive nor negative.

As with integers, we can think of negative real numbers as represented by *owing*. For example, -35.47 can be represented by owing $35.47 on a credit card.

FIGURE 2.6

Relationships
among the
Number
Systems

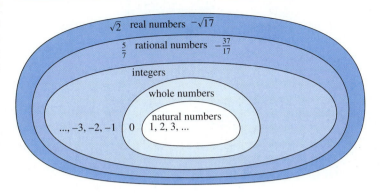

RELATIONSHIPS AMONG THE NUMBER SYSTEMS

Notice the following relationships:

- Every counting number is a whole number.

- Every whole number is an integer.

- Every integer is a rational number because if A is an integer, then you can also write it as $\frac{A}{1}$, which shows that it is a rational number.

- Every rational number is a real number because, if you have a fraction $\frac{A}{B}$, you can divide B into A to get a decimal representation of the number. (We will discuss why this makes sense in Chapter 7.)

  But none of these kinds of numbers are the same:

- Some real numbers are not rational numbers. For example, $\sqrt{2}$ is a real number that is not a rational number. (This is not at all obvious. We discuss this in Section 12.6.)

- Some rational numbers are not integers. For example, $\frac{1}{2}$ is a rational number, but it is not an integer.

- Some integers are not whole numbers. For example, -1 is an integer, but it is not a whole number.

- 0 is a whole number that is not a counting number.

Figure 2.6 shows the relationships between the various sets of numbers. All the numbers that we will be working with—whole numbers, integers, rational numbers (fractions) and real numbers (decimal numbers)—are real numbers. So the real numbers unify all the kinds of numbers we have discussed.

PRACTICE PROBLEMS FOR SECTION 2.1

Answers to these practice problems begin on page 22.

1. Define the following terms:
   counting numbers
   ordinals
   cardinals

   whole numbers
   integers
   rational numbers
   real numbers

2. Is 3 a rational number?

3. Give an example of a rational number that is not an integer.

4. Give an example of an integer that is not a whole number.

5. Is $\frac{2}{3} \times 6\frac{1}{5}$ a rational number?

ANSWERS TO PRACTICE PROBLEMS FOR SECTION 2.1

1. See text.

2. Yes, 3 is a rational number because you can also write it as $\frac{3}{1}$.

3. $\frac{2}{3}$ is an example of a rational number that is not an integer.

4. -2 is an example of an integer that is not a whole number.

5. Yes, $\frac{2}{3} \times 6\frac{1}{5}$ is a rational number because you can also write it as $\frac{62}{15}$, as shown in the following:

$$\frac{2}{3} \times 6\frac{1}{5} = \frac{2}{3} \times \frac{31}{5}$$

$$= \frac{2 \cdot 31}{3 \cdot 5}$$

$$= \frac{62}{15}$$

2.2 The Decimal System and Place Value

All the numbers we will study in this book are within the set of real numbers and can be written in decimal notation. Remember that a number such as 541, in which the decimal point has been dropped, is still considered a decimal number. In this section, we will look at the meaning of a number written as a decimal, and we will consider the origins of writing numbers as decimals. Although you are certainly familiar with writing decimal numbers, so much so that you probably take it for granted, it is actually a remarkably efficient and clever way to write numbers. The decimal system also took many thousands of years to develop, which is probably why it still takes children many years to understand what written numbers represent.

THE ORIGINS OF THE DECIMAL SYSTEM

decimal system

base-ten system

The **decimal system** is the way of writing numbers as decimal numbers. Sometimes this system is called the **base-ten system** for writing numbers. How did this way of writing numbers come about?

The earliest humans lived in small groups and gathered their food daily. These humans probably did not have much need for numbers. But eventually, some people began to live in larger groups. Then people began to specialize in their daily tasks and to trade goods among each other. As trade progressed and societies became more complex, people needed a way to record numbers.

For example, think back to a shepherd living thousands of years ago, long before the invention of writing. How did the shepherd keep track of his sheep every day as he led his flock from place to place? He may have made notches on a piece of wood or bone—one notch for each sheep. But if the shepherd had many sheep, and if he did not organize his notches, he might have had a hard time determining whether he really had the same number of sheep on one day as on the next. (See the left side of Figure 2.7.) The shepherd might have come upon

FIGURE 2.7

A Shepherd's
Tally of Sheep

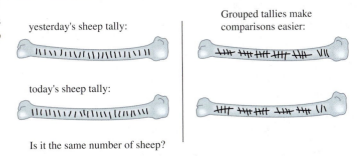

yesterday's sheep tally:

today's sheep tally:

Is it the same number of sheep?

Grouped tallies make
comparisons easier:

the idea of organizing his notches, as seen in the right side of Figure 2.7. This way of grouping tallies is a first step toward an efficient way of recording quantities by writing numbers.

As civilization progressed, some people began to live in cities, and trade became more sophisticated. Then people needed to work with larger numbers, so they replaced grouped tally marks, such as those on the right of Figure 2.7, with symbols. Different cultures used different symbols. For example, you are probably familiar with the Roman numeral V, which represents 5. To record 50 sheep, a person long ago might have written

VVVVVVVVVV.

However, it is hard to read all those Vs, so the Romans devised new symbols: X for 10, L for 50, C for 100, and M for 1000. These symbols are fine for representing numbers up to a few thousand, but what about representing 10,000? Once again,

MMMMMMMMMM

is difficult to read. What about 100,000, or even 1,000,000? To represent these, you'd want yet more new symbols.

Eventually, the practice of using a new symbol for a large number became cumbersome, and early mathematicians developed a flexible system in which we can (at least theoretically) write any number of any size without requiring new symbols. The key innovation of the decimal system is that, rather than using new symbols to represent larger and larger numbers, the decimal system uses *place value*.

CLASS ACTIVITY   NOW TURN TO CLASS ACTIVITIES MANUAL

2A Developing Place Value p. 5

PLACE VALUE

place value   The decimal system uses **place value**, which means that the meaning of a digit in a decimal number depends on the location of the digit in the decimal number. For example, the 2 in 237 stands for two *hundreds*, whereas the 2 in 14.28 stands for two *tenths*. Specifically, each position or place in a decimal number corresponds to one of the values indicated in Table 2.1.

TABLE 2.1

Values of Places
in Decimal
Numbers

VALUES OF PLACES IN A DECIMAL NUMBER			
Proceeding to the left of the decimal point		Proceeding to the right of the decimal point	
1	one		
10	ten	$\frac{1}{10}$	tenth
100	hundred	$\frac{1}{100}$	hundredth
1000	thousand	$\frac{1}{1000}$	thousandth
10,000	ten thousand	$\frac{1}{10,000}$	ten thousandth
100,000	hundred thousand	$\frac{1}{100,000}$	hundred thousdandth
1,000,000	million	$\frac{1}{1,000,000}$	millionth
⋮		⋮	

FIGURE 2.8

The Values
Represented by
the Digits in a
Decimal

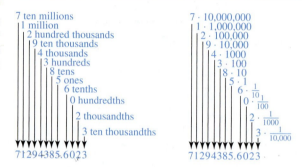

A digit in a decimal number stands for that many of its place's value. For example, Figure 2.8 shows the meaning of each digit in the decimal number 71294385.6023.

A decimal number stands for the total amount represented by all its places taken together. For example, 3618.95 stands for 3 thousands, 6 hundreds, 1 ten, 8 ones, 9 tenths, and 5 hundredths. We can put this way of expressing the number 3618.95 in a more symbolic form by writing one of the following three expressions:

$$3 \text{ thousands} + 6 \text{ hundreds} + 1 \text{ ten} + 8 \text{ ones} + 9 \text{ tenths} + 5 \text{ hundredths}$$

or

$$3 \cdot 1000 + 6 \cdot 100 + 1 \cdot 10 + 8 \cdot 1 + 9 \cdot \frac{1}{10} + 5 \cdot \frac{1}{100}$$

or

$$3000 + 600 + 10 + 8 + \frac{9}{10} + \frac{5}{100}$$

expanded form

When a decimal number is rewritten as one of these last three kinds of expressions, we say that the decimal number is in **expanded form**. The expanded form of a decimal number shows the amount represented by the number more explicitly than just the decimal number itself because the expanded form shows the values of the places. One reason that children have difficulty learning what written numbers represent is that they must keep the values of the places in mind. Because the values of the places in a decimal number are not shown explicitly, even interpreting written numbers requires a certain level of abstract thinking.

The following are some additional examples of decimal numbers written in expanded form:

$$94,152 = 9 \cdot 10,000 + 4 \cdot 1000 + 1 \cdot 100 + 5 \cdot 10 + 2 \cdot 1$$

$$12.438 = 1 \cdot 10 + 2 \cdot 1 + 4 \cdot \frac{1}{10} + 3 \cdot \frac{1}{100} + 8 \cdot \frac{1}{1000}$$

VALUES OF PLACES IN DECIMALS AND POWERS OF TEN

The decimal system for writing numbers uses place value, but what are the values of the places? In the decimal system, the value of each place is ten times the value of the place to its immediate right.

The values of the places in a decimal number can therefore be described by repeatedly multiplying by 10, as shown in Figure 2.9.

A convenient notation for writing an expression such as

$$10 \times 10 \times 10 \times 10 \times 10 \times 10$$

which is six tens multiplied together, is

$$10^6$$

So,

$$1,000,000 = 10 \times 10 \times 10 \times 10 \times 10 \times 10 = 10^6$$

The expression 10^6 is read "ten to the sixth power" or just "ten to the sixth," and we can refer **power of ten** to an expression like 10^6 as a **power of ten**. The number 6 in 10^6 is called the **exponent** of **exponent** 10^6. Notice the correlation between the six zeros in 1,000,000 and the 6 in the expression 10^6.

FIGURE 2.9

Each Place's Value Is Ten Times the Value of the Place to Its Right

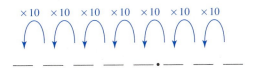

one	1	$= 1 \times 10$
ten	10	$= 10 \times 10$
hundred	100	$= 10 \times 10 \times 10$
thousand	1,000	$= 10 \times 10 \times 10 \times 10$
ten thousand	10,000	$= 10 \times 10 \times 10 \times 10 \times 10$
hundred thousand	100,000	$= 10 \times 10 \times 10 \times 10 \times 10 \times 10$
million	1,000,000	$= 10 \times 10 \times 10 \times 10 \times 10 \times 10$

TABLE 2.2

Powers of Ten

Ten	$10 = 10^1$
Hundred	$100 = 10 \times 10 = 10^2$
Thousand	$1000 = 10 \times 10 \times 10 = 10^3$
Ten thousand	$10,000 = 10 \times 10 \times 10 \times 10 = 10^4$
Hundred thousand	$100,000 = 10 \times 10 \times 10 \times 10 \times 10 = 10^5$
Million	$1,000,000 = 10 \times 10 \times 10 \times 10 \times 10 \times 10 = 10^6$
Ten million	$10,000,000 = 10^7$
Hundred million	$100,000,000 = 10^8$
Billion	$1,000,000,000 = 10^9$
Ten billion	$10,000,000,000 = 10^{10}$
Hundred billion	$100,000,000,000 = 10^{11}$
Trillion	$1,000,000,000,000 = 10^{12}$

Similarly, the expression

$$10^{17}$$

stands for 17 tens multiplied together,

$$10^{17} = 10 \times 10 \times 10 \times 10 \times 10 \times 10 \times 10 \times 10 \times 10 \times 10 \times 10 \times 10 \times 10 \times 10 \times 10 \times 10 \times 10$$

The exponent of 10^{17} is 17, and the decimal representation of 10^{17} is a 1 followed by 17 zeros:

$$10^{17} = 100,000,000,000,000,000$$

Table 2.2 shows powers of ten from 10^1 to 10^{12}.

Notice the pattern in Table 2.2: the number of zeros behind the 1 (in 1,000,000, for example) is the same as the exponent on the ten. For example, 10,000 is written as a one followed by four zeros; it also can be written as 10^4, where the exponent on the ten is 4.

By using exponents, we can show more clearly the structure of the values of the places in decimal numbers.

We can also use exponents in expanded forms:

$$17,843 = 1 \cdot 10^4 + 7 \cdot 10^3 + 8 \cdot 10^2 + 4 \cdot 10^1 + 3 \cdot 1$$

What about decimal numbers with places to the right of the decimal point? As before, there is special notational to write one tenth, one hundredth, one thousandth, and so on, as powers of 10. This time, the powers are *negative*, as in the following:

$$\frac{1}{10} = \frac{1}{10^1} = 10^{-1}$$
$$\frac{1}{100} = \frac{1}{10^2} = 10^{-2}$$
$$\frac{1}{1000} = \frac{1}{10^3} = 10^{-3}$$
$$\frac{1}{10,000} = \frac{1}{10^4} = 10^{-4}$$
$$\vdots \qquad \vdots \qquad \vdots$$

Notice that since $1/10 = .1$ and $1/100 = .01$ and $1/1000 = .001$, and so on, the negative exponents fit the following pattern of positive exponents:

$$
\begin{aligned}
10,000 &= 10^4 \\
1,000 &= 10^3 \\
100 &= 10^2 \\
10 &= 10^1 \\
1 &= 10^0 \\
.1 &= 10^{-1} \\
.01 &= 10^{-2} \\
.001 &= 10^{-3} \\
.0001 &= 10^{-4}
\end{aligned}
$$

Notice that it makes sense to *define* 10^0 to be the number one—that clearly fits with the pattern of moving the decimal point one place to the left (moving down the left hand column of numbers), and lowering the exponent on the 10 by one (moving down the column on the right).

Using the notation of exponents, we can also write 12.438 in expanded form as follows:

$$12.438 = 1 \times 10 + 2 \times 1 + 4 \times 10^{-1} + 3 \times 10^{-2} + 8 \times 10^{-3}$$

CLASS ACTIVITY NOW TURN TO CLASS ACTIVITIES MANUAL

2B Showing Powers of Ten p. 7

See the website

www.aw-bc.com/beckmann

for an amazing experience of zooming in through powers of ten.

SAYING DECIMAL NUMBERS AND WRITING DECIMAL NUMBERS WITH WORDS

The decimal number 123,456 represents the amount

1 hundred thousand and

2 ten thousands and

3 thousands and

4 hundreds and

5 tens and

6 ones.

But in most circumstances, this is not how we should read this number. Instead, we should read 123,456 as

one-hundred twenty-three thousand, four-hundred fifty-six,

separating the "thousands portion" of the number from the "hundreds portion." Similarly, when a number is in the millions or billions, we read each of the "billions portion," "millions portion," "thousands portion," and "hundreds portion" separately. For example, we read

$$3,478,256,149$$

as

three-billion, four-hundred seventy-eight million, two-hundred fifty-six thousand, one-hundred forty-nine.

When there are not too many digits to the right of the decimal point, we can read the decimal point as *and*. For example, we can read

$$139.7$$

as one-hundred thirty-nine, and seven tenths. Digits to the right of the decimal point can be read as a group. For example, we can read

$$12.37$$

as twelve and thirty-seven hundredths, and we can read

$$1.358$$

as one and three-hundred fifty-eight thousandths. However, a number with many digits to the right of its decimal point, such as

$$5.678134$$

is best read as five point six seven eight one three four.

There is one special number name that many children find amusing and fascinating: a googol. A **googol** is the name for the number whose decimal representation is a 1 followed by one hundred zeros:

$$1 \underbrace{0000000000 \ldots 0000000000}_{100 \text{ zeros}}$$

googol

A googol is so large that, according to current theories in physics, it is even larger than the number of atoms in the universe. Another very large number that has a special name is a googolplex. A **googolplex** is the name for the number whose decimal representation is a 1

googolplex

followed by a googol zeros:

$$1 \underbrace{0000000000 \ldots 0000000000}_{\text{a googol zeros}}$$

It is very difficult to comprehend such a number.

PRACTICE PROBLEMS FOR SECTION 2.2

Answers to these practice problems begin below.

1. What does it mean to say that the decimal system uses place value?

2. What is the relationship between the values of adjacent places in a decimal number?

3. Write the following decimal numbers as powers of 10:

   a. $10, 000, 000, 000, 000$

   b. .1

   c. .000001

   d. 1

   e. 10

   f. the number whose decimal representation is a 1 followed by 200 zeros

   g. a googol

   h. a googolplex

4. Write the following decimal numbers in expanded form:

   a. 102.04

   b. 300,000,027

   c. 5,000.00200

5. Write the following numbers in ordinary decimal notation:

   a. One hundred thirty-seven million

   b. One million thirty-seven

   c. One million thirty-seven thousand

6. We call 10^6 a million, we call 10^9 a billion, and we call 10^{12} a trillion. What are 10^{13} and 10^{14} called? If we call 10^{15} a thousand trillion and 10^{18} a million trillion, then what are 10^{19}, 10^{20}, and 10^{21} called?

7. The children in Mrs. Watson's class made chains of small paper dolls, as pictured in Figure 2.10. A chain of 5 dolls is 1 foot long. How long would the following chains of dolls be? In each case, give your answer in either feet or miles, depending on which answer is easiest to understand. Recall that 1 mile = 5280 feet.

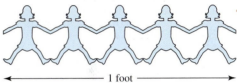

←──────── 1 foot ────────→

FIGURE 2.10

A Chain of Paper Dolls

   a. 100 dolls

   b. 1000 dolls

   c. 10,000 dolls

   d. 100,000 dolls

   e. one million dolls

   f. one billion dolls

ANSWERS TO PRACTICE PROBLEMS FOR SECTION 2.2

1. See text.

2. See text.

3. a. $10,000,000,000,000 = 10^{13}$

   b. $.1 = 10^{-1}$

   c. $.000001 = 10^{-6}$

   d. $1 = 10^0$

   e. $10 = 10^1$

f. The number whose decimal representation is a 1 followed by 200 zeros can be written 10^{200}.

g. A googol can be written 10^{100}.

h. A googolplex can be written 10^{googol} or $10^{(10^{100})}$.

4. a. $1 \cdot 100 + 2 \cdot 1 + 4 \cdot \frac{1}{100}$ or
$1 \cdot 10^2 + 2 \cdot 10^0 + 4 \cdot 10^{-2}$

b. $3 \cdot 100,000,000 + 2 \cdot 10 + 7 \cdot 1$ or
$3 \cdot 10^8 + 2 \cdot 10^1 + 7 \cdot 10^0$

c. $5 \cdot 1000 + 2 \cdot \frac{1}{1000}$ or
$5 \cdot 10^3 + 2 \cdot 10^{-3}$

5. a. $137,000,000$

b. $1,000,037$

c. $1,037,000$

6. Since 10^{12} is a trillion, 10^{13} is ten trillion and 10^{14} is a hundred trillion. Since 10^{18} is a million trillion, 10^{19} is ten-million trillion, 10^{20} is a hundred-million trillion, and 10^{21} is a billion trillion.

7. a. Because 5 dolls are 1 foot long, 10 dolls are 2 feet long. Thus, 100 dolls are 10 times as long, namely, 20 feet long.

b. 1000 dolls are 10 times as long as 100 dolls. Because 100 dolls are 20 feet long, 1000 dolls are 200 feet long.

c. 10,000 dolls are 10 times as long as 1000 dolls. Because 1000 dolls are 200 feet long, 10,000 dolls are 2000 feet long.

d. 100,000 dolls are 10 times as long as 10,000 dolls. Because 10,000 dolls are 2000 feet long, 100,000 dolls are 20,000 feet long. One mile is 5280 feet, so 20,000 feet is $20,000 \div 5280 = 3.8$ miles. So 100,000 dolls are nearly 4 miles long.

e. One million dolls are 10 times as long as 100,000 dolls. Because 100,000 dolls are 3.8 miles long, one million dolls are 38 miles long.

f. One billion dolls are 1000 times as long as one million dolls. Because one million dolls are 38 miles long, one billion dolls are 38,000 miles long. The circumference of the earth is about 24,000 miles, so one billion dolls would wrap around the earth more than $1\frac{1}{2}$ times!

PROBLEMS FOR SECTION 2.2

1. Write 100 zeros on a piece of paper and time how long it takes you. Based on the time it took you to write 100 zeros, approximately how long would it take you to write 1,000 zeros? 10,000 zeros? 100,000 zeros? one million zeros? one billion zeros? one trillion zeros? In each case, give your answer either in minutes, hours, days, or years, depending on which answer is easiest to understand. Explain your answers.

2. Explain clearly why it would not be possible for anyone to write a googol zeros. Use this to explain why no person could ever write a googolplex as a decimal number.

3. The students in Ms. Caven's class have a large poster showing a million dots. Now, the students would really like to see a billion of something. Think of at least two different ways that you might attempt to show a billion of something and discuss whether or not your methods would be feasible. Be specific and back up your explanations with calculations.

2.3   Representing Decimal Numbers

We will study two ways to represent decimal numbers. One is with physical objects. Another is with number lines.

FIGURE 2.11

Representing
1234 =
1 · 1000 +
2 · 100 +
3 · 10 + 4 · 1
with Money

dimes   pennies

1 thousand   2 hundreds   3 tens   4 ones

REPRESENTING DECIMAL NUMBERS WITH PHYSICAL OBJECTS

Writing a decimal number in its expanded form can help you understand the meaning of the number, but physical objects, while cumbersome, may be even more helpful. Appropriate physical objects can be especially helpful for developing a feel for place value.

USING MONEY TO REPRESENT DECIMAL NUMBERS

A familiar way to represent decimal numbers is with money. For example,

$$12.34 = 1 \cdot 10 \ + \ 2 \cdot 1 \ + \ 3 \cdot \frac{1}{10} \ + \ 4 \cdot \frac{1}{100}$$

can be represented by 1 ten-dollar bill, 2 one-dollar bills, 3 dimes, and 4 pennies, as shown in Figure 2.11. Notice that this same set of bills and coins can also represent

$$1234 = 1 \cdot 1000 \ + \ 2 \cdot 100 \ + \ 3 \cdot 10 \ + \ 4 \cdot 1$$

when we consider a penny as 1 *cent* rather than $\frac{1}{100}$ of a *dollar*.

Even though money is familiar, one disadvantage of using money to represent decimal numbers is that it does not show that the value of each place in a decimal number is 10 times the value of the place to the right. For example, we can't tell just by looking at a dollar bill that its value is ten times the value of a dime. Although every adult knows this, it is not an apparent relationship from the appearance of a dollar bill and a dime.

USING BUNDLED OBJECTS TO REPRESENT DECIMAL NUMBERS

One way to physically represent decimal numbers so that it is apparent that each place's value is ten times the value of the place to the right is to use bundled objects. Toothpicks are convenient since they are small, inexpensive, and readily available. (Coffee stirrers also work well if you can get enough of them.) To represent the various place's values, first make bundles of 10 toothpicks and band them together, or place 10 toothpicks in a small plastic sandwich bag. Then make bundles of 100 toothpicks by banding 10 bundles of 10. (You can put a rubber band around 10 bags that each contain 10 toothpicks.) If you have enough time and toothpicks, you can make a bundle of 1000 toothpicks by banding 10 bundles of 100 (which themselves are 10 bundles of 10). See Figure 2.12. Even though it may seem tedious, it is important to do this successive bundling because this is how you demonstrate that each place's value is 10 times the value of the place to the right. If you just put 100 loose toothpicks in a plastic bag, for

FIGURE 2.12

Bundled
Toothpicks for
Representing
Values of Places

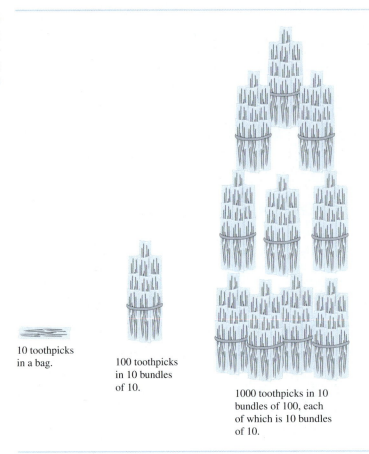

10 toothpicks
in a bag.

100 toothpicks
in 10 bundles
of 10.

1000 toothpicks in 10
bundles of 100, each
of which is 10 bundles
of 10.

instance, then you don't show the structure of the hundreds place as 10×10 (i.e., 10 groups of 10).

To show the decimal number 1234 with toothpicks, use

1 bundle of 1000 (which is 10 bundles of 100—each of which is 10 bundles of 10),

2 bundles of 100 (each of which is 10 bundles of 10),

3 bundles of 10 toothpicks, and

4 individual toothpicks,

as shown in Figure 2.13.

However, the same configuration of bundles of toothpicks can also represent

$$123.4 = 1 \cdot 100 + 2 \cdot 10 + 3 \cdot 1 + 4 \cdot \frac{1}{10}$$

by simply interpreting 1 toothpick as $\frac{1}{10}$, so that a bag of 10 toothpicks must represent 1, a bundle of 10 bags of 10 toothpicks must represent 10, and a bundle of 10 bundles of 100 (which itself is 10 bundles of 10) represents 100.

FIGURE 2.13

Representing
$1234 =$
$1 \cdot 1000 +$
$2 \cdot 100 +$
$3 \cdot 10 +$
$4 \cdot 1$ with
Bundled
Toothpicks

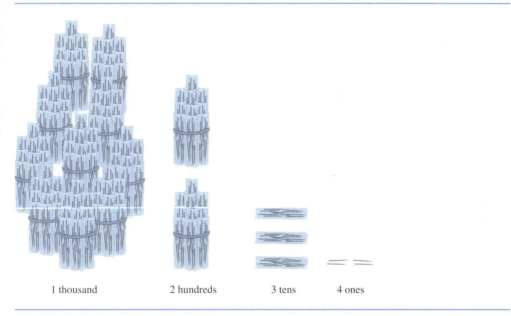

1 thousand   2 hundreds   3 tens   4 ones

Similarly, the same configuration of bundled toothpicks can represent smaller numbers, such as

$$12.34, \quad 1.234, \quad .1234, \quad .01234, \quad .001234, \quad .0001234$$

as well as larger numbers, such as

$$12{,}340, \quad 123{,}400, \quad 1{,}234{,}000, \quad 12{,}340{,}000$$

depending on what one toothpick represents.

USING BASE TEN BLOCKS TO REPRESENT DECIMAL NUMBERS

Many teachers have sets of **base ten blocks** for representing decimal numbers. Base ten blocks are sets of wooden (or plastic) blocks of several types: small cubes; "longs" that look like 10 small cubes glued together in a row; "flats" that look like 10 longs glued together to make a square of 100 small cubes; and a large cube that is constructed to look like 10 flats glued together to make a cube of 1000 small cubes. (See Figure 2.14.)

FIGURE 2.14

Base Ten Blocks

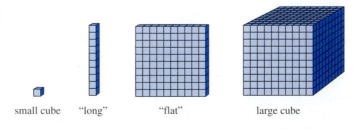

small cube   "long"   "flat"   large cube

FIGURE 2.15

Representing
$1234 =$
$1 \cdot 1000 +$
$2 \cdot 100 +$
$3 \cdot 10 + 4 \cdot 1$
with Base Ten
Blocks

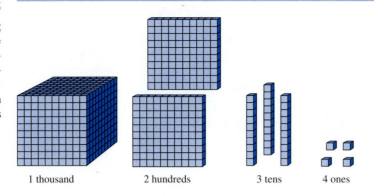

1 thousand   2 hundreds   3 tens   4 ones

To represent

$$1234 = 1 \cdot 1000 + 2 \cdot 100 + 3 \cdot 10 + 4 \cdot 1$$

with base ten blocks you can use the following:

    1 large cube (representing 1000 small cubes),

    2 flats (each representing 100 small cubes),

    3 longs (each representing 10 small cubes), and

    4 small cubes,

as shown in Figure 2.15.

As before, this same configuration of base ten blocks can represent other decimal numbers by changing the interpretation of 1 small cube. For example, it can represent the decimal number .1234 by considering 1 small block to be $\frac{1}{10,000}$, so that a long represents $\frac{1}{1000}$, a flat represents $\frac{1}{100}$, and a large cube represents $\frac{1}{10}$.

One disadvantage of using base ten blocks is that it is not always apparent to students that for successive blocks, the larger one is made of 10 times as many small cubes as the smaller one.

To summarize, Figure 2.16 shows the different ways to represent

$$1234 = 1 \cdot 1000 + 2 \cdot 100 + 3 \cdot 10 + 4 \cdot 1$$

with physical objects that we have discussed in this section.

CLASS ACTIVITY   NOW TURN TO CLASS ACTIVITIES MANUAL

2c Representing Decimal Numbers with Physical Objects p. 10

REPRESENTING DECIMAL NUMBERS ON NUMBER LINES

Bundled toothpicks show nicely that the value of a place in a decimal number is 10 times the value of the place to its right. Starting at any place in a decimal number, you can let 1 toothpick stand for the value of that place. Moving to the left in the decimal number, the values of

FIGURE 2.16

Representing
$1234 =$
$1 \cdot 1000 +$
$2 \cdot 100 +$
$3 \cdot 10 + 4 \cdot 1$

(a) money

dimes   pennies

1 thousand   2 hundreds   3 tens   4 ones

(b) bundled toothpicks

1 thousand   2 hundreds   3 tens   4 ones

(c) base ten blocks

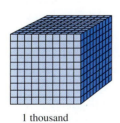 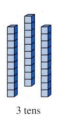

1 thousand   2 hundreds   3 tens   4 ones

FIGURE 2.17

Stage 1:
Representing
Integers on a
Number Line

subsequent places are shown by repeatedly bundling bundles. As long as you have enough toothpicks, you can keep going. But what about starting at a place in a decimal number and moving to the *right*? A good way to show the decreasing values of places as one moves to the right in a decimal number is with number lines.

number line   A **number line** is a line on which one location has been chosen as 0, and another location, to the right of 0, has been chosen as 1. Number lines stretch infinitely far in both directions, although in practice, only a small portion of a number line can be shown (and that portion may **unit**   or may not include 0 and 1). The distance from 0 to 1 is called a **unit**, and the choice of a unit is **scale**   called the **scale** of the number line. Once choices for the locations of 0 and 1 have been made, every decimal number is represented by a location on the number line, and every location on the number line corresponds to a decimal number.

DECIMAL NUMBERS "FILL IN" THE NUMBER LINE

Decimal numbers "fill in" the locations on the number line between the integers. You can think of plotting decimal numbers on the number line in successive stages. At the first stage, the integers are placed on a number line so that consecutive integers are one unit apart. The positive integers are to the right of 0, and the negative integers are to the left of zero. (See Figure 2.17.)

At the second stage, the decimal numbers that have entries in the tenths place, but no smaller place, are spaced equally between the integers, breaking each interval between consecutive integers into 10 smaller intervals, each $\frac{1}{10}$ unit long. See the stage-two number lines shown at the top of Figures 2.18 and 2.19. Notice that although the interval between consecutive integers is broken into 10 intervals, there are only 9 *tick marks* for decimal numbers in the interval, one for each of the 9 nonzero entries, 1 through 9, that go in the tenths places.

At the third stage, the decimal numbers that have entries in the hundredths place, but no smaller place, are spaced equally between the previously plotted decimal numbers, breaking each interval between previously plotted consecutive decimal numbers into 10 smaller intervals, each $\frac{1}{100}$ unit long. See the stage-three number lines in the middle of Figures 2.18 and 2.19. Notice that the number lines shown in Figures 2.18 and 2.19 all have different scales, so that they can show clearly the locations of the decimal numbers. These locations also exist in the first number line at the top, but they are not shown due to space limitations.

At each stage in the process of filling in the number line, we plot new decimal numbers in between previously plotted decimal numbers. The tick marks for these new decimal numbers should be smaller than the tick marks of the decimal numbers plotted at the previous stage, as shown in Figures 2.18 and 2.19. We use smaller tick marks to distinguish among the stages.

Imagine the process of repeatedly breaking intervals into 10 smaller intervals continuing forever. It turns out that all decimal numbers, including the ones that have infinitely many nonzero entries to the right of the decimal point, will have a location on the number line. The idea that even decimal numbers with infinitely many nonzero entries to the right of the decimal point have a location on the number line, may seem murky; in fact, it is a very subtle point, whose details were not fully worked out by mathematicians until the late 19th century.

FIGURE 2.18

Decimal Numbers "Fill In" Number Lines

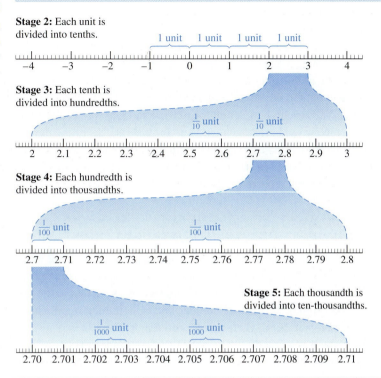

FIGURE 2.19

Negative Decimal Numbers "Fill In" Number Lines

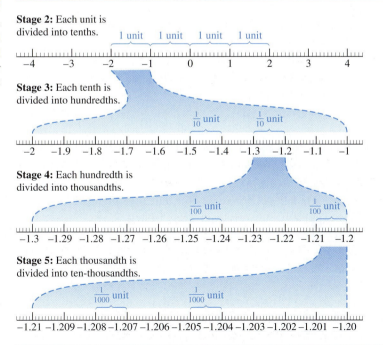

The idea of "filling in" locations on the number line, starting from the integers, is like starting with the notion of dollars as a denomination, deciding that a smaller denomination is needed, and therefore creating dimes. Ten dimes make a dollar and a dime is worth $\frac{1}{10}$ of a dollar, just as the intervals on a number line between consecutive integers are broken into 10 intervals, each $\frac{1}{10}$ of a unit long. If dimes don't make a small enough denomination, pennies are created. Ten pennies make a dime and a penny is worth $\frac{1}{100}$ of a dollar, just as the intervals between consecutive decimal numbers with no entries to the right of the tenths place are broken into 10 smaller intervals, each $\frac{1}{100}$ of a unit long. If we felt that we needed a smaller denomination than the penny, we would create a new coin. In this case, 10 of the new coins would make a penny. In theory, we could keep going, creating smaller and smaller denominations of money, just as the values of places in decimal numbers get smaller and smaller as one moves to the right, and just as we can repeatedly break intervals on a number line into 10 smaller intervals.

DECIMAL NUMBERS REPRESENT DISTANCES FROM 0

You can also view the location of a decimal number on a number line as the distance between that number and zero. A positive decimal number N is located to the right of 0, at a distance of N units away from 0. A negative decimal number $-N$ is located to the left of 0, at a distance of N units away from 0. It is as if a number line is an infinite ruler, except that number lines also show negative numbers, whereas rulers do not.

For example, 3 is located to the right of 0 at a distance of 3 units from 0. The negative decimal number -2 is located to the left of 0 at a distance of 2 units from 0. To plot a decimal number that is not an integer, such as 1.3, we can use the expanded form. Since

$$1.3 = 1 \cdot 1 + 3 \cdot \frac{1}{10}$$

locate this decimal number 1 unit and an additional 3 tenths of a unit to the right of zero. (See Figure 2.20.)

To plot a negative decimal number, such as -2.41, we can work with the expanded form of the decimal number obtained by dropping the minus sign, 2.41.

$$2.41 = 2 \cdot 1 + 4 \cdot \frac{1}{10} + 1 \cdot \frac{1}{100}$$

Therefore, -2.41 is located to the left of 0 at a distance of 2 units plus 4 tenths of a unit plus 1 hundredth of a unit away from 0. So, to locate this decimal number on the number line, you start at 0, move 2 units to the left, then move another 4 tenths of a unit to the left, and finally move yet another 1 hundredth of a unit to the left. (See Figure 2.21.) In this way, the decimal number -2.41 is plotted a full 2.41 units away from 0.

FIGURE 2.20

Plotting $1.3 = 1 \cdot 1 + 3 \cdot \frac{1}{10}$

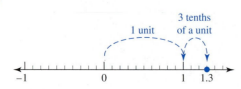

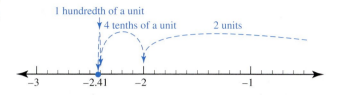

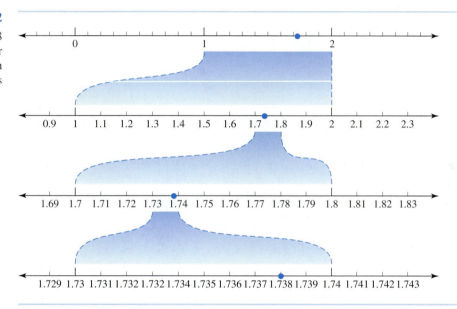

FIGURE 2.21

Plotting −2.41

FIGURE 2.22

Plotting 1.738
on Number
Lines with
Different Scales

USING NUMBER LINES OF DIFFERENT SCALES

Where you plot a decimal number on a number line depends on the scale of the number line. For example, Figure 2.22 shows the location of 1.738 on number lines of different scales.

DECIMAL NUMBERS WITH INFINITELY MANY NONZERO ENTRIES

Some decimals extend infinitely far to the right. For example, it turns out that the decimal representation of $\sqrt{2}$,

$$\sqrt{2} = 1.41421356237\ldots$$

goes on forever, never ending in zeros. *Every* decimal, even a decimal that extends infinitely to the right, has a definite location on a number line. However, such a decimal number will never fall *exactly* on a tick mark, no matter what the scale of the number line. Figure 2.23 shows where $\sqrt{2}$ is located on number lines of various scales.

CLASS ACTIVITY   NOW TURN TO CLASS ACTIVITIES MANUAL

2D Decimal Numbers on Number Lines p. 12

FIGURE 2.23

Plotting $\sqrt{2} =$
1.41421356237
. . . on a Number
Line

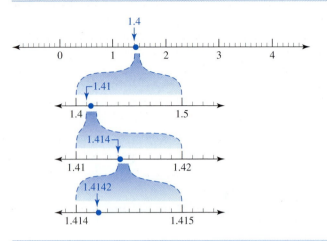

PRACTICE PROBLEMS FOR SECTION 2.3

Answers to these practice problems begin on page 41.

1. We can use bundles of toothpicks and money to represent decimal numbers. In what way are bundles of toothpicks better at representing decimal numbers than money is?

2. Describe how to represent .0278 with bundles of toothpicks. In this case, what does one toothpick represent?

3. List at least three different decimal numbers that the bundled toothpicks in Figure 2.24 could represent. In each case, state the value of one toothpick.

FIGURE 2.24

Which Decimal Numbers Can This Represent?

4. Give examples of decimal numbers that cannot be represented with bundles of toothpicks—even if you had as many toothpicks as you wanted.

5. Draw number lines like the ones in Figure 2.25. Label the tick marks on your number lines with appropriate decimal numbers.

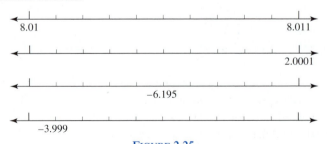

FIGURE 2.25

Label the Tick Marks

6. Draw number lines like the ones in Figure 2.26. Label the tick marks on your three number lines in three different ways. In each case, your labeling should fit with the fact that the tick marks at the ends of the number lines are longer than the other tick marks. You may further lengthen the tick marks at either end as needed.

FIGURE 2.26

Label the Tick Marks in Three Different Ways

7. Draw number lines like the ones in Figure 2.27. Label the tick marks on your three number lines in three different ways. In each case, your labeling should fit with the fact that the tick marks at the ends of the number lines are longer than the other tick marks. You may further lengthen the tick marks at either end as needed.

FIGURE 2.27

Label the Tick Marks in Three Different Ways

8. Draw a number line like the one in Figure 2.28. Label the tick marks on your number line so that the decimal numbers 3.482179 and 3.482635 can both be plotted visibly and distinctly. The distance between adjacent tick marks should be a power of ten, such as 0.1, 0.01, 0.001, and so on. The numbers do not necessarily have to fall on tick marks. Plot the two numbers.

FIGURE 2.28

A Number Line

9. Draw a number line like the one in Figure 2.28. Label the tick marks on your number line so that the decimal numbers −7.65 and −3 can both be plotted visibly and distinctly. The distance between adjacent tick marks should be a power of ten, such as 0.1, 0.01, 0.001, and so on. The numbers do not necessarily have to fall on tick marks. Plot the two numbers.

ANSWERS TO PRACTICE PROBLEMS FOR SECTION 2.3

1. See page 31.

2. Represent .0278 as 2 bundles of 100 toothpicks (each of which is 10 bundles of 10), 7 bundles of 10 toothpicks, and 8 individual toothpicks. In this case, each individual toothpick must represent $\frac{1}{10,000}$, since the expanded form of .0278 is

$$2 \cdot \frac{1}{100} + 7 \cdot \frac{1}{1000} + 8 \cdot \frac{1}{10,000}$$

3. If one toothpick represents 1, then Figure 2.24 represents 214. The table below shows several other possibilities.

If 1 toothpick represents	Then Figure 2.24 represents
100	21,400
10	2140
1	214
$\frac{1}{10}$	21.4
$\frac{1}{100}$	2.14
$\frac{1}{1000}$	.214

4. Realistically, we would be hard pressed to represent numbers with more than 4 nonzero digits with toothpicks. Even 999 would be difficult to represent. However, there are some numbers whose expanded form *can't* be represented by toothpicks (in the manner described in the text)—even if you had as many toothpicks

as you wanted. For example, consider

$$.333333333\ldots$$

where the 3s go on forever. Which place's value would you pick to be represented by 1 toothpick? If you did pick such a place, you would have to represent the values of the places to the right by tenths of a toothpick, hundredths of a toothpick, thousandths of a toothpick, and so on, forever, in order to show the expanded form of this number.

As an aside, here is a surprising fact: we *can* represent .333333333... with toothpicks, but in a different way (not by bundling so as to show the places in the decimal number). Namely, we can represent .333333333... by $\frac{1}{3}$ of a toothpick, because it so happens that $\frac{1}{3} = .333333333\ldots$ (which you can see by dividing 1 by 3).

5. See Figure 2.29.

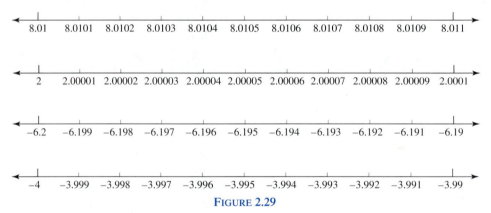

Figure 2.29

Labeled Number Lines

6. See Figure 2.30 for one way to label the tick marks. Notice that in the second and third number lines, the tick marks at the number 8 have been lengthened.

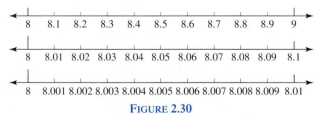

Figure 2.30

Labeled Number Lines

7. See Figure 2.31 for one way to label the tick marks. Notice that in the second and third number lines, the tick marks at the number 8 have been lengthened.

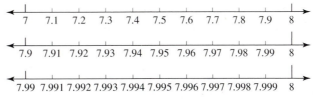

FIGURE 2.31

Labeled Number Lines

8. See Figure 2.32.

FIGURE 2.32

Choosing an Appropriate Scale

9. See Figure 2.33.

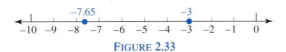

FIGURE 2.33

Choosing an Appropriate Scale

PROBLEMS FOR SECTION 2.3

1. List at least three different decimal numbers that the bundled toothpicks pictured in Figure 2.34 could represent. In each case, state the value of one toothpick.

FIGURE 2.34

Which Decimal Numbers Can These
Bundled Toothpicks Represent?

2. List at least three different decimal numbers that the base ten blocks pictured in Figure 2.35 could represent. In each case, state the value of one small block.

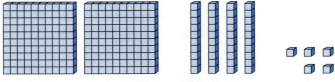

FIGURE 2.35

Which Decimal Numbers Can These Base Ten Blocks Represent?

3. Draw rough pictures of bundled toothpicks or base ten blocks to show how to represent the following decimal numbers. In each case, list two other decimal numbers that your picture could represent. Explain how to interpret your picture as representing these other decimal numbers.

   a. .26   b. 13.4   c. 1.28   d. 0.000032

4. Explain why the bagged and loose toothpicks pictured in Figure 2.36 do not represent a decimal number in the manner described in the text. Describe how to alter the appearance of these bagged and loose toothpicks so that the same total number of toothpicks *does* represent a decimal number.

FIGURE 2.36

Why Do These Not Represent a Decimal Number?

---

5. Draw three copies of the number line in Figure 2.37. Use your number lines to show three different ways to label the tick marks in Figure 2.37. In each case, your labeling should fit with the fact that the tick marks at the ends of the number lines are longer than the other tick marks. You may further lengthen the tick marks at either end as needed.

FIGURE 2.37

How to Label the Tick Marks

6. Draw three copies of the number line in Figure 2.38. Use your number lines to show three different ways to label the tick marks in Figure 2.38. In each case, your labeling should fit with the fact that the tick marks at the ends of the number lines are longer than the other tick marks. You may further lengthen the tick marks at either end as needed.

FIGURE 2.38

How to Label the Tick Marks

7. Draw three copies of the number line in Figure 2.39. Use your number lines to show three different ways to label the tick marks in Figure 2.39. In each case, your labeling should fit with the fact that the tick marks at the ends of the number lines are longer than the other

tick marks. You may further lengthen the tick marks at either end as needed.

FIGURE 2.39

How to Label the Tick Marks

8. Draw three copies of the number line in Figure 2.40. Use your number lines to show three different ways to label the tick marks in Figure 2.40. In each case, your labeling should fit with the fact that the tick marks at the ends of the number lines are longer than the other tick marks. You may further lengthen the tick marks at either end as needed.

FIGURE 2.40

How to Label the Tick Marks

9. Jerome says that the unlabeled tick mark on the number line in Figure 2.41 should be 7.10. Why might Jerome think this? Explain to Jerome why he is not correct. Describe how you could use physical objects to help Jerome determine the correct answer.

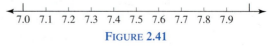

FIGURE 2.41

How to Label the Tick Mark on the Right

10. Suppose you are teaching children about decimal numbers. Tell the children what the decimal numbers 1.25 and 2.4 mean. Describe different ways that you could represent or show the children these decimal numbers. Draw on all the ways of representing and showing decimal numbers that were discussed in this section.

11. Suppose you are teaching children about decimal numbers. Tell the children about number lines and about how decimal numbers are plotted on number lines.

12. In the decimal system, every whole number can be expressed as a decimal number in only one way without a decimal point. If we used a different system for writing numbers, this might not be true. Let's create another way to write numbers that we'll call the "money numeration system." In the money numeration system, there will generally be more than one way to write a number. To write a number using the money numeration system, simply record a way to make that many cents out of coins and bills. The following examples show how the money numeration system works:

$$23(\text{cents}) = 2(\text{dimes}) + 3(\text{cents})$$

Also, $23(\text{cents}) = 1(\text{dimes}) + 2(\text{nickels}) + 3(\text{cents})$ (There are still more ways to write 23.)

$5,175(\text{cents}) = 5(\text{ten dollars}) + 1(\text{dollar}) + 3(\text{quarters}).$

a. Write $2,742$ (cents) in the money numeration system in at least 3 different ways.

b. Write 27 (cents) in the money numeration system in all possible ways. Explain how you know you've found all the ways. Is it surprising how many ways there are?

c. Which numbers can only be written in one way in the money numeration system?

d. Why is it convenient to have a system of money where most amounts can be made in many ways? Why is it convenient to have a system for writing numbers (the decimal system) where every whole number can be written only in one way (without a decimal point)?

2.4 Comparing Sizes of Decimal Numbers

Given two decimal numbers, they are either equal or one is greater than the other. We will compare decimal numbers from two different points of view. First we will consider decimal numbers as representing amounts in order to compare them. Then, we will use number lines to compare decimal numbers.

COMPARING NUMBERS BY VIEWING THEM AS AMOUNTS

greater than   If A and B are numbers that are not negative, then we say that A is **greater than** B, and we write

$$A > B$$

less than   if A represents a larger quantity than B. Similarly, we say that A is **less than** B, and we write

$$A < B$$

if A represents a smaller quantity than B. A good way to remember which symbol to use is to notice that the "wide side" faces the larger number. Some teachers have their students draw alligator teeth on the symbols, as in Figure 2.42, to help them: alligators are hungry and always want to eat the larger amount.

For example,

$$0.154 < .2$$

FIGURE 2.42

Alligator Teeth
on a "Less
Than" Symbol

$2 \lll 5$

FIGURE 2.43

Showing That
0.154 Is Less
Than 0.2

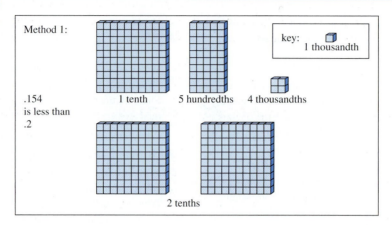

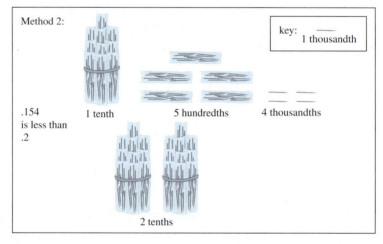

because

$$0.154 = 1 \cdot \frac{1}{10} + 5 \cdot \frac{1}{100} + 4 \cdot \frac{1}{1000},$$

$$0.2 = 2 \cdot \frac{1}{10},$$

and, as you can see in Figure 2.43, the combined amount of 5 hundredths and 4 thousandths make less than 1 tenth. In fact, extrapolating from Figure 2.43, and considering how much larger the tenths place is than the hundredths and thousandths places, we see that even 0.199, which has 9 hundredths and 9 thousandths is still less than 0.2. Even if there were nonzero values in the ten-thousdandths, hundred-thousandths, and millionths places, and so on, any decimal number of the form $0.1 * * * * \ldots * * * *$ is still less than 0.2,

$$0.1 * * * * \ldots * * * * < 0.2$$

(There is one technical exception to this rule, which we will discuss later.)

Informally speaking, as we see in Figure 2.43, places of larger value count more than all the places of lower value combined. Therefore, in order to compare the sizes of decimal numbers, we should compare numbers by comparing like places and by starting at the place of largest value in which at least one of the numbers has a nonzero entry.

For example,

$$1234 > 789$$

because 1234 has a 1 in the thousands place and 789 has 0 in the thousands place (even though this zero is not written).

$$
\begin{array}{c}
1234 \\
\updownarrow \\
0789
\end{array}
$$

Similarly,

$$.89 < 1.2$$

because 1.2 has a 1 in the ones place, but .89 has a 0 in the ones place (even though this zero is not written).

$$
\begin{array}{c}
0.89 \\
\updownarrow \\
1.2
\end{array}
$$

Finally,

$$1.2378 < 1.24$$

because both decimal numbers have a 1 in the ones place and a 2 in the tenths place, but 1.24 has a larger digit, namely, 4, in the hundredths place than does 1.2378, which only has a 3.

$$
\begin{array}{c}
1.2\,378 \\
\updownarrow \updownarrow \updownarrow \\
1.24
\end{array}
$$

Notice that determining the greater decimal number is similar to putting words in alphabetical order, except that for decimal numbers, we compare values in *like places*, whereas for words, we compare letters starting from the *beginning* of each word. Maybe this is why a decimal number such as 1.01 might at first glance look as if it is less than .998.

Since a place counts more than all the places of lower value combined, we get the following general rule for comparing two nonnegative decimal numbers:

HOW TO COMPARE DECIMAL NUMBERS

Starting from the place of largest value represented in the decimal numbers (the leftmost place), compare the digits of both decimal numbers in that place.

- If the digits in that place are not equal, then the decimal number with the larger digit is greater than the other decimal number.
- If the digits in that place are equal, keep moving one place to the right, comparing the digits of the decimal numbers in like places, until one of the decimal numbers has a larger digit than the other. The decimal number with this larger digit is greater than the other decimal number.

A TECHNICAL EXCEPTION TO THE RULE FOR COMPARING DECIMAL NUMBERS

There is a rare exception to the preceding method for determining which of two decimal numbers is greater. The exception occurs when one of the decimal numbers has an infinitely repeating 9, such as

$$37.569999999999\ldots$$

In this case, the method above would lead us to say that

$$37.57 > 37.569999999999\ldots$$

However, this is *not true*. In fact, as we will see in Section 12.6,

$$37.57 = 37.569999999999\ldots$$

USING NUMBER LINES TO COMPARE DECIMAL NUMBERS

It is especially easy to compare decimal numbers when they are plotted on a number line. When we use a number line, we can compare negative decimal numbers as well as positive ones.

If A and B are decimal numbers plotted on a number line, then the following statements are true:

COMPARING DECIMAL NUMBERS

$A < B$ provided that A is to the left of B on the number line.

$A > B$ provided that A is to the right of B on the number line.

These interpretations of "greater than" and "less than" are consistent with the interpretation we developed in the previous section for positive decimal numbers, where we viewed positive decimal numbers as representing quantities. Why? If a positive decimal number A represents a larger quantity than another positive decimal number B, then A will be plotted farther from 0 than B. Since both are plotted to the right of 0, A will be to the right of B. Likewise, if A represents a smaller quantity than B, then we will plot A closer to 0 than B. Since both are to the right of 0, A will be to the left of B.

With the number line interpretation of $>$ and $<$, it is easy to compare negative decimal numbers. For example, we plot -2.7 to the left of -1.5, as in Figure 2.44; therefore,

$$-2.7 < -1.5$$

FIGURE 2.44

Showing That
−2.7 Is Less
Than −1.5

COMPARING NEGATIVE DECIMAL NUMBERS BY VIEWING THEM AS OWED AMOUNTS

In addition to explaining why $-2.7 < -1.5$ by using a number line, we can also explain why $-2.7 < -1.5$ by considering these negative decimal numbers as amounts owed. We can think of -2.7 as owing \$2.70 and -1.5 as owing \$1.50, in other words, to have $-\$2.70$ is to owe \$2.70, and to have $-\$1.50$ is to owe \$1.50. If you owe \$2.70, you have less than if you owe \$1.50, so therefore

$$-2.7 < -1.5$$

In general, the more you owe, the less you have, and the less you owe, the more you have. In other words,

$$\text{If } A > B, \quad \text{then} \quad -A < -B.$$
$$\text{If } A < B, \quad \text{then} \quad -A > -B.$$

For example,

$$7 > 4$$

Therefore,

$$-7 < -4$$

Notice that the minus signs cause the $>$ symbol to reverse.

CLASS ACTIVITY   NOW TURN TO CLASS ACTIVITIES MANUAL

2E Finding Smaller and Smaller Decimal Numbers p. 17

2F Finding Decimals between Decimals p. 18

2G Decimals between Decimals on Number Lines p. 19

2H "Greater Than" and "Less Than" with Negative Decimal Numbers p. 21

PRACTICE PROBLEMS FOR SECTION 2.4

1. What's wrong with a grocery store sign that says, "Squash, .99 cents per pound"?

2. Which is greater, 0.01 or 0.0099999999999?

3. Draw a picture like one of the pictures in Figure 2.43 demonstrating that $1.2 > .89$.

4. Children who have heard of a googol will often think it's the largest number. Is it? What about a googolplex—is that the largest number? Is there a largest number?

5. Find a decimal number between $\sqrt{2}$ and 1.41422, and plot all three numbers visibly and distinctly on a number line in which the distance between adjacent tick marks is a power of ten, such as 0.1, 0.01, or 0.001. The numbers need not land on tick marks.

6. Name two different numbers that are between 3.456 and 3.457.

7. Use a number line to show that $1.2 > .89$.

8. Use a number line to show that $-1.2 < -.89$.

9. Explain why $-1.2 < -.89$ without using a number line.

10. The smallest integer that is greater than zero is 1. Is there a smallest *real number* that is greater than zero?

ANSWERS TO PRACTICE PROBLEMS FOR SECTION 2.4

1. Notice that .99 cents is less than one cent. It's hard to imagine squash, or any other vegetable, selling for less than 1 cent per pound. The store probably means that the price is $.99 per pound, or in other words, 99 cents per pound.

2. The number .01 is greater because it has a 1 in the hundredths place, whereas .0099999999999 has a 0 in the hundredths place.

3. See Figure 2.45.

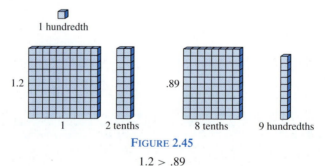

FIGURE 2.45

$1.2 > .89$

4. Notice that a googolplex is greater than a googol. The number googolplex plus one, for example, is greater than a googolplex. There is no largest number because no matter what number you choose as candidate for the largest number, that number plus one is a larger number.

5. The decimal representation of $\sqrt{2}$ is 1.414213 ..., so 1.414215 is one example of a decimal number that is in between $\sqrt{2}$ and 1.41422. There are many other examples. See Figure 2.46.

FIGURE 2.46

The Number .1414215 Is Between $\sqrt{2}$ and 1.4122

6. The numbers 3.4563 and 3.4567 both lie in between 3.456 and 3.457.

7. See Figure 2.47.

FIGURE 2.47

1.2 > .89 Because 1.2 Is to the Right of .89

8. See Figure 2.48.

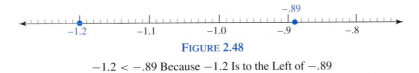

FIGURE 2.48

−1.2 < −.89 Because −1.2 Is to the Left of −.89

9. To explain why −1.2 < −.89, we can consider having −1.2 as owing $1.20 and having −.89 as owing $.89. If you owe $1.20 then you have less than if you only owe $.89. Therefore, having −1.2 means you have less than if you have −.89. So, −1.2 < −.89.

10. No, there is no smallest real number that is greater than 0. Consider the following list of numbers:

.1
.01
.001
.0001
.00001
.000001
⋮

Imagine the list continuing forever. The numbers in this list get smaller and smaller, getting ever closer to 0 without ever reaching 0. No matter what number you choose that is greater than 0, when you write it as a decimal number, it will have a nonzero entry somewhere. Based on where the first nonzero entry in your chosen number is, we can pick a decimal number from the list above that is even smaller than your chosen number. Therefore there can be no smallest positive number.

PROBLEMS FOR SECTION 2.4

1. Draw a picture like either picture in Figure 2.43 demonstrating that 1.1 > .99.

2. Mary says that the numbers in the following list are getting bigger:

4.1, 4.2, 4.3, 4.4, 4.5, 4.6, 4.7, 4.8, 4.9, 4.10, 4.11, 4.12, 4.13, 4.14, 4.15, . . . , 4.18, 4.19, 4.20, 4.21, 4.22, . . . , 4.29, 4.30, 4.31, . . . 4.98, 4.99, 4.100, 4.101, . . .

Is Mary right or not? If she's not right, then which numbers on the list are immediately to the left of a smaller number?

3. Arthur says that 6.10 > 6.9. Explain to Arthur in two different ways why his statement is not true.

4. Mark says that .178 is greater than .25. Describe several ways to convince him that it's not.

5. Explain in two different ways why −8 < −5.

6. Explain in two different ways why −3.25 < −1.4.

7. Find a number between 3.24 and 3.241 if there is one. If there is no number between them, explain why not.

8. Is there more than one number between 8.45 and 8.47? If so, find two such numbers.

---

9. a. Find a number between 3.8 and 3.9, and plot all three numbers visibly and distinctly on a number line like the one in Figure 2.49, which has a set of longer tick marks and a set of shorter tick marks. The distance between adjacent tick marks should be a power of ten, such as 0.1, 0.01, or 0.001. Label all the longer tick marks.

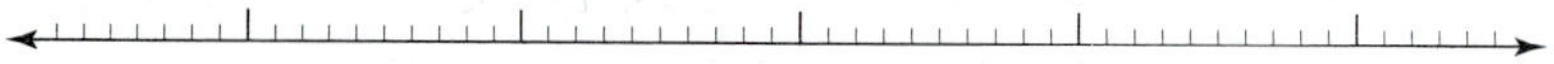

FIGURE 2.49

A Number Line

b. Describe how to use money to find a number between 3.8 and 3.9.

---

10. Find a number between 2.981 and 2.982, and plot all three numbers visibly and distinctly on a number line like the one in Figure 2.49, which has a set of longer tick marks and a set of shorter tick marks. The distance between adjacent tick marks should be a power of ten, such as 0.1, 0.01, or 0.001. Label all the longer tick marks.

11. Find a number between −34.9714 and −34.9835, and plot all three numbers visibly and distinctly on a number line like the one in Figure 2.49, which has a set of longer tick marks and a set of shorter tick marks. The distance between adjacent tick marks should be a power of ten, such as 0.1, 0.01, or 0.001. Label all the longer tick marks. The numbers need not land on tick marks.

12. Find a number between −7.834 and −7.83561, and plot all three numbers visibly and distinctly on a number line like the one in Figure 2.49, which has a set of longer tick marks and a set of shorter tick marks. The distance between adjacent tick marks should be a power of ten, such as 0.1, 0.01, or 0.001. Label all the longer tick marks. The numbers need not land on tick marks.

13. Find a number between 13 and 12.9999, and plot all three numbers visibly and distinctly on a number line like the one in Figure 2.49, which has a set of longer tick marks and a set of shorter tick marks. The distance between adjacent tick marks should be a power of ten, such as 0.1, 0.01, 0.001, and so on. Label all the longer tick marks. The numbers need not land on tick marks.

14. Find a number between 13 and 13.0001, and plot all three numbers visibly and distinctly on a number line like the one in Figure 2.49, which has a set of longer tick marks and a set of shorter tick marks. The distance between adjacent tick marks should be a power of ten, such as 0.1, 0.01, 0.001, and so on. Label all the longer tick marks. The numbers need not land on tick marks.

15. Describe an infinite list of decimal numbers, all of which are greater than 3.514, but get closer and closer to 3.514.

16. The smallest integer that is greater than 2 is 3. Is there a smallest *real number* that is greater than 2? Explain why or why not in your own words.

2.5 Rounding Decimal Numbers

Decimal numbers are often used to describe the size of actual quantities. The distance between two locations might be described as 2.7 miles; the population of a city might be given as 87,000; a tax rebate might be described as $1.3 trillion. These numbers are not the *exact* amounts: there aren't exactly 87,000 people living in the city, and the tax rebate isn't $1.3 trillion on the nose. Instead, we understand that these numbers have been *rounded*. Sometimes we round because we don't know a quantity exactly, and sometimes we round because we want to convey the approximate size of a quantity. For example, a company might know that it spent \$174,586.74 on some equipment, but in a report to management, an employee might describe the expenditure as \$175,000 because this round number is more quickly and easily grasped and is close enough to the actual expenditure for the discussion at hand.

HOW TO ROUND

round   To **round** a decimal number means to find a nearby decimal number that has fewer (or no more) nonzero digits. We can round to the nearest 100, to the nearest 10, to the nearest 1, to the nearest $\frac{1}{10}$, to the nearest $\frac{1}{100}$, and so on. For any power of ten, 10^n, we can round to the nearest 10^n. To round a given decimal number to the nearest 100, we must find the decimal number that is closest to our given decimal number and has only zeros in places smaller than the hundreds place. To round a given decimal number to the nearest $\frac{1}{10}$, we we must find the decimal number that is closest to our given decimal number and has only zeros in places smaller than the tenths place. In general, to round a given decimal number to the nearest 10^n, we must find the decimal number that is closest to our given decimal number and that has only zeros in places smaller than the 10^ns place.

To understand rounding, think about zooming in or zooming out on a number line so that you can see the tick marks that fit with the desired type of rounding. When rounding to the nearest hundred, zoom out so that you see large tick marks spaced 100 apart, as in Figure 2.50. The decimal numbers at the large tick marks are exactly those decimal numbers that have only zeros in places smaller than the hundreds place. So, when you round to the nearest hundred, you are finding the large tick mark that is closest to the location of the number you are rounding. Thus, to round 3367 to the nearest hundred, notice that 3367 lies between 3300 and 3400. But 3367 is closer to 3400 than to 3300 because 3367 is greater than 3350, which is half way between 3300 and 3400. Therefore, round 3367 to 3400.

When rounding to the nearest tenth, zoom in on the number line so that you see large tick marks spaced $\frac{1}{10}$ apart, as in Figure 2.51. The decimal numbers at these large tick marks are

FIGURE 2.50

Rounding to the Nearest Hundred

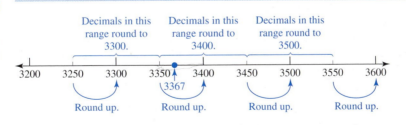

FIGURE 2.51

Rounding to the Nearest Tenth

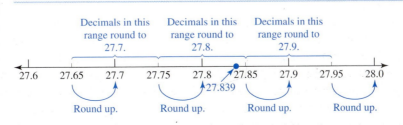

exactly those decimal numbers that have only zeros in places smaller than the tenths place. So, when you round to the nearest tenth, you are finding the large tick mark that is closest to the location of the number you are rounding. Thus, to round 27.839 to the nearest tenth, notice that 27.839 lies between 27.8 and 27.9. But 27.839 is closer to 27.8 than to 27.9 because 27.839 is less than 27.85, which is half way between 27.8 and 27.9. Therefore, round 27.839 to 27.8.

What do we do about decimal numbers that are exactly half way in between two tick marks? For example, how do we round 3450 when we are rounding to the nearest hundred? How do we round 27.45 when we are rounding to the nearest tenth? We need a convention to break ties like these. The most common convention, and the one you should use unless otherwise specified, is to *round up* when there is a tie. So when rounding 3450 to the nearest hundred, round to 3500, as shown in Figure 2.50. When rounding 27.85 to the nearest tenth, round to 27.9, as shown in Figure 2.51. Another convention for rounding, one that is often used by engineers, is to *round to the nearest even* when there is a tie. With this convention, when rounding to the nearest hundred, 3450 would be rounded to 3400, rather than 3500, because 4 is even. Similarly, when rounding to the nearest tenth, 27.85 would be rounded to 27.8, rather than 27.9, because 8 is even.

WORKING WITH DECIMAL NUMBERS THAT REPRESENT ACTUAL QUANTITIES

When we use a decimal number to describe the size of an actual quantity, such as a distance or a population, we generally assume that this number has been rounded. Furthermore, we assume that *the way a decimal number is written indicates the rounding that has taken place.* For example, when the distance between two locations is described as 6.2 miles, we assume from the presence of a digit in the tenths place, but no smaller place, that the actual distance has been rounded to the nearest tenth of a mile. Therefore, the actual distance could be anywhere between 6.15 and 6.25 miles (but less than 6.25 miles). Similarly, if the population of a city is described as 93,000, we assume from the nonzero entry in the thousands place and the zeros in all lower places that the actual population has been rounded to the nearest thousand. Therefore, the actual population could be anywhere between 92,500 and 93,500 (but less than 93,500).

Because of rounding, there is a slight difference in the meaning of decimal numbers in the abstract and decimal numbers when they are used to describe the size of an actual quantity. In the abstract,

$$8.00 = 8$$

However, when we write 8 for the weight of an object in kilograms, it has a different meaning than when we write 8.00 for the weight of an object in kilograms. Writing 8 indicates that the actual weight has been rounded to the nearest one, whereas writing 8.00 means that the actual weight has been rounded to the nearest hundredth. If the weight is reported as 8.00 kilograms, then the actual weight could be anywhere between 7.995 and 8.005 kilograms. If, however, the weight is reported as 8 kilograms, then the actual weight could be anywhere between 7.5 and 8.5 kilograms. Therefore, writing 8.00 conveys that the weight is known much more accurately than if the weight is reported as 8 kilograms.

When you solve a problem that involves real or realistic quantities, you should round your answer so that it does not appear to be more accurate than it actually is. Your answer cannot be any more accurate than the numbers you started with, so you should round your answer to fit with the rounding of your initial numbers. For example, suppose that the population of a city is given as 1.6 million people, and that after some calculations, you project that the city will have 1.95039107199 million people in 10 years. Although the decimal number 1.95039107199 may be the exact answer to your calculations, you should not report your answer this way, because it makes your answer appear to be more accurate than it actually is. We must assume that the initial number 1.6 is rounded to the nearest tenth. Therefore we should also round the answer, 1.95039107199, to the nearest tenth and report the projected population in 10 years as 2.0 million.

CLASS ACTIVITY NOW TURN TO CLASS ACTIVITIES MANUAL

2I Explaining Why a Procedure for Rounding Makes Sense p. 22

2J Can We Round This Way? p. 24

PRACTICE PROBLEMS FOR SECTION 2.5

1. Round 6.248 to the nearest tenth. Explain in words why you round the decimal number the way you do. Use a number line to support your explanation.

2. Round 173.465 to the nearest hundred, to the nearest ten, to the nearest one, to the nearest tenth, and to the nearest hundredth.

3. The distance between two cities is described as 1500 miles. Should you assume that this is the exact distance between the cities? If not, what can you say about the exact distance between the cities?

ANSWERS TO PRACTICE PROBLEMS FOR SECTION 2.5

1. The number 6.248 is between 6.2 and 6.3. But because of the 4 in the hundredths place, 6.248 is less than 6.25, which is halfway between 6.2 and 6.3. Therefore, 6.248 is closer to 6.2 than to 6.3, so 6.248 rounded to the nearest tenth is 6.2.

   The number line in Figure 2.52 illustrates what was just stated.

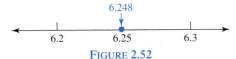

6.248

FIGURE 2.52

Rounding 6.248 to the Nearest Tenth

2. To the nearest hundred: 200
   To the nearest ten: 170
   To the nearest one: 173
   To the nearest tenth: 173.5
   To the nearest hundredth: 173.47

3. Because the reported distance has zeros in the tens and ones places, you should assume that the distance has been rounded to the nearest hundred. Therefore, the exact distance between the two cities is probably not 1500 miles; it could be anywhere between 1450 miles and 1550 (but less than 1550).

PROBLEMS FOR SECTION 2.5

1. Round 2.1349 to the nearest hundredth. Explain in words why you round the decimal number the way you do. Use a number line to support your explanation.

2. Round 9995.2 to the nearest ten. Explain in words why you round the decimal number the way you do. Use a number line to support your explanation.

3. Adam has made up his own method of rounding. Starting at the rightmost place in a decimal number, he keeps rounding to the value of the next place to the left until he reaches the place to which the decimal number was to be rounded. For example, Adam would use the following steps to round 11.3524 to the nearest tenth:

$$11.3524 \rightarrow 11.352 \rightarrow 11.35 \rightarrow 11.4$$

Is Adam's method a valid way to round? Explain why or why not.

4. The weight of an object is reported as 12,000 pounds. Should you assume that this is the exact weight of the object? If not, what can you say about the exact weight of the object? Explain.

5. In a report, a population is given as 2700. Should you assume that this is the exact population? If not, what can you say about the exact population? Explain.

Chapter 3

Fractions

I n this chapter, we continue our discussion of numbers, now focusing on fractions and percent. Much of the information we receive in daily life is presented in terms of fractions or percent. A news report might begin, "Nearly $\frac{3}{4}$ of the people surveyed...," or a sign in a store might report, "35% off." To function effectively in today's society, we must understand fractions and percents. In this chapter we will study the meaning of fractions and how to represent fractions. We will see why different fractions can be used to represent the same amount. We will also consider some of the difficulties in understanding fractions and learn key aspects to notice in order to overcome these difficulties.

3.1 The Meaning of Fractions

This section shows the meaning of (positive) fractions, some of the common difficulties in understanding the meaning of fractions, and how to use simple pictures to represent fractions.

First, consider how fractions are used in the following ordinary situations:

$\frac{1}{8}$ of a pizza

$\frac{2}{3}$ of the houses in the neighborhood

$\frac{9}{10}$ of the profit

$\frac{5}{4}$ of a cup of water

$\frac{4}{100}$ of $20,000

All these examples use the word *of*, and all the fractions above represent part *of* some object, collection of objects, or quantity.

If *A* and *B* are whole numbers, and *B* is not zero, then the **fraction**

$$\frac{A}{B}$$

of an object, a collection, or a quantity is the amount formed by *A* parts (or *A* copies of parts) when the object, collection, or quantity is divided into *B* equal parts.

numerator   If $\frac{A}{B}$ is a fraction, then *A* is called the **numerator** and *B* is called the **denominator**.
denominator   These names make sense because the word *numerator* comes from *number*, and the numerator tells you the number of parts. The word *denominator* comes from *name* and is related to *denomination*, which tells what type something is. For example, in money, the denomination of a bill tells you what type of bill it is, in other words, what value it has. In religion, a denomination is a group with the same type of religious belief and practice. The denominator of a fraction tells you what type of parts are being used. These parts can be halves, thirds, fourths, fifths, and so on. Hence, a fraction tells you how many of what type of parts.

A   ←   numerator: the number of parts
__
B   ←   denominator: the type or name of the parts

Figure 3.1 illustrates fractions of objects or collections of objects. In each case, the fraction $\frac{A}{B}$ is represented by shading *A* parts when the whole object or collection of objects has been

FIGURE 3.1

Fractions of
Objects,
Collections, and
Quantities

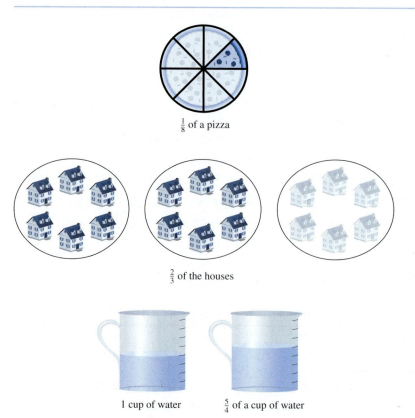

$\frac{1}{8}$ of a pizza

$\frac{2}{3}$ of the houses

1 cup of water $\frac{5}{4}$ of a cup of water

divided into B equal parts. In the pizza example, $\frac{1}{8}$ of the pizza is shown within the whole pizza. Likewise, in the houses example, $\frac{2}{3}$ of the houses are shown shaded within the whole collection of houses. However, since $\frac{5}{4}$ cup of water is more than a whole cup of water, it is shown separate from the whole cup of water.

CLASS ACTIVITY   NOW TURN TO CLASS ACTIVITIES MANUAL

3A Fractions of Objects p. 25

A FRACTION IS ASSOCIATED WITH A WHOLE

Notice the crucial word *of* in the examples of fractions of objects, fractions of collections of objects, and fractions of quantities described in the following:

$$\frac{2}{5} \text{ of the land}$$
$$\frac{2}{3} \text{ of the cars on the road}$$
$$\frac{1}{10} \text{ of the water in a lake}$$

Fractions are defined *in relation to a whole*, and this whole can be just one object, or it can be a collection of objects, such as the cars on the road, or 24 houses. The whole can also be a quantity, such as a quantity of water, or it can simply be the number 1.

When working with fractions, always keep in mind that there is an associated whole— whether or not that whole has been made explicit or drawn to your attention. Students from elementary school through college can correct many mistakes in their work with fractions if they can identify the whole associated with a fraction. That is, they need to understand what the fraction is *"of."*

One elementary school teacher, Patti Huberty, a mathematics specialist, begins her fraction lessons by asking children to find fractions of numbers, such as "What is half of 10?" Huberty finds that this helps children focus on the associated whole.

CLASS ACTIVITY   NOW TURN TO CLASS ACTIVITIES MANUAL

3B The Whole Associated with a Fraction p. 28

EQUAL PARTS

CLASS ACTIVITY   NOW TURN TO CLASS ACTIVITIES MANUAL

3C Is the Meaning of Equal Parts Always Clear? p. 31

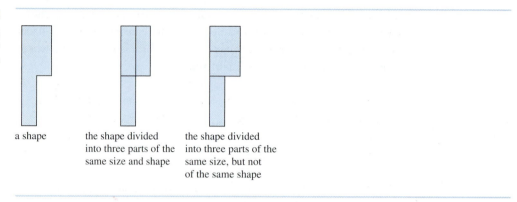

a shape | the shape divided into three parts of the same size and shape | the shape divided into three parts of the same size, but not of the same shape

If you did Class Activity 3C, then you saw that the meaning of equal parts is not always completely clear. If you want to divide a collection of toys into 4 equal parts, you might not simply put the same number of toys in each part: all the toys may not be considered equal, some may be more desirable than others. If you are asked to divide a shape, such as the one in Figure 3.2, into 3 equal parts, you might not know whether the parts should be equal in size and shape or if they only need to be the same size.

In realistic situations, it is seldom a problem to determine what constitutes equal parts. In most cases, equal parts will be defined by dollar value, by length, by area, by volume, or by number. To divide the items in an estate into equal parts, you would probably make the parts of equal dollar value. For rope, parts of the same length constitute equal parts. For land, parts of the same area would generally constitute equal parts. For a pile of gravel, parts of the same volume constitute equal parts. On the other hand, when talking about $\frac{2}{3}$ of the stamps in a collection or $\frac{2}{3}$ of the cars on the road or $\frac{2}{3}$ of the people in the room, we treat each stamp, car, or person as equal, even though some stamps may be more valuable than others, some cars may be more valuable or bigger than others, and some people are bigger than others.

UNDERSTANDING IMPROPER FRACTIONS

proper fraction A fraction of whole numbers $\frac{A}{B}$ (where B is not zero) is called a **proper fraction** if the numerator
improper fraction A is smaller than the denominator B. Otherwise, it's called an **improper fraction**. For example,

$$\frac{2}{5}, \quad \frac{3}{8}, \quad \frac{15}{16}$$

are proper fractions, whereas

$$\frac{6}{5}, \quad \frac{11}{8}, \quad \frac{27}{16}$$

are improper fractions.

Many students at all levels, from elementary school through college, find it difficult to understand improper fractions. How do we make sense of the notion of $\frac{3}{2}$ of a candy bar? According to the definition, $\frac{3}{2}$ of a candy bar is the amount formed by 3 parts (or 3 copies of parts) when the candy bar is divided into 2 equal parts. In order to conceptualize $\frac{3}{2}$ of a candy bar, one must be able to conceive of a third part that is a copy of the two equal parts that make up the whole candy bar. One must therefore be able to think of the candy bar and its parts as replicable. For example, we must be able to see 5 candy bars as 5 copies of 1 candy bar. So, two ideas are needed to interpret $\frac{3}{2}$ of a candy bar: the idea that half of a candy bar is an entity in its own right and the idea that half a candy bar can be replicated 3 times, even though only 2 of those copies are present in one full candy bar.

In some realistic situations, where an object is unique, an improper fraction of that object may not make sense. This probably contributes to the difficulty of understanding improper fractions. If Mama bakes one special pie that she's never baked before and may never bake again, does it really make sense to talk about $\frac{9}{8}$ of Mama's pie? On the other hand, if we go to a pie shop where there are many identical cherry pies, the idea of $\frac{9}{8}$ of a cherry pie does make sense. When it is easy to imagine many identical copies of an object, or large amounts of a quantity, such as water, then it becomes easier to understand improper fractions. The idea of $\frac{5}{4}$ of a cup of water may be easier to understand than $\frac{5}{4}$ of a pie because we all experience a nearly limitless supply of water coming from our taps, and one cup of water is virtually identical to any other cup of water.

In Section 3.3 we will see that an excellent way to make sense of improper fractions is with a number line.

CLASS ACTIVITY   NOW TURN TO CLASS ACTIVITIES MANUAL

3D Understanding Improper Fractions p. 32

PRACTICE PROBLEMS FOR SECTION 3.1

1. Show $\frac{1}{8}$ of the combined amount in the 3 pies in Figure 3.3 and explain why your answer is correct.

FIGURE 3.3

Three Pies

2. Hermione has a potion recipe that calls for 4 drams of snake liver oil. Hermione wants to make $\frac{2}{3}$ of the potion recipe. Rather than calculate $\frac{2}{3}$ of the number 4, Hermione measures $\frac{2}{3}$ of a dram of snake liver oil 4 times and uses that amount of snake liver oil in her potion. Use pictures and the meaning of fractions to explain why Hermione's method is valid.

3. This rectangle of x's is $\frac{3}{4}$ of another (original) rectangle of x's.

Show the original rectangle.

4. This rectangle of x's is $\frac{8}{3}$ of another (original) rectangle of x's.

x	x	x	x	x	x	x	x	x	x	x	x	x	x	x	x
x	x	x	x	x	x	x	x	x	x	x	x	x	x	x	x
x	x	x	x	x	x	x	x	x	x	x	x	x	x	x	x

Show the original rectangle.

5. Simone got $\frac{3}{4}$ of a full bar of chocolate. Hank was supposed to get $\frac{1}{4}$ of the full chocolate bar, but he has to get his share from Simone. What fraction of Simone's chocolate should Hank get? Draw a picture that helps you solve this problem. Use your picture to help you explain your solution. For each fraction in this problem, and in your solution, describe the whole associated with this fraction. In other words, describe what each fraction is *of*.

6. If one serving of juice gives you $\frac{3}{2}$ of your daily value of vitamin C, how much of your daily value of vitamin C will you get in $\frac{2}{3}$ of a serving of juice? Draw a picture that helps you solve this problem. Use your picture to help you explain your solution. For each fraction in this problem, and in your solution, describe the whole associated with this fraction. In other words, describe what each fraction is *of*.

7. Explain why you can use the fractions in (a), (b), (c), and (d) to describe the shaded region in Figure 3.4. How can four different numbers describe the same shaded region? How must you interpret the shaded region in each case?

a. $\frac{3}{4}$ b. $\frac{3}{8}$ c. $\frac{3}{1}$ d. $\frac{3}{2}$

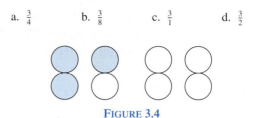

FIGURE 3.4

What Fraction Describes the
Shaded Region?

8. Divide the shaded shape in Figure 3.5 into 4 equal parts in two different ways: one where all parts have the same size and shape, and one where all parts have the same area, but some parts do not have the same shape.

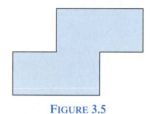

FIGURE 3.5

Divide into 4 Equal Parts

ANSWERS TO PRACTICE PROBLEMS FOR SECTION 3.1

1. If you shade $\frac{1}{8}$ of each pie individually, as in Figure 3.6, then collectively, the 3 shaded parts are $\frac{1}{8}$ of the 3 pies combined. Why? Because 8 of these combined parts make up the whole three pies.

FIGURE 3.6

$\frac{1}{8}$ of Three Pies

2. If we represent the 4 drams of snake liver oil with drawings of 4 jars, as at the top of Figure 3.7, then the shaded regions in the middle of Figure 3.7 represent $\frac{1}{3}$ of the 4 drams of snake liver oil because 3 copies of these shaded regions make the full 4 drams. Therefore, 2 copies of the shaded regions in the middle of Figure 3.7 form $\frac{2}{3}$ of the 4 drams of snake liver oil, as shown at the bottom of Figure 3.7.

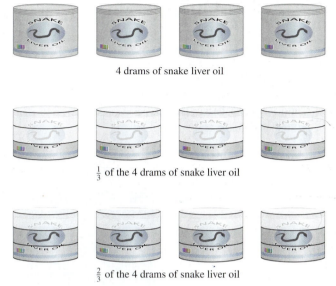

4 drams of snake liver oil

$\frac{1}{3}$ of the 4 drams of snake liver oil

$\frac{2}{3}$ of the 4 drams of snake liver oil

FIGURE 3.7

Snake Liver Oil

3. Because the rectangle shown in the practice problem is $\frac{3}{4}$ of another rectangle, it must consist of 3 parts of the other rectangle, when the other rectangle is divided into 4 equal parts. So the rectangle must consist of 3 equal parts, as shown in the following diagram:

x	x	x	x
x	x	x	x
x	x	x	x
x	x	x	x
x	x	x	x
x	x	x	x

The original rectangle that we are looking for must consist of 4 of these parts:

```
x   x   x   x
x   x   x   x
x   x   x   x
x   x   x   x
x   x   x   x
x   x   x   x
x   x   x   x
x   x   x   x
```

4. Because the rectangle shown in the practice problem is $\frac{8}{3}$ of another rectangle, it must consist of 8 parts of the other rectangle, when the other rectangle is divided into 3 equal parts. So the rectangle shown in the practice problem must consist of 8 equal parts, as shown in the following diagram:

x x	x x	x x	x x	x x	x x	x x	x x
x x	x x	x x	x x	x x	x x	x x	x x
x x	x x	x x	x x	x x	x x	x x	x x

The original rectangle must consist of 3 of these parts as shown in the following diagram:

$$x \quad x \quad x \quad x \quad x \quad x$$
$$x \quad x \quad x \quad x \quad x \quad x$$
$$x \quad x \quad x \quad x \quad x \quad x$$

5. As Figure 3.8 shows, we can divide the full chocolate bar into 4 equal parts, and Simone has 3 of those 4 parts. One of Simone's 3 parts should go to Hank. Therefore Hank should receive $\frac{1}{3}$ of Simone's chocolate.

full bar of chocolate

Simone's chocolate

$\frac{1}{4}$ of the full bar
of chocolate is $\frac{1}{3}$
of Simone's chocolate

FIGURE 3.8

$\frac{1}{4}$ of a Bar of Chocolate Is $\frac{1}{3}$ of
Simone's Chocolate

Notice that the 1 part that is to go to Hank is $\frac{1}{4}$ of the full bar of chocolate, but it is also $\frac{1}{3}$ of Simone's chocolate. In the problem, the full bar of chocolate is the whole associated with the fractions $\frac{3}{4}$ and $\frac{1}{4}$; in other words, both fractions are *of* the full bar of chocolate. But Simone's chocolate is the whole associated with the answer, $\frac{1}{3}$; in other words, the fraction $\frac{1}{3}$ is *of* Simone's chocolate.

6. Because 1 serving of juice provides $\frac{3}{2}$ of the daily value of vitamin C, one serving of juice represents 3 parts, when the daily value of vitamin C is divided into 2 equal parts. The daily value of vitamin C is represented in 2 of those parts, as shown in Figure 3.9. If you drink $\frac{2}{3}$ of a serving of juice, you will get those 2 parts of the daily value of vitamin C, so you will get the full daily value of vitamin C in $\frac{2}{3}$ of a serving of juice.

There are two different wholes associated with the fractions in this problem: a serving of juice and the daily value of vitamin C. The whole associated with the $\frac{3}{2}$ in the practice problem is the "daily value of vitamin C." The whole associated with the $\frac{2}{3}$ in the practice problem is "a serving of juice."

1 serving of juice

daily value of vitamin C

FIGURE 3.9

One Serving of Juice Provides $\frac{3}{2}$ of the
Daily Value of Vitamin C

7. We can use different numbers to describe the shaded part in Figure 3.4 by choosing different wholes. If you take the whole to be 4 circles in Figure 3.4, then because the whole consists of 4 equal parts and because 3 of those parts are shaded, the shaded part is $\frac{3}{4}$ of the chosen whole.

If you take the whole to be 8 circles, then because the whole consists of 8 equal parts and because 3 of those parts are shaded, the shaded part is $\frac{3}{8}$ of the chosen whole.

If you take the whole to be 1 circle, then because the whole consists of 1 part and because 3 (copies) of those parts are shaded, the shaded part is $\frac{3}{1}$ of the chosen whole.

If you take the whole to be 2 circles, then because the whole consists of 2 equal parts and because 3 (copies) of those parts are shaded, the shaded part is $\frac{3}{2}$ of the chosen whole.

8. See Figure 3.10.

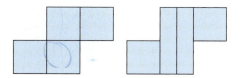

FIGURE 3.10

Two Ways to Divide the Shape
into 4 Equal Parts

PROBLEMS FOR SECTION 3.1

1. Michael says that the dark marbles in Figure 3.11 can't represent $\frac{1}{3}$ because "there are 5 marbles and 5 is more than 1 but $\frac{1}{3}$ is supposed to be less than 1." In a short paragraph, discuss the source of Michael's confusion and discuss what Michael must learn about fractions in order to overcome his confusion.

FIGURE 3.11

Dark and Light Marbles

2. a. Draw a copy of the two rectangles in Figure 3.12. Shade $\frac{1}{5}$ of the combined area of the two rectangles. Use the meaning of fractions to explain why your shaded portion is $\frac{1}{5}$ of the combined area of the two rectangles.

   b. Draw a copy of the two rectangles in Figure 3.12. Shade $\frac{2}{5}$ of the combined area of the two rectangles. Use the meaning of fractions to explain why your

shaded portion is $\frac{2}{5}$ of the combined area of the two rectangles.

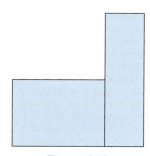

FIGURE 3.12

Two Rectangles

3. When Jean was asked to say what the 3 in the fraction $\frac{2}{3}$ means, Jean said that the 3 is the whole. Explain why it is not completely correct to say that "3 is the whole." What is a better way to say what the 3 in the fraction $\frac{2}{3}$ means?

4. Harry wants to make $\frac{3}{4}$ of a recipe of a potion. The full recipe calls for 2 vials full of newt blood. Instead of calculating $\frac{3}{4}$ of 2, Harry measures $\frac{3}{4}$ of a vial of newt

blood 2 times and uses this amount to make $\frac{3}{4}$ of his potion recipe. Is Harry's method valid? Using the meaning of fractions, explain why or why not.

5. Figure 3.13 shows a picture of three pies. Draw a copy of the three pies. Shade $\frac{3}{8}$ of the combined amount in the three pies and use the meaning of fractions to explain why your answer is correct.

FIGURE 3.13

Three Pies

6. This rectangle of x's is $\frac{4}{5}$ of another (original) rectangle of x's.

$$
\begin{array}{cccc}
x & x & x & x \\
x & x & x & x \\
x & x & x & x \\
x & x & x & x \\
x & x & x & x
\end{array}
$$

Show the original rectangle. Explain why your solution is correct.

7. This rectangle of x's is $\frac{6}{5}$ of another (original) rectangle of x's.

$$
\begin{array}{ccccc}
x & x & x & x & x \\
x & x & x & x & x \\
x & x & x & x & x \\
x & x & x & x & x \\
x & x & x & x & x \\
x & x & x & x & x
\end{array}
$$

Show the original rectangle. Explain why your solution is correct.

8. Kaitlyn gave $\frac{1}{2}$ of her candy bar to Arianna. Arianna gave $\frac{1}{3}$ of the candy she got from Kaitlyn to Cameron. What fraction of a candy bar did Cameron get? Draw a picture that helps you solve this problem. Use your picture to help you explain your solution. For each fraction in this problem and in your solution, describe the whole associated with this fraction. In other words, describe what each fraction is *of*.

9. One fourth of the beads in Maya's bead collection are blue. Of the blue beads, $\frac{1}{2}$ are small. What fraction of

Maya's bead collection consists of small, blue beads? Draw a picture that helps you solve this problem. Use your picture to help you explain your solution. For each fraction in this problem and in your solution, describe the whole associated with this fraction. In other words, describe what each fraction is *of*.

10. You were supposed to use $\frac{2}{3}$ of a cup of cocoa to make a batch of cookies, but you only have $\frac{1}{3}$ of a cup of cocoa. What fraction of the cookie recipe can you make with your $\frac{1}{3}$ cup of cocoa (assuming you have enough of the other ingredients)? Draw a picture that helps you solve this problem. Use your picture to help explain your solution. For each fraction in this problem and in your solution, describe the whole associated with this fraction. In other words, describe what each fraction is *of*.

11. Susan was supposed to use $\frac{5}{4}$ of a cup of butter in her recipe but she only used $\frac{3}{4}$ of a cup of butter. What fraction of the butter that she should have used did Susan actually use? Draw a picture that helps you solve this problem. Use your picture to help you explain your solution. For each fraction in this problem and in your solution, describe the whole associated with this fraction. In other words, describe what each fraction is *of*.

12. If $\frac{3}{4}$ of a cup of a snack food gives you your daily value of calcium, then what fraction of your daily value of calcium is in 1 cup of the snack food? Draw a picture that helps you solve this problem. Use your picture to help you explain your solution. For each fraction in this problem and in your solution, describe the whole associated with this fraction. In other words, describe what each fraction is *of*.

13. A container is filled with $\frac{5}{2}$ of a cup of cottage cheese. What fraction of the cottage cheese in the container should you eat if you want to eat 1 cup of cottage cheese? Draw a picture that helps you solve this problem. Use your picture to help you explain your solution. For each fraction in this problem and in your solution, describe the whole associated with this fraction. In other words, describe what each fraction is *of*.

14. Give three different fractions that you can legitimately use to describe the shaded region in Figure 3.14. For each fraction, explain why you can use that fraction to describe the shaded region. Write an unambiguous question about the shaded region in Figure 3.14 that can be answered by naming a fraction. Explain why your question is not ambiguous.

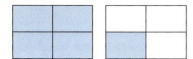

FIGURE 3.14

Describe the Shaded Part
with a Fraction

15. Give three different fractions that you can legitimately use to describe the shaded region in Figure 3.15. For each fraction, explain why you can use that fraction to describe the shaded region. Write an unambiguous question about the shaded region in Figure 3.15 that can be answered by naming a fraction. Explain why your question is not ambiguous.

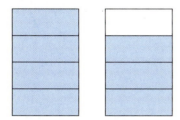

FIGURE 3.15

Describe the Shaded Part
with a Fraction

16. Marquez says that the shaded region in Figure 3.16 represents the fraction $\frac{9}{12}$. Carmina says the shaded region in Figure 3.16 represents the fraction $\frac{9}{4}$. Meili says the shaded region in Figure 3.16 represents 9. Explain why each of the three children's answers can be considered correct. Then write an unambiguous question about the shaded region in Figure 3.16 that can be answered by naming a fraction. Explain why your question is not ambiguous.

FIGURE 3.16

A Shaded Region

17. Make up a story problem or situation where *one* object (or collection, or quantity) is *both* $\frac{1}{2}$ of something and $\frac{1}{3}$ of something else.

18. Discuss which of the items on the list that follows are good, and which are not as good, for showing improper fractions. Explain your choices.

string

a cake

a box of cereal and some cup measures

apples

something else—your choice

19. NanHe made a design that used hexagons, rhombuses, and triangles like the ones shown in Figure 3.17. NanHe counted how many of each shape she used in her design and determined that $\frac{4}{11}$ of the shapes she used were hexagons, $\frac{5}{11}$ were rhombuses, and $\frac{2}{11}$ were triangles. You may combine your answers to parts (a) and (b).

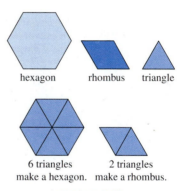

hexagon   rhombus   triangle

6 triangles   2 triangles
make a hexagon.   make a rhombus.

FIGURE 3.17

Pattern Blocks

a. Even though $\frac{4}{11}$ of the shapes NanHe used were hexagons, does this mean that the hexagons in NanHe's design take up $\frac{4}{11}$ of the area of her design? If not, what fraction of the area of NanHe's design do the hexagons take up?

b. Write a short paragraph describing how the notion of "equal parts" relates to your answer in part (a) of this problem.

20. A county has two elementary schools, both of which have an after school program for the convenience of working parents. In one school, $\frac{1}{3}$ of the children attend the after school program; in the other school, $\frac{2}{3}$ of the children attend the after school program. There are a total of 1500 children in the two schools combined.

a. If the first school has 600 students and the second school has 900 students, then what is the fraction of elementary school students in the county who attend the after school program? What if it's the other way around and the first school has 900 students and the second has 600 students?

b. Using the data in the problem statement, make up two more examples of numbers of students in each school. (These don't have to be entirely realistic numbers.) For each of your examples, determine the fraction of elementary school children in the county who attend the after school program.

c. Jamie says that if $\frac{1}{3}$ of the children in the first school attend the after school program, and if $\frac{2}{3}$ of the children in the second school attend the after school program, then $\frac{1}{2}$ of the children from the two schools combined attend the after school program. Jamie arrives at this answer because she says $\frac{1}{2}$ is half-way between $\frac{1}{3}$ and $\frac{2}{3}$. Is Jamie's answer of $\frac{1}{2}$ *always* correct? Is Jamie's answer of $\frac{1}{2}$ *ever* correct, if so, under what circumstances?

3.2 Equivalent Fractions

When we write a whole number in decimal notation, there is only one way to do so without the use of a decimal point. The only way to write 1234 as a decimal number without a decimal point is 1234. But in the case of fractions, every fraction is equal to infinitely many other fractions. For example,

$$\frac{1}{2} = \frac{2}{4} = \frac{3}{6} = \frac{4}{8} = \frac{5}{10} = \cdots$$

In general,

$$\frac{A}{B} = \frac{A \cdot 2}{B \cdot 2} = \frac{A \cdot 3}{B \cdot 3} = \frac{A \cdot 4}{B \cdot 4} = \frac{A \cdot 5}{B \cdot 5} = \cdots$$

In this section we will see why every fraction is equal to infinitely many other fractions, and we will study some consequences of this fact.

CLASS ACTIVITY   NOW TURN TO CLASS ACTIVITIES MANUAL

3E Equivalent Fractions p. 33

3F Misconceptions about Fraction Equivalence p. 35

WHY EVERY FRACTION IS EQUAL TO INFINITELY MANY OTHER FRACTIONS

To explain why every fraction is equal to infinitely many other fractions, let's start with an example. Why is

$$\frac{3}{4}$$

of a pie the same amount of pie as

$$\frac{3 \cdot 5}{4 \cdot 5} = \frac{15}{20}$$

of the same pie? We can use the meaning of fractions of objects to explain this equivalence. First, $\frac{3}{4}$ of the pie is the amount formed by 3 parts when the pie is divided into 4 equal parts.

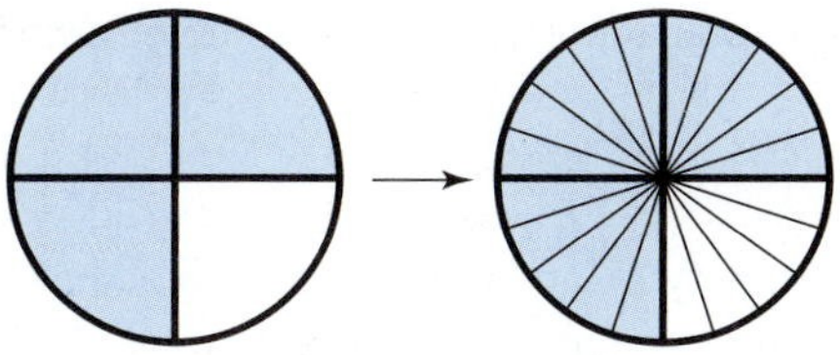

FIGURE 3.18

Subdivide Each
Part into 5 Parts

Divide each of those 4 equal parts into 5 small, equal parts, as shown in Figure 3.18. Now, the pie consists of a total of $4 \cdot 5 = 20$ small parts. The 3 parts representing $\frac{3}{4}$ of the pie have each been subdivided into 5 equal parts, therefore these 3 parts have become $3 \cdot 5 = 15$ smaller parts, as shown in the shaded portions of Figure 3.18. So 3 of the original 4 parts of pie is the same amount of pie as $3 \cdot 5$ smaller parts of the total $4 \cdot 5$ smaller parts. Therefore

$$\frac{3}{4}$$

of a pie is the same amount of pie as

$$\frac{3 \cdot 5}{4 \cdot 5} = \frac{15}{20}$$

of the pie.

There wasn't anything special about the numbers 3, 4, and 5 used here—the same reasoning will apply to other counting numbers substituted for these numbers. Therefore, in general, if N is any counting number,

$$\frac{A}{B}$$

of a pie (or any other object or collection of objects) is the same amount of pie (or object or collection) as

$$\frac{A \cdot N}{B \cdot N}$$

of the pie (or object or collection). In other words,

$$\frac{A}{B} = \frac{A \cdot N}{B \cdot N}$$

Another way to explain why

$$\frac{A}{B} = \frac{A \cdot N}{B \cdot N}$$

is to multiply by 1 in the form of $\frac{N}{N}$:

$$\frac{A}{B} = \frac{A}{B} \cdot 1$$

$$= \frac{A}{B} \cdot \frac{N}{N}$$

$$= \frac{A \cdot N}{B \cdot N}$$

This explanation for why $\frac{A}{B} = \frac{A \cdot N}{B \cdot N}$ relies on the conceptually more complex concept of fraction multiplication, whereas the previous explanation relies only on the meaning of fractions.

COMMON DENOMINATORS

Because every fraction is equal to infinitely many other fractions, you have a lot of flexibility when you work with fractions. When you start with a fraction, say, $\frac{3}{4}$, you know that it is always equal to many other fractions with larger denominators—for example,

$$\frac{3}{4} = \frac{3 \cdot 7}{4 \cdot 7} = \frac{21}{28}$$

$$\frac{3}{4} = \frac{3 \cdot 25}{4 \cdot 25} = \frac{75}{100}$$

The fractions $\frac{21}{28}$ and $\frac{75}{100}$ are just two of infinitely many fractions equal to $\frac{3}{4}$.

In some cases, a fraction is equal to another fraction with a smaller denominator:

$$\frac{8}{12} = \frac{2 \cdot 4}{3 \cdot 4} = \frac{2}{3}$$

$$\frac{170}{190} = \frac{17 \cdot 10}{19 \cdot 10} = \frac{17}{19}$$

common
denominators

When working with two fractions simultaneously, it is often desirable to give them **common denominators**, which just means the *same* denominators. When we give fractions common denominators, we describe the fractions in terms of like parts. The fractions $\frac{3}{5}$ and $\frac{2}{7}$ are in terms of fifths and sevenths, respectively. We can write these two fractions with the common denominator $5 \cdot 7 = 35$ as follows:

$$\frac{3}{5} = \frac{3 \cdot 7}{5 \cdot 7} = \frac{21}{35}$$

$$\frac{2}{7} = \frac{2 \cdot 5}{7 \cdot 5} = \frac{10}{35}$$

As you see in Figure 3.19, when we give $\frac{3}{5}$ and $\frac{2}{7}$ common denominators, we subdivide the fifths of $\frac{3}{5}$ and the sevenths of $\frac{2}{7}$ into thirty-fifths, so that both fractions are described with like parts.

The fractions $\frac{3}{5}$ and $\frac{2}{7}$ have many other common denominators as well, such as 70 and 105, but 35 is the smallest one.

For any two fractions, multiplying the denominators always produces a common denominator; however, it may not be the least common denominator. The number 24 is a common denominator for the fractions $\frac{3}{8}$ and $\frac{5}{6}$, whereas $8 \cdot 6 = 48$. We will discuss least common denominators again in Section 12.2.

CLASS ACTIVITY   NOW TURN TO CLASS ACTIVITIES MANUAL

3G Common Denominators p. 36

3H Solving Problems by Changing Denominators p. 37

FIGURE 3.19

FIGURE 3.19

Common
Denominators
Give Fractions
Like Parts

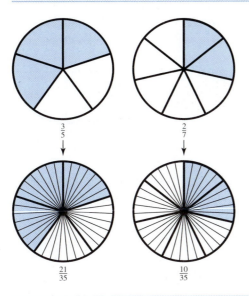

$$\frac{3}{5} \qquad \frac{2}{7}$$

$$\frac{21}{35} \qquad \frac{10}{35}$$

THE SIMPLEST FORM OF A FRACTION

Every fraction is equal to infinitely many other fractions, but in a collection of fractions that are equal to each other, there is one that is the simplest.

simplest form   A fraction of whole numbers $\frac{A}{B}$ (where B is not zero) is said to be in **simplest form** (or in **lowest terms**) if there is no whole number other than 1 that divides both A and B evenly.

The fraction

$$\frac{3}{4}$$

is in simplest form because no whole number other than 1 divides both 3 and 4 evenly. The fraction

$$\frac{30}{35}$$

is not in simplest form because the number 5 divides both 30 and 35 evenly. Notice, however, that we can put the fraction $\frac{30}{35}$ in simplest form as follows:

$$\frac{30}{35} = \frac{6 \cdot 5}{7 \cdot 5} = \frac{6}{7}$$

To put a fraction in simplest form, we essentially reverse the process of giving the fraction a new, larger denominator so that it will have a denominator in common with another fraction. For example, if we wanted to change the fraction

$$\frac{6}{7}$$

so that is has a denominator in common with the fraction $\frac{4}{5}$, then we would rewrite $\frac{6}{7}$ as follows:

$$\frac{6}{7} = \frac{6 \cdot 5}{7 \cdot 5} = \frac{30}{35}$$

FIGURE 3.20

Putting $\frac{18}{24}$ in
Simplest Form

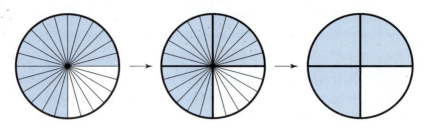

Notice that this process of creating an equivalent fraction with a larger denominator is the reverse of the process described earlier that puts $\frac{30}{35}$ in simplest form.

In general, if

$$\frac{A}{B}$$

is a fraction of whole numbers that is not in simplest form, then we can put it in simplest form as follows: First, we find the largest whole number N that divides both A and B evenly. We write A and B in the form $C \cdot N$ and $D \cdot N$, respectively, where C and D are whole numbers. Then

$$\frac{A}{B} = \frac{C \cdot N}{D \cdot N} = \frac{C}{D}$$

and $\frac{C}{D}$ is the simplest form of $\frac{A}{B}$.

To write $\frac{18}{24}$ in simplest form, notice that 6 is the largest whole number that divides both 18 and 24 evenly. Because

$$18 = 3 \cdot 6$$

and

$$24 = 4 \cdot 6$$

we have

$$\frac{18}{24} = \frac{3 \cdot 6}{4 \cdot 6} = \frac{3}{4}$$

Thus, $\frac{3}{4}$ is the simplest form of $\frac{18}{24}$.

If you think in terms of pies, putting a fraction in simplest form is like joining together small pie pieces to make larger pie pieces, as shown in Figure 3.20. In other words, putting a fraction in simplest form reverses the procedure of subdividing pie pieces that was demonstrated earlier in Figure 3.18.

We can also use several steps to put a fraction in simplest form. For example, to put $\frac{18}{24}$ in simplest form, we don't have to know right away that 6 is the largest whole number that divides both 18 and 24 evenly. We might notice instead that 18 and 24 are both evenly divisible by 2 and, therefore, that

$$\frac{18}{24} = \frac{9 \cdot 2}{12 \cdot 2} = \frac{9}{12}$$

Then we notice that 9 and 12 are both evenly divisible by 3, so

$$\frac{9}{12} = \frac{3 \cdot 3}{4 \cdot 3} = \frac{3}{4}$$

Now, 3 and 4 are not evenly divisible by any whole number except 1, so once again, we have shown that the simplest form of $\frac{18}{24}$ is $\frac{3}{4}$.

We will discuss putting fractions in simplest form again in Section 12.2 when we discuss greatest common factors.

CLASS ACTIVITY   NOW TURN TO CLASS ACTIVITIES MANUAL

3I   Simplifying Fractions p. 39

3J   When Can We "Cancel" to Get an Equivalent Fraction? p. 40

PRACTICE PROBLEMS FOR SECTION 3.2

1. Use the meaning of fractions to explain why

$$\frac{4}{5} = \frac{4 \cdot 3}{5 \cdot 3} = \frac{12}{15}$$

2. Write $\frac{3}{8}$ and $\frac{5}{6}$ with common denominators in three different ways.

3. Kelsey has $\frac{3}{5}$ of a chocolate bar. Kelsey wants to give some of her chocolate bar to Janelle. What fraction of her chocolate bar should Kelsey give Janelle so that Kelsey will be left with $\frac{1}{2}$ of the original chocolate bar?

   a. Draw pictures to help you solve this problem. Explain why your answer is correct.

   b. In solving this problem, how do $\frac{3}{5}$ and $\frac{1}{2}$ appear in different forms?

   c. What are the different wholes associated with the fractions in this problem? In other words, for each fraction in this problem and its solution, what is the fraction *of*?

4. Ken ordered $\frac{3}{4}$ of a ton of gravel. Ken wants $\frac{1}{4}$ of his order of gravel delivered now and $\frac{3}{4}$ delivered later. What fraction of a ton of gravel should Ken get delivered now?

   a. Draw pictures to help you solve this problem. Explain why your answer is correct.

b. In solving the problem, how does $\frac{3}{4}$ appear in a different form?

c. What are the different wholes associated with the fractions in this problem? In other words, for each fraction in this problem and its solution, what is the fraction *of*?

5. Use the diagram in Figure 3.21 to help you explain why the following equations that put $\frac{10}{15}$ in simplest form make sense:

$$\frac{10}{15} = \frac{2 \cdot 5}{3 \cdot 5} = \frac{2}{3}$$

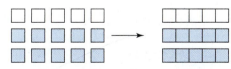

FIGURE 3.21
Explaining Why $\frac{10}{15} = \frac{2}{3}$

6. Put the following fractions in simplest form:

$$\frac{45}{72}, \quad \frac{24}{36}, \quad \frac{56}{88}$$

1. If you have a pie that is divided into 5 equal parts, and 4 are shown shaded, then if you divide each of the 5 parts into 3 smaller parts, the shaded amount will then consist of $4 \cdot 3$ small parts, and the whole pie will consist of $5 \cdot 3$ small parts. Thus, the shaded part of the pie can be described both as $\frac{4}{5}$ of the pie and as

    $$\frac{4 \cdot 3}{5 \cdot 3} = \frac{12}{15}$$

    of the pie.

2. There are infinitely many ways to write $\frac{3}{8}$ and $\frac{5}{6}$ with common denominators. Here are three:

    $$\frac{3}{8} = \frac{3 \cdot 6}{8 \cdot 6} = \frac{18}{48},$$
    $$\frac{5}{6} = \frac{5 \cdot 8}{6 \cdot 8} = \frac{40}{48}.$$

    $$\frac{3}{8} = \frac{3 \cdot 3}{8 \cdot 3} = \frac{9}{24},$$
    $$\frac{5}{6} = \frac{5 \cdot 4}{6 \cdot 4} = \frac{20}{24}.$$

    $$\frac{3}{8} = \frac{3 \cdot 9}{8 \cdot 9} = \frac{27}{72},$$
    $$\frac{5}{6} = \frac{5 \cdot 12}{6 \cdot 12} = \frac{60}{72}.$$

3. a. See Figure 3.22. If the original chocolate bar is divided into 10 equal parts, then the amount of chocolate that Kelsey has is $\frac{6}{10}$ of the original chocolate. The amount that Kelsey wants to keep is $\frac{5}{10}$ of the original chocolate, so Kelsey should give 1 part out of her 6 parts to Janelle. Therefore Janelle should get $\frac{1}{6}$ of Kelsey's chocolate.

    original chocolate bar
    Kelsey's chocolate

    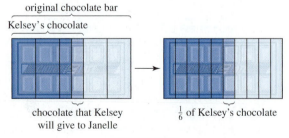

    chocolate that Kelsey $\frac{1}{6}$ of Kelsey's chocolate
    will give to Janelle

    FIGURE 3.22

    What Fraction of Kelsey's Chocolate
    Will Janelle Get?

b. In solving this problem, $\frac{3}{5}$ becomes

    $$\frac{3 \cdot 2}{5 \cdot 2} = \frac{6}{10}$$

    and $\frac{1}{2}$ becomes

    $$\frac{1 \cdot 5}{2 \cdot 5} = \frac{5}{10}$$

c. The chocolate bar is one of the wholes in this problem—Kelsey's chocolate is $\frac{3}{5}$ of this whole, as is the $\frac{1}{2}$ that Kelsey wants to keep. Kelsey's chocolate is also a whole in this problem because we describe the amount of chocolate that Janelle will get as a fraction of this whole.

4. a. In Figure 3.23, 1 ton of gravel is represented by a circle, and Ken's order is represented by $\frac{3}{4}$ of a circle. On the right, $\frac{1}{4}$ of Ken's order is represented by dividing each of the 3 parts in the $\frac{3}{4}$ into 4 equal small pieces and by picking one small piece from each of the 3 parts. These small pieces are shown shaded darkly. There are 16 small pieces in the full circle, and the 3 shaded darkly represent $\frac{1}{4}$ of Ken's order. Therefore, $\frac{1}{4}$ of Ken's order is $\frac{3}{16}$ of a ton.

    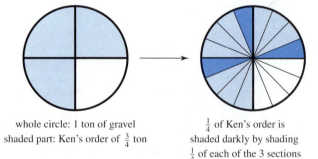

    whole circle: 1 ton of gravel $\frac{1}{4}$ of Ken's order is
    shaded part: Ken's order of $\frac{3}{4}$ ton   shaded darkly by shading
                                           $\frac{1}{4}$ of each of the 3 sections

    FIGURE 3.23

    What Fraction of a Ton Should
    Ken Get Now?

b. In solving this problem, the $\frac{3}{4}$ of a ton of gravel also appears as

    $$\frac{3 \cdot 4}{4 \cdot 4} = \frac{12}{16}$$

    of a ton of gravel.

c. A full ton of gravel is one of the wholes in this problem. Ken's order of gravel ($\frac{3}{4}$ of a ton) is another whole, because in this problem we determine what $\frac{1}{4}$ of this whole is as a fraction of 1 ton of gravel.

5. If we consider the collection of squares on the left of Figure 3.21, we see that 10 out of 15 of these squares are shaded. Therefore, $\frac{10}{15}$ of the squares on the left are shaded. The squares are in groups of 5. Viewing the squares in groups, we see that $2 \cdot 5$ out of $3 \cdot 5$ squares are shaded, so we can also say that $\frac{2 \cdot 5}{3 \cdot 5}$ of the squares are shaded. If we move the squares together horizontally as shown on the right of Figure 3.21, we see that the squares form 3 groups and that 2 of those groups are shaded. So $\frac{2}{3}$ of the squares are shaded. Therefore, we have three ways of writing the fraction of squares that are shaded, and these three ways must all be equal, so

$$\frac{10}{15} = \frac{2 \cdot 5}{3 \cdot 5} = \frac{2}{3}$$

6.
$$\frac{45}{72} = \frac{5 \cdot 9}{8 \cdot 9} = \frac{5}{8}$$

$$\frac{24}{36} = \frac{2 \cdot 12}{3 \cdot 12} = \frac{2}{3}$$

$$\frac{56}{88} = \frac{7 \cdot 8}{11 \cdot 8} = \frac{7}{11}$$

Notice that we could also simplify each of these fractions in several steps. For example,

$$\frac{45}{72} = \frac{15 \cdot 3}{24 \cdot 3} = \frac{15}{24} = \frac{5 \cdot 3}{8 \cdot 3} = \frac{5}{8}$$

PROBLEMS FOR SECTION 3.2

1. Use the meaning of fractions to explain why

$$\frac{3}{4} = \frac{3 \cdot 3}{4 \cdot 3}$$

Do not use multiplication by 1. Draw a picture to support your explanation.

2. Use the meaning of fractions to explain why

$$\frac{6}{5} = \frac{6 \cdot 2}{5 \cdot 2}$$

Do not use multiplication by 1. Draw a picture to support your explanation.

3. Explain in two different ways why

$$\frac{2}{3} = \frac{2 \cdot 4}{3 \cdot 4}$$

4. Explain in two different ways why

$$\frac{5}{6} = \frac{5 \cdot 2}{6 \cdot 2}$$

5. Using the fractions $\frac{2}{3}$ and $\frac{3}{4}$, describe how to give two fractions common denominators. In terms of pictures, what are you doing when you give fractions common denominators?

6. Using the fractions $\frac{5}{6}$ and $\frac{3}{8}$, describe how to give two fractions common denominators. In terms of pictures, what are you doing when you give fractions common denominators?

7. Using the example

$$\frac{6}{9}$$

describe how to put a fraction in simplest form and explain why this procedure makes sense. In terms of pictures, what are you doing when you put a fraction in simplest form?

8. Using the example

$$\frac{12}{16} = \frac{3}{4} \cdot \frac{4}{4} =$$

describe how to put a fraction in simplest form and explain why this procedure makes sense. In terms of pictures, what are you doing when you put a fraction in simplest form?

9. Ted put $\frac{3}{4}$ of a cup of chicken stock in his soup. Later, Ted added another $\frac{2}{3}$ of a cup of chicken stock to his soup. All together, what fraction of a cup of chicken stock did Ted put into his soup?

   a. Draw pictures to help you solve this problem. Explain why your answer is correct.

b. In solving the problem, how do $\frac{3}{4}$ and $\frac{2}{3}$ appear in different forms?

c. What are the different wholes associated with the fractions in this problem? In other words, for each fraction in this problem and its solution, what is the fraction *of*?

10. First you poured $\frac{3}{4}$ of a cup of water into an empty bowl. Then you scooped out $\frac{1}{3}$ of a cup of water. How much water is left in the bowl?

a. Draw pictures to help you solve this problem. Explain why your answer is correct.

b. In solving the problem, how do $\frac{3}{4}$ and $\frac{1}{3}$ appear in different forms?

c. What are the different wholes associated with the fractions in this problem? In other words, for each fraction in this problem and its solution, what is the fraction *of*?

11. First you poured $\frac{3}{4}$ of a cup of water into an empty bowl. Then you scooped out $\frac{1}{2}$ of the water in the bowl. How much water is left in the bowl?

a. Draw pictures to help you solve this problem. Explain why your answer is correct.

b. In solving the problem, how does $\frac{3}{4}$ appear in a different form?

c. What are the different wholes associated with the fractions in this problem? In other words, for each fraction in this problem and its solution, what is the fraction *of*?

12. You want to make a recipe that calls for $\frac{2}{3}$ of a cup of flour, but you only have $\frac{1}{2}$ of a cup of flour left. Assuming you have enough of the other ingredients, what fraction of the recipe can you make?

a. Draw pictures to help you solve this problem. Explain why your answer is correct.

b. In solving the problem, how do $\frac{2}{3}$ and $\frac{1}{2}$ appear in different forms?

c. What are the different wholes associated with the fractions in this problem? In other words, for each fraction in this problem and its solution, what is the fraction *of*?

13. Ken got $\frac{2}{3}$ of a ton of gravel even though he ordered $\frac{3}{4}$ of a ton of gravel. What fraction of his order did Ken get?

a. Draw pictures to help you solve this problem. Explain why your answer is correct.

b. In solving this problem, how do $\frac{2}{3}$ and $\frac{3}{4}$ appear in different forms?

c. What are the different wholes associated with the fractions in this problem? In other words, for each fraction in this problem and its solution, what is the fraction *of*?

14. Two thirds of a cup of Healthy SnackOs provides your full daily value of vitamin B. You ate $\frac{1}{2}$ of a cup of Healthy SnackOs. What fraction of your daily value of vitamin B did you get in the Healthy SnackOs you ate?

a. Draw pictures to help you solve this problem. Explain why your answer is correct.

b. In solving this problem, how do $\frac{2}{3}$ and $\frac{1}{2}$ appear in different forms?

c. What are the different wholes associated with the fractions in this problem? In other words, for each fraction in this problem and its solution, what is the fraction *of*?

15. Your cookie recipe calls for $\frac{2}{3}$ of a cup of butter for a batch of cookies. You decide that you only want to make $\frac{1}{3}$ of a batch of cookies. How much butter will you need?

a. Draw pictures to help you solve this problem. Explain why your answer is correct.

b. In solving the problem, how does $\frac{2}{3}$ appear in a different form?

c. What are the different wholes associated with the fractions in this problem? In other words, for each fraction in this problem and its solution, what is the fraction *of*?

16. One serving of DietMuck is $\frac{2}{3}$ of a cup. Jane wants to eat $\frac{2}{3}$ of a serving of DietMuck. What fraction of a cup of DietMuck should Jane eat?

a. Draw pictures to help you solve this problem. Explain why your answer is correct.

b. In solving the problem, how does $\frac{2}{3}$ appear in a different form?

c. What are the different wholes associated with the fractions in this problem? In other words, for each fraction in this problem and its solution, what is the fraction *of*?

17. If one serving of DietDelights provides $\frac{1}{3}$ of your daily value of vitamin C, then what fraction of your daily value of vitamin C is in $\frac{3}{4}$ of a serving of DietDelights?

a. Draw pictures to help you solve this problem. Explain why your answer is correct.

b. In solving the problem, how does $\frac{1}{3}$ appear in a different form?

c. What are the different wholes associated with the fractions in this problem? In other words, for each fraction in this problem and its solution, what is the fraction *of* ?

18. Becky moves into an apartment with two friends on August 1. Becky's friends have been in the apartment since July 1. The electric bill comes every two months, and the next one will be for the electricity used in July and August. The bill is not broken down by month. What fraction of the July/August electric bill should Becky pay and what fraction should her two friends pay if they want to divide the bill fairly?

a. Solve Becky's electric bill problem in the case where the apartment has a large communal area that is frequently used by all the friends, the friends typically eat meals together, and not much electricity is used in the separate sleeping areas. Explain your reasoning.

b. Solve Becky's electric bill problem in the case where the friends spend most of their time in their separate rooms, don't eat meals together, and don't spend much time together in their communal area. Explain your reasoning.

3.3   Fractions as Numbers

So far, we have only discussed fractions of objects, collections, or quantities, but we haven't yet looked at fractions abstractly, as numbers in their own right, which is our objective in this section.

We create the notion of the whole numbers by abstracting from our experiences with objects. For example,

$$2 \text{ apples}, \quad 7 \text{ balls}, \quad 25 \text{ people}, \ldots$$

is really like saying (somewhat awkwardly)

$$2 \textit{ of } \text{apple}, \quad 7 \textit{ of } \text{ball}, \quad 25 \textit{ of } \text{person}, \ldots$$

which abstracts to the following notion of number:

$$2, \quad 7, \quad 25, \ldots$$

In the same way, we create the following notion of fractions as numbers by abstracting from fractions of objects:

$$\frac{2}{3} \textit{ of } \text{a pie}, \quad \frac{11}{10} \textit{ of } \text{an acre of land}, \quad \frac{7}{8} \textit{ of } \text{the population of the United States}, \ldots$$

These abstract to

$$\frac{2}{3}, \quad \frac{11}{10}, \quad \frac{7}{8}, \ldots$$

But even when fractions are viewed abstractly as numbers, they are still "of a whole." Just as 5 is "five ones," so, too, $\frac{3}{4}$ is "$\frac{3}{4}$ of 1."

By viewing fractions as locations on number lines we will be able to see even more clearly how fractions are numbers.

FIGURE 3.24

FIGURE 3.24

Plotting $\frac{5}{3}$ on a
Number Line

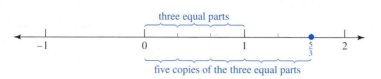

FIGURE 3.25

Showing $\frac{5}{3}$ with
Pieces of
Rectangles

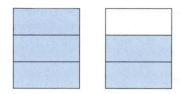

FRACTIONS ON NUMBER LINES

We will now see how fractions can be plotted on a number line in a natural way.

Any fraction $\frac{A}{B}$, where A and B are whole numbers and B is not zero can be located on a number line according to the rule for locating decimal numbers on number lines given in Section 2.3. Thus, $\frac{A}{B}$ is located to the right of zero at a distance of $\frac{A}{B}$ units from 0. Therefore, $\frac{A}{B}$ is at the right end of a line segment whose left end is at 0 and whose length is $\frac{A}{B}$ of one unit.

Where is $\frac{5}{3}$ on a number line? The fraction $\frac{5}{3}$ is at the right end of a line sement whose left end is at 0 and whose length is $\frac{5}{3}$ of a unit. To construct such a line segment, use the meaning of $\frac{5}{3}$: divide the line segment from 0 to 1 into 3 equal parts, and make a line segment consisting of 5 copies of those parts. Place this five-part line segment on the number line so that its left endpoint is at 0, as shown in Figure 3.24. Then the right endpoint of the five-part line segment is at the fraction $\frac{5}{3}$.

Number lines are especially good for showing improper fractions. When we show improper fractions with pieces of pie, or pieces of some other object, it is often unclear which whole the fraction refers to. For example, does the shaded region in Figure 3.25 represent $\frac{5}{3}$ or $\frac{5}{6}$? We can't say unless we know what the fraction is supposed to be *of*. The shaded region is $\frac{5}{3}$ of the three-piece rectangle, but it is also $\frac{5}{6}$ of the two rectangles. In contrast, on number lines, the whole is always the line segment between 0 and 1, so no ambiguity arises.

CLASS ACTIVITY   NOW TURN TO CLASS ACTIVITIES MANUAL

3K Improper Fractions Revisited p. 41

3L Plotting Fractions on Number Lines p. 42

3M Fractions of Line Segments p. 43

DECIMAL REPRESENTATIONS OF FRACTIONS

Because fractions can be plotted on number lines, and because every location on a number line corresponds to a decimal number, it must be possible to represent fractions as decimal numbers. In fact, it is easy to write a fraction $\frac{A}{B}$ as a decimal number. Just divide A by B as follows:

$$\frac{A}{B} = A \div B$$

We will see why it makes sense to write a fraction in decimal notation by dividing in Chapter 7. So,

$$\frac{5}{16} = 5 \div 16 = .3125$$

$$\frac{1}{12} = 1 \div 12 = .083333\ldots$$

$$\frac{2}{7} = 2 \div 7 = .285714285714\ldots$$

Since a fraction can be written as decimal number by dividing the numerator by the denominator, it has become common practice to write a fraction when one actually means division. In other words, it is common to write

$$\frac{A}{B} \quad \text{or} \quad A/B$$

to mean

$$A \div B$$

In fact, some calculators use the fraction symbol / instead of $\div$.

PRACTICE PROBLEMS FOR SECTION 3.3

1. According to the text, why are number lines especially good for showing improper fractions?

2. Plot 0, 1, and $\frac{8}{7}$ on a number line like the one in Figure 3.26 in such a way that each number falls on a tick mark. Lengthen the tick marks of whole numbers.

FIGURE 3.26

Plot Fractions on This Number Line

3. Plot 1, $\frac{5}{3}$, and $\frac{7}{4}$ on a number line like the one in Figure 3.26 in such a way that each number falls on a tick mark. Lengthen the tick marks of whole numbers.

4. Plot 1, .9, and $\frac{5}{4}$ on a number line like the one in Figure 3.26 in such a way that each number falls on a tick mark. Lengthen the tick marks of whole numbers.

5. The line segment in Figure 3.27 has length $\frac{3}{7}$ unit. Explain how to subdivide the line segment and possibly add pieces to the line segment to create a line segment

of length $\frac{1}{7}$ unit. Do this without first creating a segment of length 1 unit. Explain how you know the resulting line segment is the correct length.

$\frac{3}{7}$ units

FIGURE 3.27

Create a Line Segment of Length $\frac{1}{7}$ Units Given That This Line Segment Has Length $\frac{3}{7}$ Units

6. The line segment in Figure 3.28 has length $\frac{9}{8}$ unit. Explain how to subdivide the line segment and possibly add pieces to the line segment to create a line segment of length $\frac{3}{4}$ unit. Do this without first creating a segment of length 1 unit. Explain how you know the resulting line segment is the correct length.

$\frac{9}{8}$ units

FIGURE 3.28

Create a Line Segment of Length $\frac{3}{4}$
Units Given That This Line Segment
Has Length $\frac{9}{8}$ Units

7. The line segment in Figure 3.29 has length $\frac{3}{5}$ unit. Explain how to subdivide the line segment and possibly add pieces to the line segment to create a line segment of length $\frac{2}{3}$ unit. Do this without first creating a segment of length 1 unit. Explain how you know the resulting line segment is the correct length.

$\frac{3}{5}$ units

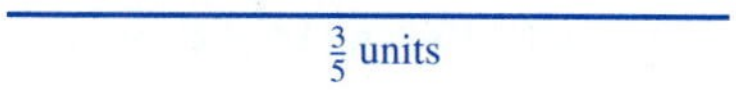

FIGURE 3.29

Create a Line Segment of Length $\frac{2}{3}$
Units Given That This Line Segment
Has Length $\frac{3}{5}$ Units

8. Plot $\frac{11}{8}$ on a number line like the one in Figure 3.30.

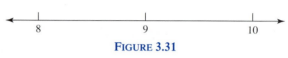

FIGURE 3.30

A Number Line

9. Plot $\frac{29}{3}$ on a number line like the one in Figure 3.31.

FIGURE 3.31

A Number Line

ANSWERS TO PRACTICE PROBLEMS FOR SECTION 3.3

1. See page 78.

2. To plot $\frac{8}{7}$, we need to work with sevenths. Therefore we must divide the 1-unit segment from 0 to 1 into 7 equal parts, as shown in Figure 3.32.

FIGURE 3.32

Plotting 0, 1, and $\frac{8}{7}$

3. Because

$$\frac{5}{3} = \frac{5 \cdot 4}{3 \cdot 4} = \frac{20}{12}$$

$$\frac{7}{4} = \frac{7 \cdot 3}{4 \cdot 3} = \frac{21}{12}$$

and

$$1 = \frac{12}{12}$$

the numbers $\frac{5}{3}$, $\frac{7}{4}$ and 1 can be plotted as shown in Figure 3.33.

FIGURE 3.33

Plotting 1, $\frac{5}{3}$, and $\frac{7}{4}$

4. Because

$$.9 = \frac{9}{10} = \frac{9 \cdot 2}{10 \cdot 2} = \frac{18}{20}$$

$$\frac{5}{4} = \frac{5 \cdot 5}{4 \cdot 5} = \frac{25}{20}$$

and

$$1 = \frac{20}{20}$$

the numbers 1, .9 and $\frac{5}{4}$ can be plotted as shown in Figure 3.34.

FIGURE 3.34

Plotting 1, .9, and $\frac{5}{4}$

---

5. Because the line segment is $\frac{3}{7}$ of a unit long, it must consist of 3 parts, each of which is $\frac{1}{7}$ of a unit long. Therefore, if the line segment is divided into 3 equal parts, one of those parts will be $\frac{1}{7}$ of a unit long, as shown in Figure 3.35.

FIGURE 3.35

From $\frac{3}{7}$ Units to $\frac{1}{7}$ Units

6. Because the original line segment is $\frac{9}{8}$ units long, it consists of 9 equal parts, each of which is $\frac{1}{8}$ of a unit long.

Since

$$\frac{3}{4} = \frac{3 \cdot 2}{4 \cdot 2} = \frac{6}{8}$$

we should create a line segment consisting of 6 of the $\frac{1}{8}$ unit long parts, as shown in Figure 3.36.

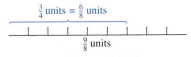

FIGURE 3.36

From $\frac{9}{8}$ Units to $\frac{3}{4}$ Units

---

7. Because

$$\frac{3}{5} = \frac{3 \cdot 3}{5 \cdot 3} = \frac{9}{15}$$

and

$$\frac{2}{3} = \frac{2 \cdot 5}{3 \cdot 5} = \frac{10}{15}$$

the original line segment of length $\frac{3}{5}$ consists of 9 equal parts, each of which is $\frac{1}{15}$ of a unit long, and the desired line segment of length $\frac{2}{3}$ units will consist 10 copies of those parts, as shown in Figure 3.37.

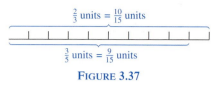

FIGURE 3.37

From $\frac{3}{5}$ Units to $\frac{2}{3}$ Units

8. To plot $\frac{11}{8}$ on a number line, divide the 1-unit segment from 0 to 1 into 8 equal parts, and measure off 11 of those parts, as shown in Figure 3.38.

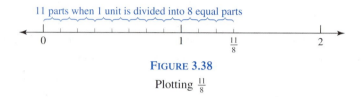

FIGURE 3.38

Plotting $\frac{11}{8}$

9. The fraction $\frac{29}{3}$ is located 29 parts away from 0, where each part is $\frac{1}{3}$ of a unit long (i.e., 3 parts are 1 unit long). Nine sets of 3 parts will use up 27 parts, and these 27 parts will end at 9 on the number line. Two more parts will get to the location of $\frac{29}{3}$ on the number line, as shown in Figure 3.39.

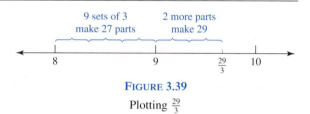

FIGURE 3.39

Plotting $\frac{29}{3}$

PROBLEMS FOR SECTION 3.3

1. Discuss why it can be confusing to show an improper fraction such as $\frac{7}{3}$ with pieces of pie or the like. What is another way to show the fraction $\frac{7}{3}$?

2. Draw a number line like the one in Figure 3.40. Then plot 0, 1, $\frac{4}{3}$, and $\frac{3}{2}$ on your number line in such a way that each number falls on a tick mark. Lengthen the tick marks of whole numbers.

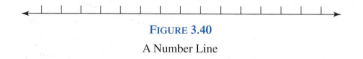

FIGURE 3.40

A Number Line

3. Draw a number line like the one in Figure 3.40. Then plot 0, $\frac{3}{8}$, and $\frac{1}{3}$ on your number line in such a way that each number falls on a tick mark. Lengthen the tick marks of whole numbers.

4. Draw a number line like the one in Figure 3.40. Then plot $\frac{1}{2}$, $\frac{3}{5}$, and $\frac{5}{8}$ on your number line in such a way that each number falls on a tick mark. Lengthen the tick marks of whole numbers (if there are any).

5. Draw a number line like the one in Figure 3.40. Then plot 0, 0.3, and $\frac{3}{5}$ on your number line in such a way

that each number falls on a tick mark. Lengthen the tick marks of whole numbers.

6. Draw a number line like the one in Figure 3.40. Then plot 1, $\frac{5}{6}$, and $\frac{8}{7}$ on your number line in such a way that each number falls on a tick mark. Lengthen the tick marks of whole numbers.

7. The next line segment has length $\frac{2}{7}$ unit. Draw a line segment like this one. Explain how to subdivide your line segment and possibly add pieces to your line segment to create a line segment of length $\frac{1}{3}$ unit. Do this

without first creating a segment of length 1 unit. Explain how you know your line segment is the correct length.

_____

$\frac{2}{7}$ unit

8. The next line segment has length $\frac{4}{5}$ unit. Draw a line segment like this one. Explain how to subdivide your line segment and possibly add pieces to your line segment to create a line segment of length $\frac{1}{2}$ unit. Do this without first creating a segment of length 1 unit. Explain how you know your line segment is the correct length.

_____

$\frac{4}{5}$ unit

9. The line segment below has length $\frac{3}{5}$ unit. Draw a line segment like this one. Explain how to subdivide your line segment and possibly add pieces to your line segment to create a line segment of length $\frac{1}{2}$ unit. Do this without first creating a segment of length 1 unit. Explain how you know your line segment is the correct length.

_____

$\frac{3}{5}$ unit

10. Draw a number line like the one in Figure 3.41. Plot $\frac{13}{3}$ on your number line.

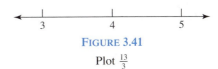

FIGURE 3.41

Plot $\frac{13}{3}$

11. Draw a number line like the one in Figure 3.42. Plot $\frac{47}{6}$ on your number line.

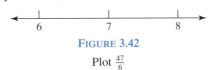

FIGURE 3.42

Plot $\frac{47}{6}$

12. Erin says the tick mark indicated in Figure 3.43 should be labeled 2.2. Is Erin right or not? If not, why not, and how can she label the tick mark properly?

Erin says to label this tick mark 2.2.

FIGURE 3.43

How to Label the Tick Mark

13. Liam says the tick mark indicated in Figure 3.44 should be labeled 1.7. Is Liam right or not? If not, why not, and how can the tick mark be labeled properly?

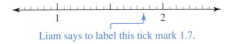

Liam says to label this tick mark 1.7.

FIGURE 3.44

How to Label the Tick Mark

3.4 Comparing Sizes of Fractions

Given two numbers in decimal representation, we can determine which one is greater by comparing the digits of the decimal numbers. However, comparing the sizes of fractions is more complicated because every fraction is equal to infinitely many other fractions. Unlike whole numbers, we can't always tell just by comparing digits of fractions whether or not they are equal, or whether one is greater than the other. You will probably recognize that a familiar fraction such as $\frac{1}{2}$ is equal to $\frac{3}{6}$ or $\frac{50}{100}$, but can you tell right away, just by looking, that

$$\frac{411}{885} \quad \text{and} \quad \frac{548}{1180}$$

are equal? It is not obvious.

There are three standard methods for determining whether two fractions are equal, or if not, which one is greater. The three methods for comparing fractions are as follows:

1. Converting to decimal numbers

2. Using common denominators

3. Cross-multiplication

COMPARING FRACTIONS BY CONVERTING TO DECIMAL NUMBERS

Every fraction can be converted to a decimal number by dividing the denominator into the numerator. Therefore, we can compare two fractions simply by converting both fractions to decimal numbers and comparing the decimal numbers.

Is $\frac{17}{35}$ equal to $\frac{43}{87}$, or if not, which is greater?

$$\frac{17}{35} = 17 \div 35 = .4857\ldots$$

$$\frac{43}{87} = 43 \div 87 = .4942\ldots$$

Since

$$.4942\ldots > .4857\ldots$$

it follows that

$$\frac{43}{87} > \frac{17}{35}$$

COMPARING FRACTIONS BY USING COMMON DENOMINATORS

When two fractions have the same denominator, you can determine whether they are equal, or if not, which is greater, just by looking at them. The fractions

$$\frac{784}{953} \quad \text{and} \quad \frac{621}{953}$$

have the same denominator. Since

$$784 > 621$$

it follows that

$$\frac{784}{953} > \frac{621}{953}$$

In general, if two fractions have the same denominator, then the one with the greater numerator is greater; if the numerators and denominators are equal, then the fractions are equal.

Why does this rule for comparing fractions with like denominators make sense? Say we have two fractions that have the same denominator. If we think of the two fractions as fractions of two identical pies, then each pie is divided into the same number of pieces. Therefore, each piece of pie in the two pies is the same size. The numerator describes the number of pieces we have of the pie. Since the pie pieces are all the same size, if we have more pieces of one pie, then we have more of that pie. See Figure 3.45. If we have the same number of pieces of each pie, then we have the same amount of each pie. Therefore, when two fractions have the same denominators, the fraction with the greater numerator is greater; and when two fractions have the same denominator and the same numerator, the fractions are equal.

But what if you have two fractions that don't have the same denominator? How can you find out if they are equal, or if they are not, how can you find out which fraction is greater? You can give the two fractions a common denominator. It can be any common denominator—it

FIGURE 3.45

Comparing
Fractions with
Equal
Denominators

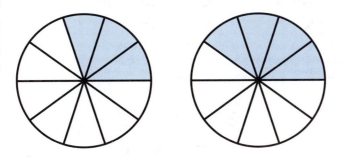

does not have to be the least one. A common denominator that always works is obtained by multiplying the two denominators. Consider the previous two fractions,

$$\frac{411}{885} \quad \text{and} \quad \frac{548}{1180}$$

We can give these fractions the common denominator

$$885 \times 1180$$

Then

$$\frac{411}{885} = \frac{411 \cdot 1180}{885 \cdot 1180} = \frac{484980}{1044300}$$

and

$$\frac{548}{1180} = \frac{548 \cdot 885}{1180 \cdot 885} = \frac{484980}{1044300}$$

The two fractions $\frac{411}{885}$ and $\frac{548}{1180}$ are equal because when both are written with the denominator $885 \cdot 1180 = 1044300$, they have the same numerator.

What are we really doing when we give two fractions common denominators in order to compare the sizes of the fractions? As we saw in Section 3.2, when we give fractions common denominators, we create *like parts*. For example, to compare the fractions

$$\frac{4}{9} \quad \text{and} \quad \frac{3}{5}$$

we can give both fractions the common denominator of $9 \cdot 5 = 45$ as follows:

$$\frac{4}{9} = \frac{4 \cdot 5}{9 \cdot 5} = \frac{20}{45}$$

$$\frac{3}{5} = \frac{3 \cdot 9}{5 \cdot 9} = \frac{27}{45}$$

As shown in Figure 3.46, the process of giving the fractions common denominators corresponds to subdividing ninths and fifths into pieces of the same size, namely forty-fifths. The denominators are now the same, and the numerator 27 is greater than the numerator 20. Therefore,

$$\frac{27}{45} > \frac{20}{45}$$

FIGURE 3.46

Comparing $\frac{4}{9}$ and $\frac{3}{5}$

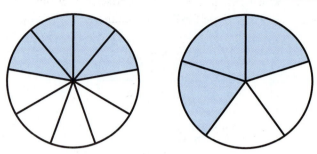

When we give fractions common denominators,
we subdivide to create like parts:

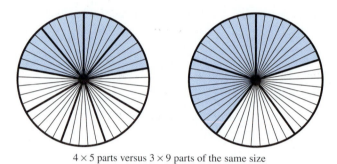

4×5 parts versus 3×9 parts of the same size

so that

$$\frac{3}{5} > \frac{4}{9}$$

COMPARING FRACTIONS BY CROSS-MULTIPLYING

Notice that to show that the two fractions $\frac{411}{885}$ and $\frac{548}{1180}$ are equal, all we needed to check was that

$$411 \cdot 1180 = 548 \cdot 885$$

because $411 \cdot 1180$ and $548 \cdot 885$ were the numerators when we gave $\frac{411}{885}$ and $\frac{548}{1180}$ a common de-nominator. So, all we had to do was to multiply the numerator of each fraction $\frac{411}{885}$ and $\frac{548}{1180}$ by the denominator of the other fraction and check whether those two resulting numbers were equal.

Similarly, to show that

$$\frac{3}{5} > \frac{4}{9}$$

all we needed to check was that

$$3 \cdot 9 > 4 \cdot 5$$

because $3 \cdot 9$ and $4 \cdot 5$ are the numerators when we give $\frac{4}{9}$ and $\frac{3}{5}$ the common denominator of $9 \cdot 5$.

Generalizing the above, if $\frac{A}{B}$ and $\frac{C}{D}$ are two fractions of whole numbers (where neither B nor D is zero), then we can give these fractions the common denominator $B \cdot D$:

$$\frac{A}{B} = \frac{A \cdot D}{B \cdot D}$$

$$\frac{C}{D} = \frac{C \cdot B}{D \cdot B}$$

To compare $\frac{A}{B}$ and $\frac{C}{D}$, we can compare

$$\frac{A \cdot D}{B \cdot D}$$

and

$$\frac{C \cdot B}{D \cdot B}$$

instead. Because the denominators $D \cdot B$ and $B \cdot D$ are equal, we only need to compare the numerators

$$A \cdot D \quad \text{and} \quad C \cdot B$$

We conclude that

- The fractions $\frac{A}{B}$ and $\frac{C}{D}$ are equal exactly when $A \cdot D$ and $C \cdot B$ are equal.
- $\frac{A}{B}$ is greater than $\frac{C}{D}$ exactly when $A \cdot D$ is greater than $C \cdot B$.
- $\frac{A}{B}$ is less than $\frac{C}{D}$ exactly when $A \cdot D$ is less than $C \cdot B$.

Stated symbolically,

$$\frac{A}{B} = \frac{C}{D} \quad \text{exactly when} \quad A \cdot D = C \cdot B$$

$$\frac{A}{B} > \frac{C}{D} \quad \text{exactly when} \quad A \cdot D > C \cdot B$$

$$\frac{A}{B} < \frac{C}{D} \quad \text{exactly when} \quad A \cdot D < C \cdot B$$

cross-multiplying   This method for checking if two fractions are equal, or if not, which one is greater, is often called the **cross-multiplying** method. Notice that it is just a way to check if the numerators are equal when the two fractions are given the common denominator obtained by multiplying the two original denominators.

USING OTHER REASONING TO COMPARE FRACTIONS

The standard methods just described for comparing fractions are useful because they are efficient and they always work, but sometimes it is possible to use other reasoning to compare fractions. Why would we want to use other reasoning when we already have several good methods? In some cases, other reasoning can be more efficient. Even more importantly, however, reasoning in other ways can help us think about the meaning of fractions and can help us understand fractions better.

Consider the case of determining whether $\frac{3}{5}$ and $\frac{4}{9}$ are equal, and if not, which is larger. We can reason that $\frac{4}{9}$ is less than $\frac{1}{2}$,

$$\frac{4}{9} < \frac{1}{2}$$

because if we divide an object into 9 equal pieces, then it would take four and a half pieces to make half of the object, so 4 pieces is certainly less than half. Similarly, we can say that

$$\frac{1}{2} < \frac{3}{5}$$

because if we divide an object into 5 equal pieces, then two and a half of them would make half the object, so three pieces is more than half the object. Since $\frac{4}{9}$ is less than a half, and $\frac{3}{5}$ is more than a half, $\frac{3}{5}$ is the larger fraction.

Which fraction is greater,

$$\frac{5}{8} \quad \text{or} \quad \frac{5}{9}?$$

Both fractions represent 5 parts, but $\frac{5}{8}$ is 5 parts when the whole is divided into 8 parts, whereas $\frac{5}{9}$ is 5 parts when the whole is divided into 9 parts. Now, if an object is divided into 8 equal parts then each part is larger than if the object is divided into 9 equal parts—fewer parts making up the same whole means each part has to be larger. So 5 of 8 equal parts must be greater than 5 of 9 equal parts. Thus,

$$\frac{5}{8} > \frac{5}{9}$$

as shown in Figure 3.47.

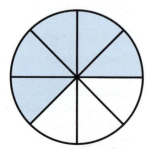 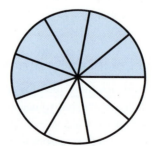

Both have 5 shaded parts, but 8ths are bigger than 9ths.

> ## CLASS ACTIVITY   NOW TURN TO CLASS ACTIVITIES MANUAL
>
> **3N** Can We Compare Fractions This Way? p. 45
>
> ---
>
> **3O** Comparing Fractions by Reasoning p. 47
>
> ---
>
> **3P** Rules For Comparing Fractions p. 48

PRACTICE PROBLEMS FOR SECTION 3.4

1. Compare the sizes of the following pairs of fractions in two ways: by cross-multiplying and by giving the fractions common denominators. How are these two methods related?

$$\frac{2}{3} \text{ and } \frac{3}{5}$$

$$\frac{8}{13} \text{ and } \frac{13}{21}$$

$$\frac{15}{20} \text{ and } \frac{6}{8}$$

$$\frac{1}{4} \text{ and } \frac{3}{8}$$

$$\frac{4}{14} \text{ and } \frac{5}{15}$$

2. Complete the following to make true statements about comparing fractions:

   a. If two fractions have the same denominators, then the one with .... is greater.

   b. If two fractions have the same numerator, then the one with .... is greater.

   Use the meaning of fractions to explain why your answers make sense.

3. For each of the following pairs of fractions, determine which is larger. Use reasoning other than finding common denominators, cross-multiplying, or converting to decimal representation in order to explain your answer.

   a. $\frac{6}{11}$ versus $\frac{6}{13}$

   b. $\frac{5}{8}$ versus $\frac{7}{12}$

   c. $\frac{5}{21}$ versus $\frac{7}{24}$

   d. $\frac{21}{22}$ versus $\frac{56}{57}$

   e. $\frac{97}{100}$ versus $\frac{35}{38}$

4. Explain why

   $$\frac{A}{B} > \frac{C}{D} \quad \text{exactly when} \quad A \cdot D > C \cdot B$$

5. Without using a calculator, order the following numbers from largest to smallest: $\frac{41}{20}, 2, \frac{11}{5}, -2, -2.3, -\frac{19}{10}$. Explain your reasoning.

6. Find a decimal number between $\frac{21}{34}$ and $\frac{34}{55}$ and plot all three numbers visibly and distinctly on a number line in which the distance between adjacent tick marks is a power of ten, such as 0.1, 0.01, 0.001, and so on. The numbers need not land on tick marks.

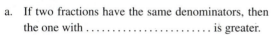

ANSWERS TO PRACTICE PROBLEMS FOR SECTION 3.4

1.

FRACTIONS	COMMON DENOMINATOR	CROSS-MULTIPLYING	CONCLUSION
$\frac{2}{3}$ $\frac{3}{5}$	$\frac{2 \cdot 5}{3 \cdot 5} > \frac{3 \cdot 3}{5 \cdot 3}$	$2 \cdot 5 > 3 \cdot 3$	$\frac{2}{3} > \frac{3}{5}$
$\frac{8}{13}$ $\frac{13}{21}$	$\frac{8 \cdot 21}{13 \cdot 21} < \frac{13 \cdot 13}{21 \cdot 13}$	$8 \cdot 21 < 13 \cdot 13$	$\frac{8}{13} < \frac{13}{21}$
$\frac{15}{20}$ $\frac{6}{8}$	$\frac{15 \cdot 8}{20 \cdot 8} = \frac{6 \cdot 20}{8 \cdot 20}$	$15 \cdot 8 = 6 \cdot 20$	$\frac{15}{20} = \frac{6}{8}$
$\frac{1}{4}$ $\frac{3}{8}$	$\frac{1 \cdot 8}{4 \cdot 8} < \frac{3 \cdot 4}{8 \cdot 4}$	$1 \cdot 8 < 3 \cdot 4$	$\frac{1}{4} < \frac{3}{8}$
$\frac{4}{14}$ $\frac{5}{15}$	$\frac{4 \cdot 15}{14 \cdot 15} < \frac{5 \cdot 14}{15 \cdot 14}$	$4 \cdot 15 < 5 \cdot 14$	$\frac{4}{14} < \frac{5}{15}$

Notice that to compare $\frac{15}{20}$ and $\frac{6}{8}$ with the common denominator method, you could choose a smaller common denominator than $20 \cdot 8$. For instance, you could choose 40 as the common denominator. In this case, you compare the numerators $2 \cdot 15$ and $5 \cdot 6$ rather than comparing $8 \cdot 15$ and $20 \cdot 6$, which is what you compare with the cross-multiplying method. You can use a similar procedure with $\frac{1}{4}$ and $\frac{3}{8}$. Of course, it's easiest just to use the common denominator 8 in that case.

When we give two fractions the common denominator obtained by multiplying the two denominators, the numerators are exactly the quantities we compare when we cross multiply. Therefore, the cross-multiplying method is really just a shortcut for comparing two fractions by giving them the common denominator that is the product of the two denominators.

2. a. If two fractions have the same denominators, then the one with the greater numerator is greater. Think of the two fractions as fractions you have eaten of two identical pies. Since the two fractions have the same denominator, the number of pieces that both pies were divided into is the same. Because the pie pieces are the same size, you have eaten more of the pie from which you ate the largest number of pieces. Therefore, you have eaten the greater fraction of pie for the fraction that has the greater numerator.

   b. If two fractions have the same numerator, then the one with the smaller denominator is greater. Think of the two fractions as fractions you have eaten of two identical pies. The fraction with the smaller denominator corresponds to the pie that was divided into fewer pieces. In the pie that was divided into fewer pieces, each piece is bigger than in the other pie. For example, the pieces in a 6-piece pie are bigger than the pieces in an 8-piece pie—fewer pieces making up the same size pie means that each piece must be bigger, as seen in Figure 3.48. Because the numerators of the two fractions are the same, you have eaten the same number of pieces from each pie. So, you have eaten the most pie from the pie that was divided into fewer pieces. Therefore, you have eaten the greater fraction of pie for the fraction that has the smaller denominator.

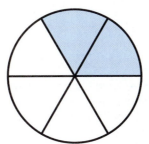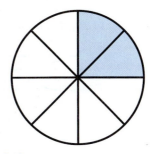

FIGURE 3.48

Two Pieces Are Eaten in Each, but the Pies
Have Different Size Pieces

---

3. a. $\frac{6}{11} > \frac{6}{13}$ If you divide a pie into 13 pieces, then each piece will be smaller than if you divide that same pie into 11 pieces. So, 6 pieces of a pie that is divided into 11 pieces will be more than 6 pieces of a pie that is divided into 13 pieces.

    b. $\frac{5}{8}$ is one piece more than half ($\frac{4}{8}$), and $\frac{7}{12}$ is one piece more than half ($\frac{6}{12}$). However, when a pie is divided into 8 pieces, each piece is larger than when an identical pie is divided into 12 pieces—fewer pieces making up the same amount means each piece must be larger. The eighths pieces are larger than the twelfths pieces, and since each fraction, $\frac{5}{8}$ and $\frac{7}{12}$, is one piece more than one half, it follows that $\frac{5}{8} > \frac{7}{12}$.

    c. Notice that both fractions are close to $\frac{1}{4}$. The fractions $\frac{1}{4}$ and $\frac{5}{20}$ are equal; therefore, $\frac{5}{21}$ is a little less than $\frac{1}{4}$. (See the reasoning in the previous answer.) The fractions $\frac{1}{4}$ and $\frac{6}{24}$ are equal; therefore, $\frac{7}{24}$ is a little bigger than $\frac{1}{4}$ (See the reasoning in the previous answer.) So, $\frac{5}{21} < \frac{7}{24}$.

    d. Notice that each fraction is "one piece less than 1 whole." The fraction $\frac{21}{22}$ is $\frac{1}{22}$ less than a whole, and $\frac{56}{57}$ is $\frac{1}{57}$ less than a whole. But $\frac{1}{22}$ is bigger than $\frac{1}{57}$ because if you divide a pie into 22 pieces, each piece will be bigger than if you divide an identical pie into 57 pieces. Therefore $\frac{21}{22} < \frac{56}{57}$ because $\frac{21}{22}$ is a bigger piece away from a whole than is $\frac{56}{57}$.

    e. The same reasoning used in the answer to the previous part applies here, too. This time, each fraction is 3 pieces away from 1 whole.

4. To compare two fractions $\frac{A}{B}$ and $\frac{C}{D}$, we can give them the common denominator $B \cdot D = D \cdot B$. Then

$$\frac{A}{B} = \frac{A \cdot D}{B \cdot D}$$

and

$$\frac{C}{D} = \frac{C \cdot B}{D \cdot B}$$

Since the fractions $\frac{A \cdot D}{B \cdot D}$ and $\frac{C \cdot B}{D \cdot B}$ have the same denominator, the fraction $\frac{A \cdot D}{B \cdot D}$ is greater than $\frac{C \cdot B}{D \cdot B}$ exactly when the numerator $A \cdot D$ is greater than the numerator $C \cdot B$. This in turn means that the original fraction $\frac{A}{B}$ is greater than $\frac{C}{D}$ exactly when $A \cdot D$ is greater than $C \cdot B$. In other words,

$$\frac{A}{B} > \frac{C}{D}$$

exactly when

$$A \cdot D > C \cdot B$$

5. Every positive number is greater than every negative number, so we know that $\frac{41}{20}$, 2, and $\frac{11}{5}$ are greater than -2, -2.3, and $-\frac{19}{10}$. We can give $\frac{41}{20}$, 2, and $\frac{11}{5}$ the common denominator of 20 by letting

$$2 = \frac{2}{1} = \frac{2 \cdot 20}{1 \cdot 20} = \frac{40}{20}$$

and

$$\frac{11}{5} = \frac{11 \cdot 4}{5 \cdot 4} = \frac{44}{20}$$

Comparing the numerators, we see that

$$\frac{11}{5} > \frac{41}{20} > 2$$

Since $\frac{19}{10} = 1.9$, and since

$$2.3 > 2 > 1.9$$

it follows that

$$-1.9 > -2 > -2.3$$

Therefore,

$$\frac{11}{5} > \frac{41}{20} > 2 > -1.9 > -2 > -2.3$$

6. The decimal representation of $\frac{21}{34}$ is .6176... and the decimal representation of $\frac{34}{55}$ is .6181...; so .618 is one example of a decimal number that is in between $\frac{21}{34}$ and $\frac{34}{55}$. There are many other examples. See Figure 3.49.

$$
\begin{array}{cccc}
.617 & \begin{matrix}\frac{21}{34}\\ = .61764...\end{matrix} & .618 \quad \begin{matrix}\frac{34}{55}\\ = .61818...\end{matrix} & .619
\end{array}
$$

FIGURE 3.49

The Number .618 Is Between $\frac{21}{34}$ and $\frac{34}{55}$

PROBLEMS FOR SECTION 3.4

1. Julie says that the picture in Figure 3.50 shows that

$$\frac{1}{4} > \frac{1}{2}$$

by comparing the areas. Explain carefully the error in Julie's reasoning.

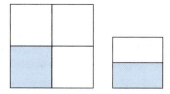

FIGURE 3.50

Is $\frac{1}{4} > \frac{1}{2}$?

2. In your own words, explain in detail why we can determine which of two fractions is greater by giving the two fractions common denominators. What is the rationale behind this method? What are we really doing when we give the fractions common denominators?

3. In your own words, explain in detail why we can determine which of two fractions is greater by using the cross-multiplying method. What is the rationale behind

this method? What are we really doing when we cross-multiply in order to compare fractions?

4. Which fraction is greater, $\frac{43}{75}$ or $\frac{43}{74}$? To explain your answer, use the meaning of fractions along with reasoning other than finding common denominators, cross-multiplying, or converting to decimal numbers.

5. Which fraction is greater, $\frac{35}{109}$ or $\frac{36}{104}$? To explain your answer, use the meaning of fractions along with reasoning other than finding common denominators, cross-multiplying, or converting to decimal numbers.

6. Which fraction is greater, $\frac{15}{31}$ or $\frac{23}{47}$? To explain your answer, use the meaning of fractions along with reasoning other than finding common denominators, cross-multiplying, or converting to decimal numbers.

7. Which fraction is greater, $\frac{19}{20}$ or $\frac{27}{28}$? To explain your answer, use the meaning of fractions along with reasoning other than finding common denominators, cross-multiplying, or converting to decimal numbers.

8. Without using a calculator, order the following numbers from smallest to largest: $\frac{11}{10}$, 1.2, $\frac{9}{8}$, -1.1, -0.98, -1. Explain your reasoning.

9. Find a number between $\frac{23}{84}$ and $\frac{29}{98}$ and plot all three numbers visibly and distinctly on a number line like the one in Figure 3.51, which has a set of longer tick marks and a set of shorter tick marks. The distance between adjacent tick marks should be a power of ten, such as 0.1, 0.01, or 0.001. Label all the longer tick marks. The numbers need not land on tick marks.

FIGURE 3.51

A Number Line

10. Find a number between $\frac{78}{134}$ and $\frac{124}{213}$ and plot all three numbers visibly and distinctly on a number line like the one in Figure 3.51, which has a set of longer tick marks and a set of shorter tick marks. The distance between adjacent tick marks should be a power of ten, such as 0.1, 0.01, or 0.001. Label all the longer tick marks. The numbers need not land on tick marks.

11. Give two different methods for solving the following problem: Find two different fractions in between $\frac{3}{5}$ and $\frac{2}{3}$ whose numerators and denominators are whole numbers.

12. Give two different methods for solving the following problem: Find two different fractions between $\frac{5}{7}$ and $\frac{6}{7}$ whose numerators and denominators are whole numbers.

13. A student says that $\frac{1}{5}$ is halfway between $\frac{1}{4}$ and $\frac{1}{6}$. Use a carefully drawn number line to show that this is not correct. What fraction is halfway between $\frac{1}{4}$ and $\frac{1}{6}$? Explain your reasoning.

14. Minju says that fractions that use bigger numbers are greater than fractions that use smaller numbers. Make up two problems for Minju to help her reconsider her ideas. For each problem, explain how to solve it, and explain why you chose that problem for Minju.

15. Malcolm says that

$$\frac{8}{11} > \frac{7}{10}$$

because

$$8 > 7 \quad \text{and} \quad 11 > 10$$

Even though it is true that $\frac{8}{11} > \frac{7}{10}$, is Malcolm's reasoning correct? If Malcolm's reasoning is correct, clearly explain why. If Malcolm's reasoning is not correct, give Malcolm two examples that show why not.

16. You may combine your answers to all three parts of this problem.

   a. Is it valid to compare

$$\frac{30}{70} \quad \text{and} \quad \frac{20}{50}$$

by "cancelling" the 0s and comparing

$$\frac{3}{7} \quad \text{and} \quad \frac{2}{5}$$

instead? Explain your answer.

   b. Is it valid to compare

$$\frac{15}{25} \quad \text{and} \quad \frac{105}{205}$$

by "cancelling" the 5s and comparing

$$\frac{1}{2} \quad \text{and} \quad \frac{10}{20}$$

instead? Explain your answer.

   c. Write a paragraph discussing the distinction between your answer in (a) and your answer in (b).

17. Consider the following list of fractions:

$$\frac{1}{1}, \quad \frac{2}{1}, \quad \frac{3}{2}, \quad \frac{5}{3}, \quad \frac{8}{5}, \quad \ldots$$

You do not have to explain your answers to the following parts:

   a. Describe a pattern in the list of fractions and use your description to find the next 5 entries in the list after $\frac{8}{5}$. You will now have the first 10 entries in the list of fractions.

   b. Use either the cross-multiplying method or the common denominator method to compare the sizes of the 1st, 3rd, 5th, 7th, and 9th fractions in the list. Describe a pattern in the sizes of these fractions. Describe a pattern that occurs when you compare the fractions.

   c. Use either the cross-multiplying method or the common denominator method to compare the sizes of the 2nd, 4th, 6th, 8th, and 10th fractions in the list. Describe a pattern in the sizes of these fractions. Describe a pattern that occurs when you compare the fractions.

   d. Convert the 10 fractions on your list to decimal numbers, and plot them on a number line. Zoom in on portions of your number line, as in Figure 2.22, so that you can show clearly where each decimal number is plotted relative to the others.

   e. If you could find more and more entries in the list of fractions, and plot them on a number line, in what region of the number line would they be located?

Do you think these numbers would get closer and closer to a particular number?

18. Suppose you start with a proper fraction and you add 1 to both the numerator and the denominator. For example, if you started with $\frac{2}{3}$, then you'd get a new fraction $\frac{2+1}{3+1} = \frac{3}{4}$.

   a. Give at least 5 examples of proper fractions $\frac{A}{B}$. In each example, compare the sizes of $\frac{A}{B}$ and $\frac{A+1}{B+1}$. What do you notice? Make sure you are working with proper fractions (where the numerator is less than the denominator).

   b. Frank says that if $\frac{A}{B}$ is a proper fraction, then $\frac{A+1}{B+1}$ is always greater than $\frac{A}{B}$ because $\frac{A+1}{B+1}$ has more parts. Regardless of whether or not Frank's conclusion is correct, discuss whether or not Frank's reasoning is valid. Did Frank give a convincing explanation that $\frac{A+1}{B+1}$ is greater than $\frac{A}{B}$? If not, what objections could you make to Frank's reasoning?

   c. Explain the phenomenon you discovered in part (a). Compare the sizes of $\frac{A}{B}$ and $\frac{A+1}{B+1}$. Explain why the fraction you say is larger really *is* larger.

3.5 Percent

Almost any time we open a newspaper, walk into a store, or listen to a sports broadcast we encounter percentages. A percentage is a number that expresses a relationship between two other numbers. In this way, percentages are like fractions: just as a fraction is *of* an associated whole, a percentage is *of* some number. In fact, percentages can be viewed as special kinds of fractions, namely those with denominator 100. Every fraction has a decimal representation, therefore every percentage also has a decimal representation.

THE MEANING OF PERCENT

percent   The word **percent**, which is usually represented using the symbol %, means "of each hundred" (per = "of each" or "for each"; cent = "hundred"). For example, 76% means 76 of each hundred. Therefore, to find 76% of 1280, we should see how many hundreds are in 1280, and multiply that number by 76. The number of hundreds in 1280 is

$$1280 \div 100$$

Therefore, 76% of 1280 is

$$76 \times (1280 \div 100)$$

which can also be expressed as

$$\frac{76}{100} \cdot 1280$$

or even

$$.76 \cdot 1280$$

So, 76% of 1280 is 972.8.

Another example of finding a percent of a number follows:

$$3\% \text{ of } 234 = \frac{3}{100} \cdot 234$$
$$= .03 \cdot 234$$
$$= 7.02$$

In general, $P\%$ of a quantity Q is

$$\frac{P}{100} \cdot Q$$

Because $P\%$ of a number is $\frac{P}{100}$ times that number, it makes sense to say that

$$P\% = \frac{P}{100}$$

so that

$$76\% = \frac{76}{100} = .76$$

and

$$125\% = \frac{125}{100} = 1.25$$

We can therefore interpret the word percent to mean *hundredths*, and we can view percents as decimal numbers, or as special kinds of fractions, namely ones that have denominator 100. When we consider percents as fractions, we can use the meaning of fractions, so that 85% of some object means the amount formed by 85 parts when the object is divided into 100 equal parts.

PERCENTS, FRACTIONS, AND PICTURES

Because we frequently encounter certain percents, we highlight them in Table 3.1 along with their equivalent fractions expressed in simplest form. We also represent these percentages in Figure 3.52.

TABLE 3.1

Common Percentages Expressed as Fractions

$$25\% = \frac{25}{100} = \frac{1}{4}, \quad 50\% = \frac{50}{100} = \frac{1}{2}, \quad 75\% = \frac{75}{100} = \frac{3}{4}$$

$$10\% = \frac{10}{100} = \frac{1}{10}, \quad 20\% = \frac{20}{100} = \frac{1}{5}, \quad 5\% = \frac{5}{100} = \frac{1}{20}$$

FIGURE 3.52

Pictures of Percentages

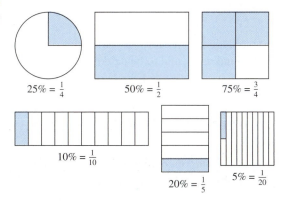

$25\% = \frac{1}{4}$ $50\% = \frac{1}{2}$ $75\% = \frac{3}{4}$

$10\% = \frac{1}{10}$ $20\% = \frac{1}{5}$ $5\% = \frac{1}{20}$

If you are familiar with the fraction representations of the percentages in Table 3.1, you can easily extend them to nearby percentages and fractions as follows:

55% is a little more than $\frac{1}{2}$ and is halfway between 50% and 60%,

or

15% is 10% less than $\frac{1}{4}$.

Some of the class activities, practice problems, and section problems ask you to use pictures to solve percent problems. Although drawing pictures is usually not an efficient way of solving percent problems, pictures can help you develop a better understanding of percents. Just as it would not be reasonable to pull out bundled toothpicks whenever we work with decimal numbers, it would also not be reasonable to do all percent calculations with pictures. However, both physical objects and pictures are valuable learning aids.

CLASS ACTIVITY   NOW TURN TO CLASS ACTIVITIES MANUAL

3Q Pictures and Percentages p. 49

3R Calculating Percentages by Using Nearby Common Fractions p. 51

THREE TYPES OF PERCENT PROBLEMS

A basic relationship involving percents can be put in the form

$$P \text{ percent of } Q \text{ is } R.$$

In equation form, we have

$$P\% \cdot Q = R$$

or

$$\frac{P}{100} \cdot Q = R.$$

Therefore, there are three basic kinds of percent problems: one for each of the cases where one of the amounts P, Q, or R is unknown and to be determined, and the other two amounts are known.

FINDING THE RESULT R WHEN THE PERCENT P AND THE INITIAL QUANTITY Q ARE KNOWN

The most straightforward kind of problem involving percentages is one in which the percent P and the quantity Q are given, and the resulting amount R must be calculated:

Susie must pay 6% tax on her purchase of $43.95. How much tax must Susie pay?

To solve this, we substitute the known values into the percent equation as follows:

$$6\% \cdot \$43.95 = R$$

Thus, Susie must pay

$$.06 \cdot \$43.95 = \$2.64$$

in tax.

FINDING THE PERCENT P WHEN THE INITIAL QUANTITY Q AND THE RESULT R ARE KNOWN

In some problems, the initial amount Q and the resulting amount R are known, but the percent P must be calculated:

John earned $47,600 last year and gave $375 to charity. What percent of his income did John give to charity?

In this problem, we know that $P\%$ of 47,600 is 375 and we want to find the percent P. So we solve the following equation for P:

$$P\% \cdot 47,600 = 375$$

By solving, we find that

$$P\% = \frac{375}{47,600} = .00788 = .788\%$$

which means that John gave about .8% (8 tenths of one percent) of his income to charity.

If Nellie leaves a $9 tip on a meal that cost $57.45, what percent was Nellie's tip? If $P\%$ is this percentage, then

$$P\% \cdot 57.45 = 9 \cdot$$

So,

$$P\% = \frac{9}{57.45} = 9 \div 57.45 = .157 = 15.7\%$$

Therefore, Nellie left a tip of nearly 16%.

Notice that to solve this type of problem, you just divide the resulting amount R by the initial amount Q. Then, multiply by 100 to write the answer as a percentage.

FINDING THE ORIGINAL QUANTITY Q WHEN THE PERCENT P AND THE RESULTING AMOUNT R ARE KNOWN

In another basic type of percent problem, we know the percent P and the resulting amount R, and we must calculate the initial quantity Q:

A store gave $15,000 to schools in the community. This $15,000 represents 3% of the store's annual profit. What was the store's annual profit?

3%	$\longrightarrow$	$15,000
1%	$\longrightarrow$	$15,000 \div 3 = $5,000
100%	$\longrightarrow$	$100 \times $5,000 = $500,000

TABLE 3.2
Calculating 100% of an Amount if 3% of the Amount is $15,000

Let A stand for the store's annual profit. Then

$$\frac{3}{100} \cdot A = 15{,}000$$

so

$$.03 \cdot A = 15{,}000$$
$$A = 15{,}000 \div .03$$
$$A = 500{,}000$$

Therefore, the store's annual profit was $500,000. Another way to solve this problem is to calculate 1% of the store's annual profit first and then calculate 100% of the store's annual profit. This line of reasoning is as follows; it is also summarized in Table 3.2:

$$3\% \text{ of annual profit is } $15{,}000$$

So

$$1\% \text{ of annual profit is } $15{,}000 \div 3 = $5{,}000$$

Therefore,

$$100\% \text{ of annual profit is } 100 \times $5{,}000 = $500{,}000.$$

CLASS ACTIVITY   NOW TURN TO CLASS ACTIVITIES MANUAL

3S Calculations with Percents p. 53

PRACTICE PROBLEMS FOR SECTION 3.5

1. For each of the shapes shown in Figure 3.53, determine what percent of the shape is shaded. Give your answer rounded to the nearest multiple of 5 (i.e., 5, 10, 15, 20, . . .).

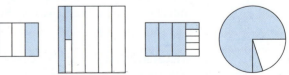

FIGURE 3.53
What Percent of Each Shape Is Shaded?

2. A restaurant server received a $7.00 tip on a meal he served. If this tip represents 20% of the cost of the meal, then how much did the meal cost? Solve this problem by using a common fraction. Then use a diagram to solve this problem. Finally, solve the problem numerically.

3. If $12.3 million is 75% of the budget, then how much is the full budget? First solve the problem by drawing a picture. Explain how your picture helps you solve the problem. Then solve the problem numerically.

4. There were 4800 gallons of water in a tank. Some of the water was drained out, leaving 65% of the original amount of water in the tank. How many gallons of water are in the tank? First solve the problem by drawing a picture. Explain how your picture helps you solve the problem. Then solve the problem numerically.

5. If your daily value of vitamin C is 60 milligrams, then how many milligrams is 95% of your daily value of vitamin C? First solve the problem by drawing a picture. Explain how your picture helps you solve the problem. Then solve the problem numerically.

6. George was given 9 grams of medicine, but the full dose that he was supposed to receive is 20 grams. What percent of his full dose did George receive? First solve the problem by drawing a picture. Explain how your picture helps you solve the problem. Then solve the problem numerically.

7. a. 1067 is 136% of what number?

   b. What percent of 843 is 1255?

8. At one point, 15% of the US budget was needed to pay interest on the federal debt, and the interest on the federal debt was $241 billion. What was the US budget at that time?

9. There are 200 marbles in a bucket. Of the 200 marbles, 80% have swirled colors and 20% are solid colored. How many swirled marbles must be removed so that 75% of the remaining marbles are swirled?

ANSWERS TO PRACTICE PROBLEMS FOR SECTION 3.5

1. See Figure 3.54.

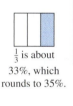

$\frac{1}{3}$ is about 33%, which rounds to 35%.

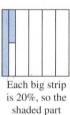

Each big strip is 20%, so the shaded part is 15%.

Each big strip is 25%. The small strips are $\frac{1}{5}$ of that, so 5%. So 80% is shaded.

The large shaded portion is 75%. The small shaded piece must be about 5% because 10% would be close to half of the remaining 25% of the pie. So about 80% is shaded.

FIGURE 3.54

Percent Shaded

---

2. Twenty percent is equal to $\frac{1}{5}$. If $7.00 represents one fifth of the cost of the meal, then the meal must cost 5 times $7.00, which is $35.00.

   See Figure 3.55 for a pictorial solution to the problem.

To solve the problem numerically, note that 20% of the cost of the meal is $7.00. So

$$20\% \text{ of } ? = 7$$
$$.20 \times ? = 7$$
$$? = 7 \div .20$$
$$? = 35$$

Therefore, the meal cost $35.

FIGURE 3.55

Calculating 100% from 20%

3. Because

$$75\% = \frac{3}{4}$$

we can think of the $12.3 million that make up 75% of the budget as distributed equally among the 3 parts of $\frac{3}{4}$, as shown in Figure 3.56. Each of those 3 parts must, therefore, contain $4.1 million. The full budget is made of 4 of those parts. Thus, the full budget must be $16.4 million.

To solve the problem numerically, notice that we are given that 75% of the full budget is $12.3 million. Therefore,

$$75\% \times \text{full budget} = \$12.3 \text{ million}$$
$$.75 \times \text{full budget} = \$12.3 \text{ million}$$
$$\text{full budget} = (\$12.3 \div .75) \text{ million}$$
$$= \$16.4 \text{ million}$$

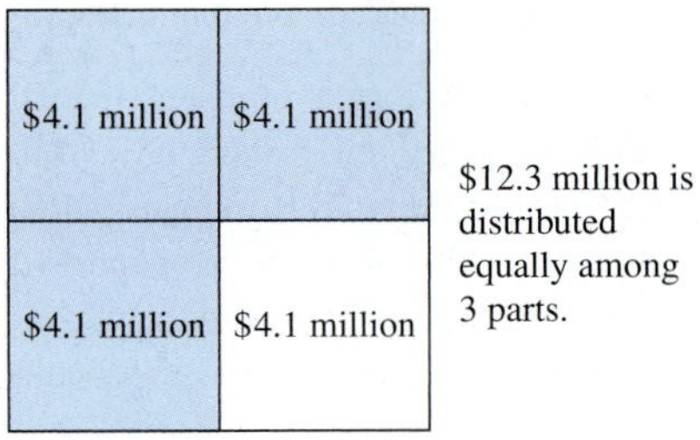

FIGURE 3.56

Calculating the Full Budget

4. As Figure 3.57 shows, we can think of 65% as 50% + 10% + 5%. Now, 50% of 4800 is $\frac{1}{2}$ of 4800, which is 2400. Since 10% of 4800 is $\frac{1}{10}$ of 4800, which is 480, it follows that 5% is half of 480, which is 240. Therefore, 65% of 4800 is 2400 + 480 + 240, which is 3120, so 3120 gallons of water are left in the tank.

To solve the problem numerically, notice that we are given that 65% of 4800 gallons are left in the tank. Therefore,

$$65\% \times 4800 = \text{gallons remaining in tank}$$
$$.65 \times 4800 = \text{gallons remaining in tank}$$
$$3120 = \text{gallons remaining in tank}$$

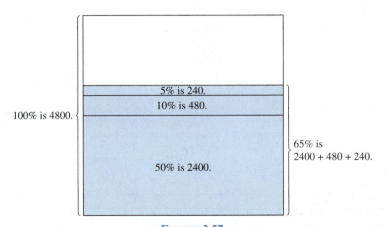

FIGURE 3.57

Calculating 65% of 4800

5. As Figure 3.58 shows, 95% is 5% less than 100%. The 10 vertical strips in Figure 3.58 are each 10%, and half of one of these vertical strips is 5%. Because 10% of 60 milligrams is $\frac{1}{10}$ of 60 milligrams which is 6 milligrams, 5% of 60 milligrams is half of 6 milligrams which is 3 milligrams. So, 95% of 60 milligrams is 3 milligrams less than 60 milligrams, which is 57 milligrams.

To solve the problem numerically, calculate 95% of 60:

$$95\% \times 60 = .95 \times 60$$
$$= 57$$

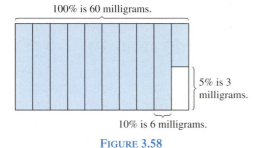

100% is 60 milligrams.

5% is 3 milligrams.

10% is 6 milligrams.

FIGURE 3.58

What Is 95% of 60 Milligrams?

6. One way to determine what percent 9 grams is of 20 grams is to note that 9 grams is a little less than $\frac{1}{2}$, or 50%, of 20 grams, as shown in Figure 3.59. Now, 50% of 20 grams is 10 grams. Nine grams is 1 gram less than 10 grams. So, if we can figure out what percent 1 gram is of 20 grams, then we can subtract this percentage from 50% to determine what percent 9 grams is of 20 grams. Because 2 grams is $\frac{1}{10}$, or 10%, of 20, it follows that 1 gram is half of 10%, namely 5%, of 20 grams. So, 9 grams is

$$50\% - 5\% = 45\%$$

of 20 grams.

Another way to determine what percent 9 grams is of 20 grams is shown in Figure 3.60. If a rectangle representing a full dose of 20 grams of medicine is divided into 10 equal parts, then each of those parts represents 10% of a full dose of medicine. Each part also must represent 2 grams of medicine. So 9 grams of medicine is represented by 4 full parts and half of a 5th part. Therefore, 9 grams of medicine is 45%.

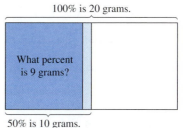

100% is 20 grams.

What percent is 9 grams?

50% is 10 grams.

FIGURE 3.59

Nine Grams Is What Percent
of 20 Grams?

To solve the problem numerically, notice that if $P\%$ is the percent of a full dose that George received, then $P\%$ of 20 grams is 9 grams, according to the information we are given. Therefore,

$$P\% \times 20 = 9$$
$$P\% = 9 \div 20$$
$$P\% = .45 = 45\%$$

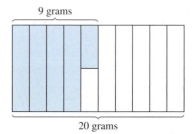

9 grams

20 grams

FIGURE 3.60

Nine Grams Is What Percent
of 20 Grams?

7. a. Stating the question numerically, we have

$$\left(\frac{136}{100}\right) \times ? = 1067$$

In other words,

$$1067 \div 1.36 = ?$$

The answer is (approximately) 784.56.

b. Because

$$1255 \div 843 = 1.489 = \frac{148.9}{100} = 148.9\%$$

(approximately), 1255 is 148.9% of 843.

8. From the information in the practice problem,

$$\left(\frac{15}{100}\right) \times (\text{U.S. budget}) = 241 \text{ billion}$$

Therefore, the U.S. budget was

$$(\$241 \text{ billion}) \div .15 = \$1,606.7 \text{ billion}$$

which is about 1.6 trillion dollars.

9. Because 80% of the 200 marbles are swirled and 20% are solid colored, 160 marbles are swirled and 40 are solid colored. We want these 40 solid colored marbles to be 25% of the remaining marbles. So, if M is the number of remaining marbles,

$$25\% \times M = 40$$

and

$$M = \frac{40}{0.25} = 160$$

Therefore, we must remove 40 swirled marbles.

PROBLEMS FOR SECTION 3.5

1. If your daily value of carbohydrates is 300 grams, then how many grams is 95% of your daily value of carbohydrates? First solve the problem by drawing a picture. Explain how your picture helps you solve the problem. Then explain how to solve the problem numerically.

2. A road crew ordered $\frac{5}{2}$ tons of gravel, but they only received $\frac{3}{2}$ tons of gravel. What percent of their order did the road crew receive? First solve the problem by drawing a picture. Explain how your picture helps you solve the problem. Then explain how to solve the problem numerically.

3. If your daily value of dietary fiber is 25 grams and if you only ate 80% of your daily value of dietary fiber, then how many grams of dietary fiber did you eat? First solve the problem by drawing a picture. Explain how your picture helps you solve the problem. Then explain how to solve the problem numerically.

4. If 36,000 people make up 15% of a population, then what is the total population? First solve the problem by drawing a picture. Explain how your picture helps you solve the problem. Then explain how to solve the problem numerically.

5. If 180 milligrams of potassium constitute 5% of your daily value of potassium, then how many milligrams is your full daily value of potassium? First solve the problem by drawing a picture. Explain how your picture helps you solve the problem. Then explain how to solve the problem numerically.

6. In Happy Valley the average rainfall in July is 5 inches, but this year, only 3.5 inches of rain fell in July. What percent of the average July rainfall did Happy Valley receive this year? First solve the problem by drawing a picture. Explain how your picture helps you solve

the problem. Then explain how to solve the problem numerically.

7. If a $\frac{3}{4}$-cup serving of cereal provides your full daily value of vitamin B6, then what percentage of your daily value of vitamin B6 will you receive in $\frac{1}{2}$ of a cup of the cereal? First solve the problem by drawing a picture. Explain how your picture helps you solve the problem. Then explain how to solve the problem numerically.

8. If a $\frac{2}{3}$-cup serving of cereal provides your full daily value of folic acid, then what percentage of your daily value of folic acid will you receive in $\frac{1}{2}$ of a cup of the cereal? First solve the problem by drawing a picture. Explain how your picture helps you solve the problem. Then explain how to solve the problem numerically.

9. The Biggo Corporation hopes that 95% of its 4,600 employees will participate in the charity fund drive. How many employees does the Biggo Corporation hope will participate in the fund drive? First solve the problem by drawing a picture. Explain how your picture helps you solve the problem. Then explain how to solve the problem numerically.

10. Malcolm ran 75% as far as Susan. How far did Susan run as a percentage of Malcolm's running distance? Draw a picture or diagram to help you solve the problem. Use your picture to help explain your answer.

11. Frank ran 80% as far as Denise. How far did Denise run as a percentage of Frank's running distance? Draw a picture or diagram to help you solve the problem. Use your picture to help explain your answer.

12. Andrew ran 40% as far as Marcie. How far did Marcie run as a percentage of Andrew's running distance? Draw a picture or diagram to help you solve the problem. Use your picture to help explain your answer.

13. GrandMart sells 115% as much soda as BigMart. How much soda does BigMart sell, calculated as a percentage of GrandMart's soda sales?

14. Connie and Benton paid for identical plane tickets, but Benton spent more than Connie (and Connie's ticket was not free).

    a. If Connie spent 75% as much as Benton, then did Benton spend 125% as much as Connie? If not, then what percentage of Connie's ticket price did Benton spend?

    b. If Benton spent 125% as much as Connie, then did Connie spend 75% as much as Benton? If not, then what percentage of Benton's ticket price did Connie spend?

15. At a newstand, 75% of the items sold are newspapers. Does this mean that 75% of the newstand's income comes from selling newspapers? Why or why not? Write a paragraph discussing this. Include examples to illustrate your points.

16. Suppose that 200 pounds of freshly picked cucumbers are 99% water by weight. After several days, some of the water from the cucumbers has evaporated, and the cucumbers are now 98% water. How much do the cucumbers weigh now? Explain your solution.

17. At a store, there is a display of 300 cans of beans. Of the 300 cans, 60% are brand A and 40% are brand B. How many cans of brand B beans must be removed so that 75% of the remaining cans are brand A? Explain your solution.

18. A company produces two types of handbags and sells a total of 500 handbags per day. Of the 500 bags sold per day, 30% are style A and 70% are style B. Suppose the company begins to sell additional style A bags every day, but does not sell any additional style B bags. How many more style A bags would the company have to sell such that 50% of their handbag sales are style A bags?

Chapter 4

Addition and Subtraction

*I*n this chapter our focus is on addition and subtraction. We begin by looking at the standard procedures used to add and subtract and then explain why these procedures make sense. We will add and subtract whole numbers, negative numbers, decimals, fractions, and percents. After looking at the standard procedures, we turn to the idea of solving addition and subtraction problems mentally. In doing so, we use two key properties of addition: the commutative property and the associative property.

4.1 Interpretations of Addition and Subtraction

The most familiar ways to think of addition and subtraction are as *combining* and *taking away*. But as we'll see, we can also view addition and subtraction on number lines, which will allow us to understand addition and subtraction of negative numbers. Subtraction is also used to compare numbers. Shopkeepers use this view of subtraction to make change.

ADDITION AND SUBTRACTION AS COMBINING AND TAKING AWAY

sum   Typically, if A and B are two numbers, then the **sum**

$$A + B$$

basic meaning of addition   represents the total number of objects you will have if you start with A objects and then get B more objects. The numbers A and B in a sum are called **terms**, **addends**, or **summands**.

You can represent the sum

$$149 + 85$$

as the total number of toothpicks you will have if you start with 149 toothpicks and you get 85 more toothpicks.

You can represent the sum

$$34.5 + 7.89$$

as the total number of dollars you will have if you start with $34.5 = 34.50$ dollars and then get 7.89 more dollars.

You can represent the sum

$$\frac{3}{4} + \frac{2}{3}$$

as the total number of cups of flour in a batch of dough if you start with $\frac{3}{4}$ cup of flour in the dough and add another $\frac{2}{3}$ cup of flour.

difference   Typically, if A and B are two numbers, the **difference**

$$A - B$$

represents the total number of objects you will have if you start with A objects and take away **basic meaning of** B of those objects. The numbers A and B in a difference can be called **terms**. The number A **subtraction** is sometimes called the **minuend** and the number B is sometimes called the **subtrahend**.

You can represent

$$142 - 83$$

as the number of toothpicks you have left if you start with 142 toothpicks and give away 83 toothpicks.

You can represent

$$319 - 148.2$$

as the number of dollars you have left if you start with $319.00 and spend $148.20.

You can represent

$$\frac{3}{4} - \frac{1}{8}$$

as the number of acres of land you have left if you start with $\frac{3}{4}$ acre and sell $\frac{1}{8}$ acre.

These interpretations of addition and subtraction are the most basic and simple interpretations, but notice that they don't really make sense for certain kinds of numbers. For example, the meaning of sums and differences like

$$\sqrt{2} + \sqrt{3}$$
$$2 + (-3)$$

and

$$1 - 5$$

seem murky with these interpretations. For example, how can you "take away" 5 objects when you only have 1? One way to interpret this situation is in terms of number of objects *owed*, but another way is to use number lines.

ADDITION AND SUBTRACTION ON NUMBER LINES

Addition and subtraction of all numbers, whether positive or negative, and whether represented by decimals or by fractions, can be interpreted easily using number lines.

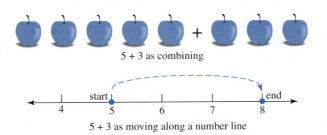

5 + 3 as combining

5 + 3 as moving along a number line

Our interpretation of addition on number lines must fit with our interpretation of addition as combining. So, consider the situation of starting with 5 apples and getting 3 more. Then we have $5 + 3$ apples. The natural way to translate this "addition as combining" situation to a number line is to start at the point 5 on the number line and to move 3 units to the right, ending at 8, as shown in Figure 4.1. We can now interpret addition not only as combining, but also as movement along a number line. In this way, we can make sense of addition problems involving negative numbers, such as $2 + (-5)$ and addition problems such as $\sqrt{2} + \sqrt{3}$, for which the combining interpretation seems strange.

If A and B are any two numbers, then they are represented by points on a number line. The sum

$$A + B$$

number line meaning of addition

corresponds to the point on the number line that is located as follows:

1. Start at A on the number line.
2. Move a distance equal to the number of units that B is away from 0.
   - Move to the right if B is positive.
   - Move to the left if B is negative.
3. The resulting point is the location of $A + B$.

Where is $2 + (-3)$ on a number line? As shown in Figure 4.2, start at 2 and move 3 units to the left. Move left because -3 is negative. You end up at -1, so

$$2 + (-3) = -1$$

Our interpretation of subtraction on number lines must fit with our interpretation of subtraction as taking away. So consider the situation of starting with 5 apples and taking away 3. Then we have $5 - 3$ apples. The natural way to translate this "subtraction as taking away" situation to a number line is to start at the point 5 on the number line and to move 3 units to the left, ending at 2, as shown in Figure 4.3. We can now interpret subtraction not only as

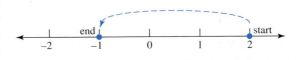

FIGURE 4.3

Subtraction as
Taking Away
Versus
Subtraction on a
Number Line

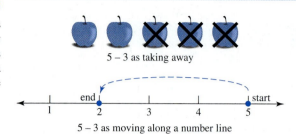

taking away, but also as movement along a number line. In this way, we can make sense of subtraction problems involving negative numbers, such as $2 - (-5)$ and subtraction problems such as $\sqrt{3} - \sqrt{2}$, for which the taking away interpretation seems strange.

If A and B are any two numbers, the difference

$$A - B$$

number line meaning of subtraction

corresponds to the point on the number line that is located as follows:

1. Start at A on the number line.

2. Move a distance equal to the number of units that B is away from 0.

   - Move to the *left* if B is positive.

   - Move to the *right* if B is negative.

3. The resulting point is the location of $A - B$.

Where is $(-2) - (-4)$ on a number line? As shown in Figure 4.4, start at -2 and move 4 units to the *right*. Move right because -4 is negative and we are subtracting. You end up at 2, so $(-2) - (-4) = 2$.

Although the combining and taking away interpretations of addition and subtraction fit with our common sense and intuition, they do not work well with certain numbers. The number line interpretation of addition and subtraction, although more abstract, applies to *all* numbers.

RELATING ADDITION AND SUBTRACTION

Every statement about subtraction corresponds to a statement about addition. Namely, to say that

$$A - B = C$$

is equivalent to saying that

$$A = C + B$$

FIGURE 4.4

Using a Number
Line to Show
That $(-2)-$
$(-4) = 2$

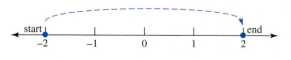

FIGURE 4.5

A Subtraction
Statement
Corresponds to
an Addition
Statement

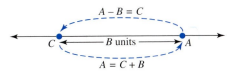

We can see why these two statements are equivalent by showing them on a number line, as in Figure 4.5.

SUBTRACTION AS COMPARISON

Although we commonly think of subtraction as taking away, we can also interpret subtraction as *comparison*. For example, we can interpret

$$7 - 5$$

as the number of blocks that remain from 7 blocks when 5 blocks are taken away. But, as seen in Figure 4.6, we can also interpret

$$7 - 5$$

as *how many more blocks* 7 blocks are than 5 blocks or, in other words, how many blocks you have to add to 5 blocks to get 7 blocks.

Notice that even the word *difference* that we use to describe the result of subtraction, $A - B$, contains the idea of comparison.

The comparison interpretation of subtraction is commonly used by shopkeepers when giving change after a purchase. The shopkeeper returns money to the patron, adding these returned amounts to the purchase price until he arrives at the amount of money that the patron gave him.

CLASS ACTIVITY NOW TURN TO CLASS ACTIVITIES MANUAL

4A The Shopkeepers' Method of Making Change p. 55

FIGURE 4.6

Two Views of
Subtraction:
Taking Away
and Comparison

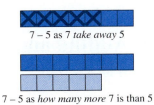

7 − 5 as 7 *take away* 5

7 − 5 as *how many more* 7 is than 5

PRACTICE PROBLEMS FOR SECTION 4.1

1. Use a number line to calculate $0 - (-2)$.

2. Use a number line to calculate $-3 - (-3)$.

ANSWERS TO PRACTICE PROBLEMS FOR SECTION 4.1

1. See Figure 4.7.

2. See Figure 4.8.

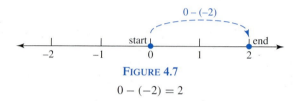

FIGURE 4.7

$0 - (-2) = 2$

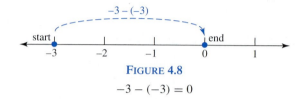

FIGURE 4.8

$-3 - (-3) = 0$

PROBLEMS FOR SECTION 4.1

1. a. Use a number line to calculate $-1 + (-2)$.

   b. Use a number line to calculate $1 - (-2)$.

   c. Use a number line to calculate $-1 - 2$.

   d. Use a number line to calculate $-1 - (-2)$.

2. Use a number line to calculate $3.8 + 1.9$.

3. Use a number line to calculate $1.2 - 2.5$.

4. When a patron of a store gives a shopkeeper $\$A$ for a $\$B$ purchase, we can think of the change owed to the patron as what is left from $\$A$ when $\$B$ are taken away. In contrast, the shopkeeper might make change by starting with $\$B$ and then proceeding to hand the customer money, adding on the amounts until they reach $\$A$.

   a. Describe in detail how a shopkeeper could use the previous method to give a patron change from a $20 bill on a $17.63 purchase.

   b. In general, explain why a patron will always get the correct amount of change when the shopkeepers'

method for making change is used. In particular, reconcile the method used by shopkeepers with the view that change is what is left from $\$A$ when $\$B$ are taken away.

5. Describe two situations other than change-making where subtraction is interpreted as comparison.

6. Use a number line like the one in Figure 4.9 to help you compare and contrast the following expressions:

$$-A + B$$
$$-(A + B)$$
$$-A - B$$
$$-(A - B)$$

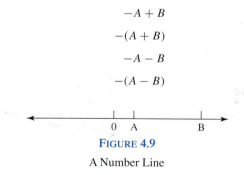

FIGURE 4.9

A Number Line

4.2 Why the Standard Algorithms for Adding and Subtracting Decimal Numbers Work

In school, we learned the standard paper-and-pencil methods for adding, subtracting, multiplying, and dividing numbers. But have you ever wondered why they work? All these standard methods were developed by people, independently, in many different places throughout the world, and at many different times. How did mathematicians know that their clever methods would yield correct results? Just as you can't bake a cake by randomly combining flour, sugar, butter, and eggs, you also can't come up with the correct answer to an addition problem by randomly combining the numbers involved. The originators of the methods we use in arithmetic today had to think deeply to derive those methods. We can now benefit from the methods, but as a teacher, you should not take the standard methods of arithmetic for granted. You should know why these methods work and the reasoning that lies behind them. Only when you understand that mathematics is based on sensible reasoning will you be able to teach mathematics so that it makes sense.

WHAT IS AN ALGORITHM?

algorithm   An **algorithm** is a method or a procedure for carrying out a calculation.

When you bake a cake, you probably follow a recipe. A cake recipe is a kind of algorithm: a step-by-step procedure taking various ingredients and turning them into a cake. In the same way, we have step-by-step procedures for adding numbers, subtracting numbers, multiplying numbers, and dividing numbers. Just as there is mystery in how flour, eggs, butter, and sugar can combine to make cake, there is mystery in how the standard algorithms result in correct answers to arithmetic problems. Somehow, we throw numbers in, mix them up in a specific way, and out comes—cake? No, but out comes the correct answer to the addition problem. In this book, we won't examine the mysteries of cake baking, but we will uncover the mysteries of the algorithms of arithmetic. In fact, they aren't mysteries at all, but very clever, efficient ways of calculating that make complete sense once you know how to look at them.

THE ADDITION ALGORITHM

When we use the standard algorithm to add decimal numbers, such as $34.5 + 7.89$ or $149 + 85$, we put the numbers one under the other, lining up the decimal points, and then we add column by column, *regrouping* as needed. When adding decimal numbers this way, if the digits in a column add to 10 or more, we write down the ones digit of the sum and shift the remaining 10 to become a 1 in the next place to the left. We usually write this 1 at the top of the next column to the left. This process of shifting a 10 from the sum of the digits in one place to a 1 in the

regrouping   next place to the left is called **regrouping**. Regrouping is also called **trading** or **carrying**.

Regrouping has taken place in the following examples:

$$
\begin{array}{r}
\overset{1\ 1}{34.50} \\
+\ \ 7.89 \\
\hline
42.39
\end{array}
\qquad
\begin{array}{r}
\overset{1\ 1}{149} \\
+\ \ 85 \\
\hline
234
\end{array}
$$

Notice that even though you don't see any decimal points in $149 + 85$, they are still lined up because we could also write this addition problem as $149.0 + 85.0$. Why do we line up the decimal points like that? Why do we add column by column? Why do we regroup? The answers to these questions lie in the interpretation of addition as combining and in the idea of place value.

WORKING WITH PHYSICAL OBJECTS TO UNDERSTAND THE ADDITION ALGORITHM FOR WHOLE NUMBERS

Consider the sum $149 + 85$. We can represent this sum as the total number of toothpicks when we combine 149 toothpicks with 85 toothpicks. In order to analyze the addition algorithm, represent the decimal numbers 149 and 85 with bundles as described in Section 2.3. Represent 149 toothpicks as 1 bundle of one hundred (which is ten bundles of ten), 4 bundles of ten, and 9 individual toothpicks, as shown in Figure 4.10. Similarly, represent 85 toothpicks as 8 bundles of ten and 5 individual toothpicks. When all these toothpicks are combined, how many are there? Adding like bundles, we have the following:

14 individual toothpicks (9 plus 5 more)

12 bundles of 10 toothpicks (4 bundles of ten plus 8 more bundles of ten)

1 bundle of 100 toothpicks

Symbolically, we can write this as follows:

$$\begin{array}{r} 1(100) + 4(10) + 9(1) \\ + 8(10) + 5(1) \\ \hline = 1(100) + 12(10) + 14(1) \end{array}$$

So there are

$$1(100) + 12(10) + 14(1)$$

toothpicks, but this expression is not the expanded form of a number in ordinary decimal notation because there are 12 tens and 14 ones. This is where we regroup our bundles. *The regrouping of bundles of toothpicks is the physical representation of the regrouping process in the addition algorithm.*

To regroup the bundled toothpicks, convert the 14 individual toothpicks into 1 bundle of ten toothpicks and 4 individual toothpicks. This 1 bundle of ten is the small "carried" 1 we write above the 4 in the standard procedure. The small 1 above the tens in the following example really stands for 1 ten:

$$\begin{array}{r} {\scriptstyle 1\ 1} \\ 149 \\ + 85 \\ \hline 234 \end{array}$$

Likewise, regroup the 12 bundles of 10 toothpicks into 1 bundle of 100 (namely 10 bundles of ten) and 2 bundles of ten. This 1 bundle of 100 is the small "carried" 1 we write above the 1 in the standard procedure. That small 1 really stands for 1 hundred. When you collect like bundles, you will have 2 bundles of a hundred, 3 bundles of ten, and 4 individual toothpicks.

FIGURE 4.10

Regrouping
149 + 85
Toothpicks

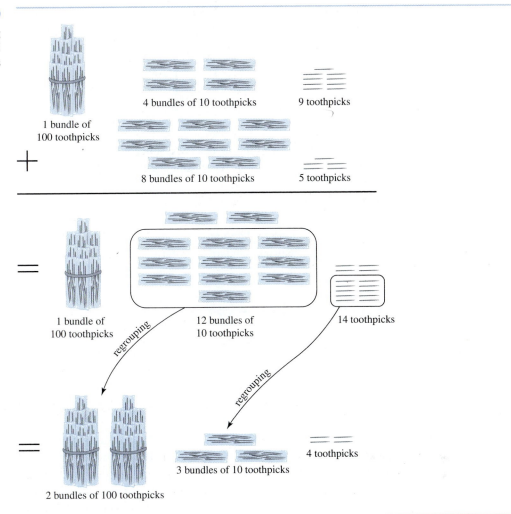

Here is the same example written symbolically:

$$1(100) +\ \ 4(10) +\ \ 9(1)$$
$$+\ \ 8(10) +\ \ 5(1)$$
$$= 1(100) +\ \ \cancel{12(10)} +\ \ \cancel{14(1)}$$
$$+\ \ 1(10) +\ \ 4(1) \quad \text{14 ones become 1 ten and 4 ones.}$$
$$+ 1(100) +\ \ 2(10) \qquad\quad \text{12 tens become 1 hundred and 2 tens.}$$
$$= 2(100) +\ \ 3(10) +\ \ 4(1)$$

We can also write this regrouping in equation form:

$$1(100) + 12(10) + 14(1) = \quad 1(100) \quad + 10(10) + 2(10) + 10(1) + 4(1)$$

$$= 1(100) + 1(100) + 2(10) + 1(10) + \quad 4(1)$$

$$= \quad 2(100) \quad + \quad 3(10) \quad + \quad 4(1)$$

$$= \quad 234$$

(The diagonal arrows show you how the 10 tens become 1 hundred and the 10 ones become 1 ten.) Notice that these equations represent symbolically the *physical actions* of regrouping the toothpicks.

The key point is that the standard addition algorithm is just a way to condense the information in equations like the ones shown and, therefore, is a way to quickly and efficiently record the physical action of adding and regrouping actual objects, such as toothpicks. This example demonstrates why the standard addition algorithm gives us correct answers to addition problems.

THE ADDITION ALGORITHM FOR DECIMALS

When we use the standard addition algorithm to add decimals, we follow the same procedure as for whole numbers. The first step in the algorithm is to line up the decimal points. Although the decimal points are usually not shown in whole numbers, this step also occurs when adding whole numbers because the places where the decimal points would be are lined up. Why do we line up the decimal points? We can see why this makes sense by working with bundled toothpicks. The key lies in place value.

As we saw in Section 2.3, bundled toothpicks can be used to represent (finite) decimals, as long as the meaning of 1 toothpick is interpreted suitably. To represent the sum $.834 + 6.7$ with toothpicks, let 1 toothpick represent $\frac{1}{1000}$. Then a bundle of 10 toothpicks represents $\frac{1}{100}$, a bundle of 100 toothpicks (a bundle of 10 tens) represents $\frac{1}{10}$ and a bundle of 1000 toothpicks (a bundle of 10 hundreds) represents 1. So .834 is then represented by

8 bundles of 100 toothpicks,

3 bundles of 10 toothpicks, and

4 individual toothpicks,

while 6.7 is represented by

6 bundles of 1000 toothpicks and

7 bundles of 100 toothpicks.

Notice that we chose 1 toothpick to represent $\frac{1}{1000}$ in *both* .834 and 6.7. In this way, when we use the bundled toothpicks to represent the sum $.834 + 6.7$, thousandths will be added to thousandths, hundredths to hundredths, tenths to tenths, and ones to ones. If 1 toothpick were to represent $\frac{1}{1000}$ in .834 but $\frac{1}{10}$ in 6.7, then we would add thousandths to tenths and hundredths to ones, which wouldn't make any sense. This would be like treating a penny as $\frac{1}{100}$ of a dollar in one setting and $\frac{1}{10}$ of a dollar in another setting. *The consistent choice for the meaning of 1 toothpick when representing both decimal numbers has the same effect as lining up the decimal points of the two decimal numbers. Decimal points must be lined up so that like terms will be added*—tens to tens, ones to ones, tenths to tenths, hundredths to hundredths, and so on.

If you are adding decimal numbers that have no digits in places lower than the hundredths place, then you can think of your addition problem in terms of money. For example, we can think of the addition problem

$$3.7 + 0.45$$

as

$$\$3.70 + \$0.45$$

When we add these numbers in a column we must line up the decimal points so that we add pennies to pennies, dimes to dimes, and dollars to dollars. If we didn't line up the decimal points, then we would add 7 dimes to 5 pennies and erroneously conclude that the result is 12 pennies.

Once the decimal points have been lined up, the addition algorithm proceeds in the same way as if the decimal points weren't there, and the explanation for why the algorithm works is the same as for whole numbers. You can write 0's for the blank places if you like, as shown here on the right:

$$
\begin{array}{r}
^{1}\\
.834\\
+\;6.7\\
\hline
\end{array}
\qquad
\begin{array}{r}
^{1}\\
0.834\\
+\;6.700\\
\hline
\end{array}
$$

THE SUBTRACTION ALGORITHM

Now let's turn to subtraction. When we subtract decimal numbers, such as $319 - 148.2$ or $142 - 83$, using the standard paper-and-pencil algorithm, we first put the numbers one under the other, lining up the decimal points. Then we subtract column by column, *regrouping* as needed so that we can subtract the numbers in a column without a negative number resulting. When subtracting numbers this way, if the digit at the top of a column is less than the digit below it, we cross out the digit at the top of the next column to the left, changing this digit to one that is 1 less, and we replace the digit at the top of our original column with 10 plus the original digit. (For example, we replace a 2 with 12.) If we can't carry out these steps because there is a 0 in the next column to the left, then we keep moving to the left, crossing out 0's until we come to a digit that is greater than 0. We cross this nonzero digit out and replace it with the digit that is 1 less, we replace all the intervening 0s with 9s, and, as before, we replace the digit at the top of the original column with 10 plus this digit. This process of changing digits so as regrouping   to be able to subtract in a column is called **regrouping**. Regrouping is also called **trading** or **borrowing**.

Regrouping is used in the following examples:

$$
\begin{array}{r}
319.0\\
-\;148.2\\
\hline
\end{array}
\;\rightarrow\;
\begin{array}{r}
{}^{2\,11\,8\ 10}\\
\cancel{3}\cancel{1}\cancel{9}.\cancel{0}\\
-\;148.2\\
\hline
170.8
\end{array}
$$

$$
\begin{array}{r}
142\\
-\;83\\
\hline
\end{array}
\;\rightarrow\;
\begin{array}{r}
{}^{0\,13\,12}\\
\cancel{1}\cancel{4}\cancel{2}\\
-\;83\\
\hline
59
\end{array}
$$

$$
\begin{array}{r}
100.2 \\
-\quad 5.3 \\
\hline
\end{array}
\quad \rightarrow \quad
\begin{array}{r}
\overset{0\,9\,9\ \ 12}{\cancel{1}\cancel{0}\cancel{0}.\cancel{2}} \\
-\quad 5.3 \\
\hline
94.9
\end{array}
$$

As with the addition algorithm, we want to understand why this procedure makes sense. We will explain why by interpreting subtraction as taking away and by considering place value.

WORKING WITH PHYSICAL OBJECTS TO UNDERSTAND THE SUBTRACTION ALGORITHM

As with addition, we can model the subtraction algorithm, particularly the regrouping process, with bundles of toothpicks. Consider the difference $142 - 83$. To represent this with bundled toothpicks, start with 142 toothpicks in 1 bundle of a hundred, 4 bundles of ten, and 2 individual toothpicks. How many toothpicks will be left when we take 83 toothpicks away? When we try to take 3 individual toothpicks away from the 142 toothpicks, we first need to do some unbundling. *This unbundling is a physical representation of the regrouping process.* This process is illustrated in Figure 4.11. One of the 4 bundles of ten can be unbundled and added to the individual toothpicks. The result is 1 bundle of a hundred, 3 bundles of ten, and 12 individual toothpicks. Using equations, we have

$$
\begin{aligned}
142 &= 1(100) + \quad 4(10) \quad + \quad 2(1) \\
&= 1(100) + 3(10) + 1(10) + \quad 2(1) \\
&\qquad\qquad\qquad\qquad\qquad \searrow \\
&= 1(100) + \quad 3(10) \quad + 10(1) + 2(1) \\
&= 1(100) + \quad 3(10) \quad + \quad 12(1)
\end{aligned}
$$

Now there won't be any problem taking 3 individual toothpicks away, but what about taking away 80, namely 8 bundles of ten? There are only 3 bundles of ten. So we must unbundle the 1 bundle of a hundred as 10 bundles of 10 and combine these 10 bundles of ten with the 3 bundles of ten to make 13 bundles of ten. We continue from the previous equations to get

$$
\begin{aligned}
142 &= 1(100) + \quad 3(10) \quad + 12(1) \\
&\qquad\quad \searrow \\
&= \qquad\quad 10(10) + 3(10) + 12(1) \\
&= \qquad\qquad 13(10) \quad + 12(1)
\end{aligned}
$$

After regrouping, there is no difficulty taking 3 individual toothpicks away from the 12 individual toothpicks and 8 bundles of ten away from the 13 bundles of ten. Notice that the physical actions of moving the toothpicks correspond exactly to the regrouping process that takes place in the standard subtraction algorithm. In expanded form, the subtraction problem $142 - 83$ can now be rewritten as

$$
\begin{array}{r}
13(10) + 12(1) \\
- \ [8(10) + \ 3(1)] \\
\hline
= \ 5(10) + \ 9(1) \ = 59
\end{array}
$$

FIGURE 4.11

Regrouping 142
Toothpicks

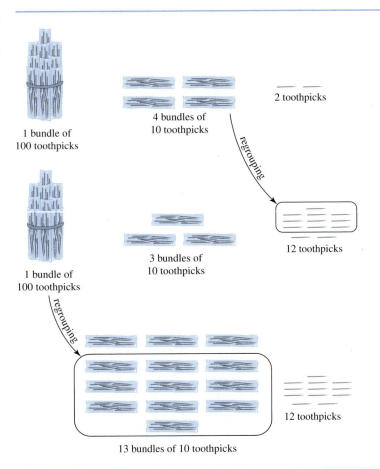

The key point is that the standard subtraction algorithm is just a way to condense the information in equations like those given previously. Therefore, it is a quick and efficient way to record the physical actions of regrouping and then taking away actual objects, such as toothpicks. This is why the standard subtraction algorithm gives us correct answers to subtraction problems.

UNDERSTANDING THE SUBTRACTION ALGORITHM FOR DECIMALS

As with the addition algorithm for decimals, we can generally still represent decimal subtraction problems with bundles of toothpicks by an appropriate interpretation of 1 toothpick. As before, decimal points are lined up so that like terms are subtracted—tens from tens, ones from ones, tenths from tenths, hundredths from hudredths, and so on.

CLASS ACTIVITY   NOW TURN TO CLASS ACTIVITIES MANUAL

4B Understanding the Standard Addition Algorithm p. 56

4C Understanding the Standard Subtraction Algorithm p. 58

4D Subtracting across Zeros p. 60

4E Regrouping with Dozens and Dozens of Dozens p. 62

4F Regrouping with Seconds, Minutes, and Hours p. 63

PRACTICE PROBLEMS FOR SECTION 4.2

1. Define the following terms:

   algorithm

   sum

   difference

   regroup (in addition and in subtraction)

2. Describe how to use bundled toothpicks to explain why the standard procedure for adding $1.8 + .67$ makes sense. Explain the regrouping process. Write corresponding equations with the numbers in expanded form.

3. Write equations with numbers in expanded form showing how to regroup the number 104 so that 69 can be subtracted from it.

4. Why do we line up decimal points before adding or subtracting decimal numbers?

5. Ellie solves the subtraction problem $2.5 - .13$ with toothpicks. She represents 2.5 with 2 bundles of 10 toothpicks and 5 individual toothpicks, and she represents .13 with 1 bundle of 10 toothpicks and 3 individual toothpicks. Ellie gets the answer 1.2. Is she right? Explain your answer.

6. A store buys action figures in boxes. Each box contains 50 bags, and each bag contains 6 action figures. At the beginning of the month, the store has

   7 unopened boxes, 15 unopened bags, and
   3 individual action figures.

   At the end of the month, the store has

   2 unopened boxes, 37 unopened bags, and
   5 individual action figures.

   How many action figures did the store sell during the month (assuming they got no additional shipments of action figures)? Write your answer in terms of boxes, bags, and individual action figures. *Work with boxes, bags, and individuals in a sort of expanded form and use regrouping to solve this problem.*

ANSWERS TO PRACTICE PROBLEMS FOR SECTION 4.2

1. See text.

2. Represent 1.8 by 1 bundle of 100 toothpicks, 8 bundles of 10 toothpicks and 0 individual toothpicks. Similarly, represent 0.67 by 6 bundles of 10 toothpicks and 7 individual toothpicks. We first combine the 0 individual toothpicks from 1.8 with the 7 individual toothpicks from 0.67 to make 7 individual toothpicks. Then we combine 8 bundles of 10 toothpicks from 1.8 with 6

bundles of 10 toothpicks from 0.67 to make 14 bundles of 10 toothpicks. We can regroup 10 of these bundles of 10 and bundle them together to make 1 bundle of 100. There are still 4 bundles of 10 toothpicks remaining. The new 1 bundle of 100 toothpicks is now combined with the 1 bundle of 100 toothpicks from 1.8 to make 2 bundles of 100 toothpicks. We have 2 bundles of 100 toothpicks, 4 bundles of 10 toothpicks, and 7 individual toothpicks, representing 2.47.

Working symbolically with expanded forms, the corresponding equations are

$$\begin{aligned}
1(1) +\ & 8\left(\tfrac{1}{10}\right) + 0\left(\tfrac{1}{100}\right) \\
+\ & 6\left(\tfrac{1}{10}\right) + 7\left(\tfrac{1}{100}\right) \\
\hline
= 1(1) +\ & 14\left(\tfrac{1}{10}\right) + 7\left(\tfrac{1}{100}\right)
\end{aligned}$$

and

$$\begin{aligned}
1(1) + 14\left(\tfrac{1}{10}\right) + 7\left(\tfrac{1}{100}\right) & \\
= \quad 1(1) \quad & + 10\left(\tfrac{1}{10}\right) + 4\left(\tfrac{1}{10}\right) + 7\left(\tfrac{1}{100}\right) \\
= 1(1) + 1(1) + \quad & \qquad\qquad 4\left(\tfrac{1}{10}\right) + 7\left(\tfrac{1}{100}\right) \\
= \quad 2(1) \quad + \quad & \qquad\qquad 4\left(\tfrac{1}{10}\right) + 7\left(\tfrac{1}{100}\right)
\end{aligned}$$

3. Here's the regrouping process, with equations in expanded form:

$$\begin{aligned}
104 = 1(100) + & \quad 0(10) \quad + \quad 4(1) \\
= & \quad 10(10) + 0(10) + \quad 4(1) \\
= & \quad 10(10) \quad + \quad 4(1) \\
= & \quad 9(10) + 1(10) + \quad 4(1) \\
= & \quad 9(10) \quad + 10(1) + 4(1) \\
= & \quad 9(10) \quad + \quad 14(1)
\end{aligned}$$

4. See text.

5. No, Ellie's answer is not correct. 1 toothpick must represent the same amount when representing both 2.5 and

.13. Ellie should think of 1 toothpick as representing $\frac{1}{100}$ in both cases. Then 2.5 is represented by 2 bundles of 100 toothpicks and 5 bundles of 10 toothpicks, while .13 is represented by 1 bundle of 10 toothpicks and 3 individual toothpicks. Now Ellie should be able to see that she'll need to regroup in order to subtract. It might help Ellie to think in terms of money: 2.5 and .13 can be represented by $2.50 and $.13. Ellie's way of using the toothpicks would be like saying that a dime is equal to a penny.

6. We must solve the following problem:

$$\begin{aligned}
7 \text{ boxes} + 15 \text{ bags} + 3 \text{ individual} \\
- (2 \text{ boxes} + 37 \text{ bags} + 5 \text{ individual}) \\
\hline
\end{aligned}$$

We can solve this by first regrouping the 7 boxes, 15 bags, and 3 individual action figures. If we open one of the bags, then there is one less bag, but 6 more individual figures, so there are 7 boxes, 14 bags, and 9 individual action figures.

If we open one of the boxes, then there is one less box, but 50 more bags of action figures, so there are 6 boxes, 64 bags, and 9 individual action figures.

It's still the same number of action figures, they are just arranged in a different way. In equation form we can write this as

$$\begin{aligned}
7 \text{ boxes} + 15 \text{ bags} &+ 3 \text{ individual} \\
= 7 \text{ boxes} + \quad 14 \text{ bags} &+ (6+3) \text{ individual} \\
= 6 \text{ boxes} + (50+14) \text{ bags} &+ 9 \text{ individual} \\
= 6 \text{ boxes} + \quad 64 \text{ bags} &+ 9 \text{ individual}
\end{aligned}$$

Now we are ready to subtract the 2 boxes, 37 bags, and 5 individual action figures:

$$\begin{aligned}
6 \text{ boxes} + 64 \text{ bags} + 9 \text{ individual} \\
- (2 \text{ boxes} + 37 \text{ bags} + 5 \text{ individual}) \\
\hline
4 \text{ boxes} + 27 \text{ bags} + 4 \text{ individual}
\end{aligned}$$

So a total of 4 boxes, 27 bags, and 4 individual action figures were sold during the month.

PROBLEMS FOR SECTION 4.2

 1. Describe how to use bundled toothpicks (or bundles of other objects) to explain regrouping in the addition problem $167 + 59$. Draw (simplified) pictures to aid your explanation.

2. Describe how to use bundled toothpicks (or bundles of other objects) to explain regrouping in the subtraction problem $231 - 67$. Draw (simplified) pictures to aid your explanation.

3. In order to add 153 and 87, Josh wants to line the numbers up this way:

$$\begin{array}{r} 153 \\ + \ 87 \\ \hline \end{array}$$

Explain to Josh why his method doesn't work, and explain why the correct way of lining up the numbers makes sense.

4. Allie solves the subtraction problem $304 - 9$ as follows:

$$\begin{array}{r} \overset{2}{}\overset{14}{} \\ \cancel{3}0\cancel{4} \\ - \quad 9 \\ \hline 205 \end{array}$$

Explain to Allie what is wrong with her method, and explain why the correct method makes sense.

5. On a space shuttle mission, a certain experiment is started 2 days, 14 hours, and 30 minutes into the mission. The experiment takes 1 day, 21 hours, and 47 minutes to run. When will the experiment be completed? Give your answer in days, hours, and minutes into the mission. Work with a sort of expanded form. In other words, work with

$$2(\text{days}) + 14(\text{hours}) + 30(\text{minutes})$$

and

$$1(\text{day}) + 21(\text{hours}) + 47(\text{minutes})$$

and *regroup among days, hours, and minutes* to solve this problem.

6. We can write dates and times in a sort of expanded form. For example, October 4th, 6:53 P.M. can be written as

$$4(\text{days}) + 18(\text{hours}) + 53(\text{minutes})$$

(In some circumstances you might want to include the month and the year, too.) How long is it from 3:27 P.M. on October 4 to 7:13 A.M. on October 19? Give your answer in days, hours, and minutes. Work in the type of expanded form described above and *regroup among days, hours, and minutes* to solve this problem.

7. Erin wants to figure out how much time elapsed between 9:45 A.M. and 11:30 A.M. Erin does the following:

$$\begin{array}{r} \overset{0}{}\overset{12}{}1 \\ 1\cancel{1}:\cancel{3}0 \\ - \quad 9:45 \\ \hline 1:85 \end{array}$$

and says the answer is 1 hour and 85 minutes.

a. Is Erin right? If not, explain what is wrong with her method and show how to *modify* her method of regrouping to make it correct. (Do not start from scratch.)

b. Solve the problem of how much time elapsed between 9:45 A.M. and 11:30 A.M. in another way than your modification of Erin's method. Explain your method.

8. The standard subtraction algorithm described in the text is not the only correct subtraction algorithm. Some people use the following algorithm instead: Line up the numbers as in the standard algorithm, and subtract column by column, proceeding from right to left. The only difference between this new algorithm and our standard one is in regrouping. To regroup with the new algorithm, move one column to the left and cross out the digit at the *bottom* of the column, replacing it with that digit *plus 1*. (So replace an 8 with a 9, replace a 9 with a 10, etc.) Then, as in the standard algorithm, replace the digit at the top of the original column with the original number plus 10. The following example shows the steps of this new algorithm:

$$\begin{array}{r} 132 \\ - \ 79 \\ \hline \end{array} \rightarrow \begin{array}{r} \overset{12}{13\cancel{2}} \\ - \ \overset{8}{\cancel{7}}9 \\ \hline 3 \end{array} \rightarrow \begin{array}{r} \overset{13}{1}\overset{12}{3\cancel{2}} \\ - \ \overset{1}{\cancel{0}}\overset{8}{\cancel{7}}9 \\ \hline 53 \end{array}$$

a. Use the new algorithm to solve $524 - 198$ and $1003 - 95$. Verify that the new algorithm gives correct answers.

b. Explain why it makes sense that this new algorithm gives correct answers to subtraction problems. What is the reasoning behind this new algorithm?

c. What are some advantages and disadvantages of this new algorithm compared to our standard one?

9. The standard subtraction algorithm described in the text is not the only correct subtraction algorithm. The next subtraction algorithm is called *adding the complement*. For a 3-digit whole number, N, the **complement** of N is $999 - N$. For example, the complement of 486 is

$$999 - 486 = 513$$

Notice that regrouping is never needed to calculate the complement of a number. To use the adding-the-complement algorithm to subtract a 3-digit whole number, N, from another 3-digit whole number, you start by adding the complement of N rather than subtracting N.

For example, to solve

$$723 - 486$$

first add the complement of 486:

$$
\begin{array}{r}
723 \\
+\ 513 \\
\hline
1236
\end{array}
$$

Then cross out the 1 in the thousands column, and add 1 to the resulting number:

$$\cancel{1}236 \quad \rightarrow \quad 236 + 1 = 237$$

Therefore, according to the *adding-the-complement* algorithm, $723 - 486 = 237$.

a. Use the adding-the-complement algorithm to calculate $301 - 189$ and $295 - 178$. Verify that you get the correct answer.

b. Explain why the adding-the-complement algorithm gives you the correct answer to any three digit subtraction problem. In order to explain this, focus on the relationship between the original problem and the addition problem in adding the complement. For example, how are the problems $723 - 486$ and $723 + 513$ related? Work with the *complement* relationship, $513 = 999 - 486$, and notice that $999 = 1000 - 1$.

c. What are some advantages and disadvantages of the *adding-the-complement* subtraction algorithm compared to the standard subtraction algorithm described in the text?

4.3 Adding and Subtracting Fractions

In the last section, we learned the reasoning behind the standard addition and subtraction algorithms for whole numbers and decimals. There are also standard algorithms for adding and subtracting fractions, and in this section, we will study why these make sense. Because they are defined in terms of addition, we will also study mixed numbers (numbers such as $2\frac{3}{4}$) and the algorithm for converting a mixed number to an improper fraction. Similarly, because we can always express a finite decimal as a sum of fractions by writing the decimal in expanded form, we will see how to write finite decimals as fractions. Finally in this section, we will consider a source of errors in fraction addition and subtraction: the failure to work with like wholes.

ADDING AND SUBTRACTING FRACTIONS WITH LIKE DENOMINATORS

The easiest situation to work with when adding fractions is adding fractions that have the same denominator. If two fractions have the same denominator, then you can add or subtract these fractions by adding or subtracting the numerators and leaving the denominator unchanged. For example,

$$\frac{4}{15} + \frac{7}{15} = \frac{4+7}{15} = \frac{11}{15}$$

Why does this make sense? We can interpret the sum

$$\frac{4}{15} + \frac{7}{15}$$

as the total amount of pie you will have if you start with $\frac{4}{15}$ of a pie and get $\frac{7}{15}$ more of the pie. But $\frac{4}{15}$ of a pie is represented by 4 pieces when the pie is divided into 15 equal pieces, and $\frac{7}{15}$ is represented by 7 of those pieces. So, if you have 4 pieces of pie and you get 7 more, then you will have $4 + 7 = 11$ pieces all together, as you see in Figure 4.12. What kind of pieces are they? All 11 pieces are fifteenths of a pie. Therefore, you have $\frac{11}{15}$ of a pie in all.

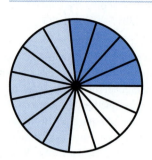

ADDING AND SUBTRACTING FRACTIONS WITH UNLIKE DENOMINATORS BY FINDING COMMON DENOMINATORS

How do we add or subtract fractions that have different denominators, such as $\frac{5}{6} + \frac{3}{8}$? We can interpret the sum

$$\frac{5}{6} + \frac{3}{8}$$

as the total distance walked if you first walk $\frac{5}{6}$ of a mile and then walk $\frac{3}{8}$ of a mile. (See Figure 4.13.) In this case, the first distance walked is described in terms of *sixths* of a mile and the second distance walked is described in terms of *eighths* of a mile. Sixths of a mile and eighths of a mile are different-size distances, so we can't just add the 5 in $\frac{5}{6}$ to the 3 in $\frac{3}{8}$. We need a common distance with which to describe both $\frac{5}{6}$ of a mile and $\frac{3}{8}$ of a mile. In other words, we need to break *sixths* and *eighths* into *like parts*.

As in comparing fractions, the process of "breaking into like parts" is achieved numerically by giving the fractions $\frac{5}{6}$ and $\frac{3}{8}$ a common denominator. *Any* common denominator will do; it does not have to be the least one. You can always produce a common denominator by

FIGURE 4.13

Using a
Common
Denominator to
Add

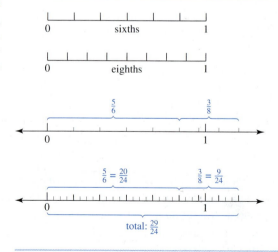

multiplying the two denominators. So, to add

$$\frac{5}{6} \quad \text{and} \quad \frac{3}{8}$$

we can use the common denominator $6 \cdot 8$, which is 48. If we want to work with smaller numbers, we can use 24, which is a common denominator because $24 = 6 \cdot 4$ and $24 = 8 \cdot 3$. Once we have common denominators, our fractions are described in terms of like parts, so we simply add the numerators to determine the total number of parts:

$$\frac{5}{6} + \frac{3}{8} = \frac{5 \cdot 8}{6 \cdot 8} + \frac{3 \cdot 6}{8 \cdot 6} = \frac{40}{48} + \frac{18}{48} = \frac{58}{48}$$

or

$$\frac{5}{6} + \frac{3}{8} = \frac{5 \cdot 4}{6 \cdot 4} + \frac{3 \cdot 3}{8 \cdot 3} = \frac{20}{24} + \frac{9}{24} = \frac{29}{24}$$

Since

$$\frac{58}{48} = \frac{29 \cdot 2}{24 \cdot 2} = \frac{29}{24}$$

the two ways just shown of adding $\frac{5}{6}$ and $\frac{3}{8}$ produce equal results. When we use the common denominator 24, the resulting sum $\frac{29}{24}$ is in simplest form. If the least common denominator is not used when adding fractions, then the resulting sum will not be in simplest form.

Figure 4.13 shows the addition of $\frac{5}{6}$ and $\frac{3}{8}$ with the common denominator 24, using a number line. In this case, each sixth is broken into 4 parts and each eighth is broken into 3 parts to produce twenty-fourths.

In general, suppose

$$\frac{A}{B} \quad \text{and} \quad \frac{C}{D}$$

are two fractions (where $B \neq 0$ and $D \neq 0$). Then

$$\frac{A}{B} = \frac{A \cdot D}{B \cdot D}$$

and

$$\frac{C}{D} = \frac{C \cdot B}{D \cdot B}$$

and the latter two fractions

$$\frac{A \cdot D}{B \cdot D} \quad \text{and} \quad \frac{C \cdot B}{D \cdot B}$$

both have denominators equal to $B \cdot D$. Both fractions are expressed in terms of like parts when they have common denominators. Therefore, we can add or subtract these fractions by adding or subtracting the number of parts, (i.e., the numerators). Therefore, in general,

$$\frac{A}{B} + \frac{C}{D} = \frac{A \cdot D + C \cdot B}{B \cdot D}$$

and

$$\frac{A}{B} - \frac{C}{D} = \frac{A \cdot D - C \cdot B}{B \cdot D}$$

WRITING MIXED NUMBERS AS IMPROPER FRACTIONS

mixed number   A **mixed number** or **mixed fraction** is a number that is written in the form

$$A\frac{B}{C}$$

where A, B, and C are whole numbers, and $\frac{B}{C}$ is a proper fraction (i.e., the numerator is less than the denominator). So

$$2\frac{3}{4} \quad \text{and} \quad 5\frac{7}{8}$$

are mixed numbers.

The mixed number

$$A\frac{B}{C}$$

stands for

$$A + \frac{B}{C}$$

A whole number A can also be written as a fraction, namely, $\frac{A}{1}$. Therefore we can write the mixed number $A\frac{B}{C}$ as follows:

$$A\frac{B}{C} = A + \frac{B}{C} = \frac{A}{1} + \frac{B}{C}$$

Now we can use the way we add fractions to show that every mixed number can be written as an improper fraction:

$$A\frac{B}{C} = \frac{A \cdot C + B}{C}$$

So,

$$2\frac{3}{4} = \frac{2 \cdot 4 + 3}{4} = \frac{11}{4}$$

WRITING FINITE DECIMALS AS FRACTIONS

Every finite decimal stands for a finite sum of fractions. We can see this representation by writing the decimal in its expanded form—for example,

$$2.7 = 2 + \frac{7}{10}$$

$$32.85 = 30 + 2 + \frac{8}{10} + \frac{5}{100}$$

$$0.491 = \frac{4}{10} + \frac{9}{100} + \frac{1}{1000}$$

By giving the fractions in the expanded form of a finite decimal a common denominator, we can write the decimal as a fraction with a denominator that is a power of 10. For example, we have

$$2.7 = 2 + \frac{7}{10}$$

$$= \frac{20}{10} + \frac{7}{10}$$

$$= \frac{27}{10}$$

or

$$0.491 = \frac{4}{10} + \frac{9}{100} + \frac{1}{1000}$$

$$= \frac{400}{1000} + \frac{90}{1000} + \frac{1}{1000}$$

$$= \frac{491}{1000}$$

Observe that this method for writing decimals as fractions only applies to finite decimals. It does not apply to a decimal such as

$$0.4444444\ldots$$

where the 4s continue forever. In fact, it turns out that some numbers, such as $\sqrt{2} = 1.4142\ldots$ cannot be written as fractions of whole numbers.

WHEN IS COMBINING NOT ADDING?

The old admonishment "You can't add apples and oranges" is especially important to remember when adding and subtracting fractions, because sometimes fractions of quantities that are to be combined refer to *different wholes*.

Whenever we add two numbers, such as

$$3 + 4 \quad \text{or} \quad \frac{2}{3} + \frac{3}{4}$$

both summands and the answer all refer to the same whole. Similarly, whenever we subtract two numbers, such as

$$4 - 3 \quad \text{or} \quad \frac{3}{4} - \frac{2}{3}$$

the minuend, the subtrahend, and the answer all refer to the same whole.

For example, we can interpret the sum

$$3 + 4$$

as the total number of apples when 3 apples are combined with 4 apples. But if you interpret $3 + 4$ by combining 3 bananas and 4 blocks, as in Figure 4.14, then the best you can say is that you have 7 *objects* in all.

Similarly, we can interpret the sum

$$\frac{1}{3} + \frac{1}{4}$$

FIGURE 4.14

3 Bananas and 4
Blocks

FIGURE 4.14

3 Bananas and 4
Blocks

FIGURE 4.15

$\frac{1}{3}$ of a Small Pie
and $\frac{1}{4}$ of a Big
Pie Does Not
Make $\frac{1}{3} + \frac{1}{4}$ of
a Pie

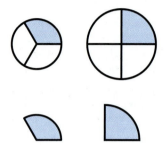

as the fraction of a pie we will have in all if we have $\frac{1}{3}$ of the pie and then get another $\frac{1}{4}$ *of the same pie or an equivalent pie.* The $\frac{1}{3}$, the $\frac{1}{4}$, and the sum $\frac{1}{3} + \frac{1}{4} = \frac{7}{12}$ *all refer to the same pie,* or to copies of an equivalent pie.

But suppose we have a small pie and a large pie, that the small pie is divided into 3 pieces and the large pie is divided into 4 pieces, as shown in Figure 4.15. If we get one piece of each pie, does this represent the sum

$$\frac{1}{3} + \frac{1}{4}?$$

No, it does not, because the $\frac{1}{3}$ and the $\frac{1}{4}$ refer to *different wholes* that are not equivalent. All we can say is that we have 2 pieces of pie. We can't say that we have $\frac{7}{12}$ of a pie—because what pie would the $\frac{7}{12}$ refer to?

So when working with fraction addition or subtraction story problems, be especially careful that the fractions in question refer to the same underlying wholes. In some cases, fractional amounts that are to be combined or taken away refer to different wholes; therefore, the story problem is not a fraction addition or subtraction problem.

CLASS ACTIVITY   NOW TURN TO CLASS ACTIVITIES MANUAL

4G Are These Story Problems for $\frac{1}{2} + \frac{1}{3}$? p. 64

4H Are These Story Problems for $\frac{1}{2} - \frac{1}{3}$? p. 65

4I Fraction Addition and Subtraction p. 66

4J Mixed Numbers p. 68

PRACTICE PROBLEMS FOR SECTION 4.3

1. Using the example $\frac{1}{4} + \frac{5}{6}$, explain why we must give fractions a common denominator in order to add them.

2. When we add or subtract fractions, *must* we use the least common denominator? Are there any advantages to using the least common denominator?

3. The usual procedure for converting a mixed number, such as $7\frac{1}{3}$, into an improper fraction is this:

$$7\frac{1}{3} = \frac{7 \cdot 3 + 1}{3} = \frac{22}{3}$$

Explain the logic behind this procedure.

4. You showed Tommy the diagram in Figure 4.16 to explain why $2\frac{1}{4} = \frac{9}{4}$, but Tommy says that it shows $\frac{9}{12}$, not $\frac{9}{4}$. What must you clarify?

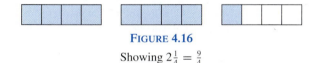

FIGURE 4.16

Showing $2\frac{1}{4} = \frac{9}{4}$

5. We usually call the number 0.43 "forty-three hundredths" and not "four tenths and three hundredths." Why do these two mean the same thing?

6. Jessica says that

$$\frac{1}{2} + \frac{2}{3} = \frac{3}{5}$$

and shows the picture in Figure 4.17 to prove it. What is the problem with Jessica's reasoning? Why is it not correct? Don't just explain how to do the problem correctly; explain what is faulty with Jessica's reasoning.

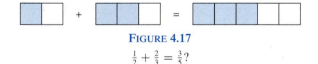

FIGURE 4.17

$\frac{1}{2} + \frac{2}{3} = \frac{3}{5}$?

7. Which of the following problems can be solved by adding $\frac{1}{2} + \frac{1}{3}$? For those problems that can't be solved by adding $\frac{1}{2} + \frac{1}{3}$, solve the problem in another way if there is enough information to do so, or explain why the problem cannot be solved.

   a. In Ms. Dock's class, $\frac{1}{2}$ of the boys want pizza for lunch and $\frac{1}{3}$ of the girls want pizza for lunch. What fraction of the children in Ms. Dock's class want pizza for lunch?

   b. $\frac{1}{2}$ of Jane's shirts have the color red on them somewhere. $\frac{1}{3}$ of Jane's shirts have the color pink on them somewhere. What fraction of Jane's shirts have either red or pink on them somewhere?

   c. $\frac{1}{2}$ of Jane's shirts have the color red on them somewhere. Of the shirts that belong to Jane and do not have red on them somewhere, $\frac{1}{3}$ have the color pink on them somewhere. What fraction of Jane's shirts have either red or pink on them somewhere?

8. Can the following story problem be solved by subtracting $\frac{1}{2} - \frac{1}{3}$? If not, explain why not, and solve the problem in a different way if there is enough information to do so.

   Story problem: There is $\frac{1}{2}$ of a pie left over from yesterday. Pratima eats $\frac{1}{3}$ of the leftover pie. How much pie is left?

9. What is wrong with the following story problem? Give two different ways to restate the problem. Explain how to solve your restated problems.

   Story problem: Liat has $\frac{3}{4}$ cup of juice. Sumin has $\frac{1}{3}$ less. How much juice does Sumin have?

ANSWERS TO PRACTICE PROBLEMS FOR SECTION 4.3

1. When we add $\frac{1}{4}$ and $\frac{5}{6}$, we first give the fractions a common denominator so they have *like parts*. We can interpret $\frac{1}{4} + \frac{5}{6}$ as the total amount of pie you have if you start with $\frac{1}{4}$ of a pie and then get $\frac{5}{6}$ of another equivalent (same size) pie. But fourths and sixths are pieces of different size, so we must express both kinds of pie pieces

in terms of like pieces. If we divide each of the 4 fourths of a pie into 3 pieces and each of the 6 sixths of the other pie into 2 pieces, then $\frac{1}{4}$ of the first pie becomes $\frac{3}{12}$ and $\frac{5}{6}$ of the second pie becomes $\frac{10}{12}$. Now both fractions of pie are expressed in terms of twelfths of a pie, as seen in Figure 4.18. Therefore, the total amount of pie you have consists of $3 + 10 = 13$ pieces, and these pieces are twelfths of a pie. Thus, you have $\frac{13}{12}$ of a pie, which is $1\frac{1}{12}$ of a pie.

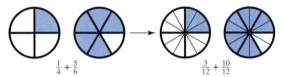

FIGURE 4.18

Finding Common Denominators to Add $\frac{1}{4} + \frac{5}{6}$

Numerically, this process of subdividing pieces of pie can be expressed succinctly with the following equations:

$$\frac{1}{4} + \frac{5}{6} = \frac{1 \cdot 3}{4 \cdot 3} + \frac{5 \cdot 2}{6 \cdot 2} = \frac{3}{12} + \frac{10}{12} = \frac{13}{12} = 1\frac{1}{12}$$

2. To add fractions, we do not need to use the least common denominator. Any common denominator will do. In practice problem 1, we could have used the common denominator 24 instead of 12 to add $\frac{1}{4} + \frac{5}{6}$. However, there are some advantages to using the least common denominator. The least common denominator is smaller than other common denominators and therefore may be easier to work with. If we use a common denominator that is not the least common denominator, then the resulting sum will not be in simplest form. For example, if we use the common denominator 24 to add $\frac{1}{4} + \frac{5}{6}$, then the resulting sum is

$$\frac{1}{4} + \frac{5}{6} = \frac{6}{24} + \frac{20}{24} = \frac{26}{24}$$

which is not in simplest form.

3. Remember that $7\frac{1}{3}$ stands for $7 + \frac{1}{3}$, and that $7 = \frac{7}{1}$. To calculate

$$7\frac{1}{3} = \frac{7}{1} + \frac{1}{3}$$

as a fraction, we need to first find a common denominator:

$$
\begin{aligned}
7\frac{1}{3} &= 7 + \frac{1}{3} \\
&= \frac{7}{1} + \frac{1}{3} \\
&= \frac{7 \cdot 3}{1 \cdot 3} + \frac{1}{3} \\
&= \frac{7 \cdot 3 + 1}{3} \\
&= \frac{22}{3}.
\end{aligned}
$$

Notice that the next-to-last step shows the procedure for turning a mixed number into an improper fraction. This procedure is really just a shorthand way to give 7 and $\frac{1}{3}$ a common denominator and add them.

4. You must identify the whole in the diagram. Tommy is taking the whole to be the full collection of squares, but to interpret the diagram as representing $2\frac{1}{4}$, we must take a strip of 4 squares to be the whole. Without identifying the whole in the diagram, we cannot interpret it unambiguously.

5. Notice that four tenths is the same as forty hundredths, so four tenths and three hundredths is forty hundredths and three hundredths, or forty-three hundredths. We can express this reasoning with the following equations:

$$
\begin{aligned}
.43 &= \frac{4}{10} + \frac{3}{100} \\
&= \frac{40}{100} + \frac{3}{100} \\
&= \frac{43}{100}.
\end{aligned}
$$

6. Jessica's pictures show that $\frac{1}{2}$ of the two-block bar combined with $\frac{2}{3}$ of the three-block bar does indeed make up $\frac{3}{5}$ of a five-block bar. The problem is that Jessica is using three different *wholes*. The fraction $\frac{1}{2}$ refers to a two-block whole, the fraction $\frac{2}{3}$ refers to a 3-block whole, and the fraction $\frac{3}{5}$ refers to a 5-block whole. When we add fractions, such as $\frac{1}{2} + \frac{2}{3}$, both fractions in the sum and the sum itself should all refer to the same whole.

7. None of the problems can be solved by adding $\frac{1}{2} + \frac{1}{3}$.

   a. In this problem, the number of *boys* in Ms. Dock's class is the whole associated with the fraction $\frac{1}{2}$, whereas the number of *girls* in Ms. Dock's class is the whole associated with the fraction $\frac{1}{3}$. These two wholes are different. When we add any two numbers, the numbers must refer to the same wholes. There is not enough information to determine what fraction of the children in Ms. Dock's class want pizza for lunch because we do not know if there is the same number of girls as boys in Ms. Dock's class.

b. In this problem, the fractions $\frac{1}{2}$ and $\frac{1}{3}$ refer to the same whole, namely Jane's shirts, but some of the shirts that have pink on them may also have red on them somewhere. So, if we start with the $\frac{1}{2}$ of Jane's shirts that have red on them, and if we then want to add on the remaining shirts that have pink on them, we do not know what fraction of Jane's shirts these remaining shirts are. We can't solve this problem because we don't know what fraction of Jane's shirts have both red and pink on them.

c. In this problem, the wholes that the fraction $\frac{1}{2}$ and $\frac{1}{3}$ refer to are not the same. The fraction $\frac{1}{2}$ refers to all of Jane's shirts, whereas the fraction $\frac{1}{3}$ refers to the half of Jane's shirts that don't have red on them. We can solve this problem, however, because the shirts that have pink but not red on them are $\frac{1}{3}$ of $\frac{1}{2}$ of Jane's shirts, which is $\frac{1}{6}$ of Jane's shirts. (You can see this by drawing a diagram.) Therefore, the fraction of Jane's shirts that have either red or pink on them is

$$\frac{1}{2} + \frac{1}{6} = \frac{4}{6} = \frac{2}{3}$$

8. The problem cannot be solved by subtracting $\frac{1}{2} - \frac{1}{3}$ because the fractions $\frac{1}{2}$ and $\frac{1}{3}$ refer to different wholes. The $\frac{1}{2}$ refers to the whole pie, whereas the $\frac{1}{3}$ refers to the pie that is left over. We can solve this problem because when Pratima eats $\frac{1}{3}$ of the $\frac{1}{2}$ pie that is left over, she is eating $\frac{1}{6}$ of the pie (as you can see by drawing a diagram). So we know the following:

$$\frac{1}{2} + \frac{1}{6} = \frac{4}{6} = \frac{2}{3}$$

of the pie has been eaten. Therefore, $\frac{1}{3}$ of the pie is left.

9. The problem is with the statement "Sumin has $\frac{1}{3}$ less." Does this mean that Sumin has $\frac{1}{3}$ *cup* less juice than Liat, or does it mean that the *amount of juice* Sumin has is $\frac{1}{3}$ less than the amount of juice Liat has? It is not clear which meaning is intended. To correct the problem, replace the statement, "Sumin has $\frac{1}{3}$ less." either with "Sumin has $\frac{1}{3}$ cup less juice than Liat." or with "Sumin has $\frac{1}{3}$ less juice than Liat." In the first case, the problem is solved by calculating

$$\frac{3}{4} - \frac{1}{3} = \frac{9}{12} - \frac{4}{12} = \frac{5}{12}$$

so that Sumin has $\frac{5}{12}$ cup of juice. In the second case, the problem is solved by first calculating $\frac{1}{3}$ of $\frac{3}{4}$ cup of juice, which is $\frac{1}{4}$ cup of juice. (There are 3 equal parts in $\frac{3}{4}$, and each part is $\frac{1}{4}$.) Therefore, Sumin has $\frac{1}{4}$ cup less juice than Liat, so Sumin has $\frac{3}{4} - \frac{1}{4} = \frac{1}{2}$ cup of juice.

PROBLEMS FOR SECTION 4.3

1. Using the example $\frac{2}{3} + \frac{3}{4}$, explain why we add fractions the way we do. In other words, what is the logic behind this procedure? Explain your answer.

2. Using the example $3\frac{5}{6}$, describe the procedure for turning a mixed number into an improper fraction and explain in your own words why this procedure makes sense. What is the logic behind this procedure? Draw pictures to support your explanation.

3. a. For each of the following decimals, show how to write the decimal as a fraction by first putting the decimal in expanded form:

   i. 2.34

   ii. 124.5

   iii. 7.938

b. Based on your results in part (a), describe a quick way to rewrite a finite decimal as a fraction. Illustrate with the example 2748.963.

4. Show how to solve $5\frac{3}{4} + 1\frac{2}{3}$ in two different ways. In each case, express your answer as a mixed number. Explain why both of your methods are legitimate.

5. Show how to solve $4\frac{2}{3} - 1\frac{3}{4}$ in two different ways. In each case, express your answer as a mixed number. Explain why both of your methods are legitimate.

6. Find two *different*, positive fractions whose sum is $\frac{1}{11}$.

7. John says $\frac{2}{3} + \frac{2}{3} = \frac{4}{6}$ and uses the picture in Figure 4.19 as evidence. Discuss what is wrong with John's reasoning. What underlying misconception does John have?

Don't just explain how to do the problem correctly; explain why John's reasoning is faulty.

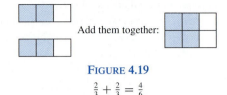

Add them together:

FIGURE 4.19

$\frac{2}{3} + \frac{2}{3} = \frac{4}{6}$

8. Denise says that $\frac{2}{3} - \frac{1}{2} = \frac{1}{3}$ and gives the reasoning indicated in Figure 4.20 to support her answer. Is Denise right? If not, what is wrong with her reasoning and how could you help her understand her mistake and fix it? *Don't* just explain how to solve the problem correctly; explain where Denise's reasoning is flawed.

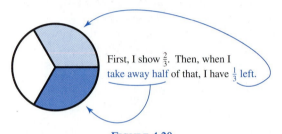

First, I show $\frac{2}{3}$. Then, when I take away half of that, I have $\frac{1}{3}$ left.

FIGURE 4.20

Denise's Idea for $\frac{2}{3} - \frac{1}{2}$

9. Can the following story problem about voters be solved by adding $\frac{1}{2} + \frac{1}{3}$? If so, explain why, if not, explain why not and give a different story problem that can be solved by adding $\frac{1}{2} + \frac{1}{3}$.

> Story problem about voters: In Kneebend county, $\frac{1}{2}$ of the women voters and $\frac{1}{3}$ of the male voters voted for a certain referendum. Altogether, what fraction of the voters voted for the referendum in Kneebend county?

10. Can the following story problem about a bird feeder be solved by subtracting $\frac{3}{4} - \frac{1}{2}$? If so, explain why. If not, explain why not and give a different story problem that can be solved by subtracting $\frac{3}{4} - \frac{1}{2}$.

> Story problem about a bird feeder: A bird feeder was filled with $\frac{3}{4}$ of a full bag of bird seed. The birds ate $\frac{1}{2}$ of what was in the bird feeder. What fraction of a full bag of bird seed did the birds eat?

11. Can the following story problem about Sarah's bead collection be solved by adding $\frac{1}{4} + \frac{1}{5}$? If so, explain why.

If not, explain why not and give a different story problem that can be solved by adding $\frac{1}{4} + \frac{1}{5}$.

> Story problem about Sarah's bead collection: One-fourth of the beads in Sarah's collection are pink. One-fifth of the beads in Sarah's collection are oblong. What fraction of the beads in Sarah's collection are either pink or oblong?

12. Can the following story problem about Jim's medicine be solved by subtracting $\frac{1}{3} - \frac{1}{4}$? If so, explain why. If not, explain why not and give a different story problem that can be solved by subtracting $\frac{1}{3} - \frac{1}{4}$.

> Story problem about Jim's medicine: Jim has a small container filled with one tablespoon of medicine. Jim poured out $\frac{1}{3}$ tablespoon of medicine. Then Jim poured out $\frac{1}{4}$ tablespoon of medicine. Now how much medicine is left in the container?

13. a. Write a story problem for $\frac{3}{4} + \frac{2}{3}$.

  b. Write a story problem for $\frac{3}{4} - \frac{2}{3}$.

14. a. Write a story problem for $2\frac{1}{2} + 1\frac{1}{3}$.

  b. Write a story problem for $2\frac{1}{2} - 1\frac{1}{3}$.

15. Suppose you start with a fraction and you add 1 to both the numerator and the denominator. For example, if you started with $\frac{2}{3}$, then you'd get a new fraction $\frac{2+1}{3+1} = \frac{3}{4}$. Is this procedure of adding 1 to the numerator and the denominator the same as adding the number 1 to the original fraction? (For example, is $\frac{2+1}{3+1}$ equal to $\frac{2}{3} + 1$?) Explain your answer.

16. In the first part of the season, the Bluejays play 18 games and win 7, while the Robins play 3 games and win 1. In the second part of the season, the Bluejays play 2 games and win 1, while the Robins play 17 games and win 8.

  a. Which team won a larger fraction of its games in the first part of the season?

  b. Which team won a larger fraction of its games in the second part of the season?

  c. Which team won a larger fraction of its games overall?

  d. Compare your answers in parts (a), (b), and (c). You may be surprised at the comparison. (If you are not surprised, then you should probably rethink your work in part (c).)

  e. Is the fraction of games won overall by the Bluejays equal to the sum of the fractions of games won by the Bluejays in the first and second parts of the season?

f. What are the different *wholes* that are associated with the fractions in this problem?

g. Write a paragraph on the relationship between your answers to (e) and (f). Refer to the discussion in the text on *combining* versus *adding*.

4.4 When Do We Add Percentages?

Every percentage can be expressed as a decimal or a fraction. So, our explanations of addition of decimals and fractions applies to the addition of percentages as well. In this brief section, we will consider percent problems in which *combining* takes place, but addition of percentages does not.

How can there be a percent problem in which combining takes place, but addition of the percentages is not appropriate? This is the very same problem that occurs for fractions, discussed in the last section. As with fractions, when you encounter a percent problem that involves combining, be aware of the *wholes* that the percentages refer to. *Adding percentages is not appropriate when the wholes involved are not the same.*

CLASS ACTIVITY   NOW TURN TO CLASS ACTIVITIES MANUAL

4K Should We Add These Percentages? p. 69

PRACTICE PROBLEMS FOR SECTION 4.4

1. Each of the pictures in Figure 4.21 shows two adjacent lots of land, lot A and lot B. In each case, 20% of lot A is shown shaded and 40% of lot B is shown shaded. What percent of the *combined amount* of lot A and lot B is shaded in each case?

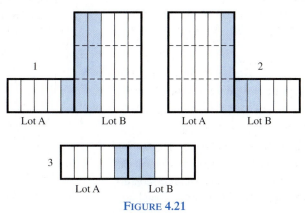

FIGURE 4.21

What Percent of the Combined Amount of
Lots A and B Is Shaded?

ANSWERS TO PRACTICE PROBLEMS FOR SECTION 4.4

1. See Figure 4.22.

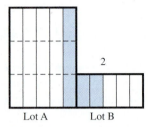

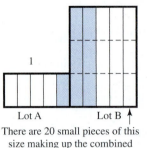

Lot A   Lot B

There are 20 small pieces of this
size making up the combined
lots A and B. Therefore, each such
piece is $\frac{1}{20}$, which is 5%. Since
7 are shaded, that means 35% of
the combined amount is shaded.

Lot A   Lot B

Here, 5 pieces are shaded. Each
piece is 5% of the total.
Therefore, 25% of the total
is shaded.

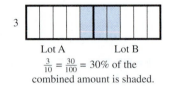

Lot A   Lot B

$\frac{3}{10} = \frac{30}{100} = 30\%$ of the
combined amount is shaded.

FIGURE 4.22

Various Percents of the Combined Amount of Lots A and B Are
Shaded

PROBLEMS FOR SECTION 4.4

1. Broad Street divides Popperville into an east side and a west side. On the east side of Popperville, 20% of the children qualify for a reduced-price lunch. On the west side of Popperville, 30% of the children qualify for a reduced-price lunch. Is it correct to calculate the percentage of children in all of Popperville who qualify for reduced-price lunch by adding 20% and 30% to get 50%? If the answer is no, why not? Explain in detail and calculate the correct percentage in at least two examples.

2. Anklescratch County and Kneebend County are two adjacent counties. In Anklescratch County, 30% of the miles of road have bike lanes, whereas in Kneebend County, only 10% of the miles of road have bike lanes.

    a. Elliot says that in the two-county area, 40% of the miles of road have bike lanes. Elliot found 40% by adding 30% and 10%. Is Elliot correct or not? Explain.

    b. What percent of miles of road in the two-county area have bike lanes if Anklescratch County has 200 miles of road and Kneebend County has 800 miles of road? What if it's the other way around and Anklescratch County has 800 miles of road, whereas Kneebend County has 200 miles of road?

    c. Ming says that 20% of the miles of road in the two-county area have bike lanes because 20% is the average of 10% and 30%. Explain why Ming's answer could be either correct or incorrect. Under what circumstances will Ming's answer be correct?

3. There are two elementary schools in the town of South Elbow. In the first school, 20% of the children are Hispanic. In the second school, 30% of the children are Hispanic. Write a brief essay about what you can and cannot tell from this data alone. Include examples to support your points.

4.5 Percent Increase and Percent Decrease

When a quantity increases or decreases, determining the amount of change is a simple matter of subtraction. However, in many situations, the actual value of the increase or decrease is less informative than the *percent* that this increase or decrease represents. For example, suppose that this year, there are 50 more children at Barrow Elementary School than there were last year. If Barrow Elementary only had 100 children last year, then 50 additional children is a very large increase. On the other hand, if Barrow Elementary had 500 children last year, an increase of 50 children is less significant. In this section, we will study increases and decreases in quantities as *percents* rather than as fixed values.

percent increase   If the value of a quantity goes up, then the increase in the quantity, figured as a percent of the original, is the **percent increase** of quantity. If a piece of furniture cost $279 last week, and this week the same piece of furniture costs $319, then the price went up by $40. The percent increase of the cost of the furniture is the percent represented by this $40 increase over the original cost, $279. What percent of $279 is $40? Let $P\%$ be this percentage. Then

$$P\% \cdot 279 = 40$$

So,

$$P\% = \frac{40}{279} = .143 = 14.3\%$$

Therefore, $40 is about 14% of $279, so the price of the furniture increased by about 14%.

percent decrease   When a quantity decreases in value, there is the notion of a **percent decrease**. It is the decrease, figured as a percent of the original. If Lower Heeltoe got 215 inches of rainfall in 2000 and only 193 inches of rainfall in 2001, then Lower Heeltoe got 22 inches less rain in 2001 than in 2000. What percent of 215 inches is 22 inches? Let $P\%$ be this percentage. Then

$$P\% \cdot 215 = 22$$

So,

$$P\% = \frac{22}{215} = .102 = 10.2\%$$

Therefore, 22 inches is 10.2% of 215 inches. So the amount of rainfall in Lower Heeltoe decreased by 10.2% from 2000 to 2001.

TWO METHODS FOR CALCULATING PERCENT INCREASE OR DECREASE

In this section we describe two methods for finding a percent increase or decrease when we know the original and final amounts.

THE FIRST METHOD

Our first method for calculating a percent increase or percent decrease uses the meaning of these terms. We used this method in the previous examples. In general, this method works as follows: Suppose a quantity changes from an amount A—the reference amount—to an amount B:

$$A \qquad \rightarrow \qquad B$$
reference amount   changed amount

To calculate the percent increase or decrease in the quantity, we use the following procedure:

1. Calculate the *change*, C, in the quantity, namely either $B - A$ or $A - B$, whichever is positive (or 0).

2. Calculate the percent that the change C is of the reference amount A. If we call this percent $P\%$, then

$$P\% \cdot A = C$$

So,

$$P\% = \frac{C}{A}$$

When you divide C by A, you will get this percentage as a decimal number. To write it as a percent, multiply by 100:

$$100 \times \frac{C}{A}$$

The result is the percent increase or decrease of the quantity from A to B.

THE SECOND METHOD

Our second method for calculating percent increase or decrease is usually more efficient. To derive this method, suppose that a quantity increases from an amount A to a larger amount B. If you calculate B as a percent of A, you will find that it is more than 100%. This makes sense because B is more than A, so B must represent more than 100% of A. *The amount by which B, calculated as a percent of A, exceeds 100%, is the percent increase from A to B.* On the left of Figure 4.23, we see that B is 135% of A. Therefore, the increase from A to B is 35%, as represented by the distance from A to B.

The following steps describe the calculation illustrated on the left of Figure 4.23:

1. Calculate B as a percent of A. You can do this by dividing B by A and multiplying the result by 100:

$$100 \times \frac{B}{A}$$

2. Subtract 100%. The result is the percent increase of the quantity from A to B.

FIGURE 4.23

Calculating Percent Increase and Decrease

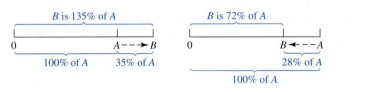

For example, suppose that the population of a city increases from 35,000 people to 44,000 people. What is the percent increase in the population?

$$44{,}000 \div 35{,}000 = 1.257 = 125.7\%$$

Subtracting 100%, we determine that the population increased by 25.7%.

Similarly, if a quantity decreases from an amount A to a smaller amount B, then if you calculate B as a percent of A, you will find that it is less than 100%. This makes sense because B is less than A, so B must represent less than 100% of A. *The amount that B, calculated as a percent of A, is under 100%, is the percent decrease from A to B.* On the right of Figure 4.23, we see that B is 72% of A. Therefore, the decrease from A to B is 28% (namely, $100\% - 72\%$), as represented by the distance from A to B.

The following steps describe the calculation illustrated on the right of Figure 4.23:

1. Calculate B as a percent of A. You can do this by dividing B by A and multiplying the result by 100:

$$100 \times \frac{B}{A}$$

2. Subtract this percent from 100%. The result is the percent decrease of the quantity from A to B.

For example, suppose that the price of a computer dropped from $899 to $825. What is the percent decrease in the price of the computer?

$$825 \div 899 = .918 = 91.8\%$$

$$100\% - 91.8\% = 8.2\%$$

The price of the computer decreased by about 8%.

TWO METHODS FOR CALCULATING AMOUNTS WHEN THE PERCENT INCREASE OR DECREASE IS GIVEN

In this section, we describe two methods for calculating a new amount when we know the original amount and the percent increase or decrease. Our second method will also allow us to calculate the original amount if we know the final amount and the percent increase or decrease.

THE FIRST METHOD

When an amount and its percent increase are given, you can calculate the new amount by calculating the increase and adding this increase to the original amount. Similarly, when an amount and its percent decrease are given, you can calculate the new amount by calculating the decrease and subtracting this decrease from the original amount.

If a table costs $187 and if the price of the table goes up by 6%, then what will the new price of the table be? We find that 6% of $187 is

$$.06 \times \$187 = \$11$$

Thus, the new price of the table will be

$$\$187 + \$11 = \$198$$

On the other hand, if a table costs $187 and the price of the table goes down by 6%, then the new price of the table will be

$$\$187 - \$11 = \$176$$

THE SECOND METHOD

Our second method for calculating amounts when percent increases or decreases are given uses the same idea as the second method for calculating percent increases and decreases. Unlike the first method, we can use this method to calculate the *initial amount* when we know the percent increase or decrease and the *final amount*.

On the left of Figure 4.24, we see that if an amount A increases by 35% to an amount B, then

$$B \text{ is } 135\% \text{ of } A$$

In general, if an amount A increases by $P\%$ to an amount B, then

$$B \text{ is } (100 + P)\% \text{ of } A$$

So, if a table costs $187 and if the price of the table goes up by 6%, then the new price of the table will be $(100 + 6)\% = 106\%$ of $187, which is

$$1.06 \times \$187 = \$198$$

The new price of the table after the 6% increase is $198.

Notice that we can also solve the following type of problem, where the percent increase and the final amount are known, and the initial amount is to be calculated: If the price of gas went up 5% and is now $2.15 per gallon, then how much did gas cost before this increase? If A was the initial price per gallon of gas before the increase, then 105% of A is $2.15. Therefore,

$$1.05 \times A = \$2.15$$

So,

$$A = \frac{\$2.15}{1.05} = \$2.05$$

FIGURE 4.24

Calculating
Amounts from
the Percent
Increase or
Percent
Decrease

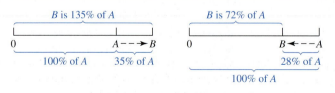

TABLE 4.1

Calculating the Initial Price of Gas before a 5% Increase by Calculating 1% of the Initial Price and then 100% of the Initial Price

			After a 5% increase, the price of gas is $2.15.
105%	$\longrightarrow$	$2.15	Therefore, 105% of the initial price is $2.15.
1%	$\longrightarrow$	$2.15 ÷ 105 = $0.0205	So 1% of the initial price is $0.0205.
100%	$\longrightarrow$	100 × $0.0205 = $2.05	Thus, 100% of the initial price of gas is $2.05.

Hence, gas cost $2.05 per gallon before the 5% increase. Table 4.1 shows another way to find the initial price of gas, using the calculation method described in Section 3.5 on page 98 and in Table 3.2.

The situation is similar when the percent decrease is given. On the right of Figure 4.24, we see that if an amount A decreases by 28% to an amount B, then

$$B \text{ is } 72\% \text{ of } A$$

In general, if an amount A decreases by $P\%$ to an amount B, then

$$B \text{ is } (100 - P)\% \text{ of } A$$

So, if a table costs $187 and if the price of the table goes down by 6%, then the new price of the table will be $(100 - 6)\% = 94\%$ of $187, which is

$$.94 \times \$187 = \$176$$

Therefore, after the 6% decrease in price, the new price of the table is $176.

Once again, we can also solve the following type of problem, where the percent decrease and the final amount are known, and the initial amount is to be calculated: If the price of gas went down 5% and is now $2.15 per gallon, then how much did gas cost before this decrease? If A was the initial price of gas before the decrease, then 95% of A is $2.15. Therefore,

$$.95 \times A = \$2.15$$

So,

$$A = \frac{\$2.15}{.95} = \$2.26$$

Then gas cost $2.26 per gallon before the 5% decrease. Table 4.2 shows another way to find the initial price of gas. As in the example of Table 4.1, this calculation uses the calculation method described in Section 3.5 on page 98 and in Table 3.2.

TABLE 4.2			After a 5% decrease, the price of gas is $2.15.
Calculating the Initial Price of Gas before a 5% Decrease by Calculating 1% of the Initial Price and then 100% of the Initial Price	95%	→ $2.15	Therefore, 95% of the initial price is $2.15.
	1%	→ $2.15 ÷ 95 = $0.0226	So 1% of the initial price is $0.0226.
	100%	→ 100 × $0.0226 = $2.26	Thus, 100% of the initial price of gas is $2.26.

THE IMPORTANCE OF THE REFERENCE AMOUNT

When you find a percent increase or decrease, make sure you calculate the increase or decrease as a percent *of the reference amount*, namely the amount for which you want to know the percent increase or decrease. Let's say the cost of a Dozey-Chair increases from $200 to $300. Then the new price is 50% more than the old price, but the old price is only 33% less than the new price. In both percent calculations, the change is $100. The different percentages come about because of the different reference amounts used. In the first case, the reference amount is $200, and $100 is 50% of $200. In the second case, the reference amount is $300, and $100 is only 33% of $300.

The *reference amount* plays the same role in percentages as the *whole* does in fractions: a percent is *of* some quantity, just as a fraction is *of* some whole.

CLASS ACTIVITY   NOW TURN TO CLASS ACTIVITIES MANUAL

4L Percent Increase and Decrease p. 71

4M Percent *Of* Versus Percent Increase or Decrease p. 73

PRACTICE PROBLEMS FOR SECTION 4.5

1. Last year, Ken had 2.5 tons of sand in his sand pile. This year, Ken has 3.5 tons of sand in his sand pile. By what percent did Ken's sand pile increase from last year to this year? First solve the problem by drawing a picture. Explain how your picture helps you solve the problem. Then solve the problem numerically.

2. Last year's profits were $16 million, but this year's profits are only $6 million. By what percent did profits decrease from last year to this year? First solve the problem by drawing a picture. Explain how your picture helps you solve the problem. Then solve the problem numerically.

3. If sales taxes are 6%, then how much should you charge for an item so that the total cost, including tax, is $35?

4. A pair of shoes has just been reduced from $75.95 to $30.38. Fill in the blanks:
   a. The new price is _____% less than the old price.
   b. The new price is _____% of the old price.
   c. The old price is _____% higher than the new price.
   d. The old price is _____% of the new price.

5. John bought a piece of land adjoining the land he owns. Now John has 25% more land than he did originally. John plans to give 20% of his new, larger amount of land to his daughter. Once John does this, how much land will John have in comparison to the amount he had originally? Draw a picture to help you solve this problem. Then solve the problem numerically, assuming, for example, that John starts with 100 acres of land.

6. The population of a certain city increased by 2% from 1996 to 1997 and then decreased by 2% from 1997 to 1998. By what percent did the population of the city change from 1996 to 1998? Did the population increase, decrease, or stay the same? Make a guess first, then calculate the answer carefully.

ANSWERS TO PRACTICE PROBLEMS FOR SECTION 4.5

1. Using a picture, we see that each strip in Figure 4.25 represents 20% of last year's sand pile. Since two additional strips have been added since last year, that is a 40% increase.

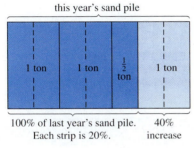

this year's sand pile

100% of last year's sand pile.   40%
Each strip is 20%.   increase

FIGURE 4.25

Last Year's and This Year's Sand Pile

Calculating the percent increase numerically gives

$$3.5 \div 2.5 = 1.4 = 140\%$$

Subtracting 100%, we see that the percent increase is 40%.

2. Using a picture, we see that each strip in Figure 4.26 represents $2 million. As the picture shows, the decrease in profits is $\frac{1}{8}$ more than 50%. Since $\frac{1}{8} = 12.5\%$, the profits decreased by 50% + 12.5%, namely by 62.5%.

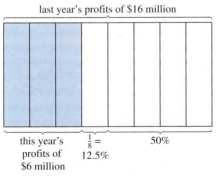

last year's profits of $16 million

this year's $\frac{1}{8} =$ 50%
profits of   12.5%
$6 million

FIGURE 4.26

Last Year's and This Year's Profits

Calculating the percent decrease numerically,

$$6 \div 16 = .375 = 37.5\%$$

Therefore, the percent decrease is $100\% - 37.5\%$, which is 62.5%.

3. The sales tax increases the amount that the customer pays by 6%. Therefore, if P represents the price of the item, 106% of P must equal $35. Thus,

$$1.06 \times P = \$35$$

So,

$$P = \$35 \div 1.06 = \$33.02$$

If the price of the item is $33.02, then with a 6% sales tax, the total cost to the customer is $35.

4. a. The new price is 60% less than the old price.
   b. The new price is 40% of the old price.
   c. The old price is 150% higher than the new price.
   d. The old price is 250% of the new price.

5. See Figure 4.27 on p. 140. If John starts with 100 acres of land and gets 25% more, then he will have 125 acres. We find that 20% of 125 acres is 25 acres, so if John gives 20% of his new amount of land away, he will have $125 - 25 = 100$ acres of land, which is the amount he started with.

6. Although you might have guessed that the population stayed the same, it actually decreased by .04%. Try this on a city of 100,000, for example. You can tell that the population will have to decrease because when the population goes back down by 2%, this 2% is *of a larger number* than the original population.

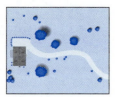

1. John's original plot of land

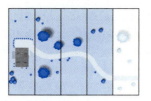

2. Now John has 25% more land.

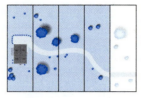

3. 20% of John's new, larger amount of land is represented by the unshaded part.

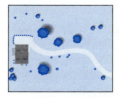

4. When John gives 20% of his larger plot of land away, he's left with the amount he started with.

FIGURE 4.27

Solving the Land Problem with a Picture

PROBLEMS FOR SECTION 4.5

1. Last year, the population of South Skratchankle was 48,000. This year, the population is 60,000. By what percent did the population increase? First solve the problem by drawing a picture. Explain how your picture helps you solve the problem. Then solve the problem numerically.

2. Last year's sales were $7.5 million. This year's sales are only $6 million. By what percent did sales decrease from last year to this year? First solve the problem by drawing a picture. Explain how your picture helps you solve the problem. Then solve the problem numerically.

3. Jayna and Lisa are comparing the prices of two boxes of cereal of equal weight. Brand A costs $3.29 per box and brand B costs $2.87 per box. Jayna calculates that

$$\frac{\$3.29}{\$2.87} = 1.15$$

Lisa calculates that

$$\frac{\$2.87}{\$3.29} = .87$$

a. Use Jayna's calculation to make *two* correct statements comparing the prices of brand A and brand B with percentages.

b. Use Lisa's calculation to make *two* correct statements comparing the prices of brand A and brand B with percentages.

4. Are the following two problems solved in the same way? Are the answers the same? Solve both and compare your solutions.

a. A television that originally cost $500 is marked down by 25%. What is its new price?

b. Last week a store raised the price of a television by 25%. The new price is $500. What was the old price?

5. How much should Swanko Jewelers charge now for a necklace if they want the necklace to cost $79.95 when they reduce their prices by 60%? Explain the reasoning behind your method of calculation.

6. If sales taxes are 7%, then how much should you charge for an item if you want the total cost, including tax, to be $15? Explain the reasoning behind your method of calculation.

7. Connie and Benton bought identical plane tickets, but Benton spent more than Connie (and Connie did not get her ticket for free).

a. If Connie spent 25% less than Benton, then did Benton spend 25% more than Connie? If not, then what percent more than Connie did Benton spend?

b. If Benton spent 25% more than Connie, then did Connie spend 25% less than Benton? If not, then what percent less than Benton did Connie spend?

8. Of the following 5 statements, which have the same meaning? In other words, which of these statements could be used interchangeably (in a news report, for example)? Explain your answers.

a. The price increased by 53%.

b. The price increased by 153%.

c. The new price is 153% of the old price.

d. The new price is higher than the old price by 53%.

e. The old price was 53% less than the new price.

9. Write a paragraph explaining the difference between a 150% increase in an amount and 150% of an amount. Give examples to illustrate.

10. The SuperDiscount store is planning a "35%-off sale" in two weeks. This week, a pair of pants costs $59.95.

a. Suppose SuperDiscount raises the price of the pants by 35% this week, and then two weeks from now, lowers the price by 35%. How much will the pants cost two weeks from now? Explain your method of calculation. Explain why it makes sense that the pants won't return to their original price of $59.95 two weeks from now. (A picture may help you explain.)

b. By what percent does SuperDiscount need to raise the price of the pants this week, so that two weeks from now, when they lower the price by 35%, the pants will return to the original price of $59.95? Explain your method of calculation.

11. Every week, DollarDeals lowers the price of items they have in stock by 10%. Suppose that the price of an item has been lowered twice, each time by 10% of that week's price. Explain why it makes sense that the total discount on the item is *not* 20%, even though the price has been lowered twice by 10% each time. A picture may help you explain. What percent is the total discount?

12. The following information about two different snack foods is taken from their packages:

SNACK	SERVING SIZE	CALORIES IN ONE SERVING	TOTAL FAT IN ONE SERVING
small crackers	28 g	140 calories	6 g
chocolate hearts	40 g	220 calories	13 g

Suppose we want to compare the amount of fat in these foods. One way to compare the fat is to compare the amount of fat in one serving of each food; another way is to compare the amount of fat in some fixed number of calories of the foods, such as 1 calorie or 100 calories; yet another way is to compare the amount of fat in a fixed number of grams of the foods, such as 1 gram or 100 grams. Solve the following problems by comparing the small crackers and chocolate hearts in different ways.

a. Explain how to interpret the information in the table so that the chocolate hearts have 117% more fat than the small crackers.

b. Explain how to interpret the information in the table so that the chocolate hearts have 52% more fat than the small crackers.

c. Explain how to interpret the information in the table so that the chocolate hearts have 38% more fat than the small crackers.

13. According to the 2000 Census, from 1990 to 2000 the population of Clarke County, Georgia, increased by 15.86% and the population of adjacent Oconee County increased by 48.85%.

   a. Can we calculate the percent increase in the total population of the two-county Clarke/Oconee area from 1990 to 2000 by adding 15.86% and 48.85%? Why or why not?

   b. Use the census data in the following table to calculate the percent increase in the total population of the two-county Clarke/Oconee area from 1990 to 2000:

COUNTY	1990 POPULATION	2000 POPULATION
Clarke	87,594	101,489
Oconee	17,618	26,225

14. In 1999, Washington County had a total population—urban and rural populations combined—of 200,000. From 1999 to 2000, the rural population of Washington County went up by 4%, and the urban population of Washington County went up by 8%.

   a. Based on the information above, make a reasonable guess for the percent increase of the total population of Washington County from 1999 to 2000. Based on your guess, what do you expect the total population of Washington County to have been in 2000?

   b. Make up three very different examples for the rural and urban populations of Washington County in 1999 (i.e., pick pairs of numbers that add to 200,000). For each example, calculate the total pop-

ulation in 2000, and calculate the percent increase in the total population of Washington County from 1999 to 2000. Compare these answers to your answers in part (a).

   c. If you only had the data given at the beginning of the problem (the 4% and 8% increases and the total population of 200,000 in 1999), would you be able to say exactly what the total population of Washington County was in 2000? Could you give a range for the total population of Washington County in 2000? In other words, could you say that the total population must have been between certain numbers in 2000? If so, what is this range of numbers?

15. One year on *National Equal Pay Day*, news reports stated that on average, women earn 74 cents for every dollar men earn.

   a. How might this "74 cents for every dollar" have been calculated? Make a reasonable guess and describe specifically how to do the calculation. (Different people may have different ideas about this calculation.)

   b. Assuming that the "74 cents for every dollar" statement is true, how much longer must the average woman work than the average man in order to make the same amount of money? Give a "ballpark" answer first, then find the exact answer.

   c. According to the description above, women make 26% less than men on average. Does it follow that the average woman has to work 26% longer in order to make the same amount of money as the average man? How does your answer to part (b) compare with this statement?

4.6 The Commutative and Associative Properties of Addition and Mental Math

In this section, we will study some techniques you can use to solve addition and subtraction problems mentally. Some of these techniques come from basic properties that we assume to be true for addition. These properties of addition are not only helpful for mental calculations, but they are also fundamental to the study of algebra.

> **CLASS ACTIVITY**   NOW TURN TO CLASS ACTIVITIES MANUAL
>
> **4N** Mental Math p. 74

In our study of the properties of addition, and later, in our study of other properties of arithmetic, we will consider mathematical expressions involving three or more numbers. Sometimes we will want to group the numbers in an expression in certain ways. In order to do this, we will need to use parentheses.

PARENTHESES IN EXPRESSIONS WITH THREE OR MORE TERMS

A sum

$$A + B + C$$

with 3 terms means the sum of $A + B$ and C. In other words, according to the meaning of $A + B + C$, this sum is to be calculated by first adding A and B, and then adding C. Likewise, if there are 4 or more terms in a sum, the sum stands for the result obtained by adding from left to right.

But what if we want to indicate the sum of A with $B + C$? We can show this with parentheses. In ordinary writing, parentheses are used for asides, but in mathematical expressions, parentheses are used to group numbers and operations $(+, -, \times, \div)$. To indicate the sum of A with $B + C$, simply write

$$A + (B + C)$$

So,

$$17 + (18 + 2) = 17 + 20$$
$$= 37$$

Notice that we can express the meaning of the sum

$$A + B + C$$

by using parentheses:

$$(A + B) + C$$

THE ASSOCIATIVE PROPERTY OF ADDITION

How can you make the problem

$$7384 + 999 + 1$$

easy to solve mentally? Rather than adding from left to right, it is easier to first add 999 and 1 to make 1000, and then add 7384 and 1000 to get 8384. In other words, rather than adding from left to right, calculating

$$(7384 + 999) + 1$$

according to the meaning of the sum $7384 + 999 + 1$, it is easier to group the 999 with the 1 and calculate

$$7384 + (999 + 1)$$

instead. Why is it legitimate to switch the way the numbers in the sum are grouped? Because, according to the *associative property of addition*, both ways of calculating the sum give equal results; in other words,

$$(7384 + 999) + 1 = 7384 + (999 + 1)$$

associative property of addition

The **associative property of addition** tells us that when we add any three numbers, it doesn't matter whether we add the first two and then add the third, or whether we add the first number to the sum of the second and the third—either way we will always get the same answer. In other words, the associative property of addition says that if A, B, and C are any three numbers, then

$$(A + B) + C = A + (B + C)$$

That is, the sum of $A + B$ and C is equal to the sum of A and $B + C$.

It might help you to understand the terminology *associative property* by remembering that to *associate* with someone means to keep company with them. In

$$(7384 + 999) + 1 = 7384 + (999 + 1)$$

the number 999 can either *associate* with 7384, or it can *associate* with 1.

We assume that the associative property of arithmetic is true for all numbers, but we can use an example to see why this property of addition makes sense. Suppose that Joe has 2 marbles, Sue has 3 marbles, and Terrell has 4 marbles. Sue and Terrell will give all their marbles to Joe, but let's compare two different ways this might happen:

First way: Sue gives her marbles to Joe and then Terrell gives his marbles to Joe.

	JOE	SUE	TERRELL
Beginning	oo	ooo	oooo
Middle	oo + ooo	Sue's go to Joe	oooo
End	(oo + ooo) + oooo		Terrell's go to Joe

According to this way of giving Joe the marbles, Joe has

$$(2 + 3) + 4$$

marbles in all. Why are the parentheses placed this way? At the last step, when Terrell gives Joe his 4 marbles, Joe already has the $2 + 3$ marbles from Sue, so the 4 marbles are added to the *combined amount* of $2 + 3$.

Second way: Terrell gives his marbles to Sue and then Sue gives the combined collection of Terrell's and her marbles to Joe.

	JOE	SUE	TERRELL
Beginning	oo	ooo	oooo
Middle	oo	ooo + oooo	Terrell's go to Sue
End	oo + (ooo + oooo)	Sue's go to Joe	

According to this way of giving Joe the marbles, Joe has

$$2 + (3 + 4)$$

marbles in all. Why are the parentheses placed this way? This time, Joe gets the *combined amount* of $3 + 4$ marbles, which is added to the 2 marbles that he already has.

Of course, Joe will have the same number of marbles in all, no matter which way he gets the marbles. We can express this by writing

$$(2 + 3) + 4 = 2 + (3 + 4)$$

The same kind of relationship would hold if other numbers of marbles replaced the numbers 2, 3, and 4. The associative property of addition states that this kind of relationship always holds, no matter what numbers are involved.

THE COMMUTATIVE PROPERTY OF ADDITION

How can you make the addition problem

$$2997 + 569 + 3$$

easy to solve mentally? Adding 2997 and 569 is hard, but if we combine the 2997 with the 3 to make 3000, then it's easy to add the 569 to make 3569. In other words, rather than adding the numbers from left to right in the order they appear, it's easier to switch the order of the 569 and the 3, and calculate

$$2997 + 3 + 569$$

instead. Why is it legitimate to switch the order of 569 and the 3 in the original sum? Because

$$569 + 3 = 3 + 569$$

and this is true by virtue of the *commutative property of addition*.

commutative property of addition

The **commutative property of addition** states that for all real numbers A and B,

$$A + B = B + A$$

To illustrate the commutative property, assume you start with 3 apples and you get 2 more. Then you have

$$3 + 2$$

FIGURE 4.28

The
Commutative
Property:
$3 + 2 = 2 + 3$

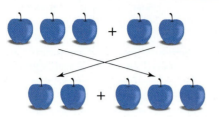

apples. But what if you had started instead with 2 apples and then got 3 more? In this case you'd have

$$2 + 3$$

apples. Common sense tells us it doesn't matter in which order you got the apples, either way you'll have the same number of apples in all, as we see in Figure 4.28. Another way to say that you have the same amount of apples either way is to say

$$3 + 2 = 2 + 3$$

The commutative property of addition says that no matter what the numbers A and B are, even if they are fractions or decimals, if you start with A objects and get B more objects, you have the same number in all as if you started with B objects and got A more. In other words,

$$A + B = B + A$$

for all numbers A and B.

In this setting, the word *commute* means *to change places with*. In going from $3 + 2$ to $2 + 3$, the 2 and the 3 change places. The term **turnarounds** is sometimes used with children instead of *commutative property* because it seems to be easier to understand.

Students sometimes get confused between the commutative and associative properties. Notice that the commutative property involves changing the order of *two* terms, while the associative property involves changing the location of parentheses when *three* terms are involved.

WRITING EQUATIONS THAT CORRESPOND TO A METHOD OF CALCULATION

We have just seen that the associative and commutative properties of addition can sometimes be used to make mental addition problems easier. In order to show a line of reasoning for a calculation, it is useful to write sequences of equations that correspond to the line of reasoning.

An equation is a mathematical statement saying that two numbers or expressions are equal. The familiar equal sign, $=$, shows this equivalence. Remember that the equal sign means one and only one thing: $=$ means *equals* or *is equal to*.

An equation is actually just a sentence, and you should read it and interpret it as a sentence. The equation

$$7 + (6 + 4) = 17$$

is equivalent to the following sentence:

7 plus the sum of 6 and 4 is equal to 17.

In order to show a method of calculation, we often string several equations together. Rather than writing

$$198 + 357 + 2 = 198 + 2 + 357$$
$$198 + 2 + 357 = 200 + 357$$
$$200 + 357 = 557$$

it is neater and easier to follow this sequence of steps:

$$198 + 357 + 2 = 198 + 2 + 357$$
$$= 200 + 357$$
$$= 557$$

When several equations are strung together, it is for the purpose of concluding that the *first* expression written is equal to the *last* expression written. The purpose of the previous equations is to conclude that $198 + 357 + 2$ is equal to 557. This uses the following general property of equality, which we assume holds true:

If A is equal to B and if B is equal to C, then A is also equal to C.

A COMMON ERROR IN USING $=$

It is common for students at all levels (including college) to make careless, incorrect use of the equal sign. Because the proper use of the equal sign is essential in algebra, and because elementary school mathematics lays the foundation for learning algebra, it is especially important for you to use the equal sign correctly.

When you use an equal sign, *make sure that the quantities before and after the equal sign really are equal to each other*. It is easy to make a careless error with the equal sign when solving a problem in several steps. Suppose you calculate

$$499 + 165$$

by taking 1 from 165, adding this 1 to 499 to make 500, and then adding on the remaining 164 get 664. How can you show this method of calculation with equations? It is easy to make the following mistake:

$$\boxed{\text{Warning: incorrect}} \quad 499 + 1 = 500 + 164$$
$$= 664$$

These equations are incorrect because $499 + 1$ *is not equal to* $500 + 164$. Instead, you can write the following correct equations:

$$499 + 165 = 499 + 1 + 164$$
$$= 500 + 164$$
$$= 664$$

USING THE ASSOCIATIVE AND COMMUTATIVE PROPERTIES TO SOLVE ADDITION PROBLEMS

Both children and adults use the associative and commutative properties of addition when solving addition problems mentally. You may use these properties and not even be aware that you are doing so. In fact, it is good to develop an easy and fluent use of the properties, because they are essential in algebra.

Children will use the commutative property of addition if they are ready to do so. Young children (first grade or so) often solve addition problems by "counting on." To solve $8 + 2$, a child could start with 8 and count: "9, 10." If the child is given the problem of solving $2 + 8$, the child might start with 2 and count on 8 more. Or the child might use the commutative property—even if he does not know this term—and replace the problem $2 + 8$ with the problem $8 + 2$, and count on from 8 instead of from 2.

Children also use the associative property of addition spontaneously. When Arianna, a kindergartner, was asked how to solve $8 + 7$, she described her reasoning this way, referring to a little jingle about $8 + 8$ that she had learned earlier:

> I know that *eight and eight is sixteen, dip your nose in ice cream*. So 8 and 7 is one less, which is 15.

The following equations correspond to Arianna's method:

$$
\begin{aligned}
8 + 7 &= 8 + (8 - 1) \\
&= (8 + 8) - 1 \\
&= 16 - 1 \\
&= 15
\end{aligned}
$$

We can see by the way that the parentheses were switched at the second equals sign that Arianna used the associative property, even though she certainly didn't know that term. Actually, since the associative property refers to *addition* and not *subtraction*, and since subtracting 1 is the same as adding -1, it is better to use the following equations to demonstrate the use of the associative property:

$$
\begin{aligned}
8 + 7 &= 8 + (8 - 1) \\
&= 8 + (8 + (-1)) \\
&= (8 + 8) + (-1) \\
&= 16 + (-1) \\
&= 15
\end{aligned}
$$

We can also use the associative property of addition to calculate the sum of two numbers by *shifting* a part of one number to the other number. For example, to solve

$$489 + 73$$

mentally, you could use this reasoning:

> If I take 11 of the 73 and add it to 489, the result is 500. This 500, plus the remaining 62, gives 562.

FIGURE 4.29

Using the
Associative
Property to Shift
11 Toothpicks,
Changing
$489 + 73$ into
$500 + 62$

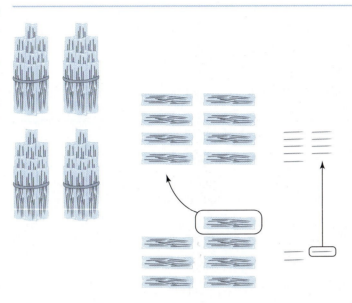

You can see why this method makes sense by imagining that you are combining 489 toothpicks with 73 toothpicks. How many toothpicks do you have in all? If you move 11 of the 73 toothpicks (1 ten and 1 one) to the pile of 489 toothpicks, then the large pile now has 500 toothpicks, and the small pile has 62 toothpicks. All together that's 562 toothpicks. Figure 4.29 illustrates this moving of toothpicks. When we write equations corresponding to this method of solution, we see that we used the associative property of addition:

$$489 + 73 = 489 + (11 + 62)$$
$$= (489 + 11) + 62$$
$$= 500 + 62$$
$$= 562$$

THE COMMUTATIVE AND ASSOCIATIVE PROPERTIES OF ADDITION IN ALGEBRA

The commutative and associative properties of addition allow for great flexibility in calculating sums: you can rearrange the terms any way you like, and you can combine the terms with each other any way you like. This flexibility is valuable not only for mental calculations, but is also an essential skill in algebra. When we combine like terms in the expression

$$3x + 5 + 2x^2 - 4 + 8x - 5x^2$$

in order to rewrite it as

$$-3x^2 + 11x + 1$$

we have used the commutative and associative properties of addition.

CLASS ACTIVITY   NOW TURN TO CLASS ACTIVITIES MANUAL

4O Using Properties of Addition in Mental Math p. 75

4P Using Properties of Addition to Aid Learning of Basic Addition Facts p. 77

OTHER MENTAL METHODS OF ADDITION AND SUBTRACTION

In addition to the associative and commutative properties of addition, there are other ways to use reasoning to solve addition and subtraction problems mentally. Ultimately, all these methods also rely on the commutative and associative properties of addition, but we will not emphasize that these properties were used.

ROUNDING AND COMPENSATING

Sometimes an addition or subtraction problem involves a number that is close to a round number that is easy to work with. In this case, either we can use the associative property to "shift" numbers (as in Figure 4.29), or we can round the numbers and compensate for that rounding after adding or subtracting.

To solve $376 + 199$ we can round and compensate:

Suppose we add 200 to 376 instead of adding 199 to 376. This makes 576. But we added 1 too many, so we must take 1 away from 576. Therefore $376 + 199 = 575$.

We can write corresponding equations as follows:

$$376 + 199 = 376 + 200 - 1$$
$$= 576 - 1$$
$$= 575$$

We can also round and compensate in subtraction problems. Consider the problem $684 - 295$. The number 295 is close to 300. So, to calculate

$$684 - 295$$

we can reason as follows:

Taking 300 objects away from 684 objects leaves 384 objects. But when we took 300 away, we took away 5 more than we should have (because 300 is 5 more than 295). This means that we must *add* 5 to 384 to get the answer, 389.

The following equations correspond to this line of reasoning:

$$684 - 295 = 684 - 300 + 5$$
$$= 384 + 5$$
$$= 389$$

SUBTRACTING BY ADDING ON

Another way to solve $684 - 295$ is to view the difference $684 - 295$ as the amount we must add to 295 to make 684. This method interprets subtraction as *comparison*, as was discussed in Section 4.1. With this point of view, we start with 295 and keep adding numbers until we reach 684:

$$295 + 5 = 300$$

$$300 + 300 = 600$$

$$600 + 84 = 684$$

or

$$295 + 5 + 300 + 84 = 684$$

All together, starting with 295, we added

$$5 + 300 + 84 = 389$$

to reach 684. Therefore, $684 - 295 = 389$. This is the method that shopkeepers frequently use to make change.

WHY SHOULD WE BE ABLE TO USE REASONING TO ADD AND SUBTRACT?

In the previous examples, we saw how to use the associative and commutative properties and other ways of reasoning in order to solve addition and subtraction problems. Why should we be able to use these methods when we already have standard algorithms that always work? Although it might be easier to teach only the standard algorithms, the nonstandard methods encourage reasoning that is essential in the study of algebra. The mathematics that you will teach in elementary school will lay the foundation for algebra. This foundation must be strong, so that your students will be able to progress in mathematics. Currently, most high schools require at least two years of math to graduate.

A strong foundation in mathematics will be a wise economic investment for your students. According to the Bureau of Labor Statistics of the U.S. Department of Labor, on average, workers with more education have higher earnings and a lower unemployment rate. (See [41].) The U.S. Department of Education's paper "Mathematics Equals Opportunity" (see [40]) reported that students who take rigorous mathematics courses are more likely to go on to college, that algebra is a "gateway" course to more advanced mathematics courses in high school, and that low-income students who take Algebra I and Geometry are much more likely to attend college than those who do not. It is not an exaggeration to say that a strong foundation in mathematics can have a lifelong positive impact.

In addition, according to the National Center for Education Statistics, workers who had scored in the top quartile on the Armed Services Vocational Aptitude Battery test in mathematics earned, on average, over 35% more per hour than workers with lower scores. (See [19].) Some of the fastest growing jobs, such as those involving computers or health care, often require a good ability in math. Some jobs that seem like they wouldn't need math actually do. For example, according to the Bureau of Labor Statistics (see [42]), foresters and conservation scientists require advanced levels of theoretical math. Landscape architects, aircraft pilots, and veterinarians need to know applied math. Sheet-metal workers, bricklayers, and jewelers need to know practical "shop" math, and arithmetic is important for ticket agents and stock clerks. Of course, there are many excellent jobs that don't require much math, but education should equip students with skills that prepare them for a broad range of opportunities.

CLASS ACTIVITY   NOW TURN TO CLASS ACTIVITIES MANUAL

4Q Writing Correct Equations p. 80

4R Writing Equations That Correspond to a Method of Calculation p. 81

4S Other Ways to Add and Subtract p. 82

4T A Third Grader's Method of Subtraction p. 84

PRACTICE PROBLEMS FOR SECTION 4.6

1. Give an example that demonstrates how you can use the associative property of addition to make a problem easier to do mentally. Describe in words how you can solve the problem mentally. Write equations that show where you used the associative property.

2. The sequence of equations that follows shows a way of using properties of addition to calculate a sum. Say specifically which properties of addition were used, and where.

$$27 + 89 + 13 = 27 + (89 + 13)$$
$$= 27 + (13 + 89)$$
$$= (27 + 13) + 89$$
$$= 40 + 89$$
$$= 129$$

3. For each of the following addition problems, write equations that correspond to a mental method for calculating the sum that uses the associative and/or the commutative properties of addition. Say specifically which properties of addition were used, and where.

   a. $993 + 2389$

   b. $398 + (76 + 2)$

4. Each arithmetic problem in this exercise has a description for how to solve the problem. In each case, write a sequence of equations that correspond to the given description.

   a. Problem: $23 + 45$

   Solution: 23 plus 40 is 63, and then 5 more makes 68.

   b. Problem: $800 - 297$

   Solution: $800 - 300 = 500$, but subtracting 300 subtracts 3 too many, so we must add 3 back, making 503.

5. Nancy writes the following equations in order to solve $37 + 14$:

$$30 + 10 = 40 + 7 = 47 + 4 = 51$$

   Write correct equations that solve $37 + 14$ and that incorporate Nancy's solution strategy.

6. Jim writes the following equations in order to solve $85 - 15$:

$$85 - 10 = 75 - 5 = 70$$

   Write correct equations that solve $85 - 15$ and that incorporate Jim's solution strategy.

7. Find ways to solve the next set of addition and subtraction problems *other than* using the standard addition or subtraction algorithms. In each case, explain your reasoning and also write equations that incorporate your thinking.

a. $786 - 47$

b. $427 + 28$

c. $999 + 999$

d. $1002 - 986$

e. $237 - 40$

ANSWERS TO PRACTICE PROBLEMS FOR SECTION 4.6

1. To calculate $49 + 37$ mentally, we can think of moving 1 from the 37 to the 49, so that the sum becomes $50 + 36$, which we can easily see is 86. The corresponding equations show that the associative property of addition was used:

$$49 + 37 = 49 + (1 + 36)$$
$$= (49 + 1) + 36$$
$$= 50 + 36$$
$$= 86$$

The associative property of addition is used at the second equals sign to switch the placement of parentheses from grouping 1 and 36 together, to grouping 49 and 1 together.

2. The associative property of addition was used at the first equals sign to say that $(27 + 89) + 13 = 27 + (89 + 13)$. The commutative property of addition was used at the second equals sign to change $89 + 13$ to $13 + 89$. The associative property of addition was used at the third equals sign to say that $27 + (13 + 89) = (27 + 13) + 89$.

3. a. $993 + 2389 = 993 + (7 + 2382) = (993 + 7) + 2382 = 1000 + 2382 = 3382$. The associative property of addition was used in rewriting $993 + (7 + 2382)$ as $(993 + 7) + 2382$.

   b. $398 + (76 + 2) = 398 + (2 + 76) = (398 + 2) + 76 = 400 + 76 = 476$. The commutative property of addition was used to rewrite $76 + 2$ as $2 + 76$. The associative property of addition was used to rewrite $398 + (2 + 76)$ as $(398 + 2) + 76$

4. a.
$$23 + 45 = 23 + (40 + 5)$$
$$= (23 + 40) + 5$$
$$= 63 + 5$$
$$= 68$$

Note that the associative property of addition is used at the second equals sign in order to switch the placement parentheses. This has the effect of grouping

the 40 with the 23 instead of with the 5 (where it was part of 45).

b.
$$800 - 297 = 800 - 300 + 3$$
$$= 500 + 3$$
$$= 503$$

5.
$$37 + 14 = 30 + 10 + 7 + 4$$
$$= 40 + 7 + 4$$
$$= 47 + 4$$
$$= 51$$

6.
$$85 - 15 = 85 - 10 - 5$$
$$= 75 - 5$$
$$= 70$$

7. Here are equations you could write:

a.
$$786 - 47 = 786 - 50 + 3$$
$$= 736 + 3$$
$$= 739$$

b.
$$427 + 28 = 427 + 3 + 25$$
$$= 430 + 25$$
$$= 455$$

c.
$$999 + 999 = 999 + 1000 - 1$$
$$= 1999 - 1$$
$$= 1998$$

d.
$$986 + 4 = 990$$
$$990 + 10 = 1000$$
$$1000 + 2 = 1002$$

so,
$$986 + 4 + 10 + 2 = 1002$$

so,
$$1002 - 986 = 4 + 10 + 2$$
$$= 16$$

e.
$$237 - 40 = 240 - 40 - 3$$
$$= 200 - 3$$
$$= 197$$

PROBLEMS FOR SECTION 4.6

1. Many teachers have a collection of small cubes that can be snapped end-to-end to make "trains" of cubes. The cubes come in various colors.

   a. Describe in detail how you could use snap-together cubes in different colors to demonstrate the commutative property of addition.

   b. Describe in detail how you could use snap-together cubes in different colors to demonstrate the associative property of addition.

2. Figure 4.30 indicates a "make-a-10" strategy for adding $6 + 8$.

   a. Write equations that correspond to the make-a-10 strategy for adding $6 + 8$ depicted in Figure 4.30. Your equations should make careful and appropriate use of parentheses. Which property of arithmetic do your equations and the picture of Figure 4.30 illustrate?

   b. Draw a picture for $7 + 5$ that illustrates a make-a-10 strategy. Write equations that correspond to the strategy indicated in your picture, making careful and appropriate use of parentheses.

   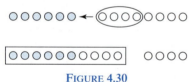

   FIGURE 4.30

   A Dot Picture for $6 + 8$

3. Tomaslav has learned the following facts well:

   • All the sums of whole numbers that add to 10 or less— Tomaslav knows these facts "forwards and backwards." For example, he knows not only that $5 + 2$ is 7, but also that 7 breaks down into $5 + 2$ or $2 + 5$.

   • $10 + 1, \ 10 + 2, \ 10 + 3, \ \ldots, \ 10 + 10$.

   • the *doubles* $1 + 1, \ 2 + 2, \ 3 + 3, \ \ldots, \ 10 + 10$.

   For each sum in this problem, describe at least three different ways that Tomaslav could use reasoning, together with the facts he knows well, to determine the sum. In each case, write equations that correspond to the strategies you describe. Take care to use parentheses appropriately and as needed.

   a. $8 + 7$

   b. $6 + 7$

   c. $8 + 9$ (Try to find four or five different ways to solve this.)

4. When we use the standard subtraction algorithm to subtract

   $$\begin{array}{r} 52 \\ -\ 19 \\ \hline \end{array}$$

   we must regroup. Which property or properties of arithmetic do we use when we regroup? Answer this question by writing 52 in expanded form and by using parentheses carefully to show regrouping.

5. To solve $159 - 73$, a student writes the following equations:

   $$160 - 70 = 90 - 3 = 87 - 1 = 86$$

   Although the student has a good idea for solving the problem, his equations are not correct. In words, describe the student's solution strategy and discuss why the strategy makes sense; then write a correct sequence of equations that correspond to this solution strategy. Write your equations in the following form:

   $$159 - 73 = \text{some expression}$$
   $$= \text{some expression}$$
   $$= \vdots$$
   $$= 86$$

6. To solve $201 - 88$, a student writes the following equations:

   $$88 + 2 = 90 + 10 = 100 + 100 = 200 + 1 = 201$$
   $$2 + 10 = 12 + 100 = 112 + 1 = 113$$

   Although the student has a good idea for solving the problem, his equations are not correct. In words, describe the student's solution strategy and explain why it makes sense; then write correct equations that correspond to this solution strategy.

7. *Problem:* If I have 12 eggs and use 5 in a recipe, how many eggs will I have left?

   Arianna's solution: There are 7 eggs left, because if I take 5 away from 10, there will be 5 left. So I then count 6 for 11 and 7 for 12.

   Write a coherent sequence of equations that solve $12 - 5$ and that show why Arianna's reasoning is valid.

8. David and Ashley want to solve $8.27 - 2.98$ by first solving $8.27 - 3 = 5.27$. David says that they must *subtract* .02 from 5.27, but Ashley says that they must *add* .02 to 5.27.

a. Draw a number line (which need not be perfectly to scale) to help you explain who is right and why. Do not just say which answer is numerically correct; use the number line to help you explain why the answer must be correct.

b. Explain in another way who is right and why.

9. Tylishia says that she can solve $324 - 197$ by adding 3 to both numbers and solving $327 - 200$ instead.

a. Draw a number line (which need not be perfectly to scale) to help you explain why Tylishia's method is valid.

b. Explain in another way why Tylishia's method is valid.

c. Could you adapt Tylishia's method to other subtraction problems, such as to the problem $183 - 49$? If so, give at least two more examples and show how to apply Tylishia's method in each case.

10. To solve $512 - 146$, a student writes the following:

$$\begin{array}{rr} 512 & 400 \\ -\ 146 & -\ 30 \\ \hline -4 & 370 \\ -30 & -4 \\ 400 & \hline 366 \end{array} \qquad 512 - 146 = 366$$

Describe the student's solution strategy and discuss why the strategy makes sense. Expanded forms may be helpful to your discussion.

11. Here's how Mo solved the subtraction problem $635 - 813$:

$$\begin{array}{r} 635 \\ -\ 813 \\ \hline -\ 222 \end{array}$$

Mo did this by working from right to left, saying

$$5 - 3 = 2$$
$$3 - 1 = 2$$
$$6 - 8 = -2$$

a. Is Mo's answer right?

b. Solve

$$\begin{array}{r} 6(100) + 3(10) +\ 5(1) \\ -\ [8(100) + 1(10) + 3(1)] \\ \hline \end{array}$$

by working with expanded forms. Discuss how Mo's work compares with your work in expanded forms.

12. Describe a way to solve $304 - 81$ mentally, by using reasoning other than the standard subtraction algorithm. Then write a coherent sequence of equations that correspond to your reasoning.

13. Give an example of an arithmetic problem that can be made easy to solve mentally by using the associative property of addition. Write equations that show your use of the associative property of addition. Your use of the associative property must genuinely make the problem easier to solve.

14. Give an example of an arithmetic problem that can be made easy to solve mentally by using the commutative property of addition. Write equations that show your use of the commutative property of addition. Your use of the commutative property must genuinely make the problem easier to solve.

15. Is there an *associative property of subtraction*? In other words, is

$$A - (B - C) = (A - B) - C$$

true for all real numbers A, B, C? Explain your answer. If the equation is always true, explain why; if the equation is not always true, find another expression that *is* equal to the expression

$$A - (B - C)$$

and explain why the two expressions are equal.

16. Is there a *commutative property of subtraction*? Explain why or why not. In your answer, include a statement of what a *commutative property of subtraction* would be if there were such a property.

Chapter 5

Multiplication

I n this chapter, we will study the meaning of multiplication, ways of representing multiplication, the properties of multiplication, and the procedures we use to multiply whole numbers. Because we use multiplication so often, we may take the process for granted, and we may think that it is easy. However, even though it is easy for adults to carry out the procedure of multiplying, the underlying concept of multiplication is much more subtle. As we will see, the procedures we use when we multiply are based on very clever uses of the properties of arithmetic.

5.1 The Meaning of Multiplication and Ways to Show Multiplication

In this section we will discuss the meaning of multiplication. We will also study various ways of picturing and thinking about multiplication of whole numbers; some people call these multiplication models. These models will provide us with different ways to organize information to make the multiplicative structure in a given situation visible. By working with these different ways of displaying multiplication in a variety of situations, you will deepen your understanding of the concept of multiplication. Part of having a deep understanding of multiplication is to know not only *that* you multiply to solve a certain problem, but to know *why* the problem calls for multiplication, and not some other operation, such as addition. In order to understand how and when to use multiplication, you will need to focus on the meaning of multiplication.

THE MEANING OF MULTIPLICATION

What does multiplication mean? If A and B are nonnegative numbers, then

$$A \times B \quad \text{or} \quad A \cdot B$$

meaning of multiplication

product

which we read as "A times B," means the total number of objects in A groups if there are B objects in *each* group. The result $A \times B$ is called the **product** of A and B, and the numbers A

factors

and *B* are called **factors**. So if there are 237 bags and each bag contains 46 potatoes, then the total number of potatoes is 237×46, which turns out to be 10, 902 potatoes.

multiple

Given a whole number *A*, a whole number *B* that can be written as a whole number times *A* is called a **multiple** of *A*. For example, 15 is a multiple of 5 because $15 = 3 \times 5$ and the numbers

$$4, \ 8, \ 12, \ 16, \ 20, \ 24, \ldots$$

are multiples of 4.

SHOWING MULTIPLICATIVE STRUCTURE BY GROUPING

grouping

The simplest way to show that a situation requires multiplication is to use **grouping**. When we arrange a collection of objects into *A* groups with *B* objects in each group, then we know that according to the meaning of multiplication, there are $A \times B$ objects in all. Figure 5.1 shows that there are a total of

$$3 \times 4$$

dots pictured because the dots are arranged into 3 groups of 4.

SHOWING MULTIPLICATIVE STRUCTURE WITH ORGANIZED LISTS

organized list

Another way to show that a situation involves multiplication is with an **organized list**, which is a list of information that is organized to show structure. Suppose you want to represent the different ways to order a burger and fries at a restaurant. Let's say there are two types of burgers from which to choose: cheeseburger C and regular burger B. Then let's say there are three types of fries from which to choose: small S, large L, and jumbo J. Now we can make the organized list in Table 5.1 to show all the possible combinations of a burger and fries.

By analyzing an organized list we can show *why* we can multiply to determine the number of choices for the combination of a burger and fries. As Figure 5.2 shows, the organized list consists of 2 natural groups with 3 items in each group (where each item in the group consists

TABLE 5.1

An Organized List
of Choices of
Burgers and Fries

B S	regular burger and small fries
B L	regular burger and large fries
B J	regular burger and jumbo fries
C S	cheeseburger and small fries
C L	cheeseburger and large fries
C J	cheeseburger and jumbo fries

FIGURE 5.2

Analyzing an
Organized List
to Show
Multiplicative
Structure

of a *combination* of burger and fries). Therefore, according to the meaning of multiplication, there are

$$2 \times 3$$

ways to choose the combination of a burger and fries.

Now suppose that, in addition to the two type of burgers (C, B) and the three types of fries (S, L, J), you also have a choice of two types of cola: regular (R) and diet (D). How many different ways are there to get a combination of a cola, a burger, and fries? Table 5.2 shows all possible combinations of soda, burger, and fries in an organized list.

As Figure 5.3 shows, the organized list in Table 5.2 consists of 2 natural groups. When we look in closer detail, we see that each of the 2 groups can be broken into 2 smaller groups, with 3 entries in each of these smaller groups. Therefore, according to the meaning of multiplication, there are

$$2 \times (2 \times 3)$$

ways to choose the combination of a cola, a burger, and fries, which works out to 12 different ways. Notice that the parentheses in the expression $2 \times (2 \times 3)$ are used to show that there are 2 large groups and that each group contains 2×3 items.

SHOWING MULTIPLICATIVE STRUCTURE WITH ARRAY DIAGRAMS

Consider once again the situation of two types of burgers (regular burger B, and cheeseburger C) and three types of fries (small S, large L, and jumbo J) to choose from at a restaurant. How many ways are there to choose the combination of a burger and fries? We can use an *array*

TABLE 5.2

An Organized List
of Choices of
Soda, Burger, and
Fries

R B S	regular cola, regular burger, and small fries
R B L	regular cola, regular burger, and large fries
R B J	regular cola, regular burger, and jumbo fries
R C S	regular cola, cheeseburger, and small fries
R C L	regular cola, cheeseburger, and large fries
R C J	regular cola, cheeseburger, and jumbo fries
D B S	diet cola, regular burger, and small fries
D B L	diet cola, regular burger, and large fries
D B J	diet cola, regular burger, and jumbo fries
D C S	diet cola, cheeseburger, and small fries
D C L	diet cola, cheeseburger, and large fries
D C J	diet cola, cheeseburger, and jumbo fries

FIGURE 5.3

Analyzing an
Organized List
to Show
Multiplicative
Structure

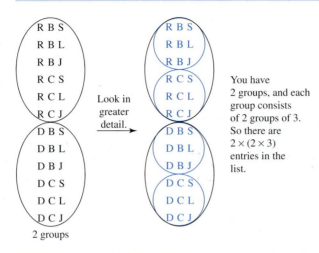

2 groups

TABLE 5.3

An Array
Diagram for the
Ways to Choose a
Burger and Fries

B, S	B, L	B, J
C, S	C, L	C, J

array diagram

diagram to display all the ways of choosing a burger and fries. An **array diagram** is a display of information that is arranged into rows and columns.

Each entry in the array in Table 5.3 corresponds to a choice of burger and fries. For example, the entry C, L corresponds to the choice of a cheeseburger and large fries. If we consider each row in the array to be a group, then the array consists of 2 groups with 3 items in each group. Therefore, according to the meaning of multiplication, there are

$$2 \times 3$$

items in the array, and thus 2×3 ways to choose the combination of a burger and fries. Notice that if we choose our groups to be the columns of the array rather than the rows, then we would say that there are

$$3 \times 2$$

items in the array, because there are 3 groups of 2 in this case.

Now consider the situation of two types of cola, regular cola R and diet cola D, in addition to the two types of burgers (B, C) and three types of fries (S, L, J). How many ways are there to choose the combination of cola, a burger, and fries? We can show all possible choices by showing two array diagrams, one for each type of cola, as in Table 5.4. Once again, we can see the multiplicative structure in this situation because the choices for a cola, burger, and fries are divided into 2 groups with 2×3 items in each group. Therefore, according to the meaning of multiplication, there are

$$2 \times (2 \times 3)$$

ways of choosing the combination of a cola, burger and fries.

TABLE 5.4

Using Two Arrays to Show the Ways to Choose a Soda, Burger, and Fries

REGULAR COLA			DIET COLA		
B, S	B, L	B, J	B, S	B, L	B, J
C, S	C, L	C, J	C, S	C, L	C, J

SHOWING MULTIPLICATIVE STRUCTURE WITH TREE DIAGRAMS

Consider once again the situation of two types of cola (regular cola R and diet cola D), two types of burgers (regular burger B and cheeseburger C) and three types of fries (small S, large L, and jumbo J) to choose from at a restaurant. How many ways are there to choose the combination of a cola, a burger, and fries? Before we organize these choices into a tree diagram, consider first the modification of the organized list in Table 5.2 shown in Table 5.5.

When we modify the way the information in Table 5.5 is displayed, we obtain the *tree diagram* pictured in Figure 5.4.

tree diagram   A **tree diagram** is a diagram consisting of line segments, called branches, that connect pieces of information. To "read" a tree diagram, start at the far left and follow branches *all the way across to the right.*

The sequence of branches that you travel along going across the tree diagram in Figure 5.4 corresponds to a choice of a soda, burger, and fries. For example, starting at the left, you might pick the R (regular cola) branch, followed by the C (cheeseburger) branch, and then followed by the L (large fries) branch, as shown in Figure 5.4. This choice of branches all the way across corresponds to the entry RCL in the organized list in Table 5.2.

As with the other ways of displaying this information, the tree diagram shows multiplicative structure. There are 2 primary branches (R and D), and each of those two primary branches has

TABLE 5.5

A Modified Organized List Is a Step toward a Tree Diagram.

regular cola	regular burger	small fries
		large fries
		jumbo fries
	cheeseburger	small fries
		large fries
		jumbo fries
diet cola	regular burger	small fries
		large fries
		jumbo fries
	cheeseburger	small fries
		large fries
		jumbo fries

FIGURE 5.4

A Tree Diagram:
Each Choice of
Cola, Burger,
and Fries Is
Represented by
a Sequence of
Branches.

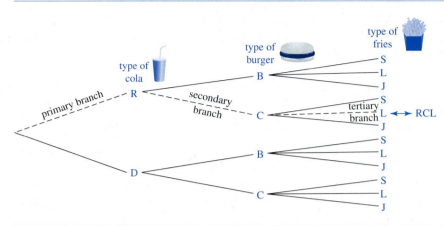

2 secondary branches (B and C). Since there are 2 groups of 2 secondary branches, there are 2×2 secondary branches according to the meaning of multiplication. For each of the 2×2 secondary branches, there are 3 third-level (tertiary) branches (S, L, and J). Therefore, there are 2×2 groups of 3 third-level branches, and according to the meaning of multiplication, there are

$$(2 \times 2) \times 3$$

third-level branches. Since each third-level branch corresponds to a choice of cola, burger, and fries, there are also $(2 \times 2) \times 3 = 12$ choices for the combination of a cola, burger, and fries.

USING ORGANIZED LISTS, ARRAY DIAGRAMS, AND TREE DIAGRAMS

Even though organized lists, arrays, and tree diagrams are relatively efficient ways to record data, they can become tedious to draw in their entirety when you are multiplying many numbers or large numbers. However, when many numbers or large numbers are involved, you can often show the multiplicative structure in a situation by *starting* an organized list or tree diagram, and drawing only a portion of it. Even if you can't finish the diagram completely, if you show the *structure* of the full diagram, you will be able to explain why multiplication is the appropriate operation to use in the situation.

CLASS ACTIVITY   NOW TURN TO CLASS ACTIVITIES MANUAL

5A Showing Multiplicative Structure p. 85

PRACTICE PROBLEMS FOR SECTION 5.1

1. How many different three-digit numbers can you write using only the digits 1, 2, and 3 if you do not repeat any digits (so that 121 and 332 are not counted)? Show how to solve this problem with an organized list and with a tree diagram. Use the meaning of multiplication to explain why this problem can be solved by multiplying.

2. How many different three-digit numbers can be made using only the digits 1, 2, and 3, where repeated digits are allowed (so that 121 and 332 are counted)? Show how to solve this problem with an organized list and with a tree diagram. Use the meaning of multiplication to explain why this problem can be solved by multiplying.

3. How many different keys can be made if there are ten places along the key that will be notched and if each notch will be one of eight depths?

4. Annette buys a wardrobe of 3 skirts, 3 pants, 5 shirts, and 3 sweaters, all of which are coordinated so that she can mix and match them any way she likes. How many different outfits can Annette create from this wardrobe? (Every day Annette wears either a skirt or pants, a shirt, and a sweater.)

5. A delicatessen offers four types of bread, twenty types of meats, fifteen types of cheese, a choice of mustard, mayonnaise, both or neither, and a choice of lettuce, tomato, both or neither for their sandwiches. How many different types of sandwiches can the deli make with one meat and one cheese? Should the deli's advertisement read "hundreds of sandwiches to choose from," or would it be better to substitute thousands or even millions for hundreds?

6. A pizza parlor offers ten toppings to choose from. How many ways are there to order a large pizza with exactly two different toppings? (For example, you could order pepperoni and mushroom, but not double pepperoni.)

ANSWERS TO PRACTICE PROBLEMS FOR SECTION 5.1

1. Figure 5.5 shows a tree diagram and an organized list displaying all such 3-digit numbers. Both show a structure of 3 big groups, one group for each possibility for the first digit. Once a first digit has been chosen, there are two choices for the second digit (since repeats aren't allowed). This means there are 3 groups of 2 possibilities for the first two digits. So by the meaning of multiplication, there are $3 \times 2 = 6$ possibilities for the first two digits. Once the first two digits are chosen, there is only one choice for the third digit—it must be the remaining unused digit. So there are $3 \times 2 = 6$ three-digit numbers using only the digits 1, 2, 3, with no digit repeated.

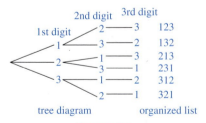

tree diagram   organized list

FIGURE 5.5

A Tree Diagram and an Organized List Showing All Three-Digit Numbers Using 1, 2, and 3 with No Digit Repeated

2. The tree diagram and the organized list in Figure 5.6 on p. 164 both show 3 big groups. Each of those 3 big groups has 3 groups of 3. Therefore, according to the meaning of multiplication, there are

$$3 \times (3 \times 3) = 27$$

three-digit numbers that use only the digits 1, 2, and 3.

3. Figure 5.7 on p. 164 shows part of a tree diagram for all such keys. There are 8 primary branches corresponding to the 8 choices for the first notch. For each of those 8 choices on the first notch, there are 8 choices for the second notch. Therefore, there are 8 groups of 8 choices for the first two notches, and so, by the meaning of multiplication, there are 8×8 choices for the first two notches. For each of the 8×8 choices for the first two notches, there are 8 choices for the third notch. Therefore, there are $8 \times 8 \times 8$ choices for the first three notches. And so on. When all 10 notches are taken into account, there are $8 \times 8 \times 8 \times 8 \times 8 \times 8 \times 8 \times 8 \times 8 \times 8 = 8^{10} = 1,073,741,824$ that can be made, which is more than a billion keys!

4. Annette can make $6 \times 5 \times 3 = 90$ different outfits.

5. There are $4 \times 20 \times 15 \times 4 \times 4 = 19,200$ different sandwiches that can be made with a meat and a cheese. The deli should substitute thousands for hundreds.

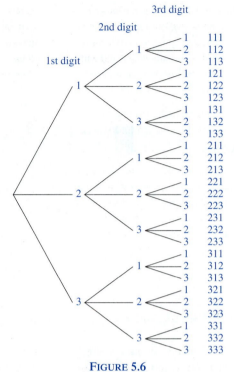

FIGURE 5.6

A Tree Diagram and an Organized
List Showing All Three-Digit Numbers
Using 1, 2, and 3

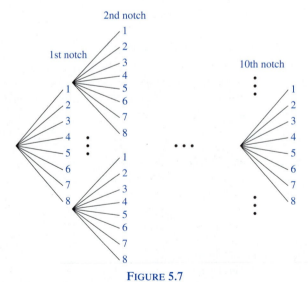

FIGURE 5.7

A Partial Tree Diagram Showing All Keys
with 10 Notches of 8 Depths

6. This is just like the clinking glasses problem of Chapter 1. For each of the 10 choices for the first topping, there are 9 choices for the second topping (there are only 9 because repeats are not allowed). This forms 10 groups of 9, so there are 10×9 choices of double toppings.

However, when they are counted this way, pepperoni and mushrooms counts as different from mushrooms and pepperoni. Therefore, we must divide by 2 so as not to double count pizzas. So there are actually only $\frac{10 \times 9}{2} = 45$ ways to order a pizza with two different toppings.

PROBLEMS FOR SECTION 5.1

1. Use the meaning of multiplication to explain why each of the following problems can be solved by multiplying:

   a. There are 3 feet in a yard. If a rug is 5 yards long, how long is it in feet?

   b. There are 5280 feet in a mile. How long in feet is a 4-mile-long stretch of road?

   c. One inch is 2.54 centimeters. Hillary is 5 feet, 3 inches tall, which is 63 inches tall. How tall is Hillary in centimeters?

   d. A kilometer is .6 miles (approximately). How many miles is a 10-kilometer run?

   e. Will is driving 65 miles per hour. If he continues driving at that speed, how far will he drive in $2\frac{1}{2}$ hours?

2. Find a tree or a large bush that has more leaves (or needles) on it than you can count. Use multiplication to estimate the number of leaves (or needles) on the tree. Describe your method and use the meaning of multiplication to explain why the problem can be solved by multiplying.

3. John, Trey, and Miles want to know how many two-letter combinations there are that don't have a repeated letter. For example, they want to count combinations such as BA and AT, but they don't want to count combinations such as ZZ or XX. John says there are $26 + 25$ because you don't want to use the same letter twice, that's why the second number is 25. Trey says he thinks it should be times, not plus: 26×25. Miles says the number is $26 \times 26 - 26$ because you need to take away the double letters. Discuss the boys' ideas. Which answers are correct, which are not, and why? Explain your answers clearly and thoroughly, drawing on the meaning of multiplication.

4. Allie and Betty want to know how many three-letter combinations, such as BMW, or DDT are possible (letters are allowed to repeat, as in DDT or BOB). Allie thinks there can be $26 + 26 + 26$ three-letter combinations, whereas Betty thinks the number is $26 \times 26 \times 26$. Which girl, if either, is right and why? Explain your answers clearly and thoroughly, drawing on the meaning of multiplication.

5. Explain your answers to the following:

   a. How many nine-digit numbers are there that use only the digits $1, 2, 3, \ldots, 8, 9$? (Repetitions are allowed, so, for example, 123211114 is allowed.)

   b. How many nine-digit numbers are there that use each of the digits $1, 2, 3, \ldots, 8, 9$ exactly once?

6. In all three parts in this problem, explain your method so that someone who is not in your class could understand why your answer is correct. In particular, if you use an operation such as multiplication or addition, explain why you use that operation as opposed to some other operation.

   a. How many whole numbers are there that have exactly 10 digits and that can be written using only the digits 8 and 9?

   b. How many whole numbers are there that have at most 10 digits and that can be written using only the digits 0 and 1? (Is this situation related to part (a)?)

   c. How many whole numbers are there that have exactly 10 digits and that can be written using only the digits 0 and 1?

7. Most Georgia car license plates currently use the format of three numbers followed by three letters (such as 123 ABC). How many different license plates can be made this way? Explain your method so that someone who is not in your class could understand why your answer is correct. In particular, if you use an operation such as multiplication or addition, explain why you use that operation as opposed to some other operation.

8. a. How many ways are there of picking a president and vice-president from a club of 40 people?

b. How many ways are there of picking a pair of co-presidents from a club of 40 people?

c. Are the answers to parts (a) and (b) the same or different? Explain why they are the same or why they are different.

9. A dance club has ten women and ten men. In each of the following parts, give an explanation clear enough that someone who is not in your class could understand why your method gives the correct answer:

a. How many ways are there to choose one woman and one man to demonstrate a dance step?

b. How many ways are there to pair each woman with a man (so that all ten women and all ten men have a dance partner at the same time)?

10. *A pizza parlor problem.* How many ways are there to order a large pizza with no double toppings? (For example, mushroom, pepperoni, and sausage is one possibility, so is mushroom and sausage, but not double mushroom and sausage. Your count should also include the case of no toppings.) You can't answer the question yet because you don't know how many toppings the pizza parlor offers.

a. Determine the number of different pizzas when there are exactly three toppings to choose from (such as mushroom, pepperoni and sausage).

b. Determine the number of different pizzas when there are exactly four toppings to choose from.

c. Determine the number of different pizzas when there are exactly five toppings to choose from.

d. Look for a pattern in your answers in parts (a), (b), and (c). Based on the pattern you see, predict the number of different pizzas when there are ten toppings to choose from.

e. Now find a different way to determine the number of different pizzas when there are ten toppings to choose from. This time think about the situation in the following way: pepperoni can be either on or off, mushrooms can be either on or off, sausage can be either on or off, and so on, for all ten toppings. Explain clearly how to use this idea to answer the question and why this method is valid.

11. A pizza parlor offers ten different toppings to choose from. How many ways are there to order a large pizza if double toppings, but not triple toppings or more are allowed? For example, double pepperoni and (single) mushroom is one possibility, as is double pepperoni and double mushroom, but triple pepperoni is not allowed. Notice that you can choose each topping in one of three ways: off, as a single topping, or as a double topping.

12. Draw a tree diagram and write a corresponding organized list to show all the ways of making 31 cents out of quarters, dimes, nickels, and pennies. Unlike the other tree diagrams in this section, this tree diagram won't have the same number of branches coming from each branch at the same stage, and therefore won't show a multiplicative structure.

13. Draw a tree diagram and write a corresponding organized list to show all the ways of presenting 234 toothpicks in bundles of hundreds, bundles of tens, and individual toothpicks. For example, 2 bundles of 100 toothpicks, 3 bundles of 10 toothpicks, and 4 individual toothpicks is one way to show 234 toothpicks, and another way is 2 bundles of 100 toothpicks and 34 individual toothpicks. As in the previous problem, this tree diagram won't have the same number of branches coming from each branch at the same stage, so it won't show a multiplicative structure.

5.2 Why Multiplying Decimal Numbers by 10 Is Easy

Why is multiplying by 10 and by powers of 10 so easy in the decimal system? To multiply a decimal number by 10, all we do is move every digit one place to the left. We can also think of this as moving the decimal point one place to the right because this procedure has the same effect. Similarly, to multiply by 100, move every digit two places to the left. To multiply by 1000, move every digit three places to the left, and so on.

Multiplying by 10 and by powers of 10 is easy because of the special structure of place value in the decimal system: in decimal numbers, the value of each place is 10 times the value of the place to its immediate right. Consider what happens when we multiply the number 34 by 10. The number 34 stands for 3 tens and 4 ones, and can be represented by 3 bundles of 10 toothpicks and 4 individual toothpicks, as shown in Figure 5.8. Then 10×34 stands for the total number of toothpicks in 10 groups of 34 toothpicks. As Figure 5.9 shows, when we form 10 groups of 34 toothpicks, each of the 3 original groups of 10 toothpicks becomes bundled into 1 group of 100. Therefore, when we multiply by 10, the 3 in the tens place moves one place over to the hundreds place. *Notice that this shifting occurs precisely because the value of the hundreds place is 10 times the value of the tens place.* Similarly, when we multiply 34 by 10, each of the 4 original individual toothpicks is bundled into 1 group of 10. Therefore, when we multiply by 10, the 4 in the ones place moves one place over to the tens place. Once again, notice that this shifting occurs precisely because the value of the tens place is 10 times the value of the ones place.

The situation is similar for other decimal numbers. We can think of multiplying

$$10 \times 400{,}000.007$$

FIGURE 5.8

The Number 34 Represented by 3 Bundles of 10 and 4 Individual Toothpicks

FIGURE 5.9

Ten Groups of 34 Is Bundled into 3 Hundreds and 4 Tens

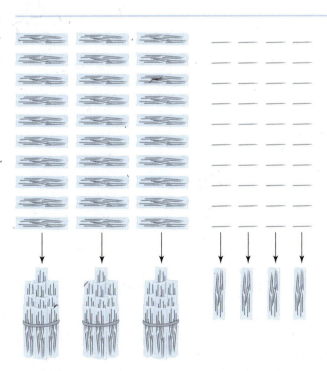

as forming 10 groups of bundled objects representing 400,000.007. As before, the following occur:

The 4 hundred thousands move one place to the left to become 4 million because the value of the millions place is 10 times the value of the hundred-thousands place.

The 7 thousandths move one place to the left to become 7 hundredths because the value of the hundredths place is 10 times the value of the thousandths place.

What about multiplying by 100, or 1000, or 10000, and so on? Because

$$100 = 10 \times 10$$

multiplying by 100 has the same effect as multiplying by 10 twice. Therefore, multiplying by 100 moves each digit in a decimal number two places to the left. Similarly, because

$$1000 = 10 \times 10 \times 10$$

multiplying by 1000 has the same effect as multiplying by 10 three times. Therefore, multiplying by 1000 moves each digit in a decimal number three places to the left.

CLASS ACTIVITY   NOW TURN TO CLASS ACTIVITIES MANUAL

5B If We Wrote Numbers Differently, Multiplying by 10 Might Not Be So Easy p. 89

5C Multiplying by Powers of 10 Explains the Cycling of Decimal Representations of Fractions p. 91

PRACTICE PROBLEM FOR SECTION 5.2

1. Using the example 10×3.4, explain why we move the digits in a decimal number one place to the left when we multiply by 10.

ANSWER TO PRACTICE PROBLEM FOR SECTION 5.2

1. Using bundled toothpicks, the decimal number 3.4 can be represented as shown in Figure 5.8, as long as 1 toothpick represents $\frac{1}{10}$ and a bundle of 10 toothpicks repre-

sents 1. Now use the explanation that is given in the text and that accompanies Figure 5.9.

PROBLEMS FOR SECTION 5.2

1. Using the example 10×23, explain why we move the digits in a decimal number one place to the left when we multiply by 10.

2. Mary says that $10 \times 3.7 = 3.70$. Why might Mary think this? Explain to Mary why her answer is not correct and why the correct answer is right. If you tell Mary a procedure, be sure to tell Mary why the procedure makes sense.

3. Once we understand that multiplying a decimal number by 10 shifts the digits one place to the left, explain how we can deduce that multiplying by 100 shifts the digits two places to the left and multiplying by 1000 shifts the digits 3 places to the left. How should we think about the numbers 100 and 1000 in order to make these deductions?

4. a. Find the decimal representation of $\frac{1}{37}$ to at least 6 places (or as many as your calculator shows). Notice the repeating pattern.

   b. Now find the decimal representations of $\frac{10}{37}$ and of $\frac{26}{37}$ to at least 6 places. Compare the repeating patterns to each other and to the decimal representation of $\frac{1}{37}$. What do you notice?

   c. Write $10 \times \frac{1}{37} = \frac{10}{37}$ and $100 \times \frac{1}{37} = \frac{100}{37}$ as mixed numbers (with a whole number part and a fractional part).

   d. What happens to the decimal representation of a number when it is multiplied by 10? By 100? Use your answer, and part (c), to explain the relationships you noticed in part (b).

5. a. Find the decimal representation of $\frac{1}{41}$ to at least 8 decimal places (preferably 10 or more). Notice the repeating pattern.

   b. Use your answer in part (a) to find the decimal representations of the numbers

   $$10 \times \frac{1}{41}, \quad 100 \times \frac{1}{41}, \quad 1000 \times \frac{1}{41},$$
   $$10{,}000 \times \frac{1}{41}, \quad 100{,}000 \times \frac{1}{41}$$

   without a calculator.

   c. Write the numbers

   $$10 \times \frac{1}{41} = \frac{10}{41}, \quad 100 \times \frac{1}{41} = \frac{100}{41},$$
   $$1000 \times \frac{1}{41} = \frac{1000}{41},$$
   $$10{,}000 \times \frac{1}{41} = \frac{10{,}000}{41},$$
   $$100{,}000 \times \frac{1}{41} = \frac{100{,}000}{41}$$

   as mixed numbers (a whole number part and a fractional part).

   d. Use your answers in part (b) to find the decimal representations of the fractional parts of the mixed numbers you found in part (c). Do not use your calculator or do long division; use part (b). Explain your reasoning.

5.3 Multiplication and Areas of Rectangles

In this section, we will examine the natural relationship between multiplication and area of rectangles. You are probably familiar with the length times width, $L \times W$, formula for areas of rectangles. We will see that we can explain why this formula is valid by using the meaning of multiplication.

Area is usually measured in square units. Depending on what we are describing—the page of a book, the floor of a room, a football field, a parcel of land—we usually measure area in square inches, square feet, square yards, square miles, square centimeters, square meters, or square kilometers. A **square inch**, often written 1 in², is the area of a square that is 1 inch wide and 1 inch long, as is the square in Figure 5.10. Similarly, one **square foot**, often written 1 ft², is the area of a square that is 1 foot wide and 1 foot long. In general, for any unit of length, one **square unit** is the area of a square that is 1 unit wide and 1 unit long.

square inch

square foot

square unit

FIGURE 5.10

A 1-Inch-by-1-Inch Square Has Area 1 in²

FIGURE 5.11

A 2-Inch-by-3-Inch Rectangle

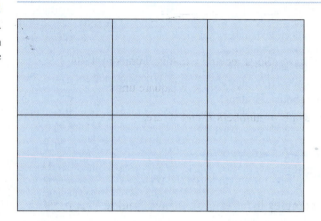

The rectangle in Figure 5.11 is three inches wide and two inches long (or high). Why can its area be calculated by multiplying its length times its width? When we subdivide the rectangle into 1 inch × 1 inch squares we have two rows of these squares, and each row has three squares. In other words, the rectangle is made up of 2 groups with 3 squares in each group. Therefore, according to the meaning of multiplication, the rectangle is made up of 2 × 3, or 6 squares. Because each 1-inch-by-1-inch square has area one square inch, the area of the whole rectangle is 2 × 3 square inches. We can use the same reasoning to determine areas of other rectangles.

In general, if L and W are any whole numbers, then a rectangle that is L units long and W units wide can be subdivided into L rows of 1-unit-by-1-unit squares, with W squares in each row, as indicated in Figure 5.12. In other words, the rectangle consists of L groups, with W squares in each group. Therefore, according to the meaning of multiplication, the rectangle is made of

$$L \times W$$

1-unit-by-1-unit squares. Because each 1-unit-by-1-unit square has area 1 square unit, the entire L-unit-by-W-unit rectangle has area

$$L \times W \quad \text{square units}$$

This explanation for why the length-times-width formula is valid applies not only when the length and width of the rectangle are whole numbers, but even when the length and width

FIGURE 5.12

A Rectangle That Is *L* Units Long and *W* Units Wide

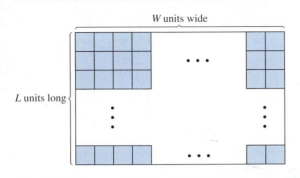

W units wide

L units long

· · ·

are fractions or decimals. So, a rectangle that is *L* units by *W* units has area

$$L \times W \text{ square units}$$

regardless of what kinds of numbers *L* and *W* are.

CLASS ACTIVITY NOW TURN TO CLASS ACTIVITIES MANUAL

5D Areas of Rectangles in Square Yards and Square Feet p. 94

5E Using Multiplication to Estimate How Many p. 95

PRACTICE PROBLEMS FOR SECTION 5.3

1. Use the meaning of multiplication to explain why the area of a carpet that is 20 feet long and 12 feet wide is 20×12 square feet.

2. One mile is 1760 yards. Does this mean that 1 square mile is 1760 square yards? If not, how many square yards are in a square mile? Explain carefully.

ANSWERS TO PRACTICE PROBLEMS FOR SECTION 5.3

1. You can think of a carpet that is 20 feet long and 12 feet wide as made up of 20 rows of squares with 12 squares in each row, each square being 1 foot wide and 1 foot long. In other words, the carpet consists of 20 groups, with 12 squares in each group. Therefore, according to the meaning of multiplication, the carpet is made of 20×12 squares. The area of a square that is 1 foot wide and 1 foot long is 1 square foot; thus, the area of the carpet is 20×12 square feet.

2. One square mile is the area of a square that is 1 mile long and 1 mile wide. Such a square is 1760 yards long and 1760 yards wide, so we can mentally decompose this square into 1760 rows, each having 1760 squares that are 1 yard by 1 yard. Therefore, according to the meaning of multiplication, there are

$$1760 \times 1760 = 3,097,600$$

1 yard by 1 yard squares in a square mile. This means that 1 square mile is 3,097,600 square yards and not 1760 square yards.

PROBLEMS FOR SECTION 5.3

1. Explain why it makes sense to multiply 4×6 to determine the area of a 4-ft-by-6-ft rug in square feet.

2. One foot is 12 inches. Does this mean that 1 square foot is 12 square inches? Draw a picture showing how many square inches are in a square foot. Use the meaning of multiplication to explain why you can calculate the area of 1 square foot in terms of square inches by multiplying.

3. A room has a floor area of 48 square yards. What is the area of the room in square feet? Solve this problem in two different ways, each time referring to the meaning of multiplication.

4. One kilometer is 1000 meters. Does this mean that 1 square kilometer is 1000 square meters? If not, what is 1 square kilometer in terms of square meters? Explain your answer in detail, referring to the meaning of multiplication.

5. A standard piece of paper is 11 inches long and $8\frac{1}{2}$ inches wide. Use the meaning of multiplication to explain why the area of a standard piece of paper is

$$11 \times 8\frac{1}{2} \text{ in}^2$$

6. Ms. Dunn's class wants to estimate the number of blades of grass in a lawn that is shaped roughly like a 30-foot-by-40-foot rectangle. The children cut 1-inch-by-1-inch squares out of sheets of paper and place these square holes over patches of grass. Each child cuts the grass from their 1-square-inch patch and counts the number of blades of grass they cut. Inside, the class calculates the average number of blades of grass they cut. This average is 37. Using this data, determine approximately how many blades of grass are in the lawn. Explain your reasoning.

7. Imagine that you are standing on a sandy beach, like the one in Figure 5.13, gazing off into the distance. How many grains of sand might you be looking at? To solve this problem, make reasonable assumptions about how wide the beach is and how far down the length of the beach you can see. Make a reasonable assumption about how many grains of sand are in a very small piece of beach and explain why your assumption is reasonable. Based on your assumptions, make a calculation that will give you a fairly good estimate of the number of grains of sand that you can see. Explain your reasoning.

FIGURE 5.13

How Many Grains of Sand Do We See Looking Down a Sandy Beach?

8. a. Explain how Figure 5.14 shows that

$$1 + 2 = \frac{1}{2}(2 \times 3)$$

and

$$1 + 2 + 3 = \frac{1}{2}(3 \times 4)$$

 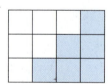

FIGURE 5.14

Pictures to Help Find Sums

b. Draw a similar picture for $1 + 2 + 3 + 4$ and explain how to generalize the picture to help you calculate other sums, such as $1 + 2 + 3 + \cdots + 10$.

c. Use your answer to part (b) to calculate $1 + 2 + 3 + \cdots + 500$.

9. The Browns need new carpet for a room with a rectangular floor that is 35 feet wide and 43 feet long. To save money, they will install the carpet themselves. The carpet comes on a large roll that is three *yards* wide. The carpet store will cut any length of carpet they like, but the Browns must buy the full three yards in width.

a. Draw clear, detailed pictures showing two different ways the Browns could lay their carpet.

b. For each way of laying the carpet, find how much carpet the Browns will need to buy from the carpet store. Which way is less expensive for the Browns?

10. The Smiths will be carpeting a room in their house. In one store, they see a carpet they like that costs $35 per square yard. Another store has a similar carpet for $3.95 per square foot. Is this more or less expensive than the carpet at the first store? Explain your reasoning.

11. One acre is 43,560 square feet. If a square piece of land is 3 acres, then what is the length and width of this piece of land in feet? (Remember that the length and width of a square are the same.) Explain your reasoning.

5.4 The Commutative Property of Multiplication

In Chapter 4 we studied some of the properties of addition, including the commutative property. In this section, we will study the commutative property of multiplication. We will explain why this property makes sense for counting numbers by working with areas of rectangles. We will also explain when to use the commutative property of multiplication.

commutative property of multiplication

The **commutative property of multiplication** says that if A and B are any real numbers, then

$$A \times B = B \times A$$

For example,

$$297 \times 43 = 43 \times 297$$

and

$$1.3 \times 5.7 = 5.7 \times 1.3$$

We assume that this property always holds for *any* pair of numbers, but we can explain why this property makes sense for counting numbers.

If you think of the commutative property of multiplication in terms of groupings, it can seem somewhat mysterious: why do 3 groups of 5 marbles have the same total number of marbles as 5 groups of 3 marbles? Why do 297 baskets of potatoes with 43 potatoes in each basket contain the same total number of potatoes as 43 baskets of potatoes with 297 potatoes in each basket? Of course, we can calculate 3×5 and 5×3 and see that they are equal, and we can calculate 297×43 and 43×297 and see that they are equal, but by simply calculating, we don't see why the commutative property should hold in other situations as well. A conceptual explanation that will show us why this property makes sense will help us understand its meaning better. We can give a conceptual explanation by working with arrays or with areas of rectangles.

Suppose you have 3 groups with 5 marbles in each group. According to the meaning of multiplication, this is a total of

$$3 \times 5$$

marbles. You can arrange the 3 groups of marbles so that they form the 3 rows of an array, as pictured on the left in Figure 5.15. But if you now choose the *columns* of the array to be the groups, as on the right in Figure 5.15, then there are 5 groups with 3 marbles in each group. Therefore, according to the meaning of multiplication, there is a total of

$$5 \times 3$$

FIGURE 5.15

Grouping
Marbles to
Show That
$3 \times 5 = 5 \times 3$

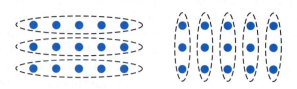

marbles. But the total number of marbles is the same, either way you count them. There-fore,

$$3 \times 5 = 5 \times 3$$

By imagining pictures like Figure 5.15, it's easy to see why the commutative property of multiplication should hold for counting numbers other than 3 and 5. Even though it would be ridiculously tedious to draw 297 horizontal rows with 43 potatoes in each row, we can imagine such a picture as similar to Figure 5.15. Such a picture, and the argument about changing groups from rows to columns, explains why

$$297 \times 43 = 43 \times 297$$

In these situations there is nothing special about the numbers 3, 5, 297, and 43. So, if A and B are any counting numbers, there will be a picture similar to Figure 5.15 illustrating that

$$A \times B = B \times A$$

We can also explain why the commutative property of multiplication makes sense by using areas of rectangles. For example, a rug that is 3 feet by 5 feet can be thought of as being made of 3 rows with 5 squares, each of area 1 square foot, in each row. Therefore, the area of the rug is

$$3 \times 5 \text{ square feet}$$

On the other hand, the rug can be thought of as made of 5 columns with 3 squares, each of area 1 square foot, in each column, as shown in Figure 5.16. Therefore, the area of the rug is

$$5 \times 3 \text{ square feet}$$

But the area is the same either way you calculate it; therefore,

$$3 \times 5 = 5 \times 3$$

As before, the same line of reasoning will work when other counting numbers replace 3 and 5.

FIGURE 5.16

Using the Area
of a Rectangle
to Show That
$3 \times 5 = 5 \times 3$

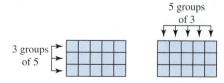

It is easy to use the commutative property of multiplication without even being consciously aware that you are doing so. Suppose you are asked to solve the following problem mentally: what are fifteen twos? Rather than counting by twos fifteen times, you might spontaneously turn this problem into: what are two fifteens? The new problem is easy to solve mentally by adding 15 and 15 to get 30. When you change the original problem to the new problem, you use the commutative property of multiplication. This is because "fifteen twos" means 15 groups of 2, which is 15×2, and "two fifteens" means 2 groups of 15, which is 2×15. You can substitute the problem 2×15 for the problem 15×2 because, according to the commutative property of multiplication, their results are equal. In other words,

$$15 \times 2 = 2 \times 15$$

CLASS ACTIVITY   NOW TURN TO CLASS ACTIVITIES MANUAL

5F Explaining the Commutative Property of Multiplication p. 97

5G Using the Commutative Property of Multiplication p. 98

PRACTICE PROBLEMS FOR SECTION 5.4

1. Use the meaning of multiplication and a picture to give a conceptual explanation for why $2 \times 4 = 4 \times 2$. Your explanation should be general enough to remain valid if other counting numbers were to replace 2 and 4.

2. Give an example to show how to use the commutative property of multiplication to make a mental math problem easier to do.

ANSWERS TO PRACTICE PROBLEMS FOR SECTION 5.4

1. See Figure 5.17. It shows that the total number of objects in 2 groups with 4 objects in each group (2×4 objects) can also be thought of as made out of 4 groups with 2 objects in each group (4×2 objects). But you have the same number of objects either way you count them; therefore,

$$2 \times 4 = 4 \times 2$$

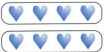

two groups of 4   four groups of 2

FIGURE 5.17

$2 \times 4 = 4 \times 2$

2. If sponges come 2 to a package and if you have 13 packages of sponges, then the total number of sponges you have is 13×2. To solve 13×2 you can instead solve 2×13, according to the commutative property of multiplication. Two groups of 13 is just $13 + 13$ which is easy to calculate mentally as 26.

1. There are 31 envelopes with 3 stickers in each envelope. Seyong calculates the total number of stickers by counting by 3s 31 times. Natasha adds $31 + 31 + 31 = 93$ instead. Explain why Natasha's strategy is valid.

2. Here is Amy's explanation for why the commutative property of multiplcation is true for counting numbers:

    Whenever I take two counting numbers and multiply them, I always get the same answer as when I multiply them in the reverse order. For example,

    $$6 \times 8 = 48$$
    $$8 \times 6 = 48$$
    $$9 \times 12 = 108$$
    $$12 \times 9 = 108$$
    $$3 \times 15 = 45$$
    $$15 \times 3 = 45$$

    It always works that way, no matter which numbers you multiply, you will get the same answer either way you multiply them.

    Explain why Amy's discussion might not convince a skeptic that the commutative property should always be true for any pair of counting numbers.

3. Sue and Tonya started the same job at the same time and earned identical salaries. After one year, Sue got a 5% raise and Tonya got a 6% raise. The following year, the situation was reversed: Sue got a 6% raise and Tonya got a 5% raise. After the first year, Tonya's salary was higher than Sue's, of course, but whose salary was higher after both raises? Solve this problem using the fact that if someone's salary goes up by 5%, then their new salary is 1.05 times their old salary (similarly for a 6% raise, of course). Using this method for calculating the women's salaries, explain how the commutative property of multiplication is relevant to comparing the salaries after both raises.

4. Suppose that the sales tax is 7% and suppose that some towels are on sale at a 20% discount. When you buy the towels, you pay 7% tax on the discounted price. What if you were to pay 7% tax on the full price, but you got a 20% discount on the price *including* the tax? Would you pay more, less, or the same amount? Explain how the commutative property of multiplication is relevant to this question.

5. A dress is marked down 25% and then it is marked down 20% from the discounted price.

    a. By what percent is the dress marked down after both discounts?

    b. Does the dress cost the same, less, or more than if the dress was marked down 45% from the start? Explain how you can determine the answer to this question without doing any calculating.

    c. If the dress was marked down 20% first and then 25%, do you get a different answer to part (a)? Explain how the commutative property of multiplication is relevant to this question.

5.5   Multiplication and Volumes of Boxes

Not only are *areas* naturally related to multiplication, but *volumes* are as well, as we'll see in this section. In the next section, we will use volumes to discuss the associative property of multiplication.

Depending on what you want to measure—a dose of liquid medicine, the size of a compost pile, the volume of coal in a mountain—volume can be measured in cubic inches, cubic feet, cubic yards, cubic miles, cubic centimeters, cubic meters, or cubic kilometers. (Volumes of liquids are also commonly measured in liters, milliliters, gallons, quarts, cups, and fluid ounces.)

One cubic centimeter, often written 1 cm^3, is the volume of a cube that is one centimeter high, one centimeter deep, and one centimeter wide. A drawing of such a cube is shown in Figure 5.18, along with a cube of volume 1 cubic inch, 1 inch^3. One cubic yard, often written 1 yd^3, is the volume of a cube that is one yard high, one yard deep, and one yard wide. In general, for any unit of length, one **cubic unit**, often written 1 unit^3, is the volume of a cube that is 1 unit high, 1 unit deep, and 1 unit wide.

cubic unit

FIGURE 5.18

Cubes of
Volume 1 inch³
and 1 cm³

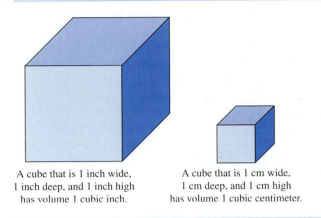

A cube that is 1 inch wide,
1 inch deep, and 1 inch high
has volume 1 cubic inch.

A cube that is 1 cm wide,
1 cm deep, and 1 cm high
has volume 1 cubic centimeter.

We will be working with volumes of boxes and box shapes (these shapes are also called rectangular prisms). The volume, in cubic units, of a box or box shape is just the number of 1-unit-by-1-unit-by-1-unit cubes that it would take to fill the box or make the box shape. You may remember a formula for the volume of a box, but if so, assume for a moment that you don't know this formula. We will see how to derive the formula for the volume of a box from the meaning of multiplication.

Suppose you have a box that is 4 inches high, 2 inches deep, and 3 inches wide, as pictured in Figure 5.19. What is the volume of this box? If you have a set of building blocks that are all one inch high, one inch deep, and one inch wide, then you can use the building blocks to build the box. The number of blocks needed is the volume of the box in cubic inches. We can use multiplication to describe the number of blocks needed by considering the box to be made of 4 horizontal layers, as shown in Figure 5.20. Each layer consists of two rows of three blocks, so according to the meaning of multiplication, each layer contains 2×3 blocks. There are 4 layers, each containing 2×3 blocks. So, according to the meaning of multiplication, there are

$$4 \times (2 \times 3)$$

blocks making up the box. Therefore, the box has a volume of

$$4 \times (2 \times 3) = 4 \times 6 = 24 \text{ cubic inches}$$

The reasoning used in this example applies generally. Suppose you have a box that is H units high, D units deep, and W units wide. If H, D, and W are whole numbers, then as before, you can build such a box out of 1-unit-by-1-unit-by-1-unit cubes. How many cubes does it

FIGURE 5.19

A 4-Inch-High,
2-Inch-Deep,
3-Inch-Wide
Box

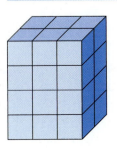

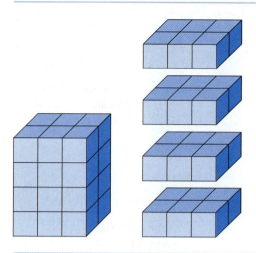

take? If you consider the box to be made of horizontal layers, then there are *H* layers. Each layer is made up of *D* rows of *W* blocks (or *W* rows of *D* blocks) and therefore contains $D \times W$ blocks, according to the meaning of multiplication. There are *H* layers with $D \times W$ blocks in each layer. Therefore, according to the meaning of multiplication, the box is made out of

$$H \times (D \times W)$$

1-unit-by-1-unit-by-1-unit cubes. Consequently, the box has volume

$$H \times (D \times W) \text{ cubic units}$$

Notice that the order in which the letters H, D, and W appear and the way the parentheses are placed in the expression $H \times (D \times W)$ corresponds to the way the box was divided into groups.

As with areas, it turns out that this height-times-depth-times-width formula for the volume of a box remains valid even when the height, depth, and width of the box are not whole numbers. The volume of a box that is *H* units high, *D* units deep, and *W* units wide is always $H \times (D \times W)$ cubic units.

In the next section we will examine the use of parentheses in expressions like $H \times (D \times W)$.

CLASS ACTIVITY   NOW TURN TO CLASS ACTIVITIES MANUAL

5H Ways to Describe the Volume of a Box with Multiplication p. 99

5I Volumes of Boxes in Cubic Yards and Cubic Feet p. 102

5J How Many Gumdrops? p. 103

PRACTICE PROBLEMS FOR SECTION 5.5

1. Use the meaning of multiplication to explain why a box that is 3 feet wide, 2 feet deep, and 4 feet high has volume $4 \times (2 \times 3)$ cubic feet.

2. How many cubic feet of mulch will you need to cover a garden that is 10 feet wide and 4 yards long with 2 inches of mulch? Can you find the amount of mulch you need by multiplying $2 \times (10 \times 4)$?

3. One cubic foot of water weighs about 62 pounds. How much will the water in a swimming pool that is 20 feet wide, 30 feet long, and 4 feet deep weigh?

ANSWERS TO PRACTICE PROBLEMS FOR SECTION 5.5

1. See the explanation in the text. Substitute "feet" where the text says "inches."

2. We know that 1 yard = 3 feet and 1 foot = 12 inches. Thus, 4 yards = 12 feet, and 2 inches = $\frac{1}{6}$ feet. So you can think of the mulch as forming a box shape that is $\frac{1}{6}$ feet high, 10 feet deep (or wide), and 12 feet wide (or deep). This box has volume

$$\frac{1}{6} \times (10 \times 12) \text{ cubic feet } = 20 \text{ cubic feet}$$

so you'll need 20 cubic feet of mulch.

You can't find the amount of mulch you need by multiplying $2 \times 10 \times 4$ because each number refers to a different unit of length.

3. Because the water in the pool is in the shape of a box that is 4 feet high, 20 feet wide, and 30 feet deep, it can be thought of as

$$4 \times (20 \times 30) = 2400$$

1-foot-by-1-foot-by-1-foot cubes of water. Each of those cubes of water weighs 62 pounds, so the water in the pool will weigh

$$2400 \times 62 \text{ pounds } = 148,800 \text{ pounds}$$

PROBLEMS FOR SECTION 5.5

1. Use the meaning of multiplication to explain why a box that is 5 inches high, 4 inches wide, and 3 inches deep has a volume of

$$5 \times (4 \times 3) \text{ cubic inches}$$

Explain the parentheses in the expression $5 \times (4 \times 3)$.

2. Figure A in Figure 5.21 on p. 180 shows a 5-unit-high, 3-unit-wide, and 2-unit-deep box made out of blocks. Figures B through G of Figure 5.21 show different ways of subdividing the box into natural groups of blocks. For each of these ways of subdividing the box, use the meaning of multiplication as we have described it to write an expression for the total number of blocks in the box. Each expression should involve the numbers 5, 3, and 2, the multiplication symbol, and parentheses. Explain why you write your expressions as you do.

3. One foot is 12 inches. Does this mean that one cubic foot is 12 cubic inches? Describe how to use the meaning of multiplication to determine what 1 cubic foot is in terms of cubic inches.

4. How much mulch will you need to cover a rectangular garden that is 20 feet by 30 feet with a 3-inch layer of mulch? Explain.

5. A lot of gumballs are in a glass container. The container is shaped like a box with a square base. When you look down on the top of the container, you see about 50 gumballs at the surface. When you look at one side of the container, you see about 60 gumballs up against the glass. You also notice that there are about 9 gumballs against each vertical edge of the container. Given this information, estimate the total number of gumballs in the container. Explain your reasoning.

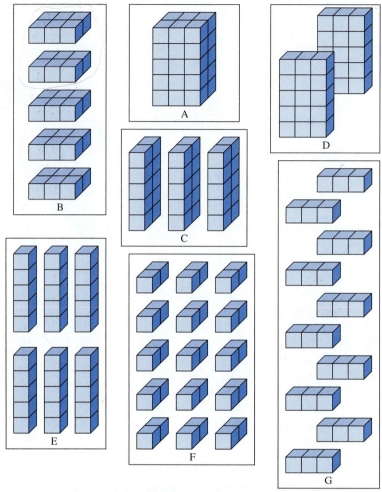

FIGURE 5.21

Different Ways to Subdivide a Box

6. Estimate how many neatly stacked hundred-dollar bills you could fit in a briefcase that is 20 inches long, 11 inches wide, and 2 inches thick on the inside. Describe your method and explain why it gives a good estimate.

7. Figure 5.22 shows a grocery store display of cases of sodas. The display consists of a large box shape with a "staircase" on top. The display is 7 cases wide and 5 cases deep; it is 4 cases tall in the front, and 8 cases tall at the back. How many cases of sodas are there in the display? Solve this problem in at least two different ways and explain your method in each case.

8. A cube that is 10 inches wide, 10 inches deep, and 10 inches high is made out of smaller cubes that are each 1 inch wide, 1 inch deep, and 1 inch high. The large cube is then painted on the outside.

a. How many of the smaller cubes that make up the large cube have paint on them? Explain.

b. How many of the smaller cubes have paint on two sides? Explain.

c. How many of the smaller cubes have paint on three sides? Explain.

9. Investigate the following two questions and explain your conclusions:

a. If you make a rectangular garden that is twice as wide and twice as long as a rectangular garden that you already have, how will the area of the larger

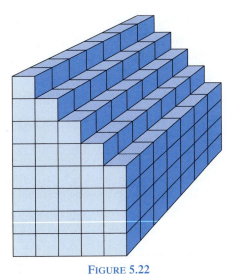

FIGURE 5.22

A Display of Cases of Soda

garden compare to the area of the original garden? (Will the larger garden be twice as big, 3 times as big, 4 times as big, etc.?)

b.   If you make a cardboard box that is twice as wide, twice as tall and twice as deep as a cardboard box that you already have, how will the volume of the larger box compare to the volume of the original box? (Will the larger box be twice as big, 3 times as big, 4 times as big, etc.?)

10. The Better Baking Company is introducing a new line of reduced-fat brownies in addition to its regular brownies. The batter for the reduced-fat brownies contains $\frac{1}{3}$ less fat than the batter for the regular brownies. Both types of brownies will be baked in the same size rectangular pan, which is 24 inches wide and 30 inches long. The regular brownies are cut from this pan by dividing the width into 12 equal segments and by dividing the length into 10 equal segments, so that each regular brownie is 2 inches by 3 inches. Each regular brownie contains 6.3 grams of fat. In addition to using a reduced-fat batter, the Better Baking Company would like to further reduce the amount of fat in their new brownies by making these brownies smaller than the regular ones. You have been contacted to help with this task. Present two different ways to divide the length and width of the pan to produce smaller brownies. In each case, calculate the amount of fat in each brownie and explain the basis for your calculation. The brownies should be of a reasonable size and there should be no waste left over in the pan after cutting the brownies. The length and width of the brownies do not necessarily have to be whole numbers of inches.

5.6 The Associative Property of Multiplication

Recall that the associative property of *addition* says that for all real numbers A, B, and C,

$$(A + B) + C = A + (B + C)$$

As we have seen, the associative property of addition gives us flexibility in calculating sums. Likewise, the associative property of *multiplication* will give us flexibility in calculating products.

associative property of multiplication   The **associative property of multiplication** says that if A, B, and C are any real numbers, then

$$(A \times B) \times C = A \times (B \times C)$$

We assume that this property holds for all real numbers, but as we'll see in this section, we can explain why it makes sense for counting numbers. We will also see how to use the associative property of multiplication.

WHY DOES THE ASSOCIATIVE PROPERTY OF MULTIPLICATION MAKE SENSE?

Why is the associative property valid? For example, why is

$$(4 \times 2) \times 3 = 4 \times (2 \times 3)?$$

In other words, why is the quantity (4×2) times 3 equal to 4 times the quantity (2×3)? In this example, because specific numbers are involved, we can evaluate the quantities on both sides of the equals sign and see that they really are equal. We get

$$(4 \times 2) \times 3 = 8 \times 3 = 24$$

and

$$4 \times (2 \times 3) = 4 \times 6 = 24$$

Therefore,

$$(4 \times 2) \times 3 = 4 \times (2 \times 3)$$

But when we evaluate the expressions, we only show that the associative property holds in this one special case. Why does it make sense that the associative property of multiplication holds for *all* real numbers?

To explain why the associative property of multiplication makes sense for all counting numbers, we will develop a conceptual explanation for why

$$(4 \times 2) \times 3 = 4 \times (2 \times 3)$$

This explanation will be general in the sense that it will also explain why the equation is true if we were to replace the numbers 4, 2, and 3 with other counting numbers. Our explanation is based on calculating the volume of a box in two different ways.

In the previous section, we decomposed a 4-inch-high, 2-inch-deep, 3-inch-wide box shape into 4 groups of blocks, with 2×3 blocks in each group, as shown on the left of Figure 5.23. Viewed that way, the total number of blocks in the box shape is

$$4 \times (2 \times 3)$$

On the other hand, the box shape can be decomposed into 4×2 groups, with 3 blocks in each group, as shown on the right of Figure 5.23. According to the meaning of multiplication, there are

$$(4 \times 2) \times 3$$

blocks in the box shape. There are 4×2 groups because the groups of 3 blocks can themselves be arranged into 4 groups of 2. But the total number of blocks in the box shape is the same either way we count them. Therefore,

$$(4 \times 2) \times 3 = 4 \times (2 \times 3)$$

FIGURE 5.23

Showing that
$4 \times (2 \times 3) =$
$(4 \times 2) \times 3$

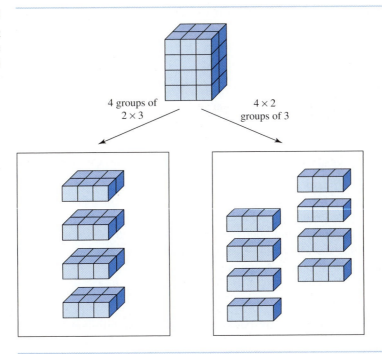

4 groups of
2×3

4×2
groups of 3

which is what we wanted to establish. Notice that the same argument works when the numbers 4, 2, and 3 are replaced with other counting numbers—only the size of the box will change. This reasoning explains why the associative property of multiplication makes sense for all counting numbers.

USING THE ASSOCIATIVE AND COMMUTATIVE PROPERTIES OF MULTIPLICATION

The associative and commutative properties of multiplication allow for great flexibility in carrying out multiplication problems.

When we multiply many numbers together, it is common not to use any parentheses, and the associative property of multiplication allows us to leave off parentheses without creating ambiguity. Rather than writing an expression such as

$$(17 \times 25) \times 4$$

or

$$17 \times (25 \times 4)$$

we can simply write the expression

$$17 \times 25 \times 4$$

without any parentheses, and we are free to multiply adjacent numbers in whatever order we choose. In this case, it's easiest to first multiply the 25 with the 4 to get 100 and then multiply

17 with 100 to get 1700. The following sequence of equations correspond to this strategy:

$$17 \times 25 \times 4 = 17 \times (25 \times 4)$$
$$= 17 \times 100$$
$$= 1700$$

Rather than writing an expression such as

$$((41 \times 2) \times 3) \times 5$$

according to the associative property of multiplication, we can simply write

$$41 \times 2 \times 3 \times 5$$

and we are free to multiply adjacent numbers in any order we please. Furthermore, we can invoke the commutative property of multiplication and switch the order in which numbers appear. In this case, it would be convenient to multiply the 2 by the 5, so we can switch the placement of the 3 and the 5 so that the 2 is adjacent to the 5:

$$41 \times 2 \times 5 \times 3$$

The following equations correspond to this strategy and show how to complete the calculation:

$$41 \times 2 \times 3 \times 5 = 41 \times 2 \times 5 \times 3$$
$$= 41 \times 10 \times 3$$
$$= 410 \times 3$$
$$= 1230$$

By using the associative and commutative property of multiplication, we can often make mental multiplication problems easier to solve.

CLASS ACTIVITY   NOW TURN TO CLASS ACTIVITIES MANUAL

5K Explaining and Using the Associative Property p. 104

5L Different Ways to Calculate the Total Number of Objects p. 107

PRACTICE PROBLEMS FOR SECTION 5.6

1. Explain how you use the associative property of multiplication when you calculate 7×600 mentally.

2. Use the meaning of multiplication and the idea of decomposing a box shape in two different ways to explain why $4 \times (2 \times 3) = (4 \times 2) \times 3$.

3. Describe how to use the design in Figure 5.24 to explain why

$$5 \times (2 \times 2) = (5 \times 2) \times 2$$

Your explanation should be general, so that it explains why the previous equation is true when you replace the numbers 5, 2, and 2 with other counting numbers, and the design in Figure 5.24 is changed accordingly.

one spiral

FIGURE 5.24

A Design of Spirals

ANSWERS TO PRACTICE PROBLEMS FOR SECTION 5.6

1. When we calculate 7×600 mentally, we first calculate 7×6 (a "basic fact") and then we multiply by 100. When we do this, we take the 6 from the 600, which is 6×100, and group the 6 with the 7 instead of with 100. In equation form, we can express this rearrangement as follows:

$$7 \times 600 = 7 \times (6 \times 100)$$
$$= (7 \times 6) \times 100$$
$$= 42 \times 100$$
$$= 4200$$

The associative property of multiplication is used at the second equal sign.

2. See text.

3. See Figure 5.25. On the one hand, we can think of the design as being made up of 5 clusters of spirals, with 2 groups of 2 spirals in each cluster. In this arrangement there are $5 \times (2 \times 2)$ spirals in the design. On the other hand, we can think of the design as made up of 5×2

groups of spirals with 2 spirals in each group. There are 5×2 groups because there are 5 sets of 2 groups. This means there are $(5 \times 2) \times 2$ spirals in the design. There are the same number of spirals, no matter how they are counted. Therefore,

$$5 \times (2 \times 2) = (5 \times 2) \times 2$$

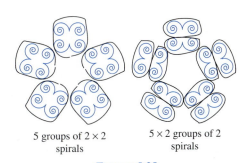

5 groups of 2×2 spirals 5×2 groups of 2 spirals

FIGURE 5.25

Using Grouping to Explain Why
$5 \times (2 \times 2) = (5 \times 2) \times 2$

PROBLEMS FOR SECTION 5.6

1. Write three different expressions for the total number of curlicues in Figure 5.26. Each expression should involve only the following: the numbers 2, 3, 5, and 8; the mutiplication symbol $\times$ or $\cdot$; and parentheses. For each expression, use the meaning of multiplication as we have described it to explain why your expression represents the total number of curlicues in Figure 5.26.

2. Suppose you have 60 pennies arranged into 12 stacks with 5 pennies in each stack. The 12 stacks are arranged

into 3 rows with 4 stacks in each row, as shown in Figure 5.27. For each of the following expressions, explain how to see a grouping of the pennies so that the expression describes the total number of pennies according to that grouping:

a. $3 \times (4 \times 5)$

b. $(3 \times 4) \times 5$

c. $5 \times (3 \times 4)$

d. $4 \times (3 \times 5)$

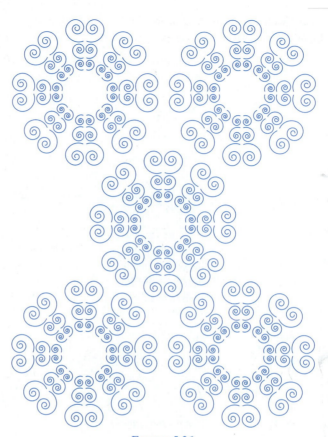

FIGURE 5.26

How Many Curlicues?

FIGURE 5.27

Stacks of Pennies

---

3. To calculate 3×80 mentally, we can just calculate $3 \times 8 = 24$ and then put a zero on the end to get the answer, 240. Use the picture in Figure 5.28 to help you explain why this method of calculation is valid.

4. Write equations to show how the commutative and associative properties of multiplication are involved when you calculate 40×800 mentally by relying on basic

multiplication facts (such as 4×8). Write your equations in the form

$$40 \times 800 = \text{some expression}$$
$$= \vdots$$
$$= \text{some expression}$$

Indicate specifically where the commutative and associative properties of multiplication are used.

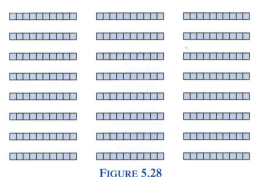

FIGURE 5.28

Explaining a Mental Method
for Calculating 3×80

5. Explain how to use the associative property of multiplication to make 16×25 easy to calculate mentally. Write equations that show why your method is valid and show specifically where you have used the associative property of multiplication. Write your equations in the form

$$16 \times 25 = \text{some expression}$$
$$= \vdots$$
$$= \text{some expression}$$

6. Use the associative property of multiplication to make the problem 32×0.25 easy to solve mentally. Write equations to show your use of the associative property of multiplication. Explain how your solution method is related to solving 32×0.25 by thinking in terms of money.

7. Explain how to make the following product easy to calculate mentally (there are five 2s and five 5s in the product):

$$2 \cdot 2 \cdot 2 \cdot 2 \cdot 2 \cdot 5 \cdot 5 \cdot 5 \cdot 5 \cdot 5$$

8. Julia says that it's easy to multiply a number by 4 because you just "double the double". Explain Julia's

idea and explain why it uses the associative property of multiplication.

9. Carmen says that it's easy to multiply even numbers by 5 because you just take half of the number and put a zero on the end. Write equations that incorporate Carmen's method and that demonstrate why her method is valid. Use the case 5×22 for the sake of concreteness. Write your equations in the following form:

$$5 \times 22 = \text{some expression}$$
$$= \text{some expression}$$
$$\vdots$$
$$= 110$$

10. Use the facts that

$$1 \text{ mile} = 1760 \text{ yards}$$
$$1 \text{ yard} = 3 \text{ feet}$$
$$1 \text{ foot} = 12 \text{ inches}$$

in order to calculate the number of inches in a mile. Do this in two different ways to illustrate the associative property of multiplication.

5.7 The Distributive Property

The associative and commutative properties apply in situations when we are adding only (as in Chapter 4) or multiplying only (as in Sections 5.4 and 5.6). In this section we introduce a property that applies to both addition and multiplication, namely the distributive property of multiplication over addition. The distributive property is the most important and computationally powerful tool in all of arithmetic. It allows for tremendous flexibility in performing mental calculations, and, as we will see in Section 5.9, the distributive property is the foundation of the standard longhand multiplication technique.

Before we discuss the distributive property of multiplication over addition, we will discuss the conventions for interpreting expressions involving both multiplication and addition.

EXPRESSIONS INVOLVING BOTH MULTIPLICATION AND ADDITION

Expressions that involve both addition and multiplication must be interpreted suitably, according to the conventions developed by mathematicians. This situation is entirely unlike the situation where only addition, or only multiplication, is involved in an expression. In those situations, parentheses can be dropped safely and adjacent numbers can be combined at will. However, when both multiplication and addition are present in an expression, parentheses cannot generally be dropped without changing the value of the expression. For example,

$$7 + 5 \times 2$$

is *not equal* to

$$(7 + 5) \times 2$$

To properly interpret an expression such as

$$7 + 5 \times 2$$

or

$$5 \times 17 + 9 \times 10^2 - 12 \times 94 + 20 \div 5$$

we need to use the following conventions:

- All powers are calculated first.
- Next, multiplications and divisions are performed from left to right.
- Finally additions and subtractions are performed from left to right.
- Expressions inside parentheses are always evaluated first, using the previous conventions.

Therefore,

$$7 + 5 \times 2 = 7 + 10$$
$$= 17$$

whereas

$$(7 + 5) \times 2 = 12 \times 2$$
$$= 24$$

Similarly,

$$5 \times 17 + 9 \times 10^2 - 12 \times 94 + 20 \div 5$$
$$= 5 \times 17 + 9 \times 100 - 12 \times 94 + 20 \div 5$$
$$= 85 + 900 - 1128 + 4$$
$$= -139$$

CLASS ACTIVITY   NOW TURN TO CLASS ACTIVITIES MANUAL

5M Order of Operations p. 109

WHY DOES THE DISTRIBUTIVE PROPERTY MAKE SENSE?

distributive   The distributive property of multiplication over addition says that for all real numbers, A, B,
property   and C,

$$A \times (B + C) = A \times B + A \times C$$

Notice the use of parentheses to group the B and C: the expression

$$A \times (B + C)$$

means A times the *quantity* $B + C$.

As with the other properties of arithmetic that we have studied, we assume that the distributive property holds for all real numbers. However, we can explain why the distributive property makes sense for counting numbers by decomposing arrays of objects or by decomposing rectangles.

Consider an array of dots consisting of 4 horizontal rows, with 7 dots in each row, as shown in Figure 5.29. The shading of the dots shows one way to decompose this array of dots into two smaller arrays. There are two ways of expressing the total number of dots in the array that correspond to this way of decomposing the array. On the one hand, there are 4 rows with $5 + 2$ dots in each row, and therefore the total number of dots is

$$4 \times (5 + 2)$$

Notice that parentheses are needed to say 4 times the *quantity* 5 plus 2. On the other hand, the shading decomposes the array of dots into two arrays. The array on the left consists of 4 rows with 5 dots in each row, and thus contains 4×5 dots; the array on the right consists of 4 rows with 2 dots in each row and hence contains 4×2 dots. Therefore, combining these two arrays of dots, there is a total of

$$4 \times 5 + 4 \times 2$$

FIGURE 5.29

An Illustration of the Distributive Property: $4 \times (5 + 2) = 4 \times 5 + 4 \times 2$

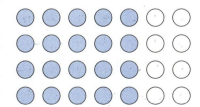

dots. But the total number of dots is the same, either way you count them. Therefore,

$$4 \times (5 + 2) = 4 \times 5 + 4 \times 2$$

This explains why the distributive property makes sense *in this case*. But notice that the reasoning is general in the sense that if we were to replace the numbers 4, 5, and 2 with other counting numbers, and if we were to correspondingly adjust the array of dots, then our argument would still hold. Therefore, the distributive property makes sense for all counting numbers.

VARIATIONS ON THE DISTRIBUTIVE PROPERTY

Several useful variations on the distributive property are listed next. All these variations can be obtained from the original distributive property by using the commutative property of multiplication, by using the distributive property repeatedly, or by using the fact that

$$B - C = B + (-C)$$

THE ORIGINAL DISTRIBUTIVE PROPERTY

$$A \times (B + C) = A \times B + A \times C$$

for all real numbers A, B, and C.

VARIATION 1

$$(A + B) \times C = A \times C + B \times C$$

for all real numbers A, B, and C.

VARIATION 2

$$A \times (B + C + D) = A \times B + A \times C + A \times D$$

for all real numbers A, B, C, and D.

VARIATION 3

$$A \times (B - C) = A \times B - A \times C$$

for all real numbers A, B, and C.

Collectively, we will refer to the original distributive property and all its variations simply as "the distributive property."

FOIL

If you have studied algebra, then you probably learned the FOIL method for multiplying expressions of the form

$$(A + B) \cdot (C + D)$$

FOIL   **FOIL** stands for *First, Outer, Inner, Last,* in order to remind us that

$$(A + B) \cdot (C + D) = A \cdot C + A \cdot D + B \cdot C + B \cdot D$$

where $A \cdot C$ is *First*, $A \cdot D$ is *Outer*, $B \cdot C$ is *Inner*, and $B \cdot D$ is *Last*.

We can derive FOIL by using the distributive property several times. Hence, FOIL is an extension of the distributive property. To derive FOIL, let's first treat $C + D$ as a single entity; think of $C + D$ as c. Then by the distributive property,

$$(A + B) \cdot c = A \cdot c + B \cdot c$$

Therefore,

$$(A + B) \cdot \overbrace{(C + D)}^{c} = A \cdot \overbrace{(C + D)}^{c} + B \cdot \overbrace{(C + D)}^{c}$$

Now we can apply the distributive property again, this time to $A \cdot (C + D)$ and to $B \cdot (C + D)$. Stringing all these equations together, we have

$$\begin{aligned}
(A + B) \cdot (C + D) &= A \cdot (C + D) + B \cdot (C + D) \\
&= A \cdot C + A \cdot D + B \cdot C + B \cdot D
\end{aligned}$$

thereby proving that FOIL is valid.

We can also see why FOIL makes sense by subdividing arrays or rectangles. Figure 5.30 shows a rectangle made of $10 + 3$ rows with $10 + 4$ small squares in each row. According to the meaning of multiplication, the total number of small squares in the rectangle is

$$(10 + 3) \cdot (10 + 4)$$

As we see by the shading in Figure 5.30, we can subdivide the rectangle into 4 natural parts: 10 groups of 10 small squares, 10 groups of 4 small squares, 3 groups of 10 small squares, and 3 groups of 4 small squares. Therefore, the total number of small squares in the rectangle is

$$10 \cdot 10 + 10 \cdot 4 + 3 \cdot 10 + 3 \cdot 4$$

But the total number of small squares is the same either way you count them. Therefore,

$$(10 + 3) \cdot (10 + 4) = 10 \cdot 10 + 10 \cdot 4 + 3 \cdot 10 + 3 \cdot 4$$

FIGURE 5.30

A Subdivided Rectangle to Show that $(10 + 3) \cdot (10 + 4) = 10 \cdot 10 + 10 \cdot 4 + 3 \cdot 10 + 3 \cdot 4$

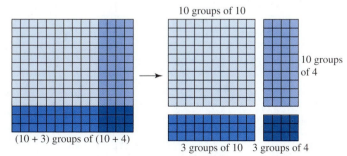

10 groups of 10

10 groups of 4

$(10 + 3)$ groups of $(10 + 4)$

3 groups of 10   3 groups of 4

which is the FOIL equation. The same reasoning will apply when the numbers 10, 3, 10, and 4 are replaced with other numbers, and the rectangle is changed accordingly. Thus, we see why FOIL must be valid for all counting numbers.

USING THE DISTRIBUTIVE PROPERTY

Just like the other properties of arithmetic that we have studied, we can often use the distributive property to make mental arithmetic problems easier to solve.

For example, what is an easy way to calculate

$$41 \times 25$$

mentally? We can use the following strategy:

40 times 25 is 1000, plus one more 25 is 1025.

The following sequence of equations corresponds to this strategy and uses the first variation on the distributive property at the second equals sign:

$$
\begin{aligned}
41 \times 25 &= (40 + 1) \times 25 \\
&= 40 \times 25 + 1 \times 25 \\
&= 1000 + 25 \\
&= 1025
\end{aligned}
$$

What is an easy way to calculate

$$39 \times 25$$

mentally? We can use a strategy similar to the previous one, this time taking away a group of 25 rather than adding it:

40 groups of 25 makes 1000; take away a group of 25 and we are left with 975.

The following sequence of equations corresponds to this strategy and uses the third variation on the distributive property (varied by the first variation) at the second equals sign.

$$
\begin{aligned}
39 \times 25 &= (40 - 1) \times 25 \\
&= 40 \times 25 - 1 \times 25 \\
&= 1000 - 25 \\
&= 975
\end{aligned}
$$

CLASS ACTIVITY   NOW TURN TO CLASS ACTIVITIES MANUAL

5N Explaining and Using the Distributive Property p. 110

5O The Distributive Property and FOIL p. 113

5P Why Isn't 23 × 23 Equal to 20 × 20 + 3 × 3? p. 115

5Q Squares and Products Near Squares p. 116

PRACTICE PROBLEMS FOR SECTION 5.7

1. Does the expression

$$3 \times 4 + 2$$

   have a different meaning than the expression

$$3 \times \quad 4 + 2$$

   which has a big space between the × and the 4? Explain.

2. Dana and Sandy are working on

$$8 \times 5 + 20 \div 4$$

   Dana says the answer is 15, but Sandy says the answer is 45. Who's right, who's wrong, and why?

3. There are 15 goodie bags. Each goodie bag contains 2 pencils, 4 stickers, and 3 wiggly worms. Write an expression using the numbers 15, 2, 4, and 3; the symbols × (or ·) and +; and parentheses, if needed, for the total number of objects in the 15 goodie bags. If you use parentheses, explain why you need them; if you do not use parentheses, explain why you do not need them.

4. Mr. Greene has a bag of plastic spiders to give out. After putting 4 spiders in each of 23 bags, he has 8 spiders left. Write an expression using the numbers 4, 23, and 8; the symbols × (or ·) and +; and parentheses, if needed, for the total number of spiders Mr. Greene had to give out. If you use parentheses, explain why you need them; if you do not use parentheses, explain why you do not need them.

5. Compute 97346 × 142349 + 2654 × 142349 mentally.

6. Write equations that correspond to the following reasoning for determining 5 × 7:

   I know 2 × 7 is 14. Then another 2 × 7 makes 28. And one more 7 makes 35.

   Which properties of arithmetic are used? Draw an array that corresponds to the reasoning.

7. Draw an array that shows why

$$20 \times 19 = 20 \times 20 - 20 \times 1$$

   Also, use this equation to help you calculate 20 × 19 mentally.

8. Draw a subdivided array to show that

$$(10 + 2) \times (10 + 3)$$
$$= 10 \times 10 + 10 \times 3 + 2 \times 10 + 2 \times 3$$

   Then write equations that use properties of arithmetic to show why the preceding equation is true.

ANSWERS TO PRACTICE PROBLEMS FOR SECTION 5.7

1. The expressions

    $$3 \times 4 + 2$$

    and

    $$3 \times \ \ 4 + 2$$

    have the same meaning. According to the convention on order of operations, multiplication is performed before addition, no matter how big a space there is following the multiplication symbol. If you want to write 3 times the *quantity* $4 + 2$ then you must use parentheses and write

    $$3 \times (4 + 2)$$

2. Sandy is right that the answer is 45. Dana, like many students, just worked from left to right. She did not use the conventions on order of operations: do multiplication and division first, then addition and subtraction. According to the conventions on the order of operations,

    $$8 \times 5 + 20 \div 4 = 40 + 5$$
    $$= 45$$

3. Since there are 15 goodie bags and each goodie bag has a total of $2 + 4 + 3$ objects in it, the total number of objects in all the goodie bags is

    $$15 \cdot (2 + 4 + 3)$$

    We need parentheses in this expression to show that 15 multiplies the entire quantity $2 + 4 + 3$.

4. Since there are 23 bags with 4 spiders in each bag, the total number of spiders in bags is 23×4. But there are also 8 more spiders, so the total number of spiders that Mr. Greene has is 8 more than 23×4 which is

    $$23 \times 4 + 8$$

    We do not need parentheses because according to the convention on order of operations, we calculate the expression above by multiplying 23 times 4 and then adding 8 to that quantity, which is exactly what we want to express.

5. By the distributive property,

    $$97,346 \times 142,349 + 2,654 \times 142,349$$
    $$= (97,346 + 2,654) \times 142,349$$
    $$= 100,000 \times 142,349$$
    $$= 14,234,900,000$$

6. The distributive property is used at the second equals sign in the following equations:

    $$5 \times 7 = (2 + 2 + 1) \times 7$$
    $$= 2 \times 7 + 2 \times 7 + 1 \times 7$$
    $$= 14 + 14 + 7$$
    $$= 28 + 7 = 35$$

    The following subdivided array corresponds to the arithmetic because it shows 2 sevens, another 2 sevens, and then a single seven, to make a total of 5 sevens.

    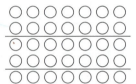

7. Figure 5.31 shows 20 rows of 20 squares. If you take away 20 rows of 1 square—namely, a vertical strip of squares—then you are left with 20 rows of 19 squares, thus illustrating that

    $$20 \times 19 = 20 \times 20 - 20 \times 1$$

    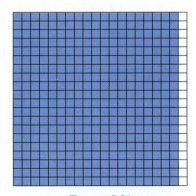

    FIGURE 5.31

    An Array Showing that
    $20 \times 19 = 20 \times 20 - 20 \times 1$

8. The array in Figure 5.32 shows $10 + 2$ groups of $10 + 3$ small squares. According to the meaning of multiplication, there are a total of $(10 + 2) \times (10 + 3)$ small squares in this array. The array is broken into four smaller arrays: 10 groups of 10, 10 groups of 3, 2 groups of 10, and 2 groups of 3. But it's the same number of small squares no matter how you count them, therefore

$$(10 + 2) \times (10 + 3) = 10 \times 10 + 10 \times 3 + 2 \times 10 + 2 \times 3$$

We can also explain this equation using the distributive property:

$$(10 + 2) \times (10 + 3) = (10 + 2) \times 10 + (10 + 2) \times 3$$
$$= 10 \times 10 + 2 \times 10 + 10 \times 3 + 2 \times 3$$

The distributive property was used at both equal signs. In fact, it was used twice at the second equals sign.

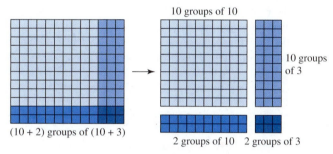

10 groups of 10

10 groups of 3

$(10 + 2)$ groups of $(10 + 3)$

2 groups of 10   2 groups of 3

FIGURE 5.32

Subdividing an Array to Show that
$(10 + 2) \times (10 + 3) = 10 \times 10 + 10 \times 3 + 2 \times 10 + 2 \times 3$

PROBLEMS FOR SECTION 5.7

1. Ben and Charles are working on

$$4 + 3 \times 2 \times 10.$$

Ben says the answer is 64. Charles says the answer is 140. Who's right, who's wrong, and why?

2. a. There are 6 cars traveling together. Each car has 2 people in front and 3 in the back. Write an expression using the numbers 6, 2, and 3; the symbols $\times$ (or $\cdot$) and $+$; and parentheses, if needed, for the total number of people riding in the 6 cars. If you use parentheses, explain why you need them; if you do not use parentheses, explain why you do not need them.

   b. Write a story problem for the expression

$$6 \times 2 + 3$$

3. The children in Mrs. Black's class are arranged as follows: 7 tables have 4 children sitting at each of them; 2 tables have 3 children sitting at each of them. Write an expression using the numbers 7, 4, 2, and 3; the symbols $\times$ (or $\cdot$) and $+$; and parentheses, if needed, for the total number of children in Mrs. Black's class. If you use parentheses, explain why you need them; if you do not use parentheses, explain why you do not need them.

4. Write one story problem for the expression

$$8 \times (4 + 2)$$

and another story problem for the expression

$$8 \times 4 + 2$$

Make clear which is which and why each story problem fits with its expression.

5. Draw arrays to help you show why the following equations are true without evaluating any of the products or sums. Explain briefly.

   a. $(5 + 1) \cdot 7 = 5 \cdot 7 + 1 \cdot 7$

   b. $6 \cdot (5 + 2) = 6 \cdot 5 + 6 \cdot 2$

   c. $(10 - 1) \cdot 6 = 10 \cdot 6 - 1 \cdot 6$

6. Josh consistently remembers that $7 \times 7 = 49$, but he keeps forgetting 7×8.

   a. Explain to Josh how 7×7 and 7×8 are related. Draw an array to help you show this relationship.

   b. Write an equation relating 7×8 to 7×7. Which property of arithmetic does your equation use? Explain.

7. Explain how to use the distributive property to make 31×25 easy to calculate mentally. Write an equation that corresponds to your strategy. Without drawing all the detail, draw a rough picture of an array that illustrates this calculation strategy.

8. Explain how to calculate 29×20 mentally using 30×20. Write an equation that uses subtraction and the distributive property and that corresponds to your strategy. Without drawing all the detail, draw a rough picture of an array that illustrates this calculation strategy.

9. Suppose that the sales tax where you live is 6%. Compare the total amount of sales tax you would pay if you bought a pair of pants and a shirt at the same time, versus if you first bought the pants and then went back to the store and bought the shirt. Write equations to show why the distributive property is relevant to this problem.

10. a. Use the distributive property several times to show why

   $$(10 + 2) \cdot (10 + 4)$$
   $$= 10 \cdot 10 + 10 \cdot 4 + 2 \cdot 10 + 2 \cdot 4$$

   b. Draw an array to show why

   $$(10 + 2) \cdot (10 + 4)$$
   $$= 10 \cdot 10 + 10 \cdot 4 + 2 \cdot 10 + 2 \cdot 4$$

   c. Relate the steps in your equations in part (a) to your array in part (b).

11. Ted thinks that because $10 \times 10 = 100$ and $2 \times 5 = 10$, he should be able to calculate 12×15 by adding $100 + 10$ to get 110. Explain to Ted in two different ways that even though his method is not correct, his

calculations can be part of a correct way to calculate 12×15:

   a. by drawing an array;

   b. by writing equations that use the distributive property.

12. Clint and Sue went out to dinner and had a nice meal that cost \$64.82. With a 7% tax of \$4.54, the total came to \$69.36. They want to leave a tip of approximately 15% of the cost of the meal (before the tax). Describe a way that Clint and Sue can mentally figure the tip.

13. Your favorite store is having a 10%-off sale, meaning that the store will take 10% off the price of each item you buy. When the clerk rings up your purchases, she takes 10% off the total (before tax), rather than 10% off each item. Will you get the same discount either way? Is there a property of arithmetic related to this? Explain!

14. SunJae is working on the multiplication problem 21×34. SunJae says that he can take 1 from the 21 and put it with the 34 to get 20×35. SunJae says that this new multiplication problem, 20×35, should have the same answer as 21×34. Is SunJae right? How might you convince SunJae that his reasoning is or is not correct in a way other than simply showing him the answers to the two multiplication problems?

15. On a television broadcast on tennis, a commentator said,

   She has made 30 out of 33 first serves. That's better than 90%.

   Explain how the commentator could figure this out without using a calculator or doing long division.

16. Frank's Jewelers runs the following advertisement: "Come to our 40%-off sale on Saturday. We're not like the competition, who raise prices by 30% and then have a 70%-off sale!"

   a. If two items start off with the same price, which gives you the lower price in the end: taking off 40% or raising the price by 30% and then taking off 70% (of the raised price)?

   b. Consider the same problem more generally, with other numbers. For example, if you raise prices by 20% and then take off 50% (of the raised price), how does this compare to taking 30% off of the original price? If you raise a price by 30% and then lower the raised price by 30%, how does that compare to the original price? Try at least 2 other pairs of percentages by which to raise and then lower a price. Describe what you observe.

Predict what happens in general: if you raise a price by $A\%$ and then take $B\%$ off of the raised price, does that have the same result as if you'd lowered the original price by $(B - A)\%$? If not, which produces the lower final price?

c. Use the distributive property or FOIL to explain the pattern you discovered in part (b). Remember that to raise a price by 15%, for example, you multiply the price by $1 + .15$, whereas to lower a price by 15%, you multiply the price by $1 - .15$.

17. a. Use an ordinary calculator to calculate $666,666,666 \times 999,999,999$. Based on the calculator's display, guess how to write the answer in ordinary decimal notation (showing *all* digits).

b. Now think some more, and determine how to write the product $666,666,666 \times 999,999,999$ in ordinary decimal notation, showing all its digits, without multiplying longhand or using a calculator or computer.

18. Calculate the product $9,999,999,999 \times 9,999,999,999$ without using a calculator or computer and without simply multiplying longhand. Give the answer in ordinary decimal notation, showing all its digits. (Both numbers in the product have 10 nines.) Explain your method.

19. Check the following:

$$11 - 2 = 3^2$$

$$1111 - 22 = 33^2$$

$$111111 - 222 = 333^2$$

Continue to find at least three more in the pattern. Does the pattern continue? Now explain why there is such a pattern. Hint: notice that $1111 - 22 = 11 \times 101 - 11 \times 2$.

20. Determine which of the following two numbers is larger and explain your reasoning:

a. $1,000,000 \times (1 + 2 + 3 + 4 + \cdots + 1,000,001)$

b. $1,000,001 \times (1 + 2 + 3 + 4 + \cdots + 1,000,000)$

21. The **square** of a number is just the number times itself. For example, the square of 4 is $4 \cdot 4 = 16$.

a. Find the squares of the numbers $15, 25, 35, 45, \ldots,$ $95, 105, 115, 125, \ldots, 195, 205,$ and three other whole numbers that end in 5.

b. Find some patterns in the answers to part (a). Specifically, in all of your answers to part (a), what do you notice about the last two digits, and what do you notice about the number formed by deleting the last two digits?

c. Use what you discovered in part (b) to predict the squares of 2,005 and 10,005.

d. Every whole number ending in 5 must be of the form $10A + 5$ for some whole number A. Find the square of $10A + 5$, namely, calculate $(10A + 5) \times (10A + 5)$.

e. How does your answer to part (d) explain the pattern you found in part (b)?

22. The **square** of a number is just the number times itself. For example, the square of 4 is 16.

a. Calculate the squares of $1, 2, \ldots, 9$ and many other whole numbers, including $17, 34, 61, 82, 99, 123,$ $255, 386,$ and 728. Record the ones digits in each case. What do you notice? Do any of your squares have a ones digit of 7, for example? Are any other digits missing from the ones digits of squares?

b. Which digits can never occur as the ones digit of a square of a whole number? Explain why some digits cannot occur as the ones digit of a square of a whole number.

c. Based on what you've discovered, could the number

$$139,787,847,234,329,483$$

be the square of a whole number? Why or why not?

5.8 Mental Math, Properties of Arithmetic, and Algebra

We have seen that we can use the commutative and associative properties of addition and multiplication and the distributive property individually to help us make some mental math problems easy to solve. In this section we will study some mental math strategies that rely on combinations of these properties of arithmetic. We will also study in greater depth the link between mental math strategies and algebra. We will see that we can take a mental calculation strategy that is expressed in words and translate it into a string of equations. Such strings of equations are the same kinds of equations that are produced in algebra but without the *x*s.

WRITING EQUATIONS THAT CORRESPOND TO A MENTAL CALCULATION STRATEGY

We will now analyze a fourth grader's reasoning and then see how to write equations that correspond to the reasoning.

In the following conversation, a fourth grader skillfully uses the distributive property to calculate mentally with fractions. The distributive property holds for all real numbers, therefore we can use it with fractions. The exchange is taken from *Developing Children's Understanding of the Rational Numbers: A New Model and an Experimental Curriculum* by Joan Moss and Robbie Case [36, page 135]. *"Experimental S1"* is one of the fourth grade students who participated in an experimental curriculum described in the article.

Experimenter: Another student told me that 7 is $\frac{3}{4}$ of 10. Is it?

Experimental S1: No, because of one half of 10 is 5. One half of 5 is
2 and $\frac{1}{2}$. So if you add $2\frac{1}{2}$ to 5, that would be
$7\frac{1}{2}$. So $7\frac{1}{2}$ is $\frac{3}{4}$ of 10, not 7.

Why is the student's reasoning valid? Notice that the student adds half of 10 and half of half of ten to get 7 and a half. In other words, the student says that $\frac{3}{4}$ of 10 is half of ten plus half of half of ten, or

$$\frac{3}{4} \cdot 10 = \frac{1}{2} \cdot 10 + \frac{1}{2} \cdot \frac{1}{2} \cdot 10$$

We can use the following two facts to explain why the student's method is correct:

- Three fourths is a half plus half of a half, or in other words,

$$\frac{3}{4} = \frac{1}{2} + \frac{1}{2} \cdot \frac{1}{2}$$

as we see in Figure 5.33.

- The distributive property.

FIGURE 5.33

Finding $\frac{3}{4}$ of 10

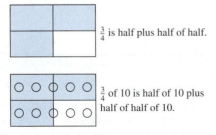

$\frac{3}{4}$ is half plus half of half.

$\frac{3}{4}$ of 10 is half of 10 plus half of half of 10.

Observe how the following equations use these two facts, and show why the student's reasoning is valid:

$$\frac{3}{4} \cdot 10 = \left(\frac{1}{2} + \frac{1}{2} \cdot \frac{1}{2}\right) \cdot 10$$

$$= \frac{1}{2} \cdot 10 + \frac{1}{2} \cdot \frac{1}{2} \cdot 10$$

$$= 5 + 2\frac{1}{2}$$

$$= 7\frac{1}{2}$$

The distributive property is used at the second equals sign. Notice that the student also used the associative property of multiplication by calculating $\frac{1}{2}$ of 10 and then calculating $\frac{1}{2}$ of 5. In other words, instead of calculating

$$\left(\frac{1}{2} \cdot \frac{1}{2}\right) \cdot 10$$

the student calculates

$$\frac{1}{2} \cdot \left(\frac{1}{2} \cdot 10\right)$$

EQUATIONS AND PROPERTIES OF ARITHMETIC ARE STEPPING STONES TO ALGEBRA

This section and several others in the book have emphasized the writing of equations and the use of properties of arithmetic to solve arithmetic problems. In some cases, we used properties of arithmetic to help us solve a problem mentally, but in other cases, especially when we write long strings of equations, it may seem that we are only making the problem longer and harder. So why bother, especially when there are efficient algorithms for carrying out the calculations? Writing equations and using properties of arithmetic to calculate are emphasized here *because these are fundamental skills in algebra.* For example, one standard type of problem in algebra is to simplify an expression such as

$$9x^2 + 4x(7 - x)$$

We can simplify this expression by writing the following string of equations:

$$9x^2 + 4x(7 - x) = 9x^2 + 4x \cdot 7 - 4x \cdot x$$
$$= 9x^2 + 28x - 4x^2$$
$$= 9x^2 - 4x^2 + 28$$
$$= 5x^2 + 28$$

In making this simplification, we used the distributive property as well as the associative and commutative properties of multiplication and addition.

The National Council of Teachers of Mathematics advocates that all children learn algebra. (See [44, p. 37].) This ambitious goal can only succeed if children develop a strong foundation to support the learning of algebra. Research shows that the leap from arithmetic to algebra is a difficult one for students because algebra uses unknown quantities. (See [25].) However, research also shows that some of the difficulties in students' algebra learning can already be found when students work with equations in arithmetic. (See [32].) It therefore makes sense that elementary school children should learn to calculate flexibly by using properties of arithmetic (even if they do not know the formal names of these properties and are not aware that they are using properties of arithmetic) and should learn to work with equations.

Developing the ability to calculate flexibly by using properties of arithmetic lays a foundation for algebra. But according to the National Council of Teachers of Mathematics, [44, p. 35], there is another benefit:

> Researchers and experienced teachers alike have found that when children in the elementary grades are encouraged to develop, record, explain, and critique one another's strategies for solving computational problems, a number of important kinds of learning can occur...

Such experiences can help children strengthen their understanding of place value and develop better number sense. (See [29].)

Curriculum frameworks or guidelines of various states recognize the importance of using properties of arithmetic by recommending or mandating that children learn to use these properties. The Georgia Quality Core Curriculum (QCC) [23] expects a third grader to master the following objective:

> Uses properties of addition and multiplication (including commutative, associative, and properties of zero and one).

The Mathematics Framework for California Public Schools [9] identifies the following as a key standard for Grade 2:

> Use the commutative and associative rules to simplify mental calculations and to check results.

CLASS ACTIVITY   NOW TURN TO CLASS ACTIVITIES MANUAL

5R Using Properties of Arithmetic to Aid the Learning of Basic Multiplication Facts p. 119

5S Solving Arithmetic Problems Mentally p. 123

5T Which Properties of Arithmetic Do These Calculations Use? p. 124

5U Writing Equations That Correspond to a Method of Calculation p. 126

5V Showing the Algebra in Mental Math p. 128

PRACTICE PROBLEMS FOR SECTION 5.8

1. A child is having difficulty remembering 8×8. Draw two arrays showing how 8×8 is related to other possibly easier facts involving smaller numbers. For each array, write a corresponding equation relating 8×8 to other multiplication facts. Which properties of arithmetic do you use?

2. The string of equations that follows corresponds to a mental method for calculating 45×11. Explain in words why the method of calculation makes sense. Which properties of arithmetic are used, and where are they used?

$$45 \times 11 = 45 \times (10 + 1)$$
$$= 45 \times 10 + 45 \times 1$$
$$= 450 + 45$$
$$= 495$$

3. Each arithmetic problem in this exercise has a description for solving the problem. In each case, write a string of equations that corresponds to the given description. Identify properties of arithmetic that are used. Write your equations in the following form:

original $=$ some expression

$$= \vdots$$

$= $ some expression

a. Problem: What is 6×40?
    Solution: 6 times 4 is 24, then you multiply by 10 and get 240.

b. Problem: What is 110% of 62?
    Solution: 100% of 62 is 62 and 10% of 62 is 6.2, so all together it's 68.2.

c. Problem: Calculate the 7% tax on a purchase of $25.
    Solution: 10% of $25 is $2.50, so 5% is $1.25. One percent of $25 is 25 cents, so 2% is 50 cents. This means 7% is $1.25 plus 50 cents, which is $1.75.

d. Problem: What is 45% of 300?
    Solution: 50% of 300 is 150. Ten percent of 300 is 30 and half of that is 15, so the answer is 135.

e. Problem: Find $\frac{3}{4}$ of 72.
    Solution: Half of 72 is 36. Half of 36 is 18. Then, to add 36 and 18, I did 40 plus 14, which is 54.

f. Problem: Calculate $59 \cdot 70$.
    Solution: 6 times 7 is 42 so 60 times 70 is 4200. Therefore, 59 times 70 is 70 less, which is 4130.

4. For each of the problems that follow, use the distributive property to help make the problem easy to solve mentally. In each case, write a string of equations that correspond to your strategies. Write your equations in the following form:

original $=$ some expression

$$= \vdots$$

$=$ some expression

a. Calculate 51% of 140.
b. Calculate 95% of 60.
c. Calculate $\frac{5}{8} \times 280$.
d. Calculate 99% of 80.

5. For each of the following arithmetic problems, use properties of arithmetic to make the problem easy to solve mentally. Write a string of equations that corresponds to your method. Say which properties of arithmetic you use and where you use them.

a. 25×84
b. 49×6
c. 486×5

6. Joey, a second grader, used the following method to mentally calculate the number of minutes in a day. First, Joey calculated 25 times 6 by finding 4 times 25 plus two times 25, which is 150. Then he subtracted 6 from 150 to get 144. Then he multiplied 144 by 10 to get the answer: 1440. Write a sequence of equations that corresponds to Joey's method and that show why his method is legitimate. What properties of arithmetic are involved?

ANSWERS TO PRACTICE PROBLEMS FOR SECTION 5.8

1. See Figure 5.34 for the arrays. The array on the left breaks 8 groups of 8 stars into 2 groups of 4×8. The corresponding equation is

$$8 \times 8 = 2 \times (4 \times 8)$$

This equation uses the associative property of multiplication because the first 8 is broken into 2×4 and this 4 is switched from associating with the 2 to associating with the 8 on the right. We could also write the equation

$$8 \times 8 = 4 \times 8 + 4 \times 8$$

which uses the distributive property instead of the associative property. Notice that if the child knows the fact $4 \times 8 = 32$, then he might be able to add $32 + 32$ quickly to get the correct answer to 8×8.

FIGURE 5.34

Relating 8×8 to Other Multiplication Facts

The array on the right of Figure 5.34 breaks 8 groups of 8 stars into 5 groups of 8 and 3 groups of 8. The corresponding equation is

$$8 \times 8 = 5 \times 8 + 3 \times 8$$

Notice that if the child knows 5×8 and 3×8 then she might be able to add $40 + 24$ quickly to get the correct answer to 8×8.

2. The product 45×11 stands for the total number of objects in 45 groups that have 11 objects in each group. These objects can be broken into 45 groups of 10 and another 45 groups of 1. The number of objects in 45 groups of 10 is 45×10, which is 450, and the number of objects in 45 groups of 1 is 45. In all that makes $450 + 45$ which is 495 objects. The distributive property is used to say that $45(10 + 1) = 45 \times 10 + 45 \times 1$.

3. a.
$$6 \times 40 = 6 \times (4 \times 10)$$
$$= (6 \times 4) \times 10$$
$$= 24 \times 10$$
$$= 240$$

The associative property of multiplication is used at the second equal sign in order to switch the placement of the parentheses and group the 4 with the 6 instead of with the 10.

b.
$$110\% \times 62 = (100\% + 10\%) \times 62$$
$$= 100\% \times 62 + 10\% \times 62$$
$$= 62 + 6.2$$
$$= 68.2$$

The distributive property is used at the second equal sign.

c.
$$7\% \times 25 = (5\% + 2\%) \times 25$$
$$= 5\% \times 25 + 2\% \times 25$$
$$= \frac{1}{2} \times 10\% \times 25 + 2 \times 1\% \times 25$$
$$= \frac{1}{2} \times 2.5 + 2 \times .25$$
$$= 1.25 + .50$$
$$= 1.75$$

The distributive property is used at the second equals sign. (Although it is hidden, technically speaking, the associative property of multiplication is actually used twice at the fourth equals sign in order to calculate 10% of 25 and 1% of 25 first, and then multiply those by $\frac{1}{2}$ and 2, respectively.)

d.
$$45\% \times 300 = (50\% - 5\%) \times 300$$
$$= 50\% \times 300 - 5\% \times 300$$
$$= 150 - \frac{1}{2} \times 10\% \times 300$$
$$= 150 - \frac{1}{2} \times 30$$
$$= 150 - 15$$
$$= 135$$

The distributive property is used at the second equals sign. (Although it is hidden, technically speaking, the associative property of multiplication is used at the fourth equals sign to calculate 10% of 300 first, and then find half of that.)

e.

$$\frac{3}{4} \times 72 = \left(\frac{1}{2} + \frac{1}{2} \times \frac{1}{2}\right) \times 72$$

$$= \frac{1}{2} \times 72 + \frac{1}{2} \times \frac{1}{2} \times 72$$

$$= 36 + \frac{1}{2} \times 36$$

$$= 36 + 18$$

$$= 36 + (4 + 14)$$

$$= (36 + 4) + 14$$

$$= 40 + 14$$

$$= 54$$

The distributive property is used at the second equals sign. The associative property of addition was used at the sixth equals sign in order to change the placement of the parentheses. (Although it is hidden, technically speaking, the associative property of multiplication is used at the third equals sign to find half of 72 first. Then we take half of that.)

f.

$$59 \cdot 70 = (60 - 1) \cdot 70$$

$$= 60 \cdot 70 - 1 \cdot 70$$

$$= (6 \cdot 10) \cdot (7 \cdot 10) - 70$$

$$= (6 \cdot 7) \cdot (10 \cdot 10) - 70$$

$$= 42 \cdot 100 - 70$$

$$= 4200 - 70 = 4130$$

The distributive property is used at the second equals sign. The commutative and associative properties of multiplication are used at the fourth equals sign. Notice that even though the stated solution to the problem did not explicitly mention multiplying 10 by 10 to get 100 and then multiplying 42 by 100, these calculations were used implicitly. So the previous equations actually expand on the stated solution, filling in the implied details.

4. a. We can decompose 51% as 50% plus 1%. Fifty percent of 140 is half of 140, which is 70. One percent of 140 is 1.4. So 51% of 140 is 70 plus 1.4, which is 71.4. Using equations, we have

$$.51 \times 140 = (.50 + .01) \times 140$$

$$= .50 \times 140 + .01 \times 140$$

$$= 70 + 1.4 = 71.4$$

The distributive property was used at the second equals sign.

b. We can decompose 95% as 100% minus 5%. Now 10% of 60 is 6, so 5% of 60 is half of that, which

is 3. Therefore, 95% of 60 is $60 - 3$, which is 57. Using equations, we have

$$.95 \times 60 = (1 - .05) \times 60$$

$$= 1 \times 60 - .05 \times 60$$

$$= 60 - \frac{1}{2} \times .10 \times 60$$

$$= 60 - \frac{1}{2} \times 6$$

$$= 60 - 3 = 57$$

The distributive property was used at the second equals sign.

c. Use the fact that $\frac{5}{8} = \frac{4}{8} + \frac{1}{8} = \frac{1}{2} + \frac{1}{8}$.

$$\frac{5}{8} \times 280 = \left(\frac{1}{2} + \frac{1}{8}\right) \times 280$$

$$= \frac{1}{2} \times 280 + \frac{1}{2} \times \frac{1}{2} \times \frac{1}{2} \times 280$$

$$= 140 + 35 = 175$$

The distributive property was used at the second equals sign.

d.

$$.99 \times 80 = (1 - .01) \times 80$$

$$= 1 \times 80 - .01 \times 80$$

$$= 80 - .8$$

$$= 79.2$$

The distributive property was used at the second equals sign.

5. a.

$$25 \times 84 = 25 \times (4 \times 21)$$

$$= (25 \times 4) \times 21$$

$$= 100 \times 21$$

$$= 2100$$

The associative property of multiplication was used to rewrite $25 \times (4 \times 21)$ as $(25 \times 4) \times 21$, in other words, to multiply the 4 with the 25 instead of with the 21.

b.

$$49 \times 6 = (50 - 1) \times 6$$

$$= 50 \times 6 - 1 \times 6$$

$$= 300 - 6$$

$$= 294$$

The distributive property was used to rewrite $(50 - 1) \times 6$ as $50 \times 6 - 1 \times 6$.

c.
$$486 \times 5 = (243 \times 2) \times 5$$
$$= 243 \times (2 \times 5)$$
$$= 243 \times 10$$
$$= 2430$$

The associative property of multiplication was used to rewrite $(243 \times 2) \times 5$ as $243 \times (2 \times 5)$.

6. The following sequence of equations shows in detail why Joey's method is valid and how it uses properties of arithmetic:

$$24 \times 60 = 24 \times (6 \times 10)$$
$$= (24 \times 6) \times 10$$
$$= [(25 - 1) \times 6] \times 10$$
$$= (25 \times 6 - 1 \times 6) \times 10$$
$$= [6 \times 25 - 6] \times 10$$
$$= [4 \times 25 + 2 \times 25 - 6] \times 10$$
$$= [100 + 50 - 6] \times 10$$
$$= (150 - 6) \times 10$$
$$= 144 \times 10$$
$$= 1440$$

The associative property of multiplication is used at the second equals sign.

The distributive property is used at the fourth and sixth equals signs.

The commutative property of multiplication is used at the fifth equals sign.

The associative property of addition was essentially used at the eighth equals sign.

PROBLEMS FOR SECTION 5.8

1. Demarcus knows his $1 \times$, $2 \times$, and $3 \times$ multiplication tables. He also knows 4×1, 4×2, 4×3, 4×4, and 4×5.

   a. Describe how the three arrays in Figure 5.35 provide Demarcus with three different ways to determine 4×6 from multiplication facts that he already knows. In each case, write an equation that corresponds to the array and that shows how 4×6 is related to other multiplication facts.

   b. Draw arrays showing two different ways that Demarcus could use the multiplication facts he already knows to determine 4×7. In each case, write an equation that corresponds to the array and that shows how 4×7 is related to other multiplication facts.

   c. Draw arrays showing two different ways that Demarcus could use the multiplication facts he already knows to determine 4×8. In each case, write an equation that corresponds to the array and that shows how 4×8 is related to other multiplication facts.

FIGURE 5.35

Different Ways to Think about 4×6

2. Suppose that a child has learned the following basic multiplication facts:

- The $\times 1$, $\times 2$, $\times 3$, $\times 4$, and $\times 5$ tables—that is,

$1 \times 1 = 1$	$1 \times 2 = 2$	$1 \times 3 = 3$	$1 \times 4 = 4$	$1 \times 5 = 5$
$2 \times 1 = 2$	$2 \times 2 = 4$	$2 \times 3 = 6$	$2 \times 4 = 8$	$2 \times 5 = 10$
$3 \times 1 = 3$	$3 \times 2 = 6$	$3 \times 3 = 9$	$3 \times 4 = 12$	$3 \times 5 = 15$
$4 \times 1 = 4$	$4 \times 2 = 8$	$4 \times 3 = 12$	$4 \times 4 = 16$	$4 \times 5 = 20$
$5 \times 1 = 5$	$5 \times 2 = 10$	$5 \times 3 = 15$	$5 \times 4 = 20$	$5 \times 5 = 25$
$6 \times 1 = 6$	$6 \times 2 = 12$	$6 \times 3 = 18$	$6 \times 4 = 24$	$6 \times 5 = 30$
$7 \times 1 = 7$	$7 \times 2 = 14$	$7 \times 3 = 21$	$7 \times 4 = 28$	$7 \times 5 = 35$
$8 \times 1 = 8$	$8 \times 2 = 16$	$8 \times 3 = 24$	$8 \times 4 = 32$	$8 \times 5 = 40$
$9 \times 1 = 9$	$9 \times 2 = 18$	$9 \times 3 = 27$	$9 \times 4 = 36$	$9 \times 5 = 45$

- The squares $1 \times 1 = 1$, $2 \times 2 = 4$, $3 \times 3 = 9, \ldots, 9 \times 9 = 81$.

For each of the following multiplication problems, find at least two different ways that some of the facts above, together with properties of arithmetic, could be used to mentally calculate the answer to the problem. Explain your answers, drawing arrays to illustrate how the following problems are related to facts in the given lists:

a. 6×7 b. 7×8 c. 6×8

3. For each of the multiplication problems in this exercise, describe a way to make the problem easy to solve mentally. Then write equations that correspond to your method of calculation. Write your equations in the following form:

$$24 \times 25 = \text{some expression}$$
$$= \text{some expression}$$
$$\vdots \ \vdots$$

In each case, state which properties of arithmetic you used and indicate where you used those properties.

a. 24×25 b. 25×48 c. 51×6

4. The following exchanges are taken from *Developing Children's Understanding of the Rational Numbers: A New Model and an Experimental Curriculum* by Joan Moss and Robbie Case [36, page 135]. "*Experimental S1*" and "*Experimental S3*" are two of the fourth-grade students who participated in an experimental curriculum described in the article.

Experimenter: What is 65% of 160?

Experimental S1: Fifty percent [of 160] is 80. I figure 10%, which would be 16. Then I I divided by 2, which is 8 [5%] then 16 plus 8 um . . . 24. Then I do 80 plus 24, which would be 104.

$$\vdots$$

Experimental S3: Ten percent of 160 is 16; 16 times 6 equals 96. Then I did 5%, and that was 8, so . . . , 96 plus 8 equals 104.

For each of the two students' responses, write strings of equations that correspond to the student's method for calculating 65% of 160. State which properties of arithmetic were used and where (be specific).

Write your string of equations in the following form:

$$65\% \times 160 = \text{some expression}$$
$$= \vdots$$
$$= \text{some expression}$$

5. Here is Marco's method for solving 38×60:

> 4 times 6 is 24, so 40 times 60 is 2400. Then 2 times 6 is 12, so it's $2400 - 120$, which is 2280.

Write equations that incorporate Marco's method and that also show why his method is valid. Write your equations in the following form:

$$38 \times 60 = \text{some expression}$$
$$= \text{some expression}$$
$$\vdots$$
$$= 2280$$

Which properties of arithmetic did Marco use (knowingly or not), and where?

6. Jenny uses the following method to find 28% of 60,000 mentally:

> 25% is $\frac{1}{4}$, and $\frac{1}{4}$ of 60 is 15, so 25% of 60,000 is 15,000. One percent of 60,000 is 600, and that times 3 is 1800. So the answer is $15,000 + 1,800$, which is 16,800.

Write a string of equations that calculates 28% of 60,000 and that incorporate Jenny's ideas. Write your equations in the following form:

$$28\% \times 60,000 = \text{some expression}$$
$$= \text{some expression}$$
$$\vdots$$
$$= 16,800$$

7. Use properties of arithmetic to calculate 35% of 440 mentally. Describe your strategy in words and write a string of equations that corresponds to your strategy. Indicate which properties of arithmetic you used, and where. Be specific. Write your equations in the following form:

$$35\% \times 440 = \text{some expression}$$
$$= \vdots$$
$$= \text{some expression}$$

8. Use the distributive property to make it easy for you to calculate 30% of 240 mentally. Then use the associative property of multiplication to solve the same problem. In each case, explain your strategy in words and then write equations that correspond to your strategy. Write your equations in the following form:

$$30\% \times 240 = \text{some expression}$$
$$= \text{some expression}$$
$$\vdots$$

9. Tamar calculated 41×41 as follows:

> Four 4s is 16, so four 40s is 160 and forty 40s is 1600. Then forty-one 40s is another 40 added on, which is 1640. So forty-one 41s is 41 more, which is 1681.

a. Explain briefly why it makes sense for Tamar to solve the problem the way she does. What is the idea behind her strategy?

b. Write equations that incorporate Tamar's work and that show clearly why Tamar's method calculates the correct answer to 41×41. Which properties of arithmetic did Tamar use (knowingly or not) and where? Be thorough and be specific. Write your equations in the following format:

$$41 \times 41 = \text{some expression}$$
$$= \text{some expression}$$
$$\vdots \quad \vdots$$
$$= 1681$$

10. Here is how Nya solved the problem $\frac{3}{4} \cdot 72$:

> Half of 72 is 36. Half of 36 is 18. Then, to add 36 and 18, I did 40 plus 14, which is 54.

Write a string of equations that incorporate Nya's ideas. Which properties of arithmetic did Nya use (knowingly or not) and where? Be thorough and be specific. Write your equations in the following format:

$$\frac{3}{4} \cdot 72 = \text{some expression}$$
$$= \text{some expression}$$
$$\vdots$$
$$= 54$$

11. a. Lindsay calculates two-fifths of 1260 using the following strategy: First, Lindsay finds half of 1260, which is 630. Then Lindsay subtracts one tenth of 1260, which is 126, from 630 and gives the answer, 504. Discuss the ideas behind Lindsay's strategy. Then write a string of equations that incorporate Lindsay's strategy and that show why the strategy is valid. What property of arithmetic is involved? Write your equations in the following form:

$$\frac{2}{5} \times 1260 = \text{some expression}$$

$$= \text{some expression}$$

$$\vdots$$

$$= 504$$

b. Terrell calculates two-fifths of 1260 in the following way: First he multiplies 1260 by 2 to get 2520. Then he multiplies 2520 by 2 to get 5040 and divides this by 10 to get 504. Discuss the idea behind Terrell's strategy. Then write a string of equations that incorporate Terrell's strategy, and that show why the strategy is valid. Write your equations in the form shown previously.

12. Maria is working on the multiplication problem 38×25. Maria says,

4 times 25 is 100, so 40 times 25 is 1000. Now take away 2, so the answer is 998.

Is Maria's method correct or not? If it is correct, write equations that incorporate Maria's work and that show why it's correct. If Maria's reasoning is not correct, work with portions that are right to correct Maria's work, and write a string of equations that corresponds to your corrected method for calculating 38×25. Write your equations in the following form:

$$38 \times 25 = \text{some expression}$$

$$= \text{some expression}$$

$$\vdots$$

13. There is an interesting mental technique for multiplying certain pairs of numbers. The following examples will illustrate how it works:

- To calculate 32×28, notice that the two factors 28 and 32 are both 2 away—in opposite directions—from 30. To calculate 32×28, do the following:

$$30 \times 30 - 2 \times 2 = 900 - 4$$

$$= 896$$

Notice that you can do this calculation in your head.

- Similarly, to calculate 59×61, notice that both factors are 1 away—in opposite directions—from 60. Then 59×61 is

$$60 \times 60 - 1 \times 1 = 3600 - 1$$

$$= 3599$$

Once again, notice that you can do this mentally.

a. Use the method just shown to calculate 83×77, 195×205, and one other multiplication problem like this that you make up.

b. Now explain why this method works. (*Hint:* A diagram might be helpful. Another approach is to notice that the technique applies to multiplication problems of the form $(A + B) \times (A - B)$.)

14. Try out this next mathematical magic trick. Do the following on a piece of paper:

a. Write the number of days a week you would like to go out (from 1 to 7).

b. Multiply the number by 2.

c. Add 5.

d. Multiply by 50.

e. If you have already had your birthday this year, add 1754 if it is 2004. (Add 1755 if it is 2005, add 1756 if it is 2006, and so on.) If you have not yet had your birthday this year, add 1753 if it is 2004. (Add 1754 if it is 2005, add 1755 if it is 2006, and so on.)

f. Finally, subtract the four-digit year you were born. You should now have a three-digit number. If not, try again. If you have a three-digit number, continue with the following:

The first digit of your answer is your original number (i.e., how many times you want to go out each week). The second two digits are your current age.

Is it magic, or is it math? Explain why the trick works.

15. In order to subtract 197 from 384 you need to regroup. The following string of equations corresponds to the regrouping process, when it is shown in detail with expanded forms:

$$384 = 3(100) + 8(10) + 4(1) \qquad (5.1)$$
$$= [2(100) + 1(100)] + [7(10) + 1(10)] + 4(1) \qquad (5.2)$$
$$= [2(100) + 10(10)] + [7(10) + 10(1)] + 4(1) \qquad (5.3)$$
$$= [2(100) + 10(10)] + 7(10) + [10(1) + 4(1)] \qquad (5.4)$$
$$= 2(100) + [10(10) + 7(10)] + [10(1) + 4(1)] \qquad (5.5)$$
$$= 2(100) + 17(10) + 14(1) \qquad (5.6)$$

Which properties of arithmetic are used, and where are they used? Be thorough and be specific.

16. In order to subtract .36 from .84 you need to regroup. The following string of equations corresponds to the regrouping process, when it is shown in detail with expanded forms:

$$.84 = 8\left(\frac{1}{10}\right) + 4\left(\frac{1}{100}\right) \qquad (5.7)$$
$$= \left[7\left(\frac{1}{10}\right) + 1\left(\frac{1}{10}\right)\right] + 4\left(\frac{1}{100}\right) \qquad (5.8)$$
$$= \left[7\left(\frac{1}{10}\right) + 10\left(\frac{1}{100}\right)\right] + 4\left(\frac{1}{100}\right) \qquad (5.9)$$
$$= 7\left(\frac{1}{10}\right) + \left[10\left(\frac{1}{100}\right) + 4\left(\frac{1}{100}\right)\right] \qquad (5.10)$$
$$= 7\left(\frac{1}{10}\right) + 14\left(\frac{1}{100}\right) \qquad (5.11)$$

Which properties of arithmetic are used in this, and where are they used? Be thorough and be specific.

5.9   Why the Procedure for Multiplying Whole Numbers Works

The standard longhand procedure for multiplying multiple-digit whole numbers is an efficient paper-and-pencil method of calculation. This method is useful because it converts a multiplication problem with numbers that have several digits to many multiplication problems with one-digit numbers. Therefore, someone who has memorized the one-digit multiplication tables (from $1 \times 1 = 1$ to $9 \times 9 = 81$) can multiply any pair of whole numbers by using the longhand multiplication algorithm. But why does this clever method give the correct answer to multiplication problems? What makes it work?

Before we continue, let's first make sense of the questions at the end of the previous paragraph. Notice the distinction between the *meaning* of multiplication and the longhand *procedure* for multiplying. If you have 58 bags of widgets and there are 764 widgets in each bag, then you can ask how many widgets you have in all. There is some specific total number of widgets. But what is this number? According to the meaning of multiplication, the total number of widgets is

$$58 \times 764$$

The longhand multiplication procedure consists of carrying out a number of steps as follows:

Multiply 8×4, write the 2, carry the 3; multiply 8×6, add the carried 3, write the 1, carry the 5, etc.

What is the connection between the meaning of 58×764 and the procedure for calculating 58×764? We need to address the following question:

Why does the standard longhand multiplication procedure for calculating 58×764 give the actual total number of widgets in 58 bags if there are 764 widgets in each bag?

The answer will be explained in this section.

THE PARTIAL-PRODUCTS MULTIPLICATION ALGORITHM

partial-products multiplication algorithm

In order to understand the standard longhand multiplication algorithm, we will work with the **partial-products multiplication algorithm**. The partial-products algorithm is essentially the same as the standard multiplication algorithm, but it has the virtue of showing more steps, so it will make our analysis easier. A few examples will illustrate how the partial-products algorithm works. In the examples, the arrows and the expressions to the right of the arrows have been added to show how to carry out the algorithm. You do not have to write these arrows and expressions when you use the algorithm yourself.

STANDARD ALGORITHM	PARTIAL-PRODUCTS ALGORITHM
$\begin{array}{r} 4 \\ 38 \\ \times\ \ 6 \\ \hline 228 \end{array}$	$\begin{array}{r} 38 \\ \times\ \ 6 \\ \hline 48 \\ 180 \\ \hline 228 \end{array}$ $\begin{array}{l} \leftarrow 6 \times 8 \\ \leftarrow 6 \times 30 \\ \leftarrow \text{add} \end{array}$

STANDARD ALGORITHM	PARTIAL-PRODUCTS ALGORITHM
$\begin{array}{r} 21 \\ 274 \\ \times\ \ \ 3 \\ \hline 822 \end{array}$	$\begin{array}{r} 274 \\ \times\ \ \ 3 \\ \hline 12 \\ 210 \\ 600 \\ \hline 822 \end{array}$ $\begin{array}{l} \leftarrow 3 \times 4 \\ \leftarrow 3 \times 70 \\ \leftarrow 3 \times 200 \\ \leftarrow \text{add} \end{array}$

STANDARD ALGORITHM	PARTIAL-PRODUCTS ALGORITHM
$\begin{array}{r} 1 \\ 1 \\ 45 \\ \times\ \ 23 \\ \hline 135 \\ 900 \\ \hline 1035 \end{array}$	$\begin{array}{r} 45 \\ \times\ \ 23 \\ \hline 15 \\ 120 \\ 100 \\ 800 \\ \hline 1035 \end{array}$ $\begin{array}{l} \leftarrow 3 \times 5 \\ \leftarrow 3 \times 40 \\ \leftarrow 20 \times 5 \\ \leftarrow 20 \times 40 \\ \leftarrow \text{add} \end{array}$

RELATING THE STANDARD AND PARTIAL-PRODUCTS ALGORITHMS

Compare the way the standard and partial-products algorithms are used to calculate 58×764:

STANDARD ALGORITHM	PARTIAL-PRODUCTS ALGORITHM
$\begin{array}{r} {\scriptstyle 3\,2} \\ {\scriptstyle 5\,3} \\ 764 \\ \times \quad\;\; 58 \\ \hline 6{,}112 \quad \leftarrow 32 + 480 + 5{,}600 \\[1em] 38{,}200 \quad \leftarrow 200 + 3{,}000 + 35{,}000 \\ \hline 44{,}312 \end{array}$	$\begin{array}{r} 764 \\ \times \quad\;\; 58 \\ \hline 32 \quad \leftarrow 8 \times 4 \\ 480 \quad \leftarrow 8 \times 60 \\ 5{,}600 \quad \leftarrow 8 \times 700 \\ 200 \quad \leftarrow 50 \times 4 \\ 3{,}000 \quad \leftarrow 50 \times 60 \\ 35{,}000 \quad \leftarrow 50 \times 700 \\ \hline 44{,}312 \end{array}$

Notice that the standard algorithm condenses the steps in the partial-products algorithm. In this example, three lines of calculations in the partial-products algorithm are condensed to one line in the standard algorithm: the 6,112 produced in the standard algorithm can be obtained by adding the three lines 8×4, 8×60, and 8×700 of the partial-products algorithm; the 38,200 can be obtained by adding 50×4, 50×60, and 50×700. When you multiply and then add carried numbers in the standard algorithm, you are really just combining lines from the partial-products algorithm.

Although the partial-products algorithm is longer than the standard algorithm, it is easier to carry out overall. However, one way that the partial-products algorithm is more difficult is in having to pay attention to place value. In the standard algorithm, you don't have to pay too much attention to place value—as long as you line up the numbers carefully, the algorithm takes care of place value, and you only have to multiply pairs of one-digit numbers. In the partial-products algorithm, on the other hand, you do have to pay attention to place value. For example, in the last line, you have to be able to calculate 50×700, instead of just 5×7. Notice in the following sequence that relating 50×700 to 5×7 uses the commutative and associative properties of multiplication:

$$50 \times 700 = 5 \times 10 \times 7 \times 100$$
$$= 5 \times 7 \times 10 \times 100 \text{ (by the commutative property of multiplication)}$$
$$= 35 \times 1000$$
$$= 35{,}000$$

The associative property of multiplication is used tacitly at the first equals sign in dropping parentheses from $(5 \times 10) \times (7 \times 100)$. The associative property of multiplication is also used tacitly at the third equal sign because the 10 is multiplied by the 100 instead of by the 5×7.

Because the partial-products algorithm is an expanded version of the standard multiplication algorithm, we can explain why the standard algorithm correctly calculates the answers to multiplication problems by explaining why the partial-products algorithm correctly calculates the answers to multiplication problems. We can also use the partial-products algorithm to explain why we put zeros in some of the lines of the standard algorithm.

WHY WE PLACE EXTRA ZEROS ON SOME LINES IN THE STANDARD ALGORITHM

We can use the partial-products algorithm to explain why we put an initial zero in the ones column of the second line in the standard multiplication algorithm (in the example above, this is the line produced by 5×4, 5×6, and 5×7). As the partial-products algorithm reveals, that line is actually produced by *multiplying by 50 rather than by 5*. By placing the initial zero in the ones place, all digits are moved one place to the left, which has the same effect as multiplying by 10. The net effect is that this second line is now actually 50 times 764 rather than 5 times 764.

What if the multiplication problem was 258×764 instead of 58×764? In the standard algorithm, there would be a third line produced by 2×4, 2×6, and 2×7. Before beginning the calculations for this third line, you would put down two zeros. Why two zeros? Because this line should really be produced by multiplying by 200 instead of by 2. By placing zeros in the ones and tens places, all digits are moved two places to the left, which has the same effect as multiplying by 100. The net effect is that the third line is then 200 times 764 instead of 2 times 764.

WHY THE ALGORITHMS PRODUCE CORRECT ANSWERS

We will now explain why the partial-products algorithm, and, hence, the standard multiplication algorithm, calculates correct answers to multiplication problems. We will use the example 58×764 to illustrate. Think of the multiplication problem 58×764 as representing the total number of widgets in 58 rows that each contain 764 widgets. The key to explaining why the partial-products algorithm gives the correct answer to 58×764, lies in observing that each step in the algorithm counts the number of widgets in a portion of the 58 rows of 764 widgets. The algorithms come from a clever way of subdividing the full collection of widgets into smaller portions whose numbers are easy to calculate as long as you know your one-digit multiplication tables. At the end of the algorithm, when you add the lines produced by multiplying, you are really just adding the numbers of widgets in each portion to get the total number of widgets.

Figure 5.36 shows 58 rows with 764 widgets in each row, where each ♟ in the figure represents a widget. Therefore, according to the meaning of multiplication, Figure 5.36 represents a total of 58×764 widgets. Because the numbers 58 and 764 are so large, the diagram doesn't actually show all 58 rows or all 764 widgets in each row.

Figure 5.36 illustrates how to subdivide the full collection of widgets so that the number of widgets in each portion corresponds to a line in the partial-products algorithm. For instance, the line 5600 in the partial-products algorithm, which comes from 8×700, is the number of widgets in the piece in the lower left-hand corner, which represents 8 groups of 700 widgets. Similarly, the line 200, which comes from 50×4, is the number of widgets in the piece at the top right-hand corner, which represents 50 groups of 4 widgets. It's the same for all the other lines in the partial-products algorithm—each line in the algorithm corresponds to a portion of the collection of widgets. When you add up the lines in the algorithm, you get the total number of widgets. Therefore, the partial-products algorithm correctly calculates the total number of widgets; in other words, it correctly calculates 58×764. Since the standard algorithm is just a condensation of the partial-products algorithm, it too must correctly calculate answers to multiplication problems.

FIGURE 5.36

Subdividing 58
Rows of 764
Widgets

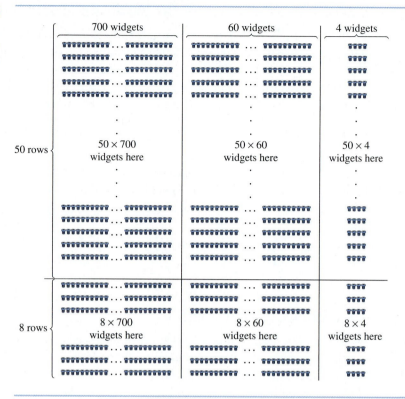

As presented in the figure, the panel is labeled with "700 widgets", "60 widgets", and "4 widgets" across the top, "50 rows" and "8 rows" on the left, with regions labeled 50×700 widgets here, 50×60 widgets here, 50×4 widgets here, 8×700 widgets here, 8×60 widgets here, and 8×4 widgets here.

Another way to explain why the partial-products algorithm, and therefore the standard algorithm, gives the correct answer to 58×764 is to put the numbers 58 and 764 in expanded forms and use properties of arithmetic to calculate 58×764:

$$
\begin{aligned}
58 \times 764 &= (50 + 8) \times (700 + 60 + 4) \\
&= 50 \times (700 + 60 + 4) + 8 \times (700 + 60 + 4) \\
&= 50 \times 700 + 50 \times 60 + 50 \times 4 + 8 \times 700 + 8 \times 60 + 8 \times 4 \\
&= 35{,}000 + 3{,}000 + 200 + 5{,}600 + 480 + 32 \\
&= 44{,}312
\end{aligned}
$$

As the preceding equations show, the six lines produced by the partial-products algorithm are exactly the following six products:

$$ 50 \times 700, \quad 50 \times 60, \quad 50 \times 4, \quad 8 \times 700, \quad 8 \times 60, \quad 8 \times 4 $$

produced by using the distributive property several times, starting with

$$ (50 + 8) \times (700 + 60 + 4) $$

CLASS ACTIVITY   NOW TURN TO CLASS ACTIVITIES MANUAL

5W The Standard versus the Partial-Products Multiplication Algorithm p. 129

5X Why the Partial-Products Algorithm Gives Correct Answers, Part I p. 130

5Y Why the Partial-Products Multiplication Algorithm Gives Correct Answers, Part II p. 132

5Z Why the Standard Multiplication Algorithm Gives Correct Answers p. 135

PRACTICE PROBLEMS FOR SECTION 5.9

1. Find the last two digits (the ones and hundreds digits) of

$$798,312,546,936 \times 74$$

2. Draw an array on graph paper and use your array to explain why the partial-products algorithm calculates the correct answer to 24×35. If graph paper is not available, draw a rectangle to represent your array rather than drawing 24 rows with 35 items in each row.

3. Relate the array you drew for Practice Problem 2 to the steps in the standard algorithm for calculating 24×35. Use this relationship to explain why the standard algorithm calculates the correct answer to 24×35.

4. Solve the multiplication problem 24×35 by writing equations that use expanded forms and the distributive property. Relate your equations to the steps in the partial-products algorithm for calculating 24×35. Use this relationship to explain why the partial-products algorithm calculates the correct answer to 24×35.

5. Relate your equations from Practice Problem 4 to the steps in the standard algorithm for calculating 24×35. Use this relationship to explain why the standard algorithm calculates the correct answer to 24×35.

6. Cameron wants to calculate 23×23. She says:

$$20 \times 20 = 400 \quad \text{and} \quad 3 \times 3 = 9,$$

so

$$23 \times 23 = 400 + 9 = 409.$$

Cameron has a good idea, but her reasoning is not completely correct. Compare Cameron's work to the steps in the partial-products algorithm for calculating 23×23. What is missing in Cameron's calculation? Draw an array to show Cameron why she must adjust her calculation.

ANSWERS TO PRACTICE PROBLEMS FOR SECTION 5.9

1. The last two digits are 64. Notice that you only have to find the last two digits of 36×74 to find this. Why? Think about doing longhand multiplication: any other contribution will be in the thousands place or higher.

2. Figure 5.37 shows an array consisting of 24 rows of small squares with 35 small squares in each row. According to the meaning of multiplication, there is a total of 24×35 small squares in the array. The shading shows how to subdivide the array into four pieces: one with 20 rows of 30, one with 20 rows of 5, one with 4 rows of 30, and one with 4 rows of 5 small squares. The number of small squares in each of these four pieces is calculated as follows:

$$20 \times 30 = 600$$
$$20 \times 5 = 100$$
$$4 \times 30 = 120$$
$$4 \times 5 = 20$$

But the number of small squares in these four pieces corresponds exactly to the four lines we produce when we use the partial-products algorithm to calculate 24×35:

$$
\begin{array}{r}
35 \\
\times\ \ 24 \\
\hline
20 \\
120 \\
100 \\
+\ 600 \\
\hline
840
\end{array}
$$

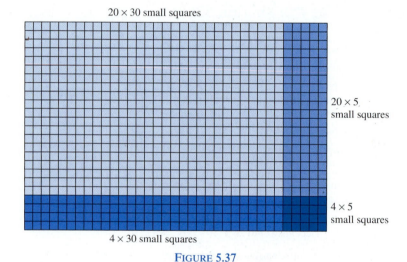

20 × 30 small squares

20 × 5 small squares

4 × 5 small squares

4 × 30 small squares

FIGURE 5.37

An Array of 24 Rows with 35 Small Squares
in Each Row Subdivided into Four Parts

So the steps in the partial-products algorithm calculate the number of squares in the four parts that make up a full array of 24 rows of 35 squares. When we add the numbers together at the end, we add up the total number of squares in all four parts, and therefore get the total number of small squares in 24 rows of 35 squares, which is 24×35 small squares.

3. The standard algorithm is a condensed version of the partial-products algorithm. The two lines we produce when we multiply 24×35 using the standard algorithm each come from combining two of the lines in the partial-products algorithm:

$$
\begin{array}{r}
\overset{1}{\underset{2}{}}\ \ \\
35 \\
\times\ \ 24 \\
\hline
140 \\
700 \\
\hline
840
\end{array}
\qquad
\begin{array}{r}
35 \\
\times\ \ 24 \\
\hline
20 \\
120 \\
100 \\
600 \\
\hline
840
\end{array}
$$

In terms of the array in Figure 5.37, the line 140 in the standard algorithm comes from the bottom 4 rows, which represent 4×35; the line 700 in the standard algorithm comes from the top 20 rows, which represent 20×35. As with the partial-products algorithm, the standard algorithm calculates the total number of small squares in an array of 24 rows with 35 small squares in each row by calculating the number of small squares in parts of the array and adding these totals.

4. $$\begin{aligned} 24 \times 35 &= (20+4) \times (30+5) \\ &= 20 \times (30+5) + 4 \times (30+5) \\ &= 20 \times 30 + 20 \times 5 + 4 \times 30 + 4 \times 5 \\ &= 600 + 100 + 120 + 20 \\ &= 840 \end{aligned}$$

The distributive property was used at the second and third equal signs. The four products, 20×30, 20×5, 4×30, and 4×5 that result from using expanded forms and the distributive property to calculate 24×35 are exactly the four products seen in the following calculation using the partial-products algorithm:

$$\begin{array}{r} 35 \\ \times \quad 24 \\ \hline 20 \\ 120 \\ 100 \\ +\ 600 \\ \hline 840 \end{array}$$

Therefore, the partial-products algorithm is really just a way of displaying the results produced by using the distributive property when the numbers 24 and 35 are put in expanded form. The partial-products algorithm produces correct calculations because these calculations come from using the distributive property.

5. The standard algorithm is a condensed version of the partial-products algorithm. The two lines we produce when we multiply 24×35 using the standard algorithm each come from combining two of the lines in the partial-products algorithm:

$$\begin{array}{r} \overset{\overset{1}{2}}{35} \\ \times \quad 24 \\ \hline 140 \\ \\ 700 \\ \hline 840 \end{array} \qquad \left\{ \begin{array}{r} 35 \\ \times \quad 24 \\ \hline 20 \\ 120 \\ 100 \\ 600 \\ \hline 840 \end{array} \right.$$

Consider the equations we write to calculate 24×35 using expanded forms and the distributive property:

$$\begin{aligned} 24 \times 35 &= (20+4) \times (30+5) \\ &= 20 \times (30+5) + 4 \times (30+5) \\ &= 20 \times 30 + 20 \times 5 + 4 \times 30 + 4 \times 5 \\ &= 600 + 100 + 120 + 20 \\ &= 840 \end{aligned}$$

The two terms in the second line of the equations, namely, $4 \times (30+5)$ and $20 \times (30+5)$, correspond to the first and second lines, respectively, produced by the standard algorithm, namely, 140 and 700. So the standard algorithm comes from breaking the bottom factor, 24, into expanded form $20 + 4$, and multiplying each of these components by the top factor, 35. The standard algorithm calculates 24×35 by calculating

$$20 \times 35 \ + \ 4 \times 35$$

which uses the distributive property.

6. When we calculate 23×23 using the partial-products algorithm, we produce four lines of products:

$$\begin{array}{r} 23 \\ \times \quad 23 \\ \hline 9 \\ 60 \\ 60 \\ +\ 400 \\ \hline 529 \end{array}$$

Cameron's calculations only produced two of those lines—the 9 and the 400—so she is missing the two lines of 60. We can see these two missing 60s in Figure 5.38, which shows 23 rows of 23 small squares, subdivided into four portions. The array shows that Cameron has left out 20×3 and 3×20, which correspond to the darkly shaded portions.

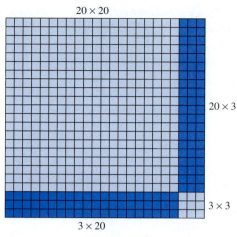

FIGURE 5.38

$$23 \times 23 = 20 \times 20 + 20 \times 3 + 3 \times 20 + 3 \times 3$$

PROBLEMS FOR SECTION 5.9

1. Solve the multiplication problem

$$\begin{array}{r} 96 \\ \times\ \ 8 \\ \hline \end{array}$$

   in three different ways: using the standard algorithm, using the partial-products algorithm, and by writing the numbers in expanded forms and using properties of arithmetic. For each of the three methods, discuss how the steps in that method are related to the steps in the other methods.

2. Solve the multiplication problem

$$\begin{array}{r} 84 \\ \times\ 76 \\ \hline \end{array}$$

   in three different ways: using the standard algorithm, using the partial-products algorithm, and by writing the numbers in expanded forms and using properties of arithmetic. For each of the three methods, discuss how the steps in that method are related to the steps in the other methods.

3. Solve the multiplication problem

$$\begin{array}{r} 237 \\ \times\ \ 43 \\ \hline \end{array}$$

   in three different ways: using the standard algorithm, using the partial-products algorithm, and by writing the numbers in expanded forms and using properties of arithmetic. For each of the three methods, discuss how the steps in that method are related to the steps in the other methods.

4. When we multiply

$$\begin{array}{r} 37 \\ \times\ 26 \\ \hline \end{array}$$

   using the standard multiplication algorithm, we start the second line by writing a zero:

$$\begin{array}{r} \overset{4}{3}7 \\ \times\ \ 26 \\ \hline 222 \\ 0 \end{array}$$

   Explain why we place this zero in the second line. What is the rationale for the procedure of placing a zero in the second line?

5. a. Use the partial-products algorithm to calculate 8×24.

   b. Draw an array for 8×24. (You may wish to use graph paper.) Subdivide the array in a natural way so that the parts of the array correspond to the steps in the partial-products algorithm.

    c.   Solve 8×24 by writing equations that use expanded forms and the distributive property. Relate your equations to the steps in the partial-products algorithm.

6. a.   Use the partial-products algorithm to calculate 19×28.

    b.   On graph paper, draw an array for 19×28. If graph paper is not available, draw a rectangle to represent the array rather than drawing 19 rows with 28 items in each row. Subdivide the array in a natural way so that the parts of the array correspond to the steps of the partial-products algorithm.

    c.   Solve 19×28 by writing equations that use expanded forms and the distributive property. Relate your equations to the steps in the partial-products algorithm.

7. a.   Use the partial-products and standard algorithms to calculate 27×28.

    b.   On graph paper, draw an array for 27×28. If graph paper is not available, then draw a rectangle to represent the array rather than drawing 27 rows with 28 items in each row. Subdivide the array in a natural way so that the parts of the array correspond to the steps in the partial-products algorithm.

    c.   On the array that you drew for part (b), show the parts that correspond to the steps of the standard algorithm.

    d.   Solve 27×28 by writing equations that use expanded forms and the distributive property. Relate your equations to the steps in the partial-products algorithm.

8. a.   Draw an array on graph paper and use your array to explain why the partial-products algorithm calculates the correct answer to 23×27. If graph paper is not available, draw a rectangle to represent the array rather than drawing 23 rows of 27 items.

    b.   Relate the array you drew for part (a) to the steps in the standard algorithm for calculating 23×27. Use this relationship to explain why the standard algorithm calculates the correct answer to 23×27.

9. Solve the multiplication problem 23×27 by writing equations that use expanded forms and the distributive property. Relate your equations to the steps in the partial-products algorithm for calculating 23×27. Use this relationship to explain why the partial-products algorithm calculates the correct answer to 23×27.

10. Draw a large rectangle to represent an array for 43×275. Show how to subdivide the rectangle in a natural way so that the parts of the rectangle correspond to the steps in calculating 43×275 using the partial-products multiplication algorithm.

11. The **lattice method** is a method that some people like to use to multiply. Figure 5.39 on p. 218 shows how to use the lattice method to multiply a 2-digit number by a 2-digit number.

    a.   Use the lattice method to calculate 38×54 and 72×83.

    b.   Use the partial-products algorithm to calculate 38×54 and 72×83.

    c.   Explain how the lattice method is related to the partial-products algorithm.

    d.   Discuss advantages and disadvantages of using the lattice method.

12. The following is a method for multiplying 21×23 that relies on repeatedly doubling numbers (by adding them to themselves):

$$
\begin{array}{rl}
23 & \\
+\ 23 & \\
\hline
46 & \quad 2 \\
+\ 46 & \qquad\qquad 21 = 16 + 4 + 1 \\
\hline
92 & \quad 4 \\
+\ 92 & \qquad\qquad\qquad\qquad 368 \\
\hline
184 & \quad 8 \qquad\qquad\qquad\quad 92 \\
+\ 184 & \qquad\qquad\qquad\quad +\ 23 \\
\hline
368 & \quad 16 \qquad\qquad\qquad 483 \\
\end{array}
$$

    The answer, 483, to the multiplication problem 21×23, is shown on the bottom of the column on the right.

    Notice that to use this repeated doubling method you only have to be able to add; you do not need to know how to multiply.

    a.   Discuss and explain the method of repeated doubling. Address the following questions in your discussion:

       i.   What is the significance of the column of the numbers 2, 4, 8, 16?

      ii.   What is the significance of writing $21 = 16 + 4 + 1$?

iii. How are the numbers 368, 92, and 23, chosen for the column on the right? Why do we add those numbers?

b. Use the method of repeated doubling to calculate 26 × 35. Show your work.

c. Use the method of repeated doubling to calculate 37 × 51. Show your work.

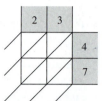
Draw a lattice as shown in the figure. Place the numbers you want to multiply on the top and down the side.

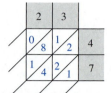
Multiply each digit along the top with each digit along the side. Write the answers in the cells as shown.

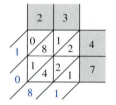
Starting at the bottom right, and moving left, then up, add numbers in diagonal strips, carrying as necessary. The answer is 1081.

FIGURE 5.39

Using the Lattice Method to Show that 23 × 47 = 1081

Chapter 6

Multiplication of Fractions, Decimals, and Negative Numbers

I n this chapter we will study multiplication of fractions, decimals, and negative numbers. We will also study the related topics of powers and scientific notation. The procedures for multiplying fractions and decimals are not hard. As a teacher, you should understand not only how to carry out those multiplication procedures, but also why those procedures are the way they are. For example, why do we put the decimal point where we do when we multiply decimals? Why do we multiply fractions by multiplying the numerators and the denominators even though we don't add fractions by adding the numerators and adding the denominators? We will be able to answer these questions by using the meaning of multiplication, as it applies to fractions and decimals.

6.1 Multiplying Fractions

In this section, we will study the meaning of multiplication of fractions and we will explain why the standard procedure for multiplying fractions gives answers to fraction multiplication problems that agree with what we expect from the meaning of multiplication.

THE MEANING OF MULTIPLICATION FOR FRACTIONS

At the beginning of Chapter 5, we defined multiplication as follows: If A and B are nonnegative numbers, then

$$A \times B \quad \text{or} \quad A \cdot B$$

represents the total number of objects in A groups if there are B objects in each group. This definition applies not only to whole numbers, but also to fractions and decimals. As we'll see,

though, when fractions are involved, we may prefer to reword the definition for the sake of clarity.

Consider the following examples:

1. $\frac{1}{2} \times 21$ means the total number of objects in $\frac{1}{2}$ groups if there are 21 objects in each group.

    Better wording: $\frac{1}{2} \times 21$ means the total number of objects in $\frac{1}{2}$ of a group if there are 21 objects in one whole group.

2. $\frac{1}{2} \times \frac{1}{2}$ means the total number of objects in $\frac{1}{2}$ groups if there are $\frac{1}{2}$ objects in each group.

    Better wording: $\frac{1}{2} \times \frac{1}{2}$ means the total number of objects in $\frac{1}{2}$ of a group if there is $\frac{1}{2}$ of an object in one whole group.

3. $3 \times \frac{2}{5}$ means the total number of objects in 3 groups if there are $\frac{2}{5}$ objects in each group. (See Figure 6.1.)

    Better wording: $3 \times \frac{2}{5}$ means the total number of objects in 3 groups if there is $\frac{2}{5}$ of an object in each group.

What do we mean by the word *object*? In a mathematical context, the word object can refer not just to a single thing, but also to a collection of things (as in "the cars in the United States") or to a quantity (as in "one cup of water" or "$20,000"). In the context of fractions, it is also natural to use the term *a whole* instead of the term object.

The wording "the total number of objects" must be understood to include the possibility of fractional objects. For example, if the object we're considering is a full jar of peanut butter, then what does the following mean?

The total number of objects in $\frac{1}{2}$ of a group when each group contains $\frac{1}{2}$ of a full jar of peanut butter.

In this case, what we mean by "the total number of objects" is really "the fraction of a full jar of peanut butter" in $\frac{1}{2}$ of $\frac{1}{2}$ of a full jar of peanut butter. This phrasing suggests alternative wording for the meaning of fraction multiplication, which you can use if you find it clearer. The following wording is most appropriate in cases where both factors are proper fractions:

$\frac{A}{B} \times \frac{C}{D}$ means the fraction of an object in $\frac{A}{B}$ of $\frac{C}{D}$ of the object.

For example,

$\frac{1}{2} \times \frac{3}{4}$ means the fraction of a container of yogurt in $\frac{1}{2}$ of $\frac{3}{4}$ of the container of yogurt. (See Figure 6.2.)

FIGURE 6.2

$\frac{1}{2}$ of $\frac{3}{4}$ of a
Container of
Yogurt

container of
yogurt

$\frac{3}{4}$ of the
container of
yogurt

$\frac{1}{2}$ of the $\frac{3}{4}$ of
the container of
yogurt

THE PROCEDURE FOR MULTIPLYING FRACTIONS

The procedure for multiplying fractions is very easy: Just multiply the numerators and multiply the denominators. In other words,

$$\frac{A}{B} \cdot \frac{C}{D} = \frac{A \cdot C}{B \cdot D}$$

For example,

$$\frac{2}{3} \cdot \frac{4}{5} = \frac{2 \cdot 4}{3 \cdot 5} = \frac{8}{15}$$

Notice that the procedure can be used even when whole numbers are involved because a whole number can always be written as a fraction by "putting it over 1". For example,

$$\frac{2}{3} \cdot 17 = \frac{2}{3} \cdot \frac{17}{1} = \frac{2 \cdot 17}{3 \cdot 1} = \frac{34}{3} = 11\frac{1}{3}$$

We can multiply mixed numbers with the fraction multiplication procedure by first converting the mixed numbers to improper fractions. For example,

$$2\frac{3}{4} \cdot 1\frac{2}{3} = \frac{11}{4} \cdot \frac{5}{3} = \frac{55}{12} = 4\frac{7}{12}$$

As with the procedure for multiplying whole numbers, we want to explain why this procedure gives correct answers to fraction multiplication problems. The meaning of fractions and the meaning of multiplication tell us what an expression of the form

$$\frac{A}{B} \cdot \frac{C}{D}$$

means. We must explain why this expression is equal to the fraction

$$\frac{A \cdot C}{B \cdot D}$$

EXPLAINING WHY THE PROCEDURE FOR MULTIPLYING FRACTIONS GIVES CORRECT ANSWERS

The procedure for multiplying fractions is easy to carry out. It also seems sensible because it involves multiplying the numerators and the denominators. However, remember that we don't add fractions by simply adding the numerators and adding the denominators. So, why is the

FIGURE 6.3

Finding $\frac{2}{7}$ of $\frac{3}{4}$
of a Rectangle

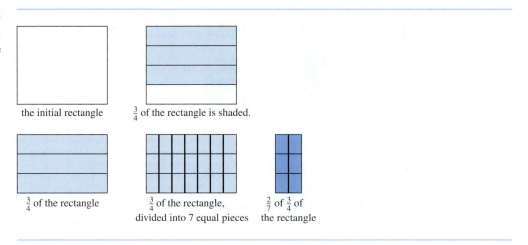

the initial rectangle $\frac{3}{4}$ of the rectangle is shaded.

$\frac{3}{4}$ of the rectangle $\frac{3}{4}$ of the rectangle, $\frac{2}{7}$ of $\frac{3}{4}$ of
divided into 7 equal pieces   the rectangle

simple procedure of multiplying the numerators and multiplying the denominators valid for fraction multiplication? We will use the meaning of fractions, the meaning of multiplication, and logical reasoning to explain why the simple procedure for multiplying fractions gives correct answers to fraction multiplication problems.

Consider the multiplication problem

$$\frac{2}{7} \cdot \frac{3}{4}$$

and consider $\frac{2}{7} \cdot \frac{3}{4}$ of a rectangle. By the meaning of multiplication,

$\frac{2}{7} \cdot \frac{3}{4}$ means the fraction of a rectangle in $\frac{2}{7}$ of a group if there is $\frac{3}{4}$ of a rectangle in each group.

In other words, $\frac{2}{7} \cdot \frac{3}{4}$ means the fraction of a rectangle in $\frac{2}{7}$ of $\frac{3}{4}$ of the rectangle.

So, starting with a rectangle, first consider $\frac{3}{4}$ of a rectangle, and then consider $\frac{2}{7}$ of the $\frac{3}{4}$ of the rectangle, as shown in Figure 6.3.

Here is the crucial point: we must identify the $\frac{2}{7}$ of the $\frac{3}{4}$ of the rectangle *as a fraction of the original rectangle*. To identify what fraction of the original rectangle is represented by $\frac{2}{7}$ of the $\frac{3}{4}$ of the rectangle, we must put it back inside the original rectangle, as shown in Figure 6.4.

Because the original rectangle was first divided into 4 equal parts, and because each of those parts was then divided into 7 equal parts, the original rectangle has therefore been

FIGURE 6.4

What Fraction is
$\frac{2}{7}$ of $\frac{3}{4}$ of a
Rectangle?

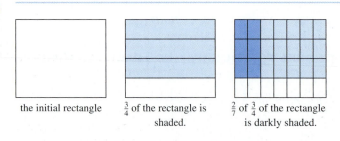

the initial rectangle $\frac{3}{4}$ of the rectangle is $\frac{2}{7}$ of $\frac{3}{4}$ of the rectangle
shaded.   is darkly shaded.

subdivided into a total of $7 \cdot 4$ small, equal parts. Notice that we multiply $7 \cdot 4$ because there are 7 columns with 4 pieces in each column. Of those $7 \cdot 4$ equal parts, the darkly shaded parts (which represent our $\frac{2}{7}$ of the $\frac{3}{4}$ of the rectangle) form $2 \cdot 3$ parts. Once again, notice that we must multiply $2 \cdot 3$ because there are 2 columns with 3 small pieces in each column that are darkly shaded. So the $\frac{2}{7}$ of the $\frac{3}{4}$ of the rectangle is represented by $2 \cdot 3$ parts out of a total of $7 \cdot 4$ equal parts making up the original rectangle. This means that $\frac{2}{7}$ of the $\frac{3}{4}$ of the rectangle is

$$\frac{2 \cdot 3}{7 \cdot 4}$$

of the rectangle. Therefore,

$$\frac{2}{7} \cdot \frac{3}{4} = \frac{2 \cdot 3}{7 \cdot 4}$$

In other words, the procedure for multiplying fractions gives the answer we expect from the meaning of fractions and the meaning of multiplication.

There wasn't anything special about the numbers 2, 3, 7, and 4; any other counting numbers could be substituted and the argument would work in the same way. Therefore, the explanation given for why

$$\frac{2}{7} \cdot \frac{3}{4} = \frac{2 \cdot 3}{7 \cdot 4}$$

also explains why we multiply other fractions by multiplying the numerators and multiplying the denominators. In other words, we have a good reason for why we multiply fractions the way we do:

$$\frac{A}{B} \cdot \frac{C}{D} = \frac{A \cdot C}{B \cdot D}$$

Notice that in order to understand fraction multiplication, we had to work with different wholes. In the example, we first took $\frac{3}{4}$ of the original rectangle. Then we took $\frac{2}{7}$ of this $\frac{3}{4}$ portion. In other words, we had to treat the $\frac{3}{4}$ of the rectangle as a new whole in its own right in order to take $\frac{2}{7}$ of it. In the end, when we found the portion that was $\frac{2}{7}$ of the $\frac{3}{4}$ of the rectangle, we had to determine what fraction that portion was of the original rectangle. In other words, we had to go back to the original whole again. As always, when we work with fractions, we must pay close attention to the underlying wholes or what the fractions are "of." When you use pictures to help you solve fraction multiplication problems, you may find it helpful to start by drawing your original reference whole as a reminder.

WHY DO WE NEED TO KNOW THIS?

As we've seen, the procedure for multiplying fractions is easy, but the explanation for why this procedure gives correct answers to fraction multiplication problems is much harder. As a teacher, it is especially important for you to have a deep conceptual understanding of the mathematics you will teach. This means not just knowing *how*, but also *why* and *when* the various procedures and formulas in mathematics work. Why is this important? Research shows that teachers who have a conceptual understanding of mathematics tend to use a conceptually directed teaching strategy. (See [34, especially pages 38–39].) These teachers teach in a way that encourages understanding. Research also shows that this kind of teaching requires deep subject-matter knowledge:

> Limited subject matter knowledge restricts a teacher's capacity to promote conceptual learning among students. Even a strong belief of "teaching mathematics for understanding" cannot remedy or supplement a teacher's disadvantage in subject matter knowledge. A few beginning teachers in the procedurally directed group wanted to "teach for understanding." They intended to involve students in the learning process, and to promote conceptual learning that explained the rationale underlying the procedure. However, because of their own deficiency in subject matter knowledge, their conception of teaching could not be realized. [34, page 36])

CLASS ACTIVITY   NOW TURN TO CLASS ACTIVITIES MANUAL

6A Writing and Solving Fraction Multiplication Story Problems p. 137

6B Misconceptions with Fraction Multiplication p. 139

6C Explaining Why the Procedure for Multiplying Fractions Gives Correct Answers p. 140

6D When Do We Multiply Fractions? p. 142

6E Multiplying Mixed Numbers p. 143

PRACTICE PROBLEMS FOR SECTION 6.1

1. What does $\frac{3}{4} \times \frac{1}{2}$ mean? Give an example of a story problem that is solved by calculating $\frac{3}{4} \times \frac{1}{2}$.

2. Which of the following problems are solved by calculating $\frac{1}{3} \times \frac{2}{3}$ and which are not?

   a. A recipe calls for $\frac{2}{3}$ cup of sugar. You want to make $\frac{1}{3}$ of the recipe. How much sugar should you use?

   b. $\frac{2}{3}$ of the cars at a car dealership have power steering. $\frac{1}{3}$ of the cars at the same car dealership have side-mounted airbags. What fraction of the cars at the car dealership have both power steering and side-mounted airbags?

   c. $\frac{2}{3}$ of the cars at a car dealership have power steering. $\frac{1}{3}$ of those cars that have power steering have side-mounted airbags. What fraction of the cars at the car dealership have both power steering and side-mounted airbags?

   d. Ed put $\frac{2}{3}$ of a bag of candies in a batch of cookies that he made. Ed ate $\frac{1}{3}$ of the batch of cookies. How many candies did Ed eat?

   e. Ed put $\frac{2}{3}$ of a bag of candies in a batch of cookies that he made. Ed ate $\frac{1}{3}$ of the batch of cookies. What fraction of a bag of candies did Ed eat?

3. Use the meaning of fractions and the meaning of multiplication to explain why

$$\frac{2}{7} \cdot \frac{3}{4} = \frac{2 \cdot 3}{7 \cdot 4}$$

4. Anthony is trying to solve $\frac{5}{7} \cdot 2$. He draws a picture as in Figure 6.5 and concludes from his picture that $\frac{5}{7} \cdot 2 = \frac{10}{14}$ because there are 14 small pieces and 10 of them are shaded. Is Anthony right or not? If not, help Anthony figure out what's wrong with his reasoning.

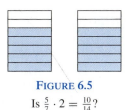

FIGURE 6.5

Is $\frac{5}{7} \cdot 2 = \frac{10}{14}$?

5. Maryann has to calculate $\frac{2}{3} \cdot 6\frac{3}{4}$. Rather than convert the $6\frac{3}{4}$ to an improper fraction, she solves the problem this way. She finds $\frac{2}{3}$ of 6, which is 4, and she finds $\frac{2}{3}$ of $\frac{3}{4}$, which is $\frac{1}{2}$. Then Maryann says the answer is $4\frac{1}{2}$. Write equations that correspond to Maryann's work and that explain why her work is correct. Which property of arithmetic is involved?

ANSWERS TO PRACTICE PROBLEMS FOR SECTION 6.1

1. $\frac{3}{4} \times \frac{1}{2}$ means the fraction of an object in $\frac{3}{4}$ of a group of $\frac{1}{2}$ of the object. A sample story problem: A recipe calls for $\frac{1}{2}$ pound of sea slugs. You decide to make $\frac{3}{4}$ of the recipe. How many pounds of sea slugs will you need?

2. a. Yes, $\frac{1}{3} \cdot \frac{2}{3}$ is the fraction of a cup of sugar you should use, since you will need $\frac{1}{3}$ of $\frac{2}{3}$ cup of sugar.

   b. No, from the information given, we don't know if $\frac{1}{3}$ of the cars that have power steering have side-mounted airbags.

   c. Yes, $\frac{1}{3}$ of $\frac{2}{3}$ of the cars at the car dealership have both power steering and side-mounted airbags.

   d. No, we can't tell how many candies Ed ate, only what fraction of a bag of candies Ed ate.

   e. Yes, Ed ate $\frac{1}{3}$ of $\frac{2}{3}$ of a bag of candies.

3. See the text.

4. No. Although Anthony has drawn a good representation of the problem, his reasoning is not completely right. 10 pieces are shaded, and these 10 pieces do represent $\frac{5}{7}$ of 2 rectangles. But Anthony must remember that each small piece represents $\frac{1}{7}$ of the original object, which we infer must have been one rectangle. So $\frac{5}{7} \cdot 2$ is the fraction of one rectangle that is shaded. This fraction is $\frac{10}{7}$. When drawing pictures such as Anthony's, it's a good idea to also draw 1 whole object somewhere as a reminder that this is your reference amount.

5.
$$\frac{2}{3} \cdot 6\frac{3}{4} = \frac{2}{3} \cdot \left(6 + \frac{3}{4}\right)$$
$$= \frac{2}{3} \cdot 6 + \frac{2}{3} \cdot \frac{3}{4}$$
$$= 4 + \frac{1}{2}$$
$$= 4\frac{1}{2}$$

The distributive property was used at the second equals sign.

PROBLEMS FOR SECTION 6.1

1. a. Anita had $\frac{1}{2}$ of a bag of fertilizer left. She used $\frac{3}{4}$ of what was left. What question about Anita's fertilizer can be answered by calculating $\frac{3}{4} \times \frac{1}{2}$?

   b. Hermione wants to make 4 batches of a potion. Each batch of potion requires $\frac{2}{3}$ cup of toober pus. What question about the potion will be answered by calculating $4 \times \frac{2}{3}$?

   c. There were 6 pieces of pizza left. Tommy ate $\frac{3}{4}$ of them. What question about the pizza can be answered by calculating $\frac{3}{4} \times 6$?

2. Paul used $\frac{3}{4}$ cup of butter in the batch of brownies he made. Paul ate $\frac{5}{16}$ of the batch of brownies. What frac-

tion of a cup of butter did Paul consume when he ate the brownies? Draw pictures to help you solve this problem. Explain in detail how your pictures help you solve the problem.

3. Which of the following are story problems for $\frac{1}{2} \times \frac{1}{3}$ and which are not? Explain your answer in each case.

   a. One-third of the children in a class have black hair. One-half of the children in the class have curly hair. How many children in the class have curly black hair?

   b. One-third of the children in a class have black hair. One-half of the children who have black hair also have curly hair. How many children in the class have curly black hair?

c. One-third of the children in a class have black hair. One-half of the children who have black hair also have curly hair. What fraction of the children in the class have curly black hair?

d. One-third of the children in a class have black hair. One-half of the children in the class have curly hair. What fraction of the children in the class have curly black hair?

4. Which of the following are story problems for $\frac{3}{4} \times 5$ and which are not? Explain your answer in each case.

a. A cake was cut into pieces of equal size. There are 5 pieces of cake left. John gets $\frac{3}{4}$ of them. What fraction of the cake does John get?

b. A cake was cut into pieces of equal size. There are 5 pieces of cake left. John gets $\frac{3}{4}$ of them. How many pieces of cake does John get?

c. A cake for a class party was cut into pieces of equal size. There are 5 pieces of cake left. Three-quarters of the class still wants cake. What fraction of the cake will be eaten?

d. A cake for a class party was cut into pieces of equal size. There are 5 pieces of cake left. Three-quarters of the class still wants cake. How many pieces does each person get?

5. Explain why it would be easy to interpret the picture in Figure 6.6 incorrectly as showing that $4 \times \frac{3}{5} = \frac{12}{20}$. Explain how to interpret the picture correctly and explain why your interpretation fits with the meaning of $4 \times \frac{3}{5}$.

FIGURE 6.6

A Picture for $4 \times \frac{3}{5}$

6. Write a story problem for

$$\frac{1}{3} \cdot \frac{1}{4}$$

Use the meaning of fractions, the meaning of multiplication, and pictures to determine the answer to the multiplication problem. Explain your answer.

7. Write a story problem for

$$2 \cdot \frac{3}{5}$$

Use the meaning of fractions, the meaning of multiplication, and pictures to determine the answer to the multiplication problem. Explain your answer.

8. Write a story problem for

$$\frac{2}{3} \cdot 5$$

Use the meaning of fractions, the meaning of multiplication, and pictures to determine the answer to the multiplication problem. Explain your answer.

9. Write a simple story problem for

$$\frac{2}{3} \cdot \frac{4}{5}$$

Use your story problem and pictures to explain why it makes sense that the answer to the fraction multiplication problem is

$$\frac{2 \cdot 4}{3 \cdot 5}$$

In particular, use your pictures to explain why we multiply the numerators and why we multiply the denominators.

10. One serving of Gooey Gushers provides 12% of the daily value of vitamin C. Tim ate $3\frac{1}{2}$ servings of Gooey Gushers.

a. Calculate $3\frac{1}{2} \times 12\%$ without using a calculator. Show your work.

b. What question about the Gooey Gushers can you answer by calculating $3\frac{1}{2} \times 12\%$?

11. a. Write a story problem for $2\frac{1}{2} \times 3\frac{1}{2}$.

b. Use pictures and the meaning of multiplication to solve the problem.

c. Use the distributive property or FOIL to calculate $2\frac{1}{2} \times 3\frac{1}{2}$ by rewriting this product as $(2 + \frac{1}{2}) \times (3 + \frac{1}{2})$.

d. Identify the four terms produced by the distributive property or FOIL [in part (c)] in a picture like the one in part (b).

e. Now write the mixed numbers $2\frac{1}{2}$ and $3\frac{1}{2}$ as improper fractions and use the standard procedure for multiplying fractions to calculate $2\frac{1}{2} \times 3\frac{1}{2}$. How do you see the product of the numerators in your picture in part (b)? How can you see the product of the denominators in your picture?

12. Manda says that

$$3\frac{2}{3} \times 2\frac{1}{5} = 3 \times 2 + \frac{2}{3} \times \frac{1}{5}$$

Explain why Manda has made a good attempt, but her answer is not correct. Explain how to work with what Manda has already written and modify it to get the correct answer. In other words, don't just start from scratch and show Manda how to do the problem, but rather take what she has already written, use it, and make it mathematically correct. Which property of arithmetic is relevant to correcting Manda's work? Explain.

13. In order to understand fraction multiplication thoroughly, we must be able to work simultaneously with different wholes. Using the example $\frac{3}{5} \times \frac{3}{4}$, explain why this is so. What are the different wholes that are associated with the fractions in the problem $\frac{3}{5} \times \frac{3}{4}$ (including the answer to the problem)?

14. In order to understand fraction multiplication thoroughly, we must understand how to divide a whole number of objects into equal parts, such as dividing 8 cookies equally among 3 people. Where do we need to understand how to divide a whole number of objects into equal parts in order to understand $\frac{2}{3} \times \frac{2}{5}$ thoroughly? Be specific.

15. Discuss why we must develop an understanding of multiplication that goes beyond seeing it as repeated addition.

16. Suppose you are teaching fraction multiplication. Make the case to your students that multiplication means the same thing whether we are multiplying fractions or whole numbers. Use story problems for 3×4 and $\frac{1}{3} \times \frac{1}{4}$ to illustrate.

17. Let's suppose that the liquid in a car's radiator is 75% water and 25% antifreeze. Suppose 30% of the radiator's liquid is drained out and replaced with pure antifreeze. Now what percent of the radiator's liquid is antifreeze? Draw pictures to help you solve this problem. Explain your answer.

18. There is a yellow flask and a red flask. The yellow flask contains more than one cup of yellow paint, and the red flask contains an equal amount of red paint. One cup of the red paint in the red flask is poured into the yellow flask and mixed thoroughly. Then one cup of the paint mixture in the yellow flask is poured into the red flask and mixed thoroughly.

   a. Without doing any calculations, which do you think will be greater, the percentage of yellow paint in the red flask or the percentage of red paint in the yellow flask? Explain your reasoning.

   b. Now use calculations to figure out the answer to the problem in part (a). Work with specific quantities of yellow and red paint. (Remember that both flasks have the same amount at the start, and both contain more than one cup.) Draw pictures to help you calculate the quantities and percentages after the mixing occurs. Do you get a different answer than you got in part (a)? If so, reconcile your answers.

6.2 Powers

In Chapter 2 we studied powers of 10, which are the foundation of the decimal system. More generally, we can consider powers of numbers other than 10. We will see that there is a precise pattern to the way powers behave when they are multiplied. In the next section, when we study the standard procedure for multiplying decimals, we will need the results from this section about multiplying powers.

CLASS ACTIVITY NOW TURN TO CLASS ACTIVITIES MANUAL

6F Multiplying Powers of 10 p. 144

POWERS OF NUMBERS OTHER THAN 10

Just as 10^5 stands for five tens multiplied together, or

$$10^5 = 10 \times 10 \times 10 \times 10 \times 10$$

the expression 2^5 stands for five 2s multiplied together:

$$2^5 = 2 \times 2 \times 2 \times 2 \times 2$$

Similarly, 3^4 stands for four 3s multiplied together:

$$3^4 = 3 \times 3 \times 3 \times 3$$

In general, if A is any real number and B is any counting number, then A^B stands for B As multiplied together:

$$A^B = \underbrace{A \times A \times \cdots \times A}_{B \text{ times}}$$

As with powers of 10, we read

$$A^B$$

exponent
as "A to the Bth power" or "A to the B," and B is called the **exponent** of A^B.

If you did Class Activity 2B in Chapter 2, then you saw that powers of 10 become very large very quickly. Powers of 2 also grow very large quickly, although they start off more slowly than powers of 10:

$$2^1 = 2$$
$$2^2 = 4$$
$$2^3 = 8$$
$$2^4 = 16$$
$$2^5 = 32$$
$$2^6 = 64$$
$$2^7 = 128$$
$$2^8 = 256$$
$$2^9 = 512$$
$$2^{10} = 1{,}024$$
$$\vdots$$
$$2^{20} = 1{,}048{,}576$$
$$\vdots$$
$$2^{30} = 1{,}073{,}741{,}824$$

exponential
growth
When a sequence of numbers consists of consecutive powers of one fixed number that is greater than 1, then we say that this sequence has **exponential growth**. The children's book *One Grain of Rice* [16], beautifully illustrates the exponential growth exhibited by powers of 2.

Exponential growth occurs in many natural systems. Population growth, whether of bacteria, animals, or people, can grow exponentially when there is enough food and space, and when there are no predators. Money left to grow in a bank account with a fixed interest rate grows exponentially. Similarly, debt, when not paid off, grows exponentially.

PRACTICE PROBLEM FOR SECTION 6.2

1. Explain why $10^A \times 10^B = 10^{A+B}$ is always true whenever A and B are counting numbers.

ANSWER TO PRACTICE PROBLEM FOR SECTION 6.2

1. The expression $10^A \times 10^B$ stands for A tens multiplied by B tens. If we multiply A tens with B tens, then all together, we multiply $A + B$ tens. Therefore,

$$10^A \times 10^B = 10^{(A+B)}$$

We can also explain this relationship with the following equations:

$10^A \times 10^B$
$= \underbrace{10 \times 10 \times \cdots \times 10}_{A \text{ times}} \times \underbrace{10 \times 10 \times \cdots \times 10}_{B \text{ times}}$
$= \underbrace{10 \times 10 \times \cdots \times 10 \times 10 \times 10 \times \cdots \times 10}_{A+B \text{ times}}$

PROBLEMS FOR SECTION 6.2

1. Is 9^5 equal to $99,999$? Explain.

2. a. Use the meaning of exponents to show how to write the expressions in (i), (ii), and (iii) as a single power of ten (i.e., in the form 10^A for some A):

   i. $(10^2)^3$ ii. $(10^3)^3$ iii. $(10^4)^2$

   b. For each of (i), (ii), and (iii) in part (a), relate the exponents in the original expression to the exponent in the answer. Describe this relationship.

   c. Explain why it is always true that $(10^A)^B = 10^{A \times B}$ whenever A and B are natural numbers.

3. Here are some mistakes that students sometimes make. In each case, explain what's wrong, discuss how the error might have come about, and write one or more correct equations that are similar to the incorrect equation.

   a. $1^{10} = 10000000000$
   b. $9^6 = 999999$
   c. $2^3 = 6$

4. a. Use a calculator to compute 2^{161}. Are you able to tell what the ones digit of this number is from the calculator's display? Why or why not?

   b. Determine the ones digits of each of the numbers $2^1, 2^2, 2^3, 2^4, 2^5, \ldots, 2^{17}$. Describe the pattern in the ones digits of these numbers.

   c. Using the pattern that you discovered in (b), predict the ones digit of 2^{161}. Explain your reasoning clearly.

6.3 Multiplying Decimals

To multiply (finite) decimals, the standard procedure is first to multiply the numbers without the decimal points, and then to place the decimal point in the answer according to a certain rule. Why is this procedure for multiplying decimals valid? The text, class activities, practice problems, and problems in this section will help you answer this question in a number of different ways.

What is the standard procedure for multiplying decimals? We first multiply the numbers without the decimal points. Then we add the number of digits to the right of the decimal points in the numbers we want to multiply, and we put the decimal point in the answer to the product computed without the decimal points that many places from the end. So, to multiply

$$\begin{array}{r} 1.36 \\ \times \ \ 3.7 \\ \hline \end{array}$$

we first multiply as if there were no decimal points:

$$\begin{array}{r} 136 \\ \times \ \ \ 37 \\ \hline 5032 \end{array}$$

Then we add the number of digits to the right of the decimal points in our two original numbers and put the decimal point that many places from the end in our answer. There are 2 digits behind the decimal point in 1.36 and another 1 digit in 3.7, for a total of $1 + 2 = 3$ digits, so the decimal point goes 3 digits from the end:

$$\begin{array}{r} 1.36 \\ \times \ \ \ 3.7 \\ \hline 5.032 \end{array}$$

Suppose you have forgotten the rule about where to put the decimal point in the answer to a decimal multiplication problem. One quick way to figure out where to put the decimal point is to think about the sizes of the numbers. For example, 1.36 is between 1 and 2 and 3.7 is between 3 and 4, so 1.36×3.7 must be between $1 \times 3 = 3$ and $2 \times 4 = 8$. So where will it make sense to put the decimal point in 5032 to get the answer to 1.36×3.7? The numbers 503.2 and 50.32 are far too big. The number 0.5032 is too small. The only number that makes sense is 5.032 because it is between 3 and 8.

How can we explain why we add the number of places behind the decimal points to determine where to place the decimal point in the answer to a decimal multiplication problem? One way is to compare the decimal multiplication problem to the multiplication problem without the decimal points. For example, how do the multiplication problems

$$0.12 \times 6.24$$

and

$$12 \times 624$$

compare? As indicated in Figure 6.7, to get from 0.12 to 12 we must multiply by ten 2 times; to get from 6.24 to 624 we must multiply by ten 2 times. Therefore, to get from

$$0.12 \times 6.24$$

to

$$12 \times 624$$

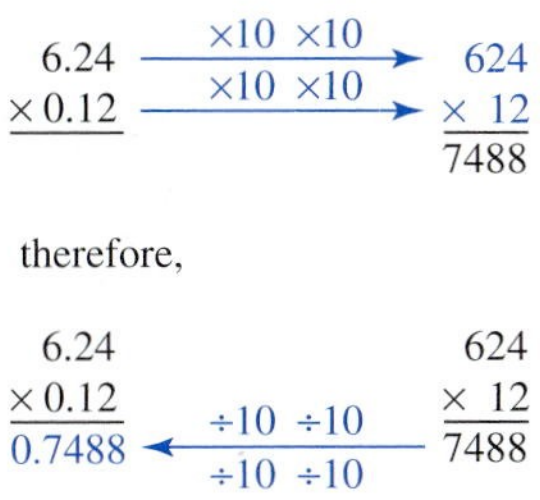

FIGURE 6.7

Comparing
0.12 × 6.24 and
12 × 624

$$\begin{array}{l} 6.24 \\ \underline{\times\, 0.12} \end{array} \xrightarrow[\times 10\ \times 10]{\times 10\ \times 10} \begin{array}{r} 624 \\ \underline{\times\ 12} \\ 7488 \end{array}$$

therefore,

$$\begin{array}{l} 6.24 \\ \underline{\times\, 0.12} \\ 0.7488 \end{array} \xleftarrow[\div 10\ \div 10]{\div 10\ \div 10} \begin{array}{r} 624 \\ \underline{\times\ 12} \\ 7488 \end{array}$$

we must multiply by ten 2 times and then another 2 times. We know that

$$12 \times 624 = 7488$$

Therefore, to get back to the answer to the original problem, 0.12×6.24, we must divide 7488 by ten 2 times and then another 2 times. When we divide by ten 2 times and then another 2 times, we shift the decimal point $2 + 2$ places to the left. So we have shown that to solve 0.12×6.24 we must move the decimal point in the answer to 12×624 to the left $2 + 2$ places, which is exactly the procedure for decimal multiplication. There wasn't anything special about the numbers we used here—the same line of reasoning applies for any other decimal multiplication problem.

Another way to explain why the standard procedure for multiplying decimals is valid is by writing the decimals as fractions. Then we can use the procedure for multiplying fractions that we have already studied. Every finite decimal can be written as a fraction with a denominator that is a power of 10. For example,

$$1.25 = \frac{125}{100}$$

$$.003 = \frac{3}{1000}$$

We can also write these denominators as products of tens:

$$1.25 = \frac{125}{100} = \frac{125}{10 \times 10}$$

$$.003 = \frac{3}{1000} = \frac{3}{10 \times 10 \times 10}$$

We will use this way of writing the denominators to explain the placement of the decimal point when we multiply decimals.

$$1.25 \times .003 = \frac{125}{10 \times 10} \times \frac{3}{10 \times 10 \times 10}$$

$$= \frac{125 \times 3}{(10 \times 10) \times (10 \times 10 \times 10)}$$

These equations tell us that to calculate the answer to $1.25 \times .003$, we should calculate the answer to 125×3 and divide by ten two times and then another 3 times. When we divide by ten 2 times and then another 3 times, we move the decimal point 2 places and then 3 places to the left. In other words, to calculate the answer to $1.25 \times .003$, we should calculate the answer to 125×3 and then move the decimal point a total of $2 + 3 = 5$ places to the left, which is exactly the standard decimal multiplication procedure.

CLASS ACTIVITY   NOW TURN TO CLASS ACTIVITIES MANUAL

6G Where Do We Put the Decimal Point When We Multiply Decimals? p. 145

6H Explaining Why We Place the Decimal Point Where We Do When We Multiply Decimals p. 146

6I Interpreting Decimal Multiplication with Pictures p. 148

PRACTICE PROBLEMS FOR SECTION 6.3

1. The product $1.35 \times 7.2 = 9.72$, but shouldn't the answer have 3 digits to the right of its decimal point? Why doesn't it?

2. Suppose you multiply a decimal number that has 5 digits to the right of its decimal point by a decimal number that has 2 digits to the right of its decimal point. Explain why you put the decimal point $5 + 2$ places from the end of the product without the decimal points.

3. Use the meaning of multiplication and a picture to explain why 2.4×1.6 has an entry in the hundredths place,

but no entries in any smaller places (thousandths, ten thousandths, etc.). Relate your explanation to the rule for placing the decimal point in 2.4×1.6.

4. Use expanded forms and properties of arithmetic or FOIL to explain why $.31 \times .024$ has digits down to the hundred-thousandths place ($\frac{1}{100,000}$), but not in any smaller places. Relate your explanation to the rule for placing the decimal point in $.31 \times .024$.

ANSWERS TO PRACTICE PROBLEMS FOR SECTION 6.3

1. The rule says to find the product without the decimal points (i.e., 135×72) and then move the decimal point in that answer $2 + 1 = 3$ places to the left. But

$$135 \times 72 = 9720$$

ends in a zero. That's why that third digit doesn't appear in the answer.

2. Let's work with a particular example to illustrate, say, 1.23456×1.23. To get from 1.23456 to 123456 we must multiply by ten 5 times. To get from 1.23 to 123 we must multiply by ten 2 times. So, to get from

$$1.23456 \times 1.23$$

to

$$123456 \times 123$$

we must multiply by ten 5 times and then another 2 times. Therefore to get from the answer to

$$123456 \times 123$$

back to the answer to

$$1.23456 \times 1.23$$

we must divide by ten 5 times and then another 2 times.

When we divide the answer to

$$123456 \times 123$$

by ten 5 times and then another 2 times, we move the decimal point to the left $5 + 2$ places. The explanation works the same way for any other decimal numbers with 5 digits and 2 digits to the right of their decimal points.

3. The product 2.4×1.6 means the total number of objects in 2.4 groups when there are 1.6 objects in each group. Starting with a square, Figure 6.8 shows 1.6 of the square, then 2 groups of 1.6 of the square and finally 2.4 groups of the 1.6 of the square. In the 2.4 groups of the 1.6 of the square, there are 2 whole squares, a bunch of "strips," each of which is $\frac{1}{10}$ of a square, and a bunch of small squares, each of which is $\frac{1}{100}$ of a square. Therefore, visibly, 2.4×1.6 has entries in the tenths and hundredths places, but not in any smaller places. According to the rule for placing the decimal point, the decimal point should go $1 + 1 = 2$ places from the right in the answer to 24×16. Thus, according to this rule, the solution to 2.4×1.6 may have entries in the tenths and hundredths places, but not in any smaller places, agreeing with the conclusion from the picture.

4. $.31 \times .024$

$$= \left(3 \cdot \frac{1}{10} + 1 \cdot \frac{1}{10^2}\right) \times \left(2 \cdot \frac{1}{10^2} + 4 \cdot \frac{1}{10^3}\right)$$

$$= 3 \cdot 2 \cdot \frac{1}{10} \cdot \frac{1}{10^2} + 3 \cdot 4 \cdot \frac{1}{10} \cdot \frac{1}{10^3}$$

$$+ 1 \cdot 2 \cdot \frac{1}{10^2} \cdot \frac{1}{10^2} + 1 \cdot 4 \cdot \frac{1}{10^2} \cdot \frac{1}{10^3}$$

$$= 6\frac{1}{10^3} + 12 \cdot \frac{1}{10^4} + 2 \cdot \frac{1}{10^4} + 4 \cdot \frac{1}{10^5}$$

We could keep on going with these equations and regroup to get the final answer, but the equations already demonstrate that there are entries down to the $\frac{1}{10^5}$ place, but no lower. According to the rule for placing the decimal point, the decimal point should go $2 + 3 = 5$ places from the end of 31×24. This placement of the decimal point will produce entries down to the $\frac{1}{10^5}$ place but no lower, agreeing with the conclusion reached by applying the distributive property or FOIL.

Thinking more generally about what will happen with other numbers, notice that the smallest place value in the product will always occur when the two smallest place values are multiplied. If you multiply $\frac{1}{10^A}$ by $\frac{1}{10^B}$ the result is $\frac{1}{10^{A+B}}$, so again this shows why we add the number of places behind the decimal points.

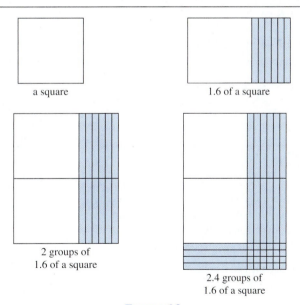

a square

1.6 of a square

2 groups of
1.6 of a square

2.4 groups of
1.6 of a square

FIGURE 6.8

Showing 2.4×1.6

PROBLEMS FOR SECTION 6.3

1. Write and solve a story problem for 14.3×1.39.

2. Write a story problem for 1.3×2.79. Solve your problem without using a calculator.

3. Leah is working on the multiplication problem 2.43×0.148. Ignoring the decimal points, Leah multiplies 243×148 and gets the answer 35964. But Leah can't remember the rule about where to put the decimal point in this answer to get the correct answer to 2.43×0.148. Explain how Leah can use reasoning about the sizes of the numbers to determine where to put the decimal point.

4. Ron used a calculator to determine that

$$0.35 \times 2.4 = 0.84$$

Ron wants to know why the rule about adding up the number of places to the right of the decimal point doesn't work in this case. Why aren't there $2 + 1 = 3$ digits to the right of the decimal point in the answer? Is Ron correct that the rule about adding the number of places to the right of the decimal points is not correct in this case? Explain.

5. When we multiply 0.48×3.9 we first multiply as if the decimal points were not there:

$$
\begin{array}{r}
3.9 \\
\times\ \ .48 \\
\hline
312 \\
1560 \\
\hline
1872
\end{array}
$$

Then we place a decimal point in 1872 to get the correct answer. Explain clearly why it makes sense to put the decimal point where we do.

6. Suppose you multiply a decimal number that has 2 digits to the right of its decimal point by a decimal number that has 3 digits to the right of its decimal point. Explain why you put the decimal point $2 + 3$ places from the end of the product calculated without the decimal points.

7. Suppose you multiply a decimal number that has 2 digits to the right of its decimal point by a decimal number

that has 4 digits to the right of its decimal point. Explain why you put the decimal point $2 + 4$ places from the end of the product calculated without the decimal points.

8. Suppose you multiply a decimal number that has M digits to the right of its decimal point by a decimal number that has N digits to the right of its decimal point. Explain why you put the decimal point $M + N$ places from the end of the product calculated without the decimal points.

9. Explain how to write 1.89 and 3.57 as improper fractions whose denominators are powers of 10. Then use fraction multiplication to explain where to place the decimal point in the solution to 1.89×3.57. Show how to use the powers of 10 in the denominators to explain why we add the number of digits behind the decimal points in 1.89 and 3.57.

10. Explain how to interpret the picture in Figure 6.9 as showing the total number of objects in 1.4 groups when there are 2.8 objects in each group (be sure to say what one object is and what one whole group is). Explain how to use this interpretation to determine the answer to 1.4×2.8. Be sure to explain how you can tell from the picture where to put the decimal point in the answer to 1.4×2.8.

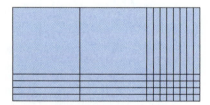

FIGURE 6.9

Showing 1.4×2.8 as 1.4 Groups with 2.8 Objects in Each Group

11. Use the meaning of multiplication and use a picture to help you explain why 1.7×3.1 has entries down to the hundredths place, but not in any place of smaller value. Be sure to explain clearly how you are interpreting your picture.

6.4 Multiplying Negative Numbers

So far, we have examined multiplication only for numbers that are not negative. What is the meaning of multiplication for negative numbers? In this section we will examine several different ways to understand multiplication of negative numbers.

What should 3×-2 mean? We know that 3×2 stands for the total number of objects in 3 groups if there are 2 objects in each group. If we try to use the same interpretation for 3×-2, then we have the following:

3×-2 is the total number of objects in 3 groups if there are -2 objects in each group.

What does this mean? One way to interpret this sensibly is to think of "-2 objects in each group" as meaning "each group *owes* 2 objects." For example, if Lakeisha, Mary, and Jayna each *owe* \$2.00, then we can think of each girl as *having* -2 dollars, and all together, the 3 girls have 3×-2 dollars. Notice that with this interpretation, it makes sense to say

$$3 \times -2 = -6$$

because all together, the 3 girls collectively owe 6 dollars, as represented by the -6. In general, if A and B are positive numbers then we can define $A \times -B$ to be $-(A \times B)$ as follows:

$$A \times -B = -(A \times B) \tag{6.1}$$

This definition is consistent with the interpretation of negative numbers as amounts owed. Equation 6.1 gives us the familiar rule:

A positive times a negative is negative.

What about a negative number times a positive number, such as

$$-2 \times 3?$$

We could perhaps interpret this product as the total number of objects in 2 owed groups with 3 objects in each group. Similarly, we could interpret a negative number times a negative number, such as

$$-4 \times -5$$

as the total number of objects in 4 owed groups with each group owing 5 objects. But these interpretations seem difficult to grasp. A better way to interpret these multiplication problems is as the change in the amount of money we will have if we give away checks or bills. (See Class Activity 6K.)

There is another way we can understand how to multiply negative numbers, which draws upon the properties of arithmetic. We have seen why the distributive property and the commutative and associative properties of addition and multiplication make sense for counting numbers. These properties of arithmetic are fundamental and describe the algebraic structure of numbers. We have seen that these properties of arithmetic underlie mental calculations and standard calculation algorithms. Because the properties of arithmetic are so important and fundamental, we should expect them to hold for any number system that is an extension of the counting numbers. In particular, the properties of arithmetic should continue to hold for all the integers, positive and negative.

Let us now assume that the properties of arithmetic we have studied hold not only for positive numbers, but for negative numbers as well. We will see that this assumption determines how we multiply negative numbers.

How do we multiply a negative number with a positive number, such as

$$-2 \times 3?$$

According to the commutative property of multiplication,

$$-2 \times 3 = 3 \times -2$$

which must be equal to -6 according to Equation 6.1 and the discussion above about how to interpret 3×-2. In general, if A and B are any positive numbers, then because we are assuming that the commutative property of multiplication holds, we must define $-A \times B$ to be $-(A \times B)$. That is,

$$-A \times B = -(A \times B) \qquad (6.2)$$

This equation gives us the following familiar rule:

A negative times a positive is negative.

Finally, how do we multiply a negative number with a negative number, as in

$$-4 \times -5?$$

We are assuming that the distributive property holds for both positive and negative numbers. Therefore,

$$\begin{aligned} -4 \times -5 + 4 \times -5 &= (-4 + 4) \times -5 \\ &= 0 \times -5 \\ &= 0 \end{aligned}$$

The preceding equations tell us that -4×-5 and 4×-5 add to zero. Since $4 \times -5 = -20$ (by Equation 6.1), -4×-5 and -20 add to zero. But 20 is the only number which, when added to -20, yields zero. Therefore, -4×-5 must equal 20. In general, if A and B are any positive numbers, then if we assume that the distributive property holds, we must define $-A \times -B$ to be $A \times B$; that is,

$$-A \times -B = A \times B \qquad (6.3)$$

This equation gives us the following familiar rule:

A negative times a negative is positive.

CLASS ACTIVITY NOW TURN TO CLASS ACTIVITIES MANUAL

6J Patterns with Multiplication and Negative Numbers p. 150

6K Using Checks and Bills to Interpret Multiplication with Negative Numbers p. 151

6L Does Multiplication Always Make Larger? p. 152

PRACTICE PROBLEMS FOR SECTION 6.4

1. Explain why it makes sense that $3 \times -2 = -6$ by interpreting negative numbers as amounts owed.

2. Given that $3 \times -2 = -6$, use a property of arithmetic to explain why it makes sense that $-2 \times 3 = 6$.

3. Given that $4 \times -5 = -20$, use a property of arithmetic to explain why it makes sense that $-4 \times -5 = 20$.

ANSWERS TO PRACTICE PROBLEMS FOR SECTION 6.4

1. See text.

2. See text.

3. See text.

PROBLEMS FOR SECTION 6.4

1. a. Explain why it makes sense that $5 \times -2 = -10$ by interpreting negative numbers as amounts owed.

   b. Given that $5 \times -2 = -10$, use a property of arithmetic to explain why it makes sense that $-2 \times 5 = -10$.

   c. Given that $5 \times -2 = -10$, use a property of arithmetic to explain why it makes sense that $-5 \times -2 = 10$.

2. For which numbers, N, is $N \times 2$ greater than 2? In other words, for which numbers, N, is

$$N \times 2 > 2$$

Investigate this question as follows:

   a. Without using a calculator, multiply 2 by the following numbers. Show your work. In each case, determine whether the resulting product is greater

than 2 or not.

$$2.85, \quad 1.25, \quad 0.95, \quad 0.03, \quad -0.7, \quad -3.5$$

$$3\frac{1}{2}, \quad 1\frac{1}{8}, \quad \frac{7}{8}, \quad \frac{2}{5}, \quad -\frac{2}{3}, \quad -2\frac{3}{4}$$

   b. Based on your answers in part (a) and on the meaning and rules of multiplication, describe the collection of all numbers, N, for which $N \times 2$ is greater than 2.

3. For which numbers, N, is $N \times -2$ greater than -2? In other words, for which numbers, N, is

$$N \times -2 > -2$$

Investigate this question as follows:

   a. Without using a calculator, multiply -2 by the following numbers. Show your work. In each case, determine whether the resulting product is greater than −2 or not.

3.15,   1.01,   0.85,   0.002,   −0.3,   −4.2

$$4\frac{3}{5}, \quad 1\frac{2}{3}, \quad \frac{9}{10}, \quad \frac{3}{5}, \quad -\frac{7}{8}, \quad -1\frac{1}{8}$$

b. Based on your answers in part (a) and on the meaning and rules of multiplication, describe the collection of all numbers, N, for which $N \times -2$ is greater than −2.

6.5 Scientific Notation

Many scientific applications require the use of very large or very small numbers: distances between stars are huge, whereas molecular distances are tiny. These kinds of numbers can be cumbersome to write in ordinary decimal notation. Therefore, a special notation, called scientific notation is often used to write such numbers. Scientific notation involves multiplying by powers of 10.

Use a calculator to multiply

$$123,456,789 \times 987,654,321$$

How is the answer displayed? The answer is probably displayed in one of the following forms:

$$1.2193263 \; 17$$

or

$$1.2193263 \; E \; 17$$

Both of these displays represent

$$1.2193263 \times 10^{17}$$

which is in scientific notation.

scientific notation   A number is in **scientific notation** if it is written as a decimal that has exactly one nonzero digit to the left of the decimal point, multiplied by a power of ten. So a number is in scientific notation if it is written in the form

$$\#.\#\#\#\#\# \times 10^{\#}$$

where the # to the left of the decimal point is not zero and where any number of digits can be displayed to the right of the decimal point. Thus, neither

$$121.93263 \times 10^{15}$$

nor

$$12.193263 \times 10^{16}$$

is in scientific notation, but

$$1.2193263 \times 10^{17}$$

is in scientific notation.

Other than 0, every real number can be expressed in scientific notation. To write a number in scientific notation, think about how multiplication by powers of 10 works. For example, how do we express

$$847,930,000$$

in scientific notation? We need to find an exponent that will make the following equation true:

$$847,930,000 = 8.4793 \times 10^?$$

The decimal point in 8.4793 must be moved 8 places to the right to get $847,930,000$; therefore, we should multiply 8.4793 by 10^8. In other words,

$$847,930,000 = 8.4793 \times 10^8$$

How do we write

$$.0000345$$

in scientific notation? We need to find an exponent that will make the following equation true:

$$.0000345 = 3.45 \times 10^?$$

The decimal point in 3.45 must be moved 5 places to the left to get .0000345; therefore, we must multiply 3.45 by

$$.00001 = \frac{1}{10^5} = 10^{-5}$$

in order to get .0000345. So

$$.0000345 = 3.45 \times 10^{-5}$$

Here are a few more examples:

Ordinary decimal notation	Scientific notation
12	1.2×10^1
123	1.23×10^2
1234	1.234×10^3
12345	1.2345×10^4
1234.5	1.2345×10^3
123.45	1.2345×10^2
12.345	1.2345×10^1
1.2345	1.2345 or 1.2345×10^0
.12345	1.2345×10^{-1}
.012345	1.2345×10^{-2}
.0012345	1.2345×10^{-3}

Although scientific notation is mainly used in scientific settings, it is common to use a variation of scientific notation when discussing large numbers in more common situations, such as when budgets or populations are concerned. For example, an amount of money may be described as

$3.5 billion,

which means

$$\$ 3.5 \times (\text{one billion})$$

or

$$\$ 3.5 \times 1{,}000{,}000{,}000$$

Two other ways to express $3.5 billion are

$$\$ 3.5 \times 10^9$$

and

$$\$3{,}500{,}000{,}000$$

The form

$$\$3.5 \text{ billion}$$

is probably more quickly and easily grasped by most people than any of the other forms.

By the way, what is a big number—or a small number? Children sometimes ask questions such as the following:

Is 100 a big number? What about 1000, is that a big number?

The answer is, "It depends." We wouldn't think of 1000 grains of sand as a lot of sand, but we might consider 10 pages of homework to be a lot.

CLASS ACTIVITY NOW TURN TO CLASS ACTIVITIES MANUAL

6M Scientific Notation versus Ordinary Decimal Notation p. 153

6N How Many Digits Are in a Product of Counting Numbers? p. 154

6O Explaining the Pattern in the Number of Digits in Products p. 155

PRACTICE PROBLEMS FOR SECTION 6.5

1. Write the following numbers in scientific notation:
   a. 153,293,043,922
   b. 0.00000321
   c. $(2.398 \times 10^{15}) \times (3.52 \times 10^9)$
   d. $(5.9 \times 10^{15}) \times (8.3 \times 10^9)$

2. The astronomical unit (AU) is used to measure distances. One AU is the average distance from the earth to the sun, which is 92,955,630 miles. We are 2 billion AU from the center of the Milky Way galaxy (our galaxy). How many miles are we from the center of the Milky Way galaxy? Give your answer in scientific notation. Ex-plain how the meaning of multiplication applies to this problem.

3. Sam uses a calculator to multiply $666,666 \times 7,777,777$. The calculator's answer is displayed as follows:

   5.1851794 E 12

   So Sam writes

   $666,666 \times 7,777,777 = 5,185,179,400,000$

   Is Sam's answer correct or not? If not, why not?

ANSWERS TO PRACTICE PROBLEMS FOR SECTION 6.5

1. a. $1.53293043922 \times 10^{11}$
    b. 3.21×10^{-6}
    c. 8.44096×10^{24}
    d. 4.897×10^{25}

2. Since we are 2 billion AU from the center of the galaxy and each AU is 92,955,630 miles, the number of miles from the earth to the center of the galaxy is the total num-ber of objects in 2 billion groups (each AU is a group) when there are 92,955,630 objects in each group (each object is 1 mile). Therefore, according to the meaning of multiplication, we are

$2 \text{ billion} \times 92,955,630 = 2 \times 10^9 \times 92,955,630$
$= 185,911,260 \times 10^9$
$= 1.86 \times 10^8 \times 10^9$
$= 1.86 \times 10^{17}$

miles from the center of the galaxy.

3. No, Sam's answer is not correct. When the calculator displays the answer to $666,666 \times 7,777,777$ as

   5.1851794 E 12

this stands for

$$5.1851794 \times 10^{12}$$

However, the calculator is forced to round its answer because it can only display so many digits on its screen.

Therefore, although it is true that

$$5.1851794 \times 10^{12} = 5,185,179,400,000$$

this is not the exact answer to $666,666 \times 7,777,777$. Instead, it is the answer to $666,666 \times 7,777,777$ rounded to the nearest hundred-thousand.

PROBLEMS FOR SECTION 6.5

1. a. The winnings of a lottery were $250 million. Write 250 million in ordinary decimal notation and in scientific notation.

   b. A company's revenues could be $15 billion. Write 15 billion in ordinary decimal notation and in scientific notation.

2. Write the following numbers in scientific notation:

   a. $201,348,761,098$

   b. 0.000000078

   c. $(2.4 \times 10^{12}) \times (8.6 \times 10^{11})$

   d. $(6.1 \times 10^{13}) \times (9.2 \times 10^{8})$

3. A calculator might display the answer to

   $$555,555 \times 6,666,666$$

   as

   $$3.7037 \text{ E } 12$$

   a. What does the calculator's display mean?

   b. What information about the solution to $555,555 \times 6,666,666$ can you obtain from the calculator's display? Can you determine the exact answer to $555,555 \times 6,666,666$ in ordinary decimal notation? Why or why not? Can you determine how many digits are in the ordinary decimal representation of the product $555,555 \times 6,666,666$? Explain.

4. Tanya says that the ones digit of 2^{59} is a 5 because her calculator's display for 2^{59} reads:

   $$5.76460752303 \text{ E } 17$$

   and there is a 5 in the ones place. Is Tanya right? Why or why not?

5. Is 2×10^7 equal to 2^7? Is 1×10^9 equal to 1^9? If not, explain the distinctions between the expressions.

6. Suppose you have a calculator that displays no more than 12 digits. Describe how to use the distributive property and the calculator to multiply

   $$8 \times 123,456,123,456$$

   showing the answer in ordinary decimal notation. Do not just calculate the product with the longhand multiplication algorithm. Make efficient use of the calculator.

7. Let's say that you want to write the answer to $179,234,652 \times 437,481,694$ as a whole number in ordinary decimal notation, showing all its digits. Find a way to use a calculator that displays no more than 12 digits to help you do this in an efficient way. Do not just multiply longhand. Explain your technique and explain why it works.

8. Suppose you multiply a 6-digit number by an 8-digit number. How many digits will the product have? The following problem will help you answer this question:

   a. When you write a 6-digit number and an 8-digit number in scientific notation, what will the exponents on the 10s be? Explain.

   b. Write

      $$(1.3 \times 10^5) \times (2.5 \times 10^7)$$

      and

      $$(7.9 \times 10^5) \times (8.3 \times 10^7)$$

      in scientific notation.

   c. Suppose you write a 6-digit number and an 8-digit number in scientific notation. If you multiply these numbers, and write the product in scientific notation, what will the exponent on the 10 be when the product is expressed in scientific notation? Explain why your answer is in an "either...or..." form. (Hint: see part (b).)

d. Use your answer to part (c) to answer the following question: When you multiply a 6-digit number by an 8-digit number, how many digits will the product have?

9. Light travels at a speed of about 300,000 kilometers per second.

   a. How far does light travel in one day? Give your answer both in scientific notation and in ordinary decimal notation. Explain your work.

   b. How far does light travel in one year? Give your answer both in scientific notation and in ordinary decimal notation, rounded to the nearest hundred-billion kilometers. Explain your work. The distance that light travels in one year is called a **light year**.

10. According to scientific theories, the solar system formed between 5 and 6 billion years ago. Light travels 186,282 miles per second. How far has the light from the forming solar system traveled in 5.5 billion years? Give your answer in scientific notation.

11. According to the Bureau of the Public Debt Online of the U.S. Department of Treasury (see www.aw-bc.com/beckmann) the total federal debt in 2003 was approximately $6.66 trillion. Assume that the debt will increase each year by 5%; that is, assume that each year's debt is 5% more than the previous year's debt. With this assumption, predict the federal debt in 2008. Give your answer rounded to the nearest hundred-billion dollars. Explain your reasoning.

12. Suppose that a laboratory has one gram of a radioactive substance that has a half-life of 100 years. "A half-life of 100 years" means that no matter what amount of the

radioactive substance one starts with, after 100 years, only half of it will be left. So, after the first 100 years, only half a gram would be left, and after another 100 years, only one quarter of a gram would be left.

   a. How many hundreds of years will it take until there is less than one hundred millionth of a gram left of the laboratory's radioactive substance? (Give a whole number of hundreds of years.)

   b. How many hundreds of years will it take until there is less than one billionth of a gram left of your radioactive substance? (Give a whole number of hundreds of years.)

13. Suppose you owe $1,000 on your credit card and that at the end of each month an additional 1.4708% of the amount you owe is added on to the amount you owe. (This is what would happen if your credit card charged an annual percentage rate of 17.6496%.) Let's assume that you do not pay off any of this debt or the interest that is added to it. Let's also assume that you don't add on any more debt (other than the interest that you are charged).

   a. How much will you owe after 6 months? after one year? after two years?

   b. Calculate $(1.014708)^{48}$ and notice that this number is just a little larger than 2. Based on this calculation, what will happen to your debt every 4 years? Explain.

   c. Suppose you didn't pay off your debt for 40 years. Using part (b), determine approximately how much you owe.

Chapter 7

Division

I n this chapter we will study division of all different kinds of numbers: whole numbers, rational numbers, real numbers, and integers. We can use division to solve a wide variety of problems, such as determining how many pencils each child will get if 100 pencils are divided equally among 22 children, determining how many servings of rice we can make from 5 cups of rice if one serving is $\frac{2}{3}$ cup of rice, or determining which paint mixture will be more yellow: one made by mixing 4 parts yellow paint with 25 parts white paint or one made by mixing 3 parts yellow paint with 16 parts white paint. On the surface, these are very different kinds of problems, but all can be solved with division.

As with multiplication, we will study the meaning of division, and we will explain why the various division procedures are valid based on the meaning of division. We will also study the link between division and fractions.

7.1 The Meaning of Division

In this section we will study the meaning of division. There are two main ways to interpret the meaning of division; both are useful and both are natural. Just as every subtraction problem can be rewritten as an addition problem, every division problem can be rewritten as a multiplication problem. We will use this connection between division and multiplication to explain why a division problem has the same answer, regardless of which of the two main interpretations of division we use. Finally, each of the two main ways of interpreting the meaning of division is subject to further interpretation, depending on whether we do or do not want to allow for a remainder.

The following are the three standard ways to denote division:

$$A \div B, \quad A/B, \quad \text{and} \quad B\overline{)A}$$

quotient

dividend

divisor

All three are read "A divided by B." In a division problem $A \div B$, the result is called the quotient, the number A is called the **dividend**, and the number B is called the **divisor**.

THE TWO INTERPRETATIONS OF DIVISION

A simple example will show that there are two distinct ways to interpret the meaning of division. How would you draw a diagram to solve $8 \div 2$? Your drawing might look like either part (a) or part (b) of Figure 7.1.

In (a) of Figure 7.1, $8 \div 2$ is represented by 8 objects divided into *groups of* 2. With this interpretation we're thinking of $8 \div 2 =?$ as asking how many groups there will be if 8 objects are divided into groups with 2 objects in each group. Whereas in (b) of Figure 7.1, $8 \div 2$ is represented by 8 objects divided into 2 *groups*. With the second interpretation, we're thinking of $8 \div 2 =?$ as asking how many objects will be in each group if 8 objects are divided equally among 2 groups.

THE "HOW MANY GROUPS?" INTERPRETATION

If A and B are nonnegative numbers, and B is not 0, then, according to the "how many groups?" interpretation of division,

$A \div B$ $A \div B$ means the number of groups when A objects are divided into groups with B objects in each group.

In other words, with this interpretation, $A \div B$ means "the number of Bs that are in A." This interpretation of division is sometimes called the measurement model of division or the subtractive model of division.

With this interpretation of division, $18 \div 6 = 3$ because if you have 18 objects and you divide these into groups of 6, then there will be 3 groups, as shown in Figure 7.2.

FIGURE 7.1

Two
Interpretations
of $8 \div 2$

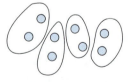 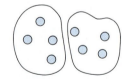

(a) How many groups?   (b) How many in each group?

FIGURE 7.2

Using the "How
Many Groups?"
Approach to
Show
$18 \div 6 = 3$

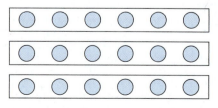

FIGURE 7.3

Using the "How Many in Each Group?" Approach to Show $18 \div 6 = 3$

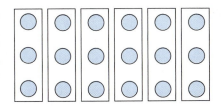

THE "HOW MANY IN EACH GROUP?" INTERPRETATION

If A and B are nonnegative numbers and B is not 0, then, according to the "how many in each group?" interpretation of division,

$A \div B$ $A \div B$ means the number of objects in each group when A objects are divided equally among B groups.

This interpretation of division is sometimes called the partitive model of division or the sharing model of division.

With this interpretation of division we conclude that $18 \div 6 = 3$ because if you divide 18 objects equally among 6 groups, then there will be 3 objects in each group, as seen in Figure 7.3.

Why do we get the same answer to division problems, regardless of which interpretation of division we use? We will answer this by relating division to multiplication.

RELATING DIVISION AND MULTIPLICATION

Consider the division problem

$$18 \div 6 = ?$$

According to the "how many groups?" interpretation of division, $18 \div 6$ is the number of groups of 6 that make 18. Therefore, according to the meaning of multiplication, that number of groups times 6 is equal to 18. In other words, with this interpretation of division,

$$18 \div 6 = ?$$

means the same as

$$? \times 6 = 18$$

In the same way, every division problem is equivalent to a multiplication problem. With the "how many groups?" interpretation of division, the problem

$$A \div B = ?$$

is equivalent to the problem

$$? \times B = A$$

On the other hand, according to the "how many in each group?" interpretation of division, $18 \div 6$ is the number of objects for which 6 groups of that many objects makes a total of 18 objects. Therefore, according to the meaning of multiplication, 6 times that number of objects is equal to 18. In other words, with this interpretation of division,

$$18 \div 6 = ?$$

means the same as

$$6 \times ? = 18$$

In the same way, every division problem is equivalent to a multiplication problem. With the "how many in each group?" interpretation of division, the problem

$$A \div B = ?$$

is equivalent to the problem

$$B \times ? = A$$

Now we can explain why a division problem will have the same answer, regardless of whether the "how many groups?" or the "how many in each group?" interpretation of division is used. The key lies in the commutative property of multiplication. According to the "how many groups?" interpretation, solving

$$A \div B = ?$$

is equivalent to solving

$$? \times B = A$$

But according to the commutative property of multiplication,

$$? \times B = B \times ?$$

Therefore, solving

$$? \times B = A$$

is equivalent to solving

$$B \times ? = A$$

which is equivalent to solving

$$A \div B = ?$$

with the "how many in each group?" interpretation of division. Thus, we will always get the same answer to a division problem, regardless of which way we interpret division.

ANSWERS WITH OR WITHOUT REMAINDERS

Consider the division problem $23 \div 4$. Depending on the context, any one of the following three answers could be most appropriate:

$$23 \div 4 = 5.75$$

$$23 \div 4 = 5\frac{3}{4}$$

$$23 \div 4 = 5, \text{ remainder } 3$$

The first two answers fit with division as we have been interpreting it so far. For example, to divide \$23 equally among 4 people, each person should get $\$23 \div 4 = \$5.75$. Or, if you have 23 cups of flour, and a batch of cookies requires 4 cups of flour, then you can make $23 \div 4 = 5\frac{3}{4}$ batches of cookies (assuming that it is possible to make $\frac{3}{4}$ of a batch of cookies). But if you have 23 pencils to divide equally among 4 children, then it doesn't make sense to give each child 5.75 or $5\frac{3}{4}$ pencils. Instead, it is best to give each child 5 pencils and keep the remaining 3 pencils in reserve. How does this answer, 5, remainder 3, fit with division as we have defined it?

In order to get the third answer, 5, remainder 3, we need to interpret division in a slightly different manner than we have so far. For each of the two main interpretations of division, there is an alternative formulation, which allows for a remainder. In these alternative formulations, we seek a quotient that is a *whole number*.

If A and B are whole numbers, and B is not zero, then

remainder
how many in
each group?

$A \div B$ is the largest whole number of objects in each group when A objects are divided equally among B groups. The **remainder** is the number of objects left over (i.e., that can't be placed in a group). This is the "**how many in each group**?" interpretation, with remainder.

remainder
how many
groups?

$A \div B$ is the largest whole number of groups that can be made when A objects are divided into groups with B objects in each group. The **remainder** is the number of objects left over (i.e., that can't be placed in a group). This is the "**how many groups**?" interpretation, with remainder.

DIVISION WITH NEGATIVE NUMBERS

So far we have defined division for nonnegative numbers. How can we make sense of division problems like

$$-18 \div 3 = ?$$

or

$$18 \div -3 = ?$$

We could perhaps talk about dividing 18 owed objects among 3 groups, but this seems strange. A better way is to rewrite a division problem with negative numbers as a multiplication problem. For example, the problem

$$-18 \div 3 = ?$$

is equivalent to

$$? \times 3 = -18$$

(if we use the "how many groups?" interpretation). Since $-6 \times 3 = -18$, and since no other number times 3 is -18, it follows that

$$-18 \div 3 = -6$$

By considering how multiplication works when negative numbers are involved, we can see that the following rules apply to division problems involving negative numbers:

$$\text{negative} \div \text{positive} = \text{negative}$$

$$\text{positive} \div \text{negative} = \text{negative}$$

$$\text{negative} \div \text{negative} = \text{positive}$$

CLASS ACTIVITY   NOW TURN TO CLASS ACTIVITIES MANUAL

7A The Two Interpretations of Division p. 159

7B Remainder, Mixed Number, or Decimal Answer to Division Problem: Which is Most Appropriate? p. 161

7C Why Can't We Divide by Zero? p. 162

7D Can We Use Properties of Arithmetic to Divide? p. 164

7E Reasoning about Division p. 165

7F Rounding to Estimate Solutions to Division Problems p. 166

PRACTICE PROBLEMS FOR SECTION 7.1

1. For each of the following story problems, write the corresponding division problem and decide which interpretation of division is involved (the "how many groups?" or the "how many in each group?" with or without remainder):

   a. There are 5237 tennis balls that are to be put into packages of 3. How many packages of balls can be made and how many will be left over?

   b. If 50 fluid ounces of mouthwash costs $4.29, then what is the price per fluid ounce of this mouthwash?

   c. If 50 fluid ounces of mouthwash costs $4.29, then how much of this mouthwash is worth $1?

   d. If you have a full 50-fluid-ounce bottle of mouthwash and you use 3 fluid ounces per day, then how many days will this mouthwash last?

   e. 1 yard is 3 feet. How long is an 80 foot long stretch of sidewalk in yards?

   f. 1 gallon is 16 cups. How many gallons are 65 cups of lemonade?

   g. If you drive 316 miles at a constant speed and it takes you 5 hours, then how fast did you go?

   h. Theresa needs to cut a piece of wood 0.33 inches thick, or just a little less thick. Theresa's ruler shows subdivisions of $\frac{1}{32}$ of an inch. How many $\frac{1}{32}$ of an inch thick should Theresa cut her piece of wood?

2. Make up two story problems for $300 \div 12$, one for each of the two interpretations of division.

3. Josh says that $0 \div 5$ doesn't make sense because if you have nothing to divide among 5 groups, then you won't be able to put anything in the 5 groups. Which interpretation of division is Josh using? What does Josh's reasoning actually say about $0 \div 5$?

4. Roland says,

   I have 23 pencils to give out to my students, but I have no students. How many pencils should each student get? There's no possible answer because I can't give out 23 pencils to my students if I have no students.

   Write a division problem that corresponds to what Roland said. Which interpretation of division does this use? What does Roland's reasoning say about the answer to this division problem?

5. Katie says,

   I have no candies to give out and I'm going to give each of my friends 0 candies. How many friends do I have? I could have 10 friends, or 50 friends, or 0 friends—there's no way to tell.

   Write a division problem that corresponds to what Katie said. Which interpretation of division does this use? What does Katie's reasoning say about the answer to this division problem?

6. Use the "how many groups?" interpretation of division to explain why $1 \div 0$ is not defined.

7. Explain why $0 \div 0$ is not defined by rewriting $0 \div 0 = ?$ as a multiplication problem.

8. Marina says that since $300 \div 50 = 6$, that means you can put 300 marbles into 6 groups with 50 marbles in each group. Marina thinks that knowing $300 \div 50 = 6$ should help her calculate $300 \div 55$. At first she tries $300 \div 50 + 300 \div 5$, but that answer seems too big. Is there some other way to use the fact that $300 \div 50 = 6$ in order to calculate $300 \div 55$?

9. According to one study, the average American produces 4.21 pounds of solid waste per day. The population of the U.S. is about 275 million.

   a. How many pounds of solid waste does the population of the U.S. produce in one year?

   b. If fifty pounds of solid waste take up one cubic foot, then how many cubic feet of solid waste does the U.S. population produce per year?

   c. If the solid waste produced by the U.S. population in a year was formed into a large cube, what would the length, width and depth of this cube be?

10. Which is a better buy: 67 cents for 58 ounces or 17 cents for 15 ounces?

11. If March 5th is a Wednesday, then why do you know right away that March 12th, 19th, 26th, and April 2nd are Wednesdays? Explain this using mathematics and knowledge of our calendar system.

12. Suppose that today is Friday. What day of the week will it be 36 days from today? What day of the week will it be 52 days from today? What about 74 days from today? Explain how division with a remainder is relevant to these questions.

ANSWERS TO PRACTICE PROBLEMS FOR SECTION 7.1

1. a. $5237 \div 3$. This uses the "how many groups?", with remainder interpretation.

   b. $\$4.29 \div 50$. This uses the "how many in each group?" interpretation (without remainder). Each ounce of mouthwash represents a group. We want to divide $4.29 equally among these groups.

   c. $50 \div 4.29$. This uses the "how many in each group?" interpretation (without remainder). Each dollar represents a group. We want to divide 50 ounces up among the 4.29 groups.

   d. $50 \div 3$. This uses the "how many groups?" interpretation, with remainder. Each 3 fluid ounces is a group. We want to know how many of these groups are in 50 fluid ounces. There will be 2 fluid ounces left over.

   e. $80 \div 3$. This uses the "how many groups?" interpretation, with or without remainder. Each 3 feet is a group (a yard). We want to know how many of these groups are in 80 feet. We can give the answer as 26 yards, 2 feet (with remainder), or as 26.7 yards (without remainder).

f. $65 \div 16$. This uses the "how many groups?" interpretation, with or without remainder. Each 16 cups is a group (a gallon). We want to know how many of these groups are in 65 cups. We can give the answer as 4 gallons and 1 cup (with remainder) or as 4.0625 gallons (without remainder). The former answer is probably more useful than the latter.

g. $316 \div 5 = 63.2$ miles per hour. This uses the "how many in each group?" interpretation, without remainder. Divide the 316 miles equally among the 5 hours. Each hour represents a group. In each hour, you drove $316 \div 5$ miles. This means your speed was 63.2 miles per hour.

h. $0.33 \div \frac{1}{32} = 10.56$ so Theresa should cut her piece of wood ten $\frac{1}{32}$ of an inch thick. This uses the "how many groups?" interpretation of division, without remainder (although in the end we round down because Theresa needs to cut her piece of wood a little less thick than the actual answer). Each $\frac{1}{32}$ of an inch represents one group. We want to know how many of these groups of $\frac{1}{32}$ of an inch are in 0.33 inches.

2. A good example for the "how many groups?" interpretation is "how many feet are in 300 inches?" Because each foot is 12 inches, this problem can be interpreted as "how many 12s are in 300?". An example for the "how many in each group?" interpretation is "300 snozzcumbers will be divided equally among 12 hungry boys. How many snozzcumbers does each hungry boy get?"

3. Josh is using the "how many in each group?" interpretation of division. Josh's statement can be reinterpreted as this: if you have 0 objects and you divide them equally among 5 groups, then there will be 0 objects in each group. Therefore $0 \div 5 = 0$.

4. The division problem that corresponds to what Roland said is this: $23 \div 0 = ?$. Roland is using the "how many in each group?" interpretation of division. Each group is represented by a student. Roland wants to divide 23 objects equally among 0 groups. But as Roland says, this is impossible to do. Therefore $23 \div 0$ is undefined.

5. The division problem that corresponds to what Katie said is this: $0 \div 0 = ?$. Katie is using the "how many groups?" interpretation of division. Each group is represented by a friend. There are 0 objects to be distributed equally, with 0 objects in each group. But

from this information, there is no way to determine how many groups there are. There could be *any* number of groups—50, or 100, or 1000. So the reason that $0 \div 0$ is undefined is there isn't *one unique answer*, in fact any number could be considered equal to $0 \div 0$.

6. With the "how many groups?" interpretation, $1 \div 0$ means the number of groups when 1 object is divided into groups with 0 objects in each group. But if you put 0 objects in each group, then there's no way to distribute the 1 object—it can never be distributed among the groups, no matter how many groups there are. Therefore, $1 \div 0$ is not defined.

7. Any division problem $A \div B = ?$ can be rewritten in terms of multiplication, namely as either $? \times B = A$ or as $B \times ? = A$. Therefore $0 \div 0 = ?$ means the same as $? \times 0 = 0$ or $0 \times ? = 0$. But *any number* times 0 is zero, so the ? can stand for any number. Since there isn't one unique answer to $0 \div 0 = ?$, we say that $0 \div 0$ is undefined.

8. Marina is right that $300 \div 50 + 300 \div 5$ won't work—it divides the 300 marbles into groups *twice*, so it gives too many groups. Also, if you get 6 groups when you put 50 marbles in each group, then you must get fewer groups when you put more marbles in each group. If you think about removing one of those 6 groups of marbles, then you could take the 50 marbles in that group and from them, put 5 marbles in each of the remaining 5 groups. Then you'd have 5 groups of 55 marbles, and you'd have 25 marbles left over. Therefore $300 \div 55 = 5$, remainder 25.

9. a. 4.23×10^{11} pounds

   b. 8.45×10^{9} cubic feet

   c. The cube would be about 2,037 feet long, wide, and deep, since $2,037 \times 2,037 \times 2,037 = 8.45 \times 10^{9}$. (You can calculate the 2,037 by taking the cube root of 8,450,000,000 on your calculator, or by taking the cube root of 8.45 on your calculator and realizing you'll have to multiply it by 10^3. Or you could calculate the 2,037 by guessing and checking.)

10. 17 cents for 15 ounces is a slightly better buy because the price for one ounce is $17 \div 15 = 1.133$ cents, whereas when you pay 67 cents for 58 ounces the price for one ounce is $67 \div 58 = 1.155$ cents.

11. Every seven days after Wednesday is another Wednesday. March 12th, 19th, 26th, and April 2nd are 7, 14, 21, and 28 days after March 5th, and 7, 14, 21, and 28 are multiples of 7, so these days are all Wednesdays too.

12. Every seven days after a Friday is another Friday. So 35 days after today (assumed to be a Friday) is another Friday. Therefore, 36 days from today is a Saturday, because it is 1 day after a Friday. Notice that we really only needed to find the remainder of 36 when divided by 7 in order to determine the answer. $52 \div 7 = 7$, re- mainder 3, so 52 days from today will be 3 days after a Friday (because the 7 groups of 7 get us to another Friday). 3 days after a Friday is a Monday. $74 \div 7 = 10$, remainder 4, so 74 days from today will be 4 days after a Friday, which is a Tuesday.

PROBLEMS FOR SECTION 7.1

1. For each of the following story problems, write the corresponding division problem, state which interpretation of division is involved (the "how many groups?" or the "how many in each group?", with or without remainder), and solve the problem.

   a. If 250 rolls are to be put in packages of 12, then how many packages of rolls can be made?

   b. If you have 500 stickers to give out equally to 23 children, then how many stickers will each child get?

   c. Given that 1 gallon is 8 pints, how many gallons of water is 45 pints of water?

   d. If your car used 12 gallons of gasoline to drive 350 miles, then how many miles per gallon did your car get?

   e. If you drove 120 miles at a constant speed and if it took you $2\frac{1}{2}$ hours, then how fast were you going?

   f. Given that 1 inch is 2.54 centimeters, how tall in inches is a man who is 186 centimeters tall? Given that 1 foot is 12 inches, how tall is the man in feet?

   g. Sue needs to cut a piece of wood 0.4 of an inch thick, or just a little less thick. Sue's ruler shows sixteenths of an inch. How many sixteenths of an inch thick should Sue cut her piece of wood?

2. Write two story problems for $38 \div 7$, one for each of the two interpretations of division.

3. Make up and solve three different story problems (word problems) for $7 \div 3$.

   a. In the first story problem, the answer should be best expressed as 2, remainder 1.

   b. In the second story problem, the answer should be best expressed as $2\frac{1}{3}$.

   c. In the third story problem, the answer should be best expressed as 2.33.

4. Explain how to solve the following problems with division. Which interpretation of division do you use?

   a. What day of the week will it be 50 days from today?

   b. What day of the week will it be 60 days from today?

   c. What day of the week will it be 91 days from today?

   d. What day of the week will it be 365 days from today?

5. a. Is $0 \div 5$ defined or not? Write a story problem for $0 \div 5$ and use your story problem to discuss whether or not $0 \div 5$ is defined.

   b. Is $5 \div 0$ defined or not? Write a story problem for $5 \div 0$ and use your story problem to discuss whether or not $5 \div 0$ is defined.

6. a. Is $0 \div 3$ defined or not? Explain your reasoning.

   b. Is $3 \div 0$ defined or not? Explain your reasoning.

7. a. Write and solve a simple story problem for $1200 \div 30$.

   b. Use the situation of your story problem in part (a) to help you solve $1200 \div 31$ without a calculator or long division by modifying your solution to $1200 \div 30$.

   c. Use the situation of your story problem in part (a) to help you solve $1200 \div 29$ without a calculator or long division by modifying your solution to $1200 \div 30$.

8. a. Write and solve a simple story problem for $630 \div 30$.

   b. Use the situation of your story problem in part (a) to help you solve $630 \div 31$ without a calculator or long division by modifying your solution to $630 \div 30$.

   c. Use the situation of your story problem in part (a) to help you solve $630 \div 29$ without a calculator or long division by modifying your solution to $630 \div 30$.

9. a.   Suppose you want to estimate

$$459 \div 38$$

by rounding 38 up to 40. Both

$$440 \div 40$$

and

$$480 \div 40$$

are easy to calculate mentally.   Use reasoning about division to determine which division problem, $440 \div 40$ or $480 \div 40$, should give you a better estimate to $459 \div 38$. Then check your answer by solving the division problems.

b.   Suppose you want to estimate

$$459 \div 42$$

by rounding 42 down to 40. As before, both $440 \div 40$ and $480 \div 40$ are easy to calculate mentally. Use reasoning about division to determine which division problem, $440 \div 40$ or $480 \div 40$, should give you a better estimate to $459 \div 42$. Then check your answer by solving the division problems.

c.   Suppose you want to estimate $632 \div 58$. What is a good way to round the numbers 632 and 58 so that you get an easy division problem which will give a good estimate to $632 \div 58$? Explain, drawing on what you learned from parts (a) and (b).

d.   Suppose you want to estimate $632 \div 62$. What is a good way to round the numbers 632 and 62 so that you get an easy division problem which will give a good estimate to $632 \div 62$? Explain, drawing on what you learned from parts (a) and (b).

10. Bob wants to estimate $1893 \div 275$. He decides to round 1893 up to 2000. Bob says that since he rounded 1893 up, he'll get a better estimate if he rounds 275 in the opposite direction, namely down to 250 rather than up to 300. Therefore, Bob says that $1893 \div 275$ is closer to 8 (which is $2000 \div 250$) than to 7, because $2000 \div 300$ is a little less than 7. This problem will help you investigate whether or not Bob's reasoning is correct.

a.   Which of $2000 \div 250$ and $2000 \div 300$ gives a better estimate to $1893 \div 275$? Can Bob's reasoning (just described) be correct?

b.   Use the meaning of division (either of the two main interpretations) to explain the following: When estimating the answer to a division problem $A \div B$,

if you round A up, you will generally get a better estimate if you also round B up rather than down. Draw diagrams to aid your explanation.

c.   Now consider the division problem $1978 \div 205$. Suppose you round 1978 up to 2000. In this case, will you get a better estimate to $1978 \div 205$ if you round 205 up to 250 (and compute $2000 \div 250$) or if you round 205 down to 200 (and compute $2000 \div 200$)? Reconcile this with your findings in part (b).

11. Explain clearly how to use the meanings of multiplication and division, as well as the following information, in order to determine how many grams 1 cup of water weighs:

1 quart	=	4 cups
1 liter	=	1.056 quarts
1 liter	weighs	1 kilogram
1 kilogram	=	1000 grams

12. Light travels at a speed of about 300,000 kilometers per second. The distance that light travels in one year is called a **light year**. The star Alpha Centauri is 4.34 light years from earth. How many years would it take a rocket travelling at 6000 kilometers per hour to reach Alpha Centauri? Solve this problem and explain how you use the meanings of multiplication and division in solving the problem.

13. Susan has a 5-pound bag of flour and an old recipe of her grandmother's calling for one kilogram of flour. She reads on the bag of flour that it weighs 2.26 kilograms. She also reads on the bag of flour that one serving of flour is about $\frac{1}{4}$ cup and that there are about 78 servings in the bag of flour.

a.   Based on the information above, how many cups of flour should Susan use in her grandmother's recipe? Solve this problem and explain how you use the meanings of multiplication and division in solving the problem.

b.   How can Susan measure this amount of flour as precisely as possible if she has the following measuring containers available: 1 cup, $\frac{1}{2}$ cup, $\frac{1}{3}$ cup, $\frac{1}{4}$ cup measures, 1 tablespoon? Remember that 1 cup = 16 tablespoons.

14. A certain experiment takes 2 days, 5 hours, and 14 minutes to perform. A lab must run this experiment 20 times. The lab is set up so that as soon as one experiment is done, the next one will start right away. How long will

it take for the 20 experiments to run? Give your answer in days, hours, and minutes. Explain why you need both multiplication and division to solve this problem.

15. You can read on the label of some soda bottles that one liter is 1 quart and 1.8 liquid ounces. Suppose gasoline sells for $1.35 per gallon. Remember that 1 gallon = 4 quarts, 1 quart = 2 pints, 1 pint = 2 cups, 1 cup = 8 liquid ounces.

    a. Mentally estimate the price of gas in dollars per liter. Explain your method.

    b. Find the exact price of gas in dollars per liter. Explain your method.

16. In 1999, February 7th and March 7th fell on the same day of the week (namely a Sunday). In 2000, February 7th and March 7th fell on different days of the week (a Monday and a Tuesday, respectively). In 2001, February 7th and March 7th fell on the same day of the week (a Wednesday). In 2002, February 7th and March 7th fell on the same day of the week (a Thursday). In 2003, February 7th and March 7th fell on the same day of the week (a Friday). Explain, using mathematics and knowledge of our calendar system, why February and March 7th fall on the same day of the week in most years, but don't fall on the same day of the week in some years.

17. November 27, 2003 was a Thursday. What day of the week is November 27, 2004? Use mathematics to solve this problem. Explain your reasoning.

18. Halloween (October 31) of 2003 was on a Friday, which was great for kids.

    a. How can you use division to determine what day of the week Halloween will be on in 2004? (The year 2004 is a leap year, so Halloween of 2004 is 366 days from Halloween of 2003.)

    b. After 2003, when are the next two times that Halloween falls on either a Friday or a Saturday? Again, use mathematics to determine this. Explain your reasoning. (The years 2004, 2008, 2012, etc. are leap years, so they have 366 days instead of 365.)

19. Must there be at least one Friday the 13th in every year? Use division to answer this question. (You may answer only for years that aren't leap years.) To get started on solving this problem, answer the following: If January 13th falls on a Monday, then what day of the week will February 13th, March 13th, etc. fall on? Use division with remainder to answer these questions. Now consider what will happen if January 13th falls on a Tuesday, a Wednesday, etc.

20. Presidents' Day is the third Monday in February. In 2003, Presidents' Day was on February 17. What is the date of Presidents' Day in 2004? Use mathematics to solve this problem without looking at a calendar. Explain your reasoning clearly.

21. I'm thinking of a number. When you divide it by 2, it has remainder 1; when you divide it by 3, it has remainder 1; when you divide by 4, 5, or 6, it always has remainder 1. The number I am thinking of could just be the number 1 (because $1 \div 2 = 0$, remainder 1; $1 \div 3 = 0$, remainder 1; $1 \div 4 = 0$, remainder 1, etc.). Find at least one other such number that is greater than 1.

22. I'm thinking of a number. When you divide it by 12, it has remainder 2; and when you divide it by 16, it also has remainder 2. The number I am thinking of could be the number 2 because $2 \div 12 = 0$, remainder 2 and $2 \div 16 = 0$, remainder 2. Find at least three other such numbers that are greater than 2. How are these numbers related?

23. Three robbers have just acquired a large pile of gold coins. They go to bed, leaving their faithful servant to guard it. In the middle of the night, the first robber gets up, gives two gold coins from the pile to the servant as hush money, divides the remaining pile of gold evenly into three parts, takes one part, and goes back to bed. A little later, the second robber gets up, gives two gold coins from the remaining pile to the servant as hush money, divides the remaining pile evenly into three parts, takes one part, and goes back to bed. A little later, the third robber gets up and does the very same thing. In the morning, when they count up the gold coins, there are 100 of them left. How many were in the pile originally? Explain your answer.

24. The United States is the second-largest country by area. (Russia is first.)

    a. Make a guess: Approximately what percent of the surface area of the earth do you think the United States covers?

    b. Now calculate the percent of the Earth's surface area that is covered by the United States. The total surface area of the Earth is about 5.1×10^8 square kilometers. The United States has an area of about 9.6×10^6 square kilometers.

25. A standard bathtub is approximately $4\frac{1}{2}$ ft long, 2 ft wide, and 1 ft deep. If water comes out of a faucet at the rate of $2\frac{1}{2}$ gallons each minute, how long will it take to fill

the bathtub $\frac{3}{4}$ full? Use the fact that 1 gallon of water occupies 0.134 cubic feet.

26. a. Use the meaning of powers of ten to show how to write the following expressions as a single power of ten (i.e., in the form $10^{\text{something}}$):

  i. $10^5 \div 10^2$ ii. $10^6 \div 10^4$ iii. $10^7 \div 10^6$

  b. In each of (i), (ii), and (iii), compare the exponents involved. In each case, what is the relationship among the exponents?

  c. Explain why it is always true that $10^A \div 10^B = 10^{A-B}$ when A and B are counting numbers and A is greater than B.

7.2 Understanding Long Division

In Section 5.9 we studied the connection between the meaning of multiplication and the standard longhand multiplication procedure. We explained why the multiplication procedure gives answers to multiplication problems that agree with the meaning of multiplication. In this section, we will do the same for division.

SOLVING DIVISION PROBLEMS WITHOUT THE STANDARD LONGHAND PROCEDURE

In order to analyze why the standard long division procedure works, we will first study how to solve division problems without the procedure and without a calculator.

In the following example, Vanessa is a student in a combined 3rd and 4th grade class ([54, p. 69]):

> *Problem 1: Jesse has 24 shirts. If he puts eight of them in each drawer, how many drawers does he use?*
> Vanessa wrote, "$24 - 8 = 16, 16 - 8 = 8, 8 - 8 = 0$," and then circled "3" for the answer.
> *Problem 2: If Jeremy needs to buy 36 cans of seltzer water for his family and they come in packs of six, how many packs should he buy?*
> This time Vanessa added, "$6 + 6 = 12, 12 + 12 = 24, 24 + 6 = 30, 30 + 6 = 36, \ldots$."

In the first problem, Vanessa starts with 24 and repeatedly *subtracts eights* until she reaches 0. In the second problem, Vanessa repeatedly *adds sixes* until she reaches 36. (The twelves are two sixes and the 24 is 4 sixes because it came from 2 sixes plus 2 sixes.) As we'll see later in this section, Vanessa's approach in the first problem is used in the standard division procedure. Vanessa's second approach is equally valid.

Consider another example, in which there are remainders. Suppose that we want to put 110 candies into packages of 8 candies each. How many packages can we make, and how many candies will be left over? To solve this problem, we must calculate $110 \div 8$, which we can do in the following way:

10 packages will use up $10 \times 8 = 80$ candies. Then there will be $110 - 80 = 30$ candies left. Three packages will use $3 \times 8 = 24$ candies. Then there will only be $30 - 24 = 6$ candies left. So we can make $10 + 3 = 13$ packages, and 6 candies will be left over.

The equations to express this reasoning are

$$110 - 10 \times 8 = 30$$

$$30 - 3 \times 8 = 6$$

These equations can be condensed to the single equation

$$110 - 10 \times 8 - 3 \times 8 = 6 \tag{7.1}$$

According to the distributive property, we can rewrite this equation as

$$110 - 13 \times 8 = 6 \tag{7.2}$$

Since 6 is less than 8,

$$110 \div 8 = 13, \text{ remainder } 6$$

To solve this problem we used the idea of repeatedly *subtracting* multiples of 8 from 110 in order to calculate $110 \div 8$. Another approach is to repeatedly *add* multiples of 8 until we get as close to 110 as possible without going over. The "repeated subtraction" reasoning is only slightly different from the "repeated addition" reasoning given in the following:

Ten packages use $10 \times 8 = 80$ candies. Three packages use $3 \times 8 = 24$ candies. So far, we have used $80 + 24 = 104$ candies in $10 + 3 = 13$ packages. Six more candies make 110, so we can make 13 packages of candies with 6 candies left over.

As before, we can write the following single equation to express this reasoning; it involves addition rather than subtraction:

$$10 \times 8 + 3 \times 8 + 6 = 110 \tag{7.3}$$

According to the distributive property, we can rewrite this equation as

$$13 \times 8 + 6 = 110 \tag{7.4}$$

This equation shows that $110 \div 8 = 13$, remainder 6.

CLASS ACTIVITY   NOW TURN TO CLASS ACTIVITIES MANUAL

7G Dividing without Using a Calculator or Long Division p. 167

STANDARD LONG DIVISION AND THE SCAFFOLD METHOD

As the example on page 256 showed, we can calculate $110 \div 8$ by repeatedly subtracting multiples of 8. Repeated subtraction is the basis of the standard long division procedure. In order to understand the standard long division procedure, we will work with a modification of it, called the **scaffold method**. The scaffold method is less efficient than the standard long division procedure, but its steps are easier to observe.

TABLE 7.1

Standard Long
Division versus
the Scaffold
Method of Long
Division

STANDARD METHOD	SCAFFOLD METHOD	
	4	← How many 7s are in 31?
	50	← How many tens of 7s are in 381?
$\underline{654}$	$\underline{600}$	← How many hundreds of 7s are in 4581?
$7\overline{)4581}$	$7\overline{)4581}$	
-42	-4200	
$\overline{38}$	$\overline{381}$	
-35	-350	
$\overline{31}$	$\overline{31}$	
-28	-28	
$\overline{3}$	$\overline{3}$	

Table 7.1 shows how to calculate $4581 \div 7$ using the standard longhand procedure and using the scaffold method. Both methods arrive at the conclusion that 4581 divided by 7 is 654 with remainder 3.

To use the scaffold method to calculate $4581 \div 7$, we begin by asking how many hundreds of sevens are in 4581. We can start with hundreds because 1000 sevens is 7000, which is already greater than 4581. There are 600 sevens in 4581 because $600 \times 7 = 4200$, but 700 sevens would be 4900, which is greater than 4581. We subtract the 600 sevens, namely 4200, from 4581, leaving 381. Now we ask how many tens of sevens are in 381. There are 50 sevens in 381 because $50 \times 7 = 350$ but $60 \times 7 = 420$ is greater than 381. We subtract the 50 sevens, namely 350, from 381, leaving 31. Finally, we ask how many sevens are in 31. There are 4, leaving 3 as a remainder. To determine the answer to $4581 \div 7$ we add the numbers at the top of the scaffold:

$$600 + 50 + 4 = 654$$

Therefore,

$$4581 \div 7 = 654 \text{ remainder } 3$$

Notice that the scaffold and standard long division methods are very similar and accomplish the same thing. The scaffold method works with entire numbers, rather than just portions of numbers (for example, 381 rather than 38). When you use the scaffold method, you have to keep track of place value. For example, in the first step, you ask, "How many *hundreds* of sevens are in 4581?" whereas, for the standard method, you only ask, "How many sevens are in 45?" The standard method is really just an abbreviated version of the scaffold method. Therefore, if we can use the meaning of division to explain why the scaffold method gives correct answers to division problems, then we will also know why the standard longhand method gives correct answers to division problems.

With the "how many groups?" interpretation of division, we can consider $4581 \div 7$ as the largest whole number of sevens in 4581, or in other words, the largest whole number that we can multiply 7 by without going over 4581. With that in mind, examine the following steps in the scaffold method:

$$
\begin{array}{r}
4 \\
50 \\
600 \\
\hline
7{\overline{\smash{\big)}\,4581}} \\
\end{array}
$$

$$
\begin{array}{rl}
7{\overline{\smash{\big)}\,4581}} & \\
-\,4200 & \leftarrow\ 600\ \text{sevens} \\
\hline
381 & \leftarrow\ \text{what is left over after subtracting } 600 \text{ sevens} \\
-\,350 & \leftarrow\ 50\ \text{sevens} \\
\hline
31 & \leftarrow\ \text{what is left over after subtracting another } 50 \text{ sevens} \\
-\,28 & \leftarrow\ 4\ \text{sevens} \\
\hline
3 & \leftarrow\ \text{what is left over after subtracting another } 4 \text{ sevens}
\end{array}
$$

In carrying out the scaffold method, we started with 4581, subtracted 600 sevens, subtracted another 50 sevens, subtracted another 4 sevens, and in the end, 3 were left over. In other words,

$$4581 - 600 \times 7 - 50 \times 7 - 4 \times 7 = 3 \tag{7.5}$$

Notice that all together, starting with 4581, we subtracted a total of 654 sevens and were left with 3. We come to the same conclusion when we first rewrite the equation

$$4581 - 600 \times 7 - 50 \times 7 - 4 \times 7 = 3$$

as

$$4581 - (600 \times 7 + 50 \times 7 + 4 \times 7) = 3$$

and then apply the distributive property to get

$$4581 - (600 + 50 + 4) \times 7 = 3$$

or

$$4581 - 654 \times 7 = 3 \tag{7.6}$$

Equation 7.6 tells us that when we take 654 sevens away from 4581, we are left with 3. Therefore we can conclude that 654 is the largest whole number of 7s in 4581, and therefore $4581 \div 7 = 654$, remainder 3.

In general, why does the scaffold method of division give correct answers to division problems, based on the meaning of division? When you solve a division problem $A \div B$ using the scaffold method, you start with the number A, and you repeatedly subtract multiples of B (numbers times B) until a number remains that is less than B. Since you subtracted as many Bs as possible, when you add the total number of Bs that were subtracted, that is the largest whole number of Bs that are in A. What's left over is the remainder. Therefore, the answer provided by using the scaffold method to calculate $A \div B$ gives the answer that we expect for $A \div B$ based on the meaning of division. Because the standard long division method is just a shorter version of the scaffold method, this also explains why the standard longhand method calculates answers to division problems that agree with the meaning of division.

CLASS ACTIVITY   NOW TURN TO CLASS ACTIVITIES MANUAL

7H Understanding the Scaffold Method of Long Division p. 169

INTERPRETING LONG DIVISION FROM THE "HOW MANY IN EACH GROUP?" VIEWPOINT

Previously, we interpreted the scaffold method of long division (and by inference also the standard method of long division) in terms of the "how many groups?" view of division. We can also interpret long division in terms of the "how many in each group?" view. In the next section, we will use the "how many in each group" view to understand how to get decimal number answers to whole number division problems.

Consider the division problem $4581 \div 7$ that we studied above. With the "how many in each group" interpretation, we can think of $4581 \div 7$ as the number of objects in each group when 4581 objects are divided equally among 7 groups. Think of dividing these 4581 objects among the 7 groups in stages. We first ask how many hundreds of objects we can put in each group, as indicated in the scaffold in Table 7.2. (We start with hundreds because if we put 1000 objects in each group we would have 7000 objects in all, which is more than 4581.) If we put 600 objects in each of the 7 groups, then we will have used up $7 \times 600 = 4200$ objects from the 4581 objects. That leaves $4581 - 4200 = 381$ objects left to distribute among the 7 groups. Now we ask how many tens of objects we can put in each of the 7 groups. If we put another 50 objects in each of the 7 groups, then we will have used up $7 \times 50 = 350$ objects, leaving $381 - 350 = 31$ objects left to be distributed among the 7 groups. Finally we ask how many individual objects we can put in each of the 7 groups. We can put another 4 objects in each group, using up $7 \times 4 = 28$ objects, and leaving $31 - 28 = 3$ objects left over. All together, each of the 7 groups got 651 objects, and 3 objects were left over.

The standard long division procedure has a nice interpretation from the "how many in each group?" viewpoint if we consider the objects that we want to divide as bundled into ones, tens, hundreds, thousands, and so on. Consider again the case of dividing 4581 objects equally among 7 groups, this time thinking of the objects as bundled into 4 thousands, 5 hundreds, 8 tens, and 1 individual object. As before, at the first step of long division we ask how many hundreds of objects we can put in each group. We can put 6 hundreds in each of the 7 groups, using 42 hundreds, as we see in the first step of the standard long division in Table 7.3. Then

---

TABLE 7.2

Using the Scaffold Method of Long Division with the "How Many in Each Group?" Viewpoint to Divide 4581 Objects Equally Among 7 Groups

4	← From 31, how many individuals can we put in each group?
50	← From 381, how many tens can we put in each group?
600	← From 4581, how many hundreds can we put in each group?
7)4581	
− 4200	← Put 600 in each group.
381	← left over after putting 600 in each group.
− 350	← Put another 50 in each group.
31	← left over after putting 50 in each group.
− 28	← Put another 4 in each group.
3	← left over after putting 4 in each group.

TABLE 7.3

Using the
Standard Method
of Long Division
with the "How
Many in Each
Group?"
Viewpoint to
Divide 4581
Objects Equally
Among 7 Groups

$$
\begin{array}{r}
654 \\
7\overline{)4581} \\
-42 \\
\hline
38 \\
-35 \\
\hline
31 \\
-28 \\
\hline
3
\end{array}
$$

← Each group gets 6 hundreds.
← 3 hundreds and 8 tens = 38 tens remain.
← Each group gets another 5 tens.
← 3 tens and 1 one = 31 ones remain.
← Each group gets another 4 ones.
← 3 ones remain.

$45 - 42 = 3$ hundreds remain. We must unbundle these 3 hundreds in order to subdivide them. Unbundled, the 3 hundreds become 30 tens. We can combine these 30 tens with the 8 tens we have in the 4581 objects. So we now have 38 tens to distribute among the 7 groups. This process of unbundling and combining corresponds to "bringing down" the 8 next to the 3 from $45 - 42$ in the long division process. From the 38 tens we now have, we can give each of the 7 groups 5 tens, leaving $38 - 35 = 3$ tens. We must unbundle these 3 tens in order to subdivide them. Unbundled, the 3 tens become 30 ones. We can combine these 30 ones with the 1 one we have in the 4581 objects. So we now have 31 ones to distribute among the 7 groups. Again, the process of unbundling and combining corresponds to "bringing down" the 1 next to the 3 from $38 - 35$ in the long division process. From the 31 ones, each group gets another 4 individual objects, leaving $31 - 28 = 3$ objects remaining. All together, each of the 7 groups got 651 objects, and 3 objects were left over, as before.

CLASS ACTIVITY   NOW TURN TO CLASS ACTIVITIES MANUAL

7I Interpreting Standard Long Division from the "How Many in Each Group?" Viewpoint p. 173

7J Zeros in Long Division p. 174

PRACTICE PROBLEMS FOR SECTION 7.2

1. Use the scaffold method to calculate $31\overline{)73125}$. Interpret the steps in the scaffold in terms of the following story problem:

   There are 73125 beads which will be put into bags with 31 beads in each bag. How many bags of beads can be made and how many beads will be left over?

2. Here is one way to calculate $239 \div 9$:

   Ten nines is 90. Another 10 nines makes 180. Five more nines makes 225. One more nine makes 234.

   Five more makes 239. All together that's 26 nines, with 5 left over. So the answer is 26, remainder 5.

   Write a single equation, in the form of Equation 7.1 or Equation 7.3, that incorporates this reasoning. Use your equation and the distributive property to write another equation, like Equation 7.2 or Equation 7.4, that shows the answer to $239 \div 9$.

3. Here is one way to calculate $2687 \div 4$:

   Five hundred fours is 2000. That leaves 687. One hundred fours is 400. That leaves 287. Another 50 fours is 200. Now 87 are left. Another 20 fours is 80. Now there are 7 left and we can get one more 4 out of that with 3 left. All together there were $500 + 100 + 50 + 20 + 1 = 671$ fours in 2687 with 3 left over, so $2687 \div 4 = 671$, remainder 3.

a. Write a scaffold that corresponds to the reasoning in problem 3.

b. Calculate $2687 \div 4$ using a scaffold with fewer steps than your scaffold for part (a).

c. Even though your scaffold in part (a) uses more steps than necessary, is it still based on sound reasoning? Explain.

4. Calculate $2950 \div 13$ without using a calculator or a long division procedure.

5. Calculate $1000 \div 27$ without using a calculator or a long division procedure.

6. Here is one way to calculate $320 \div 17$:

   20 seventeens is 340. So 19 seventeens is 17 less, which is 323. Therefore, 18 seventeens must be 306. So the answer is 18, remainder 14.

   Write a scaffold that corresponds to this reasoning.

ANSWERS TO PRACTICE PROBLEMS FOR SECTION 7.2

1.
$$
\begin{array}{r}
8 \\
50 \\
300 \\
2000 \\
\hline
31\overline{)73125} \\
-62000 \\
\hline
11125 \\
-9300 \\
\hline
1825 \\
-1550 \\
\hline
275 \\
-248 \\
\hline
27
\end{array}
$$
so $73125 \div 31 = 2358$, remainder 27.

If we make 2000 bags of beads, we will use $2000 \times 31 = 62{,}000$ beads, which will leave 11,125 beads remaining from the original 73,125. If we make another 300 bags of beads, we will use $300 \times 31 = 9300$ beads, which leaves 1825 beads remaining. If we make another 50 bags of beads, we will use 1550 beads, leaving 275 beads. Finally, if we make another 8 bags of beads, we will use 248 beads, leaving only 27 beads, which is not enough for another bag. All together, we made $2000 + 300 + 50 + 8 = 2358$ bags of beads, and 27 beads are left over.

2. $10 \times 9 + 10 \times 9 + 5 \times 9 + 1 \times 9 + 5 = 239$

So, by the distributive property,

$$(10 + 10 + 5 + 1) \times 9 + 5 = 239,$$

Therefore,

$$26 \times 9 + 5 = 239,$$

which means that

$$239 \div 9 = 26,$$

remainder 5.

3. a.
$$
\begin{array}{r}
1 \\
20 \\
50 \\
100 \\
500 \\
\hline
4\overline{)2687} \\
-2000 \\
\hline
687 \\
-400 \\
\hline
287 \\
-200 \\
\hline
87 \\
-80 \\
\hline
7 \\
-4 \\
\hline
3
\end{array}
$$
So $2687 \div 4 = 671$, remainder 3.

b.

$$
\begin{array}{r}
1 \\
70 \\
600 \\
4\overline{)2687} \\
-2400 \\
\hline
287 \\
-280 \\
\hline
7 \\
-4 \\
\hline
3
\end{array}
$$

So $2687 \div 4 = 671$, remainder 3.

c. Even though the scaffold in part (a) uses more steps than necessary, it is still based on sound reasoning. Instead of subtracting the full 600 fours from 2687, the scaffold subtracted the same number of fours in two steps instead of one: first subtracting 500 fours and then subtracting another 100 fours. If we think in terms of putting 2687 cookies into packages of 4, we are first making 500 packages and then making another 100 packages instead of making 600 packages straight away. Similarly, instead of subtracting the full 70 fours from 287, the scaffold subtracted 50 fours and then another 20 fours. All together, we are still finding the same total number of fours in 2687, just in a slightly less efficient way than in the scaffold in part (b).

4. There are many ways you could do this. Here is one way: 100 thirteens is 1300, so 200 thirteens is 2600. Taking 2600 away from 2950 leaves 350. Another 20 thirteens

is 260, leaving 90. Five thirteens make 65. Now we have 25 left. We can only get one more 13, and 12 will be left. All together we have $200 + 20 + 5 + 1 = 226$ thirteens and 12 are left, so $2950 \div 13 = 226$, remainder 12.

5. There are many ways you could do this. Here is one way: The product $10 \times 27 = 270$ and $20 \times 27 = 540$, so $30 \times 27 = 540 + 270 = 810$. Five 27s must be half of 270, which is 135. Therefore,

$$
\begin{aligned}
35 \times 27 &= (30 + 5) \times 27 \\
&= 30 \times 27 + 5 \times 27 \\
&= 810 + 135 \\
&= 945.
\end{aligned}
$$

Two 27s are 54, and adding that to 945 makes 999. Therefore, $37 \times 27 = 999$, so $37 \times 27 + 1 = 1000$, and this means that $1000 \div 27 = 37$, remainder 1.

6.

$$
\begin{array}{r}
-1 \\
-1 \\
20 \\
17\overline{)320} \\
-340 \\
\hline
-20 \\
-(-17) \\
\hline
-3 \\
-(-17) \\
\hline
14
\end{array}
$$

PROBLEMS FOR SECTION 7.2

1. a. Calculate $4215 \div 6$ and $62635 \div 32$ in two ways: with the standard division method and with the scaffold method.

   b. Compare the standard division method and the scaffold method. How are the methods alike? How are the methods different? What are advantages and disadvantages of each method?

   c. Write a single equation, like Equation 7.5, that incorporates the steps of your scaffold. Use your equation and the distributive property to write another equation, like Equation 7.6, that shows the answer to $793 \div 4$. Relate this last equation to portions in the scaffold method.

2. a. Use the scaffold method to calculate $793 \div 4$.

   b. Interpret the steps in your scaffold in terms of the following story problem: if you have 793 cookies and you want to put them in packages of 4, how many packages will there be, and how many cookies will be left over?

3. a. Use standard long division to calculate $1875 \div 8$.

   b. Interpret each step in your calculation in part (a) in terms of the following problem: You have 1875 toothpicks bundled into 1 thousand, 8 hundreds, 7 tens, and 5 individual toothpicks. If you divide these toothpicks equally among 8 groups, how many toothpicks will each group get?

4. Tamarin calculates $834 \div 25$ in the following way:

> I know that four 25s make 100, so I counted 4 for each of the 8 hundreds. This gives me 32. Then there is one more 25 in 34, but there will be 9 left. So the answer is 33, remainder 9.

   a. Explain Tamarin's method in detail and explain why her method is legitimate. (Do not just state that she gets the correct answer; explain why her method gives the correct answer.) Include equations as part of your explanation.

   b. Use Tamarin's method to calculate $781 \div 25$.

5. Felicia is working on the following problem: There are 730 balls to be put into packages of 3. How many packages can be made and how many balls will be left over? Here are Felicia's ideas:

> One hundred packages of balls will use 300 balls. After another 100 packages, we will have used up 600 balls. Another 30 packages will use another 90 balls for a total of 690 balls used. Ten more packages brings us to 720 balls used. Three more packages will bring us to 729 balls used. Then there is 1 ball left over. All together we could make $100 + 100 + 30 + 10 + 3 = 243$ packages of balls with 1 ball left over.

   Explain why the equations

   $$100 \cdot 3 + 100 \cdot 3 + 30 \cdot 3 + 10 \cdot 3 + 3 \cdot 3 + 1 = 730$$

   $$(100 + 100 + 30 + 10 + 3) \cdot 3 + 1 = 730$$

   $$243 \cdot 3 + 1 = 730$$

   correspond to Felicia's work, and explain why the last equation shows that $730 \div 3 = 243$, remainder 1.

6. Rodrigo calculates $650 \div 15$ in the following way:

$$
\begin{array}{ll}
150 \quad \leftarrow 10 & 300 \quad \leftarrow 20 \\
+\,150 \quad \leftarrow 10 & +\,300 \quad \leftarrow 20 \\
\hline
300 \quad \leftarrow 20 & 600 \quad \leftarrow 40
\end{array}
$$

$$
\begin{array}{ll}
600 \quad \leftarrow 40 & \qquad 43 \text{ R } 5 \\
+\;\;30 \quad \leftarrow \;\, 2 & \\
\hline
630 \quad \leftarrow 42 & \\
+\;\;15 & \\
\hline
645 \quad \leftarrow 43 & \\
+\;\;\;5 \quad \leftarrow \text{left} & \\
\hline
650 &
\end{array}
$$

   a. Explain why Rodrigo's method makes sense. It may help you to work with a story problem for $650 \div 15$.

   b. Write equations that correspond to Rodrigo's work and that demonstrate that $650 \div 15 = 43$, remainder 5.

7. Meili calculates $1200 \div 45$ in the following way:

$$
\begin{array}{lll}
45 & 450 \quad \leftarrow 10 & 10 + 10 + 2 + 2 + 2 \\
\times\;10 & +\,450 \quad \leftarrow 10 & \qquad\qquad = 26 \\
\hline
450 & 900 & \\
& +\;\;90 \quad \leftarrow 2 & \qquad 26 \text{ R } 30 \\
\cline{2-2}
& 990 & \\
& +\;\;90 \quad \leftarrow 2 & \\
\cline{2-2}
& 1080 & \\
& +\;\;90 \quad \leftarrow 2 & \\
\cline{2-2}
& 1170 & \\
& +\;\;30 \quad \leftarrow \text{left} & \\
\cline{2-2}
& 1200 &
\end{array}
$$

   a. Explain why Meili's strategy makes sense. It may help you to work with a story problem for $1200 \div 45$.

   b. Write equations that correspond to Meili's work and that demonstrate that $1200 \div 45 = 26$, remainder 30.

8. Use some or all of the multiplication facts

   $$2 \times 35 = 70$$
   $$10 \times 35 = 350$$
   $$20 \times 35 = 700$$

   repeatedly in order to calculate $2368 \div 35$ without using a calculator. Explain your method.

9. Use the two multiplication facts, $30 \times 12 = 360$ and $12 \times 12 = 144$, to calculate $500 \div 12$ without the use of a calculator, standard long division, or the scaffold method of division. Use both multiplication facts and explain your method. Give your answer as a whole number with a remainder. Explain your method.

10. Calculate $623 \div 8$ without using a calculator or any longhand method of division. Show your work. Then briefly describe your reasoning.

11. Calculate $2000 \div 75$ without using a calculator or any longhand method of division. Show your work. Then briefly describe your reasoning.

TABLE 7.4

A Student's
Scaffold for
$743{,}425 \div 365$

$$
\begin{array}{r}
1 \\
5 \\
10 \\
20 \\
1000 \\
\underline{1000} \\
365\,\overline{)743425} \\
\underline{-\,365000} \\
378425 \\
\underline{-\,365000} \\
13425 \\
\underline{-\,7300} \\
6125 \\
\underline{-\,3650} \\
2475 \\
\underline{-\,1825} \\
650 \\
\underline{-\,365} \\
285
\end{array}
$$

so $743425 \div 365 = 1000 + 1000 + 20 + 10 + 5 + 1$

$= 2036$ remainder 285

12. Suppose a student uses the scaffold method as shown in Table 7.4.

    a. Write another scaffold for $743{,}425 \div 365$ that uses fewer steps.

    b. Even though the student's scaffold for $743{,}425 \div 365$ uses more steps than your scaffold in part (a), is it still based on sound reasoning? Explain.

13. A student calculates $6998 \div 7$ as follows, and concludes that $6998 \div 7 = 1000 - 1$, remainder 5, which is 999, remainder 5:

$$
\begin{array}{r}
-\,1 \\
1000 \\
7\,\overline{)6998} \\
\underline{-\,7000} \\
-\,2 \\
\underline{-(-\,7)} \\
5
\end{array}
$$

Even though it is not conventional to use negative numbers in a scaffold, explain why the student's method corresponds to legitimate reasoning. Write a simple story problem for $6998 \div 7$ and use your story problem to discuss the reasoning that corresponds to the scaffold.

7.3 Fractions and Division

In Chapter 3, we stated that a fraction is equal to its numerator divided by its denominator:

$$\frac{A}{B} = A \div B$$

In fact, instead of writing the "divided by" symbol, ÷, we sometimes write $2 \div 5$ as 2/5. But the expression 2/5 also stands for the fraction $\frac{2}{5}$, so the notation we use equates fractions with division. In this section we will explain why it is legitimate to equate fractions with division. By equating fractions with division, we can explain how the remainder and mixed number answers to whole number division problems are related, and we can explain how to turn improper fractions into mixed numbers. We can also use the relationship between fractions and division to explain how to represent a fraction as a decimal. The process of turning a fraction into a decimal is the same as the process of determining the decimal answer to a whole number division problem. This process requires extending the long division procedure we studied in Section 7.2. Finally, by expressing fractions in terms of division, we can make sense of fractions that have negative numerators, negative denominators, or both negative numerators and denominators.

EXPLAINING WHY FRACTIONS CAN BE EXPRESSED IN TERMS OF DIVISION

According to our definitions of fractions and of division, the expressions

$$\frac{2}{5} \quad \text{and} \quad 2 \div 5$$

have the following different meanings:

$\frac{2}{5}$ of a pie is the amount of pie formed by 2 pieces when the pie is divided into 5 equal pieces.
$2 \div 5$ is the amount of pie each person will receive if 2 (identical) pies are divided equally among 5 people (using the "how many in each group?" interpretation).

Notice the difference: $\frac{2}{5}$ refers to *2 pieces of pie*, whereas $2 \div 5$ refers to dividing *2 pies*. But is it the same amount of pie either way? To divide 2 pies equally among 5 people you can divide each pie into 5 equal pieces and give each person one piece from each pie, as shown in Figure 7.4.

One person's share of pie consists of 2 pieces of pie, and each of those pieces is $\frac{1}{5}$ of a pie. (That is, each piece comes from a pie that has been divided into 5 equal pieces.) So each of the 5 people sharing the 2 pies gets $\frac{2}{5}$ of a pie. Thus, one person's share of pie can be described in two ways: as $2 \div 5$ of a pie and as $\frac{2}{5}$ of a pie. Therefore, both ways of describing a person's share of pie must be equal, and we have

$$2 \div 5 = \frac{2}{5}$$

FIGURE 7.4

Explaining Why $2 \div 5 = \frac{2}{5}$ by Dividing 2 Pies Equally among 5 People and Determining That One Person's Share Is $\frac{2}{5}$ of a Pie

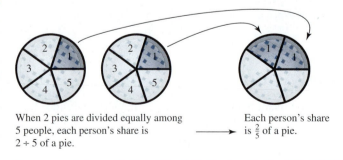

When 2 pies are divided equally among 5 people, each person's share is $2 \div 5$ of a pie.

Each person's share is $\frac{2}{5}$ of a pie.

The preceding discussion applies equally well when other whole numbers replace 2 and 5 (except that 5 should not be replaced with 0). So, in general, if A and B are whole numbers and B is not 0, then $A \div B$ really is equal to $\frac{A}{B}$.

CLASS ACTIVITY   NOW TURN TO CLASS ACTIVITIES MANUAL

7K Can Fractions Be Defined in a Different Way? p. 175

MIXED NUMBER ANSWERS TO WHOLE NUMBER DIVISION PROBLEMS

If you have a whole number division problem, how are the whole-number-with-remainder answer and the mixed number answer related? For the division problem $29 \div 6$, the remainder is 5, the divisor is 6, and the fractional part of the mixed number answer, $4\frac{5}{6}$, is $\frac{5}{6}$, which is in the form

$$\frac{\text{remainder}}{\text{divisor}}$$

This will be the case in general. Why does this make sense? The fractional part of the mixed number answer comes from dividing the remainder, as we will see in the following example:

Suppose we have 29 pizzas to be divided equally among 6 classrooms. How many pizzas does each classroom get? Since $29 \div 6 = 4$, remainder 5, we can say that each classroom will get 4 pizzas, and there will be 5 pizzas left over. Instead of just throwing away the remaining 5 pizzas, we might want to divide these 5 pizzas into 6 equal parts. When 5 pizzas are divided equally among 6 classrooms, how much pizza does each classroom get? Each classroom gets

$$5 \div 6 = \frac{5}{6}$$

of a pizza, according to the work we have just done, in which we equated fractions and division, and as we also see in Figure 7.5.

FIGURE 7.5

Dividing 5 Pizzas among 6 Classrooms

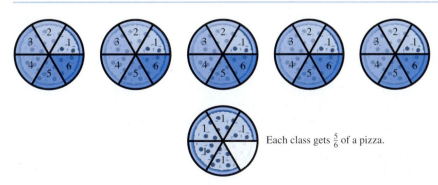

Each class gets $\frac{5}{6}$ of a pizza.

So instead of giving each class 4 pizzas and having 5 pizzas left over, we can give each class $4\frac{5}{6}$ of a pizza. In other words,

$$29 \div 6 = 4, \text{ remainder } 5$$

is equivalent to

$$29 \div 6 = 4\frac{5}{6}$$

Notice that in the mixed number version, the remainder 5 becomes the numerator of the fractional part of the answer, whereas the denominator is the divisor, 6.

The same reasoning works with other numbers. If A and B are whole numbers, then

$$A \div B = Q, \text{ remainder } R$$

is equivalent to

$$A \div B = Q\frac{R}{B}$$

In the mixed number answer to $A \div B$, the fractional part of the answer is the remainder R over the divisor, B.

USING DIVISION TO CONVERT IMPROPER FRACTIONS TO MIXED NUMBERS

In Section 4.3 we discussed how to turn mixed numbers into improper fractions by viewing a mixed number as the sum of its whole number part and its fractional part. Now we will go the other way around and turn improper fractions into mixed numbers. We do so by viewing an improper fraction in terms of division.

For example, to convert the improper fraction

$$\frac{27}{4}$$

to a mixed number, write $\frac{27}{4}$ in terms of division, and calculate the mixed number answer to the corresponding division problem:

$$\frac{27}{4} = 27 \div 4$$

$$27 \div 4 = 6 \text{ remainder } 3$$

Thus,

$$\frac{27}{4} = 27 \div 4 = 6\frac{3}{4}$$

Another way to explain why the procedure above for turning $\frac{27}{4}$ into a mixed number makes sense is to notice that $\frac{27}{4}$ stands for 27 copies of parts of an object that has been divided into 4 equal parts. Each group of 4 parts makes a whole. When we divide 27 by 4, we find that there are 6 groups of 4 and 3 left over. In terms of the fraction $\frac{27}{4}$, the 6 groups of 4 parts make 6 wholes. The 3 remaining parts make $\frac{3}{4}$ because each of the parts is one fourth.

The same reasoning works with other numbers. If A and B are whole numbers, and if

$$A \div B = Q, \text{ remainder } R$$

then

$$\frac{A}{B} = Q\frac{R}{B}$$

When we write a fraction as a mixed number, the fractional part of the mixed number is the remainder over the divisor.

CLASS ACTIVITY   NOW TURN TO CLASS ACTIVITIES MANUAL

7L Mixed Number Answers to Division Problems p. 176

USING LONG DIVISION TO CALCULATE DECIMAL ANSWERS TO WHOLE NUMBER DIVISION PROBLEMS

As we discussed at the beginning of this chapter, we can answer a whole number division problem either with a whole number and a remainder, with a mixed number, or with a decimal number. Previously, we discussed how to obtain the mixed number answer to a whole number division problem from the whole-number-with-remainder answer. To calculate the decimal answer to a whole number division problem we can use an extension of the long division procedure presented in Section 7.2. We will interpret our work to see why it makes sense to use this procedure to calculate decimal number answers to whole number division problems.

For example, how do we calculate the answer to

$$243 \div 7$$

as a decimal? We can use long division with the "how many in each group?" viewpoint, imagining, for example, that we are dividing $243 equally among 7 people:

```
      34
   7)243
   − 21    ← Each person gets 3 tens.
      33   ← 3 tens and 3 ones = 33 ones remain.
   − 28    ← Each person gets 4 ones.
       5   ← 5 ones remain.
```

Instead of stopping at this point and saying that $243 \div 7 = 34$, remainder 5, and that each person gets $34 with $5 left over, we continue to divide the remaining $5 among the 7 people. We now ask about tenths and hundredths of dollars (dimes and pennies), as indicated in Table 7.5 on page 270.

From the calculations of Table 7.5, we conclude that when we divide $243 equally among 7 people, each person gets $34.71. But there will still be 3 cents left over, which cannot be divided further because we do not have denominations smaller than one-hundredth of a dollar. In an abstract setting, or in a context where thousandths, ten-thousandths, and so on make

TABLE 7.5

Calculating the
Decimal Number
Answer to
$243 \div 7$ by
Viewing Long
Division as
Dividing $243
Equally among 7
People

$$
\begin{array}{r}
34.71 \\
7\overline{)243.000} \\
-21 \\
\hline
33 \\
-28 \\
\hline
50 \\
-49 \\
\hline
10 \\
-7 \\
\hline
3
\end{array}
$$

← Each person gets 3 tens.
← 3 tens and 3 ones = 33 ones remain.
← Each person gets 4 ones.
← 5 ones = 50 dimes (tenths) remain.
← Each person gets 7 dimes (tenths).
← 1 dime (tenth) = 10 pennies (hundredths) remain.
← Each person gets 1 penny (hundredth).
← 3 pennies (hundredths) remain.

sense, we could continue the long division process, dividing the remaining 3 hundredths to as many decimal places as we needed. We would find that

$$243 \div 7 = 34.7142\ldots$$

Notice that we obtained the digits to the right of the decimal point, $.71\ldots$, by dividing the remainder, 5, from the whole number with remainder answer to $243 \div 7$, by the divisor, 7. The decimal portion, $.71\ldots$, is another way to express $\frac{5}{7}$; in other words,

$$\frac{5}{7} = .714\ldots$$

The decimal number and mixed number answers to $243 \div 7$ are equal. They are just different ways to express the same number:

$$34\frac{5}{7} = 34.714\ldots$$

In general, for any whole number division problem, the mixed number answer and the decimal number answer are equal; they are just different ways to express the answer.

CLASS ACTIVITY   NOW TURN TO CLASS ACTIVITIES MANUAL

7M Using Long Division to Calculate Decimal Number Answers to Whole Number Division Problems p. 177

USING LONG DIVISION TO CALCULATE DECIMAL REPRESENTATIONS OF FRACTIONS

Previously, we saw how to use long division to calculate the decimal number answer to a whole number division problem. We can use this procedure to express a fraction of whole numbers as a decimal number by equating the fraction with division:

$$\frac{A}{B} = A \div B$$

For example, how can we express the fraction $\frac{5}{16}$ as a decimal? First note that

$$\frac{5}{16} = 5 \div 16$$

Then use long division to calculate $5 \div 16$ as a decimal number:

$$
\begin{array}{r}
.3125 \\
16\overline{)5.0000} \\
-48 \\
\hline
20 \\
-16 \\
\hline
40 \\
-32 \\
\hline
80 \\
-80 \\
\hline
0
\end{array}
$$

Therefore,

$$\frac{5}{16} = 0.3125$$

Similarly, we can calculate

$$\frac{1}{12} = 1 \div 12 = 0.083333333\ldots,$$

$$\frac{2}{7} = 2 \div 7 = 0.285714285714\ldots$$

The decimal representations of these last two fractions have infinitely many digits to the right of the decimal point, unlike the decimal representation of $\frac{5}{16} = 0.3125$, which only has 4 digits to the right of its decimal point.

By equating fractions with division, and by using long division, we can express any fraction of whole numbers as a decimal number.

USING DIAGRAMS TO DETERMINE DECIMAL REPRESENTATIONS OF FRACTIONS

In simple cases we can use diagrams to determine the decimal representations of fractions. These diagrams can help us better understand the relationship between decimals and fractions.

Figure 7.6 on p. 272 shows that

$$\frac{1}{4} = 0.25$$

in the following way: Consider a large square (made up of 100 small squares) as representing 1. To show $\frac{1}{4}$ of the large square we must divide the large square into 4 equal parts. One of those 4 equal parts represents $\frac{1}{4}$ of the large square. Therefore we can represent $\frac{1}{4}$ of the large square by the shaded portion of Figure 7.6, which consists of 2 vertical strips of 10 small squares, and 5 more small squares. Because each vertical strip of 10 small squares represents $\frac{1}{10}$ of the large

FIGURE 7.6

The Shaded
Area Is $\frac{1}{4}$ of the
Large Square.
Therefore,
$\frac{1}{4} = 2 \cdot \frac{1}{10} +$
$5 \cdot \frac{1}{100} = 0.25$

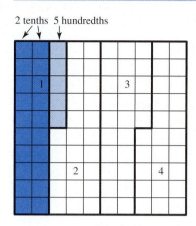

square and each small square represents $\frac{1}{100}$ of the large square, Figure 7.6 shows that

$$\frac{1}{4} = 2 \cdot \frac{1}{10} + 5 \cdot \frac{1}{100} = 0.25$$

The case of $\frac{1}{3}$ is more subtle. Once again, consider a large square made up of 100 small squares, and let the large square represent 1. To show the decimal representation of $\frac{1}{3}$ we must express $\frac{1}{3}$ of the large square in terms of strips of 10 small squares, and individual small squares. Since each vertical strip of 10 small squares represents $\frac{1}{10}$ of the large square, and since each individual small square represents $\frac{1}{100}$ of the large square, we will then have expressed $\frac{1}{3}$ in terms of tenths and hundredths, which will tell us the decimal representation of $\frac{1}{3}$ to 2 decimal places.

To divide the large square into 3 equal pieces, start by making 3 groups of 3 vertical strips of 10 squares, as in Figure 7.7. This almost divides the whole square into 3 equal pieces, except that there is a strip of 10 small squares left that must also be divided into 3 equal pieces in order to divide the whole square into 3 equal pieces. We can make 3 groups of 3 squares and

FIGURE 7.7

The Shaded
Area Is $\frac{1}{3}$ of the
Large Square.
Therefore,
$\frac{1}{3} = 3 \cdot \frac{1}{10} +$
$3 \cdot \frac{1}{100} + \ldots =$
$0.33\ldots$

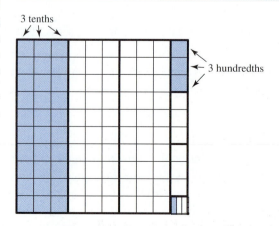

that almost divides the strip of 10 small squares into 3 equal pieces, except that there is still one small square remaining that also needs to be divided into 3 equal pieces. What does this tell us? Collecting the pieces, we see that

$$\frac{1}{3} = 3 \cdot \frac{1}{10} + 3 \cdot \frac{1}{100} + \text{a little more}$$

where the "little more" is less than $\frac{1}{100}$ (in fact it is exactly $\frac{1}{3}$ of $\frac{1}{100}$). Therefore,

$$\frac{1}{3} = 0.33\ldots$$

To determine more digits in the decimal representation of $\frac{1}{3}$ we'd need to subdivide the small remaining square into 3 equal parts, and this would lead to repeating the original procedure, as indicated by Figure 7.8 on p. 274. Therefore, the digits in the thousandths and ten-thousandths places are also 3s. If we imagine continuing this process forever, we see that all the digits in the decimal representation of $\frac{1}{3}$ are 3s:

$$\frac{1}{3} = 0.33333\ldots$$

The 3s go on forever.

CLASS ACTIVITY   NOW TURN TO CLASS ACTIVITIES MANUAL

7N Using Division to Calculate Decimal Representations of Fractions p. 178

7O Errors in Decimal Answers to Division Problems p. 181

FRACTIONS WITH NEGATIVE NUMERATORS OR DENOMINATORS

How can we make sense of fractions such as $\frac{-13}{5}$ or $\frac{13}{-5}$ or $\frac{-13}{-5}$? Earlier, we explained why, when A and B are counting numbers,

$$\frac{A}{B} = A \div B$$

We can use this relationship between fractions and division to *define* fractions that have negative numerators, negative denominators, or both. Therefore, if A and B are whole numbers and B is not zero, then

$$\frac{-A}{B} = (-A) \div B = -(A \div B) = -\frac{A}{B}$$

$$\frac{A}{-B} = A \div (-B) = -(A \div B) = -\frac{A}{B}$$

$$\frac{-A}{-B} = (-A) \div (-B) = A \div B = \frac{A}{B}$$

FIGURE 7.8

Showing the
Repeating
Nature of the
Decimal
Representation
of $\frac{1}{3} =$
$0.333333\ldots$

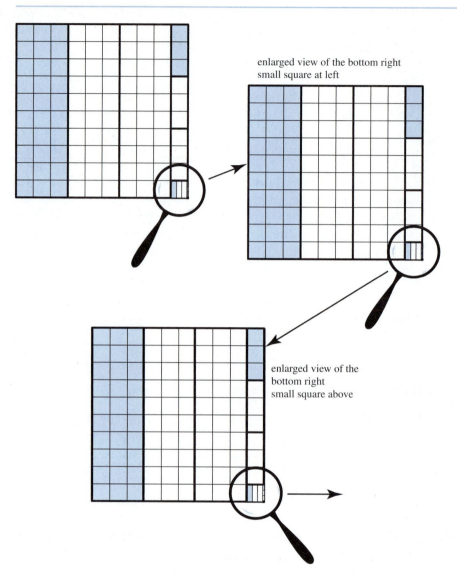

enlarged view of the bottom right
small square at left

enlarged view of the
bottom right
small square above

So,

$$\frac{-13}{5} = -\frac{13}{5}$$

$$\frac{13}{-5} = -\frac{13}{5}$$

$$\frac{-13}{-5} = \frac{13}{5}$$

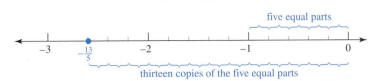

FIGURE 7.9

A Number Line
Showing
$\frac{-13}{5} = -\frac{13}{5}$

Figure 7.9 shows $\frac{-13}{5}$ plotted on a number line. According to the rules of Section 2.3 for locating numbers on number lines, because

$$\frac{-13}{5} = -\frac{13}{5}$$

$\frac{-13}{5}$ should be located to the left of 0, at a distance of $\frac{13}{5}$ units away from 0.

PRACTICE PROBLEMS FOR SECTION 7.3

1. Use the meaning of fractions and the meaning of division to explain why $\frac{3}{10} = 3 \div 10$. Your explanation should be general, in the sense that you could see why $\frac{3}{10} = 3 \div 10$ would still be true if other numbers were to replace 3 and 10.

2. Describe how the whole number with remainder and mixed number answers to $14 \div 3$ are related. Use a simple story problem to help you explain this relationship.

3. Show how to use long division (either the scaffold method or the standard method) to calculate the decimal number answer to $20 \div 11$ to the hundredths place. Interpret each step in the scaffold in terms of dividing $20 equally among 11 people.

4. Show how to use long division to write the fraction $\frac{5}{8}$ as a decimal number.

5. Use the large square in Figure 7.10 (which is subdivided into 100 small squares) to help you explain why the decimal representation of $\frac{1}{8}$ is 0.125.

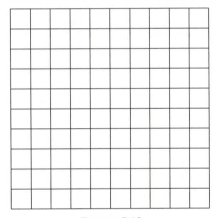

FIGURE 7.10

A Square Subdivided into 100 Smaller Squares

6. Plot $\frac{-11}{6}$ on a number line and explain why it should be plotted there.

ANSWERS TO PRACTICE PROBLEMS FOR SECTION 7.3

1. If there are 3 (identical) pies to be divided equally among 10 people, then according to the meaning of division (with the "how many in each group?" interpretation) each person will get $3 \div 10$ of a pie. To divide the pies, you can divide each pie into 10 pieces and give each person 1 piece from each of the 3 pies. One person's share is shown shaded in Figure 7.11. Each person then gets 3 pieces, where each piece is $\frac{1}{10}$ of a pie. That is, each piece is 1 part when the pie is divided into 10 equal parts. Therefore, each person gets $\frac{3}{10}$ of a pie

according to the meaning of fractions. Because each person's share can be described both as $3 \div 10$ of a pie and as $\frac{3}{10}$ of a pie, it follows that $\frac{3}{10} = 3 \div 10$.

FIGURE 7.11

The Shaded Portion Is One Person's
Share When 3 Pies Are Divided
Equally among 10 People

2. The whole-number-with-remainder answer to $14 \div 3$ is 4, remainder 2. The mixed number answer is $4\frac{2}{3}$. In general, to get the fractional part of the mixed number answer you make a fraction whose numerator is the remainder and whose denominator is the divisor. To see why this makes sense, consider an example. If you have 14 cookies to be divided equally among 3 people, then you could give each person 4 cookies and have 2 cookies left over, or you could take those 2 remaining cookies and divide them equally among the 3 people. You could do this by dividing each cookie into 3 equal parts and giving each person 2 parts (as you could show in a picture like Figure 7.5). In this way, each person would get $4\frac{2}{3}$ cookies.

3. Using the standard algorithm, we find that $20 \div 11 = 1.81\ldots$:

$$
\begin{array}{r}
1.81 \\
11\overline{)20.00} \\
\end{array}
$$

$-\ 11$ ← Each person gets \$1.
$\quad 90$ ← \$9 = 90 dimes remain.
$-\ 88$ ← Each person gets 8 dimes.
$\quad 20$ ← 2 dimes = 20 pennies remain.
$-\ 11$ ← Each person gets 1 penny.
$\quad 9$ ← 9 pennies remain.

If we think in terms of dividing \$20 equally among 11 people, then at the first step we ask how many ones we should give each person. Each of the 11 people will get \$1, leaving $\$20 - \$11 = \$9$ remaining. If we trade each dollar for 10 dimes, we will have 90 dimes. We can now give each of the 11 people 8 dimes, using a total of 88 dimes, leaving 2 dimes left. If we trade each dime for 10 pennies, we will have 20 pennies. Each person gets 1 penny, leaving $20 - 11 = 9$ pennies. Therefore, each person gets \$1.81, and $20 \div 11 = 1.81\ldots$. The decimal representation of $20 \div 11$ continues to have digits in the thousandths place, the ten-thousandths place, and so

on, but we can't use our money interpretation for these places because we don't have denominations smaller than 1 penny.

4. The following long division calculation shows that $\frac{5}{8} = 0.625$:

$$
\begin{array}{r}
.625 \\
8\overline{)5.000} \\
-\ 48 \\
\hline
20 \\
-\ 16 \\
\hline
40 \\
-\ 40 \\
\hline
0 \\
\end{array}
$$

5. By thinking of the large square as representing 1, each small square then represents $\frac{1}{100}$. To divide the large square into 8 equal pieces, first make 8 strips of 10 small squares, as indicated in Figure 7.12. Each of the 8 equal pieces into which we want to divide the large square gets one of these strips. Then there are still 20 small squares left to be divided into 8 equal pieces. You can make 8 groups of 2 small squares from these 20 small squares. Each of the 8 equal pieces gets 2 of these small squares. That leaves 4 small squares left to be divided into 8 equal pieces. Dividing each small square in half makes 8 half-squares. All together, we obtain $\frac{1}{8}$ of the big square by collecting these various parts: one strip of 10 small squares, 2 small squares, and one half of a small square, as shown in Figure 7.12. Notice that since 1 small square is $\frac{1}{100}$ of the large square, $\frac{1}{2}$ of a small square is $\frac{1}{200} = \frac{5}{1000}$ of the large square. Therefore,

$$
\begin{aligned}
\tfrac{1}{8} &= 1 \cdot \tfrac{1}{10} + 2 \cdot \tfrac{1}{100} + 5 \cdot \tfrac{1}{1000} \\
&= 0.125
\end{aligned}
$$

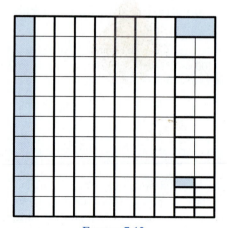

FIGURE 7.12

Showing $\frac{1}{8} = 0.125$

6. Because

$$\frac{-11}{6} = -11 \div 6 = -(11 \div 6) = -\frac{11}{6}$$

we should plot $\frac{-11}{6}$ to the left of 0, at a distance of $\frac{11}{6}$ units from 0, as shown in Figure 7.13.

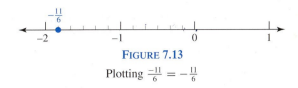

FIGURE 7.13

Plotting $\frac{-11}{6} = -\frac{11}{6}$

PROBLEMS FOR SECTION 7.3

1. Use the meaning of fractions and the meaning of division to explain in your own words why $\frac{3}{7} = 3 \div 7$. Your explanation should be general, in the sense that you could see why $\frac{3}{7} = 3 \div 7$ would still be true if other numbers were to replace 3 and 7.

2. Describe how to get the mixed number answer to $23 \div 5$ from the whole-number-with-remainder answer. Explain why your method makes sense by interpreting it in terms of a simple story problem.

3. In your own words, describe a procedure for turning an improper fraction, such as $\frac{19}{4}$, into a mixed number and explain why this procedure makes sense.

4. a. Describe how the whole-number-with-remainder and mixed number answers to $23 \div 6$ are related. Use a simple story problem to help you explain this relationship.

   b. Describe how the whole-number-with-remainder and the decimal number answer to $23 \div 6$ are related. Use a simple story problem to help you explain this relationship.

5. a. Use long division to determine the decimal number answer to $2893 \div 6$ to the hundredths place.

   b. Interpret each step in your long division calculation in part (a) in terms of dividing $2893 equally among 6 people.

6. Show how to use long division to determine the decimal representation of $\frac{1}{37}$.

7. Describe how to use either dimes and pennies or a diagram to determine the tenths and hundredths places in the decimal representation of $\frac{1}{11}$. Explain your reasoning.

8. Describe how to use either dimes and pennies or a diagram to determine the tenths and hundredths places in the decimal representation of $\frac{1}{9}$. Explain your reasoning.

9. a. Use long division to determine the decimal representation of $\frac{1}{9}$ to the ten-thousandths place.

   b. Interpret the steps to the hundredths place in part (a) in terms of dividing $1 equally among 9 people.

10. Describe how to use either dimes and pennies or a diagram to determine the tenths and hundredths places in the decimal representation of $\frac{1}{7}$. Explain your reasoning.

11. a. Use long division to determine the decimal representation of $\frac{1}{7}$ to 7 decimal places.

    b. Interpret the steps to the hundredths place in part (a) in terms of dividing $1 equally among 7 people.

12. Plot $\frac{-42}{3}$ and at least two other nearby integers on a number line and explain why you have plotted these numbers appropriately.

13. Plot $\frac{-58}{7}$ and at least two other nearby integers on a number line and explain why you have plotted these numbers appropriately.

14. Jessica calculates that $7 \div 3 = 2$, remainder 1. When Jessica is asked to write her answer as a decimal number, she simply puts the remainder 1 behind the decimal point:

$$7 \div 3 = 2.1$$

Is Jessica correct or not? If not, explain why not and explain to Jessica in a concrete way why the correct answer makes sense.

15. When you divide whole numbers using an ordinary calculator, the answer is displayed as a decimal. But what if you want the answer as a whole number with a remainder? Here's a method to determine the remainder with an ordinary calculator, illustrated with the example of $236 \div 7$.

    Use the calculator to divide the whole numbers:

$$236 \div 7 = 33.71428\ldots$$

Subtract the whole number part of the answer:

$$33.71428\ldots - 33 = .71428\ldots$$

Multiply the resulting decimal by the divisor (which was 7):

$$.71428\ldots \times 7 = 4.99999$$

Round the resulting number to the nearest whole number. This is your remainder. In this example the remainder is 5. So $236 \div 7 = 33$, remainder 5.

a. Using a calculator, solve at least two more division problems with whole numbers, and use the above method to find the answer as a whole number with a remainder. Check that your answers are correct.

b. Now solve the same division problems you did in part (a), except this time, give your answers as mixed numbers instead of as decimals.

c. Use the mixed number version to help you explain why the calculator method for determining the remainder works.

16. A year that is not a leap year has 365 days. (Leap years have 366 days and generally occur every four years.)

There are 7 days in a week and 52 whole weeks in a year. How many whole weeks are there in three years? How many whole weeks are there in seven years? Is the number of whole weeks in three years three times the number of whole weeks in one year? Is the number of whole weeks in seven years seven times the number of whole weeks in one year? Explain the discrepancy!

17. The text presented one way to explain why $\frac{A}{B} = A \div B$. This problem will help you explain in a different way why we can express fractions in terms of division.

a. Use the meaning of multiplication, the meaning of fractions, and the meaning of division to explain why

$$\frac{1}{10} \cdot 3 = 3 \div 10$$

b. Use the equation in part (a), the commutative property of multiplication, and the meaning of fractions to explain why

$$3 \div 10 = \frac{3}{10}$$

7.4 Dividing Fractions

In this section, we will discuss the two interpretations of division for fractions, and we will see why the standard "invert and multiply" procedure for dividing fractions gives answers to fraction division problems that agree with what we expect from the meaning of division.

THE TWO INTERPRETATIONS OF DIVISION FOR FRACTIONS

Let's review the meaning of division for whole numbers and see how to interpret division for fractions.

THE "HOW MANY GROUPS?" INTERPRETATION

With the "how many groups?" interpretation of division, $12 \div 3$ means the number of groups we can make when we divide 12 objects into groups with 3 objects in each group. In other words, $12 \div 3$ tells us how many groups of 3 we can make from 12.

Similarly, with the "how many groups?" interpretation of division,

$$\frac{5}{2} \div \frac{2}{3}$$

means the number of groups we can make when we divide $\frac{5}{2}$ of an object into groups with $\frac{2}{3}$ of an object in each group. In other words, $\frac{5}{2} \div \frac{2}{3}$ tells us how many groups of $\frac{2}{3}$ we can make from $\frac{5}{2}$. For example, suppose you are making popcorn balls and each popcorn ball requires

$\frac{2}{3}$ cup of popcorn. If you have $2\frac{1}{2} = \frac{5}{2}$ cup of popcorn, then how many popcorn balls can you make? In this case you want to divide $\frac{5}{2}$ cup of popcorn into groups (balls) with $\frac{2}{3}$ cup of popcorn in each group. According to the "how many groups?" interpretation of division, you can make

$$\frac{5}{2} \div \frac{2}{3}$$

popcorn balls.

THE "HOW MANY IN ONE (EACH) GROUP?" INTERPRETATION

With the "how many in each group?" interpretation of division, $12 \div 3$ means the number of objects in each group when we distribute 12 objects equally among 3 groups. In other words, $12 \div 3$ is the number of objects in one group if we use 12 objects to evenly fill 3 groups. When we work with fractions, it often helps to think of "how many in each group?" division story problems as asking "how many are in *one whole* group?", and it helps to think of *filling* groups or part of a group. So in the context of fractions, we will usually refer to the "how many in each group?" interpretation as "how many in one group?"

With the "how many in one group?" interpretation of division,

$$\frac{3}{4} \div \frac{1}{2}$$

is the number of objects in one group when we distribute $\frac{3}{4}$ of an object equally among $\frac{1}{2}$ of a group. A clearer way to say this is "$\frac{3}{4} \div \frac{1}{2}$ is the number of objects (or fraction of an object) in one whole group when $\frac{3}{4}$ of an object fills $\frac{1}{2}$ of a group." For example, suppose you pour $\frac{3}{4}$ pint of blueberries into a container and this fills $\frac{1}{2}$ of the container. How many pints of blueberries will it take to fill the whole container? In this case, $\frac{3}{4}$ pint of blueberries fills (i.e., is distributed equally among) $\frac{1}{2}$ of a group (a container). So, according to the "how many in one group?" interpretation of division, the number of pints of blueberries in one whole group (one full container) is

$$\frac{3}{4} \div \frac{1}{2}$$

One way to better understand fraction division story problems is to think about replacing the fractions in the problem with whole numbers. For example, if you have 3 pints of blueberries and they fill 2 containers, then how many pints of blueberries are in each container? We solve this problem by dividing $3 \div 2$, according to the "how many in each group?" interpretation. If we replace the 3 pints with $\frac{3}{4}$ pint and the 2 containers with $\frac{1}{2}$ container, we solve the problem in the same way as before: $3 \div 2$ now becomes $\frac{3}{4} \div \frac{1}{2}$.

Here is another way to think about the problem: Because $\frac{1}{2}$ of the container is filled, and because this amount is $\frac{3}{4}$ pint, $\frac{1}{2}$ of the number of pints in a full container is $\frac{3}{4}$ pint. In other words,

$$\frac{1}{2} \times \text{ number of pints in full container} = \frac{3}{4}$$

Therefore,

$$\text{number of pints in full container} = \frac{3}{4} \div \frac{1}{2}$$

DIVIDING BY $\frac{1}{2}$ VERSUS DIVIDING IN $\frac{1}{2}$

In mathematics, language is used much more precisely and carefully than in everyday conversation. This is one source of difficulty in learning mathematics. For example, consider the following two phrases:

dividing *by* $\frac{1}{2}$

dividing *in* $\frac{1}{2}$

You may feel that these two phrases mean the same thing; however, mathematically, they do not. To divide a number, say 5, by $\frac{1}{2}$ means to calculate $5 \div \frac{1}{2}$. Remember that we read $A \div B$ as A divided by B. We would divide 5 by $\frac{1}{2}$ if we wanted to know how many half cups of flour are in 5 cups of flour, for example. (Notice that there are 10 half cups of flour in 5 cups of flour, not $2\frac{1}{2}$.)

On the other hand, to divide a number *in* half means to find half of that number. So to divide 5 in half means to find $\frac{1}{2}$ of 5. One half of 5 means $\frac{1}{2} \times 5$. So dividing in $\frac{1}{2}$ is the same as dividing by 2.

THE "INVERT AND MULTIPLY" PROCEDURE FOR FRACTION DIVISION

Although division with fractions can be difficult to interpret, the procedure for dividing fractions is quite easy. To divide fractions, such as

$$\frac{3}{4} \div \frac{2}{3} \quad \text{and} \quad 6 \div \frac{2}{5}$$

we can use the familiar "invert and multiply" method in which we invert the divisor and multiply by it, as in

$$\frac{3}{4} \div \frac{2}{3} = \frac{3}{4} \cdot \frac{3}{2} = \frac{3 \cdot 3}{4 \cdot 2} = \frac{9}{8} = 1\frac{1}{8}$$

and

$$6 \div \frac{2}{5} = \frac{6}{1} \div \frac{2}{5} = \frac{6}{1} \cdot \frac{5}{2} = \frac{6 \cdot 5}{1 \cdot 2} = \frac{30}{2} = 15$$

reciprocal

Another way to describe this "invert and multiply" method for dividing fractions is in terms of the reciprocal of the divisor. The **reciprocal** of a fraction $\frac{C}{D}$ is the fraction $\frac{D}{C}$. In order to divide fractions, we should multiply by the reciprocal of the divisor. So, in general,

$$\frac{A}{B} \div \frac{C}{D} = \frac{A}{B} \cdot \frac{D}{C} = \frac{A \cdot D}{B \cdot C}$$

EXPLAINING WHY "INVERT AND MULTIPLY" IS VALID BY RELATING DIVISION TO MULTIPLICATION

The procedure for dividing fractions is easy enough to carry out, but why is it a valid method? Before we give a general answer to this question, consider a special case. Recall that every whole number is equal to a fraction (for example, $6 = \frac{6}{1}$). Therefore we can apply the "invert and multiply" procedure to whole numbers as well as to fractions. According to this procedure,

$$2 \div 3 = \frac{2}{1} \div \frac{3}{1} = \frac{2}{1} \cdot \frac{1}{3} = \frac{2 \cdot 1}{1 \cdot 3} = \frac{2}{3}$$

Notice that this result, $2 \div 3 = \frac{2}{3}$, agrees with our finding earlier in this chapter that we can describe fractions in terms of division, namely, $\frac{A}{B} = A \div B$.

In general, why is the "invert and multiply" procedure a valid way to divide fractions? One way to explain this is to relate fraction division to fraction multiplication. Recall that every division problem is equivalent to a multiplication problem (actually two multiplication problems). Thus,

$$A \div B = ?$$

is equivalent to

$$? \cdot B = A$$

(or $B \cdot ? = A$). So

$$\frac{3}{4} \div \frac{2}{3} = ?$$

is equivalent to

$$? \cdot \frac{2}{3} = \frac{3}{4} \tag{7.7}$$

Now remember that we want to explain why the "invert and multiply" rule for fraction division is valid. This rule says that $\frac{3}{4} \div \frac{2}{3}$ ought to be equal to

$$\frac{3 \cdot 3}{4 \cdot 2}$$

Let's verify that this fraction works in the place of the ? in Equation 7.7. In other words, let's confirm that if we multiply $\frac{3 \cdot 3}{4 \cdot 2}$ times $\frac{2}{3}$, then we really do get $\frac{3}{4}$:

$$\frac{3 \cdot 3}{4 \cdot 2} \cdot \frac{2}{3} = \frac{3 \cdot 3 \cdot 2}{4 \cdot 2 \cdot 3} = \frac{3 \cdot (3 \cdot 2)}{4 \cdot (2 \cdot 3)} = \frac{3 \cdot (3 \cdot 2)}{4 \cdot (3 \cdot 2)} = \frac{3}{4}$$

Therefore, the answer we get from the "invert and multiply" procedure really is the answer to the original division problem, $\frac{3}{4} \div \frac{2}{3}$. Notice that the preceding line of reasoning applies in the same way when other fractions replace the fractions $\frac{2}{3}$ and $\frac{3}{4}$.

It will still be valuable to explore fraction division further, interpreting fraction division directly rather than through multiplication.

CLASS ACTIVITY   NOW TURN TO CLASS ACTIVITIES MANUAL

7P Explaining "Invert and Multiply" by Relating Division to Multiplication p. 182

USING THE "HOW MANY GROUPS?" INTERPRETATION TO EXPLAIN WHY "INVERT AND MULTIPLY" IS VALID

In the previous section, we explained why the "invert and multiply" procedure for dividing fractions is valid by considering fraction division in terms of fraction multiplication. Now we will explain why the "invert and multiply" procedure is valid by working with the "how many groups?" interpretation of division.

Consider the division problem

$$\frac{2}{3} \div \frac{1}{2}$$

The following is a story problem for this division problem:

How many $\frac{1}{2}$ cups of water are in $\frac{2}{3}$ cup of water?

Or, said another way,

How many times will we need to pour $\frac{1}{2}$ cup of water into a container that holds $\frac{2}{3}$ cup of water in order to fill the container?

From the diagram in Figure 7.14 we can say right away that the answer to this problem is "one and a little more" because one half cup clearly fits in two thirds of a cup, but then a little more is still needed to fill the two thirds of a cup. But what is this "little more"? Remember the original question: How many $\frac{1}{2}$ cups of water are in $\frac{2}{3}$ cup of water? The answer should be of the form "so and so many $\frac{1}{2}$ cups of water." This means that we need to express this "little more" as *a fraction of $\frac{1}{2}$ cup of water.* How can we do that? By subdividing both the $\frac{1}{2}$ and the $\frac{2}{3}$ into common parts, namely by using common denominators.

When we give $\frac{1}{2}$ and $\frac{2}{3}$ the common denominator of 6, then, as on the right of Figure 7.14, the $\frac{1}{2}$ cup of water is made out of 3 parts (3 sixths cup of water), and the $\frac{2}{3}$ cup of water is made out of 4 parts (4 sixths cup of water), so the "little more" we were discussing in the previous paragraph is just one of those parts. Since $\frac{1}{2}$ cup is 3 parts, and the "little more" is 1 part, the "little more" is $\frac{1}{3}$ of the $\frac{1}{2}$ cup of water. This explains why $\frac{2}{3} \div \frac{1}{2} = 1\frac{1}{3}$: There's an entire $\frac{1}{2}$ cup plus another $\frac{1}{3}$ of the $\frac{1}{2}$ cup in $\frac{2}{3}$ cup of water.

To summarize, we are considering the fraction division problem $\frac{2}{3} \div \frac{1}{2}$ in terms of the story problem "how many $\frac{1}{2}$ cups of water are in $\frac{2}{3}$ cup of water?" If we give $\frac{1}{2}$ and $\frac{2}{3}$ the common denominator of 6, then we can rephrase the problem as "how many $\frac{3}{6}$ cup are in $\frac{4}{6}$ cup?" But in terms of Figure 7.14, this is equivalent to the problem "how many 3s are in 4?" which is the problem $4 \div 3 = ?$, whose answer is $\frac{4}{3} = 1\frac{1}{3}$. Notice that $\frac{4}{3}$ is exactly the same

FIGURE 7.14

How Many 1/2 Cups of Water Are in 2/3 Cup?

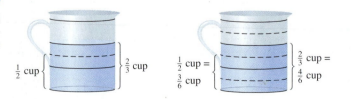

answer we get from the "invert and multiply" procedure for fraction division:

$$\frac{2}{3} \div \frac{1}{2} = \frac{2}{3} \cdot \frac{2}{1} = \frac{2 \cdot 2}{3 \cdot 1} = \frac{4}{3}$$

So the "invert and multiply" procedure gives the same answer to $\frac{2}{3} \div \frac{1}{2}$ that we get by using the "how many groups?" interpretation of division.

The same line of reasoning will work for any fraction division problem of the form

$$\frac{A}{B} \div \frac{C}{D}$$

Thinking logically, as before, and interpreting $\frac{A}{B} \div \frac{C}{D}$ as "how many $\frac{C}{D}$ cups of water are in $\frac{A}{B}$ cups of water?", we can conclude that

$$\frac{A}{B} \div \frac{C}{D} = \frac{A \cdot D}{B \cdot D} \div \frac{B \cdot C}{B \cdot D} = (A \cdot D) \div (B \cdot C) = \frac{A \cdot D}{B \cdot C}$$

The final expression, $\frac{A \cdot D}{B \cdot C}$, is the answer provided by the "invert and multiply" procedure for dividing fractions. Therefore, we know that the "invert and multiply" procedure gives answers to division problems that agree with what we expect from the meaning of division.

CLASS ACTIVITY   NOW TURN TO CLASS ACTIVITIES MANUAL

7Q "How Many Groups?" Fraction Division Problems p. 183

USING THE "HOW MANY IN ONE GROUP?" INTERPRETATION TO EXPLAIN WHY "INVERT AND MULTIPLY" IS VALID

Previously, we saw how to use the "how many groups?" interpretation of division to explain why the "invert and multiply" procedure for fraction division is valid. We can also use the "how many in one group?" interpretation for the same purpose. This interpretation, although perhaps more difficult to understand, has the advantage of showing us directly why we can multiply by the reciprocal of the divisor in order to divide fractions.

Consider the following "how many in one group?" story problem for $\frac{1}{2} \div \frac{3}{5}$:

You used $\frac{1}{2}$ can of paint to paint $\frac{3}{5}$ of a wall. How many cans of paint will it take to paint the whole wall?

This is a "how many in one group?" problem because we can think of the paint as "filling" $\frac{3}{5}$ of the wall. We can also see that this is a division problem by writing the corresponding number sentence:

$$\frac{3}{5} \cdot \text{(amount to paint the whole wall)} = \frac{1}{2}$$

Therefore,

$$\text{amount to paint the whole wall} = \frac{1}{2} \div \frac{3}{5}$$

FIGURE 7.15

The Amount of
Paint Needed
for the Whole
Wall is $\frac{5}{3}$ of the
$\frac{1}{2}$ Can Used to
Cover $\frac{3}{5}$ of the
Wall

The $\frac{1}{2}$ can of paint is divided equally among 3 parts.

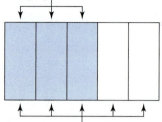

The amount of paint for the full wall is 5 times the amount in one part.

We will now see why it makes sense to solve this problem by multiplying $\frac{1}{2}$ by the reciprocal of $\frac{3}{5}$, namely, by $\frac{5}{3}$. Let's focus on the wall to be painted, as shown in Figure 7.15. Think of dividing the wall into 5 equal sections, 3 of which you painted with the $\frac{1}{2}$ can of paint. If you used $\frac{1}{2}$ can of paint to paint 3 sections, then each of the 3 sections required $\frac{1}{2} \div 3$ or $\frac{1}{2} \times \frac{1}{3}$ cans of paint. To determine how much paint you will need for the whole wall, multiply the amount you need for one section by 5. So you can determine the amount of paint you need for the whole wall by multiplying the $\frac{1}{2}$ can of paint by $\frac{1}{3}$ and then multiplying that result by 5, as summarized in Table 7.6. But to multiply a number by $\frac{1}{3}$ and then multiply it by 5 is the same as multiplying the number by $\frac{5}{3}$. Therefore, we can determine the number of cans of paint you need for the whole wall by multiplying $\frac{1}{2}$ by $\frac{5}{3}$:

$$\frac{1}{2} \cdot \frac{5}{3} = \frac{5}{6}$$

This is exactly the "invert and multiply" procedure for dividing $\frac{1}{2} \div \frac{3}{5}$. It shows that you will need $\frac{5}{6}$ can of paint for the whole wall.

TABLE 7.6

Determining How
Much Paint to
Use for a Whole
Wall if $\frac{1}{2}$ Can of
Paint Covers $\frac{3}{5}$ of
the Wall

Use	$\frac{1}{2}$ can paint	for	$\frac{3}{5}$ of the wall.
	$\downarrow \div 3$ or $\times \frac{1}{3}$		$\downarrow \div 3$ or $\times \frac{1}{3}$
Use	$\frac{1}{6}$ can paint	for	$\frac{1}{5}$ of the wall.
	$\downarrow \times 5$		$\downarrow \times 5$
Use	$\frac{5}{6}$ can paint	for	1 whole wall.
		in one step:	
Use	$\frac{1}{2}$ can paint	for	$\frac{3}{5}$ of the wall.
	$\downarrow \times \frac{5}{3}$		$\downarrow \times \frac{5}{3}$
Use	$\frac{5}{6}$ can paint	for	1 whole wall.

The preceding argument works when other fractions replace $\frac{1}{2}$ and $\frac{3}{5}$, thereby explaining why

$$\frac{A}{B} \div \frac{C}{D} = \frac{A}{B} \cdot \frac{D}{C}$$

In other words, to divide fractions, multiply the dividend by the reciprocal of the divisor.

CLASS ACTIVITY NOW TURN TO CLASS ACTIVITIES MANUAL

7R "How Many in One Group?" Fraction Division Problems p. 187

7S Are These Division Problems? p. 190

PRACTICE PROBLEMS FOR SECTION 7.4

1. Write a "how many groups?" story problem for $1 \div \frac{5}{7}$. Use the story problem and a diagram to help you solve the problem.

2. Write a "how many in one group?" story problem for $1 \div \frac{3}{4}$. Use the situation of the story problem to help you explain why the answer is $\frac{4}{3} = 1\frac{1}{3}$.

3. Annie wants to solve the division problem $\frac{3}{4} \div \frac{1}{2}$ by using the following story problem:

   I need $\frac{1}{2}$ cup of chocolate chips to make a batch of cookies. How many batches of cookies can I make with $\frac{3}{4}$ of a cup of chocolate chips?

   Annie draws a diagram like the one in Figure 7.16. Explain why it would be easy for Annie to misinterpret her diagram as showing that $\frac{3}{4} \div \frac{1}{2} = 1\frac{1}{4}$. How should Annie interpret her diagram so as to conclude that $\frac{3}{4} \div \frac{1}{2} = 1\frac{1}{2}$?

FIGURE 7.16

How Many Batches of Cookies Can We Make with $\frac{3}{4}$ Cup of Chocolate Chips if 1 Batch Requires $\frac{1}{2}$ Cup of Chocolate Chips?

4. Which of the following are solved by the division problem $\frac{3}{4} \div \frac{1}{2}$? For those that are, which interpretation of division is used? For those which are not, determine how to solve the problem, if it can be solved.

   a. $\frac{3}{4}$ of a bag of jelly worms make $\frac{1}{2}$ cup. How many cups of jelly worms are in one bag?

   b. $\frac{3}{4}$ of a bag of jelly worms make $\frac{1}{2}$ cup. How many bags of jelly worms does it take to make one cup?

   c. You have $\frac{3}{4}$ of a bag of jelly worms and a recipe that calls for $\frac{1}{2}$ cup of jelly worms. How many batches of your recipe can you make?

   d. You have $\frac{3}{4}$ cup of jelly worms and a recipe that calls for $\frac{1}{2}$ cup of jelly worms. How many batches of your recipe can you make?

   e. If $\frac{3}{4}$ pound of candy costs $\frac{1}{2}$ of a dollar, then how many pounds of candy should you be able to buy for 1 dollar?

   f. If you have $\frac{3}{4}$ pound of candy and you divide the candy in $\frac{1}{2}$, then how much candy will you have in each portion?

   g. If $\frac{1}{2}$ pound of candy costs \$1, then how many dollars should you expect to pay for $\frac{3}{4}$ pound of candy?

5. Frank, John, and David earned \$14 together. They want to divide it equally, except that David should only get a half share, since he did half as much work as either Frank or John did (and Frank and John worked equal amounts). Write a division problem to find out how much Frank should get. Which interpretation of division does this story problem use?

6. Bill leaves a tip of $4.50 for a meal. If the tip is 15% of the cost of the meal, then how much did the meal cost? Write a division problem to solve this. Which interpretation of division does this story problem use?

7. Compare the arithmetic needed to solve the following problems:

   a. What fraction of a $\frac{1}{3}$ cup measure is filled when we pour in $\frac{1}{4}$ cup of water?

   b. What is one quarter of $\frac{1}{3}$ cup?

   c. How much more is $\frac{1}{3}$ cup than $\frac{1}{4}$ cup?

   d. If $\frac{1}{4}$ cup of water fills $\frac{1}{3}$ of a plastic container, then how much water will the full container hold?

8. Use the meanings of multiplication and division to solve the following problems:

   a. Suppose you drive 4500 miles every half year in your car. At the end of $3\frac{3}{4}$ years, how many miles will you have driven?

   b. Mo used 128 ounces of liquid laundry detergent in $6\frac{1}{2}$ weeks. If Mo continues to use laundry detergent at this rate, how much will he use in a year?

   c. Suppose you have a 32 ounce bottle of weed killer concentrate. The directions say to mix two and a half ounces of weed killer concentrate with enough water to make a gallon. How many gallons of weed killer will you be able to make from this bottle?

9. The line segment pictured is $\frac{2}{3}$ of a unit long. Show a line segment that is $\frac{5}{2}$ of a unit long. Explain how this problem is related to fraction division.

_____

$\frac{2}{3}$ unit

ANSWERS TO PRACTICE PROBLEMS FOR SECTION 7.4

1. A simple "how many groups?" story problem for $1 \div \frac{5}{7}$ is "how many $\frac{5}{7}$ cup of water are in 1 cup of water?" Figure 7.17 shows 1 cup of water and shows $\frac{5}{7}$ cup of water shaded. The shaded portion is divided into 5 equal parts, and the full cup is 7 of those parts. Thus, the full cup is $\frac{7}{5}$ of the shaded part, and there are $\frac{7}{5}$ of $\frac{5}{7}$ of a cup of water in 1 cup of water, so $1 \div \frac{5}{7} = \frac{7}{5}$.

FIGURE 7.17
Showing Why $1 \div \frac{5}{7} = \frac{7}{5}$ by Considering How Many $\frac{5}{7}$ Cup of Water Are in 1 Cup of Water

2. A "how many in one group?" story problem for $1 \div \frac{3}{4}$ is "if 1 ton of dirt fills a truck $\frac{3}{4}$ full, then how many tons of dirt will be needed to fill the truck completely full?" We can see that this is a "how many in one group?" type of problem because the 1 ton of dirt fills $\frac{3}{4}$ of a group (the truck) and we want to know the amount of dirt in 1 whole group. Figure 7.18 shows a truck bed divided into 4 equal parts with 3 of those parts filled with dirt. Since the 3 parts are filled with 1 ton of dirt, each of the 3 parts must contain $\frac{1}{3}$ of a ton of dirt. To fill the truck completely, 4 parts, each containing $\frac{1}{3}$ of a ton of dirt are needed. Therefore, the truck takes $\frac{4}{3} = 1\frac{1}{3}$ tons of dirt to fill it completely, so $1 \div \frac{3}{4} = \frac{4}{3}$.

The 1 ton of dirt is divided equally among 3 parts.

4 parts are needed to fill the truck; each part is $\frac{1}{3}$ of a ton, so $\frac{4}{3}$ tons of dirt are needed to fill the truck.

truck bed

FIGURE 7.18

Showing Why $1 \div \frac{3}{4} = \frac{4}{3}$ by Considering How Many Tons of Dirt It Takes to Fill a Truck if 1 Ton Fills It $\frac{3}{4}$ Full

3. Annie's diagram shows that she can make 1 full batch of cookies from her $\frac{3}{4}$ cup of chocolate chips and that $\frac{1}{4}$ cup of chocolate chips will be left over. Because $\frac{1}{4}$ cup of chocolate chips is left over, it would be easy for Annie to misinterpret her picture as showing $\frac{3}{4} \div \frac{1}{2} = 1\frac{1}{4}$. The answer to the problem is supposed to be the number of *batches* Annie can make. In terms of batches, the remaining $\frac{1}{4}$ cup of chocolate chips makes $\frac{1}{2}$ of a batch of cookies. We can see this because 2 quarter-cup sections make a full batch, so each quarter-cup section makes $\frac{1}{2}$ of a batch of cookies. Thus, by interpreting the remaining $\frac{1}{4}$ cup of chocolate chips in terms of batches, we see that Annie can make $1\frac{1}{2}$ batches of chocolate chips, thereby showing that $\frac{3}{4} \div \frac{1}{2} = 1\frac{1}{2}$, not $1\frac{1}{4}$.

4. a. This problem can be rephrased as "if $\frac{1}{2}$ cup of jelly worms fill $\frac{3}{4}$ of a bag, then how many cups fill a whole bag?" Therefore, this is a "how many in one group?" division problem illustrating $\frac{1}{2} \div \frac{3}{4}$, not $\frac{3}{4} \div \frac{1}{2}$. Since $\frac{1}{2} \div \frac{3}{4} = \frac{1}{2} \cdot \frac{4}{3} = \frac{2}{3}$, there are $\frac{2}{3}$ cup of jelly worms in a whole bag.

   b. This problem is solved by $\frac{3}{4} \div \frac{1}{2}$, according to the "how many in one group?" interpretation. A group is a cup and each object is a bag of jelly worms.

   c. This problem can't be solved because you don't know how many cups of jelly worms are in $\frac{3}{4}$ of a bag.

   d. This problem is solved by $\frac{3}{4} \div \frac{1}{2}$, according to the "how many groups?" interpretation. Each group consists of $\frac{1}{2}$ cup of jelly worms.

   e. This problem is solved by $\frac{3}{4} \div \frac{1}{2}$, according to the "how many in one group?" interpretation. This is because you can think of the problem as saying that $\frac{3}{4}$ pound of candy fills $\frac{1}{2}$ of a group and you want to know how many pounds fills 1 whole group.

   f. This problem is solved by $\frac{3}{4} \times \frac{1}{2}$, not $\frac{3}{4} \div \frac{1}{2}$. It is dividing *in* half, not dividing *by* half.

   g. This problem is solved by $\frac{3}{4} \div \frac{1}{2}$, according to the "how many groups?" interpretation because you

want to know how many $\frac{1}{2}$ pounds are in $\frac{3}{4}$ pound. Each group consists of $\frac{1}{2}$ pound of candy.

5. If we consider Frank and John as each representing one group, and David as representing half of a group, then the \$14 should be distributed equally among $2\frac{1}{2}$ groups. Therefore, this is a "how many in one group" division problem. Each group should get

$$14 \div 2\frac{1}{2} = 14 \div \frac{5}{2} = 14 \cdot \frac{2}{5} = \frac{28}{5}$$
$$= 5\frac{3}{5} = 5\frac{6}{10} = 5.60$$

dollars. Therefore, Frank and John should each get \$5.60, and David should get half of that, which is \$2.80.

6. According to the "how many in one group?" interpretation, the problem is solved by $\$4.50 \div 0.15$ because \$4.50 fills 0.15 of a group and we want to know how much is in 1 whole group. So the meal cost

$$\$4.50 \div 0.15 = \$4.50 \div \frac{15}{100} = \$4.50 \cdot \frac{100}{15}$$
$$= \frac{\$450}{15} = \$30$$

7. Each problem, except for the first and last, requires different arithmetic to solve it.

   a. This is asking, "$\frac{1}{4}$ equals what times $\frac{1}{3}$?" We solve this by calculating $\frac{1}{4} \div \frac{1}{3}$, which is $\frac{3}{4}$. We can also think of this as a division problem with the "how many groups?" interpretation because we want to know how many $\frac{1}{3}$ cup are in $\frac{1}{4}$ cup. According to the meaning of division, this is $\frac{1}{4} \div \frac{1}{3}$.

   b. This is asking, "What is $\frac{1}{4}$ of $\frac{1}{3}$?" We solve this by calculating $\frac{1}{4} \times \frac{1}{3} = \frac{1}{12}$.

   c. This is asking, "What is $\frac{1}{3} - \frac{1}{4}$?" The answer is $\frac{1}{12}$, which happens to be the same answer as in part (b), but the arithmetic to solve it is different.

   d. Since $\frac{1}{4}$ cup of water fills $\frac{1}{3}$ of a plastic container, the full container will hold 3 times as much water, or $3 \times \frac{1}{4} = \frac{3}{4}$ of a cup. We can also think of this as a division problem with the "how many in one group?" interpretation. $\frac{1}{4}$ cup of water is put into $\frac{1}{3}$ of a group. We want to know how much is in one group. According to the meaning of division, it's $\frac{1}{4} \div \frac{1}{3}$, which again is equal to $\frac{3}{4}$.

8. a.   The number of $\frac{1}{2}$ years in $3\frac{3}{4}$ years is $3\frac{3}{4} \div \frac{1}{2}$. There will be that many groups of 4500 miles driven. So, after $3\frac{3}{4}$ years, you will have driven

$$\left(3\frac{3}{4} \div \frac{1}{2}\right) \times 4500 = \left(\frac{15}{4} \div \frac{1}{2}\right) \times 4500$$

$$= \frac{15}{2} \times 4500$$

$$= 33,750$$

miles.

b.   Since one year is 52 weeks, there are $52 \div 6\frac{1}{2}$ groups of $6\frac{1}{2}$ weeks in a year. Mo will use 128 ounces for each of those groups, so Mo will use

$$\left(52 \div 6\frac{1}{2}\right) \times 128 = \left(52 \div \frac{13}{2}\right) \times 128$$

$$= \frac{104}{13} \times 128$$

$$= 1,024$$

ounces of detergent in a year.

c.   There are $32 \div 2\frac{1}{2}$ groups of $2\frac{1}{2}$ ounces in 32 ounces. Each of those groups makes 1 gallon. So the bottle makes $32 \div 2\frac{1}{2} = 12\frac{4}{5}$ gallons of weed killer.

---

9. One way to solve the problem is to determine how many $\frac{2}{3}$ units are in $\frac{5}{2}$ units. This will tell us how many of the $\frac{2}{3}$-unit-long segments to lay end to end in order to get the $\frac{5}{2}$ unit long segment. Since $\frac{5}{2} \div \frac{2}{3} = \frac{15}{4} = 3\frac{3}{4}$, there are $3\frac{3}{4}$ segments of length $\frac{2}{3}$ units in a segment of length $\frac{5}{2}$ units. So, you need to form a line segment that is 3 times as long as the one pictured, plus another $\frac{3}{4}$ as long:

PROBLEMS FOR SECTION 7.4

1. A bread problem: If one loaf of bread requires $1\frac{1}{4}$ cups of flour, then how many loaves of bread can you make with 10 cups of flour? (Assume that you have enough of all other ingredients on hand.)

    a.   Solve the bread problem by drawing a diagram. Explain your reasoning.

    b.   Write a division problem that corresponds to the bread problem. Solve the division problem by "inverting and multiplying." Verify that your solution agrees with your solution in part (a).

2. A measuring problem: You are making a recipe that calls for $\frac{2}{3}$ cup of water, but you can't find your $\frac{1}{3}$ cup measure. You can, however, find your $\frac{1}{4}$ cup measure. How many times should you fill your $\frac{1}{4}$ cup measure in order to measure $\frac{2}{3}$ cup of water?

    a.   Solve the measuring problem by drawing a diagram. Explain your reasoning.

    b.   Write a division problem that corresponds to the measuring problem. Solve the division problem by "inverting and multiplying." Verify that your solution agrees with your solution in part (a).

3. Write a "how many groups?" story problem for $4 \div \frac{2}{3}$ and solve your problem in a simple and concrete way without using the "invert and multiply" procedure. Explain your reasoning. Verify that your solution agrees with the solution you obtain by using the "invert and multiply" procedure.

4. Write a "how many groups?" story problem for $5\frac{1}{4} \div 1\frac{3}{4}$ and solve your problem in a simple and concrete way without using the "invert and multiply" procedure. Explain your reasoning. Verify that your solution agrees with the solution you obtain by using the "invert and multiply" procedure.

5. Jose and Mark are making cookies for a bake sale. Their recipe calls for $2\frac{1}{4}$ cups of flour for each batch. They have 5 cups of flour. Jose and Mark realize that they can make two batches of cookies and that there will be some flour left. Since the recipe doesn't call for eggs, and since they have plenty of the other ingredients on hand, they decide they can make a fraction of a batch in addition to the two whole batches. But Jose and Mark have a difference of opinion. Jose says that

$$5 \div 2\frac{1}{4} = 2\frac{2}{9}$$

so he says that they can make $2\frac{2}{9}$ batches of cookies. Mark says that two batches of cookies will use up $4\frac{1}{2}$ cups of flour, leaving $\frac{1}{2}$ left, so they should be able to make $2\frac{1}{2}$ batches. Mark draws the picture in Figure 7.19 to explain his thinking to Jose.

FIGURE 7.19

Representing $5 \div 2\frac{1}{4}$ by Considering How Many $2\frac{1}{4}$ Cups of Flour are in 5 Cups of Flour

Discuss the boys' mathematics: What's right, what's not right, and why? If anything is incorrect, how could you modify it to make it correct?

---

6. Marvin has 11 yards of cloth to make costumes for a play. Each costume requires $1\frac{1}{2}$ yards of cloth.

   a. Solve the following two problems:

      i. How many costumes can Marvin make and how much cloth will be left over?

      ii. What is $11 \div 1\frac{1}{2}$?

   b. Compare and contrast your answers in part (a).

7. A laundry problem: You need $\frac{3}{4}$ cup of laundry detergent to wash one full load of laundry. How many loads of laundry can you wash with 5 cups of laundry detergent? (Assume that you can wash fractional loads of laundry.)

   a. Solve the laundry problem by drawing a diagram. Explain your reasoning.

   b. Write a division problem that corresponds to the laundry problem. Solve the division problem by "inverting and multiplying." Verify that your solution agrees with your solution in part (a).

8. Write a "how many groups?" story problem for $2 \div \frac{3}{4}$ and solve your problem in a simple and concrete way without using the "invert and multiply" procedure. Ex-

plain your reasoning. Verify that your solution agrees with the solution you obtain by using the "invert and multiply" procedure.

9. Write a "how many groups?" story problem for $\frac{1}{3} \div \frac{1}{4}$ and solve your problem in a simple and concrete way without using the "invert and multiply" procedure. Explain your reasoning. Verify that your solution agrees with the solution you obtain by using the "invert and multiply" procedure.

10. Write a "how many groups?" story problem for $\frac{1}{2} \div \frac{2}{3}$ and solve your problem in a simple and concrete way without using the "invert and multiply" procedure. Explain your reasoning. Verify that your solution agrees with the solution you obtain by using the "invert and multiply" procedure.

11. Fraction division story problems involve the simultaneous use of different wholes. Solve the following paint problem in a simple and concrete way without using the "invert and multiply" procedure. Describe how you must work simultaneously with different wholes in solving the problem.

A paint problem: You need $\frac{3}{4}$ of a bottle of paint to paint a poster board. You have $3\frac{1}{2}$ bottles of paint. How many poster boards can you paint?

12. An article by Dina Tirosh, [57], discusses some common errors in division. The following problems are based on some of the findings of this article:

   a. Tyrone says that $\frac{1}{2} \div 5$ doesn't make sense because 5 is bigger than $\frac{1}{2}$ and you can't divide a smaller number by a bigger number. Give Tyrone an example of a sensible story problem for $\frac{1}{2} \div 5$. Solve your problem and explain your solution.

   b. Kim says that $4 \div \frac{1}{3}$ can't be equal to 12 because when you divide, the answer should be smaller. Kim thinks the answer should be $\frac{1}{12}$ because that is less than 4. Give Kim an example of a story problem for $4 \div \frac{1}{3}$ and explain why it makes sense that the answer really is 12, not $\frac{1}{12}$.

13. Write a story problem for $\frac{3}{4} \times \frac{1}{2}$ and another story problem for $\frac{3}{4} \div \frac{1}{2}$. (Make clear which is which.) In each case, use elementary reasoning about the story situation to solve your problem. Explain your reasoning.

14. Sam picked $\frac{1}{2}$ gallon of blueberries. Sam poured the blueberries into one of his plastic containers and noticed that the berries filled the container $\frac{2}{3}$ full. Solve the following problems in any way you like without using a calculator, and explain your reasoning in detail:

   a. How many of Sam's containers will 1 gallon of blueberries fill? (Assume Sam has a number of containers of the same size.)

   b. How many gallons of blueberries does it take to fill Sam's container completely full?

15. A road crew is building a road. So far, $\frac{2}{3}$ of the road has been completed and this portion of the road is $\frac{3}{4}$ of a mile long. Solve the following problems in any way you like without using a calculator, and explain your reasoning in detail:

   a. How long will the road be when it is completed?

   b. When the road is 1 mile long, what fraction of the road will be completed?

16. Will has mowed $\frac{2}{3}$ of his lawn, and so far it's taken him 45 minutes. For each of the following problems, solve

the problem in two ways: (1) by using elementary reasoning about the story situation and (2) by interpreting the problem as a division problem (say whether it is a "how many groups?" or a "how many in one group?" type of problem) and by solving the division problem using standard paper and pencil methods. Do not use a calculator. Verify that you get the same answer both ways.

   a. How long will it take Will to mow the entire lawn (all together)?

   b. What fraction of the lawn can Will mow in an hour?

17. Grandma's favorite muffin recipe uses $1\frac{3}{4}$ cups of flour for one batch of 12 muffins. For each of the problems (a) through (c), solve the problem in two ways: (1) by using elementary reasoning about the story situation and (2) by interpreting the problem as a division problem (say whether it is a "how many groups?" or a "how many in one group?" type of problem) and by solving the division problem using standard paper and pencil methods. Do not use a calculator. Verify that you get the same answer both ways.

   a. How many cups of flour are in one muffin?

   b. How many muffins does 1 cup of flour make?

   c. If you have 3 cups of flour, then how many batches of muffins can you make? (Assume that you can make fractional batches of muffins and that you have enough of all the ingredients.)

18. Write a "how many in one group?" story problem for $4 \div \frac{1}{3}$, and use your story problem to explain why it makes sense to solve $4 \div \frac{1}{3}$ by "inverting and multiplying"—in other words, by multiplying 4 by $\frac{3}{1}$.

19. Write a "how many in one group?" story problem for $4 \div \frac{2}{3}$, and use your story problem to explain why it makes sense to solve $4 \div \frac{2}{3}$ by "inverting and multiplying"—in other words, by multiplying 4 by $\frac{3}{2}$.

20. Write a "how many in one group?" story problem for $9 \div \frac{3}{4}$, and use your story problem to explain why it makes sense to solve $9 \div \frac{3}{4}$ by "inverting and multiplying"—in other words, by multiplying 9 by $\frac{4}{3}$.

21. Write a "how many in one group?" story problem for $\frac{1}{2} \div \frac{3}{4}$, and use your story problem to explain

why it makes sense to solve $\frac{1}{2} \div \frac{3}{4}$ by "inverting and multiplying"—in other words, by multiplying $\frac{1}{2}$ by $\frac{4}{3}$.

22. Write a "how many in one group?" story problem for $1 \div 2\frac{1}{2}$, and use your story problem to explain why it makes sense to solve $1 \div 2\frac{1}{2}$ by "inverting and multiplying."

23. Give an example of either a hands-on activity or a story problem for elementary school children that is related to a fraction division problem (even if the children wouldn't think of the activity or problem as fraction division). Write the fraction division problem that is related to your activity or story problem. Describe how the children could solve the problem by using logical thinking aided by physical actions or by drawing pictures.

24. Buttercup the gerbil drank $\frac{2}{3}$ of a bottle of water in $1\frac{1}{2}$ days. Assuming Buttercup continues to drink water at the same rate, how many bottles of water will Buttercup drink in 5 days? Use multiplication and division to solve

this problem, explaining in detail why you can use multiplication when you do and why you can use division when you do.

25. If you used $2\frac{1}{2}$ truck loads of mulch for a garden that covers $\frac{3}{4}$ acre, then how many truck loads of mulch should you order for a garden that covers $3\frac{1}{2}$ acres? (Assume that you will spread the mulch at the same rate as before.) Use multiplication and division to solve this problem, explaining in detail why you can use multiplication when you do and why you can use division when you do.

26. If $2\frac{1}{2}$ pints of jelly filled $3\frac{1}{2}$ jars, then how many jars will you need for 12 pints of jelly? Will the last jar of jelly be completely full? If not, how full will it be? (Assume that all jars are the same size.) Use multiplication and division to solve this problem, explaining in detail why you can use multiplication when you do and why you can use division when you do.

7.5 Dividing Decimals

When we divide (finite) decimals, such as

$$2.35\overline{)3.714}$$

the standard procedure is to move the decimal point in both the divisor (2.35) and the dividend (3.714) the same number of places to the right until the divisor becomes a whole number:

$$235\overline{)371.4}$$

Then a decimal point is placed above the decimal point in the (new) dividend (371.4):

$$235\overline{)371.4}^{\,.}$$

From then on, division proceeds just as if we were doing the whole number division problem $3714 \div 235$, except that a decimal point is left in place for the answer. Why does this procedure of shifting the decimal points make sense? In this section we will explain in several different ways why we can shift the decimal points as we do. Our technique for shifting decimal points in division problems also allows us to calculate efficiently with large numbers, such as those in the millions, billions, and trillions. But first we consider the two interpretations of division for decimals.

THE TWO INTERPRETATIONS OF DIVISION FOR DECIMALS

The two interpretations that we have used for whole numbers and for fractions also apply to decimals.

THE "HOW MANY GROUPS?" INTERPRETATION

With the "how many groups?" interpretation of division, $35 \div 7$ means the number of groups we can make when we divide 35 objects into groups with 7 objects in each group. For example, if meat costs $7 per pound and you have $35, then how many pounds of meat can you buy?

Similarly, with the "how many groups?" interpretation of division,

$$14.5 \div 1.52$$

means the number of groups we can make when we divide 14.5 objects into groups with 1.52 objects in each group. For example, suppose gas costs $1.52 per gallon and you have $14.50 to spend on gas. How many gallons of gas can you buy? We want to know how many groups of $1.52 are in $14.50, so this is a "how many groups?" problem for $14.5 \div 1.52$.

THE "HOW MANY IN ONE (EACH) GROUP?" INTERPRETATION

With the "how many in one (or each) group?" interpretation of division, $35 \div 7$ means the number of objects in one group if 35 objects are divided equally among 7 groups. For example, if 7 identical action figures cost $35, then how much did one (or each) action figure cost?

Similarly, with the "how many in one group?" interpretation of division,

$$9.48 \div 5.3$$

means the number of objects in one group if 9.48 objects are divided equally among 5.3 groups. You can also think of the 9.48 objects as "filling" 5.3 groups. For example, if you bought 5.3 pounds of peaches for $9.48, then how much did 1 pound of peaches cost? The $9.48 is distributed equally among or "fills" 5.3 groups and we want to know how many dollars are in one group. So this is a "how many in one group?" problem for $9.48 \div 5.3$.

Next we turn to different methods for explaining why we move the decimal point as we do when we divide decimals.

EXPLAINING THE SHIFTING OF DECIMAL POINTS BY MULTIPLYING AND DIVIDING BY THE SAME POWER OF 10

One way to explain why it is valid to shift the decimal points in a division problem is as follows: When we shift the decimal points in both the divisor and the dividend the same number of places, we have multiplied and divided by the same power of 10, thereby replacing the original problem with an equivalent problem that can be solved by previous methods. Recall that multiplying a decimal number by a power of 10 shifts the digits in the number and therefore can be thought of as shifting the decimal point. Consider the example

$$2.35 \overline{)3.714}$$

Because

$$2.35 \times 100 = 235$$

$$3.714 \times 100 = 371.4$$

it follows that

$$
\begin{array}{cc}
\times\,100 & \div\,100 \\
\downarrow & \downarrow
\end{array}
$$

$$3.714 \div 2.35 = (3.714 \times 100) \div (2.35 \times 100)$$
$$= \quad 371.4 \div 235$$

Since we multiplied and divided by the same number, namely 100, the original problem $3.714 \div 2.35$ is equivalent to the new problem $371.4 \div 235$. In other words, the two problems have the same answer. We can give the same explanation by writing the division problems in fraction form:

$$\frac{3.714}{2.35} = \frac{3.714}{2.35} \times \frac{100}{100}$$
$$= \frac{3.714 \times 100}{2.35 \times 100}$$
$$= \frac{371.4}{235}$$

So the shifting of decimal points is really just a way to replace the problem

$$3.714 \div 2.35$$

with the equivalent problem

$$371.4 \div 235$$

EXPLAINING THE SHIFTING OF DECIMAL POINTS BY CHANGING FROM DOLLARS TO CENTS

If the decimal numbers in a division problem have no digits in places lower than the hundredths place, then we can use money to interpret the division problem. For example, we can interpret

$$1.27\overline{)4.5}$$

as follows:

$4.5 \div 1.27$ is the number of pounds of plums we can buy for \$4.50 if plums cost \$1.27 per pound.

(Note that we are using the "how many groups?" interpretation of division because we want to know how many groups of \$1.27 are in \$4.50.) If we think in terms of cents rather than dollars, then an equivalent way to rephrase the preceding statement is as follows:

$4.5 \div 1.27$ is the number of pounds of plums we can buy for 450 cents if plums cost 127 cents per pound.

But this last statement also describes the division problem

$$127\overline{)450}$$

Therefore, the problems 4.5 ÷ 1.27 and 450 ÷ 127 are equivalent. In other words, they must have the same answer. So, when we shift the decimal points of 1.27 and 4.5 two places to the right in the division problem $1.27\overline{)4.5}$ to obtain the problem $127\overline{)450}$, we have simply replaced the original problem with a problem that has the same answer and can be solved with the techniques we learned for whole number division.

EXPLAINING THE SHIFTING OF DECIMAL POINTS BY CHANGING THE UNIT

A more general way to explain why we can shift the decimal points in a decimal division problem is to think about representing the division problem with bundled objects, or with pictures representing bundled objects. For example, consider the division problem

$$0.29\overline{)1.7}$$

We can interpret this division problem as asking, "How many 0.29 are in 1.7?" from a "how many groups?" viewpoint. The diagram in Figure 7.20 gives us a way to picture this division problem, by interpreting the smallest square as representing $\frac{1}{100}$. But if we interpret the smallest square as representing 1 instead of $\frac{1}{100}$, then Figure 7.20 asks "how many 29s are in 170?" which is the division problem

$$29\overline{)170}$$

The answer must be the same, no matter which way we interpret the diagram; therefore, the division problems 1.7 ÷ 0.29 and 170 ÷ 29 are equivalent problems. Notice that from the diagram we can estimate that the answer is between 5 and 6 because 5 copies of the squares on the top will be fewer squares than on the bottom, but 6 copies of the squares on the top will be more squares than those on the bottom.

By allowing the smallest square in Figure 7.20 to represent different values, we see that this diagram can represent infinitely many different division problems, as indicated in Table 7.7. Therefore, all of these division problems must have the same answer.

FIGURE 7.20

Representing
1.7 ÷ 0.29

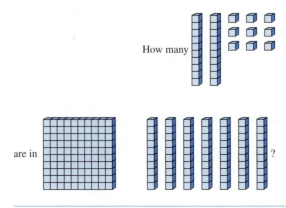

TABLE 7.7

Different Ways to Interpret Figure 7.20

IF THE SMALL SQUARE REPRESENTS	THEN FIGURE 7.20 REPRESENTS
⋮	⋮
$\frac{1}{10,000}$	$0.017 \div 0.0029$
$\frac{1}{1000}$	$0.17 \div 0.029$
$\frac{1}{100}$	$1.7 \div 0.29$
$\frac{1}{10}$	$17 \div 2.9$
1	$170 \div 29$
10	$1700 \div 290$
100	$17,000 \div 2900$
⋮	⋮
1 million	$(170 \text{ million}) \div (29 \text{ million})$
⋮	⋮

DECIDING WHERE TO PUT THE DECIMAL POINT BY ESTIMATING THE SIZE OF THE ANSWER

As a way of checking your work, or as a backup if you forget what to do about the decimal points when dividing decimals, you can often estimate to determine where to put the decimal point in the answer to a decimal division problem.

For example, suppose that you want to calculate

$$0.7\overline{)2.53}$$

but instead you calculate

$$\begin{array}{r} 36.14 \\ 7\overline{)253} \end{array}$$

How should you move the decimal point in 36.14 to get the correct answer to $2.53 \div 0.7$? By shifting the decimal point in 36.14, we can get any of the following numbers:

$$\ldots 0.03614,\ 0.3614,\ 3.614,\ 36.14,\ 361.4,\ 3614, \ldots$$

Which number in this list is a reasonable answer to $2.53 \div 0.7$? You can reason that 2.53 is close to 3 and 0.7 is close to 1, so the answer to $2.53 \div 0.7$ should be close to $3 \div 1 = 3$. Therefore, the correct answer should be 3.614.

DIVIDING NUMBERS IN THE MILLIONS, BILLIONS, AND TRILLIONS

Earlier, we explained why we can shift the decimal point in both the divisor and the dividend the same number of places to obtain an equivalent division problem. To divide decimals, we shift the decimal points to the right, thereby turning the decimal division problem into a whole number division problem. But we can also shift the decimal points to the left. By shifting decimal points to the left, we can replace a division problem involving large numbers with an equivalent division problem that involves smaller numbers. For example, to calculate

$$(170 \text{ million}) \div (29 \text{ million}) = 170{,}000{,}000 \div 29{,}000{,}000$$

we can shift the decimal points (which are not shown, but are to the right of the final 0's in each number) 6 places to the left. We thereby obtain the equivalent problem

$$170 \div 29$$

which doesn't involve so many zeros. In this way, we can make division problems involving numbers in the millions, billions, and trillions easier to solve.

What do we do if a problem mixes trillions, billions, and millions? For example, how much money will each person get if \$1.3 billion is divided equally among 8 million people? If we express 1.3 billion in terms of millions, then both numbers will be in terms of millions. Since 1 billion is 1000 million,

$$1.3 \text{ billion} = 1.3 \times 1000 \text{ million} = 1300 \text{ million}$$

Therefore,

$$(1.3 \text{ billion}) \div (8 \text{ million}) = (1300 \text{ million}) \div (8 \text{ million}) = 1300 \div 8$$

Since $1300 \div 8 = 162.5$, each person will get \$162.50.

CLASS ACTIVITY   NOW TURN TO CLASS ACTIVITIES MANUAL

7T Quick Tricks for Some Decimal Division Problems p. 191

7U Understanding Decimal Division p. 192

PRACTICE PROBLEMS FOR SECTION 7.5

1. Describe a quick way to mentally calculate $0.11 \div 0.125$ by thinking in terms of fractions.

2. Write a "how many groups?" story problem for $0.35 \div 1.45$.

3. Write a "how many in one group?" story problem for $0.35 \div 1.45$.

4. Without calculating the answers, give two different explanations for why the two division problems

$$0.65 \overline{)4.3} \qquad \text{and} \qquad 65 \overline{)430}$$

must have the same answer.

5. Explain how to calculate

$$(4 \text{ billion}) \div (2 \text{ million})$$

mentally.

6. Use the idea of multiplying and dividing by the same number to explain why the problems

$$170{,}000{,}000 \div 29{,}000{,}000$$

and

$$170 \div 29$$

are equivalent.

7. Use Figure 7.20 to explain why the problems

$$170{,}000{,}000 \div 29{,}000{,}000$$

and

$$170 \div 29$$

are equivalent.

ANSWERS TO PRACTICE PROBLEMS FOR SECTION 7.5

1. Observe that $0.125 = \frac{1}{8}$. To divide by a fraction, we can "invert and multiply." So, to divide by $\frac{1}{8}$, we multiply by 8. Therefore $0.11 \div 0.125 = 0.11 \div \frac{1}{8} = 0.11 \times 8 = 0.88$.

2. "If apples cost \$1.45 per pound, then how many pounds of apples can you buy for \$0.35?" is a "how many groups?" story problem for $0.35 \div 1.45$ because you want to know how many groups of \$1.45 are in \$0.35.

3. "If you paid \$0.35 for 1.45 liters of water, then how much does 1 liter of water cost?" is a "how many in one group?" story problem for $0.35 \div 1.45$ because \$0.35 is distributed equally among 1.45 liters and you want to know how many dollars are "in" 1 liter.

4. Explanation 1: When we shift the decimal points in the numbers 0.65 and 4.3 two places to the right, we have really just multiplied and divided the problem $4.3 \div 0.65$ by 100, thereby arriving at an equivalent division problem. In equations,

$$
\begin{array}{cc}
\times 100 & \div 100 \\
\downarrow & \downarrow \\
\end{array}
$$
$$4.3 \div 0.65 = (4.3 \times 100) \div (0.65 \times 100)$$
$$= \quad 430 \div 65$$

or

$$
\begin{aligned}
4.3 \div 0.65 &= \frac{4.3}{0.65} = \frac{4.3}{0.65} \times \frac{100}{100} \\
&= \frac{4.3 \times 100}{0.65 \times 100} \\
&= \frac{430}{65} = 430 \div 65
\end{aligned}
$$

Explanation 2: Thinking in terms of money, we can interpret $4.3 \div 0.65 = ?$ as asking, "How many groups of \$0.65 are in \$4.30?" If we phrase the question in terms of cents instead of dollars, it becomes "How many groups of 65 cents are in 430 cents?" which we can solve by calculating $430 \div 65$. The answer must be the same either way we ask the question. Therefore, $4.3 \div 0.65$ and $430 \div 65$ are equivalent problems.

We can give a third explanation by using a diagram like Figure 7.20, in which we ask how many groups of 6 ten-sticks and 5 small squares are in 4 hundred-squares and 3 ten-sticks. By interpreting the small square as $\frac{1}{100}$, the diagram represents $4.3 \div 0.65$. By interpreting the small square as 1, the diagram represents $430 \div 65$. The answer must be the same either way we interpret it, so once again, we conclude that $4.3 \div 0.65$ and $430 \div 65$ are equivalent problems.

5. Since 1 billion is 1000 million, 4 billion is 4000 million. So

$$\text{(4 billion)} \div \text{(2 million)} =$$
$$\text{(4000 million)} \div \text{(2 million)} = 4000 \div 2 = 2000$$

Mentally, we just think that 4000 million divided by 2 million is the same as 4000 divided by 2.

6. If we divide the numbers 170 million and 29 million in the division problem $170{,}000{,}000 \div 29{,}000{,}000$ by 1 million, then we will have divided and multiplied by 1 million to obtain the new problem $170 \div 29$. Since we divided and multiplied by the same number, the new problem and the old problem are equivalent. In other words, they have the same answer. In equation form,

$$\text{(170 million)} \div \text{(29 million)}$$

$$\begin{array}{cc} \times \text{ 1 million} & \div \text{ 1 million} \\ \downarrow & \downarrow \end{array}$$

$$= (170 \times 1 \text{ million}) \div (29 \times 1 \text{ million})$$
$$= 170 \div 29 = 5.86$$

In fraction form,

$$\text{(170 million)} \div \text{(29 million)} = \frac{170 \times 1 \text{ million}}{29 \times 1 \text{ million}}$$

$$= \frac{170}{29} \times \frac{1 \text{ million}}{1 \text{ million}} = \frac{170}{29} = 170 \div 29$$

7. As shown in Table 7.7, if we let the small square in Figure 7.20 represent 1 million, then Figure 7.20 represents

$$170{,}000{,}000 \div 29{,}000{,}000$$

But if we let the small square represent 1, then Figure 7.20 represents

$$170 \div 29$$

We must get the same answer either way we interpret the diagram; therefore, the two division problems $170{,}000{,}000 \div 29{,}000{,}000$ and $170 \div 29$ are equivalent.

PROBLEMS FOR SECTION 7.5

1. a. Write a "how many groups?" story problem for $5.6 \div 1.83$.

   b. Write a "how many in one group?" story problem for $5.6 \div 1.83$.

   c. Write a story problem (any type) for $0.75 \div 2.4$.

2. a. Calculate $28.3 \div 0.07$ to the hundredths place without a calculator. Show your work.

   b. Describe the standard procedure for determining where to put the decimal point in the answer to $28.3 \div 0.07$.

   c. Explain in two different ways why the placement of the decimal point that you described in part (b) is valid.

3. a. Calculate $16.8 \div 0.35$ to the hundredths place without a calculator. Show your work.

   b. Describe the standard procedure for determining where to put the decimal point in the answer to $16.8 \div 0.35$.

   c. Explain in two different ways why the placement of the decimal point that you described in part (b) is valid.

4. Ramin must calculate $8.42 \div 3.6$ longhand, but he can't remember what to do about decimal points. Instead, Ramin solves the division problem $842 \div 36$ longhand and gets the answer 23.38. Ramin knows that he must shift the decimal point in 23.38 somehow to get the correct answer to $8.42 \div 3.6$. Explain how Ramin could reason about the sizes of the numbers to determine where to put the decimal point.

5. a. Draw a diagram like Figure 7.20 and use your diagram to help you explain why the division problems $2.15 \div 0.36$ and $215 \div 36$ are equivalent.

   b. Use a diagram to help you explain why the division problems $\frac{7}{8} \div \frac{3}{8}$ and $7 \div 3$ are equivalent.

   c. Discuss how parts (a) and (b) are related.

6. A federal debt problem: If the federal debt is \$6.8 billion, and if this debt were divided equally among 290 million people, then how much would each person owe?

Describe how to estimate mentally the answer to the federal debt problem and explain briefly why your strategy makes sense.

7. A tax cut problem: If a 1.3 trillion dollar tax cut were divided equally among 290 million people over a 10-year period, then how much would each person get each year? Assume you only have a very simple calculator that cannot use scientific notation and that displays at most 8 digits. Describe how to use such a calculator to solve the tax cut problem and explain why your solution method is valid.

8. Light travels at about 300,000 kilometers per second. If Pluto is 6 billion kilometers away from us, then how long does light from Pluto take to reach us?

9. A new star was discovered. The star is about 75 trillion kilometers away from us. Light travels at about 300,000 kilometers per second. How long does light from the new star take to reach us?

10. In ordinary language, the term "divide" means "partition and make smaller," as in, "Divide and conquer."

   a. In mathematics, does dividing always make smaller? In other words, if you start with a number N and divide it by another number M, is the resulting quotient $N \div M$ necessarily less than N? Explain briefly.

   b. For which positive numbers, M, is $10 \div M$ less than 10? Determine all such positive numbers M. Use the meaning of division (either interpretation) to explain why your answer is correct.

11. When Mary converted a recipe from metric measurements to U.S. customary measurements, she discovered that she needed 8.63 cups of flour. Mary has a 1-cup measure, a $\frac{1}{2}$-cup measure, a $\frac{1}{4}$-cup measure, and a measuring tablespoon, which is $\frac{1}{16}$ cup. How should Mary use her measuring implements to measure the 8.63 cups of flour as accurately and efficiently as possible? Explain your reasoning.

12. Suppose you need to know how many thirty-secondths $\left(\frac{1}{32}\right)$ of an inch 0.685 inch is (rounded to the nearest thirty-secondth of an inch). Explain why the following method is a legitimate way to solve this problem:

   - Calculate $0.685 \times 32 = 21.92$ and round the result to the nearest whole number, namely, 22.

   - Use your result from the previous step, 22, to form the fraction $\frac{22}{32}$. Then 0.685 inch is $\frac{22}{32}$ inch rounded to the nearest thirty-secondth of an inch.

   Don't just verify that this method gives the right answer, explain why it works.

13. Sarah is building a carefully crafted cabinet and calculates that she must cut a certain piece of wood 33.33 inches long. Sarah has a standard tape measure that shows subdivisions of one sixteenth $\left(\frac{1}{16}\right)$ of an inch. How should Sarah measure 33.33 inches with her tape measure, using the closest sixteenth of an inch? (How many whole inches and how many sixteenths of an inch?) Explain your reasoning.

7.6 Ratio and Proportion

There are many situations in daily life as well as in science, medicine, and business that require the use of ratios and proportions. For example, in cooking, if we increase or decrease a recipe, we usually keep the ratios of the various ingredients the same. Any time we use percents, we are using a ratio. When pharmacists mix drugs, they must pay careful attention to the ratios of the ingredients.

Ratios are essentially just fractions, and understanding and working with ratios and proportions really just involves understanding and working with multiplication, division, and fractions.

A ratio describes a specific type of relationship between two quantities. If the ratio of flour to milk in a muffin recipe is 7 to 2, then that means that for every 7 cups of flour you use, you must use 2 cups of milk. Equivalently, for every 2 cups of milk you use, you must use 7 cups of flour. In general, to say that two quantities are in a **ratio** of A to B means that for every A units of the first quantity there are B units of the second quantity. Equivalently, for every B units of the second quantity there are A units of the first quantity.

ratio

rate   A **rate** is a ratio between two quantities that are measured in different units. For example, **speed**, which is the ratio of distance traveled to elapsed time, is a rate because distance is measured in miles, kilometers, or other similar units, whereas time is measured in hours, minutes, or seconds. If you are traveling at a speed of 60 mph, that means that you are traveling 60 miles for every hour that you travel.

To indicate a ratio we can use words, as in

the ratio of flour to milk is 7 to 2,

we can use a colon, as in

the ratio of flour to milk is 7 : 2,

or we can use a fraction, as in

the ratio of flour to milk is 7/2 $\left(\text{or } \frac{7}{2}\right)$.

Later in this section we will see why it makes sense to write ratios as fractions.

proportion   A **proportion** is a statement that two ratios are equal. For example, if the ratio of flour to milk in a muffin recipe is 7 to 2, and if the muffin recipe is to use A cups of flour (an as yet unknown amount) and 3 cups of milk, then we have the proportion

$$\frac{A}{3} = \frac{7}{2}$$

which states that the two ratios A to 3 and 7 to 2 are equal.

RATIOS AND FRACTIONS

Ratios and fractions are very closely related. In fact, ratios give rise to fractions, and fractions give rise to ratios. For this reason, ratios are sometimes written as fractions. We will now see why it makes sense to equate ratios and fractions.

Consider a bread recipe in which the ratio of flour to water is 14 to 5, so that if we use 14 cups of flour, we will need 5 cups of water. If we think of dividing those 14 cups of flour equally among the 5 cups of water, then, according to the meaning of division (the "how many in each group?" interpretation), each cup of water goes with

$$14 \div 5 = \frac{14}{5}$$

cups of flour. So in this recipe, there are $\frac{14}{5}$ cups of flour per cup of water. In this way, the ratio 14 to 5 of flour to water is naturally associated with the fraction $\frac{14}{5}$, which tells us the number of cups of flour per cup of water.

In general, a ratio A to B relating two quantities gives rise to the fraction $\frac{A}{B}$, which tells us the number of units of the first quantity there are per unit amount of the second quantity. The one exception, which doesn't even occur in ordinary circumstances, is when B is zero—in that case we cannot form the fraction $\frac{A}{B}$.

Similarly, a fraction (or other number) that tells us how much of one quantity there is per unit amount of another quantity, gives rise to a ratio between the two quantities. If we have a paint mixture in which we use $\frac{2}{3}$ cup of blue paint per cup of yellow paint, then in this mixture, the ratio of blue paint to yellow paint is

$$\frac{2}{3} \text{ to } 1$$

If we want to make the same shade of green paint using 3 cups of yellow paint, then we should use 3 times as much blue paint, or

$$3 \cdot \frac{2}{3} = 2$$

cups of blue paint. Therefore, we can also express the ratio of blue paint to yellow paint in the mixtures as

$$2 \text{ to } 3$$

In general, if there is $\frac{A}{B}$ of one quantity per unit amount of a second quantity, then the ratio of the first quantity to the second is

$$\frac{A}{B} \text{ to } 1$$

which is the same ratio as

$$B \cdot \frac{A}{B} \text{ to } B \cdot 1$$

because in the latter ratio we are using B times as much of both quantities as in the former ratio. Therefore, the fraction

$$\frac{A}{B}$$

gives rise to the ratio

$$\frac{A}{B} \text{ to } 1$$

which is equivalent to the ratio

$$A \text{ to } B$$

In this way, fractions give rise to ratios.

CLASS ACTIVITY   NOW TURN TO CLASS ACTIVITIES MANUAL

7V Ratios, Fractions, and Division p. 194

EQUIVALENT RATIOS

Consider various muffin recipes in which the ratio of flour to milk is 7 to 2, as shown in Figure 7.21 on p. 302. When the ratio of flour to milk in a muffin recipe is 7 to 2, then if we make a large batch of muffins using 14 cups of flour, which is 2 groups of 7 cups, then we must use 2 groups of 2 cups of milk, which is 4 cups of milk. If we use 21 cups of flour, which is 3

FIGURE 7.21

Different
Combinations
of Flour and
Milk That Are in
the Ratio 7 to 2

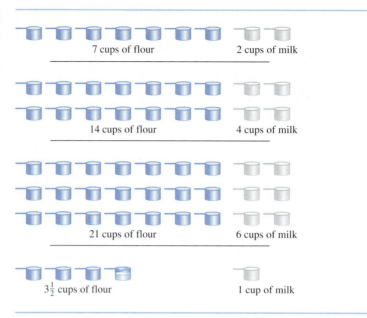

7 cups of flour   2 cups of milk

14 cups of flour   4 cups of milk

21 cups of flour   6 cups of milk

$3\frac{1}{2}$ cups of flour   1 cup of milk

groups of 7 cups, then we must use 3 groups of 2 cups of milk, which is 6 cups of milk. If we use $3\frac{1}{2}$ cups of flour, which is $\frac{1}{2}$ of a group of 7 cups, then we must use $\frac{1}{2}$ a group of 2 cups of milk, which is 1 cup of milk. Table 7.8 summarizes this simple reasoning. More generally, if we want to use

$$N \cdot 7$$

cups of flour, then we should use

$$N \cdot 2$$

cups of milk. Therefore, the ratios

$$7 \; : \; 2$$

TABLE 7.8

Determining
Different
Amounts of Flour
and Milk to Use
in Recipes in
which the Ratio
of Flour to Milk
Is 7 to 2

For	7 cups flour	use	2 cups milk.
	↓ ×2		↓ ×2
For	14 cups flour	use	4 cups milk.
For	7 cups flour	use	2 cups milk.
	↓ ×3		↓ ×3
For	21 cups flour	use	6 cups milk.
For	7 cups flour	use	2 cups milk.
	↓ ÷2		↓ ÷2
For	$3\frac{1}{2}$ cups flour	use	1 cup milk.

and

$$N \cdot 7 \, : \, N \cdot 2$$

are equivalent.

In general, using the preceding reasoning tells us that the ratio

$$A \; : \; B$$

is equivalent to the ratio

$$N \cdot A \, : \, N \cdot B$$

This is exactly the same relationship as the following is for fractions:

$$\frac{A}{B} = \frac{N \cdot A}{N \cdot B}$$

CLASS ACTIVITY   NOW TURN TO CLASS ACTIVITIES MANUAL

7W Equivalent Ratios p. 195

7X Making and Comparing Mixtures p. 197

SOLVING PROPORTIONS WITH MULTIPLICATION, DIVISION, AND SIMPLE LOGICAL REASONING

When you think of solving proportions, you may think of the method in which you set two fractions equal to each other, cross-multiply these fractions, and then solve the resulting equation. However, we can also solve proportions by using multiplication, division, and simple logical reasoning. Using the following problem, we will now study two simple methods for solving proportions.

Problem: If you want to use 4 cups of flour in a muffin recipe in which the ratio of flour to milk is 7 to 2, how much milk should you use?

Simple Solution Method 1: We first ask how much milk we would need for 1 cup of flour. If 7 cups of flour take 2 cups of milk, then we can think of the 2 cups of milk as divided equally among the 7 cups of flour. Therefore, according to the meaning of division (the "how many in each group?" interpretation), each cup of flour takes

$$2 \div 7 = \frac{2}{7}$$

of a cup of milk. Now if we want to use 4 cups of flour, then we need to use 4 groups of $\frac{2}{7}$ cup of milk. In other words, we need to use

$$4 \cdot \frac{2}{7} = \frac{8}{7} = 1\frac{1}{7}$$

cups of milk. Table 7.9 at the top of p. 304 summarizes this reasoning.

TABLE 7.9

Determining How
Much Milk to Use
for 4 Cups Flour
if the Ratio of
Flour to Milk Is 7
to 2

For	7 cups flour	use	2 cups milk.
	↓ ÷7		↓ ÷7
For	1 cup flour	use	$\frac{2}{7}$ cups milk.
	↓ ×4		↓ ×4
For	4 cups flour	use	$4 \cdot \frac{2}{7} = \frac{8}{7} = 1\frac{1}{7}$ cups milk.

TABLE 7.10

Determining How
Much Milk to Use
for 4 Cups Flour
if the Ratio of
Flour to Milk Is 7
to 2

For	7 cups flour	use	2 cups milk.
	↓ ×$\frac{4}{7}$		↓ ×$\frac{4}{7}$
For	4 cups flour	use	$\frac{4}{7} \cdot 2 = \frac{8}{7} = 1\frac{1}{7}$ cups milk.

Simple Solution Method 2: First, let's determine how many groups of 7 are in 4. According to the meaning of division (the "how many groups?" interpretation), there are

$$4 \div 7 = \frac{4}{7}$$

groups of 7 in 4. Next, for each whole group of 7, we must use 2 cups of milk, so for $\frac{4}{7}$ of a group of 7, we use

$$\frac{4}{7} \cdot 2 = \frac{8}{7} = 1\frac{1}{7}$$

cups of milk, according to the meaning of multiplication. Table 7.10 summarizes this reasoning.

CLASS ACTIVITY   NOW TURN TO CLASS ACTIVITIES MANUAL

7Y Solving Proportions with Multiplication and Division p. 199

THE LOGIC BEHIND SOLVING PROPORTIONS BY CROSS-MULTIPLYING FRACTIONS

You are probably familiar with the technique of solving proportions by cross-multiplying. Why is this a valid technique for solving proportions? We will examine this now.

Consider a light blue paint mixture made with $\frac{1}{4}$ cup blue paint and 4 cups white paint. How much blue paint will you need if you want to use 6 cups white paint and if you are using the same ratio of blue paint to white paint in order to make the same shade of light blue paint? A common method for solving such a problem is to set up the following proportion in fraction form:

$$\frac{\frac{1}{4}}{4} = \frac{A}{6}$$

Here, A represents the as yet unknown amount of blue paint you will need for 6 cups of white paint. We then cross-multiply to get

$$6 \cdot \frac{1}{4} = 4 \cdot A$$

Therefore,

$$A = \left(6 \cdot \frac{1}{4} \right) \div 4 = 1\frac{1}{2} \div 4 = \frac{3}{8}$$

so that you must use $\frac{3}{8}$ cups of blue paint for 6 cups of white paint.

Let's analyze the preceding steps. First, when we set the two fractions

$$\frac{\frac{1}{4}}{4}$$

and

$$\frac{A}{6}$$

equal to each other, why can we do that and what does it mean? If we think of the fractions as representing division—that is,

$$\frac{1}{4} \div 4$$

and

$$A \div 6$$

then each of these expressions stands for the number of cups of blue paint per cup of white paint. We want to use the same amount of blue paint per cup of white paint either way; therefore, the two fractions should be equal to each other, or

$$\frac{\frac{1}{4}}{4} = \frac{A}{6}$$

Next, why do we cross-multiply? We can cross-muliply because two fractions are equal exactly when their "cross-multiples" are equal. Recall that the method of cross-multiplying for fractions is really just a shortcut for giving the two fractions a common denominator. We can see the same phenomenon even when we work directly with the ratios. The ratio

$$\frac{1}{4} \quad \text{to} \quad 4$$

of blue paint to white paint is the same as the ratio

$$6 \cdot \frac{1}{4} \quad \text{to} \quad 6 \cdot 4$$

The ratio

$$A \quad \text{to} \quad 6$$

of blue paint to white paint is the same as the ratio

$$4 \cdot A \text{ to } 4 \cdot 6$$

Both of the ratios

$$6 \cdot \frac{1}{4} \text{ to } 6 \cdot 4$$

and

$$4 \cdot A \text{ to } 4 \cdot 6$$

refer to 24 cups of white paint ($6 \cdot 4$ and $4 \cdot 6$). Therefore, they must refer to the same amount of blue paint, which is $6 \cdot \frac{1}{4}$ cups on the one hand, and $4 \cdot A$ cups on the other hand. Thus,

$$6 \cdot \frac{1}{4} = 4 \cdot A$$

CLASS ACTIVITY   NOW TURN TO CLASS ACTIVITIES MANUAL

7Z Solving Proportions by Cross-Multiplying Fractions p. 203

7AA Can You Always Use a Proportion? p. 205

USING PROPORTIONS: THE CONSUMER PRICE INDEX

At some point you have probably heard older people wistfully recall the lower prices of days gone by, as in, "When I was young, a candy bar only cost a nickel!" As time goes by, most items become more expensive, due to inflation. (Computers are a notable exception.) So in years past, one dollar bought more than a dollar buys now. But in years past, most people also earned less than they do now. So, what is a fair way to compare costs of items across years? What is a fair way to compare wages earned in different years? The standard way to make such comparisons is with the Consumer Price Index, which is determined by the Bureau of Labor Statistics of the U.S. Department of Labor. To use the Consumer Price Index, we must work with proportions.

According to the Bureau of Labor Statistics, the Consumer Price Index (CPI) is a measure of the average change over time in the prices paid by urban consumers for a market basket of consumer goods and services. The CPI market basket is constructed from detailed expenditure information provided from surveys of families and individuals on what they actually bought. The CPI measures inflation as experienced by consumers in their day-to-day living expenses. This, and other information about the CPI, is available on the Bureau of Labor Statistics website. See www.aw-bc.com/beckmann.

Here is an example of how to use the CPI to compare salaries in different years. According to the Bureau of Labor Statistics, the average annual wages for elementary school teachers in the U.S. was $38,600 in 1998 and $41,980 in the year 2000. Now, the average wage was higher in 2000 than it was in 1998, but most items were also more expensive in 2000 than they were in 1998. So, in 2000, did elementary school teachers have more "buying power" than they

TABLE 7.11

The Consumer
Price Index for
Selected Years

THE CONSUMER PRICE INDEX (SELECTED YEARS)

year	CPI	year	CPI
1940	14.0	1994	148.2
1950	24.1	1995	152.4
1960	29.6	1996	156.9
1970	38.8	1997	160.5
1980	82.4	1998	163.0
1990	130.7	1999	166.6
1991	136.2	2000	172.2
1992	140.3	2001	177.1
1993	144.5	2002	179.9

did in 1998 or not? In other words, did elementary school teachers' salaries go up faster than inflation or not? We can use the CPI to determine the answer to this question.

The CPI assigns a number to each year (in recent history). Table 7.11 shows the CPI for some selected years.

We interpret these numbers as follows: Because the CPI was 163.0 in 1998, and 172.2 in 2000, this means that for every $163.00 you would have spent toward a "market basket of consumer goods and services" in 1998, you would have had to spend $172.20 in the year 2000. We can use this to compare the teachers' salaries in 1998 and 2000. Think of each group of $163.00 as buying a "unit" of a market basket of goods and services in 1998. So in 1998, the average elementary teacher's salary could buy

$$38,600 \div 163 = 236.81$$

units of a market basket of goods and services. To buy the same number of units of a market basket of goods and services in 2000, namely, 236.81 units, you would have needed

$$236.81 \times \$172.2 = \$40,779$$

because each unit of a market basket of goods and services cost $172.20 in 2000. In other words, $40,779 in 2000 had the same buying power as did $38,600 in 1998. Given that in the year 2000 elementary school teachers made an average of $41,980, which is more than $40,779, elementary school teachers had more buying power in 2000 than they did in 1998. Thus, from 1998 to 2000, elementary school teachers' salaries went up faster than inflation.

Here is another way to compare elementary school teachers' average salaries in 1998 and 2000 using the CPI. Because the CPI was 163.0 in 1998 and 172.2 in 2000, we can again think of $163.00 in 1998 and $172.20 in 2000 as buying a "unit" of a market basket of consumer goods and services. Now,

$$172.20 \div 163.00 = 1.05644$$

Therefore, a "unit" of a market basket of goods and services cost 1.05644 times as much in 2000 as it did in 1998. Hence, a salary of $38,600 in 1998 had the same buying power as did a salary 1.05644 times as much, namely,

$$1.05644 \times \$38,600 = \$40,779$$

in 2000. We reach the same conclusion as in the previous paragraph.

adjusting for inflation   Either of these calculations is called **adjusting for inflation**. In both calculations, we converted $38,600, the average salary for elementary teachers in 1998, to 2000 dollars, in which it became $40,779. So, a 1998 salary of $38,600 had the same buying power as a salary of $40,779 did in 2000. Notice that another way to set up the calculations to adjust for inflation would be to set up either of the following two proportions:

$$\frac{38,600}{163.0} = \frac{x}{172.2}$$

or

$$\frac{172.2}{163.0} = \frac{x}{38,600}$$

Here, x stands for the dollar amount in 2000 that had the same buying power as did $38,600 in 1998.

CLASS ACTIVITY   NOW TURN TO CLASS ACTIVITIES MANUAL

7BB The Consumer Price Index p. 207

PRACTICE PROBLEMS FOR SECTION 7.6

1. Jose mixed 3 cups of blue paint with 4 cups of red paint to make a purple paint. For each of the following fractions and division problems, interpret the fraction or the division problem in terms of Jose's paint mixture and explain why your interpretation makes sense:

   $\frac{3}{7}$ or $3 \div 7$; $\frac{4}{7}$ or $4 \div 7$; $\frac{3}{4}$ or $3 \div 4$;

   $\frac{4}{3}$ or $4 \div 3$; $\frac{7}{3}$ or $7 \div 3$; $\frac{7}{4}$ or $7 \div 4$

2. Which of the following two mixtures will be more salty?

   • 3 tablespoons of salt mixed in 4 cups of water

   • 4 tablespoons of salt mixed in 5 cups of water

   Solve this problem in two different ways, explaining why you can solve the problem the way you do.

3. Cali mixed $3\frac{1}{2}$ cups of red paint with $4\frac{1}{2}$ cups of yellow paint to make an orange paint. How many cups of red paint and how many cups of yellow paint will Cali need to make 12 cups of the same shade of orange paint? Solve this problem using only simple reasoning about multiplication or division (or both). Explain your reasoning.

4. To make a punch you mixed $\frac{1}{4}$ cup grape juice concentrate with $1\frac{1}{2}$ cups bubbly water. If you want to make the same punch using 2 cups of bubbly water, then how many cups of grape juice concentrate should you use? Solve this problem using only simple reasoning about multiplication or division (or both). Explain your reasoning.

5. In order to reconstitute a medicine properly, a pharmacist must mix 10 milliliters (mL) of a liquid medicine for every 12 mL of water. If one dose of the medicine/water mixture must contain $2\frac{1}{2}$ mL of medicine, then how many milliliters of medicine/water mixture provides one dose of the medicine? Solve this problem using only simple reasoning about multiplication or division (or both). Explain your reasoning.

6. Suppose that a logging crew can cut down five acres of trees every two days. Assume that the crew works at a steady rate. Solve problems (a) and (b) using logical thinking and using the most elementary reasoning you can. Explain your reasoning clearly.

    a. How many days will it take the crew to cut 8 acres of trees? Give your answer as a mixed number.

    b. Now suppose there are *three* logging crews that all work at the same rate as the original one. How long will it take these three crews to cut down 10 acres of trees?

7. If 3 people take 2 days to paint 5 fences, how long will it take 2 people to paint 1 fence? (Assume that the fences are all the same size and the painters are all equally good workers, and work at a steady rate.) Can this problem be solved by setting up the following proportion to find how long it will take 2 people to paint 5 fences?

$$\frac{3 \text{ people}}{2 \text{ days}} = \frac{2 \text{ people}}{x \text{ days}}$$

Solve the problem by thinking logically about the situation. Explain your reasoning clearly.

8. If a candy bar cost a nickel in 1960, then how much would you expect to have to pay for the same size and type of candy bar in 2000, according to the CPI?

9. a. What amount of money had the same buying power in 1950 as $50,000 did in 2000?

    b. What amount of money had the same buying power in 1980 as $50,000 did in 2000?

10. If an item cost $5.00 in 1990 and $6.50 in 2000, did its price go up faster, slower, or the same as inflation?

11. Suppose that from 1990 to 2000 the price of a 16-ounce box of brand A cereal went from $2.39 to $3.89. After adjusting for inflation, determine by what percent the price of Brand A cereal went up or down from 1990 to 2000.

ANSWERS TO PRACTICE PROBLEMS FOR SECTION 7.6

1. Out of the total 7 cups of purple paint, 3 cups are blue, so $\frac{3}{7}$ of the paint mixture is blue. Thinking in terms of division, if we imagine the blue paint divided equally among the 7 cups of purple paint, then there is $3 \div 7 = \frac{3}{7}$ cups of blue paint in each cup of purple paint. Similarly, 4 out of 7 cups of the purple paint are red, so $\frac{4}{7}$ of the paint is red. Imagining the red paint divided equally among the 7 cups of purple paint, there is $4 \div 7 = \frac{4}{7}$ cups of red paint in each cup of purple paint.

   Since there are 3 cups of blue paint and 4 cups of red paint in the mixture, if we think of dividing the blue paint equally among the 4 cups of red paint, then there are $3 \div 4 = \frac{3}{4}$ cups blue paint for each cup of red paint. By the same logic, there are $4 \div 3 = \frac{4}{3} = 1\frac{1}{3}$ cups red paint for each cup of blue paint.

   If we think of dividing the 7 cups of purple paint equally among the 3 cups of blue paint, then there are $7 \div 3 = \frac{7}{3} = 2\frac{1}{3}$ cups of purple paint for each cup of blue paint. By the same logic, there are $7 \div 4 = \frac{7}{4} = 1\frac{3}{4}$ cups of purple paint for each cup of red paint.

2. Method 1: If we think of the 3 tablespoons of salt in the first mixture as being divided equally among the 4 cups of water, then each cup of water contains $3 \div 4 = \frac{3}{4}$ tablespoons of salt. Similarly, each cup of water in the second mixture contains $4 \div 5 = \frac{4}{5}$ tablespoons of salt. Since $\frac{4}{5} = 0.8$ and $\frac{3}{4} = 0.75$, and since $0.8 > 0.75$, the second mixture contains more salt per cup of water. Thus, it is more salty.

   Method 2: If we make 5 batches of the first mixture and 4 batches of the second mixture, then both will contain 20 cups of water. The first mixture will contain $5 \times 3 = 15$ tablespoons of salt and the second mixture will contain $4 \times 4 = 16$ tablespoons of salt. Since both mixtures contain the same amount of water but the second mixture contains 1 more tablespoon of salt than the first, the second mixture must be more salty.

3. The $3\frac{1}{2}$ cups of red paint and $4\frac{1}{2}$ cups of yellow paint combine to make $3\frac{1}{2} + 4\frac{1}{2} = 8$ cups orange paint. Since 12 cups is $1\frac{1}{2}$ times as much as 8 cups, Cali will need $1\frac{1}{2}$ times as much red paint and yellow paint. Therefore, Cali will need

$$1\frac{1}{2} \cdot 3\frac{1}{2} = \frac{3}{2} \cdot \frac{7}{2} = \frac{21}{4} = 5\frac{1}{4}$$

cups of red paint and

$$1\frac{1}{2} \cdot 4\frac{1}{2} = \frac{3}{2} \cdot \frac{9}{2} = \frac{27}{4} = 6\frac{3}{4}$$

cups of yellow paint. This reasoning in summarized in Table 7.12.

TABLE 7.12

Determining How Much Red and Yellow Paint to Use to Make 12 Cups of Orange Paint When the Ratio of Red to Yellow Is $3\frac{1}{2}$ to $4\frac{1}{2}$

	RED PAINT		YELLOW PAINT		ORANGE PAINT
Use	$3\frac{1}{2}$ cups	and	$4\frac{1}{2}$ cups	to make	8 cups.
	$\downarrow \times 1\frac{1}{2}$		$\downarrow \times 1\frac{1}{2}$		$\downarrow \times 1\frac{1}{2}$
Use	$5\frac{1}{4}$ cups	and	$6\frac{3}{4}$ cups	to make	12 cups.

4. You can reason that if you wanted to use 3 cups of bubbly water, which is twice as much as $1\frac{1}{2}$ cups, then you should use twice as much grape juice concentrate, namely $\frac{1}{2}$ cup. So, if you wanted to use 1 cup of bubbly water, which is $\frac{1}{3}$ as much as 3 cups, then you should use $\frac{1}{3}$ of $\frac{1}{2}$ cup of grape juice concentrate, namely $\frac{1}{6}$ cups. Then to use 2 cups of bubbly water, which is twice as much as 1 cup, you should use twice as much as $\frac{1}{6}$ of a cup of grape juice concentrate, namely, $\frac{2}{6} = \frac{1}{3}$ cups. This reasoning is summarized in Table 7.13.

TABLE 7.13

Determining How Much Concentrate to Use for 2 Cups Water if the Ratio of Concentrate to Water Is $\frac{1}{4}$ to $1\frac{1}{2}$

For	$\frac{1}{4}$ cup concentrate	use	$1\frac{1}{2}$ cups bubbly water.
	$\downarrow \times 2$		$\downarrow \times 2$
For	$\frac{1}{2}$ cup concentrate	use	3 cups bubbly water.
	$\downarrow \div 3$		$\downarrow \div 3$
For	$\frac{1}{6}$ cup concentrate	use	1 cup bubbly water.
	$\downarrow \times 2$		$\downarrow \times 2$
For	$\frac{1}{3}$ cup concentrate	use	2 cups bubbly water.

5. Method 1: For 10 mL of medicine there are 12 mL of water. So if we think of dividing the 12 mL of water equally among the 10 mL of medicine, we see that there are $12 \div 10 = \frac{6}{5}$ mL of water for each milliliter of medicine. So, for $2\frac{1}{2}$ mL of medicine, there should be $2\frac{1}{2}$ times as much water, namely,

$$2\frac{1}{2} \cdot \frac{6}{5} = \frac{5}{2} \cdot \frac{6}{5} = 3$$

milliliters of water. All together, the $2\frac{1}{2}$ mL of medicine and the 3 mL of water make $5\frac{1}{2}$ mL of medicine/water mixture. This reasoning is summarized in Table 7.14.

TABLE 7.14

Determining How Much Medicine/Water Mixture Will Be Made with $2\frac{1}{2}$ mL of Medicine

	MEDICINE		WATER		MEDICINE/WATER MIXTURE
Use	10 mL	and	12 mL	to make	22 mL.
	$\downarrow \div 10$		$\downarrow \div 10$		$\downarrow \div 10$
Use	1 mL	and	$\frac{6}{5}$ mL	to make	$\frac{22}{10}$ mL.
	$\downarrow \times 2\frac{1}{2}$		$\downarrow \times 2\frac{1}{2}$		$\downarrow \times 2\frac{1}{2}$
Use	$2\frac{1}{2}$ mL	and	3 mL	to make	$5\frac{1}{2}$ mL.

Method 2: For every group of 10 mL of medicine, the pharmacist mixes in a group of 12 mL of water. How many groups of 10 mL are in $2\frac{1}{2}$ mL? You might just see right away that $2\frac{1}{2}$ mL is $\frac{1}{4}$ of a group of 10 mL. Otherwise, you can solve this using division as follows:

$$2\frac{1}{2} \div 10 = \frac{5}{2} \div \frac{10}{1} = \frac{5}{2} \cdot \frac{1}{10} = \frac{1}{4}$$

So there is $\frac{1}{4}$ of a group of 10 in $2\frac{1}{2}$. Therefore, the pharmacist should use $\frac{1}{4}$ of a group of 12 mL of water, which is 3 mL of water. All together, the $2\frac{1}{2}$ mL of medicine and the 3 mL of water make $5\frac{1}{2}$ mL of medicine/water mixture. This reasoning is summarized in Table 7.15.

TABLE 7.15

Determining How Much Medicine/Water Mixture Will Be Made with $2\frac{1}{2}$ mL of Medicine

	MEDICINE		WATER		MEDICINE/WATER MIXTURE
Use	10 mL	and	12 mL	to make	22 mL.
	$\downarrow \times \frac{1}{4}$		$\downarrow \times \frac{1}{4}$		$\downarrow \times \frac{1}{4}$
Use	$2\frac{1}{2}$ mL	and	3 mL	to make	$5\frac{1}{2}$ mL.

---

6. a. Because the crew cuts 5 acres every 2 days, it will cut half as much in 1 day, namely $2\frac{1}{2}$ acres. To determine how many days it will take to cut 8 acres, we must determine how many groups of $2\frac{1}{2}$ are in 8, which is $8 \div 2\frac{1}{2} = 3\frac{1}{5}$ days.

   b. In part (a) we saw that one crew cuts $2\frac{1}{2}$ acres per day. Therefore, three crews will cut 3 times as much per day, which is $3 \cdot 2\frac{1}{2} = 7\frac{1}{2}$ acres per day. To determine how many days it will take to cut 10 acres, we must figure out how many groups of $7\frac{1}{2}$ are in 10. This is solved by the division problem $10 \div 7\frac{1}{2}$. Because $10 \div 7\frac{1}{2} = 10 \div \frac{15}{2} = \frac{10}{1} \cdot \frac{2}{15} = \frac{20}{15} = 1\frac{5}{15} = 1\frac{1}{3}$, we conclude that it will take the 3 crews $1\frac{1}{3}$ days to cut 10 acres.

7. The proportion

$$\frac{3 \text{ people}}{2 \text{ days}} = \frac{2 \text{ people}}{x \text{ days}}$$

is not valid for this situation because, when more people are painting, it will take less time to paint a fence. It is not the case that, for each group of 3 people there are, it will take 2 days to paint a fence. Therefore, the relationship between the number of people and the number of days it takes to paint a fence is not a ratio.

Thinking logically, if 3 people take 2 days to paint 5 fences, then those 3 people will take $2 \div 5 = \frac{2}{5}$ of a day to paint just one fence (dividing the 2 days equally among the 5 fences). If just one person was painting, it would take 3 times as long to paint the fence, namely, $3 \cdot \frac{2}{5} = \frac{6}{5}$ of a day (which is $1\frac{1}{5}$ days). With 2 people painting, it will take half as much time to paint the fence, namely, $\frac{3}{5}$ of a day.

8. The CPI was 29.6 in 1960 and 172.2 in 2000. Therefore, $.05 in 1960 had the same buying power as did

$$\frac{172.2}{29.6} \cdot \$.05 = \$.29$$

in 2000.

9. a. In 2000, the CPI was 172.2, so $50,000 could buy $50{,}000 \div 172.2 = 290.36$ "units" of a market basket of goods and services. In 1950 the CPI was 24.1, so the same 290.36 units of a market basket of goods and services cost $290.36 \times \$24.10 = \$6998$ in 1950. Therefore, $6998 in 1950 had the same buying power as did $50,000 in 2000.

    b. In 2000, the CPI was 172.2; in 1980, the CPI was 82.4. Thus, a "unit" of a market basket of goods and services cost $82.4 \div 172.2 = .47851$ times as much in 1980 as it did in 2000. Therefore, $50,000 in 2000 had the same buying power as did $.47851 \times \$50{,}000 = \$23{,}926$ in 1980.

10. According to the CPI, $5.00 in 1990 had the same buying power as did

$$\frac{172.2}{130.7} \cdot \$5.00 = \$6.59$$

in 2000. So if an item cost $5.00 in 1990 and $6.50 in 2000, then its price went up a little slower than inflation.

11. In 2000, the CPI was 172.2 and in 1990, the CPI was 130.7. Therefore, adjusting for inflation, a price of $2.39 in 1990 corresponds to

$$\frac{172.2}{130.7} \cdot \$2.39 = \$3.15$$

in 2000. Thus, to make a fair comparison between the prices, we should calculate the percent increase from $3.15 to $3.89, instead of the percent increase from $2.39 to $3.89. Because

$$\frac{3.89}{3.15} = 1.23$$

the price of the cereal increased by 23%, after adjusting for inflation.

PROBLEMS FOR SECTION 7.6

1. You can make grape juice by mixing 1 can of frozen grape juice concentrate with 3 cans of water. Use simple reasoning with multiplication and division to find at least 4 other ways of mixing grape juice concentrate with water so that the result will taste the same. At least 2 of your ways should involve numbers that are not whole numbers.

2. You can make a pink paint by mixing $2\frac{1}{2}$ cups white paint with $1\frac{3}{4}$ cups red paint. Use simple reasoning with multiplication and divison to find at least 4 other ways of mixing white paint and red paint to make the same shade of pink paint.

3. Pat mixed 2 cups of blue paint with 5 cups of yellow paint to make a green paint. For each of the following fractions and division problems, interpret the fraction or the division problem in terms of Pat's paint mixture and explain why your interpretation makes sense:

    $\frac{2}{5}$ or $2 \div 5$; $\frac{5}{2}$ or $5 \div 2$; $\frac{2}{7}$ or $2 \div 7$;

    $\frac{7}{2}$ or $7 \div 2$; $\frac{5}{7}$ or $5 \div 7$; $\frac{7}{5}$ or $7 \div 5$

4. Which of the following two mixtures will have a stronger lime flavor?

    • 2 parts lime juice concentrate mixed in 5 parts water

    • 4 parts lime juice concentrate mixed in 7 parts water

    Solve this problem in two different ways, explaining why you can solve the problem the way you do.

5. Snail A moved 6 feet in 7 hours. Snail B moved 7 feet in 8 hours. Both snails moved at constant speeds. Which snail went faster? Solve this problem in two different ways, explaining why you can solve the problem the way you do.

6. You can make a soap bubble mixture by combining 2 tablespoons water with 1 tablespoon liquid dishwashing soap and 4 drops of corn syrup. How much liquid dishwashing soap and how many drops of corn syrup should you use if you want to make soap bubble mixture using 5 tablespoons of water? Solve this problem using only simple reasoning about multiplication or division (or both). Explain your reasoning.

7. You can make concrete by mixing 1 part cement with 2 parts pea gravel and 3 parts sand. How much cement and how much pea gravel should you use if you want to make concrete with 8 cubic feet of sand? Solve this problem using only simple reasoning about multiplication or division (or both). Explain your reasoning.

8. Without using a calculator, explain how you can use reasoning and mental arithmetic to determine which of the following two laundry detergents is a better buy:

   • a box of laundry detergent that washes 80 loads and costs $12.75

   • a box of laundry detergent that washes 36 loads and costs $6.75

9. Brad made some punch by mixing $\frac{1}{2}$ of a cup of grape juice with $\frac{1}{4}$ cup of sparkling water. Brad really likes his punch, so he decides to make a larger batch of it.

   a. Brad wants to make 6 cups of his punch. How much grape juice and how much sparkling water should Brad use? Use the most elementary reasoning you can to solve this problem. Explain your reasoning.

   b. Now Brad wants to make 4 cups of his punch. How much grape juice and how much sparkling water should Brad use? Use the most elementary reasoning you can to solve this problem. Explain your reasoning.

10. a. John was paid $250 for $3\frac{3}{4}$ hours of work. At that rate, how much should John make for $2\frac{1}{2}$ hours of work? Use the most elementary reasoning you can to solve this problem. Explain your reasoning.

    b. John was paid $250 for $3\frac{3}{4}$ hours of work. At that rate, how long should John be willing to work for $100? Use the most elementary reasoning you can to solve this problem. Explain your reasoning.

11. A dough recipe calls for 3 cups of flour and $1\frac{1}{4}$ cups of water. You want to use the same ratio of flour to water to make a dough with 10 cups of flour. How much water should you use?

    a. Solve this problem by setting up a proportion in which you set two fractions equal to each other.

    b. Interpret the two fractions that you set equal to each other in part (a) in terms of the recipe. Explain why it makes sense to set these two fractions equal to each other.

    c. Why does it make sense to cross-multiply the two fractions in part (a)? What is the logic behind the procedure of cross-multiplying?

d. Now solve the problem of how much water to use for 10 cups of flour in a different way, by using the most elementary reasoning you can. Explain your reasoning clearly.

12. A recipe that serves 6 people calls for $1\frac{1}{2}$ cups of rice. How much rice will you need to serve 8 people (assuming that the ratio of people to cups of rice stays the same)?

    a. Solve this problem by setting up a proportion in which you set two fractions equal to each other.

    b. Interpret the two fractions that you set equal to each other in part (a) in terms of the recipe. Explain why it makes sense to set these two fractions equal to each other.

    c. Why does it make sense to cross-multiply the two fractions in part (a)? What is the logic behind the procedure of cross-multiplying?

    d. Now solve the same problem in a different way by using logical thinking and by using the most elementary reasoning you can. Explain your reasoning clearly.

13. Marge made light blue paint by mixing $2\frac{1}{2}$ cups of blue paint with $1\frac{3}{4}$ cups of white paint. Homer poured another cup of white paint in Marge's paint mixture. How many cups of blue paint should Marge add to bring the paint back to its original shade of light blue? Use the most elementary reasoning you can to solve this problem. Explain your reasoning.

14. A batch of lotion was made at a factory by mixing 1.3 liters of ingredient A with 2.7 liters of ingredient B in a mixing vat. By accident, a worker added an extra 0.5 liters of ingredient A to the mixing vat. How many liters of ingredient B should the worker add to the mixing vat so that the ingredients will be in the original ratio? Use the most elementary reasoning you can to solve this problem. Explain your reasoning.

15. If a $\frac{3}{4}$ cup serving of snack food gives you 60% of your daily value of calcium, then what percent of your daily value of calcium is in $\frac{1}{2}$ cup of the snack food?

    a. Solve the problem with the aid of a picture. Explain how your picture helps you to solve the problem.

    b. Now solve the problem numerically. Show your work.

16. If $6,000 is 75% of a company's budget for a project, then what percent of the budget is $10,000?

a. Solve the problem with the aid of a picture. Explain how your picture helps you to solve the problem.

b. Now solve the problem numerically. Show your work.

17. Amy mixed 2 tablespoons of chocolate syrup in $\frac{3}{4}$ cup of milk to make chocolate milk. To make chocolate milk that tastes the same as Amy's, how much chocolate syrup will you need for 1 gallon of milk? Express your answer using an appropriate unit. Explain your reasoning. Recall that: 1 gallon = 4 quarts, 1 quart = 2 pints, 1 pint = 2 cups, 1 cup = 8 fluid ounces, 1 fluid ounce = 2 tablespoons.

18. A 5 gallon bucket filled with water is being pulled from the ground up to a height of 20 feet at a rate of 2 feet every 15 seconds. The bucket has a hole in it, so that water leaks out of the bucket at a rate of 1 quart ($\frac{1}{4}$ gallon) every 3 minutes. How much of the water will be left in the bucket by the time the bucket gets to the top? Solve this problem by using logical thinking and by using the most elementary reasoning you can. Explain your reasoning clearly.

19. Suppose that you have two square garden plots: one is 10 feet by 10 feet and the other is 15 feet by 15 feet. You want to cover both gardens with a one-inch layer of mulch. If the 10-by-10 garden took $3\frac{1}{2}$ bags of mulch, could you calculate how many bags of mulch you'd need for the 15-by-15 garden by setting up the proportion

$$\frac{3\frac{1}{2}}{10} = \frac{x}{15}?$$

Explain clearly why or why not. If the answer is no, is there another proportion that you could set up? It may help you to draw pictures of the gardens.

20. If you can rent 5 DVDs for 5 nights for $5, then at that rate, how much should you expect to pay to rent 1 DVD for 1 night? Solve this problem by using logical thinking and the most elementary reasoning you can. Explain your reasoning clearly.

21. If a crew of 3 people take $2\frac{1}{2}$ hours to clean a house, then how long should a crew of 2 take to clean the same house? Assume that all people in the cleaning crew work at the same steady rate. Solve this problem by using logical thinking and the most elementary reasoning you can. Explain your reasoning clearly.

22. If 6 men take 3 days to dig 8 ditches, then how long would it take 4 men to dig 10 ditches? Assume that all the ditches are the same size and take equally long to dig,

and that all the men work at the same steady rate. Solve this problem by using logical thinking and the most elementary reasoning you can. Explain your reasoning clearly.

23. If liquid pouring at a steady rate from hose A takes 15 minutes to fill a vat and liquid pouring at a steady rate from hose B takes 10 minutes to fill the same vat, then how long will it take for liquid pouring from both hose A and hose B to fill the vat? Solve this problem by using logical thinking and the most elementary reasoning you can. Explain your reasoning clearly.

24. Suppose that 400 pounds of freshly picked tomatoes are 99% water by weight. After one day, the same tomatoes only weigh 200 pounds due to evaporation of water. (The tomatoes consist of water and solids. Only the water evaporates; the solids remain.)

a. How many pounds of solids are present in the tomatoes? (Notice that this is the same when they are freshly picked as after one day.)

b. Therefore, when the tomatoes weigh 200 pounds, what percent of the tomatoes is water?

c. Is it valid to use the following proportion to solve for the percent of water, x, in the tomatoes when they weigh 200 pounds?

$$\frac{0.99}{400} = \frac{x}{200}$$

If not, why not?

25. If a loaf of bread cost $3 in 2002, then how much should a loaf of bread have cost in 1960 based on the CPI? Explain your answer.

26. a. What amount of money had the same buying power in 1970 as $30,000 did in 2000? Explain your answer.

b. What amount of money had the same buying power in 1990 as $30,000 did in 2000? Explain your answer.

27. If an item cost $3.50 in 1980 and $4.50 in 2000, did its price go up faster, slower or the same as inflation? Explain your answer.

28. Suppose that from 1995 to 2000 the price of a box of laundry detergent went from $4.85 to $5.95. After adjusting for inflation, determine by what percent the price of the box of laundry detergent went up or down from 1995 to 2000. Explain your answer.

29. Josie made $40,000 in 1995 and $50,000 in 2000. In absolute dollars, that's a 25% increase in Josie's salary from 1995 to 2000. But what is the difference in buying power? After adjusting for inflation, how much higher or lower was Josie's salary in 2000 than it was in 1995? Give your answer as a percent. Explain your reasoning.

30. Suppose that a person earned $30,000 in 1995 and got a 3% raise every year after that until the year 2000 (so that every year, her salary was 3% higher than her previous year's salary). After adjusting for inflation, how much higher or lower was the person's salary in 2000 than it was in 1995? Give your answer as a percent. Explain your reasoning.

31. According to the University of Georgia Fact Books from 1976 and 1996, expenditures on instruction at UGA were 31.1% of the total budget in 1976 and 18.7% of the total budget in 1996. In 1976 UGA spent $42,672,636 on instruction (and departmental research), whereas in 1996, UGA spent $139,672,851 on instruction. The total enrollment in 1976 was 22,879, and in 1996 it was 29,404. How can we compare expenditures on instruction in 1976 and 1996? We can take several different approaches, as in the following examples:

John says that expenditures on instruction went down by 12.4%.

Alice says that expenditures on instruction went down by 39.9%.

Richard says that expenditures on instruction went up by 227%.

Tia says that Richard's number is misleading because it doesn't adjust for inflation.

Beatrice says that the number of students should also be taken into account.

a. In what sense is John correct?

b. In what sense is Alice correct? (Compare the percentages—treat them as ordinary numbers, such as the prices of shoes. If a pair of shoes goes from $31.10 to $18.70, what percent decrease does that represent?)

c. In what sense is Richard correct?

d. Address Tia's objection. In 1976 the CPI was 56.9; in 1996 the CPI was 156.9. If you adjust for inflation and work in 1996 dollars, then by what percent did expenditures on instruction go up or down from 1976 to 1996?

e. Address Beatrice's point. Compare inflation adjusted expenditures on instruction per student in 1976 and 1996. By what percent did the per student expenditures on instruction go up or down?

Chapter 8

Geometry

Geometry is the study of space and shapes in space. The word **geometry** comes from the Greek and means measurement of the Earth (*geo = Earth, metry = measurement*). Mathematicians of ancient Greece developed fundamental concepts of geometry in order to answer such basic questions about the Earth and its relationship to the Sun, the Moon, and the planets as How big is the Earth? How far away is the Moon? and How far away is the Sun? The geometers of ancient Greece had to be able to visualize the Earth and the heavenly bodies in space. They had to extract the relevant relationships from their mental pictures, and they had to analyze these pictures mathematically. The methods the ancient Greeks developed are still useful today for solving modern problems in construction, road building, medicine, and other disciplines.

In this chapter we will study both two- and three-dimensional shapes. We will focus on visualizing shapes, on analyzing features and properties of shapes, on constructing shapes, and on relating shapes.

Traditionally, very little geometry has been taught in elementary school in the United States. Therefore, you may feel that you won't need to know much geometry in order to teach mathematics in elementary school. However, the recommendations of the National Council of Teachers of Mathematics include a strong component in geometry (see [44]):

NCTM STANDARDS

Instructional programs from prekindergarten through grade 12 should enable all students to

- *analyze characteristics and properties of two- and three-dimensional geometric shapes and develop mathematical arguments about geometric relationships;*

- *specify locations and describe spatial relationships using coordinate geometry and other representational systems;*

- *apply transformations and use symmetry to analyze mathematical situations;*

- *use visualization, spatial reasoning, and geometric modeling to solve problems.*

   For NCTM's specific geometry recommendations for grades Pre-K–2, grades 3–5, grades 6–8, and grades 9–12, see

www.aw-bc.com/beckmann.

8.1 Visualization

One important reason to study geometry is that it promotes the ability to visualize and mentally manipulate objects in space. This is a necessary skill for many professions. For example, a surgeon or dentist must be able to visualize the steps in and outcomes of an operation, a carpenter must be able to see different designs in his or her mind's eye, an architect must be able to visualize many different possibilities for a building that satisfies certain design criteria, and a clothes designer must be able to visualize how pieces of fabric will fit together to make a garment.

According to the report "What Work Requires of Schools" of the U. S. Department of Labor, [50], being able to "see things in the mind's eye" is a foundational skill for solid job performance. In order to be able to foster this in your students, you must first foster it in yourself.

This section aims to help you improve your visualization skills. For some people, visualization comes naturally; for others it is more of a struggle. No matter where you find yourself in this spectrum, you can improve your visualization skills by working at it.

In the class activities, practice problems, and problems of this section, you will encounter some basic objects and shapes studied in geometry: points, lines, planes, triangles, squares, rectangles, rhombuses, quadrilaterals, pentagons, hexagons, and circles, as well as prisms, cones, pyramids, and cylinders. Figures 8.1, 8.2, and 8.3 show examples of most of these.

FIGURE 8.1

Points, Lines, Planes; Arrow Heads Indicate That the Line or Ray Extends Indefinitely in That Direction

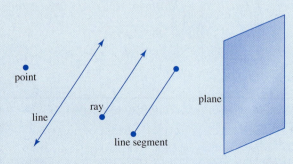

FIGURE 8.2

Examples of
Shapes

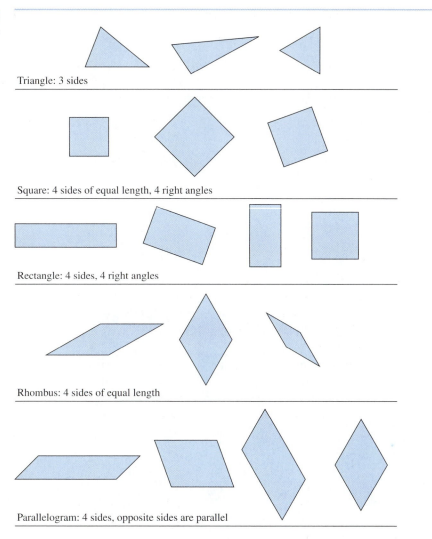

Triangle: 3 sides

Square: 4 sides of equal length, 4 right angles

Rectangle: 4 sides, 4 right angles

Rhombus: 4 sides of equal length

Parallelogram: 4 sides, opposite sides are parallel

The terms **point**, **line**, and **plane** are usually considered primitive, undefined terms; after all, you have to start somewhere. Even so, we can describe how to visualize points, lines, and planes:

- To visualize a **point**, think of a tiny dot, such as the period at the end of a sentence. A point is an idealized version of a dot, having no size or shape.

- To visualize a **line**, think of an infinitely long, stretched string that has no beginning or end. A line is an idealized version of such a string, having no thickness.

- To visualize a **plane**, think of an infinite flat piece of paper that has no beginning or end. A plane is an idealized version of such a piece of paper, having no thickness.

line segment   Related to lines are *line segments* and *rays*, as pictured in Figure 8.1. A **line segment** is the part of a line lying between two points on a line. These two points are called the **endpoints** of the line segment. You can think of a line segment as having both a beginning and an end,

FIGURE 8.3

More Examples
of Shapes

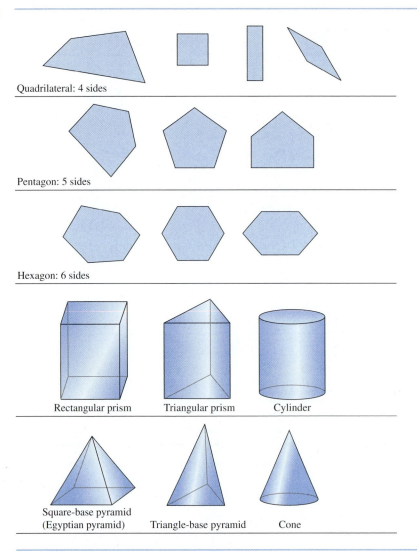

Quadrilateral: 4 sides

Pentagon: 5 sides

Hexagon: 6 sides

Rectangular prism   Triangular prism   Cylinder

Square-base pyramid
(Egyptian pyramid)   Triangle-base pyramid   Cone

ray   even though both points are called endpoints. A **ray** is the part of a line lying on one side of a point on the line. You can think of a ray as having a beginning, but no end.

Careful mathematical definitions for other shapes, pictured in Figures 8.2 and 8.3, will be given later in this chapter. For now, refer to Figures 8.2 and 8.3 so that when you encounter a shape, you can use the proper name for it.

The following activities will help you practice visualizing.

CLASS ACTIVITY NOW TURN TO CLASS ACTIVITIES MANUAL

PRACTICE PROBLEMS FOR SECTION 8.1

1. Figure 8.4 shows small versions of several patterns for shapes. Larger versions of these patterns are shown in Figure A.1 on page 711. Before you cut, fold, and tape the larger patterns in Figure A.1, visualize the folding process in order to help you visualize the final shapes. Then cut out the patterns on page 711 along the solid lines, fold down along the dotted lines, and tape sides with matching labels together. Does the shape match your visualized shape?

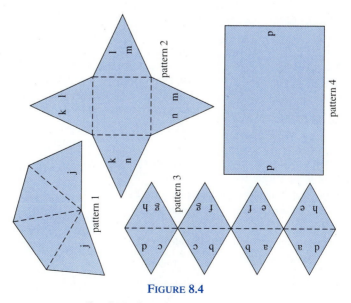

FIGURE 8.4

Small Version of Patterns for Shapes

2. Figure 8.5 shows small versions of six patterns for shapes. Larger versions of these patterns are shown in Figure A.2 on page 713. Before you cut and tape the larger patterns in Figure A.2, try to visualize the final shapes. Predict how the shapes will be alike and how they will be different. Then cut out the patterns on page 713 along the solid lines and tape sides with matching labels together. (Don't do any folding!)

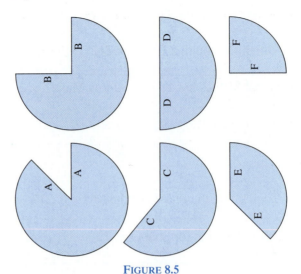

FIGURE 8.5

Small Version of Patterns for Shapes

3. What familiar parts of clothing are made from the patterns in Figure 8.6 when sides with matching labels are sewn together? Visualize!

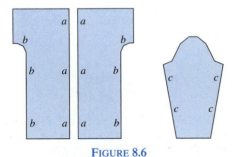

FIGURE 8.6

What Parts of Clothes Do These Patterns Make?

4. Make a paper model of a cone (like an ice cream cone). Now visualize a plane slicing through the cone. The places where the plane cuts through the cone form a shape in the plane. Describe all possible shapes in the plane that can be made this way, by slicing the cone. Use your model to help you, but also visualize each case without the use of your model.

5. Make a paper model of a cube. Now visualize a plane slicing through the cube. The places where the plane cuts through the cube form a shape in the plane. Describe how to choose a plane so that the slice forms the following shapes:

- a square
- a rectangle that is not a square
- a triangle
- a rhombus that is not a square
- a hexagon

6. Figure 8.7 shows a diagram of the Earth as seen from outer space looking down on the North Pole, which is marked N. How does the Sun appear to people at locations A, B, C, and D? Use your answers, together with the fact that the Sun rises in the east and sets in the west, to explain what time of day it is at A, B, C, and D.

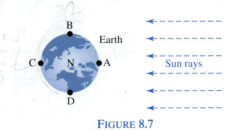

FIGURE 8.7

What Time Is It?

7. Imagine floating in outer space above the North Pole. Looking down on the Earth, which way is the Earth rotating, clockwise, or counterclockwise? Use your results in Practice Problem 6 to explain your answer.

   If you were floating above the South Pole instead of the North Pole, would the rotation look different?

8. Figure 8.8 shows four different positions of the Earth and Moon as seen from outer space, floating above the North Pole (not to scale). The Moon rotates around the Earth in the direction indicated. In which of diagrams A, B, C, and D is the Moon waxing (getting bigger), in which is it waning (getting smaller)? Explain your answers.

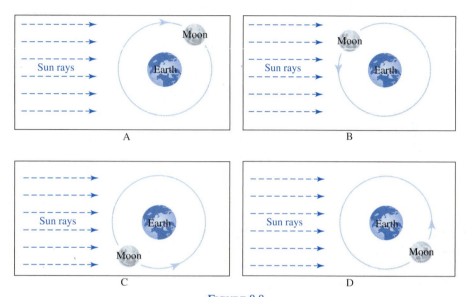

FIGURE 8.8

Waxing and Waning Moon

ANSWERS TO PRACTICE PROBLEMS FOR SECTION 8.1

1. Patterns 1 and 2 make pyramids, without and with a base, respectively. Pattern 3 makes a shape that looks like two pyramids joined at their bases. This shape is called an **octahedron**. Pattern 4 makes a cylinder.

2. All six patterns make cones. Some are shorter and wider; some are taller and narrower.

3. The two patterns on the left form a pant leg when they are sewn together. The curved portion of side *b* makes the crotch. The pattern on the right makes a sleeve. The curved portion at the top makes the arm hole. The seam at edge *c* runs straight down the arm, from the armpit to the wrist.

4. With the cone positioned as shown in Figure 8.9, horizontal planes slice the cone either in a single point or in a circle (you get a single point if the plane goes through the very bottom, small circles near the bottom, larger circles as you go up). A vertical plane slices the cone in either a V-shape or a curve called a **hyperbola**. A slanted plane slices the cone in either an **ellipse** (oval shape) or in a curve called a **parabola**. This is why circles, ellipses, hyperbolas, and parabolas are collectively referred to as **conic sections**, because they come from slicing a cone.

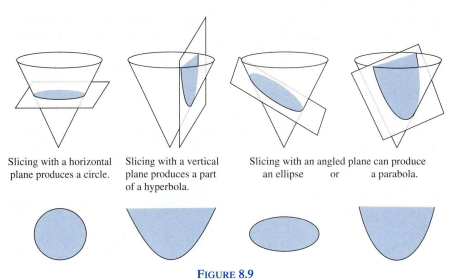

FIGURE 8.9

Slicing a Cone with a Plane

5. See Figure 8.10. Horizontal planes slice the cube in squares. A vertical plane slices the cube in either a square or a rectangle that is not a square. You can get a triangle, a rhombus, or a hexagon by slicing with angled planes. To see how to get a hexagon, cut out the two patterns in Figure A.3 on page 715 along the solid lines, fold them down along the dotted lines, and tape them to create two solid shapes. The two shapes can be put together at the hexagon to make a cube; therefore, a plane can slice a cube to create a hexagon at the slice.

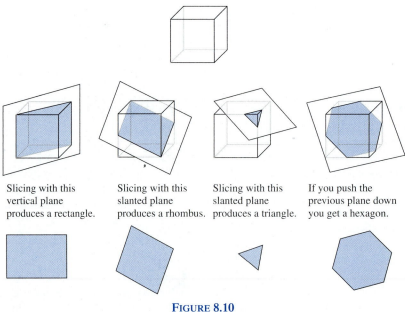

FIGURE 8.10

Slicing a Cube with a Plane

6. Picture yourself at point A in Figure 8.7. At this location, the Sun's rays are directly overhead, which means it must be about noon. Location C is not receiving any sunlight and is directly opposite the location marked noon, so it must be about midnight at point C. Now picture yourself at points B and D. At both locations, the Sun is on the horizon because the Sun's rays just graze the surface of the Earth there. Therefore, it must be sunrise or sunset at B and D. How can we tell which is which? Picture yourself standing at point D and facing north. The Sun is on your right, which is to the east. Since the Sun rises in the east, it is sunrise at point D. Now picture yourself standing at point B and facing north. Now the Sun is on your left, which is to the west. Since the Sun sets in the west, it is sunset at point B.

7. According to the results of the previous practice problem, noon, midnight, sunrise, and sunset are at the locations indicated in Figure 8.11. Because the day progresses from midnight to sunrise to noon to sunset, the Earth must be rotating counterclockwise when viewed from above the North Pole. However, when viewed

from above the South Pole, the Earth rotates in the *clockwise* direction. You can see why the sense of rotation is reversed by doing the following: Hold a small ball between your thumb and index finger. Let your index finger represent the North Pole and let your thumb represent the South Pole. Now rotate the ball counterclockwise when viewed looking down on your index finger (North Pole). Keep rotating the ball in the same way, but now look down on your thumb (South Pole). Notice that the same rotation now appears clockwise.

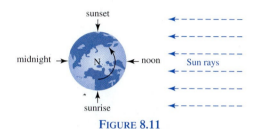

FIGURE 8.11

The Earth's Rotation as Seen Looking Down
on the North Pole

8. See Figure 8.12. The shaded portions in the Moon show what part of the Moon is visible from Earth. The portion of the Moon that is visible from Earth is the portion that is facing the Earth *and also* illuminated by the Sun. The dashed lines on the Moon help you see the part of the Moon that is illuminated by the Sun and the part of the Moon that is facing the Earth. In diagram A, the Moon is just past being full. (The Moon is full when it is opposite the Sun, with the Earth in between.) As the Moon moves from the configuration in A to the configuration shown in B and beyond, less of the Moon becomes visible. Therefore the Moon is waning in A and B. After the configuration shown in B, the Moon becomes new (when it is between the Sun and Earth) and then begins to become visible again, as in C. In D, even more of the Moon is visible than in C. So the Moon is waxing in C and D.

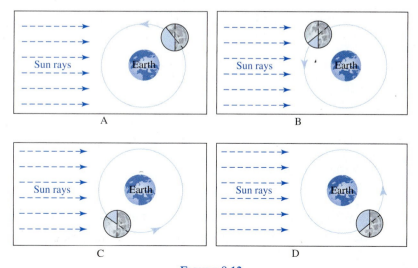

FIGURE 8.12

Is the Moon Waxing or Waning?

PROBLEMS FOR SECTION 8.1

1. Find all the different patterns for a closed 3-dimensional shape that are made out of 4 copies of the triangle in Figure 8.13 in such a way that each triangle is joined to another triangle along a whole edge, not just at a corner. (This kind of pattern is sometimes called a *net*.) For two patterns to be considered different, you should not be able to match the patterns up when they are cut out. If you cut out, fold, and tape your patterns, the shape that you make is called a *tetrahedron* or a pyramid with a triangle base.

FIGURE 8.13

A Triangle

2. a. Make three different patterns that could be cut out, folded, and taped to make an open-top 1-inch-by-1-inch-by-1-inch cube. In your patterns, each square should be joined to another square along a whole edge of the square, not just at a corner. (This kind of pattern is sometimes called a *net*.) For two patterns to be considered different, you should not be able to match the patterns up when they are cut out.

   b. How many different patterns for an open-top cube as described in part (a) are there? Find all such patterns.

3. a. Make three different patterns that could be cut out, folded, and taped to make a closed 1-inch-by-1-inch-by-1-inch cube. In your patterns, each square should be joined to another square along a whole edge of the square, not just at a corner. (This kind of pattern is sometimes called a *net*.) For two patterns to be considered different, you should not be able to match the patterns up when they are cut out.

   b. How many different patterns for a closed cube as described in part (a) are there? Find all such patterns.

4. The photo of the Earth in Figure 8.14 was taken by NASA. Identify some of the landforms shown. Use the landforms and the position of the shadow to explain how you can determine whether it is sunrise or sunset at the edge of the shadow.

FIGURE 8.14

A Photograph Taken by NASA

5. Figure 8.15 shows one possible arrangement of the Earth, Moon, and Sun as seen from outer space (not to scale). In this arrangement, what does the Moon look like to people on Earth who can see it? Is the Moon waxing or waning? Explain your answers.

FIGURE 8.15

In What Phase Is the Moon?

6. Figure 8.16 shows one possible arrangement of the Earth, Moon, and Sun as seen from outer space (not to scale). In this arrangement, what does the Moon look like to people on Earth who can see it? Is the Moon waxing or waning? Explain your answers.

FIGURE 8.16

In What Phase Is the Moon?

7. Figure 8.17 shows two possible arrangements of the Earth, Moon, and Sun as seen from outer space (not to scale). In these arrangements, what does the Moon look like to people on Earth who can see it? What is the relevance of the sentence, "Sun rays are not in the plane of the page; they come from slightly above this plane," which appears at the top of Figure 8.17? Explain your answers.

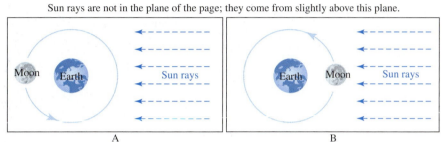

Sun rays are not in the plane of the page; they come from slightly above this plane.

FIGURE 8.17

In What Phase Is the Moon?

8. Figure 8.18 shows one possible arrangement of the Earth, Moon, and Sun as seen from outer space, shown to scale. In this arrangement, what does the Moon look like to people on Earth who can see it? Assuming that the Moon orbits counterclockwise around the Earth in the plane of the page, is the Moon waxing or waning? Explain your answers.

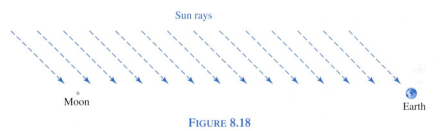

FIGURE 8.18

The Earth and Moon Shown to Scale

9. Figure 8.19 shows one possible configuration of the Earth, Moon, and Sun as seen from outer space, shown to scale. In this configuration, what does the Moon look like to people on Earth who can see it? Assuming that the Moon orbits counterclockwise around the Earth in the plane of the page, is the Moon waxing or waning? Explain your answers.

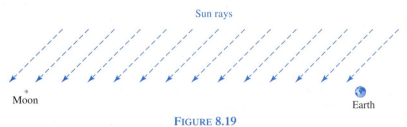

FIGURE 8.19

The Earth and Moon Shown to Scale

10. a. Figure 8.20 shows the Earth as seen from outer space, looking down on the North Pole (labeled N). What does the Sun look like to a person standing at point P? Therefore, what time of day is it at point P? Use the fact that the Sun rises in the east and sets in the west in your answer.

   b. If a person at point P in Figure 8.20 can see the Moon, and if the Moon is neither new nor full, then is the Moon waxing or is it waning? Explain how you can tell from the diagram.

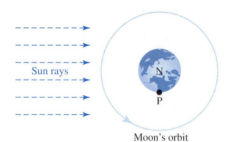

FIGURE 8.20

The Earth

11. a. Figure 8.21 shows the Earth as seen from outer space, looking down on the North Pole (labeled N). What does the Sun look like to a person standing at point P? Therefore, what time of day is it at point P? Use the fact that the Sun rises in the east and sets in the west in your answer.

   b. If a person at point P in Figure 8.21 can see the Moon, and if the Moon is neither new nor full, then

is the Moon waxing or is it waning? Explain how you can tell from the diagram.

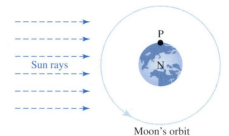

FIGURE 8.21

The Earth

12. We can often see the Moon during the day, even in the middle of the day. Mike said that he saw a full Moon at 2 P.M. Draw a diagram (like one in Figure 8.8) to help you explain why Mike couldn't be right.

13. If you go outside shortly after sunrise and you see the Moon up high in the sky, is the Moon waxing or waning? Draw a diagram (like one in Figure 8.8) to help you explain your answer.

14. Write a report on Greenwich Mean Time and the International Date Line. Explain why it is necessary to have a date line somewhere in the world.

8.2 Angles

Angles are used in two ways: to represent an amount of rotation (turning) about a fixed point, and to describe how two rays (or lines, line segments, or even planes) meet at a point. This section discusses basic facts about angles and gives some applications of angles that you might find surprising.

The two ways of thinking about angles, as amounts of rotation and as rays meeting, are closely related. It is equally valid to define angles from either point of view. Since even very young children have experience spinning around, the "rotation" point of view is perhaps more primitive. So, our first definition is that an **angle** is an amount of rotation about a fixed point. An angle at a point P is said to be **congruent** to an angle at a point Q if both represent the same amount of rotation—even though this rotation takes place around different points. (See Figure 8.22.)

Now to the other point of view about angles. Suppose there are two rays in a plane and these rays have a common endpoint P, as illustrated in Figure 8.23. In this case, the two rays

angle

congruent

FIGURE 8.22

The Same
Amount of
Rotation around
Different Points

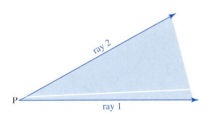

FIGURE 8.23

An Angle
Formed by Two
Rays Meeting at
Point P

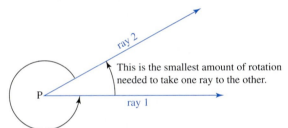

FIGURE 8.24

Amounts of
Rotation
Associated with
Two Rays

This is the smallest amount of rotation
needed to take one ray to the other.

This is another amount of rotation that takes one
ray to the other ray, but it is not the smallest possible
amount of rotation.

angle   and the region between them is called the **angle** at P formed by the two rays. This defintion of angle relates to the "rotation" definition by associating the two rays meeting at a point P to the smallest amount of counterclockwise rotation about P needed to rotate one of the rays to the position of the other ray. (See Figure 8.24.)

When there are two line segments in a plane and these line segments have a common endpoint P, then we can define the angle between these line segments in the same way as for angles between rays. (Or we can simply extend the line segments to rays having P as an endpoint and use the definition for angles between rays.)

degree   Angles are commonly measured in **degrees**. Degrees are usually indicated with a small circle: °. For example, 30 degrees is usually written $30°$. If you stand at a fixed point and rotate counterclockwise in a full circle to return to your starting position, then you will have rotated $360°$. If you stand at a point and rotate one-half of a full turn, then you will have rotated $180°$; a quarter turn is $90°$, and so on.

If two rays meet at a common endpoint Q to form a straight line, as illustrated in Figure 8.25 on page 330, then the angle at Q is $180°$. So, if you were standing at the point Q looking straight down one side of the line and rotated counterclockwise to look down the other side of the line, you would have rotated $180°$.

FIGURE 8.25

Two Rays
Forming a
Straight Line

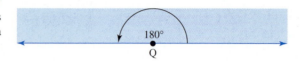

FIGURE 8.26

Some Angles

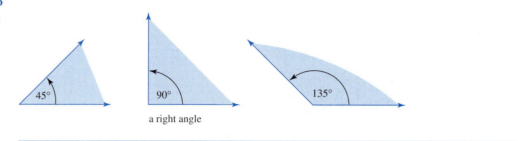

A technical point: What about clockwise rotations? Clockwise rotations give rise to negative angles. So, for example, if you stand at a point and you rotate a quarter turn clockwise, then you have rotated $-90°$. You need not be concerned with negative angles since we will not be working with them.

Notice that a small angle formed by rays has a "pointier" look than a large angle formed by rays, as seen in Figure 8.26. Angles less than $90°$ are called **acute angles**, an angle of $90°$
right angle   is called a right angle, while angles greater than $90°$ are called **obtuse** angles. Right angles are often indicated by a small square, as shown in Figure 8.27.

protractor   Angles formed by rays can be measured with a simple device called a protractor. Figure 8.28 shows a protractor measuring an angle. Protractors usually have a small hole that should be placed directly over the point where the two rays meet. This hole lies on a horizontal line through the protractor. This horizontal line should be lined up with one of the rays that form the angle to be measured. The protractor in Figure 8.28 shows that the angle it is measuring is $65°$.

Some teachers help their students learn about angles by making and using "angle explorers." An angle explorer is made by attaching two strips of cardboard with a brass fastener, as shown in Figure 8.29. By rotating the cardboard strips around, you get a feel for the relative sizes of angles.

When two lines in a plane meet, they form four angles. Figure 8.30 shows some examples.
perpendicular   When all four of the angles are $90°$, we say that the two lines are perpendicular.

Two lines in a plane that *never* meet (even somewhere far off the page they are drawn on)
parallel   are called parallel. Figure 8.31 shows examples of lines that are parallel and lines that are not parallel even though you can't see where they meet.

FIGURE 8.27

Indicating a
Right Angle

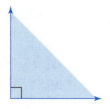

FIGURE 8.28

A Protractor
Measuring an
Angle

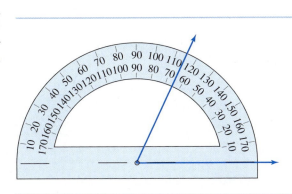

FIGURE 8.29

An "Angle
Explorer"

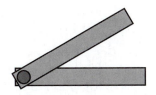

FIGURE 8.30

Two Lines
Meeting at a
Point Form 4
Angles

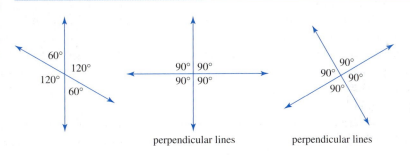

perpendicular lines   perpendicular lines

FIGURE 8.31

Parallel Lines
and Lines That
Are Not Parallel

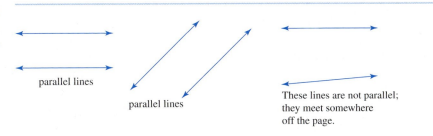

parallel lines

parallel lines

These lines are not parallel;
they meet somewhere
off the page.

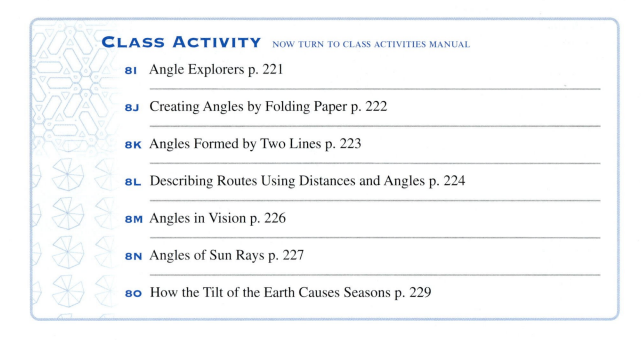

ANGLES AND REFLECTED LIGHT

When a light ray strikes a smooth reflective surface, such as a mirror, the light ray reflects in a specific way that can be described with angles.

normal line   In order to describe how light reflection works, we will need the concept of a **normal line** to a surface. The normal line at a point on a surface is the line that passes through that point and is perpendicular to the surface at that point. In order not to get too technical, you can think of "perpendicular to the surface at that point" as meaning "sticking straight out away from the surface at that point." Figure 8.32 shows a cross section of a "wiggly" surface and its normal lines at various points.

There are two fundamental physical laws that govern the reflection of light from a surface. These laws also apply to the reflection of similar radiation, such as microwaves and radio waves:

1.   The incoming and reflected light rays make the *same angle with the normal line* at the point where the incoming light ray hits the surface.

2.   The reflected light ray lies in the same plane as the normal line and the incoming light ray. The reflected light ray is not in the same location as the incoming light ray unless the incoming light ray is in the same location as the normal line.

FIGURE 8.32

Some Normal Lines to a Surface (Cross Section Shown)

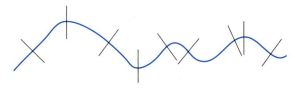

FIGURE 8.33

Light Rays
Reflecting Off
Surfaces

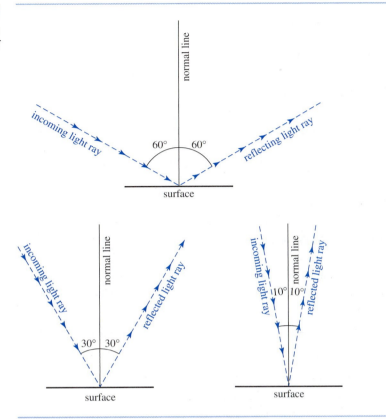

Figure 8.33 shows examples of how light rays reflect from surfaces. In each case, a cross section of the surface is shown.

You can demonstrate the reflection laws by putting a mirror on a table, shining a penlight at the mirror, and seeing where the reflected light hits the wall. (Be careful to keep the light beam away from your and others' eyes.) If you raise and lower the light, while still pointing it at the mirror, the light beam traveling toward the mirror will form different angles with the mirror. By observing the reflected light beam on the wall, you can tell that the lightbeams going toward and from the mirror make the same angle with the normal line to the mirror. (If the mirror is on a horizontal table, then the normal lines to the mirror are vertical.)

PRACTICE PROBLEMS FOR SECTION 8.2

1. My son once told me that some skateboarders can do "ten-eighties". I said he must mean 180's, not 1080's. My son was right, some skateboarders can do 1080's! What is a 1080 and why is it called that? By contrast, what would a 180 be?

2. Use a protractor to measure the angles formed by the shape in Figure 8.34.

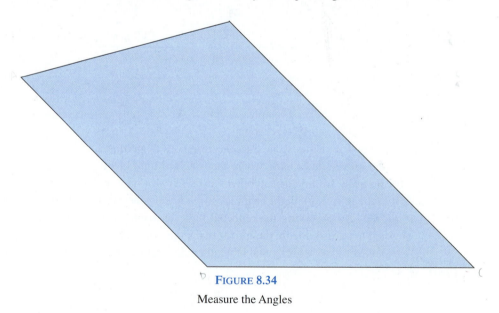

FIGURE 8.34

Measure the Angles

3. Explain why the angle at A in Figure 8.35 is not larger than the angle at B.

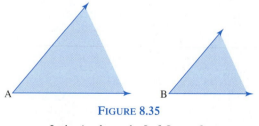

FIGURE 8.35

Is the Angle on the Left Larger?

4. Suppose that two lines in a plane meet at a point, as in Figure 8.36. Use the fact that the angle formed by a straight line is 180° to explain why $a = c$ and $b = d$.

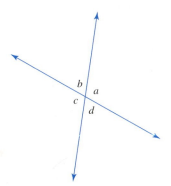

FIGURE 8.36

Lines Meeting at a Point

5. Draw pictures to show the relationship between the angle that the Sun's rays make with horizontal ground and the length of the shadow of a telephone pole.

6. Figure 8.37 shows a side view of a flashlight shining on a puppet behind a semitransparent screen. Show why the shadow of the puppet on the screen is bigger than the puppet.

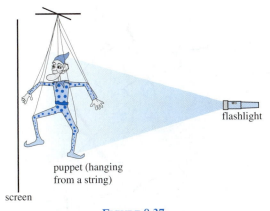

flashlight

puppet (hanging from a string)

screen

FIGURE 8.37

How Big Is the Shadow of the Puppet?

7. Dorothy walks from point A to point F along the route indicated on the map in Figure 8.38.

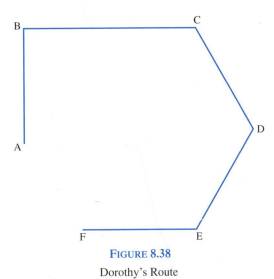

FIGURE 8.38

Dorothy's Route

a. Show Dorothy's angles of turning along her route. Use a protractor to measure these angles.

b. Use your answers to part (a) to determine Dorothy's total amount of turning along her route.

c. Now determine Dorothy's total amount of turning along her route in a different way than in part (b).

8. Figure 8.39 shows several diagrams (from the point of view of a fly looking down from the ceiling) of a person standing in a room, looking into a mirror on the wall. The direction of the person's gaze is indicated with a dashed line. What place in the room will the person see in the mirror?

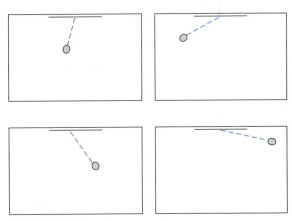

FIGURE 8.39

Looking into a Mirror

9. Figure 8.40 shows a mirror seen from the top, and a light ray hitting the mirror. Draw a copy of this picture on a blank piece of paper. Use the following paper-folding method to show the location of the reflected light ray:

- Fold and crease the paper so that the crease goes through the point where the light ray hits the mirror and so that the line labeled "mirror" folds onto itself. (This crease is perpendicular to the line labeled "mirror." You'll be asked to explain why below.)

- Keep the paper folded and now fold and crease the paper again along the line labeled "light ray."

- Unfold the paper. The first crease is the normal line to the mirror. The second crease shows the light ray and the reflected light ray.

FIGURE 8.40

Using Paper Folding to Find the Reflected Ray

a. Explain why your first crease is perpendicular to the line labeled "mirror".

b. Explain why your second crease shows the reflected light ray.

ANSWERS TO PRACTICE PROBLEMS FOR SECTION 8.2

1. A "ten-eighty" is three full rotations. This makes sense because a full rotation is 360° and

$$3 \times 360° = 1080°$$

A "180" would be half of a full rotation, which is not very impressive by comparison, although *I* certainly couldn't do it on a skateboard.

2. See Figure 8.41.

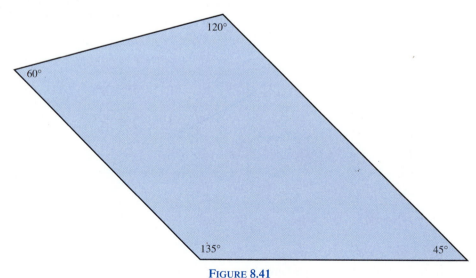

FIGURE 8.41

Angles Formed by a Shape

3. Even though the *line segments* forming the angle at A are longer, the angle at A is not larger than the angle at B. This is because the lower line segments of both angles would need to be rotated the same amount (about points A and B respectively) to get to the location of the upper line segments of the angles.

4. Since angles a and b together make up the angle formed by a straight line,

$$a + b = 180°$$

For the same reason,

$$b + c = 180°$$

So

$$a = 180° - b$$

and

$$c = 180° - b$$

Since a and c are both equal to $180° - b$, they are equal to each other (i.e., $a = c$). The same argument (with the letters changed) explains why $b = d$.

5. Figure 8.42 shows that when the Sun's rays make a smaller angle with horizontal ground, a telephone pole makes a longer shadow than when the Sun's rays make a larger angle with the ground.

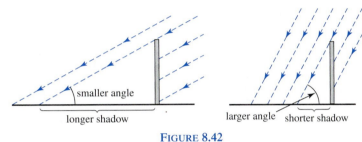

FIGURE 8.42

Sun Rays Hitting a Telephone Pole

6. Figure 8.43 shows that the shadow of the puppet is larger than the puppet itself. This occurs because the light rays from the flashlight grazing the top and bottom of the puppet are not horizontal. If the puppet were farther from the screen, or if the flashlight were closer to the puppet, the puppet's shadow would be even larger.

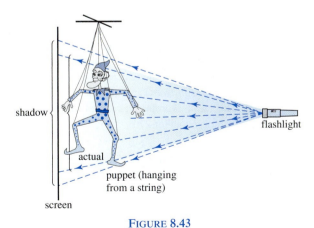

FIGURE 8.43

The Puppet's Shadow

7. a. Figure 8.44 shows Dorothy's angles of turning. At each point where the path turns, the straight arrow shows the direction that Dorothy faces before she turns. The round arrow indicates Dorothy's angle of turning.

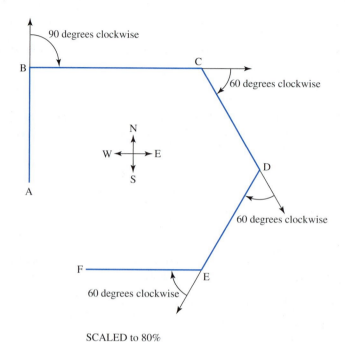

SCALED to 80%

FIGURE 8.44

Dorothy's Angles of Turning along Her Route

b. Based on the results of part (a), Dorothy turns a total of

$$90° + 60° + 60° + 60° = 270°$$

along her route.

c. Another way to determine Dorothy's total amount of turning is by considering the directions she faces as she walks along her route. If we think of Dorothy as starting out facing north, then she ends up facing west, and in between she faces all the directions clockwise from north to west. That means Dorothy makes $\frac{3}{4}$ of a full 360° turn during her walk. Since $\frac{3}{4}$ of 360° is 270°, Dorothy turns a total of 270°.

8. Figure 8.45 shows that the person will see locations A, B, C, and D, respectively, which are on various walls. Location C is in a corner. As the pictures show, the incoming light rays and their reflections in the mirror make the same angle with the normal line to the mirror at the point of reflection.

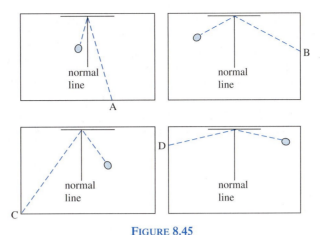

FIGURE 8.45

What a Person Sees Looking in a Mirror

---

9. a. The first crease is made so that angles a and b shown in Figure 8.46 are folded on top of each other and completely aligned. Therefore, angles a and b are equal. But angle a and angle b must add up to $180°$ because the angle formed by a straight line is $180°$. Hence, angle a and angle b must both be half of $180°$, which is $90°$. Therefore, this first crease is the normal line to the mirror at the point where the light ray hits the mirror.

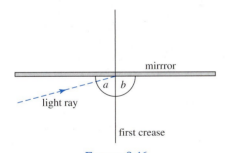

FIGURE 8.46

The First Crease

b. The second crease is made so that angles c and d, shown in Figure 8.47 are folded on top of each other and completely aligned. Therefore, angle c is equal to angle d. Since the first crease is a normal line, by the reflection laws, the second crease shows the reflected light ray.

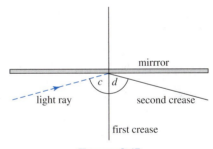

FIGURE 8.47

The Second Crease

PROBLEMS FOR SECTION 8.2

1. Figure 8.48 shows small versions of "pie wedges." Larger versions are in Figure A.4 on page 717. You can cut out these larger versions of the "pie wedges" and use them to show various angles. If available, you may want to glue these large wedges onto cardboard to make them sturdier.

Find at least two trees with low branches or bushes that are not too dense. Use your "pie wedges" to measure the angles with which limbs of the trees or bushes meet the main trunk. Also, measure the angles with which smaller branches meet main branches. (Of course you will only be able to approximate these angles because you only have so many pie wedges and because real tree branches are not as neat as straight lines.) Measure at least 6 angles for each tree or bush and record your data, noting which angles come from main limbs and which come from smaller branches. Which angles are most common? Which angles, if any, did you not find at all? Do the most common angles you find vary from tree to tree? Were the angles from the main limbs typically different from the angles from the smaller branches?

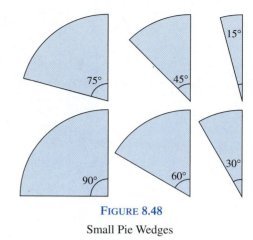

FIGURE 8.48

Small Pie Wedges

2. Amanda got in her car at point A and drove to point F along the route shown on the map in Figure 8.49.

   a. Trace Amanda's route shown in Figure 8.49; show all of Amanda's angles of turning along her route. Use a protractor to measure these angles.

   b. Determine Amanda's total amount of turning along her route by adding the angles you measured in part (a).

   c. Now describe a way to determine Amanda's total amount of turning along her route *without* measuring the individual angles and adding them up. Hint: Consider the directions that Amanda faces as she travels along her route.

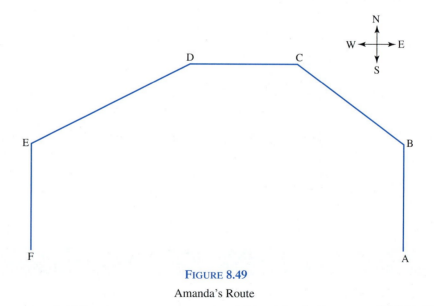

FIGURE 8.49

Amanda's Route

3. Ed is standing at point A and will walk to point D along the route shown on the map in Figure 8.50. On the map, 1 inch represents 10 of Ed's paces.

   a. Trace the map in Figure 8.50. At the points where Ed will turn, indicate his angle of turning. Use a protractor to measure these angles and mark them on your map.

b. Ed is blindfolded, so you must tell Ed exactly how to walk to point D. At points where Ed must turn, tell him how many degrees to turn, and which way.

c. Determine Ed's total amount of turning along his route by adding the angles you measured in part (a).

d. Now determine Ed's total amount of turning along his route *without* measuring the individual angles and adding them. Hint: Consider the directions that Ed faces as he travels along his route.

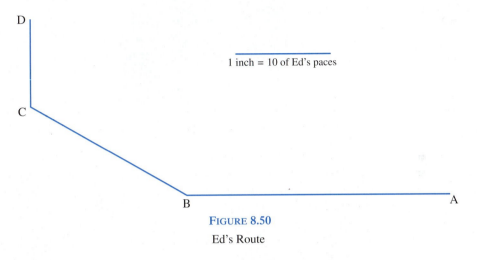

FIGURE 8.50

Ed's Route

4. Draw a map that shows the following route leading to buried treasure:

Starting at the old tree, walk 10 paces heading straight for the tallest mountain in the distance. Turn clockwise, 90°. Walk 20 paces. Turn counterclockwise, 120°. Walk 40 paces. Turn clockwise, 60°. Walk 20 paces. This is the spot where the treasure is buried.

5. Figure 8.51 on page 342 depicts Sun rays traveling toward three points on the Earth's surface, labeled A, B, and C. Assume that these Sun rays are parallel to the plane of the page.

a. Use a protractor to determine the angle that the Sun's rays make with horizontal ground at points A, B, and C.

b. Suppose there are telephone poles of the same height at locations A, B, and C. Draw pictures to show the different lengths of the shadows of these telephone poles.

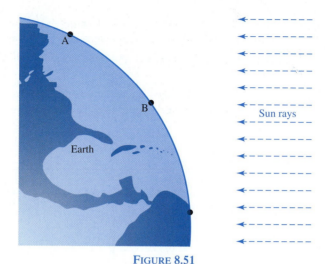

FIGURE 8.51

Sun Rays Traveling to the Earth

6. Many people mistakenly believe that the seasons are caused by the Earth's varying proximity to the Sun. In fact, the distance from the Earth to the Sun varies only slightly during the year and the seasons are caused by the tilt of the Earth's axis. As the Earth travels around the Sun during the year, the tilt of the Earth's axis causes the northern hemisphere to vary between being tilted toward the Sun to being tilted away from the Sun.

   Figure 8.52 shows the Earth as seen at a certain time of year from a point in outer space located in the plane in which the Earth rotates about the Sun. The diagram shows that the Earth's axis is tilted 23.5° from the perpendicular to the plane in which the Earth rotates around the Sun. Assume that the Sun rays are parallel to the plane of the page.

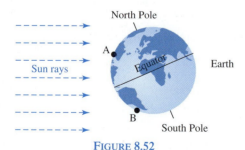

FIGURE 8.52

The Earth and Sun

a. Locations A and B in Figure 8.52 are shown at noon. Is the Sun higher in the sky at noon at location A or at location B? Explain how you can tell.

b. During the day, locations A and B will rotate around the axis through the North and South Poles. Compare the amount of sunlight that locations A and B will receive throughout the day. Which location will receive more sunlight during the day?

c. Based on your answers to parts (a) and (b), what season is it in the northern hemisphere and what season is it in the southern hemisphere in Figure 8.52? Explain.

d. At other times of year, the Earth and Sun are positioned as shown in Figure 8.53. (Assume that the Sun rays are parallel to the plane of the page.) At those times, what season is it in the northern and southern hemispheres? Why? (Notice that the second diagram in Figure 8.53 still shows the tilt of the Earth's axis.)

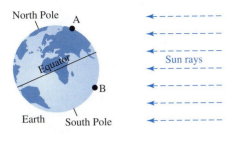

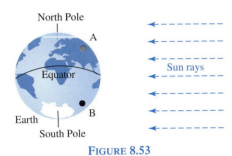

FIGURE 8.53

The Earth and Sun at Other Times of Year

e. Refer to Figures 8.52 and 8.53 and the results of the previous parts of this problem to answer the following: During which seasons are the Sun's rays most intense at the equator? Look carefully before you answer—the answer may surprise you.

7. Refer to Figures 8.52, 8.53, and 8.54 to help you answer the following: There are only certain locations on Earth where the Sun can ever be seen *directly* overhead. Where are these locations? How are these locations related to the Tropic of Cancer and the Tropic of Capricorn? Explain.

FIGURE 8.54

The Tropic of Cancer and
the Tropic of Capricorn

8. Figure 8.55 shows a cross section of Joey's toy periscope. What will Joey see when he looks in the periscope? Explain, using the laws of reflection (trace the periscope). What would be a better way to position the mirror in the telescope?

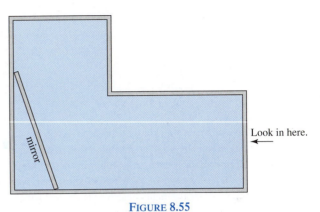

FIGURE 8.55

Joey's Periscope

9. Copy Figure 8.56 and use the laws of reflection to show how the person can look into the hand mirror and see the back of her head.

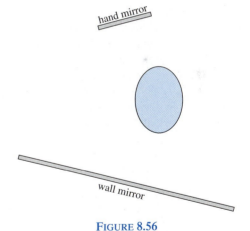

FIGURE 8.56

Using Mirrors to See the Back of One's Head

10. Department store dressing rooms often have large mirrors that actually consist of three adjacent mirrors, put together as shown in Figure 8.57 (as seen looking down from the ceiling). Use the laws of reflection to show how you can stand in such a way as to see the reflection of your back. Draw a careful picture, using an enlarged version of Figure 8.57, that shows clearly how light reflected off your back can enter your eyes. Your picture should show where you are standing, the location of your

back, and the direction of the gaze of your eyes. (You may wish to experiment with paper folding before you attempt a final drawing. See Practice Problem 9.)

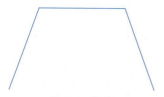

FIGURE 8.57

A Department Store Mirror

11. A **concave** mirror is a mirror that curves in, like a bowl, so that the normal lines on the reflective side of the mirror point toward each other. Makeup mirrors are often concave. The left side of Figure 8.58 shows an eye looking into a concave makeup mirror. The right side of Figure 8.58 shows an eye looking into an ordinary flat mirror. Trace the two diagrams in Figure 8.58 and, in each case, show where a woman applying eye makeup sees her eye in the mirror. Use the laws of reflection to show approximately where the woman sees the top of her reflected eye and where she sees the bottom of her reflected eye. (Assume that the woman sees light that enters the center of her eye.) Notice that the figure shows the normal lines to the concave mirror. Based on your drawings, explain why a concave mirror makes a good mirror for applying makeup. (By the way, although a concave mirror is curved in the same direction as the bowl of a spoon, the smaller amount of curvature in a concave mirror prevents it from reflecting your image upside down, as a spoon does.)

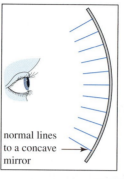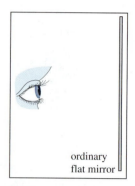

normal lines to a concave mirror →

ordinary flat mirror

FIGURE 8.58

An Eye Looking in a Concave Mirror and a Flat Mirror

12. A **convex** mirror is a mirror that curves out, so that the normal lines on the reflective side of the mirror point away from each other, as shown at the top of Figure 8.59. Convex mirrors are often used as side-view mirrors on cars and trucks. This problem will help you see why convex mirrors are useful for this purpose.

Figure 8.59 shows a bird's eye view of a cross section of a convex mirror, a cross section of a flat mirror, and eyes looking into these mirrors. Draw a copy of these mirrors and the eyes looking into them. Use the laws of reflection to help you compare how much of the surrounding environment each eye can see looking into its mirror.

Now explain why convex mirrors are often used as side-view mirrors on cars and trucks.

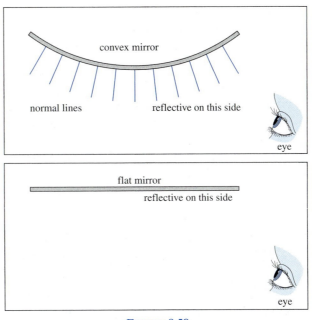

FIGURE 8.59

A Convex Mirror

13. A **convex** mirror is a mirror that curves out, so that the normal lines on the reflective side of the mirror point away from each other, as shown at the top of Figure 8.59. Convex mirrors are often used as side-view mirrors on cars and trucks, but these mirrors usually carry the warning sign "objects are closer than they appear". Explain why objects reflected in a convex mirror appear to be farther away than they actually are. Use the fact that the eye interprets a smaller image as being farther away.

8.3 Circles and Spheres

How do we define circles and spheres? To the eye, circles and spheres are distinguished by their perfect roundness. Informally, we might say that a sphere is the surface of a ball. This describes circles and spheres from an informal or artistic point of view, but there is also a mathematical point of view. As we'll see, the mathematical definitions of circles and spheres yield practical applications that cannot be anticipated by considering only the look of these shapes.

CLASS ACTIVITY NOW TURN TO CLASS ACTIVITIES MANUAL

8T Points That Are a Fixed Distance From a Given Point p. 236

DEFINITIONS OF CIRCLE AND SPHERE

circle   A **circle** is the collection of all the points in a plane that are a certain fixed distance away from a certain fixed point in the plane. This fixed point is called the **center** of the circle, and this

radius   distance is called the **radius** of the circle (plural: **radii**). So the radius is the distance from the

diameter   center of the circle to any point on the circle. The **diameter** of a circle is two times its radius. Informally, the diameter is the distance "all the way across" the circle.

For example, let's fix a point in a plane and let's call this point P, as in Figure 8.60. All the points in the plane that are 1 unit away from the point P form a circle of radius 1 unit, centered at the point P, as shown in Figure 8.60. The diameter of this circle is 2 units.

A sphere is defined in almost the same way as a circle, but in space rather than in a plane.

sphere   A **sphere** is the collection of all the points in space that are a certain fixed distance away from a certain fixed point in space. This fixed point is called the **center** of the sphere, and this distance

radius   is called the **radius** of the sphere. So the radius is the distance from the center of the sphere

diameter   to any point on the sphere. The **diameter** of a sphere is two times its radius. Informally, the diameter is the distance "all the way across" the sphere.

For example, fix a point P in space. You might want to think of this point P as located in front of you, a few feet away. Then all the points in space that are 1 foot away from the point P form a sphere of radius 1 foot, centered at P. Try to visualize this. What does it look like? It looks like the surface of a very large ball, as illustrated in Figure 8.61. The diameter of this sphere is 2 feet.

It's easy to draw almost perfect circles with the use of a common drawing tool called a

compass   **compass**, pictured in Figure 8.62. One side of a compass has a sharp point that you can stick into a point on a piece of paper. The other side of a compass has a pencil attached. To draw a circle centered at a point P and having a given radius, open the compass to the desired radius,

---

FIGURE 8.60

A Circle and a
Noncircle

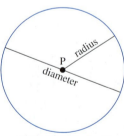

a circle of radius 1 unit
centered at P

This is NOT a circle,
but it could be a
perspective drawing
of a circle.

FIGURE 8.61

A Sphere and a
Nonsphere

a sphere of radius
1 unit, centered at P

This is NOT a sphere.

FIGURE 8.62

Drawing Circles
and Spheres

FIGURE 8.62

Drawing Circles
and Spheres

drawing a circle with a compass   a simple drawing of a sphere

FIGURE 8.63

Drawing a
Circle With a
Paper Clip

drawing a circle with a paper clip

stick the point of the compass in the point P, and spin the pencil side of the compass in a full revolution around P, all the while keeping the point of the compass at P.

It is also possible to draw good circles with simple homemade tools. For example, you can use two pencils and a paper clip to draw a circle, as shown in Figure 8.63. Put a pencil inside one end of a paper clip and keep this pencil fixed at one point on a piece of paper. Put another pencil at the other end of the paper clip and use that pencil to draw a circle.

Since a sphere is an object in space, and not in a plane, it's harder to draw a picture of a sphere. Unless you are an exceptional artist, something like the simple drawing of a sphere in Figure 8.62 will do.

CLASS ACTIVITY NOW TURN TO CLASS ACTIVITIES MANUAL

8U Circle Curiosities p. 237

8V Using Circles p. 238

WHEN CIRCLES OR SPHERES MEET

What happens when two circles meet, or two spheres meet? While this may at first seem like a topic of purely theoretical interest, it actually has practical applications.

FIGURE 8.64

The Three Ways
That Two
Circles Can
Meet

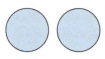 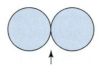

These two circles
don't meet.

These two circles
meet at a single point.

These two circles
meet at two points.

FIGURE 8.65

The Three Ways
That Two
Spheres Can
Meet

 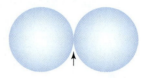 

These two spheres
don't meet.

These two spheres meet
at a single point.

These two spheres meet
along a circle.

As illustrated in Figure 8.64, there are only three possible arrangements of two distinct circles: the circles may not meet at all, the circles may meet at a single point, or the circles may meet at two distinct points.

Now what if you have two spheres, how do they meet? This is difficult, but try to visualize two spheres meeting. It might help to think about soap bubbles. When you blow soap bubbles you will occasionally see a "double bubble" which is similar (although not identical) to two spheres meeting. As with circles, two distinct spheres might not meet at all, or they might barely touch, meeting at a single point. The only other possibility is that the two spheres meet along a circle, as shown in Figure 8.65. This is a circle that is common to each of the two spheres.

CLASS ACTIVITY   NOW TURN TO CLASS ACTIVITIES MANUAL

8w The Global Positioning System (GPS) p. 240

PRACTICE PROBLEMS FOR SECTION 8.3

1. Give the (mathematical) definitions of the terms *circle* and *sphere*.

2. Points P and Q are 4 centimeters apart. Point R is 2 cm from P and 3 cm from Q. Use a ruler and a compass to draw a precise picture of how P, Q, and R are located relative to each other.

3. A radio beacon indicates that a certain whale is less than 1 kilometer away from boat A and less than 1 kilometer away from boat B. Boat A and boat B are 1 kilometer apart. Assuming that the whale is swimming near the surface of the water, draw a map showing all the places where the whale could be located.

4. Suppose that during a flight an airplane pilot realizes that another airplane is 1000 feet away. Describe the shape formed by all possible locations of the other airplane at that moment.

5. An airplane is in radio contact with two control towers. The airplane is 20 miles from one control tower

and 30 miles from another control tower. The control towers are 40 miles apart. Is this information enough to pinpoint the exact location of the airplane? Why or why not? What if you know the altitude of the airplane, do you have enough information to pinpoint the location of the airplane?

ANSWERS TO PRACTICE PROBLEMS FOR SECTION 8.3

1. See the text.

2. See Figure 8.66. Start by drawing points P and Q on a piece of paper, 4 centimeters apart. Now open a compass to 2 cm, stick its point at P, and draw a circle. Since point R is 2 cm from P it must be located somewhere on that circle. Open a compass to 3 cm, stick its point at Q, and draw a circle. Since point R is 3 cm from Q it must also be located somewhere on this circle. Thus, point R must be located at one of the two places where your two circles meet. Plot point R at either one of these two locations.

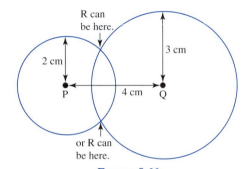

FIGURE 8.66

Locating Points P, Q, and R

3. Since the whale is less than 1 km from boat A, it must be *inside* a circle of radius 1 km, centered at boat A. Similarly, the whale is *inside* a circle of radius 1 km, centered at boat B. The places where the insides of these two circles overlap are all the possible locations of the whale, as shown in Figure 8.67.

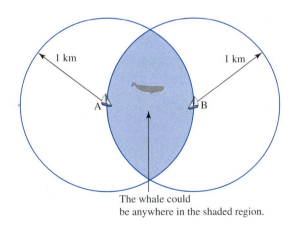

The whale could be anywhere in the shaded region.

FIGURE 8.67

Two Boats and a Whale

4. The other airplane could be at any of the points that are 1000 feet away from the airplane. These points form a sphere of radius 1000 feet.

5. The locations that are 20 miles from the first control tower form a sphere of radius 20 miles, centered at the control tower. (Some of these locations are un-

derground and therefore are not plausible locations for the airplane.) Similarly, the locations that are 30 miles from the second control tower form a sphere of radius 30 miles. The airplane must be located in a place that lies on *both* spheres (i.e., some place where the two spheres meet). The spheres must meet either at a single point (which is unlikely) or in a circle. So, most likely, the airplane could be anywhere on a circle. Therefore, we can't pinpoint the location of the airplane without more information.

Now suppose that the altitude of the airplane is known. Let's say it's 20,000 feet. The locations in the sky that are 20,000 feet from the ground form a very large sphere around the whole Earth. This large sphere and the circle of locations where the airplane might be located either meet in a single point (unlikely) or in two points. So even with this information, we still can't pinpoint the location of the airplane, but we can narrow it down to two locations.

PROBLEMS FOR SECTION 8.3

1. Smallville is 7 miles south of Gotham. Will is 8 miles from Gotham and 6 miles from Smallville. Draw a map showing where Will could be. Can you pinpoint Will to one location or not?

2. A new mall is to be built to serve the towns of Sunnyvale and Gloomington, whose centers are 6 miles apart. The developers want to locate the mall not more than 3 miles from the center of Sunnyvale and also not more than 5 miles from the center of Gloomington. Draw a simple map showing Sunnyvale, Gloomington, and *all* potential locations for the new mall based on the given information. Be sure to show the scale of your map. Explain how you determined the possible locations for the mall.

3. A radio beacon indicates that a certain dolphin is less than 1 mile from boat A and at least $1\frac{1}{2}$ miles from boat

B. Boats A and B are 2 miles apart. Draw a simple map showing the locations of the boats and *all* the places where the dolphin might be located. Be sure to show the scale of your map. Explain how you determined all possible locations for the dolphin.

4. A new Giant Superstore is being planned somewhere in the vicinity of Kneebend and Anklescratch, towns that are 10 miles apart. The developers will only say that all the locations they are considering are more than 7 miles from Kneebend and more than 5 miles from Anklescratch. Draw a map showing Kneebend, Anklescratch, and all possible locations for the Giant Superstore. Be sure to show the scale of your map. Explain how you determined all possible locations for the Giant Superstore.

8.4 Triangles

In this section we will study some basic triangle terminology and one key fact about triangles: the sum of the angles in a triangle is 180°.

triangle

closed

A **triangle** is a closed shape in a plane consisting of three line segments. The term **closed** means that every endpoint of one of the line segments meets exactly one endpoint of another line segment. Informally, we can say that a shape is closed if it has "no loose, dangling ends." Figure 8.68 shows examples of triangles and shapes that are not triangles.

Some kinds of triangles have special names. Figure 8.69 shows some examples of special kinds of triangles. A **right triangle** is a triangle that has a right angle (90°). In a right triangle, the side opposite the right angle is called the **hypotenuse**. A triangle that has three sides of the same length is called an **equilateral triangle**. A triangle that has as least two sides of the same length is called an **isosceles triangle**.

right triangle

hypotenuse

FIGURE 8.68

Triangles and
Shapes That Are
Not Triangles

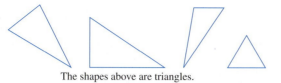

The shapes above are triangles.

The shapes below are not triangles.

This is not made out of   This is not closed.
line segments.

FIGURE 8.69

Special Kinds of
Triangles

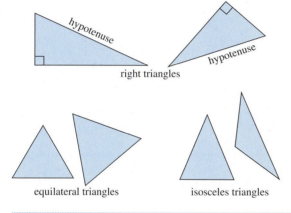

There are two especially famous theorems about triangles. One is that the sum of the angles
in a triangle is 180°, and the other is the Pythagorean theorem. In this section, we will study
the theorem about the sum of the angles in a triangle. We will study the Pythagorean theorem
in Chapter 10.

theorem   What is a theorem? A **theorem** is a mathematical statement that has been proven to be
true by the use of logical reasoning, based on previously known (or assumed) facts.

CLASS ACTIVITY   NOW TURN TO CLASS ACTIVITIES MANUAL

8X Making Right Triangles, Isosceles Triangles, and Equilateral Triangles p. 243

8Y Constructing Triangles of Specified Side Lengths p. 245

THE SUM OF THE ANGLES IN A TRIANGLE

You probably already know the theorem that the angles in a triangle add to 180°. If you stop for a moment to think about this, it is really quite remarkable. No matter what the triangle, no matter how oddly shaped it may be—as long as it's a triangle—the angles *always* add to 180°. This theorem tells us about *all* triangles, and not only that, but we can explain why this theorem is true *without checking every single triangle individually*. Many people are attracted to mathematics because the powerful reasoning of mathematics enables us to deduce general truths.

Class Activities 8Z, 8AA, and 8BB will give you several ways to think about why the angles in a triangle always add to 180°. Here is how we commonly use this fact: If you know two of the angles in a triangle, you can figure out what the third angle must be. For example, in Figure 8.70 the triangle was constructed to have an angle of 40° and an angle of 70°. These two angles already add to 110°; thus, the third angle must be 70° so that all three will add to 180°.

FIGURE 8.70

What Is the
Third Angle?

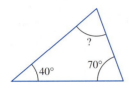

40°   70°   ?

CLASS ACTIVITY   NOW TURN TO CLASS ACTIVITIES MANUAL

8Z Seeing That the Angles in a Triangle Always Add to 180° p. 246

8AA Using Parallel Lines to Explain Why the Angles in a Triangle Add to 180° p. 247

8BB Explaining Why the Angles in a Triangle Add to 180° by Walking and Turning p. 249

8CC Angles and Shapes Inside Shapes p. 252

PRACTICE PROBLEMS FOR SECTION 8.4

1. Figure 8.71 shows a method for constructing an equilateral triangle. Explain why this method must always produce an equilateral triangle.

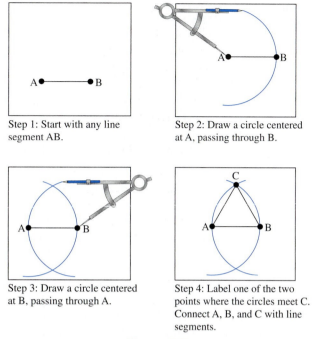

FIGURE 8.71

Constructing an Equilateral Triangle

2. Use a ruler and a compass to construct a triangle that has one side of length 3 inches, one side of length 2 inches, and one side of length 1.5 inches. Describe your method and explain why it must produce the desired triangle.

3. For each of the triangles in Figure 8.72, determine the unknown angle without measuring it.

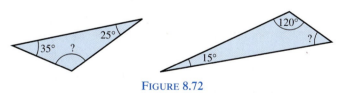

FIGURE 8.72

Find the Missing Angles

4. Use the "walking and turning" method of Class Activity 8BB to explain why the angles in a triangle add to 180°.

5. Figure 8.73 shows a square inscribed in a triangle. Since the square is *inside* the triangle, how can it be that the angles in the square add up to more degrees (360°) than the angles in the triangle (which only add to 180°)?

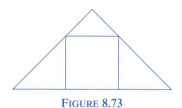

FIGURE 8.73

A Square in a Triangle

ANSWERS TO PRACTICE PROBLEMS FOR SECTION 8.4

1. The method shown in Figure 8.71 produces an equilateral triangle because of the way it uses circles. Remember that a circle consists of all the points that are the same fixed distance away from the center point. The circle drawn in step 2 consists of all points that are the same distance from A as B is; since C is on this circle, the distance from C to A is the same as the distance from B to A. The circle drawn in step 3 consists of all points that are the same distance from B as A is; since C is also on this circle, the distance from C to B is the same as the distance from A to B. Therefore, the three line segments AB, AC, and BC all have the same length, and the triangle ABC is an equilateral triangle.

2. Figure 8.74 shows the construction marks leading to the desired triangle. Starting with a line segment AB that is 3 inches long, draw (part of) a circle with center A and radius 2 inches. Then draw (part of) a circle with center B and radius 1.5 inches. (You can just as well draw the circle of radius 1.5 inches centered at A and the circle of radius 2 inches centered at B if you want to.) Let C be one of the two locations where the two circles meet. Connect A to C and connect B to C to create the desired triangle.

   The construction creates the desired triangle because it uses the defining property of circles. When you draw the circle with center A and radius 2 inches, the points on that circle are *all* 2 inches away from A. Similarly, when you draw the circle with center B and radius 1.5 inches, the points on that circle are *all* 1.5 inches away from B. So a point C where the two circles meet is *both* 2 inches from A and 1.5 inches from B. Hence, the side CA of the triangle ABC is 2 inches long and the side CB is 1.5 inches long. The original side AB was 3 inches long, so the construction creates a triangle with sides of lengths 3 inches, 2 inches, and 1.5 inches.

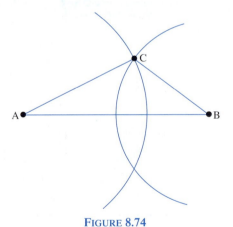

FIGURE 8.74

Constructing a Triangle with Sides of
Lengths 3 Inches, 2 Inches, and 1.5 Inches

3. Left triangle: 120°. Right triangle: 45°. In each case, use the fact that the angles in a triangle must add to 180° to determine the missing angle.

4. As seen in Class Activity 8BB, if a person were to walk around a triangle, returning to her original position, she would have turned a total of 360°. This means that $d + e + f = 360°$, where d, e, and f are the exterior angles, as shown in Figure 8.75. But since

$$a + f = 180°,$$

$$b + d = 180°, \text{ and}$$

$$c + e = 180°,$$

it follows that, on the one hand,

$$(a + f) + (b + d) + (c + e) = 180° + 180° + 180°$$
$$= 540°$$

while, on the other hand,

$$(a + f) + (b + d) + (c + e)$$
$$= (a + b + c) + (d + e + f)$$
$$= (a + b + c) + 360°$$

Because $(a + f) + (b + d) + (c + e)$ is equal to both $540°$ and $(a + b + c) + 360°$, these last two expressions must be equal to each other. That is,

$$(a + b + c) + 360° = 540°$$

Therefore,

$$a + b + c = 540° - 360° = 180°$$

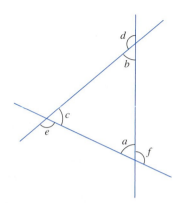

FIGURE 8.75

The Exterior Angles of a
Triangle Add to 360°

5. Unlike area, the angles in the square and the triangle don't depend on the sizes of either of these shapes. The fact that the square is inside the triangle has nothing to do with the sum of the angles in the square. We could even enlarge the square, so that the triangle fits inside the square, without changing the sum of the angles in the square.

PROBLEMS FOR SECTION 8.4

1. Figure 8.76 shows a method for making an isosceles triangle. Starting with an ordinary rectangular piece of paper, fold the paper in half either lengthwise or widthwise. With the paper still folded, use a ruler to draw a line from anywhere along the fold to anywhere along either one of the open edges of the paper that are perpendicular to the fold. Cut the folded piece of paper along the line you drew. When you unfold, you will have an isosceles triangle.

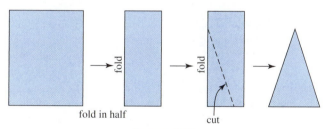

fold in half   cut

FIGURE 8.76

A Method for Making Isosceles Triangles

a. Use the method just described to make an isosceles triangle.

b. Explain why this method must always produce an isosceles triangle.

c. Use this method to make an isosceles triangle that has one side that is 5 inches long and two sides that are 10 inches long.

d. Use this method to make an isosceles triangle that has one side that is 10 inches long and two sides that are 7 inches long.

2. Figure 8.77 shows a method for constructing isosceles triangles.

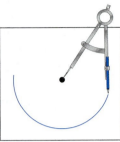

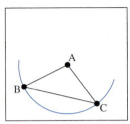

Step 1: Draw part of a circle.   Step 2: Connect the center of the
                                      circle with two points on the circle.

FIGURE 8.77

A Method for Constructing Isosceles Triangles

a. Use the method of Figure 8.77 to draw two different isosceles triangles.

b. Use the definition of circles to explain why this method will always produce an isosceles triangle.

c. Use this method to draw an isosceles triangle that has two sides of length 6 inches and one side of length 4 inches.

---

3. Draw a line segment AB that is 4 inches long. Now modify the steps of the construction shown in Figure 8.71 to produce an isosceles triangle that has sides of lengths 7 inches, 7 inches, and 4 inches. Explain why you carry out your construction as you do.

4. a. Use a ruler and compass to help you draw a triangle that has one side of length 5 inches, one side of length 3.5 inches, and one side of length 4 inches.

   b. Explain why your method of construction must produce a triangle with the required side lengths.

5. Is there a triangle that has one side of length 4 inches, one side of length 2 inches, and one side of length 1 inch? Explain.

6. In your own words, explain clearly why the sum of the angles in a triangle must always add to 180°.

7. Figure 8.78 shows a large triangle subdivided into four smaller triangles. If the angles in each of the four smaller triangles add to 180°, then why don't the angles in the larger triangle add to

$$180° + 180° + 180° + 180° = 720°$$

instead of 180°? Relate the 720° to the sum of the angles in the large triangle.

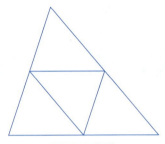

FIGURE 8.78

A Triangle Subdivided into
Four Smaller Triangles

8. Take any right triangle that has angles a, b, and 90°. Draw a line segment from the corner where the right angle is to the hypotenuse so that this line segment is perpendicular to the hypotenuse, as shown in Figure 8.79. The line segment divides the original right triangle into two smaller right triangles. Without measuring, determine the angles in these smaller right triangles. How are these angles related to the angles in the original large triangle? Explain your reasoning.

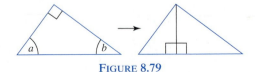

FIGURE 8.79

Subdivide a Right Triangle

8.5 Quadrilaterals and Other Polygons

In this section we will investigate properties of some special kinds of quadrilaterals and we will see the relationships among various kinds of quadrilaterals. We will also see that triangles and quadrilaterals are certain kinds of polygons. At the end of this section we use Venn diagrams, a convenient method of showing relationships among groups, to illustrate the relationships among various shapes.

quadrilateral

closed

A **quadrilateral** is a closed shape in a plane consisting of four line segments that do not cross each other. As with triangles, the term **closed** means that every endpoint of one of the line segments meets exactly one endpoint of another line segment. Figure 8.80 shows examples of quadrilaterals and shapes that are not quadrilaterals. The name *quadrilateral* makes sense because it means *four sided* (quad = four, lateral = side).

polygon

Triangles and quadrilaterals are kinds of polygons. A **polygon** is a closed shape in a plane consisting of a finite number of line segments that do not cross each other. Triangles are polygons made of 3 line segments, quadrilaterals are polygons made of 4 line segments, **pentagons** are polygons made of 5 line segments, **hexagons** are polygons made of 6 line segments, **octagons** are polygons made of 8 line segments. Figure 8.81 shows some examples. A polygon with *n* sides can be called an "*n*-gon." For example, a 13-sided polygon is a 13-gon.

pentagons

hexagons

octagons

The name *polygon* makes sense because it means *many sided* (poly = many, gon = side). Similarly, for the names *pentagon*, *hexagon*, and so on, "penta" means 5, "hexa" means 6, and so on. It would be perfectly reasonable to call triangles *trigons* and quadrilaterals *quadrigons*, although these terms are not used conventionally.

FIGURE 8.80

Quadrilaterals and Shapes That Are not Quadrilaterals

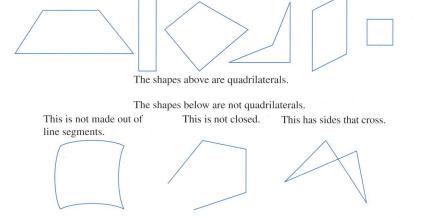

The shapes above are quadrilaterals.

The shapes below are not quadrilaterals.

This is not made out of line segments.   This is not closed.   This has sides that cross.

FIGURE 8.81

Pentagons, Hexagons, and Octagons

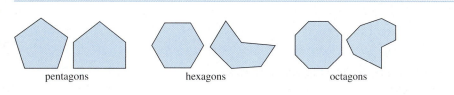

pentagons   hexagons   octagons

regular polygon   A polygon is called **regular** if all sides have the same length and all angles are equal. In Figure 8.81 the leftmost pentagon, hexagon, and octagon are all regular polygons, whereas the pentagon, hexagon, and octagon on the right are not regular polygons.

We will now study some special kinds of quadrilaterals. The most familiar special quadrilaterals are squares and rectangles. Following is a list of definitions of some special quadrilaterals (see Figure 8.82 for examples):

square—quadrilateral with four right angles whose sides all have the same length

rectangle—quadrilateral with four right angles

rhombus—quadrilateral whose sides all have the same length. The name **diamond** is sometimes used instead of rhombus.

parallelogram—quadrilateral for which opposite sides are parallel

trapezoid—quadrilateral for which at least one pair of opposite sides are parallel. (Some books define a trapezoid as a quadrilateral for which *exactly one* pair of opposite sides are parallel.)

diagonals   Some of the activities that follow explore special properties of various kinds of quadrilaterals. Several investigate diagonals of quadrilaterals. The **diagonals** of a quadrilateral are line segments connecting opposite corner points, as shown in Figure 8.83.

FIGURE 8.82

Some Special
Quadrilaterials

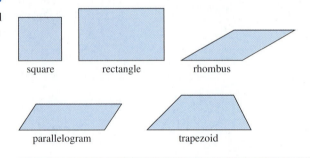

square   rectangle   rhombus

parallelogram   trapezoid

FIGURE 8.83

The Diagonals
of a
Quadrilateral

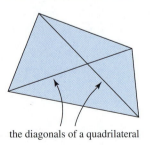

the diagonals of a quadrilateral

SHOWING RELATIONSHIPS WITH VENN DIAGRAMS

Venn diagram
set

There is a convenient way to use pictures to show relationships among collections of items. A **Venn diagram** is a picture that shows how certain sets are related. A **set** is a collection of objects.

Figure 8.84 shows a Venn diagram relating the set of mammals and the set of animals with four legs. The overlapping region represents mammals that have four legs because these animals fit in both categories: animals with four legs and mammals.

Venn diagrams are not just used in mathematics. Some teachers use Venn diagrams in language arts, for example, to compare the attributes of a child with the attributes of a story character, as in Figure 8.85.

FIGURE 8.84

A Venn Diagram Showing the Relationship between Two Sets

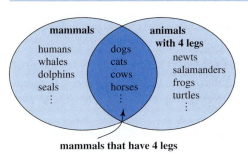

mammals that have 4 legs

FIGURE 8.85

A Language Arts Venn Diagram Comparing Matilda's and Joey's Attributes

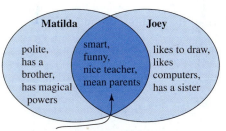

common attributes of Matilda and Joey

FIGURE 8.86

Other Venn
Diagrams
Showing the
Relationships
between Two
Sets

FIGURE 8.87

A Venn Diagram
Showing the
Relationships
among Three
Sets

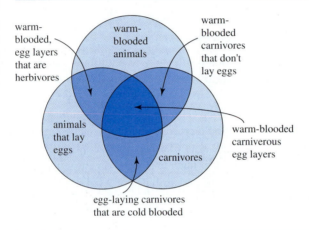

A Venn diagram of two sets doesn't have to overlap the way the previous two examples did. Figure 8.86 shows two other kinds of Venn diagrams. The first shows a Venn diagram relating the set of boys and the set of girls. Of course, these sets do not overlap. The other Venn diagram in Figure 8.86 shows that the set of whole numbers is contained within the set of rational numbers. This is so because every whole number can also be expressed as a fraction (by "putting it over 1").

When representing three or more sets, Venn diagrams can become quite complex. In general, there can be double overlaps, triple overlaps, or more (if more than three sets are involved). For example, Figure 8.87 shows a Venn diagram relating the set of warm-blooded animals, the set of animals that lay eggs, and the set of carnivorous animals. There are three double overlaps and one triple overlap.

We can use Venn diagrams to show the relationships among the various kinds of special quadrilaterals.

CLASS ACTIVITY   NOW TURN TO CLASS ACTIVITIES MANUAL

8HH Venn Diagrams Relating Quadrilaterals p. 260

The following class activities will help you discover that some kinds of quadrilaterals have special properties that other quadrilaterals don't always have.

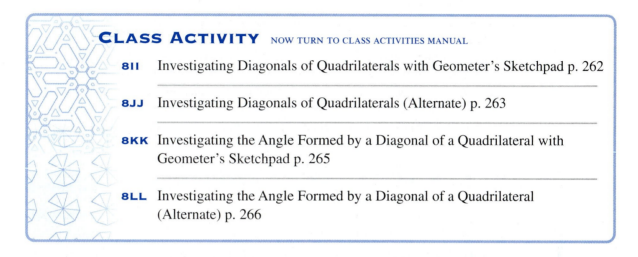

DEFINITIONS OF SHAPES VERSUS ADDITIONAL PROPERTIES THE SHAPES HAVE

If you did the class activities in this section, then you discovered that shapes often have special properties that are not readily apparent from their definitions. For example, when you construct a rhombus and look at it carefully, you will notice that opposite sides of the rhombus appear to be parallel (in fact, they are). Therefore, every rhombus should also be a parallelogram. But now look back at the definition of rhombus: a rhombus is a quadrilateral whose four sides all have the same length. There is nothing in this *definition* that tells us that opposite sides of a rhombus should be parallel. We would somehow need to *deduce* that every rhombus is in fact also a parallelogram.

Here is another example: If you did activities investigating diagonals of quadrilaterals, then you may have found that the following seem to be true for every rhombus (in fact, they are true):

- The diagonals meet at a point that is halfway across each diagonal.
- The diagonals are perpendicular.
- The diagonals have different lengths unless the rhombus is a square.
- A diagonal cuts an angle of a rhombus in half.

You may also have found that the following seem to be true for every rectangle (in fact, they are true):

- The diagonals meet at a point that is halfway across each diagonal.
- The diagonals are only perpendicular when the rectangle is a square.
- The diagonals always have the same length.

Once again, these are examples of properties of shapes that we can't tell are true just from reading the definitions of these shapes—in fact, the definitions of rhombus and rectangle don't say anything at all about diagonals.

This observation leads straight to the heart of what theoretical mathematics is all about: the theoretical study of mathematics is about *starting with some assumptions and some definitions of objects and concepts, discovering additional properties that these objects or concepts must have, and then reasoning logically to deduce that the objects or concepts do indeed have those properties.* You probably already carried out such a logical deduction, for example, if you proved that the sum of the angles in a triangle is 180°. The *definition* of triangle does not tell us that the angles in a triangle will always add to 180°, but logical reasoning allows us to *deduce* this theorem. In the foregoing examples, we took the first step of *discovering* additional properties that rhombuses and rectangles must have, beyond those properties given in the definition. In these cases, we won't carry out the next step of deducing logically that rhombuses and rectangles really do have these additional properties, but it is possible to do so. In other words, it is possible to go beyond observing that rhombuses and rectangles always seem to have certain properties, to explain why these shapes must always have those properties.

PRACTICE PROBLEMS FOR SECTION 8.5

1. Define the following terms: quadrilateral, polygon, pentagon, hexagon, octagon, 13-gon, square, rectangle, rhombus, parallelogram, and trapezoid.

2. Use the definition of circles and rhombuses to explain why the quadrilateral ABDC produced by the method of Figure 8.88 must necessarily be a rhombus.

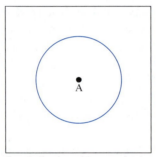

Step 1: Starting with any point A, draw a circle with center A.

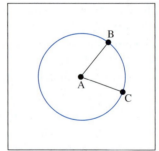

Step 2: Let B and C be any two points on the circle that are not opposite each other. Draw line segments AB and AC.

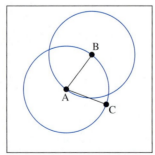

Step 3: Draw a circle centered at B and passing through A.

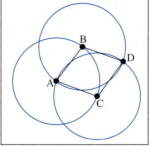

Step 4: Draw a circle centered at C and passing through A. Label the point other than A where these last two circles meet D.

FIGURE 8.88

A Method for Constructing Rhombuses

3. Draw a Venn diagram showing the relationship between the set of rectangles and the set of rhombuses. Is there an overlap? If so, what does it consist of? Explain.

4. Draw a Venn diagram showing the relationship between the set of parallelograms and the set of trapezoids. Explain.

5. Some books define trapezoids as quadrilaterals that have *exactly one* pair of parallel sides. Draw a Venn diagram showing how the set of parallelograms and the set of trapezoids are related when this alternate definition of trapezoid is used. Explain.

6. Draw a Venn diagram showing the relationship between the set of rhombuses and the set of parallelograms. In doing this, can you rely only on the information given directly in the the the definitions of rhombus and parallelogram, or would you need to deduce some additional information?

7. What special properties do diagonals of rhombuses have that diagonals of other quadrilaterals do not necessarily have?

8. What special properties do diagonals of rectangles have that diagonals of other quadrilaterals do not necessarily have?

ANSWERS TO PRACTICE PROBLEMS FOR SECTION 8.5

1. See text.

2. A circle is the set of all points that are the same fixed distance away from the center point. Therefore, because points B and C are on a circle centered at A, the distance from B to A is equal to the distance from C to A. Because the point D is on a circle centered at B and passing through A, the distance from D to B is equal to the distance from B to A. Similarly, because D is on a circle centered at C and passing through A, the distance from D to C is equal to the distance from C to A. Therefore the four line segments AB, AC, BD, and CD have the same length. Rhombuses are quadrilaterals that have four sides of the same length. So the quadrilateral ABDC formed by the line segments AB, AC, BD, and CD is a rhombus.

3. See Figure 8.89. The shapes in the overlap are those shapes that have four right angles (as rectangles do) and have four sides of the same length (as rhombuses do). According to the definition, those are exactly the shapes that are squares.

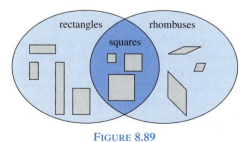

FIGURE 8.89

Venn Diagram of Rectangles and Rhombuses

4. See Figure 8.90. According to the definitions, every parallelogram is also a trapezoid because parallelograms have two pairs of parallel sides, so they can be said to have at least one pair of parallel sides.

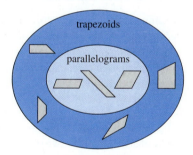

FIGURE 8.90

Venn Diagram of Parallelograms and Trapezoids

5. See Figure 8.91. According to the alternate definition, no parallelogram is a trapezoid because parallelograms have two pairs of parallel sides, so they don't have exactly one pair of parallel sides. Therefore, with this alternate definition, the set of parallelograms and the set of trapezoids do not have any overlap.

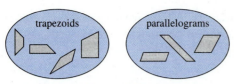

FIGURE 8.91

Venn Diagram of Parallelograms and
Trapezoids Using the Alternate Definition

FIGURE 8.92

Venn Diagram of Rhombuses and
Parallelograms

6. See Figure 8.92. Looking at rhombuses, it appears that opposite sides are parallel and therefore that every rhombus is also a parallelogram. Since the definition of rhombus does not say anything about opposite sides being parallel, we would need to deduce some additional information in order to explain why rhombuses really are parallelograms.

7. The diagonals of a rhombus are perpendicular. The two diagonals of a rhombus meet at a point that is halfway across each diagonal. A diagonal of a rhombus cuts both angles it passes through in half.

8. The two diagonals of a rectangle have the same length and meet at a point that is halfway across each diagonal.

PROBLEMS FOR SECTION 8.5

1. Create two different rhombuses, both of which have four sides of length 4 inches. You may create your rhombuses by drawing, by folding and cutting paper, or by fastening objects together. In each case, explain how you know your shape really is a rhombus.

2. The line segments AB and AC in Figure 8.93 have been constructed so that they could be two sides of a rhombus.

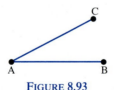

FIGURE 8.93

Two Sides of a
Rhombus

   a. Trace the angle BAC of Figure 8.93 on a piece of paper. Use a compass and straightedge to finish constructing a rhombus that has AB and AC as two of its sides.

   b. By referring to the *definition of rhombus*, explain why your construction in part (a) must produce a rhombus.

3. a. Copy the line segment AB in Figure 8.94 on dot paper or graph paper. Then draw a square that has AB as one of its sides.

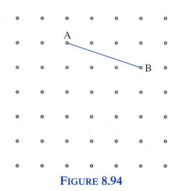

FIGURE 8.94

Make a Square on Dot Paper

   b. Copy the line segment CD in Figure 8.95 on dot paper or graph paper. Then draw a square that has CD as one of its sides.

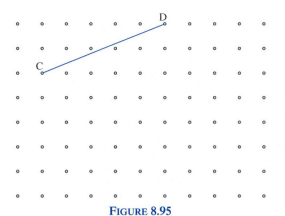

FIGURE 8.95

Make a Square on Dot Paper

c. On dot paper or graph paper draw two more line segments that are neither vertical nor horizontal and are not identical to those in parts (a) and (b). Draw two squares that have your line segments as sides.

4. a. Draw at least 3 different quadrilaterals and in each case, show how to subdivide the quadrilateral into 2 triangles.

   b. Use the fact that every quadrilateral can be subdivided into 2 triangles and the fact that the sum of the angles in a triangle is 180° to determine the sum of the angles in a quadrilateral. Explain your reasoning.

   c. Draw at least 3 different pentagons and, in each case, show how to subdivide the pentagon into 3 triangles.

   d. Use the fact that every pentagon can be subdivided into 3 triangles and the fact that the sum of the angles in a triangle is 180° to determine the sum of the angles in a pentagon. Explain your reasoning.

   e. Based on your work above, what should the sum of the angles in a hexagon be? In general, what should the sum of the angles in an n-gon be? Explain briefly.

5. Figure 8.96 A shows a pentagon with angles a, b, c, d, and e. Figure 8.96 B shows the same pentagon subdivided into triangles. Use the fact that the sum of the angles in a triangle is 180° to determine the sum,

$a + b + c + d + e$, of the angles in the pentagon. Explain your reasoning.

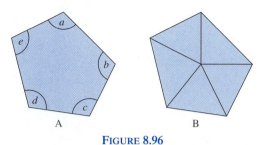

FIGURE 8.96

A Pentagon Subdivided into Triangles

6. a. Draw a Venn diagram showing the relationship between the set of rectangles and the set of parallelograms.

   b. Discuss whether the relationship between the set of rectangles and the set of parallelograms that you found in part (a) can be determined immediately from the definitions of these shapes or whether additional information that derives from the definitions but is not stated directly in the definitions is necessary. If a blind person knew only the definitions of rectangles and parallelograms, would that person immediately be able to understand the relationship you found in part (a), or did you rely on visual information about rectangles and parallelograms that is not stated in the definitions of these shapes?

   c. Give a careful, thorough explanation for why the set of rectangles and the set of parallelograms are related the way they are. Hint: To show that opposite sides of a rectangle must be parallel, explain why these sides couldn't meet if they were extended to be lines. Show that if these two lines did meet somewhere they would be part of a triangle whose angles would add to more than 180°.

7. Draw a Venn diagram that shows the relationships among the sets of quadrilaterals, squares, rectangles, parallelograms, rhombuses, and trapezoids. Explain briefly.

8. A problem that was given to 5th graders is shown in Figure 8.97. Criticize the problem on *mathematical grounds* and rewrite it to word it correctly.

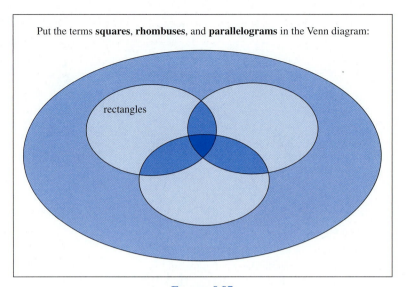

Put the terms **squares**, **rhombuses**, and **parallelograms** in the Venn diagram:

rectangles

FIGURE 8.97

A Problem Given to 5th Graders

9. Before builders lay the foundation for a rectangular house, they usually pound stakes into the ground where the 4 corners of the house will be. The builders then measure the diagonals of the quadrilateral formed by the four stakes. Suppose that these diagonals turn out to have different lengths. In this case, based on the discussion in this section about diagonals of quadrilaterals, what can you conclude about the angles at the corners? Explain. It is important for the corners of the foundation to be right angles, so that the walls above will be structurally stable. Should the builders be satisfied with the proposed foundation, or should they make adjustments to it?

10. Figure 8.98 shows a way to fold and cut a parallelogram, starting with a rectangular piece of paper of any size. (Part (2) of Class Activity 8DD describes how to make paper rectangles of different sizes.) When you unfold, you will have a parallelogram.

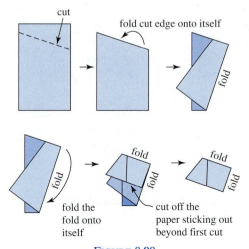

FIGURE 8.98

Folding a Parallelogram

a. Make a parallelogram using the method shown in Figure 8.98.

b. What makes this method work? Hint: What is special about the fold lines?

11. a. Use the reasoning of Class Activity 8BB and the answer to Practice Problem 4 on page 354 for triangles to determine the sum of the exterior angles of the quadrilateral in Figure 8.99. In other words, determine $e + f + g + h$. Measure with a protractor to check that your formula is correct for this quadrilateral.

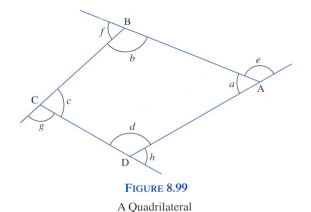

FIGURE 8.99

A Quadrilateral

b. Will there be a similar formula for the sum of the exterior angles of pentagons, hexagons, 7-gons (a.k.a. heptagons), octagons, etc.? Explain.

c. Using your formula for the sum of the exterior angles of a quadrilateral, deduce the sum of the interior angles of the quadrilateral. In other words, find $a + b + c + d$, as pictured in Figure 8.99. Explain your reasoning. Measure with a protractor to verify that your formula is correct for this quadrilateral.

d. Based on the class activity and your work above, what formula would you expect to be true for the sum of the interior angles of a pentagon? What about for hexagons? What about for a polygon with n sides?

12. a. Determine the angles in a regular pentagon. Explain your reasoning.

b. Determine the angles in a regular hexagon. Explain your reasoning.

c. Determine the angles in a regular n-gon. Explain your reasoning.

8.6 Constructions with Straightedge and Compass

In this section we will explore some easy constructions that can be done with a compass and a straightedge. We will see that these constructions work because of special properties of rhombuses. A **straightedge** is just a "straight edge," namely a ruler, except that it need not have any markings.

Constructions with straightedge and compass have their origins in the desire to make precise drawings using only simple, reliable tools. Think back to the times before there were computers. If you were studying plane shapes, and if you wanted to discover additional properties these shapes had beyond those properties given in the definitions, then you would need some way to make precise drawings. A sloppy drawing could hide interesting features, or worse, could seem to show features that aren't really there. Nowadays we have computer technology that can help us make accurate drawings easily, but we can still benefit from learning some of the old "hands-on" techniques. The slower process of drawing by hand gives us time to detect features and relationships that are lost when shapes are constructed at the speed of light with a computer.

FIGURE 8.100

Constructing a
Line That Is
Perpendicular to
a Line Segment
and Divides It in
Half

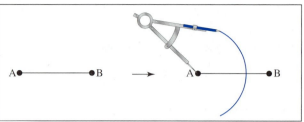

Step 1: Starting with a line segment AB, open a compass to at least half
the length of the line segment and draw (part of) a circle centered at A.

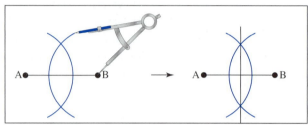

Step 2: Keep the compass opened to the same width and draw (part of) a
circle centered at B. Draw a line through the two points where the two
circles meet. This line is perpendicular to AB and divides AB in half.

DIVIDING A LINE SEGMENT IN HALF AND CONSTRUCTING A PERPENDICULAR LINE

Stop to think for a moment about how you could draw a perfect (or nearly perfect) right angle without using a previously made right angle. It's not so obvious.

Figure 8.100 shows how you can start with any line segment and construct a line that is perpendicular to the line segment *and* that passes through the point halfway across the line segment. The point halfway across a line segment is often called a **midpoint** of the line segment. So the procedure shown in Figure 8.100 actually does two things at once: It constructs a perpendicular line, and it finds a midpoint of a line segment. Instead of saying that the construction of Figure 8.100 divides the line segment *in half* we can also say that the construction **bisects** the line segment or that it constructs the *perpendicular bisector* to the line segment. Notice that the word *bisect* is similar to the word *dissect*. The former is used in geometry whereas the latter in used in biology.

midpoint

bisect (line segment)

DIVIDING AN ANGLE IN HALF

Figure 8.101 shows how you can start with any angle at a point P and divide the angle in half. In other words, given two rays that meet at a point P, the construction in Figure 8.101 produces another ray, also starting at P, that is halfway between the two original rays. Instead of saying that the construction of Figure 8.101 divides an angle in half, we can say that the construction **bisects** the angle.

bisect (angle)

FIGURE 8.101

Dividing an
Angle in Half

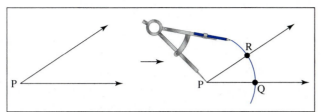

Step 1: Starting with two rays that meet at a point P, draw a circle
centered at P. Let Q and R be the points where the circle meets the rays.

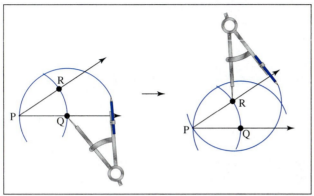

Step 2: Keep the compass opened to the same width. Draw a circle
centered at Q and another circle centered at R.

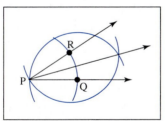

Step 3: Draw a line through point P and the other point where the
last two circles drawn meet.

CLASS ACTIVITY NOW TURN TO CLASS ACTIVITIES MANUAL

8MM Constructing a Square and an Octagon with Straightedge and
Compass p. 267

8NN Relating the Constructions to Properties of Rhombuses p. 269

PRACTICE PROBLEMS FOR SECTION 8.6

1. Draw several line segments. Use a straightedge and compass to construct lines that are perpendicular to your line segments and divide your line segments in half.

2. Draw several angles. Use a straightedge and compass to divide your angles in half.

3. a. Draw the rhombus that is naturally associated with the construction of a perpendicular bisector of a line segment shown in Figure 8.100.

   b. Explain why the quadrilateral that you identify as a rhombus in part (a) really is a rhombus, according to the definition of rhombus.

   c. Which special properties of rhombuses are related to the construction of a perpendicular bisector of a line segment shown in Figure 8.100? Explain.

ANSWERS TO PRACTICE PROBLEMS FOR SECTION 8.6

1. See Figure 8.100.

2. See Figure 8.101.

3. a. Starting with the finished construction of a perpendicular bisector shown in Figure 8.102, let C and D be the points in the construction where the two circles meet. Form the four line segments AC, BC, AD, and BD, as shown in Figure 8.103.

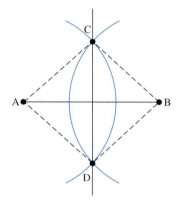

FIGURE 8.103

The Construction for a
Perpendicular Line Forms a
Rhombus

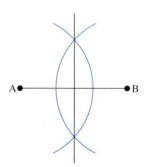

FIGURE 8.102

The Result of Using a
Straightedge and Compass
to Construct a
Perpendicular Bisector

b. Notice that AC and AD in Figure 8.103 are radii of the circle centered at A. Therefore, AC and AD have the same length. Similarly, BC and BD have the same length because they are radii of the circle centered at B. But the circles centered at A and at B have the same radius because they were constructed that way (without changing the width of the compass). Therefore, all four line segments AC, BC, AD, and BD have the same length. By definition, this means that the quadrilateral ACBD is a rhombus.

c. The original line segment AB is one of the diagonals of the rhombus ACBD of Figure 8.103. The perpendicular bisector CD is the other diagonal of the rhombus ACBD. According to the list of properties of rhombuses on page 361, the diagonals of a rhombus meet at a point that is halfway across

each diagonal, and the diagonals of a rhombus are perpendicular. These properties fit exactly with the properties of the line segments AB and CD because CD was constructed to be perpendicular to AB and to divide AB in half.

PROBLEMS FOR SECTION 8.6

1. a. Draw a ray with endpoint A. Use a straightedge and compass to carefully construct a 45° angle to your ray at A. Briefly describe your method and explain why it makes sense.

    b. On a blank piece of paper, draw a ray with endpoint A. Carefully fold your paper to create a fold line that makes a 45° angle with your ray at A. Briefly describe your method and explain why it makes sense.

2. a. Draw a ray with endpoint A. Use a straightedge and compass to carefully construct a 22.5° angle to your ray at A. Briefly describe your method and explain why it makes sense. (Hint: Notice that 22.5 is half of 45.)

    b. On a blank piece of paper, draw a ray with endpoint A. Carefully fold your paper to create a fold line that makes a 22.5° angle with your ray at A. Briefly describe your method and explain why it makes sense.

3. Use a straightedge and compass to carefully construct a square.

4. a. Draw the rhombus that is naturally associated with the straightedge and compass construction of a ray dividing an angle in half, as shown in Figure 8.101.

    b. Explain why the quadrilateral that you identify as a rhombus in part (a) really is a rhombus, according to the definition of rhombus.

    c. Which special properties of rhombuses described in Section 8.5 on page 361 are closely related to the construction of a ray dividing an angle in half that is shown in Figure 8.101? Explain.

5. Describe how to use a compass to construct the pattern of three circles shown in Figure 8.104 so that the triangle shown inside the circles is an equilateral triangle. Then explain why the triangle really is an equilateral triangle.

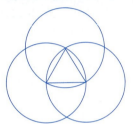

FIGURE 8.104

A Pattern of Three Circles

6. a. Use a compass to draw a pattern of circles like the one in Figure 8.105. Briefly describe how to create this pattern.

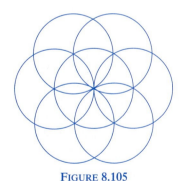

FIGURE 8.105

A Pattern of Circles

    b. Use a straightedge and compass to carefully construct a regular hexagon. Construct your hexagon by first drawing another pattern of circles like the one in Figure 8.105, but this time, draw only parts of the circles, so that you don't clutter your drawing.

7. Use a straightedge and compass to carefully construct a regular 12-gon. Briefly describe your method and explain why it makes sense. It may help you to examine Figure 8.105.

8. Draw a large circle on a piece of paper. This circle will represent a pie. Use a straightedge and compass to carefully subdivide your pie into 6 equal pie pieces. Briefly describe your method and explain why it makes sense. It may help you to examine Figure 8.105. (You may find this construction helpful in the future if you need to make pie pieces to help your students learn about fractions.)

9. Draw a large circle on a piece of paper. This circle will represent a pie. Use a straightedge and compass to carefully subdivide your pie into 12 equal pieces. Briefly describe your method and explain why it makes sense. It may help you to examine Figure 8.105. (You may find this construction helpful in the future if you need to make pie pieces to help your students learn about fractions.)

8.7   Polyhedra and Other Solid Shapes

So far in this chapter we have studied various plane shapes: circles, triangles, quadrilaterals, and other polygons. All of these shapes are flat because they lie in a plane. But plane shapes can be used to make a variety of interesting three-dimensional, solid shapes. In this section we will study some special kinds of solid shapes that can be made out of polygons as well as some related solid shapes.

polyhedron (polyhedra)

A closed shape in space that is made out of polygons is called a **polyhedron**. The plural of polyhedron is **polyhedra**. Figure 8.106 shows some polyhedra and a shape that is not a polyhedron. The polygons that make up the surface of the polyhedron are called the **faces** of the polyhedron. The place where two faces come together is called an **edge** of the polyhedron. A corner point where several faces come together is called a **vertex** or **corner** of the polyhedron. The plural of vertex is **vertices**.

faces

edge

vertex (vertices)

The name *polyhedron* comes from the Greek; it makes sense because *poly* means *many* and *hedron* means *base*, so that *polyhedron* means *many bases*.

PRISMS, CYLINDERS, PYRAMIDS, AND CONES

One special type of polyhedron is a prism. Some people have glass or crystal prisms hanging in their windows to bend the incoming light and cast rainbow patterns around the room.

From a mathematical point of view, a *right prism* is a polyhedron that, roughly speaking, can be thought of as "going straight up over a polygon," as shown in Figure 8.107. We can think of **right prisms** as formed in the following way: Take two paper copies of any polygon and lay both flat on a table, one on top of the other so that they match up. Move the top polygon *straight up* above the bottom one. If vertical rectangular faces are now placed so as to connect

right prism

FIGURE 8.106

Polyhedra and a Shape That Is Not a Polyhedron

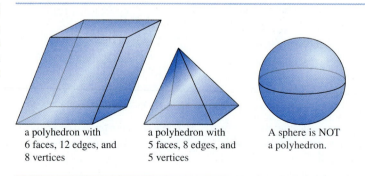

a polyhedron with 6 faces, 12 edges, and 8 vertices

a polyhedron with 5 faces, 8 edges, and 5 vertices

A sphere is NOT a polyhedron.

FIGURE 8.107

Some Right
Prisms

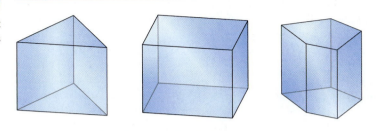

FIGURE 8.108

Oblique Prisms

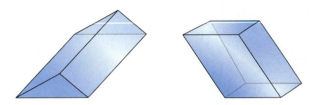

bases

corresponding sides of the two polygons, then the shape formed this way is a right prism. The two polygons that you started with are called the **bases** of the right prism.

By modifying the previous description, we get a description of *all* prisms, not just right prisms. As before, start with two paper copies of any polygon and lay both flat on a table, one on top of the other so that they match up. Move the top polygon up without twisting, away from the bottom polygon, keeping the two polygons parallel. This time, the top polygon does not need to go straight up over the bottom polygon. Instead, it can be positioned to one side, as long as it is not twisted and it remains parallel to the bottom polygon. Once again, if faces are now placed so as to connect corresponding sides of the two polygons, then the shape formed

prism

this way is a **prism**. (This time the faces will be parallelograms.) Every right prism is a prism, but Figure 8.108 shows prisms that are not right prisms. A prism that is not a right prism can

oblique prism

be called an **oblique prism**. As before, the two polygons that you started with are called the **bases** of the prism.

If a prism is moved to a different orientation in space, it is still a prism. So, for example, the polyhedra shown in Figure 8.109 are prisms, although you may need to rotate them mentally to be convinced that they really are prisms.

Prisms are often named according to the kind of polygon that make the bases of the prism.

triangular,
rectangular
prism

So, for example, a prism with a triangle base can be called a **triangular prism**, and a prism with a rectangle base can be called a **rectangular prism**.

FIGURE 8.109

A Prism Turned
Sideways is Still
a Prism

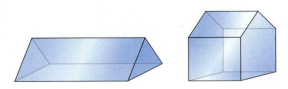

FIGURE 8.110

Cylinders

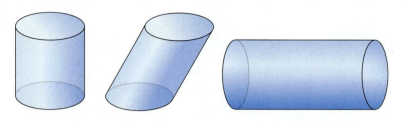

FIGURE 8.111

Pyramids

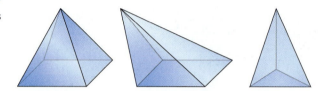

A cylinder is a kind of shape that is related to prisms. Roughly speaking, a cylinder is a tube-shaped object, as shown in Figure 8.110. The tube inside a roll of paper towels is an example of a cylinder. **Cylinders** can be described in the following way: Draw a closed curve on paper (such as a circle or an oval), cut it out, make a copy of it, and lay the the two copies on a table, one on top of the other, so that they match. Next, take the top copy and move it up without twisting, away from the bottom copy, keeping the two copies parallel. Now imagine paper or some other kind of material connecting the two curves in such a way that every line between corresponding points on the curves lies on this paper or material. The shape formed by this paper or other material is a cylinder. The regions formed by the two starting curves can again be called the **bases** of the cylinder. In some cases, you want to consider the two bases as part of the cylinder, in other cases not. Both kinds of shapes can be called cylinders.

cylinder

As with prisms, a cylinder can be a **right cylinder** or an **oblique cylinder**, according to whether one base is or is not "straight up over" the other.   The outer two cylinders in Figure 8.110 are right cylinders, whereas the middle cylinder is an oblique cylinder.

right cylinder
oblique cylinder

In addition to prisms, another special type of polyhedron is a *pyramid*. You have probably seen pictures of the famous pyramids in Egypt. Mathematical pyramids include shapes like the Egyptian pyramids as well as variations on this kind of shape, as shown in Figure 8.111. **Pyramids** can be described as follows: Start with any polygon and a separate single point that does not lie in the plane of the polygon. Now use additional polygons to connect the point to the original polygon. This should be done in such a way that all the lines that connect the point to the polygon lie on the sides of these additional polygons. These new polygons, together with the original polygon, form a pyramid. As usual, the original polygon is called the **base** of the pyramid. Figure 8.112 shows a roof that is a pyramid with an octagon base.

pyramid

As with prisms and cylinders, certain kinds of pyramids are called right pyramids. A **right pyramid** is a pyramid for which the point lies "straight up over" the center of the base. A pyramid that is not a right pyramid can be called an **oblique pyramid**. In Figure 8.111 the outer two pyramids are right pyramids, whereas the middle pyramid is an oblique pyramid. Figure 8.113 shows an oblique pyramid decorating the top of a building.

right pyramid
oblique pyramid

FIGURE 8.112

A Pyramid Roof with an Octagon Base

FIGURE 8.113

An Oblique Pyramid on a Building

Just as cylinders are shapes that are related to prisms, *cones* are shapes that are related to pyramids. Roughly speaking, cones are objects like ice-cream cones or cone-shaped paper cups, as well as related objects, as shown in Figure 8.114. As with pyramids, **cones** can be described by starting with a closed curve in a plane and a separate point that does not lie in that plane. Now imagine paper or some other kind of material that joins the point to the curve in such a way that all the lines that connect the point to the curve lie on that paper (or other material). That paper (or other material), together with the original curve, form a cone. As usual, the starting curve and the region inside it is called the **base** of the pyramid. Sometimes the base of a cone is considered a part of the cone, and sometimes it isn't. Either shape (with or without the base) can be called a cone.

As with prisms, cylinders, and pyramids, certain kinds of cones are called right cones. A **right cone** is a cone whose point lies directly over the center of its base. A cone that is not a right cone can be called an **oblique cone**. In Figure 8.114, the outer two cones are right cones, whereas the cone in the middle is an oblique cone.

cones

right cone

oblique cone

FIGURE 8.114

Cones

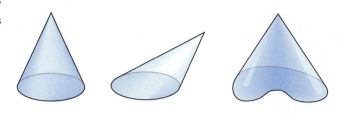

THE PLATONIC SOLIDS

Five special polyhedra are called the **Platonic solids**, in honor of Plato, who thought of these solid shapes as associated with earth, fire, water, air, and the whole universe. Examples are pictured in Figure 8.115. The following are the five Platonic solids:

Tetrahedron is made of 4 equilateral triangles, with 3 triangles coming together at each vertex.

Cube is made of 6 squares, with 3 squares coming together at each vertex.

Octahedron is made of 8 equilateral triangles, with 4 triangles coming together at each vertex.

Dodecahedron is made of 12 regular pentagons, with 3 pentagons coming together at each vertex.

Icosahedron is made of 20 equilateral triangles, with 5 triangles coming together at each vertex.

FIGURE 8.115

The Platonic Solids

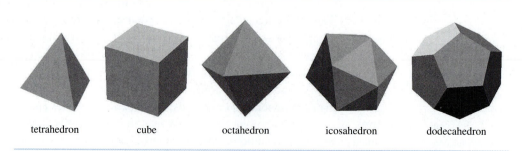

tetrahedron   cube   octahedron   icosahedron   dodecahedron

FIGURE 8.116

A Dodecahedron
Sculpture in
Jesup, Georgia

The names of the Platonic solids make sense because *hedron* comes from the Greek for "base" or "seat," while *tetra*, *octa*, *dodeca*, and *icosa* mean 4, 8, 12, and 20, respectively. So for example, *icosahedron* means *20 bases*, which makes sense because an icosahedron is made out of 20 triangular "bases."

The Platonic solids are special because each one is made of only one kind of regular polygon, and the same number of polygons come together at each vertex. In fact, the Platonic solids are the only *convex* polyhedra having these two properties. A shape in the plane or in

convex   space is **convex** if any line segment connecting two points on the shape lies entirely within the shape. So convex polyhedra do not have any protrusions or indentations.

Although cubes are commonly found in daily life (such as boxes and ice cubes), the other Platonic solids are not commonly seen. However, these shapes do occur in nature occasionally. For example, the mineral pyrite can form a crystal in the shape of a dodecahedron (this crystal is often called a pyritohedron).

The mineral fluorite can form a crystal in the shape of an octahedron, and although rare, gold can too. Some viruses are shaped like an icosahedron. Figure 8.116 shows a picture of a dodecahedron sculpture in front of a school in Jesup, Georgia. See

`www.aw-bc.com/beckmann`

for some additional examples of Platonic solids.

The real attraction of the Platonic solids is their beautiful perfection and symmetry. You can't really appreciate the Platonic solids unless you make models of these wonderful shapes, so I recommend that you construct models of them—see Class Activity 8SS and Problem 2.

CLASS ACTIVITY   NOW TURN TO CLASS ACTIVITIES MANUAL

8SS Making Platonic Solids with Toothpicks and Marshmallows p. 276

8TT Why Are There No Other Platonic Solids? p. 277

The following class activity applies to all polyhedra, not just the Platonic solids:

CLASS ACTIVITY   NOW TURN TO CLASS ACTIVITIES MANUAL

8UU Relating the Numbers of Faces, Edges, and Vertices of Polyhedra p. 280

PRACTICE PROBLEMS FOR SECTION 8.7

1. Examine the small patterns in Figure 8.117. Try to visualize the polyhedra that would result if you were to cut these patterns out on the heavy lines, fold them on the dotted lines, and tape various sides together.

   Now cut out the large versions of these patterns on page 719 and 721 along the heavy lines. Fold the shapes and tape the sides together to make polyhedra. Were your predictions correct? If not, undo your polyhedra and try to visualize again how they turn into their final shapes.

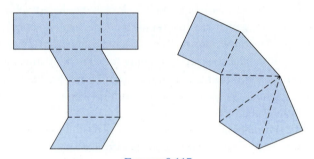

FIGURE 8.117

Small Patterns for Two Polyhedra

2. Make a pattern for a prism whose two bases are identical to the triangle in Figure 8.118. Include the bases in your pattern. (Use a straightedge and compass to make a copy of the triangle in Figure 8.118.) Label all sides that have length a, b, and c on your pattern.

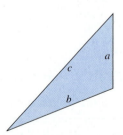

FIGURE 8.118

A Base for a Prism
and a Pyramid

3. Make a pattern for a pyramid whose base is identical to the triangle in Figure 8.118. Include the base in your pattern. (Use a straightedge and compass to make a copy of the triangle in Figure 8.118). Label all sides that have length a, b, and c on your pattern.

4. What other name can you call a tetrahedron? (See Figure 8.119.)

5. What happens if you try to make a convex polyhedron whose faces are all equilateral triangles and for which 6 triangles come together at every vertex?

6. Is it possible to make a convex polyhedron so that 7 or more equilateral triangles come together at every point? Explain.

FIGURE 8.119
A Tetrahedron

ANSWERS TO PRACTICE PROBLEMS FOR SECTION 8.7

1. The pattern on the left of Figure 8.117 makes an oblique square prism. The pattern on the right makes an oblique pyramid with a square base.

2 & 3. See Figure 8.120.

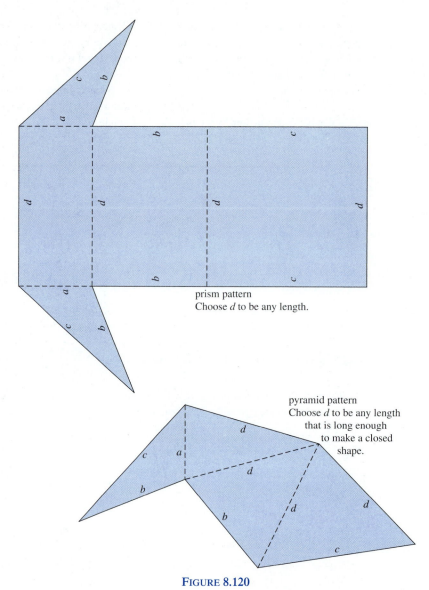

prism pattern
Choose *d* to be any length.

pyramid pattern
Choose *d* to be any length
that is long enough
to make a closed
shape.

FIGURE 8.120

Patterns for a Prism and a Pyramid with Triangle Bases

4. A tetrahedron is a special kind of pyramid with a triangle base.

5. If you put 6 equilateral triangles together so that they all meet at one point, then you will find that they make a flat hexagon, as seen in Figure 8.121. It makes sense that 6 triangles put together at a point make a flat shape because the angle at a corner of an equilateral triangle is 60°, so 6 of these angles side by side will make a full 360°. If you tried to make a convex polyhedron in such a way that 6 equilateral triangles came together at every vertex, you could never get the polyhedron to "close up"—you could only get a flat shape this way.

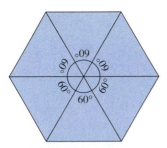

FIGURE 8.121

Six Triangles Put Together at a Point

6. If you put 7 or more equilateral triangles together so that they all meet at one point, then you will find that some triangles will have to slant up and others will have to slant down. Thus, you will have to create indentations or protrusions when you put triangles together in this way, and therefore the polyhedron will not be convex. It makes sense that 7 triangles meeting at a point will do this because the angle at a corner of an equilateral triangle is 60°, so 7 or more of these angles side by side will make more than 360°. This forces some triangles to slant up and others to slant down in order to put so many triangles together at a point.

PROBLEMS FOR SECTION 8.7

1. For each of the following solid shapes, find at least two examples of a real-world object that has that shape:

   a. prism

   b. cylinder

   c. pyramid

   d. cone

2. Referring to the descriptions of the Platonic solids on page 376, construct the five Platonic solids by cutting out the triangles, squares, and rectangles on pages 723, 725, and 727, and taping these shapes together. You may wish to copy pages 723, 725, and 727 onto card stock in order to make sturdier models that are easier to work with.

3. Use a straightedge and compass to help you make a pattern for a prism whose two bases are identical to the triangle in Figure 8.118. You may wish to draw your pattern on graph paper. Make your pattern significantly different from the one for a prism shown in Figure 8.120. Include the bases in your pattern. (Use a straightedge and compass to make a copy of the triangle in Figure 8.118.) Label all sides that have length a, b, and c on your pattern.

4. Use a straightedge and compass to help you make a pattern for a pyramid whose base is identical to the triangle in Figure 8.118. Make your pattern significantly different from the one for a pyramid shown in Figure 8.120. Include the base in your pattern. (Use a straightedge and compass to make a copy of the triangle in Figure 8.118.) Label all sides that have length a, b, and c on your pattern.

5. Use a ruler and compass to help you make a pattern for a prism whose two bases are triangles that have one side of length 2.5 inches, one side of length 2 inches, and one side of length 1.5 inches. It may help you to draw your pattern on $\frac{1}{4}$-inch graph paper. Indicate which sides of your pattern would be joined in order to make the prism.

6. Use a ruler and compass to help you make a pattern for a pyramid with a triangular base that has one side of length 2 inches, one side of length 3 inches, and one side of length 4 inches. Indicate which sides of your pattern would be joined in order to make the pyramid.

7. a. Make a pattern for an oblique cylinder with a circular base. You may leave the bases off your pattern. (Advice: Be willing to experiment first. You might start by making a pattern for a right cone and modifying this pattern.)

   b. The sleeves of most shirts and blouses are more or less in the shape of an oblique cylinder. If your pattern in part (a) were to make a sleeve, what part of your pattern would be at the shoulder? What part of your pattern would be at the armpit?

8. Make a pattern for an oblique cone with a circular base. You may leave the base off your pattern. (Advice: Be willing to experiment first. You might start by making a pattern for a right cone and modifying this pattern.)

9. Make a pattern for the "bottom portion" of a right cone, as pictured in Figure 8.122. What article of clothing is often shaped like this?

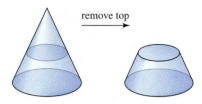

FIGURE 8.122

The Bottom Portion of a Cone

remove top

10. Answer the following without using a model:

　a. How many faces (including the bases) does a prism with a parallelogram base have? What shapes are the faces? Explain briefly.

　b. How many edges does a prism with a parallelogram base have? Explain.

　c. How many vertices does a prism with a parallelogram base have? Explain.

11. Answer the following without using a model:

　a. How many faces (including the base) does a pyramid with a rhombus base have? What shapes are the faces? Explain briefly.

　b. How many edges does a pyramid with a rhombus base have? Explain.

　c. How many vertices does a pyramid with a rhombus base have? Explain.

12. Recall that an n-gon is a polygon with n sides. For example, a polygon with 17 sides can be called a 17-gon. A triangle could be called a 3-gon. Find formulas (in terms of n) for the following, and explain why your formulas are valid:

　a. the number of faces (including the bases) on a prism that has n-gon bases

　b. the number of edges on a prism that has n-gon bases

　c. the number of vertices on a prism that has n-gon bases

　d. the number of faces (including the bases) on a pyramid that has an n-gon base

　e. the number of edges on a pyramid that has an n-gon base

　f. the number of vertices on a pyramid that has an n-gon base

13. This problem goes with Class Activity 8RR on the Magic 8 Ball. Make a closed, three-dimensional shape out of some or all of the triangles in Figure A.10. Make any shape you like—be creative. Answer the following:

　a. Do you think your shape could be the one that is actually inside the Magic 8 Ball? Why or why not?

　b. If you were going to make your own advice ball but with a (possibly) different shape inside, what would be the advantages or disadvantages of using your shape?

14. A cube is a polyhedron that has 3 square faces meeting at each vertex. What would happen if you tried to make a polyhedron that has 4 square faces meeting at each vertex? Explain.

15. The Platonic solids are convex polyhedra with faces that are either equilateral triangles, squares, or regular pentagons. If you try to make a convex polyhedron whose faces are all regular hexagons, what will happen? Explain.

16. Two gorgeous polyhedra can be created by *stellating* an icosahedron and a dodecahedron. *Stellating* means *making starlike*. Imagine turning each face of an icosahedron into a "star point," namely, a pyramid whose base is a triangular face of the icosahedron. Likewise, imagine turning each face of a dodecahedron into a "star point," namely, a pyramid whose base is a pentagonal face of the dodecahedron. A stellated icosahedron will then have 20 "star points," while a stellated dodecahedron will have 12 "star points."

　a. Make 20 copies of Figure 8.123 on card stock. Cut, fold, and tape them to make 20 triangle-based pyramid "star points." (The pattern makes star-point pyramids that don't have bases.) Tape the star points together as though you were making an icosahedron out of their (open) bases.

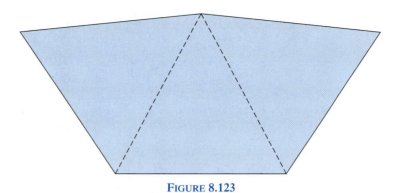

FIGURE 8.123

Pattern for a Star Point of an Icosahedron

b. Make 12 copies of Figure 8.124 on card stock. Cut, fold, and tape them to make 12 pentagon-based pyramid "star points." (The pattern makes star point pyramids that don't have bases.) Tape the star points together as though you were making a dodecahedron out of their (open) bases.

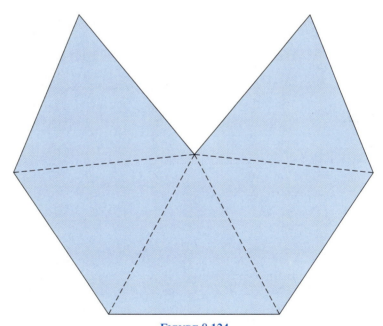

FIGURE 8.124

Pattern for a Star Point of a Dodecahedron

Chapter 9

Geometry of Motion and Change

I*n this chapter we extend our study of shapes by allowing them to move and to change size. The movements of shapes leads to the study of symmetry: by moving shapes in different ways, we can create designs with different kinds of symmetry. We will also study the extent to which a given shape can be changed. Certain shapes are rigid and inflexible, while others are "floppy" and moveable. This difference in flexibility and rigidity is related to construction practices. Finally, we will study what happens when objects change size, but otherwise retain the same shape. The theory of size change has widely used practical applications, such as the measurement of distances in land surveying.*

9.1   Reflections, Translations, and Rotations

In the last chapter, we described special properties of shapes in a plane. One especially attractive property of some shapes and designs is symmetry. In order to study symmetry we will first examine certain transformations of planes, namely translations, reflections, and rotations. We will use translations, reflections, and rotations to define what we mean by symmetry, as well as to create symmetrical designs. From this point of view, translations, reflections, and rotations are the "building blocks" of symmetry.

Roughly speaking, a **transformation** of a plane is just what the name implies: an action that changes or transforms a plane. We are interested in transformations that start with a plane and change it back into the same plane. Even though we will start and end with the same plane, the transformation will usually have caused most or all of the individual points on the plane to change location. The three kinds of transformations that we will study are *rotations*, *reflections*, and *translations*. We will also briefly consider a fourth kind of transformation: *glide-reflections*.

reflection
line of reflection
    A **reflection** (or **flip**) of a plane across a chosen line, called the **line of reflection**, is the following kind of transformation: For any point Q in the plane, imagine a line that passes

through Q and is perpendicular to the line of reflection, as shown in Figure 9.1. Now imagine moving the point Q to the other point on the line that is the same distance from the line of reflection. (If the point Q had been on the line of reflection, then it would stay in its place.) If we move each point in the plane as just described, then the end result is a reflection of the plane.

One way to think about reflections is that under a reflection, each point in the plane goes to its "mirror image" on the other side of the line of reflection (hence the name reflection).

An informal way to see the effect of a reflection is to draw a point on paper in wet ink or paint and quickly fold the paper along the desired line of reflection: the location where the wet ink rubs off onto the paper shows the final position of the point after reflecting. Another way to see the effect of a reflection is by drawing a point on a semitransparent piece of paper and flipping the paper upside down by twirling it around the desired line of reflection: you will see the location of the reflected point on the other side of the paper.

translation   A **translation** (or **slide**) of a plane by a given distance in a given direction is the end result of moving each point in the plane the given distance in the given direction. To illustrate a translation, put a flat piece of paper on a table top and slide the piece of paper in some direction (without rotating it). Now imagine doing this sliding process with a whole plane instead of a piece of paper: this would produce a translation of the plane. Figure 9.2 indicates initial and final positions of various points under a translation.

rotation   A **rotation** (or **turn**) about a point through a given angle is a transformation of a plane that is the end result of rotating all points in the plane about a fixed point, through a fixed angle. To

FIGURE 9.1

The Effect of a
Reflection on
Points P, Q,
and R

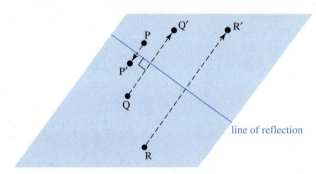

FIGURE 9.2

The Effect of a
Translation on
Points S, T, and
U. The solid
arrow represents
the distance and
direction of the
translation.

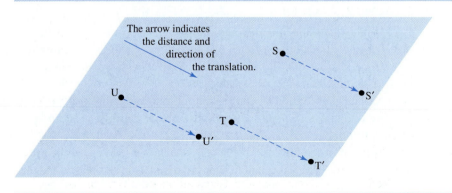

FIGURE 9.3

The Effect of a
Rotation about
Point A on
Points V, W,
and X

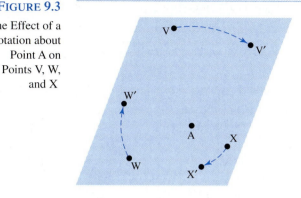

illustrate a rotation about a point, put a flat piece of paper on a table top, stick a pin through the point of rotation so as to hold that point fixed on the table, and rotate the piece of paper about that point. Now imagine rotating an infinite plane instead of a piece of paper: this movement would produce a rotation of the plane about the fixed point. Figure 9.3 shows the initial and final positions of some points in a plane under a rotation about the point A. Notice that points farther from the point A move a greater *distance* than points closer to A, even though all points rotate through the same *angle*.

glide-reflection   A **glide-reflection** is the end result of combining a reflection and then a translation in the direction of the line of reflection.

Rotations, reflections, translations, and glide-reflections are special because these transformations *don't change distances*. For example, suppose you pick two points P and Q in the plane and suppose that you now rotate the plane. Then, even though the position and orientation of the pair of points P and Q may change after rotation, *the distance between the two points* is still the same, even after they have been rotated. Similarly, translations, reflections, and glide-reflections preserve the distances between pairs of points. Furthermore, mathematicians have proven that every transformation of the plane that preserves distances between all pairs of points must be either a rotation, a translation, a reflection, or a glide-reflection.

CLASS ACTIVITY   NOW TURN TO CLASS ACTIVITIES MANUAL

PRACTICE PROBLEMS FOR SECTION 9.1

1. Match the specified transformations in A, B, C, and D of Figure 9.4 to the effects shown in 1, 2, 3, and 4.

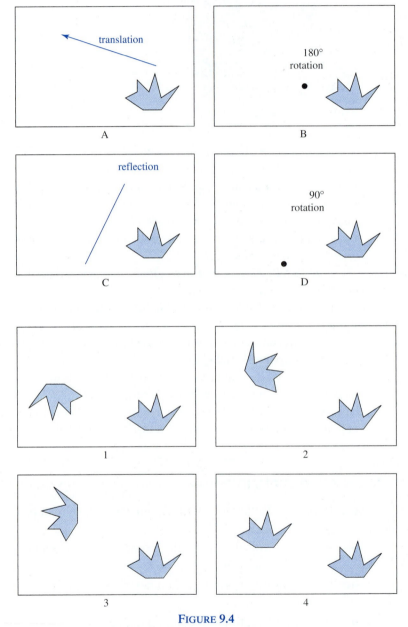

FIGURE 9.4

Various Transformations and Their Effects

2. For each picture in Figure 9.5, determine what kind of transformation (reflection, translation, or rotation) will take the initial shape to the final shape.

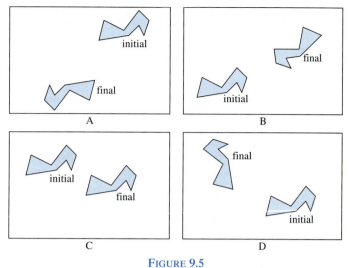

FIGURE 9.5

What Kinds of Transformations?

3. Draw the result of rotating the shaded shapes in Figure 9.6 by 90° counterclockwise around the point where the two heavy lines meet. Explain how you know where to draw your rotated shapes.

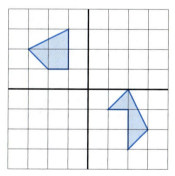

FIGURE 9.6

Determine the Locations of the
Shapes after a 90°
Counterclockwise Rotation

ANSWERS TO PRACTICE PROBLEMS FOR SECTION 9.1

1. A–4, B–1, C–2, D–3.

2. See Figure 9.7.

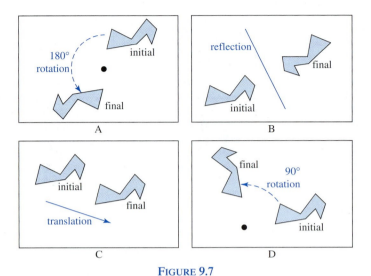

FIGURE 9.7

The Effects of Various Transformations on a Shape

3. See Figure 9.8. To determine where to draw the rotated shapes, notice that the horizontal heavy line rotates to the vertical one and the vertical heavy line rotates to the horizontal one. So the rotation takes the point A, which is 2 squares to the right of the center of rotation, to the point A′, which is 2 squares up from the center of rotation. The point B, which is 3 squares below point A must therefore rotate to the point B′, which is 3 squares to the right of point A′. Point C, which is 3 squares up and 1 square to the left of the center point will rotate to the point C′, which is 3 squares to the left and 1 square down from the center point. Similarly, by considering the location of each corner point in the shaded shapes relative to the heavy lines, we can determine where these points go when they are rotated, thereby determining the location of the rotated shapes.

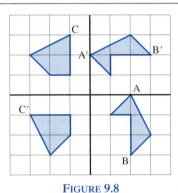

FIGURE 9.8

Shapes and Their Locations after a 90° Counterclockwise Rotation

PROBLEMS FOR SECTION 9.1

1. On graph paper, draw two shapes that are not symmetrical. Draw the result of reflecting the shapes across a grid line. Indicate the line of reflection. Explain how you know where to draw your reflected shapes.

2. a. On graph paper, draw two shapes that are not symmetrical. Draw the result of translating the shapes 5 squares to the right.

   b. On graph paper, draw two shapes that are not symmetrical. Draw the result of translating the shapes in the direction given by the arrow in Figure 9.9. Explain how you know where to draw your translated shapes.

FIGURE 9.9

An Arrow for a
Translation

3. On graph paper, draw two shapes that are not symmetrical. Draw the result of rotating the shapes 180° around a point. Indicate the center point of the rotation. Explain how you know where to draw your rotated shapes.

4. On graph paper, draw two shapes that are not symmetrical. Draw the result of rotating the shapes 90° counterclockwise around a point. Indicate the center point of the rotation. Explain how you know where to draw your rotated shapes.

5. On graph paper, draw two shapes that are not symmetrical. Draw the result of rotating the shapes 90° clockwise around a point. Indicate the center point of the rotation. Explain how you know where to draw your rotated shapes.

6. Investigate the following questions, either with Geometer's Sketchpad (preferable) or by making drawings on (graph) paper. Start by drawing a shape (or design) that is not symmetrical. What is the *net effect* if you rotate your shape about some point by 180° and then rotate the resulting shape by 180° *about some other point*? What *single* transformation (reflection, translation, or rotation) will take your initial shape to your final shape?

7. Investigate the following questions, either with Geometer's Sketchpad (preferable) or by making drawings on (graph) paper. Start by drawing a shape (or design) that is not symmetrical. What is the *net effect* if you reflect your shape across a line and then reflect the resulting shape across another line that is perpendicular to the first line? What *single* transformation (reflection, translation, or rotation) will take your initial shape to your final shape?

8. Investigate the following questions, either with Geometer's Sketchpad or by making drawings on (graph) paper. Draw a shape (or design) that is not symmetrical, draw a separate line, and draw a point on the line.

   a. If you reflect the shape across the line and then rotate the reflected shape 180° about the point, will the final position of the shape be the same as if you had *first* rotated the shape 180° and *then* reflected the rotated shape across the line?

   b. If you reflect the shape across the line and then rotate the reflected shape 90° counterclockwise about the point, will the final position of the shape be the same as if you had *first* rotated the shape 90° counterclockwise and *then* reflected the rotated shape across the original, unrotated line?

9.2 Symmetry

Symmetry is an area shared by mathematics, the natural world, and art, so it offers opportunities for cross-disciplinary study. Something about symmetry is deeply appealing to most people. Why is that so? Maybe it is because objects with symmetry seem more perfect than objects that don't have symmetry. Or maybe it is because objects with symmetry involve repetition. Is there something about human nature that causes us to enjoy repetition? For example, almost all music involves repetition of themes—different themes are often repeated throughout a piece of music in a certain pattern. Young children love to hear the same story over and over again. (Parents sometimes dread reading a favorite book for the umpteenth time!) Maybe we like repetition because it helps us learn and understand, and maybe our enjoyment of repetition makes symmetrical designs appealing.

When we look at natural objects in the world around us, we find a mix of symmetry and asymmetry. Most creatures are (mostly) symmetrical. Plants typically have symmetrical parts even if they are not symmetrical over all: leaves and flowerheads are usually symmetrical. On the other hand, geological features are rarely symmetrical. It would be surprising to see a mountain, a lake, a river, or a rock that we would describe as symmetrical, even though these objects are usually made up of smaller, repeated parts. On the other hand, some volcanoes, snowflakes, crystals, beaches, river stones, and waves are symmetrical, or nearly so.

Just as the concept of *circle* has both an informal or artistic interpretation as well as a mathematical definition, the notion of symmetry also has both an informal interpretation as well as a specific mathematical definition that applies to shapes in a plane or in space. We will now turn our attention to the mathematical definition of symmetry, which requires that we draw upon the transformations that we have just studied: rotations, reflections, and translations. There are three ways that we will use these transformations in our study of symmetry. First, we will use rotations, reflections, and translations to *define* symmetry. In other words, we will use these transformations to say what it means for a shape or a design to have specific kinds of symmetry. Second, we will use rotations, reflections, and translations to *create* shapes and designs that have symmetry. Finally, we will *analyze* symmetrical designs by looking for parts of the design that can be used to create the whole design by applying rotations, reflections, and translations.

DEFINING SYMMETRY

There are four kinds of symmetry that a shape or design in a plane can have: reflection symmetry, translation symmetry, rotation symmetry, and glide-reflection symmetry. We will use reflections, translations, and rotations to say what it means for a shape or design to have symmetry of one of these types.

reflection (mirror) symmetry
A shape or design in a plane has **reflection symmetry** or **mirror symmetry** if there is a line in the plane such that the shape or design as a whole occupies the same place in the plane both before and after reflecting across the line. This line is called a **line of symmetry**. For example, Figure 9.10 shows a design and a line of symmetry of the design. Notice that reflecting across the line of symmetry causes most points on the design to swap locations with another point on the design, but the design *as a whole* occupies the same place in the plane.

Shapes or designs can have more than one line of symmetry, as shown in Figure 9.11.

To determine whether a design or shape drawn on a semitransparent piece of paper has reflection symmetry with respect to a certain line on the paper, fold the paper along that line. If you can see that the parts of the design on the two parts of the paper match one another, then the design has reflection symmetry, and the line you folded along is a line of symmetry. Another way to see whether a design or shape has reflection symmetry is to use a mirror that has a straight edge. Place the straight edge of the mirror along the line that you think might be a line of symmetry, and hold the mirror so that it is perpendicular to the design. When you look in the mirror, is the design the same as it was without the mirror in place? If so, then the

FIGURE 9.10

A Shape with Reflection Symmetry

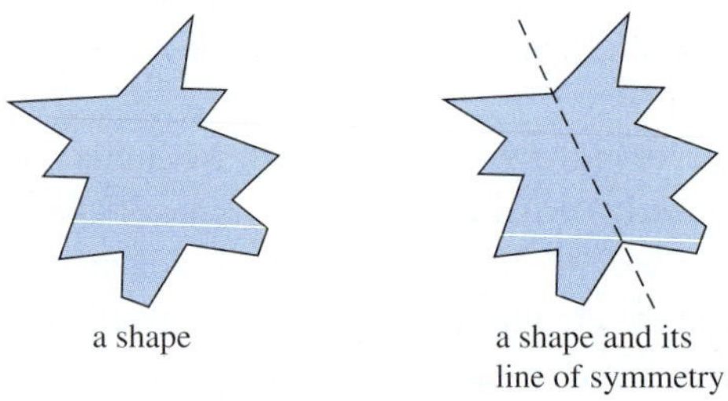

a shape

a shape and its line of symmetry

FIGURE 9.11

A Design with
Four Lines of
Symmetry

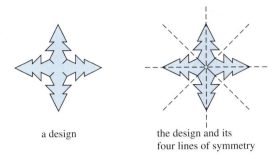

a design

the design and its
four lines of symmetry

design has reflection symmetry. These methods may help you get a better feel for reflection symmetry when you are first learning about it, but the goal is to be able to use *visualization* to determine whether or not a design has reflection symmetry.

**translation
symmetry**   A design or pattern in a plane has **translation symmetry** if there is a translation of the plane such that the design or pattern *as a whole* occupies the same place in the plane both before and after the translation. True translation symmetry occurs only in designs or patterns that take up an infinite amount of space. So when we draw a picture of a design or pattern that has translation symmetry, we can only show a small portion of it; we must imagine the pattern continuing on indefinitely. Figure 9.12 shows a pattern with translation symmetry. In fact, notice that this pattern has two independent translations that take the pattern as a whole to itself: one is "shift right", another is "shift up." Wallpaper patterns generally have translation symmetry with respect to translations in two independent directions.

Frieze patterns—often seen on narrow strips of wallpaper positioned around the top of the walls of a room—provide additional examples of designs with translation symmetry. Figure 9.13 on page 394 shows a frieze pattern that has translation symmetry. Unlike a wallpaper pattern, a frieze pattern will only have tranlation symmetry in one direction (and its "reverse") instead of in two independent directions.

**rotation
symmetry**   A shape or design in a plane has **rotation symmetry** if there is a rotation of the plane, of more than 0° but less than 360°, such that the shape or design *as a whole* occupies the same points in the plane both before and after rotation. For example, see Figure 9.14 on page 394. Rotating 72° about the center of the design takes the design as a whole to the same location in the plane,

FIGURE 9.12

A Wallpaper
Pattern with
Translation
Symmetry

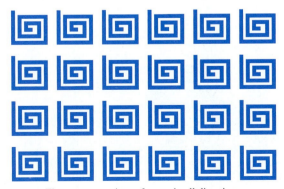

The pattern continues forever in all directions.

FIGURE 9.13

A Frieze Pattern
with Translation
Symmetry

The pattern continues forever to the right and to the left.

FIGURE 9.14

A Design with
5-fold Rotation
Symmetry

even though each individual curlicue in the design moved to the position of another curlicue. The design of Figure 9.14 is said to have **5-fold rotation symmetry** because by applying the $72° = 360° \div 5$ rotation about the center of the design 5 times, all points on the design return to their starting positions. Generally, shapes or designs can have 2-fold, 3-fold, 4-fold, etc. rotation symmetry. A design has ***n*-fold rotation symmetry** provided that there is a rotation of more than 0° but less than 360° that takes the design as a whole to the same location and such that applying this rotation n times returns every point on the design to its initial position. This is more complicated to say than to see. Figure 9.15 shows some examples of designs with various rotation symmetries.

n-fold rotation
symmetry

You can determine whether a design or shape drawn on a piece of paper has rotation symmetry: Make a copy of the design on semitransparent paper. Place the semitransparent paper over the original design so that the two designs match one another. If you can rotate the semitransparent paper less than a full turn so that the two designs match one another again, then the design has rotation symmetry. If you can keep rotating by the same amount for a total of 2, 3, 4, 5, etc., times until the design on the semitransparent paper is back to its initial position, then the design has 2-fold, 3-fold, 4-fold, 5-fold, etc., rotation symmetry, respectively. When you are first learning about rotation symmetry it may help you to physically rotate designs, however, the goal is to use *visualization* to determine if a design has rotation symmetry.

FIGURE 9.15

Designs with
Rotation
Symmetry

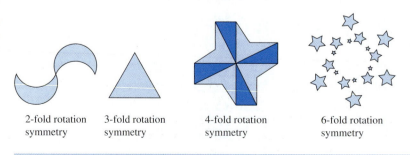

2-fold rotation   3-fold rotation   4-fold rotation   6-fold rotation
symmetry   symmetry   symmetry   symmetry

FIGURE 9.16

A Frieze Pattern
with
Glide-Reflection
Symmetry

The pattern continues forever to the right and left.

FIGURE 9.17

Understanding
Glide-Reflection
Symmetry

1. Reflect across the horizontal.

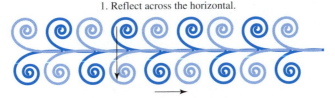

2. Translate right.

glide-reflection symmetry

Finally, a design or pattern in a plane has **glide-reflection symmetry** if there are a reflection and a translation such that, after applying the reflection followed by the translation, the *design as a whole* occupies the same location in the plane. Glide reflection symmetry is often seen on frieze patterns. For example, Figure 9.16 shows an example of a frieze pattern with glide-reflection symmetry. If this design is reflected across a horizontal line through its middle and then translated to the right (or left), the design as a whole will occupy the same location in the plane as it did originally. (See Figure 9.17.)

Notice that the frieze pattern in Figure 9.16 also has translation symmetry. Many designs have more than one type of symmetry. For example, the design shown in Figure 9.18 on p. 396 has both 2-fold rotation symmetry as well as reflection symmetry with respect to two lines: one horizontal and one vertical.

CLASS ACTIVITY NOW TURN TO CLASS ACTIVITIES MANUAL

9G Checking for Symmetry p. 289

9H Frieze Patterns p. 290

CREATING SYMMETRY

One way to create symmetrical designs is to start with any design and then either reflect it, repeatedly rotate it, or repeatedly translate the design, keeping the original and all copies of the design produced in the process. The new design, consisting of the original and all copies, will have reflection symmetry, rotation symmetry, or translation symmetry, respectively. Figure 9.19 on p. 397 indicates the process of creating symmetrical designs this way.

FIGURE 9.18

A Design on
Egyptian Fabric

To create a design with 2-fold rotation symmetry, start with any design and rotate it $\frac{360°}{2} = 180°$ about any point in the plane. The original design and the rotated design together form a new design that has 2-fold rotation symmetry. To create a design with 6-fold rotation symmetry, start with any design and rotate it $\frac{360°}{6} = 60°$ five times, keeping the original design and all the rotated designs. All six designs together form a design with 6-fold rotation symmetry. In general, to create a design with n-fold rotation symmetry, start with any design and rotate it $\frac{360°}{n}$ repeatedly until the original design gets back to its initial position. Thereafter, if you rotate the whole design, the constituent designs will cycle around, but the design as a whole occupies the same location in the plane. Since n rotations of $\frac{360°}{n}$ produce a total of 360°, which is a full rotation, the design we produce this way has n-fold rotation symmetry.

CLASS ACTIVITY   NOW TURN TO CLASS ACTIVITIES MANUAL

9I Creating Symmetrical Designs with Geometer's Sketchpad p. 291

9J Creating Symmetrical Designs (Alternate) p. 292

FIGURE 9.19

Creating
Symmetrical
Designs

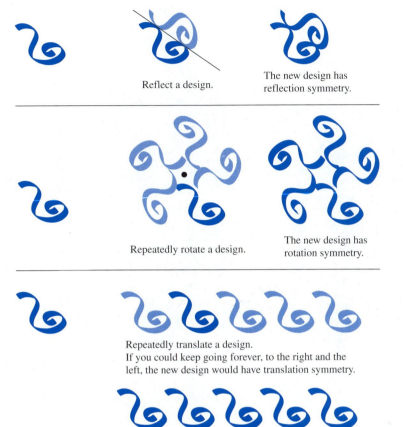

Reflect a design.

The new design has
reflection symmetry.

Repeatedly rotate a design.

The new design has
rotation symmetry.

Repeatedly translate a design.
If you could keep going forever, to the right and the
left, the new design would have translation symmetry.

The artist M. C. Escher (1898–1972) created many interesting symmetrical designs involving congruent, interlocking shapes—something like Figure 9.20 on p. 398. You can see some of Escher's artwork at

www.aw-bc.com/beckmann

There is software called TesselMania! that even young children can use to create Escher-type designs. However, if you do the next class activity, you will see that you can also use Geometer's Sketchpad to create and modify Escher-type designs. By using Geometer's Sketchpad, you can appreciate what makes these fascinating designs work.

CLASS ACTIVITY NOW TURN TO CLASS ACTIVITIES MANUAL

9K Creating Escher-Type Designs with Geometer's Sketchpad (for Fun) p. 293

FIGURE 9.20

One Kind of
Escher-Type
Design

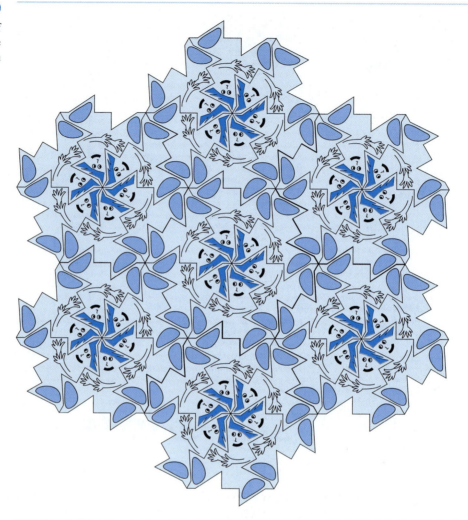

ANALYZING DESIGNS

By their very nature, symmetrical designs are made up of smaller designs, put together by rotating, reflecting, or translating the smaller designs. In some cases it is not hard to see how a symmetrical design is made up of smaller parts, such as the design in Figure 9.21. This design was made by repeatedly rotating a "curlicue."

The quilt design in Figure 9.22 is a little more complex. Copies of the small square shown at the top left of Figure 9.22 were used to create the quilt design by repeatedly reflecting or translating and rotating the small square. Figure 9.23 on p. 400 shows a way to analyze and describe the quilt's design.

For some designs or patterns, it can be challenging to find the copies of the small component shapes that create the larger pattern. For example, Figure 9.24 on p. 401 shows a wallpaper pattern at the top of the picture. If you saw this pattern on a wall, it might not be clear how it was created from a smaller design. The middle of Figure 9.24 shows a way of subdividing the pattern into smaller designs. The bottom of the picture shows how the pattern is broken

FIGURE 9.21

A Simple
Symmetrical
Design

FIGURE 9.22

A Quilt

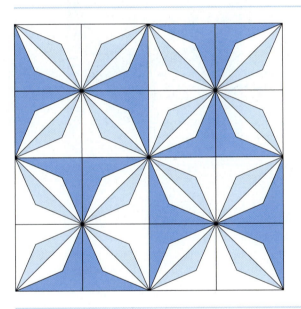

into smaller parts. This is only one of many ways to subdivide the pattern into smaller designs. The picture shows that the wallpaper pattern can be created by translating one design over and over so as to fill up a plane (or a wall).

CLASS ACTIVITY NOW TURN TO CLASS ACTIVITIES MANUAL

9L Analyzing Designs p. 294

FIGURE 9.23

Analyzing a
Quilt Design

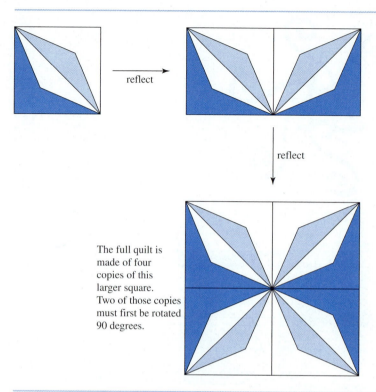

reflect

reflect

The full quilt is
made of four
copies of this
larger square.
Two of those copies
must first be rotated
90 degrees.

FIGURE 9.24

Subdividing a
Pattern into
Smaller Designs

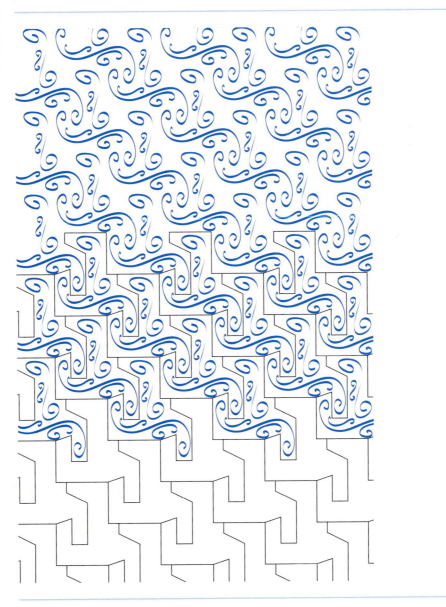

PRACTICE PROBLEMS FOR SECTION 9.2

1. Symmetrical designs are found throughout the world in all different cultures. Figure 9.25 shows a small sample of such designs. Determine the kinds of symmetry these designs have.

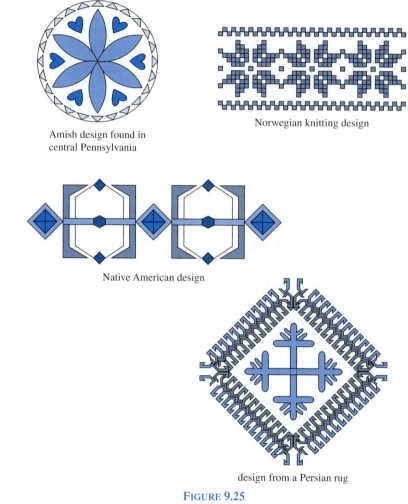

Amish design found in
central Pennsylvania

Norwegian knitting design

Native American design

design from a Persian rug

FIGURE 9.25

Some Symmetrical Designs from around the World

2. Practice Class Activities 9I and 9J.

3. Draw a design that is made out of copies of the shaded shape in Figure 9.26 and has 4-fold rotation symmetry but no reflection symmetry.

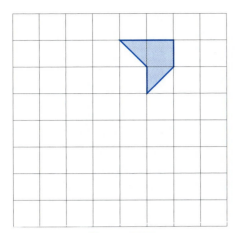

FIGURE 9.26

Create a Symmetrical Design

4. Draw a design that is made out of copies of the shaded shape in Figure 9.26 and has 2 lines of symmetry.

5. For both of the patterns in Figure 9.27 on page 404, find a piece of the pattern such that the larger pattern can be thought of as made up of repetitions of this smaller design. Show how the pattern is made up of this smaller, repeated design.

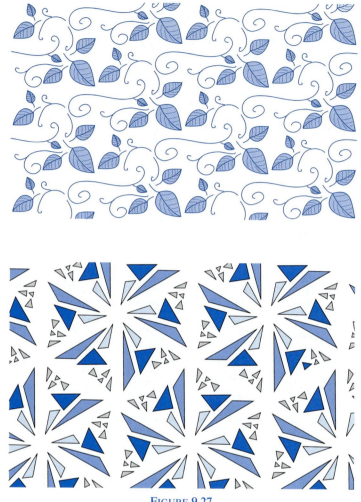

FIGURE 9.27
Analyze These Patterns

ANSWERS TO PRACTICE PROBLEMS FOR SECTION 9.2

1. The Amish design has 6-fold rotation symmetry in addition to reflection symmetry with respect to 6 different lines: 3 lines that pass through the middles of opposite flower petals and 3 lines that pass through the middles of opposite hearts.

   The Norwegian knitting design has 2-fold rotation symmetry in addition to reflection symmetry with respect to 2 different lines: one horizontal line and one vertical line. (Notice that the individual "snowflake" designs within the design have 4-fold rotation symme-try and have reflection symmetry with respect to horizontal, vertical, and two diagonal lines, but the entire Norwegian knitting design has less symmetry.)

   The Native American design has the same symmetries as the Norwegian knitting design.

   The design from a Persian rug has 4-fold rotation symmetry in addition to reflection symmetry with respect to 4 different lines: one vertical, one horizontal, and two diagonal lines.

3. See Figure 9.28.   4. See Figure 9.29.

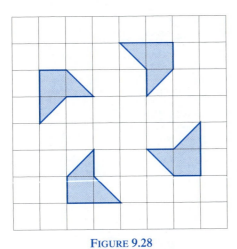

FIGURE 9.28

A Design with 4-Fold Rotation Symmetry

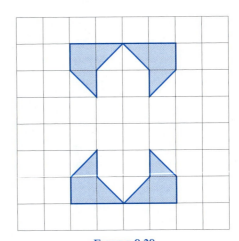

FIGURE 9.29

A Design with 2 Lines of Symmetry

5. See Figure 9.30 on page 406.

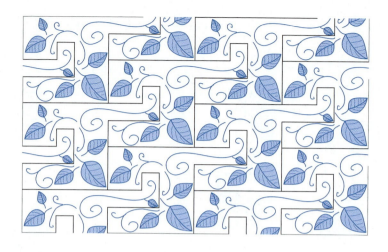

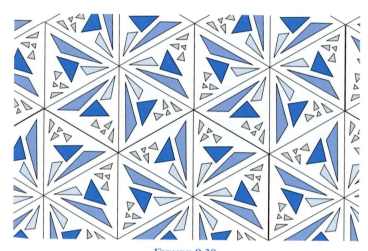

FIGURE 9.30

Analysis of Patterns

PROBLEMS FOR SECTION 9.2

1. Find examples of symmetrical designs from a modern-day culture or a historical one. Make copies of the designs (either by photocopying or by drawing them) and determine what kinds of symmetry the designs have.

2. Determine all the symmetries of Design 1 and Design 2 in Figure 9.31. Consider each design as a whole. For each design, describe all lines of symmetry (if the design has reflection symmetry) and determine whether the design has 2-fold, 3-fold, 4-fold, or other rotation symmetry. Explain your answers.

Design 7, pattern continues forever to the right and left.

Design 8, pattern continues forever to the right and left.

Design 9, pattern continues forever in all directions.

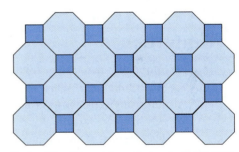

Design 10, pattern continues forever in all directions.

FIGURE 9.31

Some Symmetrical Designs

FIGURE 9.32

Some Patterns

3. Determine all the symmetries of Design 3 and Design 4 in Figure 9.31. Consider each design as a whole. For each design, describe all lines of symmetry (if the design has reflection symmetry) and determine whether the design has 2-fold, 3-fold, 4-fold, or other rotation symmetry. Explain your answers.

4. Determine all the symmetries of Design 5 and Design 6 in Figure 9.31. Consider each design as a whole. For each design, describe all lines of symmetry (if the design has reflection symmetry) and determine whether the design has 2-fold, 3-fold, 4-fold, or other rotation symmetry. Explain your answers.

5. Determine all the symmetries of Design 7 and of Design 8 in Figure 9.32. Consider each design as a whole. For each design, describe all lines of symmetry (if the design has reflection symmetry) and determine whether the design has 2-fold, 3-fold, 4-fold, or other rotation symmetry. Don't forget to look for translation symmetry and glide-reflection symmetry too. Explain your answers.

6. Determine all the symmetries of Design 9 and Design 10 in Figure 9.32. Consider each design as a whole. For each design, describe all lines of symmetry (if the design has reflection symmetry) and determine whether the design has 2-fold, 3-fold, 4-fold, or other rotation symmetry. Don't forget to look for translation symmetry and glide-reflection symmetry too. Explain your answers.

7. a. Which triangles have reflection symmetry? Explain.

   b. Which triangles have rotation symmetry? Explain.

8. a. Determine all the symmetries of a square. Explain.

b. Determine all the symmetries of a rectangle that is not a square. Explain.

c. Determine all the symmetries of a parallelogram that is not a rectangle or a rhombus. Explain.

9. Compare translation and translation symmetry. Explain how these two concepts are related.

10. Compare rotation and rotation symmetry. Explain how these two concepts are related.

11. Compare (mathematical) reflection and reflection symmetry. Explain how these two concepts are related.

12. Draw a simple asymmetrical shape or design. Then copy your shape or design so as to create a single new design that has both 2-fold rotation symmetry and translation symmetry *simultaneously*. (You may wish to use graph paper.)

13. Draw a simple asymmetrical shape or design. Then copy your shape or design so as to create a single new design that has both 4-fold rotation symmetry and translation symmetry *simultaneously*. (You may wish to use graph paper or software.)

14. Draw a simple asymmetrical shape or design. Then copy your shape or design so as to create a single new design that has both 2-fold rotation symmetry and reflection symmetry *simultaneously*. (You may wish to use graph paper.)

15. Draw a simple asymmetrical shape or design. Then copy your shape or design so as to create a single new design that has both 4-fold rotation symmetry and reflection symmetry *simultaneously*. (You may wish to use graph paper or software.)

16. Draw a simple asymmetrical shape or design. Then copy your shape or design so as to create a single new design that has rotation, reflection, and translation symmetry *simultaneously*. (You may wish to use graph paper or software.)

17. Find a symmetrical quilt, rug, or wallpaper pattern, or find a picture of a symmetrical quilt, rug or wallpaper pattern. Find component pieces of your quilt, rug, or wallpaper pattern and describe how the quilt, rug, or wallpaper pattern can be thought of as put together from these components.

9.3   Congruence

Informally, two shapes (either in a plane or in space) that are the same size and shape are called *congruent*. What information about two shapes will guarantee that they are congruent? Surprisingly, the answer for triangles differs from that for other shapes. Triangles, unlike other polygons, are structurally rigid, and a criterion for the congruence of triangles is related to this property. This distinction between triangles and other polygons explains some standard practices in building construction. So, even though the study of congruence may seem purely abstract and theoretical, it has many practical applications.

congruent   Two shapes or designs in a plane are **congruent** if there is a rotation, a reflection, a translation, or a combination of these transformations, that take one shape or design to the other shape or design. For example, Figure 9.33 shows two shaded shapes that are congruent. They are congruent because the shaded shape on the right was created by first translating the shaded shape on the left horizontally to the right, and then rotating the translated shape 90° about the point shown.

In practice, if you have two shapes or designs drawn on two semitransparent pieces of paper, you can determine whether they are congruent by superimposing one on the other. This may require sliding the pieces of paper around, rotating them, and perhaps even flipping one piece of paper over (which has the effect of reflecting the shape or design). If it is possible to move the pieces of paper around so that the shapes match one another, then the shapes are congruent; if it is not possible to match the shapes, then they are not congruent.

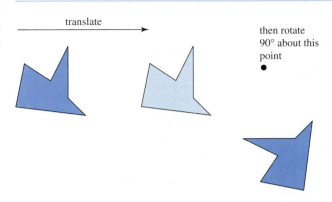

FIGURE 9.33

Congruent
Shapes

translate

then rotate
90° about this
point
•

FIGURE 9.34

A Triangle and a
Quadrilateral
Made of Straws

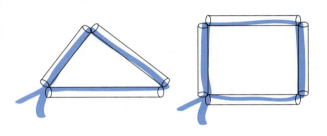

CLASS ACTIVITY   NOW TURN TO CLASS ACTIVITIES MANUAL

9M Triangles of Specified Side Lengths p. 296

9N Triangles and Quadrilaterals of Specified Side Lengths p. 297

9O Triangles with an Angle, a Side, and an Angle Specified p. 298

If you form a triangle by threading a 3-inch, a 4-inch, and a 5-inch piece of straw together and a quadrilateral by threading a 3-inch, a 4-inch, a 3-inch, and a 4-inch piece of a straw together in that order (see Figure 9.34 and the instructions in Class Activity 9N), you will find that the triangle is rigid, whereas the quadrilateral is "floppy." That is, the triangle's sides cannot be moved independently, whereas the quadrilateral's sides can be moved independently, forming many different quadrilaterals. (See Figure 9.35 on page 410.) We can interpret this difference about triangles and quadrilaterals in terms of congruence. *All triangles with sides of length 3 inches, 4 inches, and 5 inches are congruent*, whereas there are quadrilaterals with sides of length 3 inches, 4 inches, 3 inches, 4 inches (in that order) that are *not* congruent.

What if you have a triangle with sides of lengths other than 3 inches, 4 inches, and 5 inches? Imagine making this different triangle out of straws as well. Do you think it would also be rigid like the 3 inch, 4 inch, 5 inch one? In fact, any triangle, no matter what the

FIGURE 9.35

Straws Forming
a Rectangle and
the Same Straws
Forming a
Parallelogram

lengths of its sides are, is structurally rigid. This fact about triangles can be stated in terms of congruence:

Given a triangle that has sides of length a, b, and c units, it is congruent to all other triangles that have sides of length a, b, and c units.

SSS congruence

This last statement is called **side-side-side congruence** or SSS congruence for triangles; it is the mathematical way to say that triangles are structurally rigid.

The structural rigidity of triangles makes them common in building construction. For example, builders temporarily brace the frame of a house under construction with additional pieces of wood to keep the walls from falling over. (See Figure 9.36.) These additional pieces of wood create triangles; since triangles are rigid, the walls are held securely in place.

Triangles are also used in structures that need to be sturdy but relatively lightweight. For example, the crane in Figure 9.37 consists of many triangles. Because the triangles are rigid, they make the crane strong without using solid metal, which would be very heavy.

In contrast, the structural flexibility of quadrilaterals makes them useful in objects that must move and change shape. For example, the folding laundry rack in Figure 9.38 is able

FIGURE 9.36

Extra Pieces of
Wood Create
Triangles for
Stability

FIGURE 9.37

A Crane Is
Made Out of
Many Triangles

FIGURE 9.37

A Crane Is
Made Out of
Many Triangles

FIGURE 9.38

A Folding
Laundry Rack
Uses Hinged
Rhombuses in
Its Construction.
The Triangle at
the Top Keeps
the Rack from
Collapsing
When in Use.

to collapse because it is made from hinged rhombuses. The triangle at the top keeps the rack from collapsing when it is in use. When the rack is ready to be stowed, the triangle at the top is disconnected and the rhombuses collapse.

In addition to side-side-side congruence, there are other cases where all triangles with specific properties are congruent. One such case is when we specify one side of a triangle and we also specify the two angles at either end of this side. For example, given a line segment AB, which triangles have a 40° angle at A and a 60° angle at B? Figure 9.39 on page 412 shows that there are two such triangles, one triangle is "above" AB and the other is "below" AB. However, these two triangles are congruent, as you can see by reflecting one triangle across AB.

So all triangles that have AB as one side and that have a 40° angle at A and a 60° angle at B are congruent. The same will be true of other angles as well—as long as the two specified angles add to less than 180°. (Otherwise, a triangle cannot be formed, since all three positive angles in a triangle must add to 180°.) Thus, in general,

if a line segment is specified as a side of a triangle, and if two angles that add to less than 180° are specified at the two ends of the line segment, then all triangles formed from that line segment and those angles are congruent.

ASA congruence   This last statement is called **angle-side-angle congruence** or ASA congruence for triangles.

FIGURE 9.39

The Two
Triangles That
Have AB as One
Side and Have
Specified
Angles at A
and B

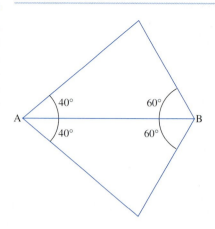

PRACTICE PROBLEMS FOR SECTION 9.3

1. What is SSS congruence?

2. What is ASA congruence?

3. Suppose someone tells you that he has a garden with four sides, two of which are 10 feet long and opposite each other, and the other two of which are 15 feet long and opposite each other. With only this information, can you determine the exact shape of the garden? Take into account that the person might like gardens in unusual shapes.

ANSWERS TO PRACTICE PROBLEMS FOR SECTION 9.3

1. See text.

2. See text.

3. No, this information alone does not allow you to determine the shape of the garden. Although most people would probably make their garden rectangular, someone with an artistic flair might make the garden in the shape of a parallelogram that is not a rectangle. You can simulate this with four strung-together straws, as in Class Activity 9O and Figure 9.34.

PROBLEMS FOR SECTION 9.3

1. Suppose that Ada, Bada, and Cada are three cities and that Bada is 20 miles from Ada, Cada is 30 miles from Bada, and Ada is 40 miles from Cada. There are straight-line roads between Ada and Bada, Ada and Cada, and Bada and Cada.

   a. Draw a careful and precise map showing Ada, Bada, and Cada and the roads between them, using a scale of 10 miles = 1 inch. Describe how to use a compass to make a precise drawing.

   b. If you were to draw another map, or if you were to compare your map to a classmate's, how would they compare? In what ways might the maps differ? In what ways would they be the same? Which criterion for triangle congruence is most relevant to these questions?

2. This problem continues the investigation of Class Activity 8Q in Section 8.2 on the size of your reflected face in a mirror. Figure 9.40 shows a side view of a person looking into a mirror. The mirror is parallel to AE. Lines BF and DG are normal lines to the mirror.

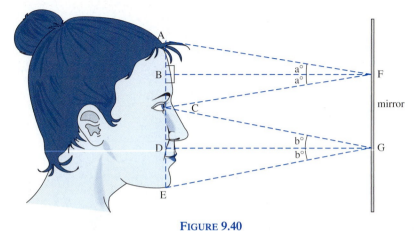

FIGURE 9.40

A Side View of a Person Looking in a Mirror

a. What is the significance of the points F and G in Figure 9.40? Explain why, referring to the laws of reflection.

b. Using the theory of congruent triangles discussed in this section, explain why triangles ABF and CBF are congruent and explain why triangles CDG and EDG are congruent.

c. Use your answer to part (b) to explain why the length of BD is half the length of AE. Therefore, explain why the reflection of your face in a mirror is half as long as your face's actual length. (Notice that BD and FG have the same length.)

---

3. Ann and Kelly are standing on a river bank, wondering how wide the river is. Ann is wearing a basball cap, so she comes up with the following idea: she lowers her cap until she sees the tip of the visor just at the opposite bank of the river. She then turns around to face away from the river, being careful not to tilt her head or cap, and has Kelly walk to the spot where she can just see Kelly's shoes. By pacing off the distance between them, Ann and Kelly figure that Kelly was 50 feet away from Ann. If the ground around the river is level, what, if anything, can Ann and Kelly conclude about how wide the river is? Relate this to triangle congruence.

9.4 Similarity

Two shapes are congruent if they have the same shape and size. This description is another way to say that two geometric objects are "equal." But how do we describe two objects if they are identical *except* for their size? Shapes, such as those in Figure 9.41, are called *similar*. The notion of similarity is used in solving many problems in geometry; it also has a wide range of practical applications, such as in surveying and mapmaking. Representational drawing uses the concept of similarity.

similar   Informally, we say that two objects that have the same shape, but not necessarily the same size, are **similar**. Another way to say this is as follows: Two objects or shapes are similar if one object represents a scaled version of the other (scaled up or down). For example, a scale model of a train is similar (at least on the outside) to the actual train on which it is modeled.

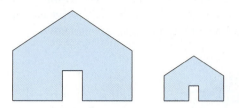

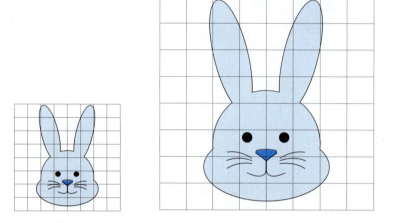

The network of streets on a street map is similar to the network of real streets it represents—at least if the streets are on flat ground.

One way to create a shape or design that is similar to another shape or design is by using grids of lines. If you enjoy doing crafts, then you may already be familiar with this technique. Draw a network of equally spaced parallel and perpendicular grid lines over the design you want to scale (or use an overlay of such grid lines), as shown in Figure 9.42. If you want the new design to be, say, twice as wide and twice as long, then make a new network of parallel and perpendicular grid lines that are spaced twice as wide as the original grid lines. Now copy the design onto the new grid lines, square by square. The new design will be similar to the original design.

A key feature of similarity is that if two objects (in a plane, or even in space) are similar, then distances between corresponding parts of the objects *all scale by the same factor*. More precisely, if two objects are similar, then there is a positive number, k, such that the distance between two points on the second object is k times as long as the distance between the corresponding points on the first object. This number k is called the **scale factor** from the first object to the second object.

scale factor

For example, the box of a toy model car might indicate that the toy car is a 24 : 1 scale model of an actual car. This means that the scale factor from the toy car to the actual car is 24, or equivalently, that the scale factor from the actual car to the toy car is $\frac{1}{24}$. In particular, the length, width, and height of the actual car are 24 times the length, width, and height, respectively, of the toy car. Similarly, for other distances, such as the width of the windshield

FIGURE 9.43

Using the "Scale Factor" Method

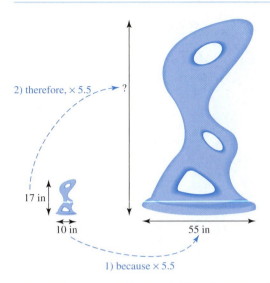

2) therefore, × 5.5 ⤑ ?

17 in

10 in

55 in

1) because × 5.5

or the length of the hood, each distance on the actual car is 24 times the corresponding distance on the toy car.

Warning: Scale factors only apply to lengths, *not* to areas or volumes. In Section 11.11 we will see how areas and volumes behave under scaling.

In many practical situations, two shapes or objects are given to be similar. If we know various lengths on one object, then we can determine all the corresponding lengths on the other object, as long as we know at least one of the corresponding lengths. There are three common ways to do this, and all use the scale factor, even if indirectly. Let us now study these three ways using a specific example.

Problem: Suppose that an artist creates a scale model of a sculpture. The scale model is 10 inches wide and 17 inches tall. If the actual sculpture is to be 55 inches wide, then how tall will the actual sculpture be?

Solutions:

scale factor method

1. *"Scale Factor" Method:* (See Figure 9.43.) Since the scale model and the actual sculpture are to be similar, there is a scale factor, k, from the model to the actual sculpture such that every length on the actual sculpture is k times as long as the corresponding length on the scale model. Therefore, the width of the actual sculpture is $k \times 10$ inches and the height of the actual sculpture is $k \times 17$ inches. Since the width of the sculpture is given as 55 inches,

$$k \times 10 \text{ inches } = 55 \text{ inches}$$

so

$$k = 55 \div 10 = 5.5$$

Thus, the height of the sculpture is

$$k \times 17 \text{ inches } = 5.5 \times 17 \text{ inches } = 93.5 \text{ inches}$$

FIGURE 9.44

Using the
"Relative Sizes"
Method

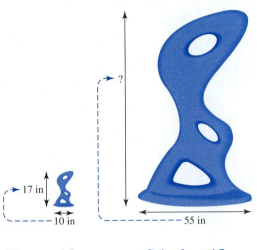

1) because × 1.7   2) therefore, × 1.7

relative sizes   **2.**   *"Relative Sizes" Method:* (See Figure 9.44.) Since the scale model is 10 inches wide
method and 17 inches tall, it is $\frac{17}{10} = 1.7$ times as tall as it is wide. The actual sculpture should,
therefore, also be 1.7 times as tall as it is wide. Since the sculpture is to be 55 inches wide,
it should be

$$1.7 \times 55 \text{ inches} = 93.5 \text{ inches}$$

tall. Why should the actual sculpture also be 1.7 times as tall as it is wide? There is
a scale factor, k, such that the width and height of the sculpture are $k \times 10$ and $k \times 17$,
respectively. Therefore,

$$\frac{\text{sculpture height}}{\text{sculpture width}} = \frac{k \times 17}{k \times 10} = \frac{17}{10} = 1.7$$

proportion   **3.**   *"Set up a Proportion" Method:* As before, since the scale model and the actual sculpture
method are to be similar, there is a scale factor, k, from the model to the actual sculpture. k can be
calculated as

$$k = \frac{\text{width of sculpture}}{\text{width of model}}$$

and

$$k = \frac{\text{height of sculpture}}{\text{height of model}}$$

Therefore,

$$\frac{\text{width of sculpture}}{\text{width of model}} = \frac{\text{height of sculpture}}{\text{height of model}}$$

All the quantities in this last equation are known, except for the height of the sculpture.
Let's let h stand for the height of the sculpture in inches. Then, substituting the known

quantities, we have

$$\frac{55 \text{ in}}{10 \text{ in}} = \frac{h \text{ in}}{17 \text{ in}}$$

But we know that we can determine whether two fractions are equal by cross-multiplying. So, because the previous equation is true,

$$55 \times 17 = 10 \times h$$

Dividing both sides of this equation by 10, we see that

$$h = \frac{55 \times 17}{10} = 93.5$$

so the sculpture is 93.5 inches tall.

Although the third method is more common, notice that the logic behind it is a little more subtle than the first two methods.

CLASS ACTIVITY   NOW TURN TO CLASS ACTIVITIES MANUAL

9P A First Look at Solving Scaling Problems p. 300

9Q A Common Misconception About Scaling p. 301

9R Enlarging and Reducing p. 302

9S Using Scaling to Understand Astronomical Distances p. 305

WHEN ARE TWO SHAPES SIMILAR?

In the situations we have encountered thus far, objects have been *given* as being similar. In other words, the very nature of the situation tells us that the objects are similar. However, there can be cases where it is not entirely obvious or clear that two shapes in question are similar. In those cases, how can we tell whether or not the shapes are similar? It is tempting to think that one can always tell "by eye," but this method is not reliable. Figure 9.45 shows two triangles that appear to be similar, but are not.

triangle similarity criterion   Here is a criterion for similarity of triangles: two triangles are similar exactly when the two triangles have the same size angles. To clarify, two triangles are similar exactly when it is possible to match each angle of the first triangle with an angle of the second triangle in such

FIGURE 9.45

These Triangles May Look Similar, but They Are Not

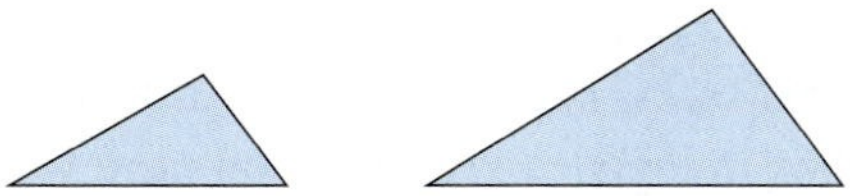

FIGURE 9.46

Similar
Triangles

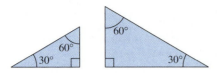

FIGURE 9.47

What Is the
Length of DE?

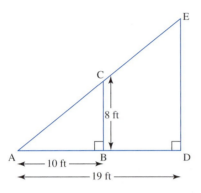

a way that matched angles have equal measures. So, for example, we can determine that the triangles in Figure 9.46 are similar because they have the same size angles.

Actually, notice that in order to determine whether two triangles are similar, you really only need to check that *two* of their angles are the same size because the sum of the angles in a triangle is always 180°. If two angles are known, the third one is determined as the angle that makes the three add to 180°. Thus, if two angles in one triangle are the same size as two respective angles in another triangle, then the third angles will automatically also be equal in size.

The next figure shows an example of how to use the criterion for triangle similarity. In Figure 9.47, what is the length of side DE? First, notice that the picture shows that triangles ABC and ADE have two angles of the same size: the angle at A and a right angle. Therefore, these triangles are similar. Now we can apply any of our three methods to determine the length of side DE. Using the "scale factor" method, the scale factor is $\frac{19}{10}$, so the length of DE is

$$\frac{19}{10} \times 8 \text{ ft} = 15.2 \text{ ft}$$

In general, a line that crosses two parallel lines makes the same angle with both parallel lines. This statement (or a variation of it) is called the *parallel postulate*. It is often taken as a basic "starting assumption" in mathematics, and we will also assume that it is true. So if two triangles ABC and ADE share an angle at A, as in Figure 9.48, and if BC and DE are parallel, then the angles at B and D are the same size, and the triangles ABC and ADE are similar.

Another common situation in which similar triangles are created is illustrated in Figure 9.49. In this case, two lines cross at a point A. If the lines BC and DE are parallel, then the triangles ABC and ADE are similar. This is because the angle at A in ABC is the same size as the angle at A in ADE, as described in Practice Problem 4 on page 335 and explained on page 337. Also, the angles at B and at D are equal in size because these are the points where the

FIGURE 9.48

Parallel Lines
Create Similar
Triangles

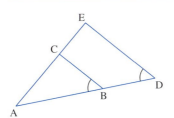

FIGURE 9.49

Parallel Lines
on Opposite
Sides of the
Point Where
Two Lines Meet
Create Similar
Triangles

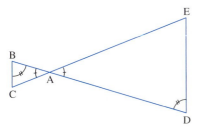

parallel line segments BC and DE meet the line segment BD. As stated previously, we simply assume this to be true, and it should seem plausible. Therefore, two out of three angles in ABC and ADE have the same size, so all three must have the same size, and the triangles are similar.

Why does the "three equal angles" criterion for triangle similarity work? Basically, it is because the process of scaling doesn't change angles. If two triangles are similar, then one triangle can be scaled to match the other, and since this scaling doesn't change angles, the triangles must have the same angles. Conversely, suppose two triangles have the same angles, say a, b, and c. Then one of the triangles can be scaled so that the sides common to angles a and b have the same length in both triangles. In the scaling process, the angles don't change, so if you match the sides that have the same length, they both have angles a and b at either end, and so these triangles are identical. Therefore, the original triangles must have been similar.

USING SIMILAR TRIANGLES TO DETERMINE DISTANCES

Similar triangles can be used in practical situations to find an unknown distance or length when several other related distances and lengths are known. Often, the tricky part in applying the theory of similar triangles is locating the similar triangles and sketching them. You will need to apply your visualization skills to help you sketch the situation, showing the relevant components. In the following examples, think carefully about why the pictures were drawn the way they were. If they had been drawn from a different perspective would you see the relevant similar triangles? (In some cases, yes, but in many cases, no.)

Here is an example of a realistic application of the theory of similar triangles: When we look at objects in the distance, they appear to be smaller than they actually are. One way to describe how big an object appears to be is to compare it with the size of your thumb by stretching your arm out straight in front of you, closing one eye, and "sighting" from your thumb to the object you are considering. This situation creates a pair of similar triangles, as shown in Figure 9.50. The similar triangles are ABC (eye, base of thumb, top of thumb) and

FIGURE 9.50

"Thumb
Sighting" a
Picture on a
Wall

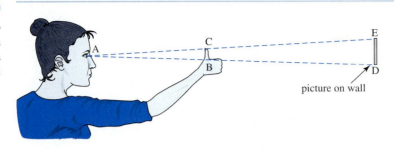

FIGURE 9.51

"Thumb
Sighting" a Man

ADE (eye, bottom of picture, top of picture). These triangles are similar because the angles at A are equal and the angles at B and D are equal if the thumb is held parallel to the picture. (See page 418 and Figure 9.48.)

You can also use this "thumb sighting" to find a distance. Let's say that I see a man standing in the distance and at this distance, he appears to be "1 thumb tall." If the man is actually 6 feet tall, then approximately how far away is he? As explained above, this "thumb sighting" creates similar triangles. (See Figure 9.51, which is *not* drawn to scale.) The distance from my sighting eye to the base of my thumb on my outstretched arm is 22 inches. My thumb is 2 inches tall. So the distance from my eye to my thumb is $\frac{22}{2}$ times as long as the length of my thumb. Therefore, according to the "relative sizes" method, the man is

$$\frac{22}{2} \times 6 \text{ feet } = 66 \text{ feet}$$

away. You might object that this tells us the distance from my eye to the man's feet, but, because the man is fairly far away, this distance is almost identical to the distance along the ground from my feet to his feet. Plus, we are only getting an *estimate* of how far away the man is because all the measurements used are very rough. So this extra bit of inaccuracy is insignificant.

How could you modify the method of "thumb sighting" to make it more useful and more accurate? You could use other objects besides your thumb for sighting—a ruler held vertically would be much more accurate, for example. And what if you didn't just hold the ruler in your hand, but had it attached to a fixed length of pole—that would also create greater accuracy. Making these sorts of improvements leads to some of the surveying equipment used today. When you see people at construction sights looking through a device on a tripod at a pole in the distance (such as in Figure 9.52), these people are surveying.

Some kinds of surveying equipment measure distances based on the theory of similar triangles, using a somewhat more elaborate method than the primitive "thumb sighting." In order to measure the distance between the surveying equipment and a pole of known height in

FIGURE 9.52

Surveying

the distance, the person surveying looks through the equipment and "sights" the pole. Instead of a thumb, there is a kind of ruler inside the surveying equipment.

CLASS ACTIVITY   NOW TURN TO CLASS ACTIVITIES MANUAL

9T Measuring Distances by "Surveying" p. 306

9U Determining the Height of a Tree p. 308

PRACTICE PROBLEMS FOR SECTION 9.4

1. Amber has drawn a simple design on a rectangular piece of paper that is 4 inches wide and 12 inches long. Amber wants to make a scaled-up version of her design on a rectangular piece of paper that is 10 inches wide. She must figure out how long to make the larger rectangle. Amber says that since the larger rectangle is 6 inches wider than the smaller rectangle (because $10 - 4 = 6$), she should make her larger rectangle 6 inches longer than the smaller rectangle. She figures that she should make the larger rectangle $12 + 6 = 18$ inches long. Will Amber be able to make a correctly proportioned scaled-up version of her design on a rectangle that is 10 inches wide and 18 inches long? If not, how long should she make her rectangle and why?

2. Hannah wants to make a scale drawing of herself standing straight with her arms down. Hannah is 4 feet 6 inches tall. Her arm is 22 inches long. If Hannah wants to make the drawing of herself 10 inches tall, then how long should she draw her arm?

   Solve this problem in two different ways: once using the *scale factor* method and once using the *relative sizes* method, explaining your reasoning both times.

3. Ryan wants to make a large model of the Great Pyramid at Giza in Egypt. The pyramid is about 481 feet tall and each of its four sides is about 756 feet long at the base. Ryan wants the sides of his model's pyramid to be 5 feet long at the base. How tall should Ryan's model pyramid be?

Use the three different methods: *scale factor*, *relative sizes*, and *set up a proportion* to solve the problem. In each case, explain the reasoning behind the method.

4. (Continuation of Practice Problem 3) Now Ryan also wants to make a model of the Sphinx that is near the Great Pyramid at Giza in Egypt. The sphinx is about 240 feet long and 66 feet high. Ryan wants to use the same scale as he did with his model pyramid. How long and how tall should his model sphinx be?

5. Ms. Bullock's class is making a display of the Sun and the planets. Each planet will be depicted as a circle. The children in Ms. Bullock's class want to show the Earth as a circle of diameter 10 cm, and they want to show the correct relative sizes of the planets and the Sun. Given the information in Table 9.1, what should the diameters of these scale drawings be? Since the Sun is so big, it wouldn't be practical to make a full model of the Sun; how could the children make a sliver of the Sun of the correct size?

TABLE 9.1

Diameters of Heavenly Bodies

HEAVENLY BODY	APPROXIMATE DIAMETER
Sun	1,392,000 km
Mercury	4,900 km
Venus	12,100 km
Earth	12,700 km
Mars	6,800 km
Jupiter	138,000 km
Saturn	115,000 km
Uranus	52,000 km
Neptune	49,500 km
Pluto	2,300 km

6. A sculptor makes a scale model for a sculpture she plans to carve out of marble that is 15 inches wide, 8 inches deep, and 27 inches high. She finds a block of marble that is 5 feet wide, 4 feet deep, and 10 feet high. What will the finished dimensions of the sculpture be if she makes the sculpture as large as possible?

7. Go to the website

   www.aw-bc.com/beckmann

   to learn how to make a **camera obscura** from a Pringles® potato chip can. A camera obscura is a fun device that projects images onto a screen through a small hole. It illustrates a fundamental idea of photography: projecting an image through a small hole. You can make a camera obscura from any tube by cutting off a piece of the tube (about 2 inches from one end), placing semi-transparent paper or plastic over the place where you cut, and taping the tube back together, with the semi-transparent paper now in the middle of the tube. Cover one end of the tube with aluminum foil and use a pin to poke a small hole in the middle of the foil. Cover the side of the tube with aluminum foil to keep light out.

Now look through the open end of the tube at a well-lit scene. You should see an upside-down projected image of the scene on the semitransparent paper.

Suppose you make a camera obscura out of a tube of diameter 3 inches, and suppose you put the semi-transparent paper or plastic 2 inches from the hole in the aluminum foil. How far away would you have to stand from a 6-foot-tall man in order to see the entire man on the camera obscura's screen?

8. Suppose that you go outside on a clear night when much of the Moon is visible and use a ruler to "sight" the Moon

(as described in Class Activity 9T). Use the information that follows to determine how big the Moon will appear to you on your ruler. (You will also need to measure the distance from your eye to a ruler that you are holding with your arm stretched out.) On a night when the Moon is visible, try this and verify it.

The distance from the Earth to the Moon is approximately 384,000 km.

The diameter of the Moon is approximately 3,500 km.

ANSWERS TO PRACTICE PROBLEMS FOR SECTION 9.4

1. Amber will not be able to make a scaled-up version of her design on paper that is 10 inches wide and 18 inches long. As we see in Figure 9.53, if Amber makes the larger rectangle 10 inches by 18 inches, it will not be proportioned in the same way that the original rectangle is. Adding the same amount to the length and width generally does not preserve the ratio of length to width in a rectangle. Instead, Amber could reason that since the original rectangle is 3 times as long as it is wide, the larger rectangle should also be 3 times as long as it is wide. So if the larger rectangle is 10 inches wide, it should be $3 \times 10 = 30$ inches wide.

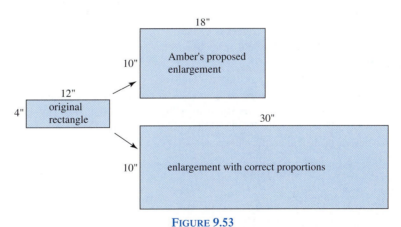

FIGURE 9.53

How Should Amber Scale Up Her 4-Inch-by-12-Inch Rectangle?

---

2. First, let's determine Hannah's height in inches. Each foot is 12 inches, so 4 feet is $4 \times 12 = 48$ inches. Therefore 4 feet 6 inches is $48 + 6 = 54$ inches, so Hannah is 54 inches tall.

Relative Sizes Method: Hannah's arm is 22 inches long, which is $\frac{22}{54}$ of her height. Therefore, in her drawing, Hannah's arm should also be $\frac{22}{54}$ of her height in the

drawing, namely,

$$\frac{22}{54} \times 10 \text{ inches} = 4 \text{ inches}$$

So Hannah should draw her arm 4 inches long.

Scale Factor Method: Because Hannah is 54 inches tall and the scale drawing of herself is to be 10 inches tall, the height of Hannah's drawing is $\frac{10}{54}$ of Hannah's actual height. In other words, the scale factor from Hannah to the drawing of herself is $\frac{10}{54}$. The length of Hannah's arm in her drawing should also be $\frac{10}{54}$ of the length of her actual arm. Therefore, the drawing of Hannah's arm should be

$$\frac{10}{54} \times 22 \text{ inches } = 4 \text{ inches}$$

long.

3. *Scale Factor Method:* Because the actual pyramid is 756 feet wide and Ryan's scale model is to be 5 feet wide, the scale factor from the pyramid to the model is the number k such that

$$k \times 756 \text{ ft } = 5 \text{ ft}$$

So

$$k = \frac{5}{756}$$

The same scale factor relates the height of the actual pyramid to the height of the scale model. Therefore, the height of the model should be

$$k \times 481 \text{ ft } = \frac{5}{756} \times 481 \text{ ft } = 3.2 \text{ ft}$$

Because 1 foot is 12 inches, .2 feet is $.2 \times 12 = 2.4$ inches. Thus, Ryan's model should be about 3 feet, $2\frac{1}{2}$ inches tall.

Relative Sizes Method: Because the pyramid is 481 feet tall and 756 feet wide at the base, the height of the pyramid is $\frac{481}{756}$ times as long as its width along the base. The same relationship should hold for the scale model. Therefore, the height of the scale model is

$$\frac{481}{756} \times 5 \text{ feet} = 3.2 \text{ feet}$$

Proportion Method: There are two ways of expressing the scale factor k (from the pyramid to the model) as a fraction. One is

$$k = \frac{5}{756}$$

because the scale factor k is the number you multiply 756 feet by to get 5 feet. The other is

$$k = \frac{(\text{height of model pyramid in ft })}{481}$$

because the scale factor k is the number you multiply 481 feet by to get the height of the model pyramid in

feet. The two fractions that are equal to the scale factor k must also be equal to each other; therefore,

$$\frac{5}{756} = \frac{(\text{height of model pyramid in ft})}{481}$$

But we know that two fractions are equal exactly when their "cross multiples" are equal. Thus,

$$5 \times 481 = 756 \times (\text{height of model pyramid in ft})$$

so

$$\text{height of model pyramid } = \frac{5 \times 481}{756} \text{ ft}$$
$$= 3.2 \text{ ft}$$

4. The *relative sizes method* provides a straightforward way to solve this problem. Because the pyramid is 756 feet wide and the Sphinx is 240 feet long, the length of the Sphinx is $\frac{240}{756}$ times as long as the width of the pyramid. Similarly, because the Sphinx is 66 feet high, the height of the Sphinx is $\frac{66}{756}$ times as long as the width of the pyramid. The same relationships should hold for the model pyramid and model Sphinx. Because the model pyramid is 5 feet wide, the length of the model Sphinx should be

$$\frac{240}{756} \times 5 \text{ feet} = 1.6 \text{ feet}$$

and the height of the model Sphinx should be

$$\frac{66}{756} \times 5 \text{ feet} = .4 \text{ feet}$$

Because 1 foot is 12 inches, .6 feet is $.6 \times 12 = 7.2$ inches and .4 feet is $.4 \times 12 = 4.8$ inches. So the model Sphinx should be about 7 inches long and 5 inches tall.

5. From the problem statement, we infer that we want the collection of circles representing the planets and Sun to be similar to cross sections of the actual planets—all with the same scale factor. If you use the *relative sizes* method, then you don't have to worry about converting kilometers to centimeters (or vice versa), and you won't have to work with huge numbers.

Because the Earth's diameter is 12,700 km and Mercury's diameter is 4,900 km, Mercury's diameter is $\frac{4,900}{12,700}$ times as long as the Earth's diameter. The same relationship should hold for the models of Mercury and the Earth. Since the model Earth is to have a diameter of 10 cm, the model Mercury should have a diameter of

$$\frac{4,900}{12,700} \times 10 \text{ cm} = 3.9 \text{ cm}$$

Table 9.2 shows the calculations for the other heavenly bodies.

TABLE 9.2
Diameters of
Scale Planets

HEAVENLY BODY	APPROXIMATE DIAMETER OF CIRCLE REPRESENTING IT
Sun	$\frac{1,392,000}{12,700} \times 10 \text{ cm} = 1096 \text{ cm} = 10.96 \text{ m}$
Mercury	$\frac{4,900}{12,700} \times 10 \text{ cm} = 3.9 \text{ cm}$
Venus	$\frac{12,100}{12,700} \times 10 \text{ cm} = 9.5 \text{ cm}$
Earth	given as 10 cm
Mars	$\frac{6,800}{12,700} \times 10 \text{ cm} = 5.4 \text{ cm}$
Jupiter	$\frac{138,000}{12,700} \times 10 \text{ cm} = 108.7 \text{ cm}$
Saturn	$\frac{115,000}{12,700} \times 10 \text{ cm} = 90.6 \text{ cm}$
Uranus	$\frac{52,000}{12,700} \times 10 \text{ cm} = 40.9 \text{ cm}$
Neptune	$\frac{49,500}{12,700} \times 10 \text{ cm} = 39 \text{ cm}$
Pluto	$\frac{2,300}{12,700} \times 10 \text{ cm} = 1.8 \text{ cm}$

The circle representing the Sun should have a diameter of about 11 meters and therefore a radius of about 5.5 meters. The children could measure a piece of string $5\frac{1}{2}$ meters long. One child could hold one end and stay in a fixed spot, while another child could attach a pencil to the other end and use it to draw a piece of a 5.5 meter radius circle representing the Sun.

6. One way to solve this problem is to think about the scale factors you might use. Since each foot is 12 inches, 5 feet is $5 \times 12 = 60$ inches, 4 feet is $4 \times 12 = 48$ inches, and 10 feet is $10 \times 12 = 120$ inches. Thinking only about the width, the scale factor from the model to the sculpture, k, should be such that

$$k \times 15 = 60$$

so

$$k = \frac{60}{15} = 4$$

Similarly, thinking only about the depth and the height, the scale factors would be

$$k = \frac{48}{8} = 6$$

and

$$k = \frac{120}{27} = 4.4$$

respectively. The artist will need to use the smallest of these scale factors, otherwise her sculpture would require more marble than she has. So the artist should use $k = 4$, in which case the dimensions of the sculpture are as follows:

$$\text{width} = 4 \times 15 \text{ in} = 60 \text{ in}$$
$$\text{depth} = 4 \times 8 \text{ in} = 32 \text{ in}$$
$$\text{height} = 4 \times 27 \text{ in} = 108 \text{ in}$$

7. Figure 9.54 shows how light from an object DE enters the pinhole (at point A) and projects onto the screen inside a camera obscura. If the object being looked at and the screen BC of the camera obscura are both vertical, then BC and DE are both vertical and are therefore parallel. Therefore, as described in the text, ABC and ADE are similar. Since these two triangles are similar, all corresponding distances on the triangle scale by the same scale factor. Since the distance from the screen of the camera obscura to the pinhole is $\frac{2}{3}$ times the length of the screen, by the *relative sizes* method, the distance from the pinhole to the man should be $\frac{2}{3}$ times the length of the man, which is $\frac{2}{3} \times 6$ feet $= 4$ feet. So you need to stand at least 4 feet from the man to see the full length of him on the camera obscura.

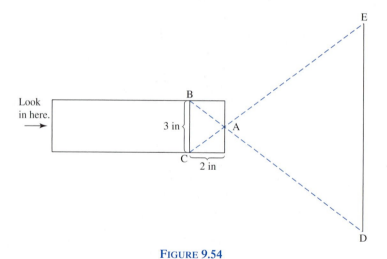

FIGURE 9.54

A Side View of a Camera Obscura

8. When you sight the Moon with a ruler you create similar triangles ABC and ADE as shown in Figure 9.55 (not to scale). Since the diameter of the Moon is 3,500 km and the distance to the Moon is 384,000 km, the Moon's diameter (DE) is $\frac{3,500}{384,000}$ times as long as the distance to the Moon (AD). The same relationship must hold for the apparent size of the Moon on the ruler (BC) and the distance from your eye to the ruler (AB): the apparent size of the Moon on the ruler must be $\frac{3,500}{384,000}$ times as long as the distance from your eye to the ruler. (This is the *relative sizes* method). Notice that $\frac{3,500}{384,000}$ is about $\frac{1}{100}$. So, if the distance from your eye to the ruler is about 22 inches, then the Moon's diameter will appear to be about .2 inches on your ruler.

FIGURE 9.55

Sighting the Moon

PROBLEMS FOR SECTION 9.4

1. Frank's dog, Fido, is 16 inches tall and 30 inches long. Frank wants to draw Fido 4 inches tall. Frank figures that because Fido's height in the drawing will be 12 inches less than Fido's actual height (because $16 - 12 = 4$), Fido's length in the drawing should also be 12 inches less than Fido's actual length. Therefore, Frank figures that he should draw Fido $30 - 12 = 18$ inches long. Will the drawing of Fido be correctly proportioned if Frank makes it 18 inches long? If not, how long should Frank draw Fido? Explain.

2. Tyler's flag problem: Tyler has designed his own flag on a rectangle that is 3 inches tall and 6 inches wide. Now Tyler wants to draw a larger version of his flag on a rectangle that is 9 inches tall. How wide should Tyler make his larger flag?

   Solve Tyler's flag problem in two ways: once using the *relative sizes* method and once using the *scale factor* method. For each method of solution, explain as clearly and concretely as you can why the method makes sense. Explain as if you were explaining the method to 4th or 5th graders who understand multiplication and division (but who do not know about setting up proportions).

3. Jasmine's flag problem: Jasmine has designed her own flag on a rectangle that is 8 inches tall and 12 inches wide. Now Jasmine wants to draw a smaller version of her flag on a rectangle that is 6 inches tall. How wide should Jasmine make her smaller flag?

   Solve Jasmine's flag problem in two ways: once using the *relative sizes* method and once using the *scale factor* method. For each method of solution, explain why the method makes sense as clearly and concretely as you can, as if you were explaining the method to 4th or

5th graders who understand multiplication and division (but who do not know about setting up proportions).

4. Kelsey wants to make a scale drawing of herself standing straight and with her arms down. Kelsey is 4 feet 4 inches tall. Her leg is 24 inches long. If Kelsey wants to make the drawing of herself 10 inches tall, then approximately how long should she draw her leg?

   Solve this problem in three different ways: once using the *scale factor* method, once using the *relative sizes* method, and once using the *set up a proportion* method, explaining your reasoning each time.

5. A painting that is 4 ft 3 in by 6 ft 4 inches will be reproduced on a card. The side that is 4 ft 3 inches long will become 3 inches long on the card. Determine how long the 6 ft 4 inches side will become on the card. Explain why your method of solution is valid.

6. A "thumb sighting" problem: Suppose you are looking down a road and you see a person ahead of you. You hold out your arm and "sight" the person with your thumb, finding that the person appears to be as tall as your thumb is long. Assume your thumb is 2 inches long, and that the distance from your sighting eye to your thumb is 22 inches. If the person is 5 feet 4 inches tall, then how far away are you from the person?

   a. Draw a picture showing that the "thumb sighting" problem involves similar triangles. Explain why the triangles are similar.

   b. Solve the "thumb sighting" problem in two different ways. In both cases, explain the logic behind the method you use.

7. Ms. Bullock's class wants to make a display of the solar system showing the correct relative distances of the planets from the Sun (in other words, showing the distances of the planets from the Sun to scale). Table 9.3 shows the approximate actual distances of the planets from the Sun.

TABLE 9.3

Distances of Planets from the Sun

PLANET	APPROXIMATE DISTANCE FROM SUN
Mercury	36,000,000 miles
Venus	67,000,000 miles
Earth	93,000,000 miles
Mars	141,000,000 miles
Jupiter	484,000,000 miles
Saturn	887,000,000 miles
Uranus	1,783,000,000 miles
Neptune	2,794,000,000 miles
Pluto	3,666,000,000 miles

If the distance from the Sun to Mercury is to be represented in the display as 1 inch, then how should the distances from the Sun to the other planets be represented? Explain your reasoning in detail for one of the planets in such a way that 5th graders who understand multiplication and division but who do not know about setting up proportions might be able to understand. Calculate the answers for the other planets without explanation.

8. Ms. Winstead's class went outside on a sunny day and measured the lengths of some of their classmates' shadows. The class also measured the length of a shadow of a tree. Inside, the children made a table like the one shown in Table 9.4.

Based on their table, the children estimated that the tree should be about 1.66 times as tall as its shadow was long. So the children figured that the tree is about 1.66 × 22 feet = 37 feet tall.

TABLE 9.4

Heights of Children and Lengths of Shadows

	Tyler	Jessica	SunJae	Lameisha	tree
shadow length	33 in	34 in	32 in	34 in	22 feet
height	53 in	57 in	52 in	58 in	?
height ÷ shadow length	1.61	1.68	1.63	1.71	

a. Explain how the reasoning used by the children in Ms. Winstead's class actually involves similar triangles. (Draw a picture to aid your explanation.)

b. Explain why the triangles you described in part (a) really are similar.

c. We studied three methods for solving scaling problems in this section. Which method did the children in Ms. Winstead's class use (even though they didn't know its name and hadn't studied it formally)? Explain your answer.

9. An art museum owns a painting that it would like to reproduce in reduced size onto a 24-inch-by-36-inch poster. The painting is 85 inches by 140 inches. Give your recommendation for the size of the reproduced painting on the poster—how wide and long do you suggest that it be? Draw a scale picture showing how you would position the reproduced painting on the poster. Explain your reasoning.

10. Most ordinary cameras produce a negative, which is a small picture of the scene that was photographed (with reversed colors). A photograph is produced from a negative by printing onto photographic paper. When a photograph is printed from a negative, one of two things happens: either the printed picture shows the full picture that was captured in the negative, or the printed picture is cropped, showing only a portion of the full picture on the negative. In either case, the rectangle forming the printed picture is similar to the rectangular portion of the negative that it comes from.

An ordinary 35-mm camera produces a negative in the shape of a $1\frac{7}{16}$-inch-by-$\frac{15}{16}$-inch rectangle. Some of the most popular sizes for printed pictures are $3\frac{1}{2}$ in $\times$ 5 in, 4 in $\times$ 6 in, and 8 in $\times$ 10 in. Can any of these size photographs be produced without either cropping the picture or leaving blank space around the picture? Why or why not? Explain your reasoning clearly.

11. Let's say that you are standing on top of a mountain looking down at the valley below, where you can see cars driving on a road. You stretch out your arm, use your thumb to "sight" a car, and find that the car appears to be as long as your thumb is wide. Estimate how far away you are from the car. Explain your method clearly. (You will need to make an assumption to solve this problem. Make a realistic assumption and make it clear what your assumption is.)

12. a. During a total solar eclipse, the Moon moves in front of the Sun, obscuring the view of the Sun from the Earth (at some locations on the Earth). Surprisingly, the Moon seems to be superimposed on the Sun during a solar eclipse, appearing to be almost identical in size as seen from the Earth. How does this situation give rise to similar triangles? Draw a sketch (it does not have to be to scale).

b. Use the data below and either the *scale factor* method or the *relative sizes* method to determine the approximate distance of the Earth to the Sun. (Do not use the *set up a proportion* method.) Explain the reasoning behind the method you use.

The distance from the Earth to the Moon is approximately 384,000 km.

The diameter of the Moon is approximately 3500 km.

The diameter of the Sun is approximately 1,392,000 km.

13. On page 420 there is a description of how the theory of similar triangles can be used to measure distances in land surveying. This problem will help you understand the theory for how the altitude above sea level of a location can be determined by surveying as well.

Suppose a person stands on the side of a hill at a point A. Using surveying equipment, the person measures the distance AB to a point B up the hill as 38 feet. Surveying equipment can also be used to measure angles. Suppose that the surveying equipment reports that the line connecting A and B makes an angle of 15 degrees with a horizontal line, AC, as shown in Figure 9.56. Suppose that, by previous surveying, the point A is already known to be 523 feet above sea level.

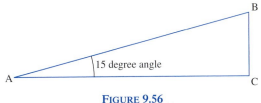

FIGURE 9.56

Surveying on a Hill

a. The triangle in Figure 9.56 is drawn to scale. Use a ruler to measure parts of this triangle. Then use these measurements, together with mathematical reasoning and the information about the given points A and B, to find the altitude of the point B above sea level.

b. Even if you measured carefully, your results won't be really accurate, so in practice you need a better method than measuring a scale drawing. A more accurate method uses trigonometry. Here is what trigonometry tells us in this situation: If we declare the distance from A to B to be 1 unit, then the distance from B to C is $\sin(15°)$ units $= .258819$ units and the distance from A to C is $\cos(15°)$ units $= .965926$ units.

Use the more accurate information from the paragraph above to find the altitude of point B above sea level *to the nearest inch.* Assume that point A is actually 523 feet and 3 inches above sea level and that the distance from point A to point B is actually 38 feet and 2 inches. Give your answer in feet and inches (for example, 682 feet and 10 inches).

14. Which TV screen appears bigger: a 50 inch screen viewed from 8 feet away or a 25 inch screen viewed from 3 feet away? Explain carefully. (TV screens are typically measured on the diagonal, so to say that a TV has a "50-inch screen" means that the diagonal of the screen is 50 inches long.)

15. A pinhole camera is a very simple camera made from a closed box with a small hole on one side. Film is put inside the camera, on the side opposite the small hole. The hole is kept covered until the photographer wants to take a picture, when light is allowed to enter the small hole, producing an image on the film. Unlike an ordinary camera, a pinhole camera does not have a view finder, so with a pinhole camera, you can't see what the picture you are taking will look like.

a. Suppose that you have a pinhole camera for which the distance from the pinhole to the opposite side (where the film is) is 3 inches. Suppose that the piece of film to be exposed (opposite the pinhole) is about 1 inch tall and $1\frac{1}{2}$ inches wide. Let's say you want to take a picture of a bowl of fruit that is about 15 inches wide and piled 6 inches high with fruit and that you want the bowl of fruit to fill up most of the picture. Approximately how far away from the fruit bowl should you locate the camera? Explain, using similar triangles.

b. Explain why the image produced on film by a pinhole camera is upside-down.

16. If the map in Figure 9.57 has a scale of 1 inch $= 15$ miles, then what is the scale of the map in Figure 9.58, which shows the same region? Explain your reasoning.

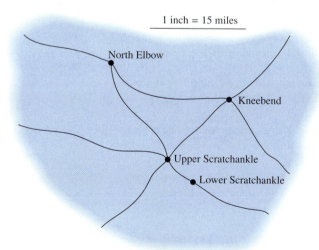

1 inch = 15 miles

North Elbow

Kneebend

Upper Scratchankle

Lower Scratchankle

FIGURE 9.57

A Map with a Scale of 1 Inch $= 15$ Miles

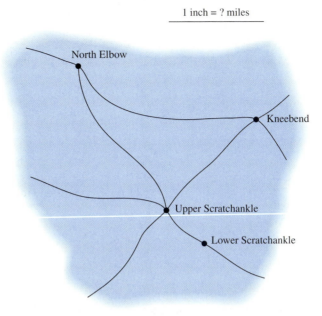

1 inch = ? miles

North Elbow

Kneebend

Upper Scratchankle

Lower Scratchankle

FIGURE 9.58

What Is the Scale of This Map?

17. Imagine that the Moon were to remain the same size as it is now, but were to orbit the Earth at *half* the distance from the Earth as it currently does. If you went outside at night and "sighted" the Moon with a ruler (as described in Class Activity 9T and Practice Problem 8), how would its new sighted diameter compare to its current sighted diameter? Use pictures (which do not need to be to scale) to help you explain your answer. (The distance from the Earth to the Moon is approximately 384,000 km, and the diameter of the Moon is approximately 3500 km.)

18. Sue has a rectangular garden. If Sue makes her garden twice as wide and twice as long as it is now, will the area of her garden be twice as big as its current area? Examine this problem *carefully* by working out some examples and drawing pictures. Explain your conclusion. If the garden is not twice as big (in terms of area), how big is it compared to the original?

Chapter 10

Measurement

In this chapter we will examine some fundamentals of measurement. We will study the standard systems of measurement that we use in the United States, conversions between different units of measurement, and reporting of measurements. All measurements of actual objects are necessarily inexact. The way we report a measurement conveys the confidence we have in the measurement's accuracy. We will also learn about the various different measurable attributes that objects can have. We can often compare the sizes of two objects with respect to several different measurable attributes, so we must take care to specify how we are comparing objects when we say that one object is larger than another. We will also begin to study the mathematically important measurable attributes of length, area, and volume.

Concerning the learning of measurement, the National Council of Teachers of Mathematics recommends the following (see [44]):

Instructional programs from prekindergarten through grade 12 should enable all students to—

NCTM STANDARDS

- understand measurable attributes of objects and the units, systems, and processes of measurement;

- apply appropriate techniques, tools, and formulas to determine measurements.

For NCTM's specific measurement recommendations for grades Pre-K–2, grades 3–5, grades 6–8, and grades 9–12, see

www.aw-bc.com/beckmann

and look for the grade band you are interested in.

10.1 The Concept of Measurement

In this section we will study the idea of measurement and the two main systems of measurement that we use today in the United States.

The concept of measurement is so familiar that you probably don't pay much attention to the underlying requirements for taking measurements. After all, we buy food by the pound, gasoline by the gallon, and electricity by the kilowatt-hour. We also measure the passage of time, the amount of land a person owns, and the distance between cities. But what does it mean to measure a quantity, and what does it take to make the careful and accurate measurements we need for everyday tasks and for scientific applications? In order to measure a quantity, we first need a fixed "reference amount" of this quantity. This "reference amount" is called a **unit**. For example, a mile is a unit of length, an acre is a unit of area, a gallon is a unit of capacity (or volume), and a kilowatt-hour is a unit of electical power. To **measure** a given quantity means to compare that quantity with a unit of the quantity. Usually, you must determine how many units of the quantity make up the given quantity.

How do we measure quantities? There are often many ways to measure quantities, and we have special devices for measuring them. But the simplest and most direct way to measure a quantity is to count how many of the units are in the quantity to be measured. For example, to measure the amount of rice in a small bag, you can simply scoop the rice out, cup by cup, and count how many cups of rice you scooped. But to measure the amount of water in your bathtub, it would be tedious to scoop it out cup by cup or gallon by gallon. So you would probably want to use an indirect method for measuring this amount of water.

The simplest and most familiar measuring device is a ruler. A ruler, whose unit is an inch, displays a number of inch-long lengths. In order to use the ruler to measure how long an object is in terms of inches, we simply read off the number of inches on the ruler, rather than repeatedly laying down an inch-long object and counting how many it took to cover the length of the object. Similarly, other measuring devices such as scales, calipers, and speedometers allow us to measure quantities indirectly, rather than making direct counts of a number of units.

What quantities can be used as units? *Any fixed* (positive) *amount* of a measurable quantity can be a unit, but for purposes of clear communication it usually makes sense to use commonly accepted standard units. For example, in order to encourage a child to develop her concept of measurement, you might have her measure the length of a piece of ribbon by laying pens end to end. She might report that the ribbon is $5\frac{1}{2}$ pens long. In this case, the unit of length is "a pen," which is not a standard unit of length. (See Figure 10.1.) On the other hand, if you want to buy ribbon from a spool at a store, you will probably use yards, feet, inches, or some combination of these to say how long a piece you want. That is, if the store is in the U.S.—if you were in France, you would use meters and centimeters to describe the length of ribbon.

There are instances in which it is natural to use a nonstandard unit. What is the area of the design shown in Figure 10.2 (a)? The easiest way to describe the area is in terms of one of the pieces that make up the design (Figure 10.2 (b)). We can decide to call the area of the piece "one squiggle," and to use one squiggle as our unit of measurement. The design in Figure 10.2 (a) is made up of 36 congruent pieces (4 rows, with $5 + 4$ pieces in each row), each of which has area one squiggle, so the total area of the design is 36 squiggles.

Our view of a standard unit varies from place to place in the world, and over time.

FIGURE 10.1

Using a
Nonstandard
Unit to
Measure: This
Desk Is 5 Pens
Wide

FIGURE 10.2

To Find the Area
of the Design in
(a), We Use the
Nonstandard
Unit "One
Squiggle"
shown in (b)

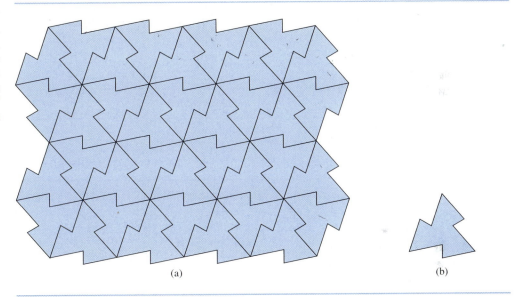

(a)   (b)

SYSTEMS OF MEASUREMENT

A system of measurement is a collection of standard units. In the United States today, two systems of measurement are in common use: the U.S. customary system of measurement and the metric system (also known as the international system of units, or the SI system).

THE U.S. CUSTOMARY SYSTEM

Table 10.1 on page 436 shows some standard units that are commonly used in the U.S. customary system of measurement.

For any unit of length, there are corresponding units of area and volume: a square unit, and a cubic unit, respectively. A square unit is the area of a square that is one unit wide and one unit long. A cubic unit is the volume of a cube that is one unit wide, one unit deep, and one unit high. For example, a square inch is the area of a square that is one inch wide and one inch long. A cubic inch is the volume of a cube that is one inch wide, one inch deep, and one inch high, as shown in Figure 10.3 on page 437.

Units in the U.S. Customary System		
Units of Length		
UNIT	**ABBREVIATION**	**SOME RELATIONSHIPS**
inch	in	
foot	ft	1 ft = 12 in
yard	yd	1 yd = 3 ft
mile	mi	1 mi = 1760 yd = 5280 ft
Units of Area		
square inch	in^2	
square foot	ft^2	$1\ ft^2 = 12^2\ in^2 = 144\ in^2$
square yard	yd^2	$1\ yd^2 = 3^2\ ft^2 = 9\ ft^2$
square mile	mi^2	
acre		$1\ acre = 43{,}560\ ft^2$
Units of Volume		
cubic inch	in^3	
cubic foot	ft^3	
cubic yard	yd^3	$1\ yd^3 = 3^3\ ft^3 = 27\ ft^3$
Units of Capacity (Volume)		
teaspoon	tsp	
tablespoon	T or tbs or tbsp	1 T = 3 tsp.
fluid ounce (or liquid ounce)	fl oz	1 fl. oz = 2 T.
cup	c	1 c = 8 fl. oz.
pint	pt	1 pt = 2 c. = 16 fl. oz.
quart	qt	1 qt = 2 pt. = 32 fl. oz.
gallon	gal	1 gal = 4 qt. = 128 fl. oz.
Units of Weight (Avoirdupois)		
ounce	oz	
pound	lb	1 pound = 16 ounces
ton	t	1 ton = 2000 pounds
Unit of Temperature		
degree Fahrenheit	° F	water freezes at 32° F
		water boils at 212° F

Capacity is essentially the same as volume, except that we use the term *capacity* for the volume of a container, the volume of liquid in a container, or the volume of space taken up by a substance filling a container, such as flour or berries.

Notice the distinction between an *ounce*, which is a unit of weight, and a *fluid ounce*, which is a unit of capacity or volume.

FIGURE 10.3

Examples of Units of Length, Area, and Volume in the U.S. Customary System. The Unit of Length Is the Inch, the Unit of Area Is the Square Inch, and the Unit of Volume Is the Cubic Inch.

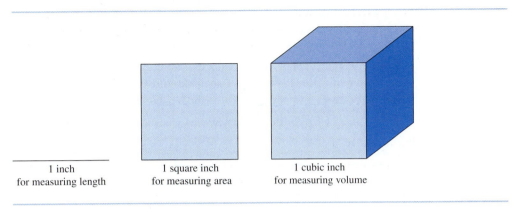

| 1 inch | 1 square inch | 1 cubic inch |
| for measuring length | for measuring area | for measuring volume |

The definitions of units have changed over time. For example, the inch was originally defined as the width of a thumb or as the length of three barleycorns. Since widths of thumbs and lengths of barleycorns can vary, these definitions are obviously not very precise. The inch is currently defined to be *exactly* 2.54 centimeters. (See [27].)

THE METRIC SYSTEM

The metric system was first used in France around the time of the French Revolution (1790). At that time, which is often called the *Age of Enlightenment*, there was a focus on rational thought, and scientists sought to organize their subjects in a rational way. The metric system is a natural outcome of this desire since it is an efficient, organized system that is designed to be compatible with the decimal system for writing numbers.

The metric system gives names to units in a uniform way. For each kind of quantity to be measured, there is a base unit (such as, meter, gram, liter). Each related unit is labeled with a prefix that indicates the unit's relationship to the base unit. For example, the prefix *kilo* means *thousand*, so a *kilometer* is a thousand meters, and a *kilogram* is a thousand grams. Many of the metric system prefixes are only used in scientific contexts, and not in everyday situations. Some of the metric system prefixes are listed in Table 10.2.

TABLE 10.2

The Metric System Uses Prefixes to Create Larger and Smaller Units from Base Units

Some Metric System Prefixes		
PREFIX	**MEANING**	
nano-	$10^{-9} = \frac{1}{1,000,000,000}$	billionth
micro-	$10^{-6} = \frac{1}{1,000,000}$	millionth
milli-	$10^{-3} = \frac{1}{1000}$	thousandth
centi-	$10^{-2} = \frac{1}{100}$	hundredth
deci-	$10^{-1} = \frac{1}{10}$	tenth
deka-	10	ten
hecto-	$10^2 = 100$	hundred
kilo-	$10^3 = 1000$	thousand
mega-	$10^6 = 1,000,000$	million
giga-	$10^9 = 1,000,000,000$	billion

Table 10.3 shows some of the most commonly used units in the metric system.

Figure 10.4 shows lines of length 1 millimeter, 1 centimeter, and 1 inch. The inch and the centimeter are related by 1 inch = 2.54 cm. A meter is about a yard and 3 inches. A kilometer is about 0.6 miles, so a bit more than half a mile.

Every unit of length has an associated unit of area and an associated unit of volume. Figure 10.5 shows a square centimeter and a cubic centimeter, which are the units of area and volume associated with the centimeter.

TABLE 10.3

Common Units of Measurement in the Metric System

Units in the Metric System		
Units of Length		
UNIT	**ABBREVIATION**	**SOME RELATIONSHIPS**
millimeter	mm	$1 \text{ mm} = \frac{1}{1000} \text{ m}$
centimeter	cm	$1 \text{ cm} = \frac{1}{100} \text{ m} = 10 \text{ mm}$
meter	m	$1 \text{ m} = 100 \text{ cm}$
kilometer	km	$1 \text{ km} = 1000 \text{ m}$
Units of Area		
square millimeter	mm^2	
square centimeter	cm^2	$1 \text{ cm}^2 = 10^2 \text{ mm}^2 = 100 \text{ mm}^2$
square meter	m^2	$1 \text{ m}^2 = 100^2 \text{ cm}^2 = 10{,}000 \text{ cm}^2$
square kilometer	km^2	$1 \text{ km}^2 = 1000^2 \text{ m}^2 = 1{,}000{,}000 \text{ m}^2$
Units of Volume		
cubic millimeter	mm^3	
cubic centimeter	cm^3 or cc	$1 \text{ cm}^3 = 10^3 \text{ mm}^3 = 1000 \text{ mm}^3$
cubic meter	m^3	$1 \text{ m}^3 = 100^3 \text{ cm}^3 = 1{,}000{,}000 \text{ cm}^3$
cubic kilometer	km^3	$1 \text{ km}^3 = 1000^3 \text{ m}^3 = 1{,}000{,}000{,}000 \text{ m}^3$
Units of Capacity		
milliliter	mL	$1 \text{ mL} = \frac{1}{1000} \text{ L}$
liter	L	$1 \text{ L} = 1000 \text{ mL}$
Units of Mass (Weight)		
milligram	mg	$1 \text{ mg} = \frac{1}{1000} \text{ g}$
gram	g	
kilogram	kg	$1 \text{ kg} = 1000 \text{ g}$
Unit of Temperature		
degree Celsius	° C	water freezes at 0° C
		water boils at 100° C

FIGURE 10.4

Comparing
Millimeters,
Centimeters,
and Inches

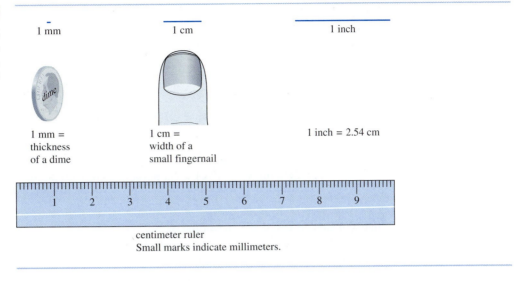

1 mm   1 cm   1 inch

1 mm =   1 cm =   1 inch = 2.54 cm
thickness   width of a
of a dime   small fingernail

centimeter ruler
Small marks indicate millimeters.

FIGURE 10.5

Examples of
Units of Length,
Area, and
Volume in the
Metric System.
The Unit of
Length Is the
Centimeter, the
Unit of Area Is
the Square
Centimeter, and
the Unit of
Volume Is the
Cubic
Centimeter.

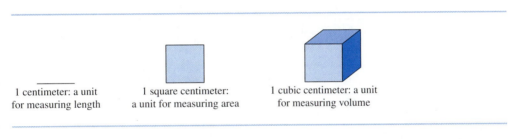

1 centimeter: a unit   1 square centimeter: 1 cubic centimeter: a unit
for measuring length   a unit for measuring area   for measuring volume

Soft drinks often come in 2-liter bottles. One liter is a little more than a quart. Doses of liquid medicines are often measured in milliliters. In this case, a milliliter is also often called a cc, for cubic centimeter.

cc (cubic centimeter)

Nutrition labels on food packages often refer to grams. Serving sizes are often given both in grams and in ounces. One ounce is about 28 grams. One kilogram is the weight of 1 liter of water, which is about 2.2 pounds. The weight of vitamins in foods is often given in terms of milligrams. Also, although 1000 kilograms should be called a megagram, it is usually called a **metric ton** or **tonne**.

metric ton

In the metric system, the units of length, capacity, and mass (weight) are related in a simple and logical way:

1 milliliter of water weighs one gram; and
1 milliliter of water fills a cube that is 1 cm wide, 1 cm deep, and 1 cm high.

The following table shows some basic relationships between the metric and U.S. customary systems of measurement:

Some Metric/U.S. Customary Relationships	
length	1 in = 2.54 cm (exact)
capacity	1 gal = 3.79 L (good approximation)
weight (mass)	1 kg = 2.2 lb (good approximation)

Table 10.4 summarizes the basics of the metric system.

Even though the metric system was designed to be more logical and organized than the old customary systems, the definitions of units in the metric system have changed over time too. As technology and scientific understanding advance, it is possible to define units with greater precision. For example, the meter was originally defined as one ten-millionth of the distance from the equator to the North Pole, whereas it is currently defined as the length of a path traveled by light in a vacuum in

$$\frac{1}{299,792,458}$$

of a second (see [27]).

TABLE 10.4

Summary of the Metric System

Metric System				
	PREFIX	**LENGTH**	**CAPACITY**	**WEIGHT**
.001 $\frac{1}{1000}$	**milli-**	millimeter	milliliter	milligram
.01 $\frac{1}{100}$	**centi-**	centimeter		centigram
.1 $\frac{1}{10}$	**deci-**			
1	**base unit**	meter	liter	gram
10	**deka-**			
100	**hecto-**			
1000	**kilo-**	kilometer		kilogram
Fundamental Relationship				
1 milliliter of water weighs 1 gram and has a volume of 1 cubic centimeter.				

CLASS ACTIVITY   NOW TURN TO CLASS ACTIVITIES MANUAL

10A Misconceptions about Measurement p. 311

10B Becoming Familiar with Centimeters and Meters p. 312

10C Becoming Familiar with Units of Area p. 313

10D Becoming Familiar with Units of Volume p. 316

PRACTICE PROBLEMS FOR SECTION 10.1

1. What does it mean to measure a quantity?

2. What is the simplest, most basic way to measure?

3. What is a unit?

4. What are the two main systems of measurement used in the U.S. today?

5. How many feet are in one mile? How many ounces are in one pound? How many pounds are in one ton?

6. What is the difference between an ounce and a fluid ounce?

7. What is special about the way units in the metric system are named?

8. How is one milliliter related to one gram and to one centimeter?

9. Use square centimeter tiles or centimeter graph paper to show or to draw three different shapes that have an area of 8 cm^2.

10. Use cubic-inch blocks (or cubic-centimeter blocks) to make a variety of solid shapes that have a volume of 24 in^3 (or 24 cm^3).

ANSWERS TO PRACTICE PROBLEMS FOR SECTION 10.1

1. See page 434.

2. See page 434.

3. See page 434.

4. See page 435.

5. There are 5280 feet in a mile, 16 ounces in a pound, and 2000 pounds in a ton.

6. An ounce is a unit of weight, whereas a fluid ounce is a unit of capacity (or volume).

7. The decimal system uses *base units* such as meter, liter, gram, and it uses *prefixes* such as milli, centi, deci, deka, hecto, and kilo to create other units such as milliliter and kilogram.

8. One milliliter of water weighs one gram and occupies a volume of 1 cubic centimeter.

9. Arrange 8 square-centimeter tiles in any way to form a shape, or draw a shape that encloses 8 squares, as in Figure 10.6 on page 442.

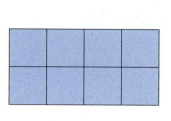 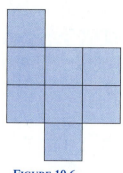 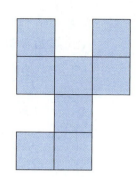

FIGURE 10.6

Three Ways to Show 8 cm²

10. Arrange 24 cubic-inch blocks (or cubic-centimeter blocks) in any way to make a solid shape.

PROBLEMS FOR SECTION 10.1

1. For each of the following metric units, give examples of two objects that you encounter in daily life whose sizes could appropriately be described using that unit:

   a. meter

   b. gram

   c. liter

   d. milliliter

   e. millimeter

   f. kilogram

   g. kilometer

2. For each of the following items, state which U.S. customary units and which metric units in common use would be most appropriate for describing the size of the item. In each case, say briefly why you chose the units.

   a. The volume of water in a full bathtub

   b. The length of a swimming pool

   c. The weight of a slice of bread

   d. The volume of a slice of bread

   e. The weight of a ship

   f. The length of an ant

3. What does it mean to say that a shape has an area of 8 square inches? Discuss your answer in as clear and direct a fashion as you can, as if you were teaching children.

4. Discuss why it is easy to give an incorrect solution to the following area problem and what you must understand about measurement in order to solve the problem correctly:

   An area problem: Draw a shape that has an area of 2 square inches.

5. Write an essay discussing the three most interesting ideas you learned from reading this section.

6. Visit a store and write down at least ten different items and their metric measurements that you find on their package labels. For each of the following units, find at least one item that is described using that unit: grams (g), kilograms (kg), milligrams (mg), liters (l), milliliters (mL).

10.2 Error and Accuracy in Measurements

When measuring actual physical quantities, it is impossible to avoid a certain amount of error and uncertainty. The way you report a measurement indicates the accuracy of that measurement. Similarly, your interpretation of a measurement depends on the digits reported in the measurement.

FIGURE 10.7

Using Rulers
with Different
Markings to
Measure

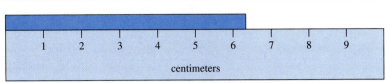

(a) With this ruler, we report the length of the strip as 6 cm.

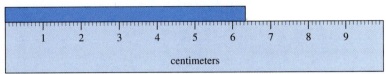

(b) With this ruler, we report the length of the strip as 6.3 cm.

Any reported measurement of an actual physical quantity is necessarily only approximate—you can never say that a measurement of a real object is exact. For example, if you use a ruler to measure the width of a piece of paper, you may report that the paper is $8\frac{1}{2}$ inches wide. But the paper isn't *exactly* $8\frac{1}{2}$ inches wide, it is just that the mark on your ruler that is closest to the edge of the paper is at $8\frac{1}{2}$ inches.

The accuracy of a measurement depends on the accuracy of the tools you use to measure with (and how well you use those tools). For example, Figure 10.7 (a) shows a centimeter ruler measuring the length of a small shaded strip. This ruler has markings at whole numbers of centimeters only. Reading from this ruler, we see that the shaded strip is between 6 and 7 centimeters long, but closer to 6 centimeters than to 7 centimeters. It looks like the strip might be 6.3 or 6.4 centimeters long, but because the ruler doesn't have finer markings, if we use only this ruler, we should report the length of the strip as 6 centimeters. Now look at the ruler in Figure 10.7 (b). This ruler is measuring an identical shaded strip, but has markings at centimeters and at tenths of centimeters (millimeters). Reading from this ruler, we see that the shaded strip is between 6.3 and 6.4 centimeters long, but closer to 6.3 centimeters than to 6.4 centimeters. In this case, we should report the length of the strip as 6.3 centimeters. Perhaps the strip is 6.34 or 6.33 centimeters long, but we can't tell using this ruler.

significant digits   In a measurement, the digits that we can accurately report are called **significant digits**. In Figure 10.7 (a), the ruler only allowed measurement to one significant digit (6), whereas in Figure 10.7 (b), the ruler allowed measurement to two significant digits (6.3). If we wanted to measure the shaded strip in Figure 10.7 to *three* significant digits then we would need a special measuring device because it would be difficult to mark, or even to see, hundredths of centimeters on a ruler. It is usually difficult to measure more than 3 significant digits.

INTERPRETING REPORTED MEASUREMENTS

The way that a measurement is reported reflects its accuracy. For example, suppose that the distance between two cities is reported as 1200 miles. Then, since there are zeros in the tens and ones places, we assume that this measurement is *rounded to the nearest hundred* (unless we are told otherwise). In other words, the actual distance between the two cities is between 1100 miles and 1300 miles, but closer to 1200 miles than to either 1100 miles or 1300 miles. Thus, the

FIGURE 10.8

Interpreting a
Reported
Distance of
1200 Miles

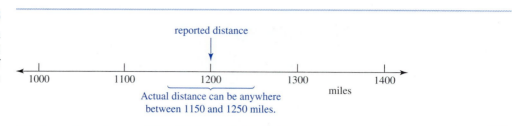

FIGURE 10.9

Interpreting a
Reported
Distance of
1230 Miles

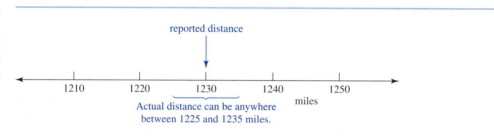

actual distance between the two cities could be anywhere between 1150 and 1250 miles, as indicated in Figure 10.8.

On the other hand, if the distance between two cities is reported as 1230 miles, then we assume (unless we are told otherwise) that this measurement is *rounded to the nearest ten*. In other words, the actual distance between the cities is between 1220 miles and 1240 miles, but closer to 1230 miles than to either 1220 miles or 1240 miles. In this case, the actual distance between the two cities could be anywhere between 1225 and 1235 miles, as indicated in Figure 10.9. Notice that in this case, we are getting a much smaller range of numbers that the actual distance could be than in the previous example. Here it's a range of 10 miles (1225 miles to 1235 miles), whereas in the previous case it's a range of 100 miles (1150 miles to 1250 miles).

Similarly, if your weight on a digital scale is reported as 130.4 pounds, then (assuming the scale's report is accurate) this is your actual weight, *rounded to the nearest tenth*. In other words, your actual weight is between 130.3 pounds and 130.5, but closer to 130.4 pounds than to either 130.3 pounds or 130.5 pounds. In this case, your actual weight is between 130.35 and 130.45 pounds.

Notice that there is a difference in reporting that the weight of an object is 130.0 pounds and reporting the weight as 130 pounds. When the weight is reported as 130.0 pounds, we assume that this weight is *rounded to the nearest tenth* of a pound. In this case, the actual weight is between 129.95 and 130.05 pounds. On the other hand, when the weight of an object is reported as 130 pounds, we assume this weight is *rounded to the nearest ten* pounds. In this case, we only know that the actual weight is between 125 and 135 pounds, which allows for a much wider range of actual weights than the range of 129.95 to 130.05 pounds (a 10-pound range versus a .1-pound range). So, when a weight is reported as 130.0 pounds, the weight is known with greater accuracy than when the weight is reported as 130 pounds.

What do we do when we know a distance more accurately than the way we write it would suggest? For example, suppose we know that a certain distance is 130,000 km, but that this

FIGURE 10.10

Calculating with
Measured
Distances

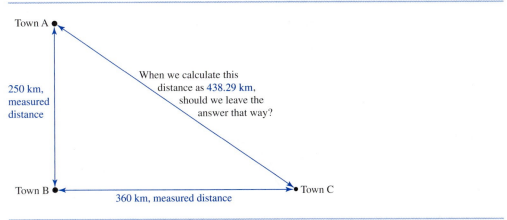

is rounded to the nearest *hundred* kilometers, not to the nearest *ten-thousand* kilometers, as the way the number is written would suggest. In this case, we can report the distance as "130,000 km to 4 significant digits."

WORKING WITH MEASUREMENTS

In later sections of this chapter we will calculate various areas and volumes of objects using lengths associated with these objects. In some cases, the lengths are taken to be *given* as opposed to actually *measured*, and the area and volume calculations are then purely theoretical. For example, we will be able to calculate the area of a circle that has radius 6.25 inches. In this case, we are imagining a circle that has radius 6.25 inches, and we are calculating the exact area that such an imagined circle would have.

In other cases, we will work with actual or hypothetical measured distances to calculate other distances, or to calculate areas or volumes. In these cases, we should assume that the distances we are working with have been rounded, as described previously. Therefore, when we determine our final answer to a calculation involving measured distances, we should also round that answer, because the answer cannot be known more accurately than the initial distances we began the calculations with. For example, we might calculate the distance from town A to town C given that the distance from town A to town B is 250 km, and the distance from town B to town C is 360 km, as shown in Figure 10.10. By calculating, we determine that the distance from town A to town C is 438.29 km. But we should not leave our answer like that, because if we do, we are communicating that we know the distance from town A to town C rounded to the nearest *hundredth* of a kilometer. Because we only know the distances from town A to town B and town B to town C rounded to the nearest *ten* kilometers, we cannot possibly know the resulting distance from town A to town C with greater accuracy. Instead, we should report the distance from town A to town C as 440 km, which fits with the rounding of the distances we were given.

Another point to keep in mind when calculating with actual or hypothetical measurements is that even though you may be rounding your final answer, you should wait until you are done with all your calculations before you round. Waiting to the end to round will give you the most accurate answer that fits with the measurements you started with.

CLASS ACTIVITY   NOW TURN TO CLASS ACTIVITIES MANUAL

10E   Reporting and Interpreting Measurements p. 317

PRACTICE PROBLEMS FOR SECTION 10.2

1. The long shaded line on the dial in Figure 10.11 indicates the weight of an object. Would it be acceptable to report the weight of the object as 2.66 kg? Why or why not? If not, how should you report the weight?

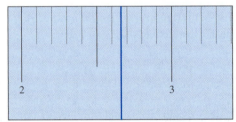

kilograms

FIGURE 10.11

The Dial of a Scale

2. What is the difference between reporting that an object weighs 2 pounds and reporting that it weighs 2.0 pounds?

3. If the distance between two cities is reported as 2500 miles, does that mean that the distance is exactly 2500 miles? If not, what can you say about the exact distance?

ANSWERS TO PRACTICE PROBLEMS FOR SECTION 10.2

1. No, you should not report the weight of the object as 2.66 kg because if you do, you are indicating that you know the weight rounded to the nearest hundredth of a kilogram. Although the weight does seem to be close to 2.66 kg, there are no markings on the dial that would allow you to say with certainty that the weight is closer to 2.66 kg than to either 2.65 kg or 2.67 kg. (In fact, the manufacturer will not put these extra markings on unless the scale is accurate enough to be able to report 3 significant digits.) Instead, you should report the weight of the object as 2.7 kg because the line indicating the weight of the object is between the 2.6 mark and the 2.7 mark, but closer to the 2.7 mark.

2. If an object is reported as weighing 2 pounds, then we assume that the weight has been rounded to the nearest whole pound. Therefore, the actual weight is between 1 and 3 pounds, but closer to 2 pounds than to either 1 pound or 3 pounds. The actual weight must be be-

tween 1.5 pounds and 2.5 pounds. On the other hand, if an object is reported as weighing 2.0 pounds, then we assume that the weight has been rounded to the nearest *tenth* of a pound. Therefore, the actual weight is between 1.9 pounds and 2.1 pounds, but closer to 2.0 pounds than to either 1.9 pounds or 2.1 pounds. The actual weight must be between 1.95 pounds and 2.05 pounds. So, when the weight is reported as 2.0 pounds, the weight is known with greater accuracy than when it is reported as 2 pounds.

3. If the distance between two cities is reported as 2500 miles, then we assume that this number is the actual distance, rounded to the nearest hundred. Therefore, the actual distance is between 2400 miles and 2600 miles, but closer to 2500 miles than to either 2400 miles or 2600 miles. The actual distance between the two cities could be anywhere between 2450 miles and 2550 miles.

PROBLEMS FOR SECTION 10.2

1. Figure 10.12 shows two rulers measuring identical shaded strips.

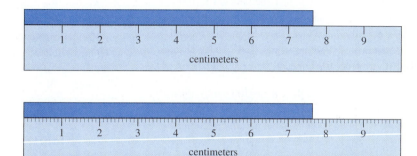

FIGURE 10.12

Two Rulers Measuring a Strip of the Same Size

   a. How should you report the length of the strip using the first ruler? Explain.

   b. How should you report the length of the strip using the second ruler? Explain.

2. The long, shaded line on the dial in Figure 10.13 indicates the weight of an object. How should you report the weight of the object? Explain why you should report the weight that way and not some other way.

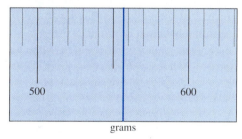

FIGURE 10.13

The Dial of a Scale

3. One source says that the average distance from the Earth to the Moon is 384,467 kilometers. Another source says that the average distance from the Earth to the Moon is 384,000 kilometers. Can both of these descriptions be correct, or must at least one of them be wrong? Explain.

4. If an object is described as weighing 6.0 grams, then is this the exact weight of the object? If not, what can you say about the weight of the object? Explain your answer in detail.

5. Tyra is calculating the distance from town A to town C. Tyra is given that the distance from town A to town B is 120 miles, the distance from town B to town C is 230 miles, that town B is due south of town A, and that town C is due east of town B. Tyra calculates that the distance from town A to town C is 259.4224 miles. Should Tyra leave her answer like that? Why or why not? If not, what answer should Tyra give? Explain. (You may assume that Tyra has calculated correctly.)

6. John has a paper square that he believes is 100 cm wide and 100 cm long, but in reality, the square is actually 1% longer and 1% wider than he believes. In other words, John's square is actually 101 cm wide and 101 cm long.

   a. Since John believes the square is 100 cm by 100 cm, he calculates that the area of his square is 10,000 cm^2. What percent greater is the actual area of the square than John's calculated area of the square? Is the actual area of John's square 1% greater than John's calculated area, or is it larger by a different percent?

   b. Draw a picture (which need not be to scale) to show why your answer in part (a) is not surprising.

   c. Answer the following based on your answers to parts (a) and (b) above. In general, if you want to know the area of a square to within 1% of its actual area, will it be good enough to know the lengths of the sides of the square to within 1% of their actual lengths? If not, will you need to know the lengths more accurately or less accurately?

7. Sally has a Plexiglas cube that she believes is 100 cm wide, 100 cm deep, and 100 cm tall, but in reality, the cube is actually 1% wider, 1% deeper, and 1% taller than she believes. In other words, Sally's cube is actually 101 cm wide, 101 cm deep, and 101 cm tall.

    a. Since Sally believes the cube is 100 cm by 100 cm by 100 cm, she calculates that the volume of her cube is 1,000,000 cm^3. What percent greater is the actual volume of the cube than Sally's calculation of the volume of the cube? Is the actual volume of Sally's cube 1% greater than her calculated volume, or is it larger by some different percent?

    b. Answer the following based on your answer to part (a) above: In general, if you want to know the volume of a cube to within 1% of its actual volume, will it be good enough to know the lengths of the sides of the cube to within 1% of their actual lengths? If not, will you need to know the lengths more accurately or less accurately?

10.3   Length, Area, Volume, and Dimension

All physical objects have several different measurable attributes which we can use to describe their size and to compare the object with other objects. For example, consider a spool of wire that you might find at a hardware store. What are the measurable quantities of this spool of wire? There is the length of wire that is wound onto the spool. There is the diameter of the wire. (The cross section of the wire is a small circle.) There are the diameter of the spool and the length of the spool. There are the weight of the wire and the weight of the wire and spool. The most commonly used measurable attributes are length, area, volume, and weight. In this section we will study length, area, volume, and the related concept of dimension.

Since the units for length, area, and volume are related, it is easy to get confused about which unit to use when. Most real objects have one or more aspects or parts that should be measured by a unit of length, another aspect or part that should be measured by a unit of area, and yet another aspect or part that should be measured by a unit of volume.

We use a unit of *length*, such as inches, centimeters, miles, or kilometers, when we want to answer one of the following kinds of questions:

How far?

How long?

How wide?

For example, "How much moulding (i.e., *how long* a piece of moulding) will I need to go around this ceiling?" The answer, in feet, is the number of foot-long segments that can be put end to end to go all around the ceiling. (See Figure 10.14.)

length   A **length** describes the size of something (or a part of something) that is one-dimensional; the length of that one-dimensional object is how many of a chosen unit of length (such as inches, cm, etc.) it takes to cover the object without overlapping. Roughly speaking, an object **one-dimensional**   is **one-dimensional** if at each location, there is only one independent direction along which to move within the object. An imaginary creature living in a one-dimensional world could only move forward and backward in that world. Living in a one-dimensional world would be like being stuck in a tunnel. Here are some examples of one-dimensional objects:

a line segment

a circle (only the outer part, not the inside)

FIGURE 10.14

A Length
Problem

How many of these
1 foot lengths ——
does it take to go around
the ceiling of this room
(i.e., around the outside)?

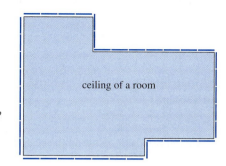

ceiling of a room

the four line segments making a square (not the inside of the square)

a curved line

the equator of the earth

the edge where two walls in a room meet

an imaginary line drawn from one end of a student's desk to the other end

perimeter   In general, the "outer edge" around a (flat) shape is one-dimensional. The **perimeter** of a shape is the distance around a shape; it is the total length of the outer edge around the shape.

We use a unit of *area*, such as square feet or square kilometers, when we want to answer questions like the following:

How much material (such as paper, fabric, or sheet metal) does it take to make this?

How much material does it take to cover this?

For example, "How much carpet do I need to cover the floor of this room?" The answer, in square feet, is the number of 1-foot-by-1-foot squares it would take to cover the floor. (See Figure 10.15.)

area   An **area** or a **surface area** describes the size of an object (or a part of an object) that
surface area   is two-dimensional; the area of that two-dimensional object is how many of a chosen unit of area (such as square inches, square centimeters, etc.) it takes to cover the object without
two-dimensional   overlapping. Roughly speaking, an object is **two-dimensional** if at each location, there are two independent directions along which to move within the object. An imaginary creature living in a two-dimensional world could only move in the forward/backward and right/left directions, as

FIGURE 10.15

An Area
Problem

How many of these
1-ft-by-1-ft squares
does it take to cover the
floor of this room?

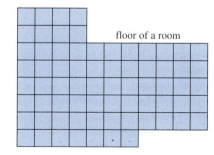

floor of a room

well as in "in between" combinations of these directions (such as diagonally), but *not* up/down. What would it be like to live in a two dimensional world? The book *Flatland* by A. Abbott [1] is just such an account. Here are some examples of two-dimensional objects:

a plane

the *inside* of a circle (this is also called a disk)

the *inside* of a square

a piece of paper (only if you think of it as having no thickness)

the *surface* of a balloon (not counting the inside)

the *surface* of a box

some farm land (only counting the surface, not the soil below)

the surface of the earth

Notice that most of the units that are used to measure area, such as

$$\text{cm}^2, \text{m}^2, \text{km}^2, \text{in}^2, \text{ft}^2, \text{mi}^2,$$

have a superscript "2" in their abbreviation. This "2" reminds us that we are measuring the size of a two-dimensional object.

We use a unit of *volume*, such as cubic yards or cubic centimeters, or even gallons or liters, when we want to answer questions like the following:

How much of a substance (such as air, water, or wood) is in this object?

How much of a substance does it take to fill this object?

For example, "How much air is in this room?" The answer, in cubic feet, is the number of 1-foot-by-1-foot-by-1-foot cubes that are needed to fill the room. (See Figure 10.16.)

volume   A **volume** describes the size of an object (or a part of an object) that is three-dimensional; the volume of that three-dimensional object is how many of a chosen unit of volume (such as cubic inches, cubic centimeters, etc.) it takes to fill the object without overlapping. Roughly speaking, **three-dimensional** an object is **three-dimensional** if at each location, there are three independent directions along which to move within the object. Our world is three-dimensional because we can move in

FIGURE 10.16

A Volume Problem

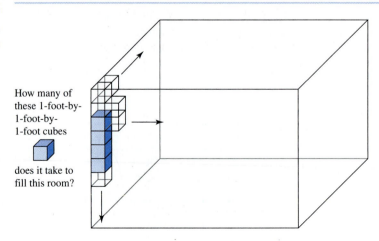

How many of these 1-foot-by-1-foot-by-1-foot cubes

does it take to fill this room?

the three independent directions: forward/backward, right/left, and up/down, as well as all "in between" combinations of these directions. Here are some examples of three-dimensional objects:

the air around us

the inside of a balloon

the inside of a box

the water in a cup

feathers filling a pillow

the compost in a wheelbarrow

the inside of the earth

Notice that many of the units that are used to measure volume, such as

$$\text{cm}^3, \text{m}^3, \text{in}^3, \text{ft}^3$$

have a superscript "3" in their abbreviation. This "3" reminds us that we are measuring the size of a three-dimensional object.

One of the difficulties with understanding whether to measure a length, an area, or a volume is that many objects have parts of different dimensions. For example, consider a balloon. Its outside *surface* is two-dimensional, whereas the air filling it is three-dimensional. Then there is the circumference of the balloon, which is the length of an imaginary one-dimensional circle drawn around the surface of the balloon. The size of each of these parts should be described in a different way: the air in a balloon is described by its volume, the surface of the balloon is described by its area, and the circumference of the balloon is described by its length.

Another example in which the same object has parts of different dimensions is the earth. The *inside* of the earth is three-dimensional, its *surface* is two-dimensional, and the equator is a one-dimensional imaginary circle drawn around the surface of the earth. Once again, the sizes of these different parts of the earth should be described in different ways: by volume, area, and length, respectively.

Another example is farmland. The size of farmland is often measured in acres, which is a measure of area—this only takes into account how big the two-dimensional top surface of the farmland is. This is a practical way to talk about how big farmland is, even though it doesn't take into account the most important part of the land—the three-dimensional soil below the surface.

For a fascinating look at various dimensions, go to the website:

www.aw-bc.com/beckmann

CLASS ACTIVITY NOW TURN TO CLASS ACTIVITIES MANUAL

10F Dimension and Size p. 319

PRACTICE PROBLEMS FOR SECTION 10.3

1. a. Use centimeter or inch graph paper to make a pattern for a closed box. The box should have six sides, and when you fold the pattern, there should be no overlapping pieces of paper.

    b. How much paper is your box made of? Be sure to use an appropriate unit in your answer.

    c. Describe one-dimensional, two-dimensional, and three-dimensional parts or aspects of your box. In each case, give the size of the part or aspect of the box using an appropriate unit.

2. Describe one-dimensional, two-dimensional, and three-dimensional parts or aspects of a water tower. In each case, name an appropriate U.S. customary unit and an appropriate metric unit for measuring or describing the size of that part or aspect of the water tower. What are practical reasons for wanting to know the sizes of these parts or aspects of the water tower?

3. Describe one-dimensional, two-dimensional, and three-dimensional parts or aspects of a store. In each case, name an appropriate U.S. customary unit and an appropriate metric unit for measuring or describing the size of that part or aspect of the store. What are practical reasons for wanting to know the sizes of these parts or aspects of the store?

ANSWERS TO PRACTICE PROBLEMS FOR SECTION 10.3

1. a. Figure 10.17 shows a scaled picture of a pattern for one possible box. Each small square represents a 1-inch-by-1-inch square.

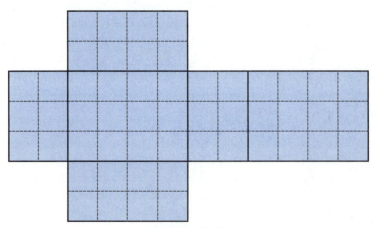

FIGURE 10.17

A Pattern for a Box

    b. The box formed by the pattern in Figure 10.17 (in its actual size) is made of 52 square inches of paper because the pattern that was used to create the box consists of 52 squares that each have area 1 square inch.

    c. The edges of the box are one-dimensional parts of the box. Depending on how the box is oriented, the lengths of these edges are the height, width, and depth of the box. These measure 2 inches, 3 inches, and 4 inches. Another one-dimensional aspect of the box is its *girth* (i.e., the distance around the box). Depending on where you measure, the girth is either 10 inches, 12 inches, or 14 inches.

       The surface of the box is a two-dimensional part of the box. Its area is 52 square inches, as discussed in part (b).

The air inside the box is a three-dimensional aspect of the box. If you filled the box with 1-inch-by-1-inch-by-1-inch cubes, there would be 2 layers with 3×4 cubes in each layer. Therefore, according to the meaning of multiplication, there would be $2 \times 3 \times 4 = 24$ cubes in the box. Thus, the volume of the box is 24 cubic inches.

2. The height of the water tower (or an imaginary line running straight up through the middle of the water tower) is a one-dimensional aspect of the water tower. Feet and meters are appropriate units for describing the height of the water tower. Some towns do not want high structures, therefore, the town might want to know the height of a proposed water tower before it decides whether or not to build it.

The surface of the water tower is a two-dimensional part of the water tower. Square feet and square meters are appropriate units for describing the size of the surface of the water tower. We would need to know the approximate size of the surface of the water tower in order to know approximately how much paint it would take to cover it.

The water inside a water tower is a three-dimensional part of a water tower. Cubic feet, gallons, liters, and cubic meters are appropriate units for measuring the volume of water in the water tower. The owner of a water tower will certainly want to know how much water the tower can hold.

3. The edge where the store meets the sidewalk is a one-dimensional aspect of the store. Its length is appropriately measured in feet, yards, or meters. This length is the width of the store front, which is one way to measure how much exposure the store has to the public eye.

The surface of the store facing the sidewalk is a two-dimensional part of the store. Its area is appropriately described in square feet, square yards, or square meters. The area of the store front is another way to measure how much exposure the store has to the public eye. (A store front that is narrow but tall might attract as much attention as a wider, lower store front.)

Another important two-dimensional part of a store is its floor. The area of the floor is appropriately measured in square feet, square yards, or square meters, or possibly even in acres if the store is very large. The area of the floor indicates how much space there is on which to place items to sell to customers.

The air inside a store is a three-dimensional aspect of the store. Its volume is appropriately measured in cubic feet, cubic yards, cubic meters, or possibly even in liters or gallons. The builder of a store might want to know the volume of air in the store in order to figure what size air conditioning units will be needed.

PROBLEMS FOR SECTION 10.3

1. Describe one-dimensional, two-dimensional, and three-dimensional parts or aspects of a soda bottle. In each case, specify an appropriate U.S. customary unit and an appropriate metric unit for measuring or describing the size of that part or aspect of the soda bottle. What are practical reasons for wanting to know the sizes of these parts or aspects of the soda bottle?

2. Describe one-dimensional, two-dimensional, and three-dimensional parts or aspects of a car. In each case, specify an appropriate U.S. customary unit and an appropriate metric unit for measuring or describing the size of that part or aspect of the car. What are practical reasons for wanting to know the sizes of these parts or aspects of the car?

10.4 Calculating Perimeters of Polygons, Areas of Rectangles, and Volumes of Boxes

Children sometimes get confused when deciding how to calculate perimeter, area, or volume. We *add* to calculate perimeter, but we *multiply* to calculate areas of rectangles and volumes of boxes (rectangular prisms). When we calculate the perimeter of a rectangle we add the lengths of all four sides (or we add the lengths of two adjacent sides and multiply by 2). On the other hand, when we calculate the area of a rectangle we multiply only two of the sides' lengths. It's hard to keep all these different methods of calculation straight unless we have a clear idea of what perimeter, area, and volume mean. In this section we will study why we calculate perimeters of polygons, areas of rectangles, and volumes of boxes (rectangular prisms) in the way we do.

CALCULATING PERIMETERS OF POLYGONS

Why do we add the lengths of the sides of a polygon in order to calculate its perimeter? The perimeter of a polygon (or other shape in the plane) is the total distance around the polygon. A hands-on way to think about perimeter is as the length of a string that wraps snugly once around the shape. When we calculate the perimeter of a polygon by adding the lengths of the sides around the polygon, it is as if we had cut the string into pieces and are determining the total length of the string by adding the lengths of the pieces. We are calculating the total distance around the shape by adding up the lengths of pieces that together encircle the shape.

CLASS ACTIVITY   NOW TURN TO CLASS ACTIVITIES MANUAL

10G   Explaining Why We Add to Calculate Perimeters of Polygons p. 320

10H   Perimeter Misconceptions p. 321

CALCULATING AREAS OF RECTANGLES

Why do we multiply the length of a rectangle by its width in order to determine the area of the rectangle? Consider a rectangle that is 4 cm long and 3 cm wide, as shown in Figure 10.18 (a). If we cover this rectangle with 1-cm-by-1-cm squares, then there will be 4 rows of squares with 3 squares in each row. Therefore, there are 4×3 squares covering the rectangle (without overlaps). Since each square has area 1 cm², the total area of the rectangle is 4×3 square centimeters, which is 12 square centimeters.

The line of reasoning for the 4-cm-by-3-cm rectangle above works generally to explain why a rectangle that is W units wide and L units long has an area of $L \times W$ square units. As Figure 10.18 (b) indicates, if we cover such a rectangle with 1-unit-by-1-unit squares, then there will be L rows with W squares in each row. Therefore, there are $L \times W$ squares covering the rectangle. Since each square has area 1 unit², the total area of the rectangle is $L \times W$ square units.

The *length times width* formula,

$$L \times W$$

for areas of rectangles is valid not only when L and W are whole numbers, but also when L or W are fractions, mixed numbers, or decimals. The previous line of reasoning works in these cases as well. For example, consider a rectangle that is $3\frac{1}{2}$ cm by $2\frac{1}{2}$ cm, as shown in Figure 10.19 on page 456. If we cover such a rectangle with 1-cm-by-1-cm squares and parts of 1-cm-by-1-cm squares, then there will be $3\frac{1}{2}$ groups of squares with $2\frac{1}{2}$ squares in each

FIGURE 10.18

Why
Multiplying the
Side Lengths of
a Rectangle
Gives the Area
of the Rectangle

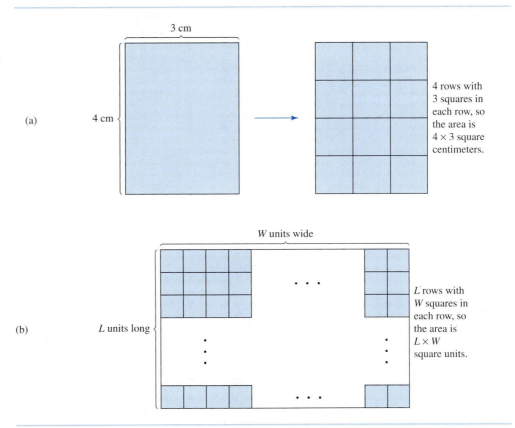

group. Each square has area 1 cm^2, so one whole group has area $2\frac{1}{2}$ cm^2. Therefore, $3\frac{1}{2}$ groups have a total area of

$$3\frac{1}{2} \times 2\frac{1}{2} \text{ cm}^2$$

Numerically, we can calculate that

$$3\frac{1}{2} \times 2\frac{1}{2} = \frac{7}{2} \times \frac{5}{2} = \frac{35}{4} = 8\frac{3}{4}$$

Hence, the area of the rectangle is $8\frac{3}{4}$ cm^2. If we count up whole squares and collect together partial squares in Figure 10.19, we also see that the rectangle consists of a total of $8\frac{3}{4}$ square centimeters.

CLASS ACTIVITY   NOW TURN TO CLASS ACTIVITIES MANUAL

1O1 Explaining Why We Multiply to Determine Areas of Rectangles p. 323

FIGURE 10.19

Why a $3\frac{1}{2}$-cm-by-$2\frac{1}{2}$-cm Rectangle Has Area $3\frac{1}{2} \times 2\frac{1}{2}$ cm^2

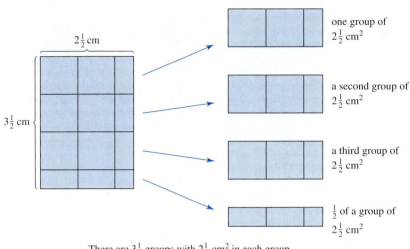

There are $3\frac{1}{2}$ groups with $2\frac{1}{2}$ cm^2 in each group, for a total area of $3\frac{1}{2} \times 2\frac{1}{2}$ cm^2.

CALCULATING VOLUMES OF BOXES

Why do we multiply the height by the width by the depth of a box (rectangular prism) in order to calculate its volume? Consider a box that is 4 units high, 3 units wide, and 2 units deep, as pictured in Figure 10.20. If we fill this box with 1-unit-by-1-unit-by-1-unit cubes, then there will be 4 groups (layers) of cubes, with 3×2 cubes in each group. Therefore, it will take

$$4 \times 3 \times 2$$

cubes to fill the box. Since each cube has volume 1 cubic unit, the total volume of the cubes, and therefore of the box, is $4 \times 3 \times 2$ cubic units. The same reasoning explains why the volume of a box that is H units high, W units wide, and D units deep is

$$H \times W \times D \text{ cubic units}$$

As with the area formula for rectangles, this volume formula is valid not only when the height, width, and depth of the box are whole numbers, but also when they are fractions, mixed numbers, or decimals. In the cases of fractions, mixed numbers, and decimals, we must consider layers that contain partial cubes, such as a layer consisting of $7\frac{1}{2}$ cubes or $8\frac{3}{4}$ cubes. We must also consider partial layers (groups). For example, if the box is $4\frac{1}{2}$ units high instead of 4 units high, then we would have $4\frac{1}{2}$ groups of cubes instead of 4 groups of cubes.

CLASS ACTIVITY   NOW TURN TO CLASS ACTIVITIES MANUAL

10J   Explaining Why We Multiply to Determine Volumes of Boxes p. 324

FIGURE 10.20

Why a
4-Unit-High,
3-Unit-Wide,
and
2-Unit-Deep
Box Has
Volume
$4 \times 3 \times 2$ Cubic
Units

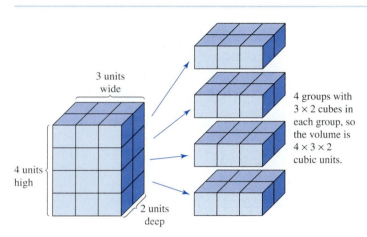

3 units wide

4 units high

2 units deep

4 groups with
3×2 cubes in
each group, so
the volume is
$4 \times 3 \times 2$
cubic units.

PRACTICE PROBLEM FOR SECTION 10.4

1. The children in a class have been learning about the perimeter and area of rectangles. The teacher notices that one child consistently gives the wrong answer to problems on the area of rectangles: the child's answer is almost always 2 times the correct answer. For example, if the correct answer is 12 square feet, the child gives the answer 24 square feet. What might be the source of this error?

ANSWER TO PRACTICE PROBLEM FOR SECTION 10.4

1. If the child had been calculating perimeters of rectangles by adding the lengths of two adjacent sides of the rectangle and multiplying this result by 2 (to account for the other two sides), then the child might be attempting to calculate area in an analogous way. Instead of just multiplying the length times width of the rectangle, the child may think that the result must be multiplied by 2, as it was for the perimeter.

PROBLEMS FOR SECTION 10.4

1. Suppose that a child in your class wants to know why we multiply only two of the lengths of the sides of a rectangle in order to determine the rectangle's area. After all, when we calculate the perimeter of a rectangle we add the lengths of the *four sides* of the rectangle, so why don't we multiply the lengths of the *four sides* in order to find the area?

   a. Explain to the child what perimeter and area mean and explain why we carry out the perimeter and area calculations for a rectangle the way we do.

   b. Describe some problems or activities that might help the child understand the calculations.

2. Suppose that a child in your class wants to know why we multiply only three of the lengths of the edges of a box in order to calculate the volume of the box. Why don't we have to multiply *all* the lengths of the edges?

   a. Explain to this child why it makes sense to calculate the volume of a box as we do.

   b. Describe some problems or activities that might help the child understand the calculation.

3. a. Explain how to see the area of the large rectangle in Figure 10.21 on page 458 as consisting of $2\frac{1}{2}$ groups with $3\frac{1}{2}$ cm^2 in each group, thereby explaining why

it makes sense to multiply

$$2\frac{1}{2} \times 3\frac{1}{2}$$

to determine the area of the rectangle in square centimeters.

b. Calculate $2\frac{1}{2} \times 3\frac{1}{2}$ without a calculator, showing your calculations. Then verify that this calculation has the same answer as when you determine the area of the rectangle in Figure 10.21 in square centimeters by counting full 1-cm-by-1-cm squares and combining partial squares.

FIGURE 10.21

Explain Why the Area is $2\frac{1}{2} \times 3\frac{1}{2}$ cm²

4. a. Explain how to see the area of the large rectangle in Figure 10.22 as consisting of $4\frac{1}{2}$ groups with $5\frac{3}{4}$ cm² in each group, thereby explaining why it makes sense to multiply

$$4\frac{1}{2} \times 5\frac{3}{4}$$

to determine the area of the rectangle in square centimeters.

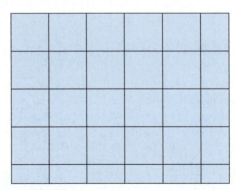

FIGURE 10.22

Explain Why the Area is $4\frac{1}{2} \times 5\frac{3}{4}$ cm²

b. Calculate $4\frac{1}{2} \times 5\frac{3}{4}$ without a calculator, showing your calculations. Then verify that this calculation has the same answer as when you determine the area of the rectangle in Figure 10.22 in square centimeters by counting full 1-cm-by-1-cm squares and combining partial squares.

5. Sarah is confused about the difference between the *perimeter* and the *area* of a polygon. Explain the two concepts and the distinction between them.

6. a. Johnny is confused about the difference between the *surface area* and the *volume* of a box. Explain the two concepts to Johnny in a way that could help

Johnny learn to distinguish between the two concepts.

b. Determine the surface area and the volume of a closed box that is 5 inches wide, 4 inches deep, and 6 inches tall. Explain in detail why you calculate as you do.

7. Tim needs a sturdy cardboard box that is 3 ft tall by 2 ft wide by 1 ft deep. Tim wants to make the box out of a large piece of cardboard that he will cut, fold and tape. The box must close up completely, so it needs a top and a bottom. Show Tim how to make such a box out of one rectangular piece of cardboard: tell Tim what size cardboard he'll need to get (how wide, how long) and explain or show how he should cut, fold, and tape the cardboard to make the box. Be sure to specify exact lengths of any cuts Tim will need to make. Include pictures where appropriate. Your instructions and box should be practical—one that Tim can actually make and use. You might want to make a scale model for your box out of paper. Note: many boxes have flaps around all four sides that one can fold down and interlock to make the top and bottom of the box. You can make this kind of top and bottom, or something else if you prefer, but make sure it will make a sturdy box that closes completely.

10.5 Comparing Sizes of Objects

In this brief section we discuss an important issue to consider when comparing sizes of objects, namely that we must be clear about which measurable attribute of the objects we are using as a basis for the size comparison.

We often say that one object is larger than another one. For example, we might say that one box of cereal is larger than another box of cereal or that one bottle of juice is larger than another bottle of juice. In the case of the juice, we are probably comparing the sizes of the bottles by the volume of juice they contain. Maybe one bottle contains a liter of juice and the other bottle contains 750 mL of juice, which is less than a liter. But in the case of the cereal, we might be comparing the volume of cereal in each box, or we might be comparing the weight of the cereal in each box. Some cereals are more dense than others, so that a smaller volume of one cereal can weigh more than a larger volume of another. Therefore, the two ways of comparing the boxes of cereal—by volume or by weight—could lead us to different conclusions about which box of cereal is larger.

In order to compare sizes of objects we must specify, either implicitly or explicitly, a measurable attribute as a basis for the comparison. So be careful when comparing sizes of objects: if there is any chance of confusion, you should specify the measurable attribute with which you are comparing the objects.

CLASS ACTIVITY NOW TURN TO CLASS ACTIVITIES MANUAL

10K The Biggest Tree in the World p. 325

10L Comparing Sizes of Boxes p. 327

PRACTICE PROBLEM FOR SECTION 10.5

1. Table 10.5 displays information about two boxes of different types of breakfast cereals. Explain in several different ways why either of the two boxes of cereal can be considered the larger of the two.

TABLE 10.5
Comparing Boxes of Cereal

	CEREAL BOX 1	CEREAL BOX 2
Weight of Cereal in Box	16 oz	11 oz
Volume of Cereal in Box	4 cups	10 cups
Servings Per Box	8	10
Calories Per Box	1600	1100

ANSWER TO PRACTICE PROBLEM FOR SECTION 10.5

1. As Table 10.5 shows, if we compare the cereal either by weight or by the number of calories per box, then Cereal Box 1 is larger (16 ounces versus 11 ounces and 1600 calories versus 1100 calories). On the other hand, if we compare the cereal by volume or by the number of servings per box, then Cereal Box 2 is larger (4 cups versus 10 cups and 8 servings per box versus 10 servings per box).

PROBLEMS FOR SECTION 10.5

1. Explain why either of the two rectangles in Figure 10.23 can be considered the larger of the two.

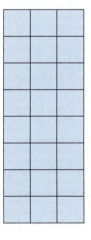

FIGURE 10.23
Two Rectangles

2. Describe one-dimensional, two-dimensional, and three-dimensional parts or aspects of the blocks in Figure 10.24. In each case, compare the sizes of the three blocks, using an appropriate unit. Use this unit to show that each block can be considered biggest of all three.

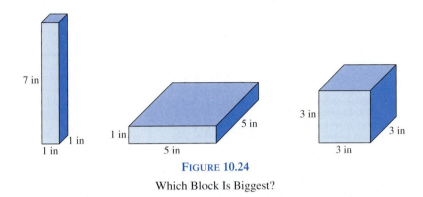

FIGURE 10.24

Which Block Is Biggest?

3. The Lazy Daze Pool Club and the Slumber-N-Sunshine Pool Club have a friendly rivalry going. Each club claims to have the bigger swimming pool. Make up realistic sizes for the clubs' swimming pools so that each club has a legitimate basis for saying that it has the larger pool. Explain clearly why each club can say its pool is biggest. Be sure to use appropriate units. Considering the function of a pool, what is a good way to compare sizes of pools? Explain.

4. Suppose there are two rectangular pools: one is 30 feet wide, 60 feet long, and 5 feet deep throughout, the other is 40 feet wide, 50 feet long, and 4 feet deep throughout. Show that each pool can be considered "biggest" by comparing the sizes of the pools in two meaningful ways other than by comparing one-dimensional aspects of the pools.

10.6 Converting from One Unit of Measurement to Another

As we've seen, we can use different units to measure the same quantity. For example, we can measure a length in miles, yards, feet, inches, kilometers, meters, centimeters, or millimeters. If we know a length in terms of one unit, how can we describe it in terms of another unit? This kind of problem is a conversion problem, which is the topic of this section.

To convert a measurement in one unit to another unit, we first need to know the relationship between the two units. For example, if we want to convert the weight of a bag of oranges given in kilograms into pounds, then we need to know the relationship between kilograms and pounds. Referring back to the table on page 440 in Section 10.1, we see that 1 kg = 2.2 lb.

Once we know how the units are related, we either multiply or divide to convert from one unit to another. How do we know which operation to use? A good way to decide is to think about the meanings of multiplication and division. For example, let's say a bag of oranges weighs 2.5 kg. What is the weight of the bag of oranges in pounds? If one kilogram is 2.2 pounds, then we can think of 2.5 kilograms as 2.5 groups with 2.2 pounds in each group, as illustrated in Figure 10.25 on page 462. Thus, according to the meaning of multiplication, 2.5 kilograms is 2.5 × 2.2 = 5.5 pounds.

Here's another example: How many gallons are there in a 25-liter container? Referring back to the table on page 440 in Section 10.1, we see that 1 gal = 3.79 L. That is, every group

FIGURE 10.25

Converting 2.5
Kilograms to
Pounds

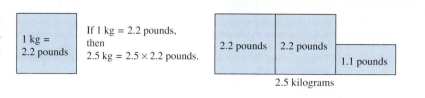

of 3.79 liters is equal to 1 gallon. Therefore, to find how many gallons are in 25 liters, we need to find out how many groups of 3.79 liters are in 25 liters. This is a division problem, using the "how many groups?" interpretation of division. So there are $25 \div 3.79 = 6.6$ gallons in 25 liters.

In some cases, the tables in Section 10.1 do not tell us directly how two units are related. For example, the table on page 440 in Section 10.1 does not tell us how kilometers relate to miles. In order to convert 3 miles to kilometers we must work with the length relationships that we do know from the tables in Section 10.1. In this case, we use the relationship 1 in = 2.54 cm, the relationships among inches, feet, and miles, and the relationships among centimeters, meters, and kilometers to convert 3 miles to kilometers. We go through the following chain of conversions:

$$\text{miles} \; \rightarrow \; \text{feet} \; \rightarrow \; \text{inches} \; \rightarrow \; \text{centimeters} \; \rightarrow \; \text{meters} \; \rightarrow \; \text{kilometers}$$

You can think of this as "ratcheting down," "going across," and then "ratcheting back up," as indicated in Figure 10.26.

Now let's convert 3 miles to kilometers. First, convert miles to feet: since 1 mile is 5280 feet, 3 miles is 3 times as many feet (3 groups of 5280 feet), so

$$3 \text{ miles} \; = 3 \times 5280 \text{ feet} \; = 15,840 \text{ feet}$$

Next, convert feet to inches: since 1 foot is 12 inches, 15,840 feet are 15,840 times as many inches (15,840 groups of 12), so

$$15,840 \text{ feet} = 15,840 \times 12 \text{ inches} \; = 190,080 \text{ inches}$$

Now convert inches to centimeters: since 1 inch is 2.54 centimeters, 190,080 inches are 190,080 times as many centimeters (190,080 groups of 2.54 centimeters), so

$$190,080 \text{ inches} \; = 190,080 \times 2.54 \text{ cm} \; = 482,803.2 \text{ cm}$$

FIGURE 10.26

Converting
from Miles to
Kilometers

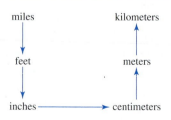

TABLE 10.6

A Calculator
Flowchart for
Converting 3
Miles to
Kilometers

enter 3,
times 5280 → 15840,
times 12 → 190080,
times 2.54 → 482803.2,
divided by 100 → 4828.032,
divided by 1000 → 4.828032.

Next, convert centimeters to meters: since 100 centimeters make 1 meter, we need to find how many groups of 100 centimeters are in 482,803.2 centimeters. According to the "how many groups?" interpretation of division, we can determine the number of meters by dividing:

$$482,803.2 \text{ cm} = (482,803.2 \div 100) \text{ m} = 4,828.032 \text{ m}$$

Finally, convert meters to kilometers: since 1000 meters make 1 kilometer, we need to find how many groups of 1000 meters are in 4,828.032 meters. According to the "how many groups?" interpretation of division, we can determine the number of kilometers by dividing:

$$4,828.032 \text{ m} = (4,828.032 \div 1000) \text{ km} = 4.828032 \text{ km}$$

Thus, 3 miles is exactly 4.828032 kilometers, which is just a little less than 5 kilometers. If you ran 3 miles and your friend ran 5 kilometers, then your friend ran farther, but not by much.

It is easy to carry out the previous calculations for converting 3 miles to kilometers on a calculator. At each step, you either multiply or divide the number currently in the calculator's display. So after each step, be sure to leave the result of your calculation in your calculator, and use this number for the next calculation. For example, when you get 15,840 feet from multiplying 3 times 5280 feet, just leave the 15,840 in your calculator for the next calculation, which is multiplying by 12. That way you don't risk making an error by punching the 15,840 back in. Table 10.6 shows a flowchart for solving the problem of converting 3 miles to kilometers using a calculator. The arrows in the table point to results calculated by the calculator.

DIMENSIONAL ANALYSIS

Many people use **dimensional analysis** to convert a measurement from one unit to another. When we use this method, we repeatedly multiply by the number 1 expressed as a fraction that relates two different units. The resulting calculations are identical to those done to convert units by using multiplication and division, as shown above.

For example, suppose that a German car is 5.1 meters long. How long is the car in feet? To solve this problem, we will use the information provided in the tables of Section 10.1 in order to carry out the following chain of conversions:

$$\text{meters} \rightarrow \text{centimeters} \rightarrow \text{inches} \rightarrow \text{feet}$$

The following equation uses dimensional analysis to convert 5.1 meters to feet:

$$5.1 \text{ m} = 5.1 \text{ m} \times \frac{100 \text{ cm}}{1 \text{ m}} \times \frac{1 \text{ in}}{2.54 \text{ cm}} \times \frac{1 \text{ ft}}{12 \text{ in}} = 16.7 \text{ ft} \qquad (10.1)$$

So, the car is 16.7 feet long.

How do we choose the fractions to use in dimensional analysis? Each fraction that we multiply by must be equal to 1. That way, when we multiply by the fraction we do not change the original amount, we just express the amount in different units. We used the following fractions in Equation 10.1:

$$\frac{100 \text{ cm}}{1 \text{ m}} = 1, \quad \frac{1 \text{ in}}{2.54 \text{ cm}} = 1, \quad \frac{1 \text{ ft}}{12 \text{ in}} = 1$$

We also choose the fractions in dimensional analysis so that all the units will "cancel" except the unit in which we want our answer expressed. In Equation 10.1, the first fraction cancels the meters but leaves us with centimeters, the second fraction cancels the centimeters but leaves us with inches, and the third fraction cancels the inches, leaving feet, which is how we want our answer expressed. Thus,

$$5.1 \text{ m} = 5.1 \text{ m} \times \frac{100 \text{ cm}}{1 \text{ m}} \times \frac{1 \text{ in}}{2.54 \text{ cm}} \times \frac{1 \text{ ft}}{12 \text{ in}} = 16.7 \text{ ft}$$

AREA AND VOLUME CONVERSIONS

Once you know how to convert from one unit of length to another, you can then convert between the related units of area and volume. However, in order to calculate area and volume conversions correctly, it's important to remember the meanings of the units for area and volume (such as square feet, cubic inches, square meters, and so on). Otherwise, it is very easy to make the following common kind of mistake:

"Since 1 yard is 3 feet, 1 square yard is 3 square feet."

Figure 10.27 will help you see why this is a mistake. One square yard is the area of a square that is 1 yard wide and 1 yard long. Such a square is 3 feet wide and 3 feet long. Figure 10.27 shows that such a square is made up of $3 \times 3 = 9$ square feet, *not* 3 square feet.

Let's say you are living in an apartment that has a floor area of 800 square feet, and you want to tell your Italian pen pal how big that is in square meters. There are several ways you might approach such an area conversion problem. One good way is this:

First determine what one (*linear*) foot is in terms of meters. Then use that information to determine what one *square* foot is in terms of *square* meters.

FIGURE 10.27

One Square Yard Is 9 Square Feet, Not 3 Square Feet

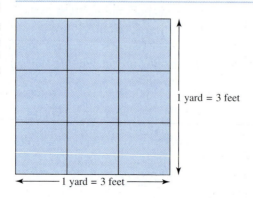

1 yard = 3 feet

1 yard = 3 feet

Here's how we can find 1 foot in terms of meters:

$$1 \text{ ft} = 12 \text{ in}$$
$$12 \text{ in} = 12 \times 2.54 \text{ cm} = 30.48 \text{ cm}$$

because each inch is 2.54 cm and there are 12 inches.

$$30.48 \text{ cm} = (30.48 \div 100) \text{ m} = .3048 \text{ m}$$

because 1 m = 100 cm, so we must find how many 100s are in 30.48 to find how many meters are in 30.48 cm.

So,

$$1 \text{ ft} = .3048 \text{ m}$$

Remember that one square foot is the area of a square that is one foot wide and one foot long. According to the previous calculation, such a square is also .3048 meters wide and .3048 meters long. Therefore, by the "length times width" formula for areas of rectangles,

$$1 \text{ ft}^2 = (.3048 \times .3048) \text{ m}^2 = .0929 \text{ m}^2$$

This means that 800 square feet is $800 \times .0929 = 74.32$ square meters, so the apartment is about 74 square meters.

Volume conversions work similarly. Suppose we want to find 13 cubic feet in terms of cubic meters, for example. We already calculated that

$$1 \text{ ft} = .3048 \text{ m}$$

Now 1 cubic foot is the volume of a cube that is 1 foot high, 1 foot deep, and 1 foot wide. Such a cube is also .3048 meters high, .3048 meters deep, and .3048 meters wide. So, by the "height times width times depth" formula, for volumes of rectangular prisms (boxes),

$$1 \text{ ft}^3 = .3048 \times .3048 \times .3048 \text{ m}^3 = .0283 \text{ m}^3$$

Therefore, 13 cubic feet is 13 times as much as $.0283 \text{ m}^3$:

$$13 \text{ ft}^3 = 13 \times .0283 \text{ m}^3 = .37 \text{ m}^3$$

APPROXIMATE CONVERSIONS AND CHECKING YOUR WORK

In this discussion, we focused on exact (or fairly exact) conversions. However, in many practical situations, you really don't need an exact answer—an estimate will do. Being able to make a quick estimate also gives you a way to check your work for an exact conversion: if your answer is far from your estimate, you've probably made a mistake. If it's not too far off, there's a good chance your work is correct.

The following approximate relationships will help you estimate conversions from one unit to another:

An inch is about $2\frac{1}{2}$ centimeters, so 2 inches is about 5 centimeters.

A meter is a little more than a yard.

A kilometer is about $\frac{6}{10}$ of a mile (so a little over half a mile).

A liter is a little more than a quart.

A kilogram is a bit more than 2 pounds.

For example, you can mentally calculate the following quickly:

A 5 kilometer run is about 3 miles.

Gas that costs $1.00 per liter in Europe costs about $1.00 per quart, in other words, $4.00 per gallon.

A 5 pound bag of oranges is about $2\frac{1}{2}$ kilograms (a little less).

CLASS ACTIVITY   NOW TURN TO CLASS ACTIVITIES MANUAL

10M Conversions: When Do We Multiply? When Do We Divide? p. 328

10N Converting Measurements With and Without Dimensional Analysis p. 330

10O Area and Volume Conversions: Which Are Correct and Which Are Not? p. 331

PRACTICE PROBLEMS FOR SECTION 10.6

1. A class needs a 175-inch-long piece of rope for a project. How long is the rope in yards?

   a. Solve the rope problem simply and explain why your method of solution makes sense.

   b. Describe a number of different correct ways to write the answer to the rope problem. Explain briefly why these different ways of writing the answer mean the same thing.

2. Solve the following conversion problems using the basic fact 1 inch = 2.54 cm in each case.

   a. A track is 100 meters long. How long is it in feet?

   b. If the speed limit is 70 miles per hour, what is it in kilometers per hour?

   c. Some farmland covers 2.4 square kilometers. How many square miles it is?

   d. Convert the volume of a compost pile, one cubic yard, to cubic meters.

   e. A man is 1.88 meters tall. How tall is he in feet?

   f. How many miles is a 10 kilometer race?

3. According to a research study, the average American consumes 63 doughnuts per year. Let's say that each doughnut is about one inch thick. If all the doughnuts consumed in the U.S. in one year were placed in one tall stack, would this stack reach to the Moon? The Moon is about 240 thousand miles away from the Earth and the population of the U.S. is about 275 million.

4. One liter is the capacity of a cube that is 10 cm wide, 10 cm deep, and 10 cm tall. One gallon is 0.134 cubic feet. One quart is one quarter of a gallon. Which is more: one quart or one liter? Use the basic fact 1 inch = 2.54 cm to figure this out.

5. One acre is 43,560 square feet. The area of land is often measured in acres.

   a. What is a square mile in acres?

   b. What is a square kilometer in acres?

6. Suppose you want to cover a football field with artificial turf. You'll need to cover an area that's bigger than the actual field—let's say you'll cover a rectangle that's 60 yards wide and 130 yards long. How many square feet (not square yards) of artificial turf will it take? The artificial turf comes in a roll that is 12 feet wide. Let's say the turf costs $8 per linear foot cut from the roll. (So, if you cut off 10 feet from the roll, it would cost $80 and you'd have a rectangular piece of artificial turf that's 10 feet long and 12 feet wide.) How much will the turf you need cost? (I've been told that preparing the field costs even more than the artificial turf itself.)

ANSWERS TO PRACTICE PROBLEMS FOR SECTION 10.6

1. a. One method for determining the length of the rope in yards is to first figure out how many inches are in a yard. One foot is 12 inches. One yard is 3 feet. Since each foot is 12 inches, 3 feet is 3 groups of 12 inches, which is $3 \times 12 = 36$ inches. To find the length of the 175-inch-long piece of rope in yards, we must figure out how many groups of 36 inches are in 175 inches. This is a "how many groups?" division problem:

    $$175 \div 36 = 4, \text{ remainder } 31$$

    so there are 4 groups of 36 inches in 175 inches with 31 inches left over. Therefore, 175 inches is 4 yards and 31 inches.

    b. We found that we can describe the length of the rope as 4 yards, 31 inches. The 31 inches can also be described as 2 feet, 7 inches, because there are 12 inches in each foot and

    $$31 \div 12 = 2 \text{ remainder } 7$$

    So we can also say that the rope is 4 yards, 2 feet, and 7 inches long. Instead of using a remainder in solving the division problem in part (a), we can give a mixed number or decimal answer to the division problem. So we can also say that the length of the rope in feet is 4.86 yards or $4\frac{31}{36}$ yards, although these answers would probably not be very useful in practice. Similarly, we could give the answer to the division problem $31 \div 12$ as a mixed number or as a decimal and say that the rope is 4 yards and $2\frac{7}{12}$ feet long or 4 yards and 2.6 feet long. Although correct, these answers would probably not be useful in practice.

2. a. $1 \text{ m} = 100 \text{ cm} = (100 \div 2.54) \text{ in} = 39.37 \text{ in} = (39.37 \div 12) \text{ ft} = 3.28$ feet. So 100 m = 328 feet.

    b. 1 mile = 5280 feet = 63,360 inches = 160934.4 cm = 1609.344 m = 1.609 km. So 70 miles per hour is 70×1.609 kilometers per hour, which is 113 kilometers per hour.

    c. $1 \text{ km} = 1000 \text{ m} = 100,000 \text{ cm} = 39370.08 \text{ in} = 3280.84 \text{ ft} = .621$ miles. Therefore, 1 square kilometer is $.621 \times .621 = .39$ square miles, and 2.4 square kilometers are $2.4 \times .39$ square miles, which is .9 square miles.

    d. 1 yd = 3 ft = 36 in = 91.44 cm = .9144 m. So, 1 cubic yard is $.9144 \times .9144 \times .9144$ cubic meters = .76 cubic meters.

    e. 1 m = 100 cm = 39.37 in = 3.28 feet. So 1.88 meters = 6.17 feet, which is 6 feet, 2 inches.

    Why is .17 feet equal to 2 inches? This is *not* explained by rounding the .17 to 2. Instead, since 1 foot = 12 inches, .17 feet = 17×12 inches = 2.04 inches, which rounds to 2 inches. Remember that you want to keep the accuracy of your answer in line with the accuracy of the initial data, so it would be misleading to report the man's height as 6 feet, 2.04 inches.

    f. $1 \text{ km} = 1000 \text{ m} = 100,000 \text{ cm} = (100,000 \div 2.54) \text{ in} = 39,379 \text{ in} = (39,379 \div 12) \text{ ft} = 3280.8 \text{ ft} = 3280.8 \div 5280 \text{ miles} = .62$ miles.

3. The stack of doughnuts would be 273,438 miles tall, which is more than the distance to the Moon.

4. 1 liter = 1000 cubic centimeters. Convert 1 cm to feet: 1 cm = 3937 in = .0328 feet. So 1 cubic centimeter is .000035314 cubic feet, and 1 liter = .035314 cubic feet. Since each gallon is .134 cubic feet, 1 liter = $(.035314 \div .134)$ gallons = .2635 gallons, which is a little more than $\frac{1}{4}$ gallon. So a liter is a little more than a quart.

5. a. 1 mile = 5280 feet, so 1 square mile = 5280×5280 square feet = 27,878,400 square feet. Since each acre is 43,560 square feet, we want to know how many 43,560 are in 27,878,400. This is a division problem. $27,878,400 \div 43,560 = 640$. So there are 640 acres in a square mile.

    b. $1 \text{ km} = 1000 \text{ m} = 100,000 \text{ cm} = (100,000 \div 2.54) \text{ in} = 39,379 \text{ in} = (39,379 \div 12) \text{ ft} = 3280.8399 \text{ ft}$. So $1 \text{ km}^2 = 3280.8399^2 \text{ ft}^2 = 10,763,910.42 \text{ ft}^2 = (10,763,910.42 \div 43,560)$ acres = 247 acres.

6. You'll need $60 \times 130 = 7,800$ square yards of turf. Each square yard is $3 \times 3 = 9$ square feet, so you'll need $7,800 \times 9 = 70,200$ square feet of turf. (Or, notice that 60 yards = 180 feet and 130 yards = 390 feet, so you'll need $180 \times 390 = 70,200$ square feet.)

    $8 buys you 12 square feet of turf, so it will cost at least

    $$8 \times (70,200 \div 12) = \$46,800$$

for the turf. But if you lay the turf in strips across the width of the field then you will need 33 strips that are 60 yards = 180 feet long, which will cost $47,520. (The 33 strips come from the fact that each strip is 4 yards wide and $130 \div 4 = 32.5$.)

PROBLEMS FOR SECTION 10.6

1. A recipe calls for 4 ounces of chocolate. If you make 15 batches of the recipe (for a large party), then how many pounds of chocolate will you need?

   a. Solve the chocolate problem simply and explain why your method of solution makes sense in a way that children could understand.

   b. Describe a number of different correct ways to write the answer to the chocolate problem. Explain briefly why these different ways of writing the answer mean the same thing.

2. A class needs 27 pieces of ribbon, each piece 2 feet, 3 inches long. How many yards of ribbon will the class need?

   a. Solve the ribbon problem simply and explain why your method of solution makes sense in a way that children could understand.

   b. Describe a number of different correct ways to write the answer to the ribbon problem. Explain briefly why these different ways of writing the answer mean the same thing.

3. The following problem could appear in a mathematics textbook for use in elementary schools:

   A rectangle is 9 feet long and 6 feet wide. Which is greater, the perimeter of the rectangle or the area of the rectangle?

   a. Calculate the perimeter and the area of the rectangle in feet and square feet, respectively. Explain briefly.

   b. Calculate the perimeter and the area of the rectangle in yards and square yards, respectively. Explain briefly.

   c. Criticize the given textbook math problem on mathematical grounds. Why is the problem not a good problem?

4. A car is 16 feet, 3 inches long. How long is it in meters? Use the fact that 1 inch = 2.54 cm to determine your answer. Give a complete and thorough explanation of the reasoning behind your calculations. Do not use dimensional analysis.

5. A car is 415 centimeters long. How long is it in feet and inches? Use the fact that 1 inch = 2.54 cm to determine your answer. Give a complete and thorough explanation of the reasoning behind your calculations. Do not use dimensional analysis.

6. The distance between two cities is described as 260 kilometers. What is this distance in miles? Make an estimate first, explaining your reasoning briefly. Then calculate your answer in two ways: (a) using dimensional analysis and (b) using logical thinking about multiplication and division. In both cases, use the fact that 1 inch = 2.54 cm to determine your answer. Remember to round your answer appropriately: The way you write your answer should reflect its accuracy. Your answer should not be reported more accurately than your starting data.

7. In Germany, people often drive 130 kilometers per hour on the Autobahn (highway). How fast are they going in miles per hour? Make an estimate first, explaining your reasoning briefly. Then calculate your answer in two ways: (a) using dimensional analysis and (b) using logical thinking about multiplication and division. In both cases, use the fact that 1 inch = 2.54 cm to determine your answer.

8. A classroom has a floor area of 600 square feet. What is the floor area of the classroom in square meters? Describe three different correct ways to solve this area problem. Each way of solving should use the fact that 1 inch = 2.54 cm. Explain your reasoning.

9. A house has a floor area of 800 square meters. What is the floor area of this house in square feet? Describe three different correct ways to solve this area problem. Each way of solving should use the fact that 1 inch = 2.54 cm. Explain your reasoning.

10. A house has a floor area of 250 square meters. Convert the house's floor area to square feet. Describe three different correct ways to solve this conversion problem.

Each way of solving should use the fact that 1 inch = 2.54 cm. Explain your reasoning.

11. A penny is $\frac{1}{16}$ of an inch thick.

   a. Suppose you had a thousand pennies. If you made a stack of these pennies, how tall would it be? Give your answer in feet and inches (for example, 7 feet, 3 inches). Explain your reasoning.

   b. Suppose you had a million pennies. If you made a stack of these pennies, how tall would it be? Give your answer in feet and inches (for example, 7 feet, 3 inches). Would the stack be more than a mile tall or not? Explain your reasoning.

   c. Suppose you had a billion pennies. If you made a stack of these pennies, how tall would it be? Give your answer in miles. Explain your reasoning.

   d. Suppose you had a trillion pennies. If you made a stack of these pennies, how tall would it be? Give your answer in miles. Would the stack reach to the Moon, which is about 240,000 miles away? Would the stack reach to the Sun, which is about 93 million miles away? Explain your reasoning.

12. A construction company has dump trucks that hold 10 cubic yards. If the company digs a hole that is 8 feet deep, 20 feet wide, and 30 feet long, then how many dump truck loads will they need to haul away the dirt they dug out from the hole? Explain your reasoning.

13. Assuming that 1 gram of gold is worth $10, how much would 1 million dollars worth of gold weigh in pounds? Explain your reasoning.

14. Imagine that all the people on Earth could stand side by side together. How much area would be needed? Would all the people on Earth fit in Rhode Island? Explain your reasoning. Rhode Island, the smallest state in terms of area, has a land area of 1000 square miles. Assume that there are 6 billion people on Earth and assume that each person needs 4 square feet to stand in.

15. Joey has a toy car that is a 1:64 scale model of an actual car. (In other words, the length of the actual car is 64 times as long as the length of the toy car.) Joey's toy car can go 5 feet per second. What is the equivalent speed in miles per hour for the actual car? Explain.

Chapter 11

More about Area and Volume

I n this chapter we will examine area and volume in greater detail than we did in Chapter 10. We will study some of the familiar area and volume formulas, explain why these formulas are valid, and see where the formulas come from. Even though children in elementary school may not study area and volume formulas, they can still learn to work informally with some of the key principles underlying these area and volume formulas. As a teacher, you should understand how these principles explain the area and volume formulas. With this foundation, you will be better able to help your students develop a strong foundation for their future learning about area and volume.

11.1 The Moving and Combining Principles about Area

In this section we will study the two most fundamental principles that are used in determining the area of a shape or a 2-dimensional aspect of an object. These principles agree entirely with our common sense. In fact, you have probably used them without being consciously aware of it.

The "moving and combining" principles about area are as follows:

1. If you move a shape rigidly without stretching it, then its area does not change.
2. If you combine (a finite number of) shapes *without overlapping* them, then the area of the resulting shape is the sum of the areas of the individual shapes.

We can use these two principles to determine the areas of many shapes.

A common way to use the combining principle to find the area of some shape is as follows: subdivide the shape into pieces whose areas are easy to determine, and then add the areas of these pieces. The resulting sum is the area of the original shape because we can think of the original shape as the combination of the pieces.

For example, what is the surface area of the swimming pool pictured on page 472 in Figure 11.1 (a) (a bird's eye view)? To answer this question, imagine dividing the surface of

FIGURE 11.1

To Determine
the Area of the
Surface of a
Swimming
Pool, Subdivide
it into 2
Rectangles

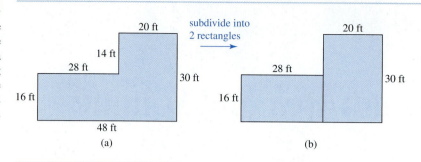

(a)   (b)

the pool into two rectangular parts, as shown in Figure 11.1 (b). We can then think of the surface of the whole pool as the combination of these two rectangular pieces. The rectangle on the left is 28 feet wide and 16 feet long, so its area is

$$28 \times 16 \text{ ft}^2 = 448 \text{ ft}^2$$

The rectangle on the right is 20 feet wide and 30 feet long, so its area is

$$20 \times 30 \text{ ft}^2 = 600 \text{ ft}^2$$

Therefore, according to the combining principle, the area of the pool is

$$448 \text{ ft}^2 + 600 \text{ ft}^2 = 1048 \text{ ft}^2$$

Another way to use the moving and combining principles to determine the area of a shape is to subdivide the shape into pieces, then move and recombine those pieces, without overlapping, to make a new shape whose area is easy to determine. By the moving and combining principles, the area of the original shape is equal to the area of the new shape.

For example, what is the area of the patio pictured in Figure 11.2 (a)? To answer this, imagine slicing off the "bump" and sliding it down to fill in the corresponding "dent," as indicated in Figure 11.2 (b). Since the "bump" was moved rigidly without stretching, and since the "bump" fills the "dent" perfectly, without any overlaps, by the moving and combining principles, the area of the resulting rectangle is the same as the area of the original patio. Since the rectangle has area

$$24 \times 60 \text{ ft}^2 = 1440 \text{ ft}^2$$

the patio also has an area of 1440 square feet.

By starting with a shape whose area we know and by subdividing the shape and recombining the pieces without overlapping to make a new shape, we can demonstrate that many different shapes can have the same area. (See Figure 11.3.)

FIGURE 11.2

Finding the
Area of a Patio
by Subdividing,
Moving, and
Recombining
Pieces

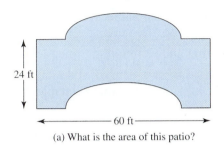

24 ft

——— 60 ft ———

(a) What is the area of this patio?

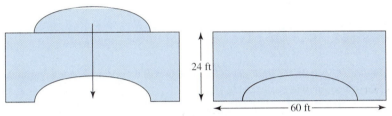

24 ft

——— 60 ft ———

(b) Subdivide, move, and recombine the pieces to form a rectangle. The area of the
rectangle is the same as the area of the original patio.

FIGURE 11.3

Children Show
35 Square
Inches in
Several Ways by
Subdividing a
Rectangle and
Recombining
the Parts
without
Overlapping

CLASS ACTIVITY   NOW TURN TO CLASS ACTIVITIES MANUAL

11A Different Shapes with the Same Area p. 333

11B Using the Moving and Combining Principles p. 335

11C Using the Moving and Combining Principles to Determine Surface Area p. 337

PRACTICE PROBLEMS FOR SECTION 11.1

1. Figure 11.4 shows the floor plan of a loft that will be getting a new wood floor. How many square feet of flooring will be needed?

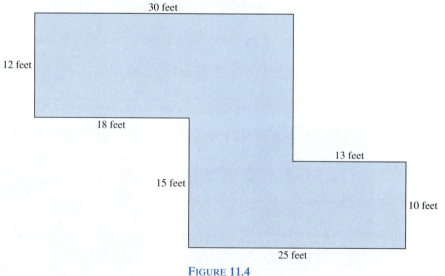

FIGURE 11.4

Floor Plan of a Loft

2. Figure 11.5 shows the floor plan of an apartment. How many square feet is the apartment? The notation 8′ stands for 8 feet.

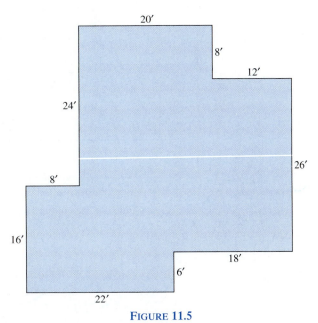

FIGURE 11.5

A Floor Plan for an Apartment

3. What is the surface area of a closed box that is 4 feet wide, 3 feet deep, and 5 feet tall? Draw a scaled-down pattern for the box to help you determine its surface area.

4. What is the area of the shape in Figure 11.6? (This is not a perspective drawing, it is a flat shape.)

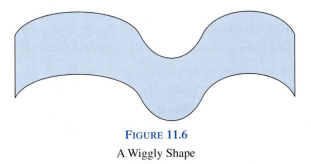

FIGURE 11.6

A Wiggly Shape

5. Use the moving and combining principles about area to determine the area of the triangle in Figure 11.7 (page 476) in *two different ways*. In both cases, do not use a formula for areas of triangles. The grid lines in Figure 11.7 are 1 cm apart.

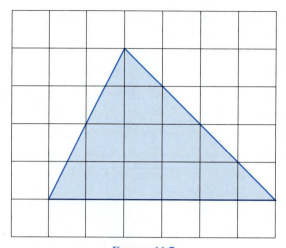

FIGURE 11.7

Find the Area of the Triangle.

6. Use the fundamental principles about area to determine the area of the octagon in Figure 11.8.

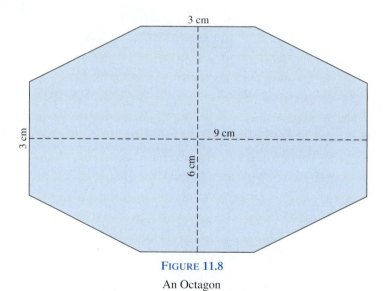

FIGURE 11.8

An Octagon

ANSWERS TO PRACTICE PROBLEMS FOR SECTION 11.1

1. First, what is the length of the unlabeled side? To answer this question, compare the "vertical" distance across the floor on the left side of the room with the right side of the room. On the left side of the room, the vertical distance across the floor is 12 feet + 15 feet = 27 feet. On the right side of the room, the vertical distance across the room is *unlabeled length* + 10 feet. The two vertical distances across the floor are equal; therefore,

$$\text{unlabeled length} + 10 = 27$$

So the unlabeled length must be 17 feet.

Notice that we can subdivide the floor into three rectangular pieces: one 30 ft by 12 ft, one 12 ft by 5 ft, and one 25 ft by 10 ft. We can think of the floor as the combination of these rectangular pieces. Therefore, according to the combining principle about area, the number of square feet of flooring needed is $30 \times 12 + 12 \times 5 + 25 \times 10 = 670$.

---

2. Figure 11.9 shows a way to subdivide the floor of the apartment into four rectangles. According to the combining principle about area, the area of the apartment is the sum of the areas of the rectangles. Therefore, the area of the apartment is

$$20 \times 8 + 32 \times 26 + 8 \times 16 + 14 \times 6 \text{ ft}^2 = 1204 \text{ ft}^2$$

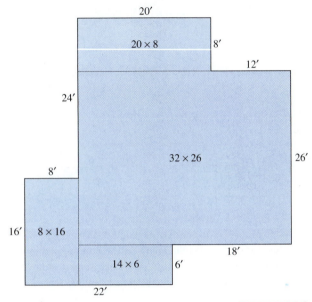

FIGURE 11.9

Subdividing the Floor into Rectangles

3. Figure 11.10 on page 478 shows a scaled down pattern for the desired box. By subdividing the pattern into rectangles and using the combining principle about areas, we see that the surface area of the box is

$$2 \times 12 + 2 \times 20 + 2 \times 15 \text{ ft}^2 = 94 \text{ ft}^2$$

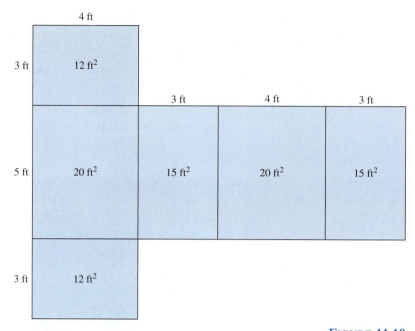

FIGURE 11.10

A Pattern for a Box

4. We can subdivide the "wiggly shape" of Figure 11.6 into pieces and recombine the pieces as shown in Figure 11.11 to make a rectangle that is 1 inch wide and $4\frac{1}{2}$ inches long. (Slice off the two "bumps" on top and the bump on the bottom, and move them to fill the corresponding "dents.") The rectangle has area $4\frac{1}{2}$ square inches, so by the moving and combining principles about area, the "wiggly shape" must also have area $4\frac{1}{2}$ square inches.

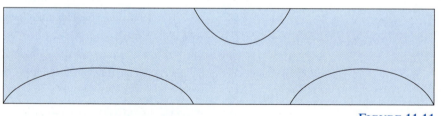

FIGURE 11.11

The Wiggly Shape Becomes a Rectangle after Subdividing and Recombining

5. Method 1: Figure 11.12 (top, left) shows how to subdivide the triangle into three pieces. Imagine rotating two of those pieces down to form a 2-cm-by-6-cm rectangle (as indicated at the top right). Since the rectangle was formed by moving portions of the triangle and recombining them without overlapping, by the moving and combining principles, the original triangle has the same area as the 2-cm-by-6-cm rectangle, namely 12 cm^2.

   Method 2: Figure 11.12 (bottom) shows how to subdivide the triangle into two smaller triangles that have areas A and B. Imagine making copies of the two triangles, rotating them, and attaching them to the original triangle to form a 4-cm-by-6-cm rectangle. Then, according to the moving and combining principles,

$$2A + 2B = 24 \text{ cm}^2$$

By the distributive property,

$$2 \cdot (A + B) = 24 \text{ cm}^2$$

Therefore,

$$A + B = 12 \text{ cm}^2$$

Since $A + B$ is the area of the original triangle, the area of the original triangle is 12 cm^2.

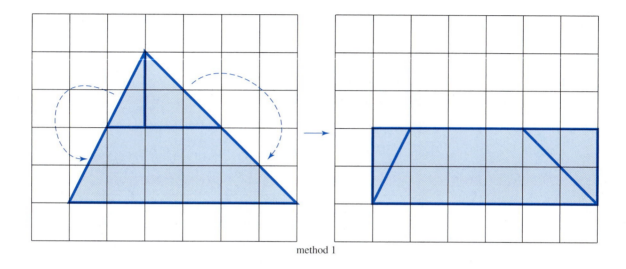

method 1

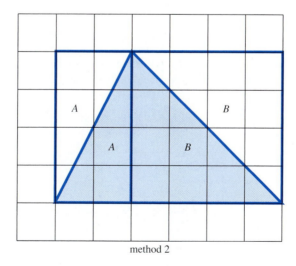

method 2

FIGURE 11.12

Two Ways to Determine the Area of the Triangle

6. The area of the octagon is 45 cm². One way to determine the area of the octagon is to think of the octagon as a 6-cm-by-9-cm rectangle with four triangles cut off, as indicated in Figure 11.13. Then, according to the combining principle about areas,

$$\text{area of octagon} + \text{area of four triangles} = \text{area of rectangle}$$

But the four triangles can be combined, without overlapping, to form two small 1.5 cm by 3 cm rectangles. So, by the previous equation, we have

$$\text{area of octagon} + 2 \times 1.5 \times 3 \text{ cm}^2 = 6 \times 9 \text{ cm}^2$$

Therefore,

$$\text{area of octagon} = 54 - 9 \text{ cm}^2 = 45 \text{ cm}^2$$

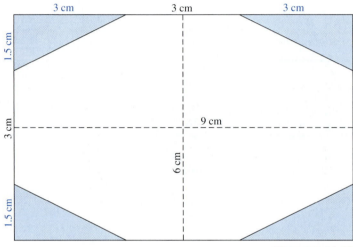

FIGURE 11.13

The Octagon is a Rectangle with Four Triangles Removed

Another way to determine the area of the octagon is to subdivide it as shown in Figure 11.14. Once again, the four triangles can be combined to make two rectangles.

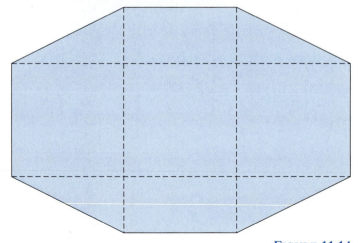

FIGURE 11.14

An Octagon

PROBLEMS FOR SECTION 11.1

1. Make a shape that has area 25 in^2 but that has no (or almost no) straight edges. Explain how you know that your shape has area 25 in^2.

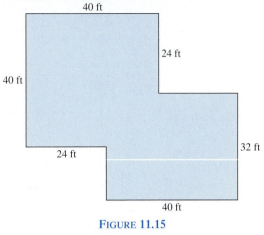

FIGURE 11.15

The Floor Plan of a House

2. If a cardboard box with a top, a bottom, and four sides is W feet wide, D feet deep, and H feet tall, then what is the surface area of this box? Find a formula for the surface area in terms of W, D, and H. Explain clearly why your formula is valid.

3. Figure 11.15 shows the floor plan for a one-story house. Calculate the area of the floor of the house. Explain your method clearly.

4. Figure 11.16 shows the floor plan for a one-story house. Calculate the area of the floor of the house. Explain your method clearly.

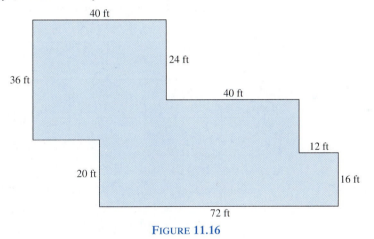

FIGURE 11.16

The Floor Plan of a House

5. Use the moving and combining principles to determine the area of the triangle in Figure 11.17 on page 482. Do not use a formula for areas of triangles. The grid lines in Figure 11.17 are 1 cm apart. Explain your reasoning.

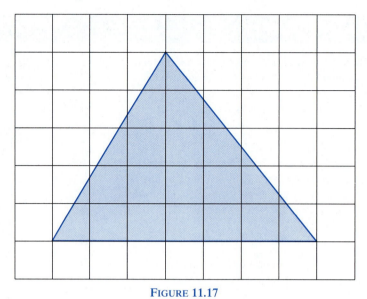

FIGURE 11.17

Determine the Area of the Triangle

6. Use the moving and combining principles to determine the area of the quadrilateral in Figure 11.18. The grid lines are 1 cm apart. Explain your reasoning.

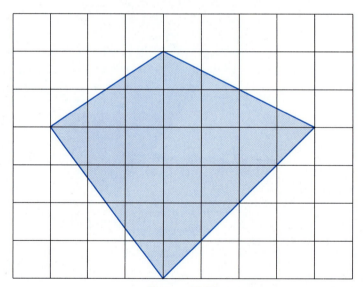

FIGURE 11.18

Determine the Area of the Quadrilateral

7. Use the moving and combining principles to determine the area of the rhombus in Figure 11.19. Adjacent dots are 1 cm apart. Explain your reasoning.

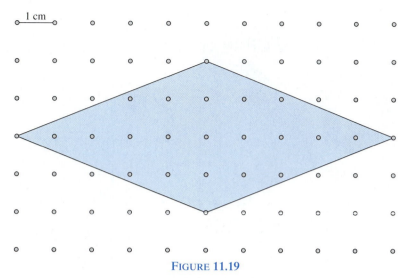

FIGURE 11.19

Determine the Area of the Rhombus

8. Use the moving and combining principles to determine the area of the rhombus in Figure 11.20. The diagonals of this rhombus are 2 inches and 5 inches long. Explain your reasoning.

FIGURE 11.20

Determine the Area of the Rhombus

9. An area problem: The Johnsons are planning to build a 5-foot-wide brick walkway around their rectangular garden, which is 20 feet wide and 30 feet long. What will the area of this walkway be? Before you solve the problem yourself, use Kaitlyn's idea:

a. Kaitlyn's idea is to "take away the area of the garden". Explain how to solve the problem about the area of the walkway using Kaitlyn's idea. Explain clearly how to apply one or both of the moving and combining principles on area in this case.

b. Now solve the problem about the area of the walkway in another way than you did in part (a). Explain your reasoning.

10. Figure 11.21 on page 484 shows a design for an herb garden, with approximate measurements. Four identical plots of land in the shape of right triangles (shown lightly shaded) are surrounded by paths (shown darkly shaded). Use the moving and combining principles to determine the area of the paths.

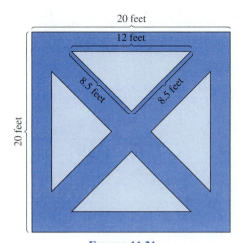

FIGURE 11.21

An Herb Garden

What must Bob have in mind? Explain why Bob's method is a legitimate way to calculate the floor area of the house, and explain clearly how one or both of the moving and combining principles on area apply in this case.

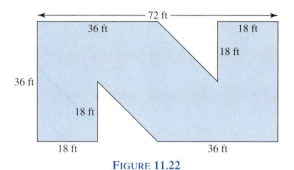

FIGURE 11.22

A Floor Plan for a Modern House

11. Figure 11.22 shows the floor plan for a modern, one-story house. Bob calculates the area of the floor of the house this way:

$$36 \times 72 - 18 \times 18 = 2268 \text{ ft}^2$$

12. a. Use the moving and combining principles to determine the areas of the 3 lightly shaded triangles in Figure 11.23. (Do not use a formula for areas of triangles.) The grid lines are 1 cm apart. Explain your reasoning.

  b. Use the moving and combining principles and your results from part (a) to determine the area of the dark shaded triangle in Figure 11.23. Explain your reasoning.

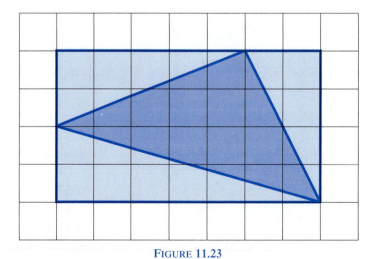

FIGURE 11.23

Four Triangles Forming a Rectangle

13. a. Use the moving and combining principles to determine the areas of the 4 lightly shaded triangles in Figure 11.24. (Do not use a formula for areas of triangles.) The grid lines are 1 cm apart. Explain your reasoning.

   b. Use the moving and combining principles and your results from part (a) to determine the area of the dark shaded quadrilateral in Figure 11.24. Explain your reasoning.

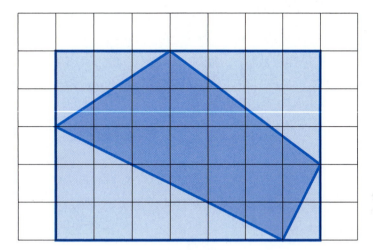

FIGURE 11.24

Four Triangles and a Quadrilateral Forming a Rectangle

14. Use the moving and combining principles about area to determine the area, in square inches, of the shaded design in Figure 11.25. The shape is a 2-inch-by-2-inch square, with a square, placed diagonally inside, removed from the middle. In determining the area of the shape, use no formulas other than the one for areas of rectangles. Explain your method clearly.

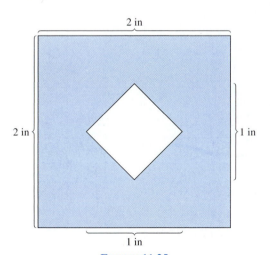

FIGURE 11.25

A Shape with a Hole

15. Use the moving and combining principles about area to determine the area, in square inches, of the shaded flower design in Figure 11.26. In determining the area of the shape, use no formulas other than the one for areas of rectangles. Explain your method clearly.

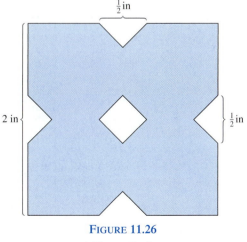

FIGURE 11.26

A Flower Design

16. Use the moving and combining principles to determine the area of the triangle in Figure 11.27. Do not use a formula for areas of triangles. Explain your method clearly. The grid lines in Figure 11.27 are 1 cm apart.

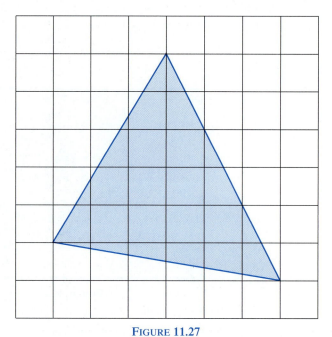

FIGURE 11.27

Determine the Area of the Triangle

17. Use the moving and combining principles to determine the area of the quadrilateral in Figure 11.28. The grid lines are 1 cm apart. Explain your reasoning.

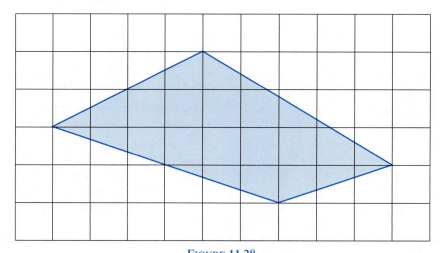

FIGURE 11.28

Determine the Area of the Quadrilateral

18. Complete the following area puzzle:

a. Trace the 8-unit-by-8-unit square in Figure 11.29.

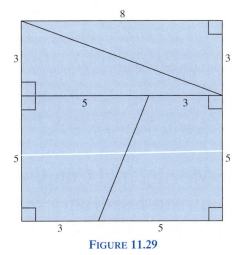

FIGURE 11.29

A Subdivided 8-by-8 Square

b. Cut out the pieces and reassemble them to form the 5-unit-by-13-unit rectangle in Figure 11.30.

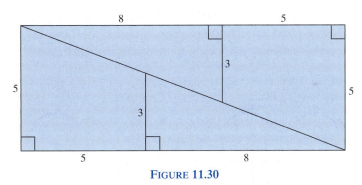

FIGURE 11.30

A Subdivided 5-by-13 Rectangle

c. Calculate the area of the square and the area of the rectangle. Is there a problem? Can the pieces from your 8-unit-by-8-unit square really fit together perfectly, without overlaps or gaps, to form a 5-unit-by-13-unit rectangle?

d. Explain the problem that you discovered in part (c). *Hints:* Can the two pieces shown in Figure 11.31 (page 488), that you cut out of the square, really fit together as shown in the rectangle to form an actual triangle? If so, wouldn't there be some similar triangles? Look at the lengths of corresponding sides of the supposedly similar triangles. Do these numbers work they way they should for similar triangles?

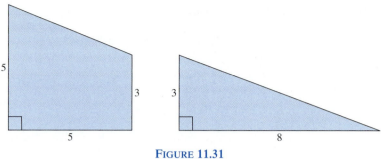

FIGURE 11.31

Two Pieces Put Together

11.2 Using the Moving and Combining Principles to Prove the Pythagorean Theorem

In this section we will use the moving and combining principles about area to prove the Pythagorean theorem. Although most children do not study the Pythagorean theorem in elementary school, the moving and combining principles, which children can learn informally in elementary school, pave the way to understanding more advanced mathematics such as the Pythagorean theorem.

The Pythagorean theorem is named for the Greek mathematician Pythagoras, who lived around 500 B.C. There is evidence, though, that this theorem was known long before that, perhaps even by the ancient Babylonians in around 2000 B.C. Pythagoras founded a secretive school, whose motto was "All is number." The members of this school, the Pythagoreans, tried to explain all of science, philosophy, and religion in terms of numbers.

The Pythagorean theorem (or Pythagoras's theorem) is a theorem about right triangles. Recall that, in a right triangle, the side opposite the right angle is called the hypotenuse. The **Pythagorean theorem** says,

Pythagorean Theorem

> In a right triangle, the square of the length of the hypotenuse is equal to the sum of the squares of the lengths of the other two sides. In other words, if c is the length of the hypotenuse of a right triangle, and if a and b are the lengths of the other two sides, then
>
> $$a^2 + b^2 = c^2 \tag{11.1}$$

For example, in Figure 11.32, the right triangle on the left has sides of lengths a, b, and c, and the hypotenuse has length c, so, according to the Pythagorean theorem,

$$a^2 + b^2 = c^2$$

The right triangle on the right of Figure 11.32 has sides of lengths d, e, and f, and the hypotenuse has length f, so, in this case, according to the Pythagorean theorem,

$$d^2 + e^2 = f^2$$

In Equation 11.1 it is understood that *all sides of the triangle are expressed in the same units*. For example, all sides can be given in centimeters, or all sides can be given in feet. If you have a right triangle where one side is measured in feet and the other in inches, for example,

FIGURE 11.32

Right Triangles

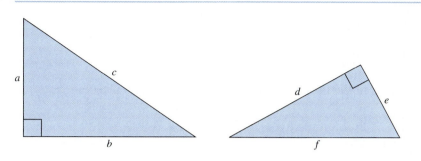

then you will need to convert both sides to inches or to feet (or to some other unit) in order to use Equation 11.1.

If you have a particular right triangle in front of you, such as those in Figure 11.32, you can measure the lengths of its sides and check that the Pythagorean theorem really is true in that case. For example, in the triangle on the left of Figure 11.32,

$$a = 3 \text{ cm}, \quad b = 4 \text{ cm}, \quad \text{and} \quad c = 5 \text{ cm}$$

so

$$a^2 + b^2 = 25, \quad \text{and} \quad c^2 = 25$$

and $a^2 + b^2$ really is equal to c^2. For the other triangle in Figure 11.32, $d = 2$ cm, $e = 4$ cm, and f is approximately 4.4 cm, so $d^2 + e^2$ is approximately 19, and f^2 is approximately 19, too. We could continue to check many right triangles to see if the Pythagorean theorem really is true in those cases. If we checked many triangles, it would be compelling evidence that the Pythagorean theorem is always true, but we could never check all the right triangles because there are an infinite number.

One of the cornerstones of mathematics, an idea that dates back to the time of the mathematicians of ancient Greece, is that a lot of evidence for a statement is not enough to know that the statement is true. *Proof* is required in order to know for sure that a statement really is true. A **proof** is a thorough, precise, logical explanation for why a statement is true, based on assumptions or facts that we already know or assume to be true. So, a proof is what establishes that a theorem is true. Proofs are one of the important aspects of this book, too, even if we don't usually call our explanations proofs.

proof

Professional and amateur mathematicians have developed literally hundreds of proofs of the Pythagorean theorem. If you do Class Activity 11F you will work through and discover one of these proofs.

CLASS ACTIVITY   NOW TURN TO CLASS ACTIVITIES MANUAL

11D   Using the Pythagorean Theorem p. 338

11E   Can We Prove the Pythagorean Theorem by Checking Examples? p. 339

11F   A Proof of the Pythagorean Theorem p. 340

PRACTICE PROBLEMS FOR SECTION 11.2

1. A garden gate that is 3 feet wide and 4 feet tall needs a diagonal brace to make it stable. How long a piece of wood will be needed for this diagonal brace? See Figure 11.33.

FIGURE 11.33
Garden Gate With
Diagonal Brace

2. A boat's anchor is on a line that is 75 feet long. If the anchor is dropped in water that is 50 feet deep, then how far away will the boat be able to drift from the spot on the water's surface that is directly above the anchor? Explain.

3. An elevator car is 8 feet wide, 6 feet deep, and 9 feet tall. What is the longest pole you could fit in the elevator? Explain.

4. Imagine a pyramid with a square base (like an Egyptian pyramid). Suppose that each side of the square base is 200 yards and that the distance from one corner of the base to the very top of the pyramid (along an outer edge) is 245 yards. How tall is the pyramid? Explain.

5. Given a right triangle with short sides of lengths a and b and hypotenuse of length c, use Figure 11.34 to explain why

$$a^2 + b^2 = c^2$$

(You may assume that all shapes that look like squares really are squares, and that in each picture, all four triangles with side lengths a, b, c are identical right triangles.)

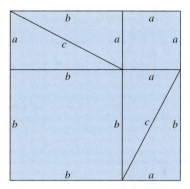 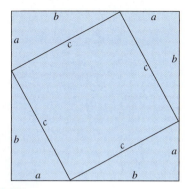

FIGURE 11.34

Shapes Forming Identical Large Squares

6. The front and back of a greenhouse have the shape and dimensions shown in Figure 11.35. The greenhouse is 40 feet long from front to back, and the angle at the top of the roof is 90°. The entire roof of the greenhouse will be covered with screening in order to block some of the light entering the greenhouse. How many square feet of screening will be needed?

6 feet

20 feet

FIGURE 11.35

A Greenhouse

ANSWERS TO PRACTICE PROBLEMS FOR SECTION 11.2

1. If we let x be the length in feet of the diagonal brace, then, according to the Pythagorean theorem,

$$3^2 + 4^2 = x^2$$

Therefore, $x^2 = 25$, so $x = 5$, which means that the diagonal brace is 5 feet long.

2. If we let x be the distance that the boat can drift away from the spot on the water's surface that is directly over the anchor, then, according to Figure 11.36 (page 492) and the Pythagorean theorem,

$$x^2 + 50^2 = 75^2$$

Therefore,

$$x^2 = 5625 - 2500 = 3125$$

so

$$x = \sqrt{3125} = 55.9$$

This tells us that the boat can drift 56 feet horizontally.

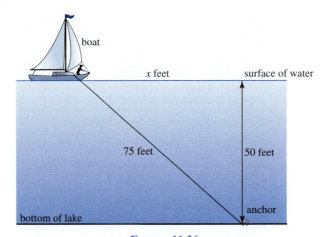

FIGURE 11.36

How Far Can an Anchored Boat Drift?

3. The length of the longest pole that will fit is the distance between point A and point C in Figure 11.37, which is about 13 feet. To determine the distance from A to C, we will use the Pythagorean theorem twice: first with the right triangle on the floor of the elevator, to determine the length of AB, and then with triangle ABC to determine the length of AC. By applying the Pythagorean theorem to the triangle on the floor of the elevator, which has short sides of lengths 8 feet and 6 feet, we conclude that $8^2 + 6^2 = AB^2$. So $AB = \sqrt{100} = 10$ feet. Notice that ABC is also a right triangle, with the right angle at B. Therefore, by the Pythagorean theorem, $AB^2 + BC^2 = AC^2$, so $AC = \sqrt{181} = 13.5$ feet. Thus, the longest pole that can fit in the elevator is about 13 feet.

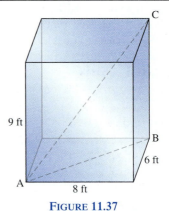

FIGURE 11.37

A Pole in an Elevator

4. The pyramid is about 200 yards tall. Here's why. Notice that the height of the pyramid is the length of BC in Figure 11.38. We will determine the length of BC by using the Pythagorean theorem twice, first with the triangle ABD, to determine the length of AB, and then with the triangle ABC, to determine the length of BC. By the Pythagorean theorem, $AD^2 + BD^2 = AB^2$, so $AB = \sqrt{20000} = 141.4$ yards. (Actually, as you'll see in the next step, we really only need AB^2, not AB, so we don't even have to calculate the square root.) The triangle ABC is a right triangle, with right angle at B. So, by the Pythagorean theorem, $141.4^2 + BC^2 = 245^2$, and it follows that $BC = \sqrt{40025} = 200$ yards. So the pyramid is about 200 yards tall.

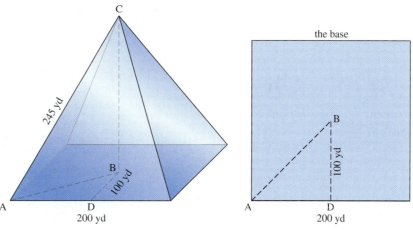

FIGURE 11.38

A Pyramid

5. One way to prove that $a^2 + b^2 = c^2$ is to imagine taking away the four triangles in each large square of Figure 11.34. Both of the large squares in Figure 11.34 have sides of length $a + b$, so both large squares have the same area. Hence, according to the moving and combining principles, if we remove the 4 copies of the right triangle from each large square, the remaining areas will still be equal. From the square on the left, two smaller squares remain, one with sides of length a and one with sides of length b. Thus, the remaining area on the left is $a^2 + b^2$. From the square on the right, a single square with sides of length c remains. Therefore, the remaining area on the right is c^2. The remaining area on the left is equal to the remaining area on the right, therefore $a^2 + b^2 = c^2$.

6. We need 1130 square feet of screening. The roof is made out of two rectangular pieces, each of which is 40 feet long and A feet wide, where A is shown in Figure 11.39. Because the angle at the top of the roof is 90°, we can use the Pythagorean theorem to determine A:

$$A^2 + A^2 = 20^2$$

Therefore, $2A^2 = 400$, so $A^2 = 200$, which means that $A = 14.14$ feet. Hence, each rectangular piece of roof needs 40×14.14 ft² = 565.69 ft² of screening. The two pieces of roof require twice as much. Rounding our answer, we see that we need about 1130 square feet of screening.

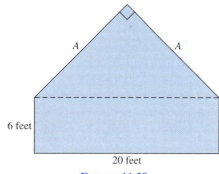

FIGURE 11.39

A Greenhouse

PROBLEMS FOR SECTION 11.2

1. Town B is 380 km due south of Town A. Town C is 460 km due east of Town B. What is the distance from Town A to Town C? Explain your reasoning. Be sure to round your answer appropriately.

2. How long a piece of ribbon will you need to stretch from the top of a 25-foot pole to a spot on the ground that is 10 feet from the bottom of the pole? Explain.

3. Rover, the dog, is on a 30-foot leash. One end of the leash is tied to Rover, who is 2 feet tall. The other end of the leash is tied to the top of a 6-foot pole. How far can Rover roam from the pole? Explain.

4. Carmina and Antone measure that the distance between the spots where they are standing is 10 feet, 7 inches. When measuring, Antone held his end of the tape measure up 4 inches higher than Carmina's end. If Carmina and Antone had measured the distance between them along the floor, would the distance be appreciably different than what they measured? Explain.

5. A company will manufacture a tent that will have a square, 15-foot × 15-foot base, four vertical walls, each 10 feet high, and a pyramid-shaped top made out of four triangular pieces of cloth, as shown in Figure 11.40. At its tallest point in the center of the tent, the tent will reach a height of 15 feet.

tent. Label your pattern with enough information so that someone making the tent would know how to measure and cut the material for the top of the tent.

6. Assuming that the Earth is a perfectly round, smooth ball of radius 4000 miles, and that 1 mile = 5280 feet, how far away does the horizon appear to be to a 5-foot-tall person on a clear day? Explain your reasoning. To solve this problem, start by drawing a picture that shows the cross section of the earth, a person standing on the surface of the earth, and the straight line of the person's gaze reaching to the horizon. (Obviously, you won't want to draw this to scale.) You will need to use the following geometric fact: If a line is *tangent* to a circle at a point P (meaning it just "grazes" the circle at the point P; it meets the circle only at that one point), then that line is *perpendicular* to the line connecting P and the center of the circle, as illustrated in Figure 11.41.

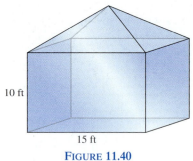

10 ft

15 ft

FIGURE 11.40

A Tent

Draw a pattern for one of the four triangular pieces of cloth that will make the pyramid-shaped top of the

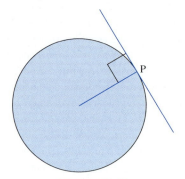

P

FIGURE 11.41

A Line Tangent to a Circle

11.3 Approximating Areas of Irregular Shapes

We can use moving and combining principles to easily find the areas of some shapes, such as those discussed, but they can't always help us find a precise area for irregular shapes. For example, how could we determine the area of the bean-shaped region in Figure 11.42? There isn't any way to subdivide and recombine the bean-shaped region into regions whose areas are easy to determine. Usually, we cannot determine the area of an irregular region exactly; instead, we must be satisfied with determining the approximate area.

One hands-on way to approximate the area of an irregular region, such as the bean-shaped region in Figure 11.42, is to cut it out, weigh it, and compare its weight to the weight of a full piece of paper. Then use proportional reasoning to find the area of the bean-shaped region. For example, if the irregular region weighs $\frac{3}{8}$ as much as a whole piece of paper, then its area is also $\frac{3}{8}$ as much as the area of the whole piece of paper.

FIGURE 11.42

A Bean-Shaped
Region

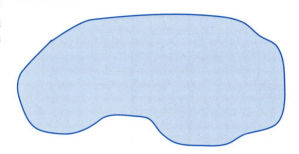

You can also determine the area of an irregular region approximately by using graph paper, as indicated in Figure 11.43. The lines on this graph paper are spaced $\frac{1}{2}$ cm apart, with heavier lines spaced 1 cm apart (so that 4 small squares make 1 square centimeter).

By counting the number of 1-cm-by-1-cm squares (each consisting of 4 small squares) inside the bean-shaped region, and by mentally combining the remaining portions of the region that are near the boundary, we can determine that the bean-shaped region has an area of about 19 square centimeters.

We can also use graph paper to find under- and overestimates for the area of an irregular region. For example, the shaded squares in Figure 11.44 lie entirely within the bean-shaped region, so their combined area must be less than the area of the region. There are 58 small squares that lie inside the bean-shaped region. Each of these small squares is $\frac{1}{2}$ cm by $\frac{1}{2}$ cm and thus has an area of

$$\frac{1}{2} \cdot \frac{1}{2} \text{ cm}^2 = \frac{1}{4} \text{ cm}^2$$

Therefore, the 58 small squares have an area of

$$58 \cdot \frac{1}{4} \text{ cm}^2 = 14\frac{1}{2} \text{ cm}^2$$

So the area of the shaded region must be greater than $14\frac{1}{2}$ cm^2.

On the other hand, the dark outline in Figure 11.44 (page 496) surrounds the squares that contain some portion of the bean-shaped region. The combined area of these squares must

FIGURE 11.43

Determining the
Area of a
Bean-Shaped
Region

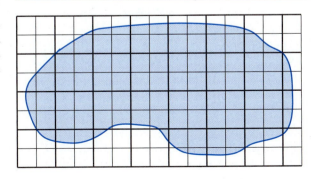

FIGURE 11.44

Determining
Under- and
Overestimates
for the Area of
the
Bean-Shaped
Region

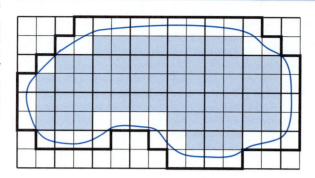

therefore be greater than the area of the region. There are 97 such small squares, each of which has area $\frac{1}{4}$ cm^2; thus, these 97 squares have a combined area of

$$97 \cdot \frac{1}{4} \ \text{cm}^2 = 24\frac{1}{4} \ \text{cm}^2$$

The area of the bean-shaped region must be less than $24\frac{1}{4}$ cm^2. So the area of the bean-shaped region must be between $14\frac{1}{2}$ cm^2 and $24\frac{1}{4}$ cm^2. If we used finer and finer graph paper we would get narrower and narrower ranges between our under- and overestimates for the area of the region.

CLASS ACTIVITY   NOW TURN TO CLASS ACTIVITIES MANUAL

11G Determining the Area of an Irregular Shape p. 343

PRACTICE PROBLEMS FOR SECTION 11.3

1. a. In Figure 11.45, the small squares in the grid are 1 cm by 1 cm. Find an underestimate and an overestimate for the area of the shaded region by considering the squares that lie entirely within the region, and the squares that contain a portion of the region (as indicated in Figure 11.44).

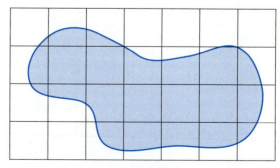

FIGURE 11.45

Find Under- and Overestimates of the Area

b. Use the grid of Figure 11.45 to determine approximately the area of the shaded region in that figure.

c. Figure 11.46 shows the same shaded region as the region in parts (a) and (b), but this time, with a finer grid. In this figure, the grid lines are $\frac{1}{2}$ cm apart. Use this finer grid to give better over- and underestimates for the area of the shaded region than you found in part (a).

d. Use the finer grid of Figure 11.46 to determine approximately the area of the shaded region.

2. Suppose that you have a map with a scale of 1 inch $= 50$ miles. You trace a state on the map onto $\frac{1}{2}$-inch graph paper. (The grid lines are spaced $\frac{1}{2}$ inch apart.) You count that the state takes up about 91 squares of graph paper. Approximately what is the area of the state? Explain.

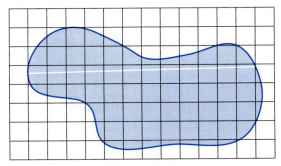

FIGURE 11.46

Find New Under- and Overestimates of the Area

ANSWERS TO PRACTICE PROBLEMS FOR SECTION 11.3

1. a. There are 5 squares that lie entirely within the region, therefore an underestimate for the area of the region is 5 square centimeters. There are 22 squares that contain some portion of the region. Therefore 22 square centimeters is an overestimate for the area of the region.

    b. Counting whole squares and combining partial squares, the area of the shaded region is about 13 square centimeters.

    c. There are 41 small squares that are contained completely within the shaded region. Each small square is $\frac{1}{2}$ cm by $\frac{1}{2}$ cm; therefore, the area of each small square is $\frac{1}{2} \cdot \frac{1}{2}$ square centimeters, which is $\frac{1}{4}$ cm². Consequently, an underestimate for the area of the region is

$$41 \cdot \frac{1}{4} \text{ cm}^2 = 10\frac{1}{4} \text{ cm}^2$$

There are 77 small squares that contain some portion of the region. Therefore, an overestimate for

the area of the region is

$$77 \cdot \frac{1}{4} \text{ cm}^2 = 19\frac{3}{4} \text{ cm}^2$$

Notice that this finer grid gives a narrower range of over- and underestimates for the area of the region.

    d. Counting whole squares and combining partial squares, we find that the area is approximately the area of 55 squares. Since each square has an area of $\frac{1}{4}$ cm², the area of the region is approximately

$$55 \cdot \frac{1}{4} \text{ cm}^2 = 13\frac{3}{4} \text{ cm}^2$$

or about 14 square centimeters.

2. Since each square on the graph paper is $\frac{1}{2}$ inch by $\frac{1}{2}$ inch, the area of each square of graph paper is $\frac{1}{2} \cdot \frac{1}{2}$ square inches, which is $\frac{1}{4}$ square inch. Therefore, 91 squares of graph paper have a combined area of

$$91 \cdot \frac{1}{4} \text{ in}^2 = 22\frac{3}{4} \text{ in}^2$$

Since the scale of the map is 1 inch = 50 miles, 1 square inch on the map represents

$$50 \cdot 50 \text{ mi}^2 = 2500 \text{ mi}^2$$

of actual land. Thus, $22\frac{3}{4}$ square inches on the map represents

$$22\frac{3}{4} \cdot 2500 \text{ mi}^2 = 56,875 \text{ mi}^2$$

or about 57,000 square miles.

PROBLEMS FOR SECTION 11.3

1. Suppose that you have a map with a scale of 1 inch = 100 miles. You trace a state on the map onto $\frac{1}{4}$-inch graph paper (the grid lines are spaced $\frac{1}{4}$ inch apart). You count that the state takes up about 80 squares of graph paper. Approximately what is the area of the state? Explain.

2. Suppose that you have a map with a scale of 1 inch = 25 miles. You cover a county on the map with a $\frac{1}{8}$-inch-thick layer of modeling dough. Then you re-form this piece of modeling dough into a $\frac{1}{8}$-inch-thick rectangle. The rectangle is $1\frac{3}{4}$ inches by $2\frac{1}{4}$ inches. Approximately what is the area of the county? Explain.

3. Suppose that you have a map with a scale of 1 inch = 30 miles. You trace a state on the map, cut out your tracing, and draw this tracing onto card stock. Using a scale, you determine that a full $8\frac{1}{2}$-inch-by-11-inch sheet of card stock weighs 10 grams. Then you cut out the tracing of the state that is on card stock and weigh this card-stock tracing. It weighs 5 grams. Approximately what is the area of the state? Explain.

11.4 Cavalieri's Principle about Shearing and Area

In addition to subdividing a shape and recombining its parts without overlapping them, there is another way, called *shearing*, to change a shape into a new shape that has the same area.

To illustrate shearing, start with a polygon, pick one of its sides, and then imagine slicing the polygon into extremely thin (really, infinitesimally thin) strips that are parallel to the chosen side. Now imagine giving those thin strips a push from the side, so that the chosen side remains in place, but the thin strips slide over, remaining parallel to the chosen side and remaining the same distance from the chosen side throughout the sliding process. Then you will have a new polygon, as indicated in Figures 11.47 and 11.48. This process of "sliding infinitesimally thin

shearing   strips" is called **shearing**.

To simulate shearing, we replace the infinitesimally thin strips with toothpicks. If you then give the stack of toothpicks a push from the side, they will slide over, as in shearing. (See Figure 11.49.)

FIGURE 11.47

Shearing a
Parallelogram

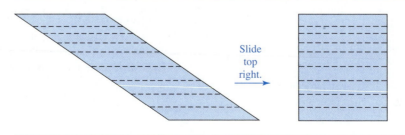

Slide
top
right.

FIGURE 11.48

Shearing a
Triangle

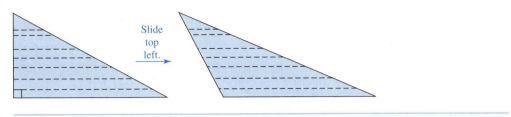

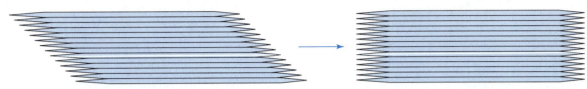

FIGURE 11.49

Shearing a Toothpick Parallelogram

Cavalieri's
principle

Cavalieri's principle for areas says that when you shear a shape as described above, the area of the original and sheared shapes are equal.

Observe the following about shearing:

1.   During shearing each point moves along a line that is *parallel* to the fixed side.

2.   During shearing, the thin strips *remain the same width and length*. The strips just slide over, they are not compressed either in width or in length.

3.   Shearing does not change the height of the "stack" of thin strips. In other words, if you think of shearing in terms of sliding toothpicks, the height of the stack of toothpicks doesn't change during shearing.

We can use shearing to determine areas of shapes. For example, to determine the area of a parallelogram, we could shear the parallelogram into a rectangle. By Cavalieri's principle, the area of the rectangle is the same as the area of the original parallelogram. So if we calculate the area of the rectangle, we will also have calculated the area of the original parallelogram.

We can also combine shearing with the moving and combining principles to determine the area of a shape. For example, to determine the area of a triangle we could first shear the triangle into a right triangle. By Cavalieri's principle, the area of the original triangle is the same as the area of the new right triangle. So if we calculate the area of the right triangle, we will also have calculated the area of the original triangle. To determine the area of the right triangle, we can combine two copies of the right triangle to make a rectangle. By the moving and combining principles, the area of the right triangle is then $\frac{1}{2}$ the area of the rectangle.

CLASS ACTIVITY   NOW TURN TO CLASS ACTIVITIES MANUAL

PRACTICE PROBLEMS FOR SECTION 11.4

1. Figure 11.50 shows a parallelogram on a pegboard. (You can think of the parallelogram as made out of a rubber band, which is hooked around four pegs.) Show two ways to move points C and D of the parallelogram to other pegs, keeping points A and B fixed, in such a way that the area of the new parallelogram is the same as the area of the original parallelogram.

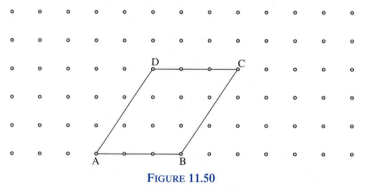

FIGURE 11.50

A Parallelogram on a Pegboard

2. Using two ordinary plastic drinking straws, cut two 4-inch pieces of straw and two 3-inch pieces of straw. Lace these pieces of straw onto a string in the following order: a 3-inch piece, a 4-inch piece, a 3-inch piece, a 4-inch piece. Tie a knot in the string so that the four pieces of straw form a quadrilateral, as pictured in Figure 11.51.

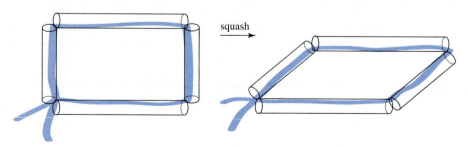

squash

FIGURE 11.51

"Squashing" a Quadrilateral Made of Straws

Put your straw quadrilateral in the shape of a rectangle. Gradually "squash" the quadrilateral so that it forms a parallelogram that is not a rectangle, as indicated in Figure 11.51. Is this "squashing" process for changing the rectangle into a parallelogram the same as the shearing process? Why or why not?

ANSWERS TO PRACTICE PROBLEMS FOR SECTION 11.4

1. See Figure 11.52. The parallelograms ABEF and ABGH are two examples of parallelograms that have the same area as ABCD. These parallelograms have the same area as ABCD by Cavalieri's principle, because they are just sheared versions of ABCD. In fact, if we move C and D anywhere along the line of pegs that goes through C and D, keeping the same distance between them, the resulting parallelogram will be a sheared version of ABCD and hence will have the same area as ABCD.

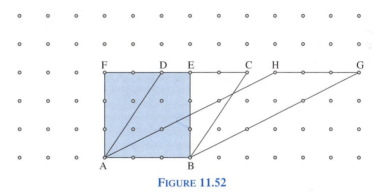

FIGURE 11.52

Three Parallelograms with the Same Area

2. No, this "squashing" process is not the same as shearing. You can tell that it's not shearing because if you made the rectangle out of very thin strips and you slid them over to make a parallelogram, they would have to become thinner in order to make the parallelogram that is formed from the straws (because the parallelogram is not as tall as the rectangle). But in the shearing process, the size of the strips does not change (either in length or in width). Therefore, "squashing" is not the same as shearing.

PROBLEMS FOR SECTION 11.4

1. Figure 11.53 on page 502 shows a triangle on a pegboard. (You can think of the triangle as made out of a rubber band which is stretched around three pegs.) Describe or show pictures of at least two ways to move point C of the triangle to another peg (keeping points A and B fixed) in such a way that the area of the new triangle is the same as the area of the original triangle. Explain your reasoning.

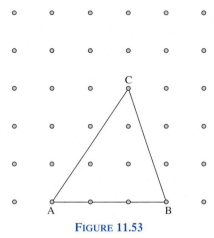

FIGURE 11.53

A Triangle on a Pegboard

2. a.   Draw a picture showing the result of shearing the parallelogram in Figure 11.54 into a rectangle. Explain how you know you have sheared the parallelogram correctly.

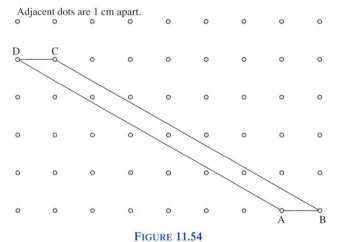

FIGURE 11.54

A Parallelogram to Shear

   b.   Use part (a) to determine the area of the parallelogram in Figure 11.54. Explain why you can determine the area of the parallelogram this way.

3. a.   Draw a picture showing the result of shearing the parallelogram in Figure 11.55 into a rectangle. Explain how you know you have sheared the parallelogram correctly. Note: shearing does not have to be horizontal.

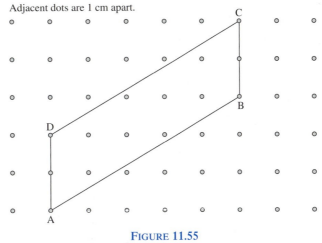

FIGURE 11.55

A Parallelogram to Shear

b. Use part (a) to determine the area of the parallelogram in Figure 11.55. Explain why you can determine the area of the parallelogram this way.

4. a. Draw a picture showing the result of shearing the triangle in Figure 11.56 into a right triangle. Explain how you know you have sheared the triangle correctly.

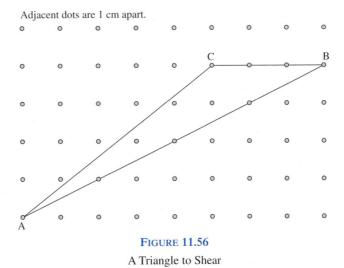

FIGURE 11.56

A Triangle to Shear

b. Use the moving and combining principles to help you determine the area of your right triangle in part (a). Explain your reasoning. (Do not use a formula for areas of triangles.)

c. Use the results of parts (a) and (b) to determine the area of the triangle in Figure 11.56. Explain why you can determine the area of the triangle this way.

5. a. Draw a picture showing the result of shearing the triangle in Figure 11.57 (page 504) into a right triangle. Explain how you know you have sheared the triangle correctly. Note: Shearing does not have to be horizontal.

Adjacent dots are 1 cm apart.

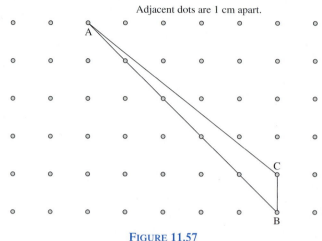

FIGURE 11.57

A Triangle to Shear

b.   Use the moving and combining principles to help you determine the area of your right triangle in part (a). Explain your reasoning. (Do not use a formula for areas of triangles.)

c.   Use the results of parts (a) and (b) to determine the area of the triangle in Figure 11.57. Explain why you can determine the area of the triangle this way.

6.   The boundary between the Johnson and the Zhang properties is shown in Figure 11.58. The Johnsons and the Zhangs would like to change this boundary so that the new boundary is one straight line segment and so that each family still has the same amount of land area. Describe a precise way to redraw the boundary between the two properties. Explain your reasoning. *Hint:* Consider shearing the triangle ABC.

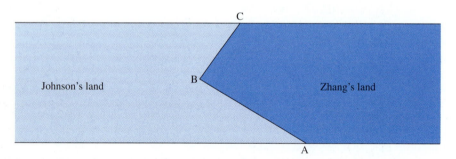

FIGURE 11.58

The Boundary between Two Properties

11.5 Areas of Triangles

So far, we have used the moving and combining principles or a combination of shearing, moving, and combining to determine the areas of various shapes, including triangles. There is another way to calculate areas of triangles namely the familiar *one half the base times the height* formula. This formula enables us to calculate areas of triangles quickly and easily. In

FIGURE 11.59

Two Ways to Select the Base (*b*) and Height (*h*) of Triangle ABC

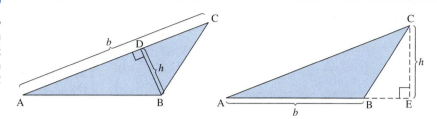

this section we will explain why this formula is valid for every triangle. For some triangles, we will be able to see why the area formula is valid by describing the triangle as "half of a rectangle" and by applying the moving and combining principles about area. For other triangles, we will use Cavalieri's principle in order to explain why the area formula is valid.

BASE AND HEIGHT FOR TRIANGLES

base

height

Before we discuss the *one half the base times the height* formula for the area of a triangle, we need to know what *base* and *height* of a triangle mean. The **base** of a triangle can be any one of its three sides. In a formula, the word *base* or a letter, such as *b*, which represents the base, really means *length of the base*.

Once a base has been chosen, the **height** is the line segment that is

1. perpendicular to the base and

2. connects the base, or an extension of the base, to the corner of the triangle that is not on the base.

In a formula, the word *height* or a letter, such as *h*, which represents the height, actually means *length of the height*.

For example, Figure 11.59 shows two copies of a triangle ABC, and two of the three ways to choose the base *b* and the height *h*. (Notice that the base and height have different lengths for the different choices.) In the second case, the height is the (dashed) line segment CE. In this case, even though the height *h* doesn't meet *b* itself, it meets an *extension* of *b*.

CLASS ACTIVITY   NOW TURN TO CLASS ACTIVITIES MANUAL

11L Choosing the Base and Height of Triangles p. 350

THE FORMULA FOR THE AREA OF A TRIANGLE

formula for the area of a triangle

The familiar **formula for the area of a triangle** with base *b* and height *h* is

$$\text{area of triangle} = \frac{1}{2}b \times h \text{ square units}$$

If a triangle has a base that is 5 inches long, and if the corresponding height of the triangle is 3 inches long, then the area of the triangle is

$$\frac{1}{2}(5 \times 3) \text{ in}^2 = 7.5 \text{ in}^2$$

In the formula

$$\frac{1}{2}b \times h$$

the base b and the height h must be described using the *same unit*. For example, both the base and the height could be in feet, or both lengths could be in centimeters. But if one length is in feet and the other is in inches, for example, then you must convert both lengths to a common unit before calculating the area. The area of the triangle is then in *square units* of whatever common unit you used for the base and height. So if the base and height are both in centimeters, then the area resulting from the formula is in square centimeters (cm^2).

WHY IS THE AREA FORMULA FOR TRIANGLES VALID?

Suppose we have a triangle, and suppose we have chosen a base b and a height h for the triangle. Why is the area of the triangle equal to one half the base times the height? In some cases, such as those shown in Figure 11.60, two copies of the triangle can be subdivided and recombined, without overlapping, to form a b-by-h rectangle. In such cases, because of the moving and combining principles about area,

$$2 \times \text{area of triangle} = \text{area of rectangle}$$

But since the rectangle has area $b \times h$,

$$2 \times \text{area of triangle} = b \times h$$

so

$$\text{area of triangle} = \frac{1}{2}b \times h$$

FIGURE 11.60

These Triangles are Half of a b-by-h Rectangle

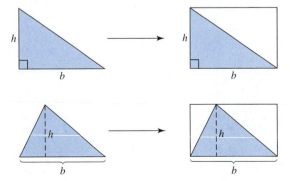

But what about the triangle in Figure 11.61? For this triangle, it's not so clear how to turn it into half of a *b*-by-*h* rectangle. However, the formula for the area of the triangle is still valid, and we can explain why by using Cavalieri's principle.

As shown in Figure 11.62, the triangle can be sheared to form a right triangle. Recall that you can think of this shearing process as slicing the triangle into (infinitesimally) thin strips and giving those strips a push to move them over. Notice that since the shearing is done parallel to the base of the triangle, the *sheared triangle still has the same base* b *and height* h *as the original triangle*.

The new right triangle has area $\frac{1}{2}b \times h$ square units since two copies of this triangle can be combined without overlapping to form a *b*-by-*h* rectangle, as shown in Figure 11.63. Remember, this is the same *b* and *h* as for the original triangle. Now, according to Cavalieri's principle, the shearing process does not change the area. So the area of the original triangle is the same as the area of the new right triangle that was formed by shearing. Therefore, the area of the original triangle in Figure 11.61 is also $\frac{1}{2}b \times h$ square units, which is what we wanted to show. Notice that although we worked with the particular triangle in Figure 11.61, there wasn't anything special about this triangle. The same argument applies to show that the $\frac{1}{2}b \times h$ formula for the area of a triangle is valid for *any* triangle.

FIGURE 11.61

Why is the Area $\frac{1}{2}b \times h$?

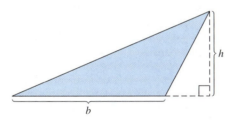

FIGURE 11.62

Shearing a Triangle

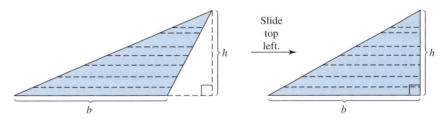

FIGURE 11.63

The Right Triangle Is Half of a Rectangle

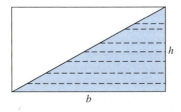

CLASS ACTIVITY   NOW TURN TO CLASS ACTIVITIES MANUAL

11M Explaining Why the Area Formula for Triangles Is Valid p. 351

PRACTICE PROBLEMS FOR SECTION 11.5

1. Show the heights of the triangles in Figure 11.64 that correspond to the bases that are labeled *b*. Then determine the areas of the triangles.

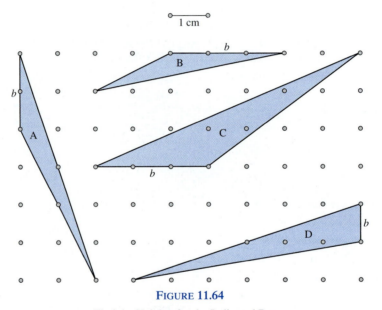

FIGURE 11.64

Find the Heights for the Indicated Bases

2. Determine the areas, in square centimeters, of the shaded regions in Figures 11.65.

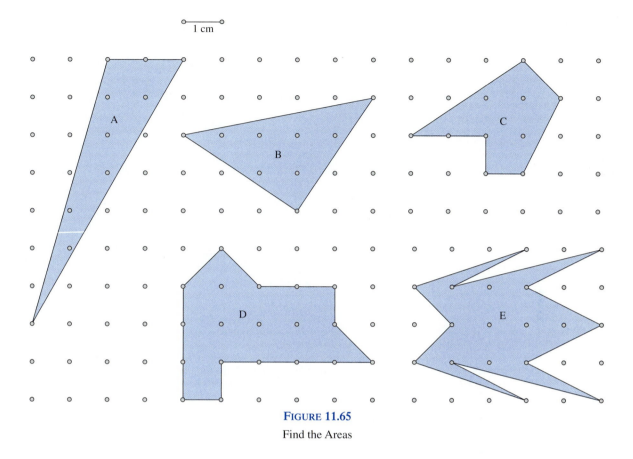

FIGURE 11.65

Find the Areas

ANSWERS TO PRACTICE PROBLEMS FOR SECTION 11.5

1. See Figure 11.66. Notice that for each of these triangles, we must extend the base in order to show the height meeting it at a right angle. Triangle A has height 2 cm and area $\frac{1}{2} \times 2 \times 2$ cm^2 = 2 cm^2. Triangle B has height 1 cm and area $\frac{1}{2} \times 3 \times 1$ cm^2 = $1\frac{1}{2}$ cm^2. Triangle C has height 3 cm and area $\frac{1}{2} \times 3 \times 3$ cm^2 = $4\frac{1}{2}$ cm^2. Triangle D has height 6 cm and area $\frac{1}{2} \times 1 \times 6$ cm^2 = 3 cm^2.

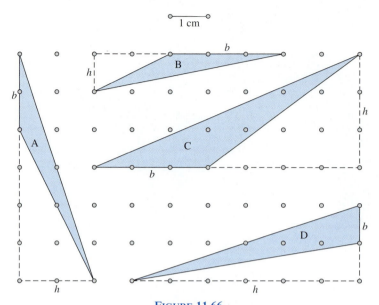

FIGURE 11.66

Bases and Heights

2. Shape A is a triangle whose base can be chosen to be the top line segment of length 2 cm. Then the height has length 7 cm, so the area of triangle A is 7 cm².

As shown in Figure 11.67, triangle B and three additional triangles combine to make a 3-cm-by-5-cm rectangle. The rectangle has area 15 cm², and the three additional triangles have areas $2\frac{1}{2}$ cm², 3 cm², and 3 cm². Therefore,

$$(\text{area of B}) + 2\frac{1}{2} + 3 + 3 \text{ cm}^2 = 15 \text{ cm}^2$$

So triangle B has area $15 - 8\frac{1}{2}$ cm² $= 6\frac{1}{2}$cm².

The areas of the remaining shaded shapes can be calculated by subdividing them into rectangles and triangles. Shape C has area $5\frac{1}{2}$ cm². Shape D has area $10\frac{1}{2}$ cm². Shape E has area 10 cm².

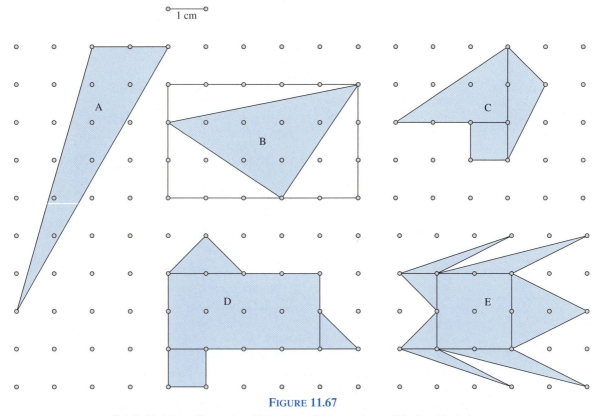

FIGURE 11.67

Subdivide These Shapes into Triangles and Rectangles to Calculate Their Areas

PROBLEMS FOR SECTION 11.5

1. For each triangle in Figure 11.68 on page 512, show the height of the triangle that corresponds to the indicated base b. Then use these bases and heights to determine the area of each triangle.

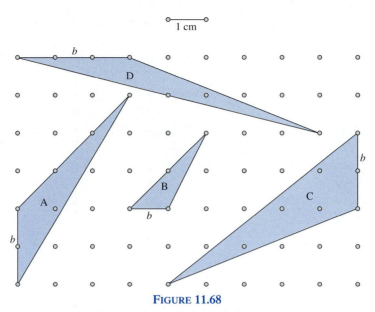

FIGURE 11.68

Determine the Heights and Areas of the Triangles

2. a. Use a ruler and compass to draw three identical triangles, having one side of length 4 inches, one side of length $2\frac{3}{4}$ inches, and one side of length $1\frac{3}{4}$ inches.

   b. For each of the three sides of the triangle you drew in part (a), let that side be the base, and draw the corresponding height of the triangle.

   c. Measure each of the three heights and use each of these measurements to determine the area of the triangle. If your three answers for the area aren't exactly the same, explain why.

3. Determine the area of the shaded shape in Figure 11.69. Explain your reasoning.

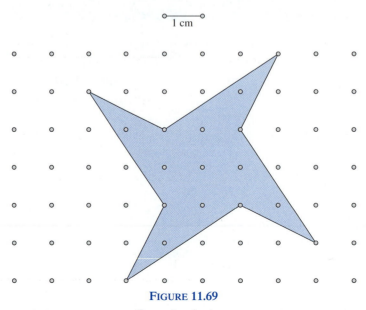

FIGURE 11.69

Determine the Area

4. Determine the area of the shaded shape in Figure 11.70. Explain your reasoning.

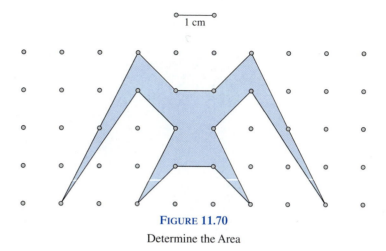

FIGURE 11.70

Determine the Area

5. Figure 11.71 shows a map of some land. Determine the size of this land in acres. Recall that one acre is 43,560 square feet.

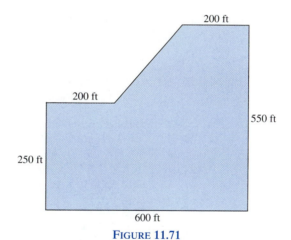

FIGURE 11.71

How Many Acres of Land?

6. Becky was asked to divide a rectangle into 4 equal pieces and to shade one of those pieces. Figure 11.72 shows Becky's solution. Is Becky right or not? Explain your answer.

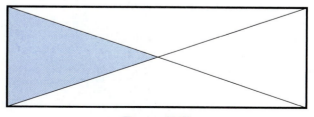

FIGURE 11.72

Four Equal Parts or Not?

7. Explain clearly in your own words why the triangle in Figure 11.73 has area $\frac{1}{2}b \times h$ for the given choice of base b and height h.

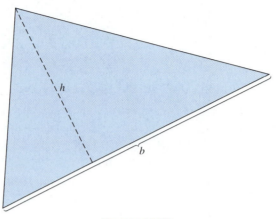

FIGURE 11.73

A Triangle

8. Explain clearly in your own words why the triangle in Figure 11.74 has area $\frac{1}{2}b \times h$ for the given choice of base b and height h.

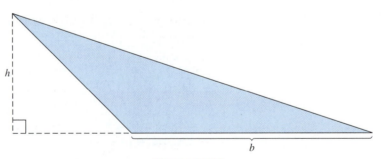

FIGURE 11.74

A Triangle

9. Class Activity 11M asked you to evaluate an incorrect explanation based on Figure 11.75 of why the area of the triangle ABC is $\frac{1}{2}bh$ for the given choice of b and h. However, it *is* possible to use Figure 11.75 and the moving and combining principles about area to give a correct explanation for why the area of the triangle is $\frac{1}{2}bh$. Use Figure 11.75 to explain why the area of the triangle ABC is $\frac{1}{2}bh$ for the given choice of b and h.

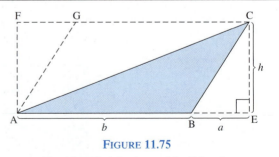

FIGURE 11.75

You *Can* Find the Area of the Triangle
by Using This Diagram

10. Use the Pythagorean theorem to help you determine the area of an equilateral triangle with sides of length 1 unit.

11. Imagine a pyramid with a square base (like an Egyptian pyramid). Suppose that the sides of the square base are all 200 yards long and that the distance from one corner of the base to the top of the pyramid (along an outer edge) is 245 yards. Determine the surface area of the pyramid (not including the base). Explain your reasoning.

11.6 Areas of Parallelograms

In this section we'll study areas of parallelograms. Since a rectangle is a special kind of parallelogram, and since the area of a rectangle is the width of the rectangle times the length of the rectangle, it would be natural to think that this same formula is true for parallelograms. But is it? Class Activity 11N examines this question.

CLASS ACTIVITY   NOW TURN TO CLASS ACTIVITIES MANUAL

11N Do Side Lengths Determine the Area of a Parallelogram? p. 353

AN AREA FORMULA FOR PARALLELOGRAMS

If you did Class Activity 11N, then you saw that unlike rectangles, it is not possible to determine the area of a parallelogram from the lengths of its sides alone. However, like triangles, there is a formula for the area of a parallelogram in terms of a base and a height.

base   As with triangles, the **base** of a parallelogram can be chosen to be any one of its four sides. In a formula, the word *base* or a letter, such as *b*, which represents the base, actually means *length of the base*.

height   Once a base has been chosen, the **height** of a parallelogram is a line segment that is

1. perpendicular to the base and

2. connects the base, or an extension of the base, to a corner of the parallelogram that is not on the base.

In a formula, the word *height* or a letter, such as *h*, which represents the height, actually means *length of the height*.

Figure 11.76 on page 516 shows a parallelogram and one way to choose a base *b* and a height *h*.

There is a very simple formula for the area of a parallelogram. If a parallelogram has a **parallelogram area formula** base that is *b* units long, and height that is *h* units long, then the area of the parallelogram is

$$b \times h \text{ square units}$$

In this formula, we assume that *b* and *h* are described with the same unit (for example, both in centimeters, or both in feet). If *b* and *h* are in different units, then you must convert them

FIGURE 11.76

One Way to
Choose the Base
and Height of
This
Parallelogram

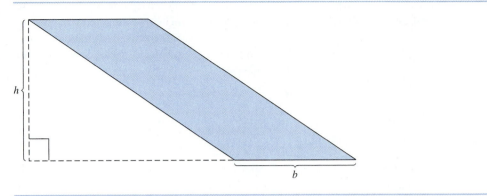

to a common unit before using the formula. For example, if a parallelogram has a base that is 2 meters long and a height that is 5 cm long, then the area of the parallelogram is

$$200 \times 5 \ \text{cm}^2 = 1000 \ \text{cm}^2$$

because 2 meters = 200 cm.

Why is the $b \times h$ formula for areas of parallelograms valid? In some cases, such as the parallelogram in Figure 11.77 (a), we can explain why the area formula is valid by subdividing the parallelogram and recombining it to form a b by h rectangle, as shown in Figure 11.77 (b). According to the moving and combining principles about area, the area of the original parallelogram and the area of the newly formed rectangle are equal. Because the newly formed rectangle has area $b \times h$ square units, the original parallelogram also has area $b \times h$ square units.

In other cases, such as in the case of the parallelogram in Figure 11.78 (a), the moving and combining principles do not apply easily for the given choice of base b and height h. In this case, as well as in all other cases, the parallelogram can be sheared to form a b-by-h rectangle, as shown in Figure 11.78 (b). According to Cavalieri's principle, the original parallelogram and the new rectangle formed by shearing have the same area. Because the rectangle has area $b \times h$ square units, the original parallelogram also has area $b \times h$ square units.

FIGURE 11.77

Subdividing and
Recombining a
Parallelogram
(a) to Make a
Rectangle (b)

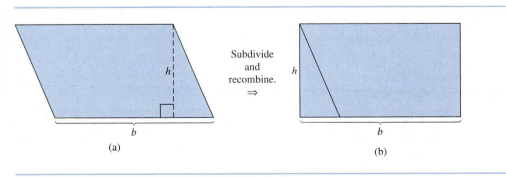

Subdivide
and
recombine.
⇒

(a)

(b)

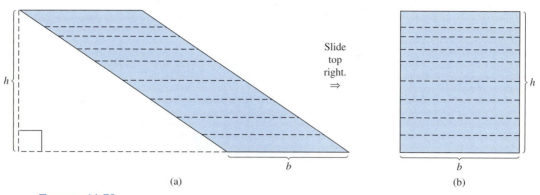

FIGURE 11.78

Shearing a Parallelogram (a) Into a Rectangle (b)

PRACTICE PROBLEMS FOR SECTION 11.6

1. Every rectangle is also a parallelogram. Viewing a rectangle as a parallelogram, we can choose a base and height for it, as for any parallelogram. In the case of a rectangle, what are other names for *base* and *height*?

2. How are the *base* × *height* formula for areas of parallelograms and the *length* × *width* formula for areas of rectangles related?

3. Why can there not be a parallelogram area formula that is expressed only in terms of the lengths of the sides of the parallelogram?

4. What is a formula for the area of a parallelogram, and why is this formula valid?

ANSWERS TO PRACTICE PROBLEMS FOR SECTION 11.6

1. In the case of a rectangle, the *base* and *height* are the *length* and *width* of the rectangle. The base can be either the length or the width.

2. The *base* × *height* formula for areas of parallelograms generalizes the *length* × *width* formula for areas of rectangles because these two formulas are the same in the case of rectangles. When a parallelogram is also a rectangle, the base can be chosen to be the length of the rectangle, and then the height is the width of the rectangle.

3. The three parallelograms in Figure 11.79 (page 518) all have sides of the same length. (See also Class Acti-

vity 11N; notice that the straw rectangle in the left photo of Figure 9.35 can be "squashed" to form the straw parallelogram in the right photo, which has a smaller area but still has the same side lengths.) If there were a parallelogram area formula that was expressed only in terms of the lengths of the sides of the parallelogram, then all three of these parallelograms would have to have the same area. But it is clear that the three parallelograms have different areas. Therefore, there can be no parallelogram area formula that is expressed only in terms of the lengths of the sides of the parallelogram.

4. See pages 515–517.

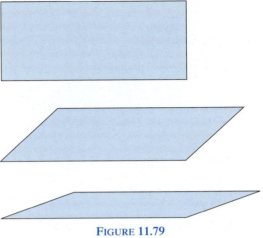

FIGURE 11.79

Three Parallelograms

PROBLEMS FOR SECTION 11.6

1. Josie has two wooden beams that are 15 feet long and two wooden beams that are 10 feet long. Josie plans to use these four beams to form the entire border around a closed garden. Josie likes unusual designs. Without any other information, what is the most you can say about the area of Josie's garden? Explain.

2. We used Cavalieri's principle to explain why a parallelogram such as the one in Figure 11.80 (I) has area $b \times h$ square units for the given choices of b and h. This problem examines some other ways to explain why the area of the parallelogram is $b \times h$ square units.

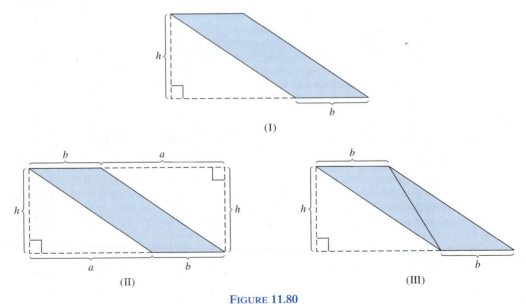

FIGURE 11.80

Why Is the Area $b \times h$?

a. Use the moving and combining principles and the large rectangle surrounding the parallelogram in Figure 11.80 (II) to explain why the area of the shaded parallelogram in Figure 11.80 (II) is $b \times h$ square units.

b. Use the area formula for triangles and the way the parallelogram is subdivided into two triangles in Figure 11.80 (III) to explain why the area of the shaded parallelogram in Figure 11.80 (III) is $b \times h$ square units.

---

3. Figure 11.81 shows a trapezoid. This problem will help you find a formula for the area of the trapezoid and explain why this formula is valid.

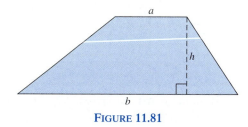

FIGURE 11.81

A Trapezoid

By subdividing the trapezoid into two triangles, as shown in Figure 11.82, find a formula in terms of a, b, and h for the area of the trapezoid and explain why your formula is valid.

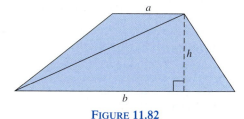

FIGURE 11.82

A Subdivided Trapezoid

4. Figure 11.81 shows a trapezoid. This problem will help you to find a formula for the area of the trapezoid, and to explain why this formula is valid.

a. Using Cavalieri's principle, explain why the trapezoid in Figure 11.81 has the same area as the new trapezoid in Figure 11.83.

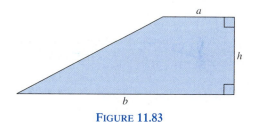

FIGURE 11.83

A Trapezoid with Two Right Angles

b. Find a formula for the area of the *new trapezoid* in Figure 11.83 in terms of a, b, and h. Explain clearly why your formula gives the area of this new trapezoid.

c. Using parts (a) and (b) of this problem, give a formula in terms of a, b and h for the area of the *original trapezoid* in Figure 11.81, and explain clearly why this formula gives the area of the original trapezoid.

11.7 Areas of Circles and the Number Pi

What is the area of a circle? You have probably seen the familiar formula

$$\pi r^2$$

for the area of a circle of radius r. The Greek letter π, pronounced "pie," stands for a mysterious number that is approximately equal to 3.14159. Since at least the time of the ancient Babylonians and Egyptians, nearly 4000 years ago, people have known about, and been fascinated by, the remarkable number π. In this section, we will discuss the number π, and then we will see why the area of a circle of radius r units is πr^2 square units.

circumference   We need some terminology first. The **circumference** of a circle is the distance around the circle. (See Figure 11.84 on page 520.) Recall that the radius of a circle is the distance from

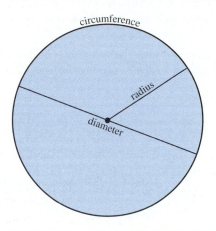

the center of the circle to any point on the circle. Recall also that the diameter of a circle is the distance across the circle, going through the center; it is twice the radius.

For any circle whatsoever, whether huge, tiny, or in between, the circumference divided by the diameter is always equal to the same number; this number is called **pi** and is written with the Greek letter π. That the circumference of a circle divided by its diameter always results in the same number can be explained by establishing that all circles are similar, so that every circle is a scaled version of one fixed circle. In scaling, both the circumference and the diameter are multiplied by the same scale factor. Therefore, when the circumference is divided by the diameter, the same value results for all circles.

pi (π)

Given any circle,

$$\text{circumference} \div \text{diameter} = \pi$$

Thus,

$$\text{circumference} = \pi \times \text{diameter}$$

Now let r stand for the radius of a circle. Because the diameter of a circle is twice its radius, we obtain the following familiar expression $2\pi r$ for the circumference of a circle of radius r:

$$\text{circumference} = \pi \times 2r = 2\pi r$$

So, for example, if you plan to make a circular garden with a radius of 10 feet, and you want to enclose the garden with a fence, then you will need

$$2\pi \times 10 \text{ ft} = 63 \text{ ft}$$

of fence.

You can do a simple experiment to determine the approximate value of π. Take a sturdy cup, can, or plate, measure the distance around the rim with a tape measure, and measure the distance across the cup, can, or plate at the widest part. When you divide these numbers you get an approximate value for π if the rim of the cup, can, or plate is a circle. But even with a big plate, an accurate tape measure, and careful measuring, you will probably only be able to determine that π is about 3.1. If you enter π on your calculator, however, you will see many more decimal places:

$$\pi = 3.14159265\ldots$$

Your calculator can only show a finite number of digits behind the decimal point, but it turns out that the decimal expansion of π goes on forever and does not have a repeating pattern. The mathematician Johann Lambert (1728–1777) first proved this in 1761.

Why is it that such a simple and perfect shape as a circle gives rise to such a mysterious and complicated number as π? Many people are attracted to mathematics because it provides a glimpse into the mysterious, the perfect, and the infinite. Although mathematics can be used to solve many practical problems, some people find the mystery and infinity in mathematics deeply appealing. It is not unlike the appeal of the greatest pieces of music, literature, and art. In this way, mathematics is not only a practical subject, but also informs us about what it means to be human, as music, literature, and art do.

How do we know the decimal expansion of the number π? There are many known formulas for π and these formulas can be used to find the decimal representation of π. One elegant formula for π comes from the following equation:

$$\frac{\pi}{4} = 1 - \frac{1}{3} + \frac{1}{5} - \frac{1}{7} + \frac{1}{9} - \frac{1}{11} + \cdots$$

This equation was discovered by Indian mathematicians in the 15th century. The ellipsis in the expression to the right of the equal sign indicates that this expression goes on forever, continuing the pattern of adding and subtracting fractions whose denominators are the odd numbers in sequence. It is beyond the scope of this book to explain why this formula is true or where it comes from.

Not only is the circumference of a circle related to the number π, but the area of a circle **circle area** is as well. A circle of radius r units has area
formula

$$\pi r^2 \text{ square units}$$

For example, suppose we want to find the area of a circular patio of diameter 30 feet. If the diameter is 30 feet, then the radius is 15 feet, and the area of the patio is

$$\pi \times 15^2 \text{ ft}^2 = \pi \times 225 \text{ ft}^2 = 707 \text{ ft}^2$$

When you use the πr^2 area formula, be sure that you only square the value of r, and not the value of πr. For example, if you multiply π times r first, and then square that result, your answer will be π times too large.

CLASS ACTIVITY   NOW TURN TO CLASS ACTIVITIES MANUAL

PRACTICE PROBLEMS FOR SECTION 11.7

1. Suppose you don't know the formula for the area of a circle, but you do know about areas of squares. What can you deduce about the area of a circle of radius r units from Figure 11.85?

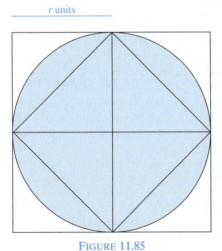

r units

FIGURE 11.85

Estimating the Area of the Circle

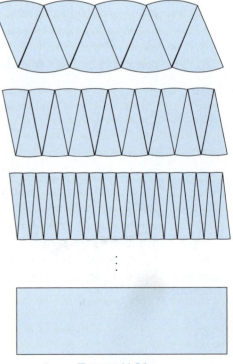

FIGURE 11.86

Rearranging a Circle

2. Using Figure 11.86 (see also Class Activity 11Q), explain why it makes sense that a circle of radius r units has area πr^2 square units, assuming we already know that a circle of radius r has circumference $2\pi r$.

3. Suppose you take a rectangular piece of paper, roll it up, and tape two ends together, without overlapping them, to make a tube. If the tube is 12 inches long and has a diameter of $2\frac{1}{2}$ inches, then what were the length and width of the original rectangular piece of paper?

4. Some trees in an orchard need to have their trunks wrapped with a special tape in order to prevent an attack of pests. Each tree's trunk is about 1 foot in diameter and must be covered with tape from ground level up to a height of 4 feet. The tape is 3 inches wide. Approximately how long a piece of tape will be needed for each tree?

5. The Browns plan to build a 5-foot-wide garden path around a circular garden of diameter 25 feet, as shown in Figure 11.87. What is the area of the garden inside the path? What is the area of the garden path? Explain your answers.

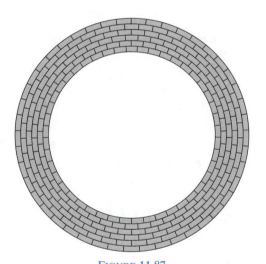

FIGURE 11.87

A Garden Path

6. Find a formula for the surface area of a cylinder of radius r units and height h units. Include the top and bottom of the cylinder.

7. What is the area of the four-petal flower in Figure 11.88? The square is 6 cm by 6 cm.

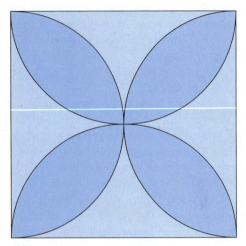

FIGURE 11.88

A Flower

8. A cone is to be made from a circle of radius 3 cm (for the base) and a quarter-circle (for the slanted portion). Determine the radius of the quarter-circle.

ANSWERS TO PRACTICE PROBLEMS FOR SECTION 11.7

1. We can conclude that the area of the circle is between $2r^2$ square units and $4r^2$ square units. The area of the circle is less than $4r^2$ because we can think of the outer square in Figure 11.85 as made up of four r-by-r squares. Since the circle lies completely inside this outer square, the area of the circle must be less than the area of the outer square. Therefore, the area of the circle is less than $4r^2$ square units. We can subdivide the inner square (di-amond) into four triangles, which we can recombine to make two r-by-r squares. Since the circle completely contains the inner square, the area of the circle must be greater than $2r^2$ square units. Just by eyeballing, the area of the circle looks to be roughly halfway in between these under- and overestimates, so the area of the circle looks to be roughly $3r^2$ (which is pretty close to the actual πr^2).

2. If you cut a circle into 8 pie pieces and rearrange them as at the top of Figure 11.86, then you get a shape that looks something like a rectangle. If you cut the circle into 16 or 32 pie pieces and rearrange them as in the middle of Figure 11.86, then you will get shapes that look even more like rectangles. If you could keep cutting the circle into more and more pie pieces, and keep rearranging them as before, you would get shapes that look more and more like the rectangle shown at the bottom of Figure 11.86.

   The height of the rectangle in Figure 11.86 is the radius r of the circle. To determine the width of the rectangle (in the horizontal direction), notice that in the rearranged circles at the top and in the middle of Figure 11.86, half of the pie pieces point up, and half point down. The circumference of the circle is divided equally between the top and bottom sides of the rectangle. The circumference of the circle is $2\pi r$, therefore the width (in the horizontal direction) of the rectangle is πr. So the rectangle is r units by πr units, and therefore has area $\pi r \times r = \pi r^2$ square units. Since the rectangle is basically a cut up and rearranged circle of radius r, the area of the rectangle ought to be equal to the area of the circle. Therefore, it makes sense that the area of the circle is also πr^2.

3. The two edges of paper that are rolled up make circles of diameter 2.5 inches. Therefore the lengths of these edges are $\pi \times 2.5$ inches, which is about 8 inches. The other two edges of the paper run along the length of the cylinder, so they are 12 inches long. Thus, the original piece of paper was about 8 inches by 12 inches.

4. One "wind" of tape all the way around a tree trunk makes an approximate circle. Because the tree trunk has diameter 1 foot, each wind around the trunk uses about π feet of tape. The tape is 3 inches wide, so it will take 4 winds for each foot of trunk height to be covered. Therefore, it will take 16 winds to cover the desired amount of trunk. This will use about $16 \times \pi$ ft $= 50$ ft of tape.

5. Since the diameter of the circular garden is 25 ft, its radius is 12.5 feet. The garden together with the path form a larger circle of radius $12.5 + 5$ feet $= 17.5$ feet. By the combining principle about areas,

   area of garden + area of path = area of larger circle

Therefore,

   area of path = area of larger circle − area of garden
   $$= \pi 17.5^2 - \pi 12.5^2 \text{ ft}^2$$
   $$= 471 \text{ ft}^2$$

6. The surface of the cylinder consists of two circles of radius r units (the top and the bottom) and a tube. Imagine slitting the tube open along its length and unrolling it, as indicated in Figure 11.89. The tube then becomes a rectangle. The height, h, of the tube becomes the length of two sides of the rectangle. The circumference of the tube, $2\pi r$, becomes the length of two other sides of the rectangle. Therefore, the rectangle has area $2\pi rh$. According to the moving and combining principles about area, the surface area of the cylinder is equal to sum of the areas of the two circles (from the top and bottom) and the area of the rectangle (from the tube), which is

$$2\pi r^2 + 2\pi rh \text{ units}^2$$

7. You can make the design by covering the square with four half-circles of tissue paper; then the flower petals are exactly the places where *two* pieces of tissue paper overlap. The remaining parts of the square are only covered with *one* layer of tissue paper. Therefore,

   area of four half-circles =
   $$\text{area of flower} + \text{area of square}$$

So that

area of flower
$$= \text{area of two circles} - \text{area of square}$$
$$= 18\pi - 36 \text{ cm}^2$$
$$= 20.5 \text{ cm}^2$$

8. Think about how the slanted portion of the cone would be attached to the base if you made the cone out of paper. The $\frac{1}{4}$ portion of the circle making the slanted portion of the cone must wrap completely around the base. Since the base has radius 3 cm, the circumference of the base is $2\pi \cdot 3$ cm. So if r is the radius of the larger circle, $\frac{1}{4}$ of $2\pi r$ must be equal to $2\pi \cdot 3$. In other words,

$$\frac{1}{4} \cdot 2\pi r = 2\pi \cdot 3$$

Therefore $r = 12$ and so the radius of the quarter-circle making the slanted portion of the cone is 12 cm.

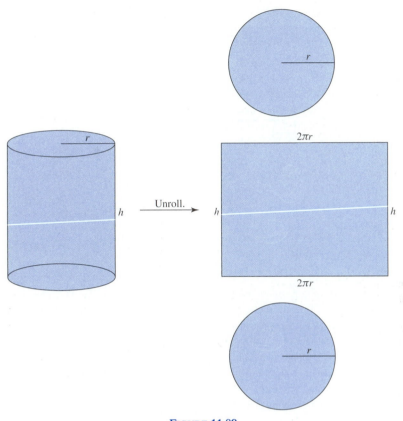

FIGURE 11.89

Taking a Cylinder Apart

PROBLEMS FOR SECTION 11.7

1. A large running track is constructed to have straight sections and two semicircular sections with dimensions given in Figure 11.90 on p. 526. Assume that runners always run on the inside line of their track. A race will consist of one full counterclockwise revolution around the track plus an extra portion of straight segment, to end up at the finish line shown. What should the distance x between the two starting blocks be in order to make a fair race?

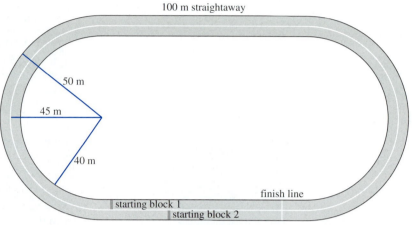

100 m straightaway

50 m

45 m

40 m

finish line

starting block 1

starting block 2

FIGURE 11.90

A Track

2. Suppose you have a large spool used for winding rope (just like a spool of thread), such as the ones shown in Figures 11.91 and 11.92. Suppose that the spool is 1 meter long and has an inner diameter of 20 cm and an outer diameter of 60 cm. Approximately how long a piece of 5-cm-thick rope can be wound onto this spool? (Assume that the rope is wound on neatly, in layers. Each layer will consist of a row of "winds," and each "wind" will be approximately a circle.)

FIGURE 11.91

A Large Spool of Wire

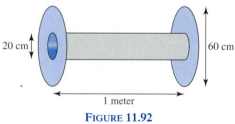

20 cm

60 cm

1 meter

FIGURE 11.92

A Large Spool

3. A city has a large cone-shaped Christmas tree that stands 20 feet tall and has a diameter of 15 feet at the bottom. The lights will be wound around the tree in a spiral, so that each "row" of lights is about 2 feet higher than the previous row of lights. Approximately how long a strand of lights (in feet) will the city need? To answer this, it might be helpful to think of each "row" (one "wind") of lights around the tree as approximated by a circle.

4. Jack has a truck that requires tires that are 26 inches in diameter. (Looking at a tire from the side of a car, a tire looks like a circle. The diameter of the tire is the diameter of this circle.) Jack puts tires on his truck that are 30 inches in diameter.

   a. A car's speedometer works by detecting how fast the car's tires are rotating. Speedometers do not detect how big a car's tires are. When Jack's speedometer reads 60 miles per hour is that accurate, or is Jack actually going slower or faster? Explain your reasoning. An exact determination of Jack's speed is not needed.

   b. Determine Jack's speed when his speedometer reads 60 mph. Explain your answer thoroughly.

5. The text gives the formula

$$\frac{\pi}{4} = 1 - \frac{1}{3} + \frac{1}{5} - \frac{1}{7} + \frac{1}{9} - \frac{1}{11} + \cdots$$

This formula can be used to give better and better approximations to π by using more and more terms from the expression to the right of the equal sign. Here's how this works. Read the symbol $\approx$ as *is approximately equal to*.

One term: $\frac{\pi}{4} = 1$, so $\pi \approx 4$

Two terms: $\frac{\pi}{4} = 1 - \frac{1}{3}$, so $\pi \approx 2.667$

Three terms: $\frac{\pi}{4} = 1 - \frac{1}{3} + \frac{1}{5}$, so $\pi \approx 3.467$

a. Use four, five, six, seven, and eight terms of the expression to the right of the equal sign above to find five more approximations to π. Also find the 20th and the 21st approximations to π.

b. Looking at the three examples given and your results in part (a), describe a pattern to the approximations to π obtained by this method. (Look at the sizes of your answers.) How do these approximations compare to the actual value of π?

c. You can get better approximations of π by taking averages of successive approximations. For example, the average of the first and second approximations to π is

$$\frac{4 + 2.667}{2} = 3.334$$

and the average of the second and third approximations to π is

$$\frac{2.667 + 3.467}{2} = 3.067$$

Find the average of the 3rd and 4th approximations to π, the 4th and 5th approximations, the 5th and 6th, the 6th and 7th and the 7th and 8th approximations to π. Also find the average of the 20th and the 21st approximations. How close is this last average to the actual value of π?

6. Tim works on the following exercise:

For each radius r, find the area of a circle of that radius:

$r = 2$ in, $r = 5$ ft, $r = 8.4$ m

Tim gives the following answers:

39.48,   246.74,   696.399

Identify the errors that Tim has made. How did Tim likely calculate his answers? Discuss how to correct the errors, including a discussion on the proper use of a calculator in solving Tim's exercise. Be sure to discuss the appropriate way to write the answers to the exercise.

7. A popular brand of soup comes in cans that are $2\frac{5}{8}$ inches in diameter and $3\frac{3}{4}$ inches tall. Each such can has a paper label that covers the entire side of the can (but not the top or the bottom).

a. If you remove a label from a soup can you'll see that it's made from a rectangular piece of paper. How wide and how long is this rectangle (ignoring the small overlap where two ends are glued together)? Use mathematics to solve this problem, even if you have a can to measure.

b. Ignoring the small folds and overlaps where the can is joined, determine how much metal sheeting is needed to make the entire can, including the top and bottom. Use the moving and combining principles about area to explain why your answer is correct. Be sure to use an appropriate unit to describe the amount of metal sheeting.

8. Let r units denote the radius of each circle in Figure 11.93. For each shaded circle portion in Figure 11.93, find a formula for its area in terms of r. Use the moving and combining principles about areas to explain why your formulas are valid.

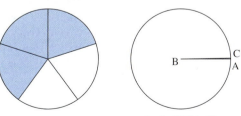

Angle ABC is 1°.   Angle DEF is 107°.

FIGURE 11.93

Parts of Circles

9. a. Make a pattern for a cone such that the slanted portion of the cone is made from a portion of a circle of radius 4 inches and such that the base of the cone is a circle of radius 3 inches. Show any relevant calculations, explaining your reasoning.

    b. Determine the total surface area (including the base) of your cone in part (a).

10. Suppose that when pizza dough is rolled out it costs 25 cents per square foot, and that sauce and cheese, when spread out on a pizza, have a combined cost of 60 cents per square foot. Let's say sauce and cheese are always spread out to within one inch of the edge of the pizza. Compare the sizes and costs of a circular pizza of diameter 16 inches and a 10-inch-by-20-inch rectangular pizza.

11. Lauriann and Kinsey are in charge of the annual pizza party. In the past, they've always ordered 12-inch-diameter round pizzas, and each 12-inch pizza has always served 6 people. This year, the jumbo 16-inch-diameter round pizzas are on special, so Lauriann and Kinsey decide to get 16-inch pizzas instead. Lauriann and Kinsey think that since a 12-inch pizza serves 6 (which is half of 12), a 16-inch pizza should serve 8 (which is half of 16). But when Lauriann and Kinsey see a 16-inch pizza, they think it ought to serve even more than 8 people. Suddenly, Kinsey realizes the flaw in their reasoning that a 16-inch pizza should serve 8. Kinsey has an idea for determining how many people a 16-inch pizza will serve. What mathematical reasoning might Kinsey be thinking of, and how many people should a 16-inch-diameter pizza serve if a 12-inch-diameter pizza serves 6? (See Figure 11.94.)

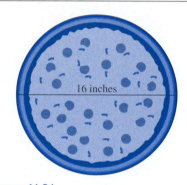

FIGURE 11.94

A 12-Inch-Diameter Pizza and a 16-Inch-Diameter Pizza

12. Penguins huddle together to stay warm in very cold weather. Suppose that a certain type of penguin has a circular cross section approximately 14 inches in diameter (so that if you looked down on the penguin from above, the shape you would see would be a circle, 14 inches in diameter). Suppose that a group of this type of penguin is huddling in a large circular cluster, about 20 feet in diameter. (All the penguins are still standing upright on the ground; they are not piled on top of each other.)

    a. Assuming that the penguins are packed together tightly, estimate how many penguins are in this cluster. (You might use areas to do this.) Is this an overestimate or an underestimate?

    b. The coldest penguins in the cluster are the ones around the circumference. Approximately how many of these cold penguins are there at any given time?

    c. So that no penguin gets too cold, the penguins take turns being at the circumference. How many minutes per hour does each penguin spend at the circumference if each penguin spends the same amount of time at the circumference?

11.8 Relating the Perimeter and Area of a Shape

If you know the distance around a shape, can you determine its area? If you know the area of a shape, can you determine the distance around the shape? We will study these questions in this section.

Recall that the perimeter of a shape is the distance around a shape. For relatively small shapes, you can determine the perimeter by placing a piece of string around the shape and cutting the string so that it goes around exactly one time. The length of the string is then the perimeter of the shape. For example, the perimeter of a circle is its circumference. The shape in Figure 11.95 has perimeter

$$4 + 6 + 2 + 3 + 5 + 1 + 3 + 2 \text{ cm} = 26 \text{ cm}$$

Note the difference between perimeter and area. The shape in Figure 11.95 has perimeter 26 cm, but it has area 21 cm^2. The perimeter of a shape is described by a unit of length, such as centimeters, whereas the area of a shape is described by a unit of area, such as square centimeters. The perimeter of the shape in Figure 11.95 is the number of 1-cm segments it takes to go all the way *around the shape*, whereas the area of the shape in Figure 11.95 is the number of 1-cm-by-1-cm squares it takes to *cover the shape*.

CLASS ACTIVITY   NOW TURN TO CLASS ACTIVITIES MANUAL

11S How Are Perimeter and Area Related? p. 361

If you did Class Activity 11S, then you discovered that for a given, fixed perimeter, there are many shapes that can have that perimeter, and these shapes can have different areas. By

FIGURE 11.95

A Shape of Perimeter 26 cm and Area 21 cm^2

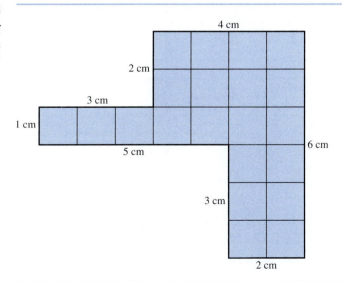

placing a loop of string on a flat surface, and by arranging the loop in different ways, you can show various shapes that have the same perimeters but different areas. (See Figure 11.96.) Therefore, perimeter does not determine area. But, for a given, fixed perimeter, which areas can occur?

Your intuition probably tells you that of all shapes having a given, fixed perimeter, the circle is the one with the largest area. For example, of all shapes with perimeter 15 inches, a circle of circumference 15 inches is the shape that has the largest area. Since the circumference of this circle is 15 inches, its radius is

$$15 \div 2\pi \text{ in} = 2.4 \text{ in}$$

and therefore its area is

$$\pi \times 2.4^2 \text{ in}^2 = 18 \text{ in}^2$$

By moving a 15-inch loop of string into various positions on a flat surface, you will probably find it plausible that *every* positive number less than 18 is the area, in square inches, of *some* shape of perimeter 15 in. So, for example, there is a shape that has perimeter 15 in and area 17.35982 in^2, and there is a shape that has perimeter 15 in and area 2.7156 in^2. Notice that we can say this *without actually finding shapes* that have those areas and perimeters, which could be quite a challenge.

The following is true in general: Among all shapes of a given, fixed perimeter P, the circle of circumference P has the largest area, and every positive number that is less than the area of that circle is the area of some shape of perimeter P. Explanations for why these facts are true are beyond the scope of this book.

What can we say about areas of *rectangles* of a given, fixed perimeter? By stretching the loop of string between four thumbtacks pinned to cardboard, and by moving the thumbtacks, you can show various rectangles that all have the same perimeter, but have different areas. If you did Class Activity 11S, then you probably discovered that of all rectangles of a given, fixed perimeter, the one with the largest area is a square. For example, among all rectangles of perimeter 24 inches, a square that has four sides of length 6 inches has the largest area, and this area is $6 \times 6 \text{ in}^2 = 36 \text{ in}^2$. By moving a 24-inch loop of string to form various rectangles,

FIGURE 11.96

Four Strings of
Equal Length
Make Shapes of
Different Areas

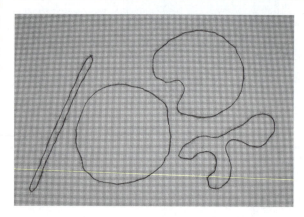

you can probably tell that *every* positive number less than 36 is the area, in square inches, of *some* rectangle of perimeter 24 inches. So, for example, there is a rectangle that has perimeter 24 in and area 35.723 in^2, and there is a rectangle that has perimeter 24 in and area 3.72 in^2, even though it would take some work to find the exact lengths and widths of such rectangles.

The following is true in general: Among all rectangles of a given, fixed perimeter P, the square of perimeter P has the largest area, and every positive number that is less than the area of that square is the area of some rectangle of perimeter P. Although we will not do this here, these facts can be explained with algebra or calculus.

CLASS ACTIVITY   NOW TURN TO CLASS ACTIVITIES MANUAL

11T Can We Determine Area by Measuring Perimeter? p. 364

PRACTICE PROBLEM FOR SECTION 11.8

1. A piece of property is described as having a perimeter of 4.7 miles. Without any additional information about the property, what is the most informative answer you can give about the area of the property? If you assume that the property is shaped like a rectangle, then what is the most informative answer you can give about its area?

ANSWER TO PRACTICE PROBLEM FOR SECTION 11.8

1. Among all shapes that have perimeter 4.7 miles, the circle with circumference 4.7 miles has the largest area. This circle has radius

$$4.7 \div 2\pi \text{ miles} = .75 \text{ miles}$$

and therefore it has area

$$\pi \times .75^2 \text{ miles}^2 = 1.8 \text{ miles}^2$$

Any area that is less than the area of this circle is a possible area for the property. So, without any additional information, the best we can say about the property is that its area is at most 1.8 square miles and the actual area could be anywhere between 0 and 1.8 square miles.

Now suppose that the property is shaped like a rectangle. Among all rectangles of perimeter 4.7 miles, the square of perimeter 4.7 miles has the largest area. This square has 4 sides of length $4.7 \div 4$ miles = 1.175 miles, and therefore this square has area 1.175 × 1.175 miles^2 = 1.4 miles^2. Any area that is less than the area of this square is a possible area for the property. So, if we know that the property is in the shape of a rectangle, then the most informative answer we can give about the area of the property is that its area is at most 1.4 square miles, and the actual area could be anywhere between 0 and 1.4 square miles.

PROBLEMS FOR SECTION 11.8

1. Anya wants to draw many different rectangles that have a perimeter of 16 centimeters. Anya draws a few rectangles, but then she stops drawing and starts looking for pairs of numbers that add to 8.

   a. Why does it make sense for Anya to look for pairs of numbers that add to 8? How is Anya likely to use these pairs of numbers in drawing additional rectangles?

   b. How could you adapt Anya's idea if you were going to draw many different rectangles that have a perimeter of 14 centimeters?

   c. Use Anya's idea to help you draw 3 different rectangles that have a perimeter of 5 inches. Label your rectangles with their lengths and widths.

2. a. On graph paper, draw 4 different rectangles that have perimeter $6\frac{1}{2}$ inches.

   b. Without using a calculator, determine the areas of the rectangles you drew in part (a). Show your calculations or explain briefly how you determined the areas of the rectangles.

3. Which of the lengths that follow could be the length of one side of a rectangle that has perimeter 7 inches? In each case, if the length is a possible side length of a rectangle of perimeter 7 inches, then determine the lengths of the other 3 sides without using a calculator. If the length is not a possible side length of such a rectangle, then explain why not. Show your calculations.

   a. $2\frac{3}{4}$ inches

   b. $5\frac{1}{2}$ inches

   c. $1\frac{7}{8}$ inches

   d. $3\frac{3}{8}$ inches

   e. $3\frac{5}{8}$ inches

4. a. Without using a calculator, find the lengths and widths of 5 different rectangles that have perimeter $4\frac{1}{2}$ inches. Show your calculations and explain them briefly.

   b. Without using a calculator, find the areas of the 5 rectangles you found in part (a). Show your calculations.

5. a. Draw 4 different rectangles, all of which have a perimeter of 8 inches. At least two of your rectangles should have side lengths that are not whole numbers (in inches). Label your rectangles with their lengths and widths.

   b. Determine the areas of each of your 4 rectangles in part (a) without using a calculator. Show your calculations, or explain briefly how you determined the areas. Then label your rectangles A, B, C, and D in decreasing order of their areas, so that A has the largest area and D has the smallest area among your rectangles.

   c. Qualitatively, how do the larger-area rectangles you drew in part (a) look different from the smaller-area rectangles? Describe how the shapes of the rectangles change as you go from the rectangle of largest area to the rectangle of smallest area.

6. a. Draw 4 different rectangles, all of which have area 4 square inches. Label your rectangles with their lengths and widths.

   b. Determine the perimeters of each of your rectangles in part (a). Then label your rectangles A, B, C, and D in increasing order of their perimeters, so that A has the smallest perimeter and D has the largest perimeter among your rectangles.

   c. Qualitatively, how do the smaller perimeter rectangles you drew in part (a) look different from the larger perimeter rectangles? Describe how the shapes of the rectangles change as you go from the rectangle of smallest perimeter to the rectangle of largest perimeter.

7. A forest has a perimeter of 210 miles, but no information is given about the shape of the forest. Justify your answers to the following (in all parts of this problem, the perimeter is still 210 miles):

   a. Is it possible that the area of the forest is 3000 square miles? Explain.

   b. Is it possible that the area of the forest is 3600 square miles? Explain.

   c. If the forest is shaped like a rectangle, then is it possible that the area of the forest is 3000 square miles? Explain.

   d. If the forest is shaped like a rectangle, then is it possible that the area of the forest is 2500 miles? Explain.

8. Bob wants to find the area of an irregular shape. He cuts a piece of string to the length of the perimeter of

the shape. He measures to see that the string is about 60 cm long. Bob then forms his string into a square on top of centimeter graph paper. Using the graph paper, he determines that the area of his string square is about 225 cm². Bob says that therefore the area of the irregular shape is also 225 cm². Is Bob's method for determining the area of the irregular shape valid or not? Explain. If the method is not valid, what can you determine about the area of the irregular shape from the information that Bob has? Explain.

9. Write a short essay discussing whether or not Nick's idea for estimating the area of an irregular shape given in Class Activity 11T provides a valid way to estimate the area of the irregular shape. If Nick's method is not valid, is there a way that he could use the string to get some sort of information about the area of the shape?

10. a. Describe a concrete way to demonstrate that many different shapes can have the same perimeter.

   b. Describe a concrete way to demonstrate that many different shapes can have the same area.

11. Consider all rectangles whose *area* is 4 square inches, including rectangles that have sides whose lengths are not whole numbers. What are the possible perimeters of these rectangles? Is there a smallest perimeter? Is there a largest? Answer this question either by pure thought or by actual examination of rectangles. Then write a paragraph describing your answer (including pictures, if relevant) and stating your conclusions clearly.

11.9   Principles for Determining Volumes

As with areas, we have fundamental principles that help us determine the volumes of various solid shapes: moving and combining principles and Cavalieri's principle. These principles for volumes are almost identical to their counterparts for areas, and we will use them in the same ways we used the principles for determining areas. We will also consider a hands-on way to determine volumes: submersing in water. When we submerse an object in water, we can get information about the volume of the object. On the other hand, we will see that when an object *floats* in water, we can get information about the *weight* of the object.

THE MOVING AND COMBINING PRINCIPLES ABOUT VOLUMES

As with shapes in a plane, there are fundamental principles about how volumes behave when solid shapes are moved or combined:

1. If you move a solid shape rigidly without stretching or shrinking it, then its volume does not change.

2. If you combine (a finite number of) solid shapes *without overlapping* them, then the volume of the resulting solid shape is the sum of the volumes of the individual solid shapes.

We have already used these principles implicitly, in explaining why we can determine the volume of a box by multiplying its height times its width times its depth. We thought of the box as subdivided into layers, and each layer as made up of 1-unit-by-1-unit-by-1-unit cubes. Each small cube has volume 1 cubic unit, and the volume of the whole box (in cubic units) is the sum of the volumes of the cubes, which is just the number of cubes.

If you have a solid lump of clay, then you can mold it into various different shapes. Each of these different shapes is made of the same volume of clay. From the point of view of the moving and combining principles, it is as if the clay had been subdivided into many tiny pieces, and then these tiny pieces were recombined in a different way to form a new shape; therefore, the new shape is made of the same volume of clay as the old shape.

Similarly, if you have water in a container and you pour the water into another container, the volume of water stays the same, even though its shape changes.

CAVALIERI'S PRINCIPLE ABOUT SHEARING AND VOLUMES

As with a shape in a plane, we can shear a solid shape and obtain a new solid shape that has the same volume.

To illustrate the shearing of a solid shape, start with a polyhedron, pick one of its faces, and then imagine slicing the polyhedron into extremely thin (really, infinitesimally thin) slices that are parallel to the chosen face—this is rather like slicing a salami with a meat slicer. Now imagine giving those thin slices a push from the side, so that the chosen slice remains in place, but so that the other thin slices slide over, remaining parallel to the chosen face, and remaining the same distance from the chosen slice throughout the sliding process. Then you will have a new solid shape, as indicated in Figure 11.97. This process of "sliding infinitesimally thin slices" is called **shearing**.

You can show shearing nicely with a stack of paper. Give the stack of paper a push from the side, so that the sheets of paper slide over as shown in Figure 11.97. In order to understand shearing of solid shapes, it may help you to think of the thin slices as made out of paper.

Note that in the shearing process, each thin slice remains unchanged: each slice is just slid over, and is not compressed or stretched.

Cavalieri's principle for volumes says that when you shear a solid shape as described previously, the volume of the original and sheared solid shapes are equal.

shearing

Cavalieri's principle

SUBMERSING AND VOLUME, FLOATING AND WEIGHT

A hands-on way to determine the volume of an object is to submerse the object in a known volume of water and measure how much the water level goes up. For example, if you have a measuring cup filled with 300 mL of water and if the water level goes up to 380 mL when you put a plastic toy that sinks in the water, then the volume of the toy is 80 mL. Recall that 1 cubic centimeter holds 1 mL of liquid, so we can also say that the volume of the toy is 80 cm^3. Why does it make sense that the volume of the toy is 80 mL or 80 cm^3? When you place the toy in the water, the toy takes up space where water had been. That water that had been where the

FIGURE 11.97

Shearing a Stack of Paper

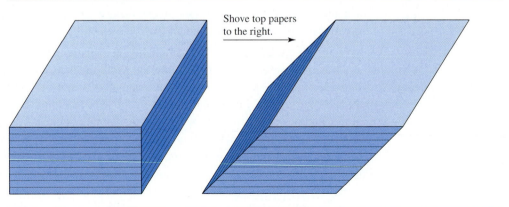

Shove top papers to the right.

toy is now, is the "extra" water at the top of the measuring cup. So the volume of this "extra" water at the top of the measuring cup is the volume of the toy.

But what if an object is not *submersed* in water but *floats* on the water instead? For example, suppose you have a measuring cup filled with 300 mL of water and suppose that when you put a toy in the water, the toy floats, and the water level goes up to 380 mL. In this case, the amount of water that the toy displaced does not have the same *volume* as the toy, but instead has the same *weight* as the toy. This fundamental physical fact, that *an object that floats displaces an amount of water that weighs as much as the object*, is called **Archimedes' principle**, in honor of Archimedes, the great mathematician and physicist who lived in ancient Greece, 287–212 B.C.

Archimedes' principle

Using Archimedes' principle, you can understand why some objects float. Think about gradually lowering an object into water. Will the object float or not? As you lower the object into the water, it displaces more and more water. At some point, the object may have displaced a volume of water that weighs as much as the object. If so, then the object will float at that point. But if you lower an object into water, and if the amount of water it displaces never weighs as much as the object, then the object will sink. Notice that floating is not just a matter of how *light* the object is: it has to do with the *shape* of the submersed part of the object, and whether this shape displaces enough water. Otherwise heavy ships made of steel would never be able to float.

CLASS ACTIVITY NOW TURN TO CLASS ACTIVITIES MANUAL

11U Determining a Volume by Submersing in Water p. 365

11V Floating versus Sinking: Archimedes' Principle p. 367

PRACTICE PROBLEMS FOR SECTION 11.9

1. A measuring cup contains 300 mL of water. When a ball is put into the water in the measuring cup, the ball sinks to the bottom and the water level rises to 400 mL.
   a. What, if anything, can you deduce about the volume of the ball?
   b. What, if anything, can you deduce about the weight of the ball?

2. Suppose you have a paper cup floating in a measuring cup that contains water. When the paper cup is empty, the water level in the measuring cup is at 250 mL. When you put some flour into the measuring cup, the cup is still floating and the water level in the measuring cup goes up to 350 mL. What information about the flour in the measuring cup can you deduce from this experiment? Explain.

1. a. Since the water level rose 100 mL, the ball displaced 100 mL of water. One hundred milliliters has a volume of 100 cubic centimeters because 1 mL has a volume of 1 cm³. So the ball has a volume of 100cm³.

    b. If the ball were *floating*, then we would be able to say that the ball weighs 100 grams, because according to Archimedes' principle, a floating body displaces an amount of water that weighs as much as the body. But because the ball is not floating, we cannot determine the exact weight of the ball from this experiment. However, we can say that the ball must weigh more than 100 grams. Here's why. Think of gradually lowering the ball into the water. When we lower the ball into the water there can never be a time when the amount of water that the ball displaces weighs as much as the ball—otherwise the ball would float according to Archimedes' principle. So the amount of water that the ball displaces as it is lowered into the water must always weigh less than the ball. Since the ball displaces 100 mL of water, and since 100 mL of water weighs 100 g, the ball must weigh more than 100 g.

2. Since the water level rose from 250 mL to 350 mL, the flour floating in the cup displaced 100 mL of water. Because the cup with the flour in it is floating, the weight of the displaced water is equal to the weight of the flour according to Archimedes' principle. Because 1 mL of water weighs 1 gram, 100 mL of water weighs 100 g, so the flour weighs 100 g.

PROBLEMS FOR SECTION 11.9

1. Suppose that you have a recipe that calls for 200 grams of flour, but you don't have a scale. Explain in detail how to apply Archimedes' principle to measure 200 grams of flour using a measuring cup that has metric markings.

2. Suppose that you have a recipe that calls for $\frac{1}{2}$ pound of flour, but you don't have a scale. Explain in detail how to apply Archimedes' principle and the fact that 1 kilogram $= 2.2$ pounds to measure $\frac{1}{2}$ pound of flour using a measuring cup that has metric markings. As a point of interest, we often say that "a pint is a pound"; therefore, you might think that 1 cup of flour (which is $\frac{1}{2}$ of a pint) weighs $\frac{1}{2}$ pound. However, the "pint is a pound" rule applies to water and to similar liquids like milk and clear juices. But ordinary flour is less dense than water, so 1 cup of flour actually weighs less than $\frac{1}{2}$ pound.

11.10 Volumes of Prisms, Cylinders, Pyramids, and Cones

In this section we will study volume formulas for prisms, cylinders, pyramids, and cones, and we will see why these formulas make sense.

VOLUME FORMULAS FOR PRISMS AND CYLINDERS

height (of prism or cylinder)

Before introducing the simple volume formula for prisms and cylinders, recall that prisms and cylinders can be thought of as formed by joining two parallel, congruent bases. The **height** of a prism or cylinder is the distance between the planes containing the two bases of the prism or cylinder, measured in the direction perpendicular to the bases, as indicated in Figure 11.98. *The height is measured in the direction perpendicular to the bases, not on the slant.*

prism and cylinder volume formula

There is a very simple formula for volumes of prisms and cylinders:

$$(\text{height}) \times (\text{area of base})$$

FIGURE 11.98

Volume of Prism
or Cylinder =
(Height) ×
(Area of Base)

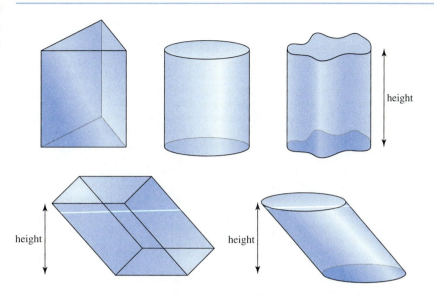

In the volume formula it is understood that if the height is measured in some unit, then the area of the base is measured in square units of the same unit. The volume of the prism or cylinder resulting from the formula is then in cubic units of the same basic unit.

For example, what is the volume of a 4-inch-tall can that has a circular base of radius 1.5 inches? The area of the base is $\pi (1.5)^2$ in^2 = 7.07 in^2; therefore, the volume of the can is 4×7.07 in^3 = 28 in^3.

CLASS ACTIVITY   NOW TURN TO CLASS ACTIVITIES MANUAL

11W Why the Volume Formula for Prisms and Cylinders Makes Sense p. 369

11X Using Volume Formulas p. 372

11Y Filling Boxes and Jars p. 373

VOLUME FORMULAS FOR PYRAMIDS AND CONES

Before introducing the volume formula for pyramids and cones, recall that pyramids and cones can be thought of as formed by joining a base with a point. The **height** of a pyramid or cone is the perpendicular distance between the point of the pyramid or cone and the plane containing the base. As indicated in Figure 11.99 on page 538, *the height is measured in the direction perpendicular to the base, not on the slant.*

height (of pyramid or cone)

pyramid and cone volume formula

The formula for volumes of pyramids and cones is

$$\frac{1}{3}(\text{height}) \times (\text{area of base})$$

FIGURE 11.99

Volume of
Pyramid or
Cone =
$\frac{1}{3}$(Height) ×
(Area of Base)

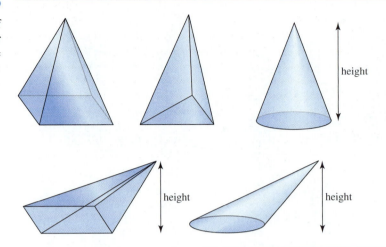

In the volume formula it is understood that if the height is measured in some unit, then the area of the base is measured in square units of the same unit. The volume of the pyramid or cone resulting from the formula is then in cubic units of the same basic unit.

For example, what is the volume of sand in a cone-shaped pile that is 15 feet high and has a radius at the base of 7 feet? According to the volume formula, the volume of sand is

$$\frac{1}{3} \times 15 \times \pi(7)^2 \text{ ft}^3$$

which is about 770 cubic feet of sand.

Where does the $\frac{1}{3}$ in the volume formula for pyramids and cones come from? Class Activities 11Z and 11AA will look at this.

CLASS ACTIVITY   NOW TURN TO CLASS ACTIVITIES MANUAL

11Z   Comparing the Volume of a Pyramid with the Volume of a Rectangular Prism p. 374

11AA   The $\frac{1}{3}$ in the Volume Formula for Pyramids and Cones p. 375

11BB   The Volume of a Rhombic Dodecahedron p. 377

PRACTICE PROBLEMS FOR SECTION 11.10

1. The water in a full bathtub is roughly in the shape of a rectangular prism that is 54 inches long, 22 inches wide, and 9 inches high. Use the fact that 1 gallon = .134 cu- bic feet to determine how many gallons of water are in the bathtub.

2. A concrete patio will be made in the shape of a 15-foot-by-15-foot square with half-circles attached at two opposite ends, as pictured in Figure 11.100. If the concrete will be 3 inches thick, how many cubic feet of concrete will be needed?

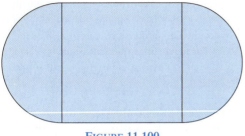

FIGURE 11.100

A Patio

3. Suppose that a tube of toothpaste contains 15 cubic inches of toothpaste and that the circular opening where the toothpaste comes out has a diameter of $\frac{5}{16}$ inch. If every time you brush your teeth you squeeze out a $\frac{1}{2}$-inch-long piece of toothpaste, how many times can you brush your teeth with this tube of toothpaste?

4. A typical ice cream cone is $4\frac{1}{2}$ inches tall and has a diameter of 2 inches. How many cubic inches does it hold (just up to the top)? How many fluid ounces is this? (One fluid ounce is 1.8 cubic inches.)

ANSWERS TO PRACTICE PROBLEMS FOR SECTION 11.10

1. There are about 46 gallons of water in the bathtub. Because the water in the tub is $54 \div 12$ ft $= 4.5$ ft long, $22 \div 12$ ft $= 1.83$ ft wide, and $9 \div 12$ ft $= .75$ ft tall, the volume of water in the tub is $4.5 \times 1.83 \times .75$ ft$^3 = 6.1875$ ft^3. Since each gallon of water is .134 ft^3, the number of gallons of water in the tub is the number of .134 ft^3 in 6.1875 ft^3, which is $6.1875 \div .134$ gallons. This is about 46 gallons.

2. You need 100 cubic feet of concrete for the patio. To explain why, notice that the patio is a kind of cylinder, so its volume can be calculated with the (height) × (area of base) formula. The base, which is the surface of the patio, consists of a square and two half-circles. Therefore, by the moving and combining principle for areas, the area of the base is the area of the square plus the area of the circle that is created by putting the two half-circles together. All together, the area of the patio (the base) is $15^2 + \pi(7.5)^2 = 401.7$ square feet. The height of the concrete patio is $\frac{1}{4}$ of a foot—notice that we must convert 3 inches to feet in order to have consistent units. Therefore, according to the (height) × (area of base)

formula, the volume of concrete needed for the patio is $\frac{1}{4} \times 401.7$ cubic feet, which is about 100 cubic feet of concrete.

3. You will be able to brush your teeth 391 times. Each time you brush, the amount of toothpaste you use is the volume of a cylinder that is $\frac{1}{2}$ inch high and has a radius of $\frac{5}{32}$ inch. According to the volume formula for cylinders, this volume is $\frac{1}{2} \times \pi(\frac{5}{32})^2$ in$^3 = .0383$ in^3. The number of times you can brush is the number of .0383 in^3 in 15 in^3, which is $15 \div .0383 = 391$ times.

4. The cone holds 4.7 cubic inches, which is 2.6 fluid ounces. The diameter of an ice cream cone is 2 in, therefore its radius is 1 in, and so the area of the base of an ice cream cone (the circular hole that holds the ice cream) is $\pi \times 1^2$ in$^2 = 3.14$ in^2. According to the volume formula for cones, the volume of an ice cream cone is $\frac{1}{3} \times 4.5 \times 3.14$ in$^3 = 4.7$ in^3. Since each fluid ounce is 1.8 in^3, the number of fluid ounces the cone holds is the number of 1.8 in^3 in 4.7 in^3, which is $4.7 \div 1.8 = 2.6$ fluid ounces.

PROBLEMS FOR SECTION 11.10

1. a. Measure how fast water comes out of some faucet of your choice. Give your answer in gallons per minute and explain how you arrived at your answer. (Notice that you do not have to fill up a gallon container to do this.)

    b. Figure 11.101 shows a bird's eye view of a swimming pool in the shape of a cross. The pool is 4 feet deep but doesn't have any water in it. There is a small faucet on one side with which to fill the pool. How long would it take to fill up this pool, assuming that the water runs out of this faucet at the same rate as water from your faucet? Use the fact that 1 gallon = .134 cubic feet.

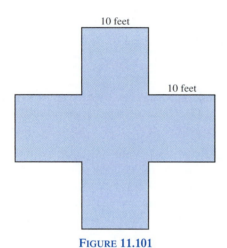

10 feet

10 feet

FIGURE 11.101

A Swimming Pool

2. Find a gallon container, a half-gallon container, a quart container, a pint container, or a one-cup measure. Measure the lengths of various parts of your chosen container and use these measurements to determine the volume of your container in cubic inches. Use your answer to estimate how many gallons are in a cubic foot. (There are about $7\frac{1}{2}$ gallons in a cubic foot.)

3. One gallon is 3.79 liters, and 1 cm³ holds 1 milliliter of liquid. Use these facts to determine the number of gallons in a cubic foot.

4. A cake recipe will make a round cake that is 6 inches in diameter and 2 inches high.

    a. If you use the same recipe but pour the batter into a round cake pan that is 8 inches in diameter, how tall will the cake be?

    b. Suppose you want to use the same cake recipe to make a rectangular cake. If you use a rectangular pan that is 8 inches wide and 10 inches long, and if you want the cake to be about 2 inches tall, then how much of the recipe should you make? (For example, should you make twice as much of the recipe, half as much, three-quarters as much, or some other amount?) Give an approximate but *practical* answer.

5. A recipe for gingerbread makes a 9-in × 9-in × 2-in pan full of gingerbread. Suppose that you want to use this recipe to make a gingerbread house. You decide that you can either use a 10-in × 15-in pan or a 11-in × 17-in pan, and that you can either make a whole recipe or $\frac{1}{2}$ of the recipe of gingerbread.

    a. Which pan should you use, and should you make the whole recipe or just half a recipe, if you want the gingerbread to be between $\frac{1}{4}$ inch and $\frac{1}{2}$ inch thick? Explain.

    b. Draw a careful diagram showing how you would cut the gingerbread into parts to assemble into a gingerbread house. Indicate the different parts of the house (front, back, roof, etc.). When assembled, the gingerbread house should look like a real 3-dimensional house—but be as creative as you like in how you design it. You do not have to use every bit of the gingerbread to make the house, but try to use as much as possible.

    c. If you want to put a solid, 2-inch-tall fence around your gingerbread house so that the fence is 6 inches away from the house, what will be the perimeter of this fence? Explain.

    d. To make the fence in part (c), how many batches of gingerbread recipe will you need, and what pan (or pans) will you use? Indicate how you will cut the fence out of the pan (or pans).

6. The front (and back) of a greenhouse have the shape and dimensions shown in Figure 11.102. The greenhouse is 40 feet long, and the angle at the top of the roof is 90°. A fungus has begun to grow in the greenhouse, so a fungicide will need to be sprayed. The fungicide is simply sprayed into the air. To be effective, one tablespoon of fungicide is needed for every cubic yard of volume in the greenhouse. How much fungicide should be used? Give your answer in terms of units that are practical.

(For example, it would not be practical to have to measure 100 tablespoons, nor would it be practical to have to measure 3.4 quarts. But it would be practical to measure 1 quart and 3 fluid ounces.)

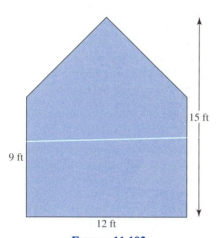

FIGURE 11.102

A Greenhouse

7. A conveyor belt dumps 2500 cubic yards of gravel to form a cone-shaped pile. How high could this pile of gravel be, and what could the circumference of the pile of gravel be at ground level? Give *two different realistically possible pairs of answers* for the height and circumference of the pile of gravel (both piles of volume 2500 cubic yards), and compare how the piles of gravel would look in the two cases.

8. A construction company wants to know how much sand is in a cone-shaped pile. The company measures that the distance from the edge of the pile at ground level to the very top of the pile is 55 feet. The company also measures that the distance around the pile at ground level is 220 feet.

   a. How much sand is in the pile? (Be sure to state the units in which you are measuring this.)

   b. The construction company has trucks that carry 10 cubic yards in each load. How many loads will it take to move the pile of sand?

9. One of the Hawaiian volcanoes is 30,000 feet high (measured from the bottom of the ocean), and has volume 10,000 cubic miles. Assuming that the volcano is shaped like a cone with a circular base, find the distance around the base of the volcano (at the bottom of the ocean). In other words, if a submarine were to travel all the way

around the base of the volcano, at the bottom of the ocean, how far would the submarine go?

10. Suppose that a gasoline-powered engine has a gas tank in the shape of an inverted cone, with radius 10 inches and height 20 inches, as shown in Figure 11.103. The gas flows out through the tip of the cone, which points down. The shaded region represents gas in the tank. The height of this "cone of gas" is 10 inches, which is half the height of the gas tank.

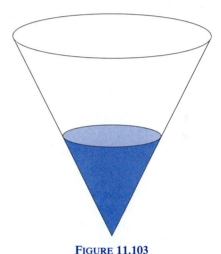

FIGURE 11.103

A Cone-Shaped Gas Tank

   a. What do you think? In terms of volume, is the tank half full, less than half full, or more than half full? Draw a picture of the gas tank and mark approximately where you think the gas would be when the tank is $\frac{3}{4}$ full, $\frac{1}{2}$ full, and $\frac{1}{4}$ full in terms of volume.

   b. Determine the radius of the "cone of gas." *Hint:* Think back to Chapter 9.

   c. Using part (b), find the volume of the gas in the tank. Is this half, less than half, or more than half the volume of the tank?

   d. Now find the volume of gas if the height of the gas was $\frac{3}{4}$ of the height of the tank. Is this more or less than half the volume of the tank? Is this surprising?

11. Make a pattern for a cone-shaped cup that will hold about 120 mL when filled to the rim. Use a pattern like the one for a cone-shaped cup in Figure A.12 on page 733. Indicate measurements on your cup's pattern, and explain why the pattern will produce a cup that holds 120 mL. A centimeter tape measure may be helpful—you may wish to cut out the one on page 731.

12. Cut out the pattern in Figure A.12 on page 733 and tape the two straight edges together to make a small cone-shaped cup. You might want to leave a small "tab" of paper on one of the edges, especially if you use glue instead of tape. In any case, make sure that the two straight edges are joined.

   a. Determine approximately how many fluid ounces the cone-shaped cup holds by filling it with a dry, pourable substance (such as rice, sugar, flour, or even sand), and pouring the substance into a measuring cup.

   b. The pattern for the cone-shaped cup indicates that the two straight edges you taped together are 10 cm long, and the curved edge that makes the rim of the cup is 25 cm long. Use these measurements to determine the volume of the cone-cup in milliliters. (Remember that 1 cubic centimeter holds 1 milliliter.) Explain your method.

   c. If you look on a can of soda, you'll see that 12 fluid ounces is 355 mL. Use this relationship, and your answer to part (b), to determine how many fluid ounces the cone-cup holds. Compare your answer to your estimate in part (a).

13. The slanted part of a cone is made from a quarter-circle. The base of the cone is a circle of radius 3 cm. What is the volume of the cone? Explain your reasoning.

14. a. Determine the volume of a cone that has a circular base of radius 6 cm and height 8 cm.

   b. Make a pattern for the slanted portion of the cone in part (a). Show all relevant calculations, explaining your reasoning.

11.11   Areas, Volumes, and Scaling

When two objects are similar, as in Figure 11.104, there is a scale factor such that corresponding *lengths* on the objects are related by multiplying by the scale factor. (See Section 9.4.) Does this mean that the *surface areas* of the objects are related by multiplying by the scale factor? Does this mean that the *volumes* of the objects are related by multiplying by the scale factor? We will examine these questions in this section.

CLASS ACTIVITY   NOW TURN TO CLASS ACTIVITIES MANUAL

11CC   Areas and Volumes of Similar Boxes p. 378

11DD   Areas and Volumes of Similar Cylinders p. 380

FIGURE 11.104
Two Similar Cups

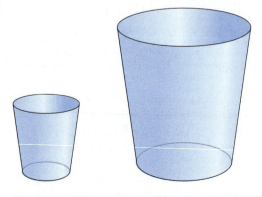

If you did Class Activities 11CC and 11DD, then you probably discovered that if an object is scaled with a scale factor k, then even though the *lengths* of various parts of the object scale by the factor k, the *surface area* of the object scales by the factor k^2, and the *volume* of the object scales by the factor k^3. In other words, if one object (object 1) is the same as another object (object 2) except that it is k times as wide, k times as deep, and k times as tall as object 2, then *the surface area of object 1 is k^2 times as big* as the surface area of object 2, and *the volume of object 1 is k^3 times as big* as the volume of object 2.

For example, if you make a scale model of a building, and if the scale factor from the model to the building is 150 (so that all lengths on the building are 150 times as long as the corresponding length on the model), then the surface area of the building is $150^2 = 22,500$ times as large as the surface area of the model, and the volume of the building is $150^3 = 3,375,000$ times as large as the volume of the model.

Notice how nicely these scale factors for surface area and volume fit with their dimensions and with the units used to measure surface area and volume: The surface of an object is 2-dimensional. It is measured in units of in^2, cm^2, ..., and when an object is scaled with scale factor k, the surface area scales by k^2. Similarly, the size of a full, 3-dimensional object can be measured by volume, which is measured in units of in^3, cm^3, ..., and when an object is scaled with scale factor k, the volume scales by k^3.

CLASS ACTIVITY   NOW TURN TO CLASS ACTIVITIES MANUAL

11EE   Determining Areas and Volumes of Scaled Objects p. 382

PRACTICE PROBLEMS FOR SECTION 11.11

1. If there were a new Goodyear blimp that was $2\frac{1}{2}$ times as long, $2\frac{1}{2}$ times as wide, and $2\frac{1}{2}$ times as tall as the current one, then how much material would be needed to make the new blimp, compared to the current one? How much gas would be needed to fill the new blimp (at the same pressure), compared to the current one? (See Figure 11.105.)

FIGURE 11.105

A Blimp and the Blimp Scaled with Scale Factor $2\frac{1}{2}$

---

2. Suppose you make a scale model for a pyramid using a scale of 1:100 (so that the scale factor from the model to the actual pyramid is 100). If your model pyramid is made out of 4 cardboard triangles, using a total of 60 square feet of cardboard, then what is the surface area of the actual pyramid?

3. If a giant was 12 feet tall, but proportioned like a typical 6-foot-tall man, about how much would you expect the giant to weigh? (See Figure 11.106.)

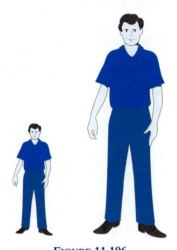

FIGURE 11.106

A 6-Foot-Tall Man and a
12-Foot-Tall Giant

4. Suppose that a larger box is 3 times as wide, 3 times as deep, and 3 times as high as a smaller box. Use a formula for the surface area of a box to explain why the larger box's surface area is 9 times the smaller box's surface area. Use a formula for the volume of a box to explain why the larger box's volume is 27 times the smaller box's volume.

5. Suppose a large cylinder has twice the radius and twice the height of a small cylinder. Use a formula for the surface area of a cylinder to explain why the surface area of the large cylinder is 4 times the surface area of the small cylinder. Use the formula for the volume of a cylinder to explain why the volume of the large cylinder is 8 times the volume of the small cylinder.

ANSWERS TO PRACTICE PROBLEMS FOR SECTION 11.11

1. According to the way surface area behaves under scaling, the new blimp's surface area would be

$$\left(2\frac{1}{2}\right)^2 = \frac{25}{4} = 6\frac{1}{4}$$

times as large as the current blimp's surface area. So the new blimp would require $6\frac{1}{4}$ times as much material to make as the current blimp. According to the way volume behaves under scaling, the new blimp's volume would be

$$\left(2\frac{1}{2}\right)^3 = \frac{125}{8} = 15\frac{5}{8}$$

times as large as the current blimp's volume. So the new blimp would require $15\frac{5}{8}$ times as much gas to fill it as the current blimp.

2. According to the way surface areas behave under scaling, the actual pyramid's surface area is 100^2, or 10,000 times as large as the surface area of the pyramid. The model is made out of 60 square feet of cardboard, so this is its surface area. Therefore, the surface area of

the actual pyramid is $10{,}000 \times 60$ square feet, which is 600,000 square feet.

3. Since the giant is twice as tall as a typical 6-foot-tall man, and since the giant is proportioned like a 6-foot-tall man, the giant should be a scaled version of a 6-foot-tall man, with scale factor 2. Therefore, the volume of the giant should be $2^3 = 8$ times the volume of the 6-foot-tall man. Assuming that weight is proportional to volume, the giant's weight should be 8 times the weight of a typical 6-foot-tall man. If a typical 6-foot-tall man weighs between 150 and 180 pounds, then the giant should weigh 8 times as much, or between 1200 and 1440 pounds.

4. A box that is w units wide, d units deep, and h units high has a volume of

$$wdh$$

cubic units. A box that is 3 times as wide, 3 times as deep, and 3 times as high is $3w$ units wide, $3d$ units deep, and $3h$ units high, so it has volume

$$(3w)(3d)(3h) = 27wdh$$

cubic units. Therefore, a box that is 3 times as wide, 3 times as deep, and 3 times as high as another smaller box has a volume that is 27 times the volume of the smaller box.

A box that is w units wide, d units deep, and h units high has a surface area of

$$2wd + 2wh + 2dh$$

square units. This is because the box has two faces that are w units by d units, two faces that are w units by h units, and two faces that are d units by h units, as you can see by looking at the pattern for a box in Figure 11.107. If another box is 3 times as wide, 3 times as deep, and 3 times as high, then this bigger box will have surface area

$$2(3w)(3d) + 2(3w)(3h) + 2(3d)(3h)$$

square units because its width, depth, and height are $3w$, $3d$, and $3h$ units, respectively. But

$$2(3w)(3d) + 2(3w)(3h) + 2(3d)(3h)$$
$$= 9(2wd + 2wh + 2dh)$$

Therefore, a box that is $3w$ units wide, $3d$ units deep, and $3h$ units high has 9 times the surface area of a box that is w units wide, d units deep, and h units high.

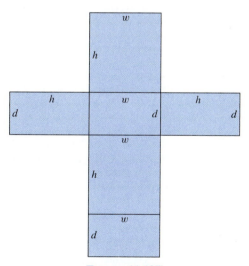

FIGURE 11.107

A Pattern for a Box of Width w,
Height h, and Depth d

5. Let's call the radius of the small cylinder r units and the height of the small cylinder h units. Then the volume of the small cylinder is

$$h\pi r^2$$

cubic units, according to the *(height)* × *(area of base)* volume formula. The larger cylinder has twice the radius and twice the height of the smaller cylinder; therefore, the larger cylinder has radius $2r$ units and height $2h$ units, so the larger cylinder has volume

$$(2h)\pi(2r)^2 = 8(h\pi r^2)$$

cubic units. Because $8h\pi r^2$ is 8 times $h\pi r^2$, the volume of the large cylinder is 8 times the volume of the small cylinder.

The surface area of the small cylinder of radius r and height h is

$$2\pi r^2 h + 2\pi rh$$

square units. (See the answer to Practice Problem 6 of Section 11.7 on page 524). Using the same formula again, but now with $2r$ substituted for r and $2h$ substituted for h, we obtain the surface area of the big cylinder as

$$2\pi(2r)^2 + 2\pi(2r)(2h) = 8\pi r^2 + 8\pi rh$$
$$= 4(2\pi r^2 + 2\pi rh)$$

square units. Because $4(2\pi r^2 + 2\pi rh)$ is 4 times $2\pi r^2 h + 2\pi rh$, the surface area of the large cylinder is 4 times the surface area of the small cylinder.

1. In the triangle ADE pictured in Figure 11.108, the point B is halfway from A to D and the point C is halfway from A to E. Compare the areas of the triangle ABC and the ADE. Explain your reasoning clearly.

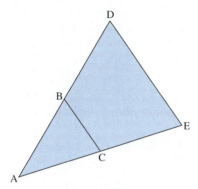

FIGURE 11.108

Triangles

2. A scale model is constructed for a domed baseball stadium using a scale of 1 foot to 100 feet. The model's dome is made out of 40 square feet of cardboard. The model contains 50 cubic feet of air. How many square feet of material will the actual stadium's dome be made out of? How many cubic feet of air will the actual stadium contain?

3. Og, a giant mentioned in the Bible, might have been 13 feet tall. If Og was proportioned like a 6-foot-tall man weighing 200 pounds, then how much would Og have weighed? Explain your reasoning.

4. An artist plans to make a large sculpture of a person out of solid marble. She first makes a small scale model out of clay, using a scale of 1 inch = 2 feet. The scale model weighs 1.3 pounds. Assuming that a cubic foot of clay weighs 150 pounds and a cubic foot of marble weighs 175 pounds, how much will the large marble sculpture weigh? Explain your reasoning.

5. According to one description, King Kong was 19 feet, 8 inches tall and weighed 38 tons. Typical male gorillas are about 5 feet, 6 inches tall and weigh between 300 and 500 pounds. Assuming that King Kong was proportioned like a typical male gorilla, does his given weight of 38 tons agree with what you would expect? Explain.

6. If you know the volume of an object in cubic inches, can you find its volume in cubic feet by dividing by 12? Discuss.

Chapter 12

Number Theory

I n this chapter we will study some of the theory of numbers, including factors, multiples, prime numbers, even and odd numbers, and some quick tests for determining if a number is divisible by another number. Most elementary school curricula do not emphasize these topics, however, some children do learn about many of these topics in the upper elementary grades. Even young children can, and often do, learn about even and odd numbers.

12.1   Factors and Multiples

In this brief section we study situations where counting numbers decompose into products of counting numbers.

If A and B are counting numbers and if $A \div B$ is a counting number with no remainder, then we say that A is **divisible** by B or **evenly divisible** by B, or that B **divides** A. For example, 12 is divisible by 4 and 5 divides 30.

If A and B are counting numbers, then we say that B is a **factor** or a **divisor** of A if there is a counting number C such that

$$A = B \times C$$

So 7 is a factor of 21 because

$$21 = 7 \times 3$$

The numbers 3, 1, and 21 are also factors of 21, and these are the only other factors of 21. Notice that the concepts of divisibility and of factor are closely linked: A counting number B is a factor of a counting number A exactly when A is divisible by B.

One way to find all the factors of a counting number is to divide the number by all the counting numbers smaller than it, to see which ones divide the number evenly, without a remainder. To make the work efficient, keep track of the quotients, because the whole number quotients will also be factors. For example, to find all the factors of 40, divide 40 by 1, by 2, by

3, by 4, by 5, and so on, recording those numbers that divide 40 and recording the corresponding quotients:

$$1,\ 40 \qquad (\text{because } 1 \times 40 = 40)$$

$$2,\ 20 \qquad (\text{because } 2 \times 20 = 40)$$

$$4,\ 10 \qquad (\text{because } 4 \times 10 = 40)$$

$$5,\ 8 \qquad (\text{because } 5 \times 8 = 40)$$

Notice that, after dividing 40 by 1 through 8 and verifying that 1, 2, 4, 5, and 8 are the only factors of 40 up to 8, you don't have to check if 9, 10, 11, 12, and so on divide 40. Why not? Any counting number larger than 8 that divides 40 would have a quotient less than 5 and would already have to appear as one of the quotients in the previous list. So we can conclude that the full list of factors of 40 is

$$1,\ 2,\ 4,\ 5,\ 8,\ 10,\ 20,\ 40$$

Notice that every counting number except 1 must have at least two distinct factors, namely, 1 and itself.

multiple   If A and B are counting numbers, then we say that A is a **multiple** of B if there is a counting number C such that

$$A = C \times B$$

So 44 is a multiple of 11 because

$$44 = 4 \times 11$$

Similarly, the numbers 55 and 66 are also multiples of 11. The list of multiples of 11 is infinitely long:

$$11, 22, 33, 44, 55, 66, \ldots$$

To find all the multiples of a counting number we simply multiply the number by 1, by 2, by 3, by 4, by 5, and so on. We can never explicitly list all the multiples of a number because the list is infinitely long.

Notice that the concepts of factors and multiples are closely linked: a counting number A is a multiple of a counting number B exactly when B is a factor of A. For this reason it's easy to get the concepts of factors and multiples confused. Remember that the factors of a number are the numbers you get from writing the original number as a product (i.e., from "breaking down" the number by dividing it). On the other hand, the multiples of a number are the numbers you get by multiplying the number by counting numbers (i.e., by "building up" the number by multiplying it).

to factor   So far we've use the word "factor" as a noun. However, the word "factor" can also be used as a verb. If A is a counting number, then **to factor** A means to write A as a product of two or more counting numbers, each of which is less than A. So we can factor 18 as 9 times 2:

$$18 = 9 \times 2$$

But notice that we can factor 18 even further, because we can factor 9 as 3 times 3, so

$$18 = 3 \times 3 \times 2$$

CLASS ACTIVITY   NOW TURN TO CLASS ACTIVITIES MANUAL

12A Factors, Multiples, and Rectangles p. 383

12B Problems about Factors and Multiples p. 384

12C Finding All Factors p. 386

12D Do Factors Always Come in Pairs? p. 387

PRACTICE PROBLEMS FOR SECTION 12.1

1. What are the factors of the number 72?

2. What are the multiples of the number 72?

3. Tanya has a large bulletin board that is 54 inches wide across the top. Tanya wants to put a repeating pattern of scrolls (such as the pattern of scrolls shown in Figure 12.1) across the top of her bulletin board in such a way that the length of each scroll is a whole number of inches and a whole number of scrolls fit perfectly across the top of the bulletin board (i.e., there aren't any partial scrolls at the end). Describe Tanya's options for the length of each scroll. In each case, how many scrolls will there be along the top of the bulletin board? Solve this problem and write a related problem about factors or multiples (whichever is most appropriate).

FIGURE 12.1

A Repeating Pattern of Scrolls

4. Tawanda has painted many dry noodles and will put strings through them to make necklaces that are completely filled with noodles, as in Figure 12.2 on page 550. Each noodle is 4 centimeters long. What are the lengths of the necklaces Tawanda can make? Solve this necklace problem and write a related problem about factors or multiples (whichever is most appropriate).

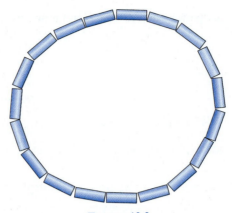

FIGURE 12.2

A Noodle Necklace

ANSWERS TO PRACTICE PROBLEMS FOR SECTION 12.1

1. The factors of the number 72 are the counting numbers that divide 72. To find these numbers, we can simply divide 72 by smaller counting numbers, checking to see which ones divide 72. So we can divide 72 by 1, by 2, by 3, by 4, by 5, and so on. In order to work efficiently, we can record the factors we find and the corresponding quotients as follows:

   1, 72   (because $1 \times 72 = 72$)

   2, 36   (because $2 \times 36 = 72$)

   3, 24   (because $3 \times 24 = 72$)

   4, 18   (because $4 \times 18 = 72$)

   6, 12   (because $6 \times 12 = 72$)

   8, 9   (because $8 \times 9 = 72$)

   Once we have checked the counting numbers up to 9 to see if they divide 72, we won't have to continue dividing 72 by counting numbers greater than 9. This is because any counting number greater than 9 that divides 72 must have a quotient that is less than 9 and therefore must already occur in the column on the right in the previous list. Hence, we can conclude that the list of factors of 72 is

   1, 2, 3, 4, 6, 8, 9, 12, 18, 24, 36, 72

2. The multiples of 72 are the numbers that can be expressed as 72 times another counting number. They are

   72, 144, 216, 288, ...

   because these numbers are

   $72 \times 1, \quad 72 \times 2, \quad 72 \times 3, \quad 72 \times 4, \quad ...$

3. Because the number of scrolls times the length of each scroll in inches is equal to 54 inches, this problem is basically the same as the following problem: find all the ways to factor 54 into a product of two counting numbers. The ways of factoring 54 into a product of two counting numbers are

   $1 \times 54, \quad 2 \times 27, \quad 3 \times 18, \quad 6 \times 9$

   and the reverse factorizations,

   $54 \times 1, \quad 27 \times 2, \quad 18 \times 3, \quad 9 \times 6$

   So Tanya can use 1 scroll that is 54 inches long, 2 scrolls that are 27 inches long, 3 scrolls that are 18 inches long, 6 scrolls that are 9 inches long, 9 scrolls that are 6 inches long, 18 scrolls that are 3 inches long, 27 scrolls that are 2 inches long, or 54 scrolls that are 1 inch long.

4. Whenever Tawanda makes a necklace from these noodles, the length of the necklace will be the number of noodles he used, times 4 centimeters. Therefore, this problem is almost the same as the problem of finding the multiples of 4. In Tawanda's case, some of the multiples of 4 would make necklaces that are too short to be practical, and some multiples of 4 would not work because Tawanda doesn't have enough noodles. The multiples of 4 are

   4, 8, 12, 16, 20, 24, ...

PROBLEMS FOR SECTION 12.1

1. Manuela is looking for all the factors of 90. So far, Manuela has divided 90 by all the counting numbers from 1 to 10, listing those numbers that divide 90 and listing the corresponding quotients. Here is Manuela's work so far:

$$1, 90 \qquad 1 \times 90 = 90$$
$$2, 45 \qquad 2 \times 45 = 90$$
$$3, 30 \qquad 3 \times 30 = 90$$
$$5, 18 \qquad 5 \times 18 = 90$$
$$6, 15 \qquad 6 \times 15 = 90$$
$$9, 10 \qquad 9 \times 10 = 90$$
$$10, 9 \qquad 10 \times 9 = 90$$

Should Manuela keep checking numbers to see if any numbers larger than 10 divide 90, or can Manuela stop dividing at this point? If so, why? What are all the factors of 90?

2. Show how to find all the factors of the following numbers in an efficient manner:

   a. 63

   b. 75

   c. 126

3. Write a problem about a realistic situation that involves the concept of factors. Explain how your problem involves the concept of factors. Solve your problem.

4. Write a problem about a realistic situation that involves the concept of multiples. Explain how your problem involves the concept of multiples. Solve your problem.

5. Johnny says that 3 is a multiple of 6 because you can arrange 3 cookies into 6 groups by putting $\frac{1}{2}$ of a cookie in each group. Discuss Johnny's idea in detail: in what way does he have the right idea about what the term *multiple* means, and in what way does he need to modify his idea?

6. Solve the next two problems, and determine whether the answers are or are not different. Explain why or why not.

   a. Josh has 1159 bottle caps in his collection. In how many different ways can Josh arrange his bottle-cap collection into groups so that the same number of bottle caps are in each group and so that there are no bottle caps left over (i.e., not in a group)?

   b. How many different rectangles can be made whose side lengths, in centimeters, are counting numbers and whose area is 1159 square centimeters?

12.2 Greatest Common Factor and Least Common Multiple

In Section 12.1, we discussed factors and multiples of individual numbers. If we have two or more counting numbers in mind, we can consider the factors that the two numbers have in common and the multiples that the two numbers have in common. By considering common multiples and common factors of two or more counting numbers, we arrive at the concepts of *greatest common factor* and *least common multiple*, which we will discuss in this section. Greatest common factors and least common multiples can be useful when working with fractions.

greatest common factor
GCF

If you have two or more counting numbers, then the **greatest common factor**, abbreviated **GCF**, or **greatest common divisor**, abbreviated **GCD**, of these numbers is the greatest counting number that is a factor of all the given counting numbers. For example, what is the GCF of 12 and 18? The factors of 12 are

$$1, \ 2, \ 3, \ 4, \ 6, \ 12$$

and the factors of 18 are

$$1, \ 2, \ 3, \ 6, \ 9, \ 18$$

Therefore, the common factors of 12 and 18 are the numbers that are common to the two lists, namely,

$$1,\ 2,\ 3,\ 6$$

The greatest of these numbers is 6; therefore, the greatest common factor of 12 and 18 is 6.

least common multiple
LCM

Similarly, if you have two or more counting numbers, then the **least common multiple**, abbreviated **LCM**, of these numbers is the least counting number that is a multiple of all the given numbers. For example, what is the LCM of 6 and 8? The multiples of 6 are

$$6,\ 12,\ 18,\ 24,\ 30,\ 36,\ 42,\ 48,\ 54,\ 60,\ 66,\ 72,\ 78, \ldots$$

and the multiples of 8 are

$$8,\ 16,\ 24,\ 32,\ 40,\ 48,\ 56,\ 64,\ 72,\ 80, \ldots$$

Therefore, the common multiples of 6 and 8 are the numbers that are common to the two lists, namely,

$$24,\ 48,\ 72, \ldots$$

The least of these numbers is 24, therefore, the least common multiple of 6 and 8 is 24.

Although there are other methods for calculating GCFs and LCMs, we can always use the definition of these concepts to calculate GCFs and LCMs.

CLASS ACTIVITY   NOW TURN TO CLASS ACTIVITIES MANUAL

12E   Situations Involving the Least Common Multiple and the Greatest Common Factor p. 388

12F   Problems Involving Greatest Common Factors and Least Common Multiples p. 390

12G   Relationships Between the GCF and the LCM p. 391

USING GCFS AND LCMS WITH FRACTIONS

We often use GCFs and LCMs when working with fractions.

To put a fraction in simplest form, we can divide the numerator and denominator of the fraction by the GCF of the numerator and the denominator. For example, to put

$$\frac{24}{36}$$

in simplest form, we can first determine that the greatest common factor of 24 and 36 is 12. Since $24 = 2 \cdot 12$ and $36 = 3 \cdot 12$, we have

$$\frac{24}{36} = \frac{2 \cdot 12}{3 \cdot 12} = \frac{2}{3}$$

So $\frac{2}{3}$ is the simplest form of $\frac{24}{36}$.

It is not necessary to determine the greatest common factor of the numerator and denominator of a fraction in order to put the fraction in simplest form. Instead, we can keep simplifying the fraction until it can no longer be simplified. For example, we could use the following steps to simplify $\frac{24}{36}$:

$$\frac{24}{36} = \frac{12 \cdot 2}{18 \cdot 2} = \frac{12}{18} = \frac{6 \cdot 2}{9 \cdot 2} = \frac{6}{9} = \frac{2 \cdot 3}{3 \cdot 3} = \frac{2}{3}$$

We can use the LCM of the denominators of two fractions to add the fractions. In order to add fractions, we must first find a common denominator. Although any common denominator will do, we may wish to work with the least common denominator. Because common denominators of two fractions must be multiples of both denominators, the least common denominator is the least common multiple of the two denominators. For example, to add

$$\frac{1}{6} + \frac{3}{8}$$

we need a common denominator which is a multiple of 6 and of 8. The least common multiple of 6 and 8 is 24. Using this least common denominator, we calculate

$$\frac{1}{6} + \frac{3}{8} = \frac{1 \cdot 4}{6 \cdot 4} + \frac{3 \cdot 3}{8 \cdot 3} = \frac{4}{24} + \frac{9}{24} = \frac{13}{24}$$

To add $\frac{1}{6} + \frac{3}{8}$ we could also have used the common denominator $6 \cdot 8 = 48$. In this case,

$$\frac{1}{6} + \frac{3}{8} = \frac{1 \cdot 8}{6 \cdot 8} + \frac{3 \cdot 6}{8 \cdot 6} = \frac{8}{48} + \frac{18}{48} = \frac{26}{48}$$

Because we did not use the least common denominator, the resulting sum is not in simplest form. We need the following additional step to put the answer in simplest form:

$$\frac{26}{48} = \frac{13 \cdot 2}{24 \cdot 2} = \frac{13}{24}$$

When we use a common denominator that is not the least common multiple of the denominators, the resulting sum will not be in simplest form. However, even if we use the least common denominator when adding two fractions, the resulting fraction may not be in simplest form. For example,

$$\frac{1}{2} + \frac{1}{6} = \frac{3}{6} + \frac{1}{6} = \frac{4}{6}$$

but $\frac{4}{6}$ is not in simplest form even though we used the least common denominator when adding.

CLASS ACTIVITY NOW TURN TO CLASS ACTIVITIES MANUAL

12H Using GCFs and LCMs with Fractions p. 394

PRACTICE PROBLEMS FOR SECTION 12.2

1. Calculate the least common multiple of 9 and 12 and explain why your answer is correct.

2. Calculate the greatest common factor of 36 and 63 and explain why your answer is correct.

3. Calculate the greatest common factor of 16 and 27.

4. If you have 100 pencils and 75 small notebooks, what is the largest group of children that you can give all the pencils and all the notebooks to so that each child gets the same number of pencils and each child gets the same number of notebooks, and so that no pencils or notebooks are left over? Explain why you can solve the problem the way you do.

5. If pencils come in packages of 12 and small notebooks come in packages of 5, what is the smallest number of pencils and notebooks you can buy so that you can match each pencil with a notebook and so that no pencils or notebooks will be left over? Explain why you can solve the problem the way you do.

ANSWERS TO PRACTICE PROBLEMS FOR SECTION 12.2

1. The multiples of 9 are

$$9, \ 18, \ 27, \ 36, \ 45, \ 54, \ 63, \ 72, \ 81, \ldots$$

and the multiples of 12 are

$$12, \ 24, \ 36, \ 48, \ 60, \ 72, \ 84, \ldots$$

Therefore, the common multiples of 9 and 12 are

$$36, \ 72, \ldots$$

The least of these is 36, which is thus the least common multiple of 9 and 12.

2. The factors of 36 are

$$1, \ 2, \ 3, \ 4, \ 6, \ 9, \ 12, \ 18, \ 36$$

and the factors of 63 are

$$1, \ 3, \ 7, \ 9, \ 21, \ 63$$

Therefore, the common factors of 36 and 63 are

$$1, \ 3, \ 9$$

The greatest of these numbers is 9; hence, the greatest common factor of 36 and 63 is 9.

3. The factors of 16 are

$$1, \ 2, \ 4, \ 8, \ 16$$

and the factors of 27 are

$$1, \ 3, \ 9, \ 27$$

The only common factor of 16 and 27 is 1, so this must be the greatest common factor of 16 and 27.

4. If each child gets the same number of pencils and no pencils are left over, then the number of children you can give the pencils to must be a factor of 100. Similarly, the number of children that you can give the notebooks to must be a factor of 75. Since we are looking for the largest number of children to give the pencils and notebooks out to, this will be the greatest common factor of 100 and 75. The factors of 100 are

$$1, \ 2, \ 4, \ 5, \ 10, \ 20, \ 25, \ 50, \ 100$$

The factors of 75 are

$$1, \ 3, \ 5, \ 15, \ 25, \ 75$$

The common factors of 100 and 75 are

$$1, \ 5, \ 25$$

Therefore, the GCF of 100 and 75 is 25, so the largest group of children you can give the pencils and notebooks out to is 25.

5. The total number of pencils you buy will be a multiple of 12, and the total number of notebooks you buy will be a multiple of 5. You want both of these multiples to be equal and to be as small as possible. Therefore, the number of pencils and notebooks you will buy will be the least common multiple of 12 and 5. The multiples of 12 are

$$12, \ 24, \ 36, \ 48, \ 60, \ 72, \ 84, \ 96, \ 108, \ldots$$

The multiples of 5 are

5, 10, 15, 20, 25, 30, 35, 40, 45, 50, 55, 60, 65, ...

The only common multiples that we see so far in these

lists is 60, so this is the LCM of 12 and 5. Therefore, you should buy 5 packs of pencils and 12 packs of notebooks for a total of 60 pencils and 60 notebooks.

PROBLEMS FOR SECTION 12.2

1. Why do we not talk about a *greatest* common multiple and a *least* common factor?

2. Determine the greatest common factor of 27 and 36, showing all details.

3. Determine the greatest common factor of 30 and 77, showing all details.

4. Determine the least common multiple of 44 and 55, showing all details.

5. Determine the least common multiple of 7 and 8, showing all details.

6. Show all the details in the following calculations:

   a. Put $\frac{90}{126}$ in simplest form by first determining the greatest common factor of 90 and 126. Show the details of determining this greatest common factor.

   b. Put $\frac{90}{126}$ in simplest form without first determining the greatest common factor of 90 and 126.

7. Show all the details in the following calculations:

   a. Add $\frac{5}{12} + \frac{1}{8}$ by using the least common denominator. Show the details of determining the least common denominator. Give the answer in simplest form.

   b. Add $\frac{5}{12} + \frac{1}{8}$ by using the common denominator $12 \cdot 8$. Give the answer in simplest form.

8. Make up a story problem that can be solved by calculating the least common multiple of 45 and 40. Solve your problem, explaining how the least common multiple is involved.

9. Make up a story problem that can be solved by calculating the greatest common factor of 45 and 40. Solve your problem, explaining how the greatest common factor is involved.

10. In a clothing factory, a worker can sew 18 Garment A seams in a minute and 30 Garment B seams per minute. If the factory manager wants to complete equal numbers of Garments A and B every minute, how many workers should she hire for each type of garment? Explain your answer.

11. Keiko has a rectangular piece of fabric that is 48 inches wide and 72 inches long. Keiko wants to cut her fabric into identical square pieces, leaving no fabric remaining. She wants the side lengths of the squares to be whole numbers of inches.

    a. Draw rough sketches indicating three different ways that Keiko could cut her fabric into squares.

    b. Keiko decides that she wants her squares to be as large as possible. How big should Keiko make her squares? Explain your reasoning. Draw a sketch showing how Keiko should cut the squares from her fabric.

12. A large gear is used to turn a smaller gear. The large gear has 60 teeth and makes 12 revolutions per minute; the small gear has 24 teeth. How fast does the small gear turn (i.e., how many revolutions per minute does it make)? Explain your reasoning.

12.3 Prime Numbers

In this section we will discuss the prime numbers, which can be considered the "building blocks" of the counting numbers.

prime numbers

primes

The **prime numbers**, or simply, the **primes**, are the counting numbers other than 1 that are divisible only by 1 and themselves. In other words, the prime numbers are the counting numbers that cannot be factored in a non-trivial way. For example, 2, 3, 5, and 7 are prime numbers, but 6 is not because $6 = 2 \times 3$.

The prime numbers are considered to be the *building blocks of the counting numbers*. Why? Because it turns out that every counting number greater than or equal to 2 is either a prime number or can be factored as a product of prime numbers. For example,

$$145 = 29 \times 5$$

$$2009 = 41 \times 7 \times 7$$

$$264,264 = 13 \times 11 \times 11 \times 7 \times 3 \times 2 \times 2 \times 2$$

In this way, all counting numbers from 2 onward are "built" from prime numbers. Furthermore, except for rearranging the prime factors, there is only one way to factor a counting number that is not prime into a product of prime numbers. In other words, it is not possible for a product of prime numbers to be equal to a product of different prime numbers. For example, it couldn't happen that $5 \times 7 \times 23 \times 29$ is equal to a product of some different prime numbers, say, involving 11 or 13 or any primes other than 5, 7, 23, and 29. The fact that every counting number greater than 1 can be factored as a product of prime numbers in a unique way as just described is not at all obvious. It is a theorem called the **Fundamental Theorem of Arithmetic**.

Fundamental Theorem of Arithmetic

The prime numbers are to the counting numbers as atoms are to matter—basic and fundamental. And just as many physicists study atoms, so too, many mathematicians study the prime numbers.

How do we find prime numbers and how can we tell if a number is prime? We will consider these questions next.

THE SIEVE OF ERATOSTHENES

Sieve of Eratosthenes

As with geometry, many fundamental contributions to knowledge about prime numbers were made by mathematicians of ancient Greece. Eratosthenes (275 – 195 B.C.) discovered a nice method for finding and listing prime numbers, called the **Sieve of Eratosthenes**. To use the Sieve of Eratosthenes, list all the whole numbers from 2 up to wherever you want to stop looking for prime numbers. The list in Table 12.1 goes from 2 to 30, so we will look for all the primes up to 30. Next, carry out the following process of circling and crossing out numbers:

1. On the list, circle the number 2, and then cross out every other number after 2. (So cross out 4, 6, 8, etc., the multiples of 2.)

2. Then circle the number 3 and cross out every third number after 3—even if it has already been crossed out. (So cross out 6, 9, 12, etc., the multiples of 3. Notice that you can do this "mechanically," by repeatedly counting, "1, 2, 3," and crossing out on "3.")

3. Circle the number 5 and cross out every fifth number after 5—even if it has already been crossed out. (So cross out 10, 15, 20, etc., the multiples of 5. Again, notice that you can do this "mechanically," by repeatedly counting, "1, 2, 3, 4, 5," and crossing out on "5.")

TABLE 12.1

Using the Sieve of Eratosthenes to Find Prime Numbers

2	3	4	5	6	7	8	9	10
11	12	13	14	15	16	17	18	19	20
21	22	23	24	25	26	27	28	29	30

4. Continue in this way, going back to the beginning of the list, circling the next number N that has not been crossed out and then crossing out every N^{th} number after it until every number in the list has either been circled or crossed out.

The circled numbers in the list are the prime numbers. Why? If a number is circled, then it was not crossed out using one of the previously circled numbers. The numbers we cross out are exactly the multiples of a circled number (beyond that circled number). For example, when we circle 3 and cross out every third number after 3, we cross out the multiples of 3 beyond 3, namely, 6, 9, 12, 15, and so on. Therefore the circled numbers are the numbers that are not multiples of any smaller number (other than 1). Hence, a circled number is divisible only by 1 and itself and so is a prime number.

CLASS ACTIVITY   NOW TURN TO CLASS ACTIVITIES MANUAL

121   The Sieve of Eratosthenes p. 395

THE TRIAL DIVISION METHOD FOR DETERMINING WHETHER A NUMBER IS PRIME

How can you tell whether or not a number, such as 239, is a prime number? You could use the Sieve of Eratosthenes to find all prime numbers up to 239, but that would be slow. A faster way is to divide your number by 2, if it's not divisible by 2, divide it by 3; if it's not divisible by 3, divide it by 5; if it's not divisible by 5, divide it by 7, and so on, dividing by consecutive prime numbers. If you ever find a prime number that divides your number, then your number is not prime; otherwise, it is prime.

trial division

This method for determining whether a number is prime is called the method of **trial division**.

Why is trial division a valid way to determine if a number is prime? Why do you only have to check *prime* numbers to see if they divide the number in question? Why don't you have to check other numbers, such as 4, 6, 8, and 9, which are not prime numbers, to see if they divide the given number? If a number is divisible by 4, for example, then it must also be divisible by 2; if a number is divisible by 6, then it must also be divisible by 2 and by 3. In general, if a number is divisible by some counting number other than 1 and itself, then, by the fundamental theorem of arithmetic, that divisor is either a prime number or can be factored into a product of prime numbers, each of those prime factors must also divide the number in question. So, to determine whether a number is prime, we only have to find out whether any prime numbers divide it.

When you use trial division to determine if a number is prime, how do you know when to stop dividing by primes? Consider the example of 283. To see if 283 is prime or not, divide it by 2, by 3, by 5, by 7, and so on, dividing by consecutive prime numbers. We find that none of these prime numbers divide 283, but how do we know when we can stop dividing? Let's look at the quotients that result when we divide 283 by prime numbers in order:

$$283 \div 2 = 141.5$$

$$283 \div 3 = 93.33\ldots$$

$$283 \div 5 = 56.6$$

$$283 \div 7 = 40.43\ldots$$

$$283 \div 11 = 25.73\ldots$$

$$283 \div 13 = 21.77\ldots$$

$$283 \div 17 = 16.65\ldots$$

Notice that as we divide by larger prime numbers, the quotients get smaller. Up to 13, the quotient is larger than the divisor, but when we divide 283 by 17, the quotient, 16.65, is smaller than the divisor. Now imagine that we were to continue dividing 283 by larger prime numbers: by 19, by 23, by 29, and so on. The quotients would get smaller and would be less than 17. So if a prime number larger than 17 were to divide 283, the corresponding quotient would be a whole number less than 17 and thus would have a prime number divisor less than 17. But we already checked all the prime numbers up to 17 and found that none of them divide 283. Therefore, we can already tell that 283 is prime.

In general, to determine whether a number is prime by the trial division method, divide the number by consecutive prime numbers starting at 2. Suppose that none of the prime numbers divide your number. If you reach a point when the quotient becomes smaller than the prime number by which you are dividing, then the previous reasoning tells us your number must be prime.

Of course, it may happen that the number you are checking is not prime. For example, consider 899. When you divide 899 by 2, by 3, by 5, by 7, and so on, dividing by consecutive prime numbers, you will find that none of these primes divide 899 until you get to 29. Then you will discover that $899 = 29 \times 31$, so 899 is not prime.

CLASS ACTIVITY   NOW TURN TO CLASS ACTIVITIES MANUAL

12J   The Trial Division Method for Determining if a Number Is Prime p. 396

FACTORING INTO PRODUCTS OF PRIME NUMBERS

The prime numbers are "building blocks" of the counting numbers because every counting number greater than 1 that is not already a prime number can be broken into a product of prime numbers. How do we write numbers as products of prime numbers?

factor tree

A convenient way to factor a counting number into a product of prime numbers is to create a factor tree. A factor tree is a diagram like the one in Figure 12.3, which shows how to factor a number, how to factor the factors, how to factor the factors of factors, and so on, until prime numbers are reached.

FIGURE 12.3

A Factor Tree
for 600

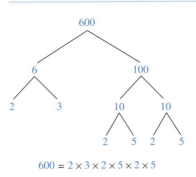

$$600 = 2 \times 3 \times 2 \times 5 \times 2 \times 5$$

Starting with a given counting number, such as 600, you can create a factor tree as follows: Factor 600 in any way you can think of—for example,

$$600 = 6 \times 100$$

Record this factorization below the number 600, as shown in Figure 12.3. Then proceed by factoring each of 6 and 100 separately, and recording these factorizations below each number. Continue until all your factors are prime numbers. Then the numbers at the bottom of your factor tree are the prime factors of your original number. From Figure 12.3, we determine that

$$600 = 2 \times 3 \times 2 \times 5 \times 2 \times 5$$

You might want to rearrange your factorization by placing identical primes next to each other:

$$600 = 2 \times 2 \times 2 \times 3 \times 5 \times 5$$

Using the notation of exponents, we can also write

$$600 = 2^3 \times 3 \times 5^2$$

Even if you thought of a different way to factor the number initially or along the way, you will still wind up with the same prime factors in the end. This is what the Fundamental Theorem of Arithmetic tells us.

What if you don't immediately see a way to factor your number? For example, how can we factor 26,741? In this case, try to divide it by consecutive prime numbers until you find a prime number that divides it evenly. We find that 26,741 is not divisible by 2, by 3, by 5, or by 7, but is divisible by 11:

$$26,741 = 11 \times 2431$$

Now we need to factor 2431. It is not divisible by 2, by 3, by 5, or by 7. (If it were, 26,741 would also be divisible by one of these.) But we still must find out whether 2431 is divisible by 11, and sure enough, it is:

$$2431 = 11 \times 221$$

FIGURE 12.4

A Factor Tree
for 26,741

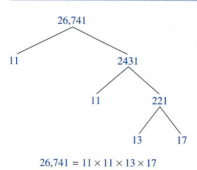

$$26{,}741 = 11 \times 11 \times 13 \times 17$$

Now determine whether 221 is divisible by 11. It isn't. So see if 221 is divisible by the next prime, 13. It is:

$$221 = 13 \times 17$$

Collecting the prime factors we have found, we conclude that

$$26{,}741 = 11 \times 11 \times 13 \times 17$$

The factor tree in Figure 12.4 records how we factored 26,741.

HOW MANY PRIME NUMBERS ARE THERE?

Mathematicians through the ages have been fascinated by prime numbers. There are many interesting facts and questions about them. For instance, does the list of prime numbers go on forever, or does it ever come to a stop? Ponder this for a moment before you read on.

Do you have a definite opinion about this or are you unsure? If you do have a definite opinion, can you give convincing evidence for it? This is not so easy.

It turns out that the list of prime numbers does go on forever. How do we know this? It certainly isn't possible to demonstrate this by listing all the prime numbers. It is not at all obvious why there are infinitely many prime numbers, but there is a beautiful line of reasoning that proves this.

In about 300 B.C., Euclid, another mathematician who lived in ancient Greece, wrote the most famous mathematics book of all time, *The Elements*. In it appears an argument that proves there are infinitely many prime numbers [18, Proposition 20 of book IX]. The following is a variation of that argument, presented in more modern language:

Suppose that

$$P_1, P_2, P_3, \ldots, P_n$$

is a list of n prime numbers. Consider the number

$$P_1 \cdot P_2 \cdot P_3 \cdot \ldots \cdot P_n + 1$$

which is just the product of the list of prime numbers with the number one added on. Then this new number is not divisible by any of the primes $P_1, P_2, P_3, \ldots, P_n$ on our list, because each of these leave a remainder of 1 when our new number is divided by any of these numbers.

Therefore, either $P_1 \cdot P_2 \cdot P_3 \cdot \ldots \cdot P_n + 1$ is a prime number that is not on our list of primes or, when we factor our number $P_1 \cdot P_2 \cdot P_3 \cdot \ldots \cdot P_n + 1$ into a product of prime numbers, each of these prime factors is a new prime number that is not on our list. Hence, no matter how many prime numbers we started with, we can always find new prime numbers, and it follows that there must be infinitely many prime numbers.

PRACTICE PROBLEMS FOR SECTION 12.3

1. What is a prime number?

2. What is the Sieve of Eratosthenes and why does it work?

3. For each of the listed numbers, determine whether it is prime. If it is not prime, factor the number into a product of prime numbers.

   a. 1081

   b. 1087

   c. 269,059

   d. 2081

   e. 1147

4. Given that $550 = 2 \cdot 5 \cdot 5 \cdot 11$, find all the factors of of 550 and explain your reasoning.

ANSWERS TO PRACTICE PROBLEMS FOR SECTION 12.3

1. See text.

2. See text.

3. a. $1081 = 23 \times 47$.

   b. 1087 is prime.

   c. $269,059 = 7 \times 7 \times 17 \times 17 \times 19$.

   d. 2081 is prime.

   e. $1147 = 31 \times 37$.

4. 1 and 550 are factors of 550. Given any other factor of 550, when this factor is factored, it must be a product of some of the primes in the list 2, 5, 5, 11. (If not, you would get a different way to factor 550 into a product of prime numbers, which would contradict the Fundamental Theorem of Arithmetic.) Therefore, the factors of 550 are

$$1, \ 550, \ 2, \ 5, \ 11, \ 2 \cdot 5 = 10, \ 2 \cdot 11 = 22,$$
$$5 \cdot 5 = 25, \ 5 \cdot 11 = 55, \ 2 \cdot 5 \cdot 5 = 50,$$
$$2 \cdot 5 \cdot 11 = 110, \ 5 \cdot 5 \cdot 11 = 275$$

PROBLEMS FOR SECTION 12.3

1. For which counting numbers, N, is there only one rectangle whose side lengths, in inches, are counting numbers and whose area, in square inches, is N? Explain.

2. For each of the numbers in (a) through (d), determine whether it is a prime number. If it is not a prime number, factor the number into a product of prime numbers.

   a. 8303

   b. 3719

   c. 3721

   d. 80,000

3. Without calculating the number $19 \times 23 + 1$, explain why this number is not divisible by 19 or by 23.

4. Following Euclid's proof that there are infinitely many primes, and starting the list with 5 and 7, you form the number $5 \times 7 + 1 = 36$, and you get the new prime numbers 2 and 3 because $36 = 2 \times 2 \times 3 \times 3$. Which new prime numbers would you get if your starting list was the following?

   a. 2, 3, 5

   b. 2, 3, 5, 7

   c. 3, 5

12.4   Even and Odd

The study of even and odd numbers is fertile ground for investigating and exploring math. This is an area of math where it's not too hard to find interesting questions to ask about what is true, to investigate these questions, and to explain why the answers are right. Even young children can do this. In this section, in addition to asking and investigating questions about even and odd numbers, we will examine the familiar method for determining if a number is even or odd.

How would you answer the following two questions?

- What does it *mean* for a whole number to be even?

- How can you *tell* if a whole number is even?

Did you give different answers? Then why do these two different characterizations describe *the same numbers*?

even   The meaning of even is this: A whole number is **even** if it is divisible by 2, in other words, if there is no remainder when you divide the number by 2. There are various equivalent ways to say that a whole number is even:

- A whole number is even if it is divisible by 2, in other words, if there is no remainder when you divide the number by 2.

- A whole number is even if it can be factored into 2 times another whole number (or another whole number times 2).

- A whole number is even if you can divide that number of things into two equal groups with none left over.

- A whole number is even if you can divide that number of things into groups of 2 with none left over. (See Figure 12.5.)

odd   If a whole number is not even, then we call it **odd**. Each of the ways previously described of saying what it means for a counting number to be even can be modified to say what it means for a counting number to be odd:

- A whole number is odd if it is not divisible by 2, in other words, if there is a remainder when you divide the number by 2.

- A whole number is odd if it cannot be factored into 2 times another whole number (or another whole number times 2).

- A whole number is odd if there is one thing left over when you divide that number of things into two equal groups.

- A whole number is odd if there is one thing left over when you divide that number of things into groups of 2. (See Figure 12.5.)

FIGURE 12.5

Even and Odd

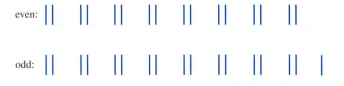

If you have a particular number in mind, such as 237,921, how do you *tell* if the number is even or odd? Chances are, you don't actually divide 237,921 by 2 to see if there is a remainder or not. Instead, you look at the ones digit. In general, if a counting number has a 0, 2, 4, 6, or 8 in the ones place, then it is even; if it has a 1, 3, 5, 7, or 9 in the ones place, then it is odd. So 237,921 is odd because there is a 1 in the ones place. *But why is this a valid way to determine that there will be a remainder when you divide 237,921 by 2?*

To answer this question, think about representing counting numbers with bundled toothpicks. For example, we represent 351 with 3 bundles of 100, 5 bundles of 10, and 1 individual toothpick; we represent 4736 with 4 bundles of 1000, 7 bundles of 100, 3 bundles of 10, and 6 individual toothpicks. A counting number is even if that number of toothpicks can be divided into groups of 2 with none left over; the number is odd if there is a toothpick left over. So consider dividing bundled toothpicks that represent a counting number into groups of 2. To divide these toothpicks into groups of 2 we could divide each of the bundles of 10, of 100, of 1000, and so on, into groups of 2. Each bundle of 10 toothpicks can be divided into 5 groups of 2. Each bundle of 100 toothpicks can be divided into 50 groups of 2. Each bundle of 1000 toothpicks can be divided into 500 groups of 2, and so on, for all the higher places. For each place from the 10s place on up, each bundle of toothpicks can always be divided into groups of 2 with none left over. Therefore, the full number of toothpicks can be divided into groups of 2 with none left over exactly when the number of toothpicks that are in the ones place can be divided into groups of 2 with none left over. This is why we only have to check the ones place of a counting number to see if the number is even or odd.

CLASS ACTIVITY   NOW TURN TO CLASS ACTIVITIES MANUAL

12K Why Can We Check the Ones Digit to Determine if a Number is Even or Odd? p. 398

12L Problems about Even and Odd Numbers p. 400

12M Questions about Even and Odd Numbers p. 401

12N Extending the Definitions of Even and Odd p. 402

PRACTICE PROBLEMS FOR SECTION 12.4

1. Explain why it is valid to determine whether a counting number is divisible by 2 by seeing if its ones digit is 0, 2, 4, 6, or 8.

2. If you add an odd number and an even number, what kind of number do you get? Explain why your answer is always correct.

3. If you multiply an even number and an even number, what kind of number do you get? Explain why your answer is always correct.

ANSWERS TO PRACTICE PROBLEMS FOR SECTION 12.4

1. See text.

2. If you add an odd number to an even number, the result is always an odd number. One way to explain why is to think about putting together an odd number of blocks and an even number of blocks, as in Figure 12.6. The odd number of blocks can be divided into groups of 2 with 1 block left over. The even number of blocks can be divided into groups of 2 with none left over. When the two block collections are joined together, there will be a bunch of groups of 2, and the 1 block that was left over from the odd number of blocks will still be left over. Therefore, the sum of an odd number and an even number is odd.

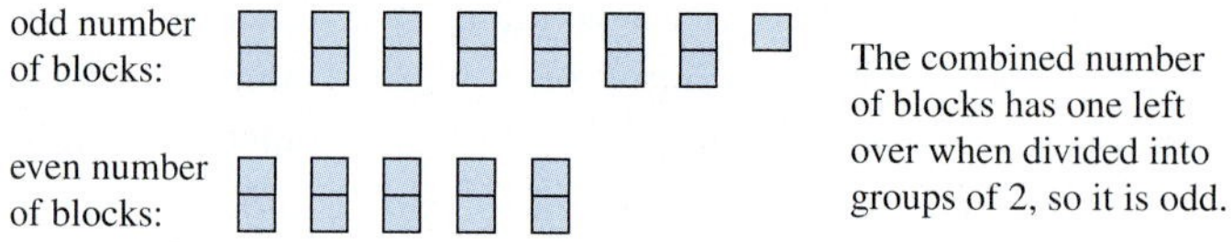

FIGURE 12.6

Adding an Odd Number to an Even Number

3. If you multiply and even number and an even number, the result is always an even number. One way to explain why is to think about creating an even number of groups of blocks, where each group of blocks contains the same even number of blocks, as in Figure 12.7. According to the meaning of multiplication, the total number of blocks is the product of the number of groups with the number of blocks in each group. Because each group of blocks can be divided into groups of 2 with none left over, the whole collection of blocks can be divided into groups of 2 with none left over. Therefore, an even number times an even number is always an even number. (Notice that this argument actually shows that any counting number times an even number is even.)

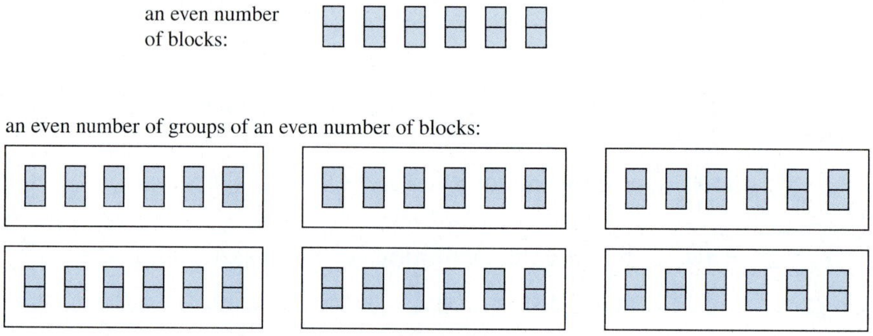

FIGURE 12.7

Multiplying an Even Number with an Even Number

PROBLEMS FOR SECTION 12.4

1. Describe a way that the children in Mrs. Verner's kindergarten class can tell if there are an even number or an odd number of children present in the class. Your method should not involve counting.

2. For each of the two designs in Figure 12.8, explain how you can tell whether the number of dots in the design is even or odd without determining the number of dots in the design.

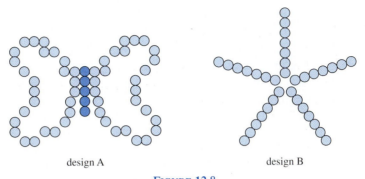

design A   design B

FIGURE 12.8

Two Dot Designs

---

3. If you add an even number and an even number, what kind of number do you get? Explain why your answer is always correct.

4. If you multiply an odd number and an odd number, what kind of number do you get? Explain why your answer is always correct.

5. If you add a number that has a remainder of 1 when it is divided by 3 to a number that has a remainder of 2 when it is divided by 3, then what is the remainder of the sum when you divide it by 3? Investigate this question by working examples. Then explain why your answer is always correct.

6. If you multiply a number that has a remainder of 1 when it is divided by 3 with a number that has a remainder of 2 when it is divided by 3, then what is the remainder of the product when you divide it by 3? Investigate this question by working examples. Then explain why your answer is always correct.

12.5 Divisibility Tests

How can you tell if one counting number is divisible by another counting number? In the previous section, we saw a way to explain why we can look at the ones digit to tell whether a counting number is even or odd. Are there easy ways to tell if a counting number is divisible by other numbers, such as 3, 4, or 5? Yes, there are, and we will study some of them in this section.

divisibility test
A method for *checking* or *testing* to see if a counting number is divisible by another counting number, without actually carrying out the division, is called a **divisibility test**.

divisibility test for 2
The **divisibility test for 2** is just the familiar test to see if a number is even or odd: check the ones digit of the number. If the ones digit is 0, 2, 4, 6, or 8, then the number is divisible by 2 (it is even), otherwise the number is not divisible by 2 (it is odd). This divisibility test was discussed in the previous section.

divisibility test for 10
A familiar divisibility test is the **divisibility test for 10**. The test is as follows: given a counting number, check the ones digit of the number. If the ones digit is 0, then the number is divisible by 10; if the ones digit is not 0, then the number is not divisible by 10. Another familiar divisibility test is the **divisibility test for 5**. The test is as follows: given a whole number, check the ones digit of the number. If the ones digit is 0 or 5, then the original number is divisible by 5, otherwise it is not divisible by 5. Problems 1 and 2 will help you explain why these divisibility tests are valid. The explanations are very similar to the explanation developed in the last section for why we can see that a number is divisible by 2 by checking its ones digit.

divisibility test for 5

The **divisibility test for 3** is as follows: given a counting number, add the digits of the number. If the sum of the digits is divisible by 3, then the original number is also divisible by 3, otherwise the original number is not divisible by 3.

This divisibility test makes it easy to determine whether a large whole number is divisible by 3. For example, is 127,358 divisible by 3? To answer this, all we have to do is add all the digits of the number:

$$1 + 2 + 7 + 3 + 5 + 8 = 26$$

Because 26 is not divisible by 3, the original number 127,358 is also not divisible by 3. Similarly, is 111,111 divisible by 3? Because the sum of the digits is 6,

$$1 + 1 + 1 + 1 + 1 + 1 = 6$$

and because 6 is divisible by 3, the original number 111,111 must also be divisible by 3.

Why does the divisibility test for 3 work? In other words, why is it a valid way to determine whether a counting number is divisible by 3 or not? As with the divisibility test for 2 (the test for even or odd), let's consider counting numbers as represented by bundled toothpicks. A counting number is divisible by 3 exactly when that number of toothpicks can be divided into groups of 3 with none left over. So consider dividing bundled toothpicks that represent a counting number into groups of 3. To divide these toothpicks into groups of 3, we could start by dividing each of the bundles of 10, of 100, of 1000, and so on into groups of 3. Each bundle of 10 can be divided into 3 groups of 3 with 1 left over. Each bundle of 100 can be divided into 33 groups of 3 with 1 left over. Each bundle of 1000 can be divided into 333 groups of 3 with 1 left over. For each place from the 10s place on up, each bundle of toothpicks can always be divided into groups of 3 with 1 left over. Now think of collecting all those left over toothpicks from each bundle and joining them with the individual toothpicks. How many toothpicks will that be? If the original number was 248, then there will be 2 leftover toothpicks from the 2 bundles of 100, as indicated in Figure 12.9. There will be 4 leftover toothpicks from the 4 bundles of 10. (Even though we could make another group of 3 from these, let's

FIGURE 12.9

Explaining the
Divisibility Test
for 3 by
Dividing
Individual
Bundles into
Groups of 3 and
Collecting the
Remainders

248
bundled
toothpicks

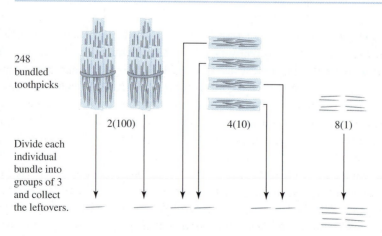

2(100)   4(10)   8(1)

Divide each
individual
bundle into
groups of 3
and collect
the leftovers.

2 + 4 + 8 toothpicks are left over after dividing each individual
bundle into groups of 3.

not, because we want to explain why we can add the digits of the number to determine if the number is divisible by 3. So we want to work with the digit 4, not 1.) If we combine the leftover toothpicks with the 8 individual toothpicks, then in all, that is

$$2 + 4 + 8$$

toothpicks. The case is similar for any other counting number: if we combine the individual toothpicks with all the leftover toothpicks that result from dividing each bundle of 10, 100, 1000, and so on into groups of 3, then the number of toothpicks we will have is given by the sum of the digits of the number. All the other toothpicks have already been put into groups of 3, so if we can put these left over toothpicks into groups of 3, then the original number is divisible by 3. If we can't put these left over toothpicks into groups of 3 with none left, then the original number is not divisible by 3. In other words, we can determine whether a number is divisible by 3 by adding its digits and seeing if that number is divisible by 3.

We can express the previous explanation for why the divisibility test for 3 is valid in a more algebraic way. Let's restrict our explanation to 4-digit counting numbers. Such a number is of the form $ABCD$. Now consider the following equations:

$$\begin{aligned} ABCD &= A \cdot 1000 + B \cdot 100 + C \cdot 10 + D \\ &= (A \cdot 999 + B \cdot 99 + C \cdot 9) + (A + B + C + D) \\ &= (A \cdot 333 + B \cdot 33 + C \cdot 3) \cdot 3 + (A + B + C + D) \end{aligned}$$

The first equation expresses $ABCD$ in its expanded form. After the second $=$ sign, we have separated the leftover toothpicks that result from dividing the bundles of 1000, 100, and 10 into groups of 3, and we have joined these leftovers with the individual toothpicks to make $A + B + C + D$ toothpicks. The expression after the third $=$ sign shows that the toothpicks other than these remaining $A + B + C + D$ can be divided evenly into groups of 3, namely into $A \cdot 333 + B \cdot 33 + C \cdot 3$ groups of 3. Therefore, the original $ABCD$ toothpicks can be divided into groups of 3 with none left over exactly when the remaining $A + B + C + D$ toothpicks can be divided into groups of 3 with none left over.

divisibility test for 9   The **divisibility test for 9** is similar to the divisibility test for 3: given a counting number, add the digits of the number. If the sum of the digits is divisible by 9, then the original number is also divisible by 9, otherwise the original number is not divisible by 9. To explain why this test is a valid way to see if a number is divisible by 9, we can use the same argument as for the divisibility test for 3, suitably modified for 9 instead of 3. You are asked to explain this divisibility test in Problem 9.

divisibility test for 4   The **divisibility test for 4** is as follows: given a whole number, check the number formed by its last two digits. For example, given 123,456,789, check the number 89. If the number formed by the last two digits is divisible by 4, then the original number is divisible by 4, otherwise, it is not. So, because 89 is not divisible by 4, according to the divisibility test for 4, the number 123,456,789 is also not divisible by 4. You are asked to explain this divisibility test in Practice Problem 2.

CLASS ACTIVITY   NOW TURN TO CLASS ACTIVITIES MANUAL

120   The Divisibility Test for 3 p. 403

PRACTICE PROBLEMS FOR SECTION 12.5

1. Use the divisibility test for 3 to determine which of the following numbers are divisible by 3:

   a. 125,389,211,464

   b. 111,111,111,111,111

   c. 123,123,123,123,123

   d. 101,101,101,101,101

2. Explain why the divisibility test for 4 described in the text is a valid way to determine if a counting number is divisible by 4.

ANSWERS TO PRACTICE PROBLEMS FOR SECTION 12.5

1. a. The sum of the digits of 125,389,211,464 is $1 + 2 + 5 + 3 + 8 + 9 + 2 + 1 + 1 + 4 + 6 + 4 = 46$, which is not divisible by 3. Therefore, 125,389,211,464 is not divisible by 3.

    b. The sum of the digits of 111,111,111,111,111 is 5×3, which is divisible by 3. Therefore, 111,111,111,111,111 is divisible by 3.

    c. The sum of the digits of 123,123,123,123,123 is 5×6, which is divisible by 3. Therefore, 123,123,123,123,123 is divisible by 3.

    d. The sum of the digits of 101,101,101,101,101 is 5×2, which is not divisible by 3. Therefore, 101,101,101,101,101 is not divisible by 3.

2. A counting number is divisible by 4 exactly when that number of toothpicks can be divided into groups of 4 with none left over. Consider representing a counting number physically with bundled toothpicks and consider dividing the bundles of toothpicks, from the hundreds place on up, into groups of 4. Each bundle of 100 toothpicks can be divided into 25 groups of 4. Each bundle of 1000 toothpicks can be divided into 250 groups of 4. Each bundle of 10,000 toothpicks can be divided into 2500 groups of 4, and so on, for all the higher places. For each place from the 100s place on up, each bundle of toothpicks can always be divided evenly into groups of 4 with none left over. Therefore, the full number of toothpicks can be divided into groups of 4 with none left over, exactly when the number of toothpicks that are together in the tens and ones places can be divided into groups of 4 with none left over.

PROBLEMS FOR SECTION 12.5

1. According to the divisibility test for 10, to determine whether a counting number is divisible by 10, you only have to check its ones digit. If the ones digit is 0, then the number is divisible by 10. Otherwise, it is not. Use the following ideas to help you give a clear and complete explanation for why this divisibility test is a valid way to determine if a number is divisible by 10:

   A counting number is evenly divisible by 10 exactly when that many toothpicks can be divided into groups of 10 with none left over. Suppose you represent the ex-panded form of a counting number with bundled toothpicks. When you divide the toothpicks into groups of 10, what can you say about each bundle of 10, each bundle of 100, each bundle of 1000, etc.? Why do we only have to check the ones digit of a number to determine whether that number is divisible by 10?

2. According to the divisibility test for 5, to determine whether a counting number is divisible by 5, you only have to check its ones digit. If the ones digit is 0 or

5, then the number is divisible by 5. Otherwise, it is not. Use the following ideas to help you give a clear and complete explanation for why this divisibility test is a valid way to determine whether a number is divisible by 5:

A counting number is divisible by 5 exactly when that many toothpicks can be divided into groups of 5 with none left over. Suppose you represent the expanded form of a counting number with bundled toothpicks. When you divide the toothpicks into groups of 5, what will happen to each bundle of 10, each bundle of 100, each bundle of 1000, etc.? Why do we only have to check the ones digit of a number to determine whether that number is divisible by 5?

3. Use the divisibility test for 3 to determine whether the following numbers are divisible by 3 or not:

   a. 7,591,348

   b. 777,777,777

   c. 157,157,157

   d. 241,241,241,241

4. Beth knows the divisibility test for 3. Beth says that she can tell just by looking, and without doing any calculations at all, that the number

   $$999,888,777,666,555,444,333,222,111$$

   is divisible by 3. How can Beth do that? Explain why it's not just a lucky guess.

5. What are all the different ways to choose the ones digit, A, in the number $572435A$, so that the number will be divisible by 3? Explain your reasoning.

6. Sam used his calculator to calculate

   $$123,123,123,123,123 \div 3$$

   Sam's calculator displayed the answer as

   $$4.1041041041\text{E}13$$

   Sam says that because the calculator's answer is not a whole number, the number $123,123,123,123$ is not evenly divisible by 3. Is Sam right? Why or why not? How do you reconcile this with Sam's calculator's display?

7. Explain how to modify the divisibility test for 3 so that you can determine the remainder of a counting number when it is divided by 3 without dividing the number by 3. Explain why your method for determining the remainder when a number is divided by 3 is valid. Illustrate

your method by determining the remainder of 8,127,534 when it is divided by 3 without actually dividing.

8. For each of the numbers in (a) through (d), verify that the divisibility test for 9 accurately predicts which numbers are divisible by 9. That is, for each number, determine whether it is divisible by 9 by using long division or a calculator. Then see if the sum of the digits of the number is divisible by 9. The two conclusions should agree.

   Example: 52,371 is divisible by 9 because $52,371 \div 9$ is the whole number 5819, with no remainder. The sum of the digits of 52,371, namely, $5 + 2 + 3 + 7 + 1 = 18$, is also divisible by 9.

   a. 18, 27, 36, 45, 54, 63, 72, 81

   b. 2,578,109

   c. 777,777

   d. 777,777,777

9. a. In your own words, explain why the divisibility test for 9 is a valid way to determine whether a three-digit counting number ABC is divisible by 9. Use the following ideas to help you:

   A whole number is divisible by 9 exactly when that many toothpicks can be divided into groups of 9 with none left over. Suppose you represent the expanded form of a counting number ABC with bundled toothpicks. Now consider dividing ABC toothpicks into groups of 9 by first dividing *each individual* bundle of 10 and *each individual* bundle of 100 into groups of 9. Determine how many toothpicks will be left over if you divide the bundles this way and if you do no further dividing into groups of 9.

   b. Relate your explanation in part (a) to the following equations:

   $$\begin{aligned} ABC &= A \cdot 100 + B \cdot 10 + C \\ &= (A \cdot 99 + B \cdot 9) + (A + B + C) \\ &= (A \cdot 11 + B \cdot 1) \cdot 9 + (A + B + C) \end{aligned}$$

10. a. What are all the different ways to choose the ones digit, A, in $271854A$ so that the number will be divisible by 9? Explain your reasoning.

    b. What are all the different ways to choose the tens digit, A, and the ones digit, B, in the number $631872AB$ so that the number will be divisible by 9? Explain your reasoning.

11. a. Find a divisibility test for 25, in other words, find a way to determine if a counting number is divisible by 25 without actually dividing the number by 25.

   b. Explain why your divisibility test for 25 is a valid way to determine whether a counting number is divisible by 25.

12. a. Find a divisibility test for 8. In other words, find a way to determine if a counting number is divisible by 8 without actually dividing the number by 8.

   b. Explain why your divisibility test for 8 is a valid way to determine whether a counting number is divisible by 8.

13. a. Is it true that a whole number is divisible by 6 exactly when the sum of its digits is divisible by 6? Investigate this by considering a number of examples. State your conclusion.

   b. How could you determine whether the number

$$111,222,333,444,555,666,777,888,999,000$$

is divisible by 6 without using a calculator or doing long division? Explain! (*Hint:* $6 = 2 \times 3$.)

14. Investigate the questions in parts (a) and (b) by considering a number of examples.

   a. If a whole number is divisible both by 6 and by 2, is it necessarily divisible by 12?

   b. If a whole number is divisible both by 3 and by 4 is it necessarily divisible by 12?

   c. Based on your examples, what do you think the answers to the questions in (a) and (b) should be?

   d. Based on your answer to part (c), determine whether

$$3,321,297,402,348,516$$

is divisible by 12 without using a calculator or doing long division. Explain your reasoning.

12.6   Rational and Irrational Numbers

In this section, we will see how the rational numbers and the real numbers are related—that is, how fractions and decimals are related. We saw in Section 7.3 that every fraction of whole numbers can be written as a decimal by dividing the denominator into the numerator. Thus, every rational number can be written as a decimal. In this section, we will see that the decimals that come from rational numbers are of two special types: they either repeat or terminate. We will then see that every decimal that either repeats or terminates can be written as a fraction, and we will see how to write such decimals as fractions. In particular, we will see that the number 0.999999..., where the 9s repeat forever, is equal to 1. Finally, in this section, we will see why numbers such as $\sqrt{2}$ and $\sqrt{3}$ are irrational. (i.e., cannot be written as a fraction of whole numbers).

DECIMAL REPRESENTATIONS OF FRACTIONS EITHER TERMINATE OR REPEAT

Recall from Section 7.3 that we can use longhand division to write a fraction of whole numbers as a decimal number. It turns out that when we write a fraction as a decimal, the decimal representation is one of two types. For example, consider the decimal representations of the fractions in Table 12.2. The fractions on the right in Table 12.2 have decimal representations that are **terminating** because these decimals only have finitely many nonzero digits. The fractions on the left in Table 12.2 have decimal representations that are **repeating** because it turns out that each of these has a single digit or a fixed string of digits that repeats forever. Notice that the repeating portion of the decimal need not begin right after the decimal point, as in the decimal representations of $\frac{1}{12}$ and $\frac{1}{888}$.

terminating

repeating

TABLE 12.2

Decimal Representations of Fractions Are of Two Types: Repeating or Terminating

$\frac{5}{9} = 0.555555555555555\ldots$	$\frac{1}{4} = 0.25$
$\frac{1}{7} = 0.142857142857142\ldots$	$\frac{5}{8} = 0.625$
$\frac{41}{333} = 0.123123123123123\ldots$	$\frac{1}{80} = 0.125$
$\frac{1}{12} = 0.083333333333333\ldots$	$\frac{41}{100} = 0.41$
$\frac{1}{888} = 0.001126126126126\ldots$	$\frac{12,345}{100,000} = 0.12345$

To indicate that a digit or a string of digits repeats forever, we can write a bar above the repeating portion. For example,

$$0.\overline{3} \quad \text{means} \quad 0.333333\ldots, \text{ where the 3s repeat forever}$$

$$0.001\overline{126} \quad \text{means} \quad 0.001126126126\ldots, \text{ where the 126s repeat forever}$$

We will now use the long division process to explain why the decimal representation of every fraction of whole numbers must either terminate or repeat. Consider a proper fraction of whole numbers, $\frac{A}{B}$. When we carry out the long division process to calculate the decimal representation of $\frac{A}{B}$, we will get various remainders along the way. For example, when we divide $1 \div 7$, we get the remainders

$$1, \ 3, \ 2, \ 6, \ 4, \ 5, \ 1, \ 3, \ldots$$

and when we divide $1 \div 8$, we get the remainders

$$1, \ 2, \ 4, \ 0$$

as shown in Table 12.3 on page 572. There are three key points to observe about the longhand division process for writing the fraction as a decimal:

1. If you get a remainder of 0, then the decimal representation terminates at that point (i.e., all subsequent digits in the decimal representation of the fraction are 0).

2. If you get a remainder that you got before, the decimal representation will repeat from there on.

3. We can only get remainders that are less that the denominator of the fraction. For example, in finding the decimal representation of $\frac{1}{7}$ we could only get the remainders 0, 1, 2, 3, 4, 5, or 6. (In fact, in that case we get all those remainders except for 0.)

Now imagine carrying out the longhand division process to find the decimal representation of a proper fraction $\frac{A}{B}$. Suppose that we have found $B - 1$ digits to the right of the decimal point. If any of the remainders were 0, then the decimal representation terminates. Now suppose that none of the remainders we found were 0. Each of the $B - 1$ digits we found in the decimal representation was obtained from a remainder by putting a 0 behind the remainder and dividing B into the resulting number. Because of item 3, and because we are assuming that we didn't get 0 as a remainder, the remainders must be among the numbers

$$1, \ 2, \ \ldots, \ B - 1$$

TABLE 12.3

Using Long Division, We See That Decimal Representations of Fractions Either Repeat or Terminate

repeats			terminates		
↓			↓		

```
        repeats                          terminates
           ↓                                 ↓
      0.1428571                          0.125
   7) 1.0000000   remainder 1        8) 1.000      remainder 1
      −7                                −8
       30          remainder 3           20         remainder 2
      −28                               −16
        20         remainder 2           40         remainder 4
       −14                              −40
        60         remainder 6            0         remainder 0
       −56                                          terminates
        40         remainder 4
       −35
        50         remainder 5
       −49
        10         remainder 1
        −7         repeats
         3         remainder 3
```

When we calculate the Bth digit to the right of the decimal point, we use the Bth remainder, which either is 0 or must also be among the numbers

$$1, \ 2, \ \ldots \ B - 1$$

If the Bth remainder is 0 then the decimal representation terminates at that point. But if the Bth remainder is not 0, then since there are only $B - 1$ distinct nonzero remainders that we can possibly get, there must be 2 equal remainders among the first B remainders. When we get a remainder that has appeared before, the decimal representation repeats. So, in this case, the decimal representation of our fraction is repeating. Therefore, for any fraction of whole numbers $\frac{A}{B}$, the decimal representation of $\frac{A}{B}$ either terminates or repeats after at most $B - 1$ places to the right of the decimal point.

We can use the conclusion we just derived to show that there are decimal numbers that are not the decimal representation of any fraction of whole numbers. For example, consider the decimal number

$$0.28228222822228222228\ldots$$

where the pattern of putting more and more 2s in between 8s continues forever. Notice that even though there is a pattern to this decimal, it does not have a *fixed string* of digits that repeats forever. Therefore this decimal number is neither terminating nor repeating, so it cannot be the decimal representation of a fraction of whole numbers. Thus, the set of real numbers, which consists of all decimal numbers, is strictly larger than the set of rational numbers, which consists of all fractions of integers.

> ### CLASS ACTIVITY   NOW TURN TO CLASS ACTIVITIES MANUAL
>
> **12P** Decimal Representations of Fractions p. 406

WRITING REPEATING AND TERMINATING DECIMALS AS FRACTIONS

As we saw previously, if you start with a fraction of whole numbers, its decimal representation is necessarily either terminating or repeating. So if a decimal is neither terminating nor repeating, then it is definitely not the decimal representation of a fraction of whole numbers. But if we have a decimal that is either terminating or repeating, is it necessarily the decimal representation of a fraction of whole numbers? The answer is yes, as we will see.

First, it is easy to write a terminating (i.e., finite) decimal as a fraction by using a denominator that is an appropriate power of 10. For example,

$$0.12345 = \tfrac{12,345}{100,000} \quad 1.2345 = \tfrac{12,345}{10,000}$$

$$0.2004 = \tfrac{2004}{10,000} \quad\quad 12.8 = \tfrac{128}{10}$$

Think of these fractions in terms of division; that is,

$$\frac{2004}{10,000} = 2004 \div 10,000$$

From that point of view, we are simply choosing a denominator so that when we divide, the decimal point will go in the correct place.

Now let's see how to write a repeating decimal as a fraction of whole numbers. By using long division we can determine the useful facts in Table 12.4. We can use these facts to write repeating decimals as fractions as shown in Table 12.5. If we have a repeating decimal whose repeating pattern doesn't start right after the decimal point, then we first shift the decimal point by dividing by a suitable power of 10. We then add on the nonrepeating part, as demonstrated in Table 12.6 on page 574.

TABLE 12.4

Facts We Can Use to Write Repeating Decimals as Fractions

$$\frac{1}{9} = 0.\overline{1} = 0.111111\ldots \qquad\qquad \frac{1}{9999} = 0.\overline{0001} = 0.00010001\ldots$$

$$\frac{1}{99} = 0.\overline{01} = 0.010101\ldots \qquad\qquad \frac{1}{99,999} = 0.\overline{00001} = 0.0000100001\ldots$$

$$\frac{1}{999} = 0.\overline{001} = 0.001001\ldots \qquad\qquad\qquad \vdots$$

TABLE 12.5

Writing Repeating Decimals as Fractions Using Facts from Table 12.4

$$0.\overline{12} = 12 \times 0.010101\ldots = 12 \times \frac{1}{99} = \frac{12}{99}$$

$$0.\overline{123} = 123 \times 0.001001001\ldots = 123 \times \frac{1}{999} = \frac{123}{999}$$

$$0.\overline{1234} = 1234 \times 0.000100010001\ldots = 1234 \times \frac{1}{9999} = \frac{1234}{9999}$$

TABLE 12.6

Writing Repeating Decimals Whose Repeating Part Doesn't Start Right after the Decimal Point as Fractions

$$0.000\overline{12} = \frac{1}{1000} \times 0.121212\ldots = \frac{1}{1000} \times \frac{12}{99} = \frac{12}{99,000}$$

$$0.987\overline{12} = 0.987 + 0.000\overline{12} = \frac{987}{1000} + \frac{12}{99,000} = \frac{97,725}{99,000}$$

CLASS ACTIVITY   NOW TURN TO CLASS ACTIVITIES MANUAL

12Q   Writing Terminating and Repeating Decimals as Fractions p. 410

ANOTHER METHOD FOR WRITING REPEATING DECIMALS AS FRACTIONS

Another nice method for writing a repeating decimal as a fraction involves shifting the decimal point by multiplying by a suitable power of 10 and then subtracting so that the resulting quantity is a whole number. For example, let N stand for the repeating decimal $0.123123123\ldots = 0.\overline{123}$, so that

$$N = 0.\overline{123} = 0.123123123\ldots$$

Then

$$1000N = 123.123123123\ldots$$

We can subtract N from $1000N$ in two different ways:

$$
\begin{array}{ll}
1000N & 123.123123123\ldots \\
\underline{-N} & \underline{-0.123123123\ldots} \\
999N & 123
\end{array}
$$

Therefore,

$$999N = 123$$

and

$$N = \frac{123}{999}$$

Thus,

$$0.\overline{123} = \frac{123}{999}$$

Notice that by multiplying N by 1000 we were able to arrange for the decimal portions to cancel when we subtracted. If we had multiplied by 10 or 100 this canceling would not have happened.

If the repeating part of the decimal does not start right after the decimal point, then we can modify the above method by shifting with additional powers of 10. For example, let

$$N = 0.98\overline{123} = 0.98123123123\ldots$$

Then

$$100,000N = 98,123.123123123\ldots$$

and

$$100N = 98.123123123\ldots$$

Now subtract $100N$ from $100,000N$ in the following two different ways:

$$\begin{array}{ll} 100,000N & 98,123.123123123\ldots \\ \underline{-100N} & \underline{-98.123123123\ldots} \\ 99,900N & 98,025 \end{array}$$

Therefore,

$$99,900N = 98,025$$

and

$$N = \frac{98,025}{99,900}$$

Thus,

$$0.98\overline{123} = \frac{98,025}{99,900}$$

Notice that by multiplying by 100,000 and by 100 we arranged for the decimal portions to cancel when we subtracted.

THE SURPRISING FACT THAT $0.99999\ldots = 1$

We can apply either of our two methods for writing a repeating decimal as a fraction to the repeating decimal

$$0.\overline{9} = 0.99999999\ldots$$

When we do this, we reach the following surprising conclusion:

$$0.\overline{9} = 9 \times 0.111111111\ldots = 9 \times \frac{1}{9} = \frac{9}{9} = 1$$

Similarly, let

$$N = 0.\overline{9} = 0.999999\ldots$$

Then

$$10N = 9.999999\ldots$$

Now subtract N from $10N$ in the following two ways:

$$
\begin{array}{rr}
10N & 9.999999\ldots \\
-N & -0.999999\ldots \\
\hline
9N & 9
\end{array}
$$

Therefore,

$$9N = 9$$

and

$$N = 1$$

Thus,

$$0.\overline{9} = 1$$

Although it may seem like $0.999999\ldots$ should be just a little bit less than 1, it is in fact *equal to* 1.

CLASS ACTIVITY   NOW TURN TO CLASS ACTIVITIES MANUAL

12R What is 0.9999…? p. 412

THE IRRATIONALITY OF THE SQUARE ROOT OF 2 AND OTHER SQUARE ROOTS

square root Given a positive number N, the **square root** of N, denoted $\sqrt{N}$, is the positive number S such that

$$S^2 = N$$

For example, $\sqrt{2}$ is the positive number such that

$$\left(\sqrt{2}\right)^2 = 2$$

Square roots of numbers provide many examples of irrational numbers—that is, numbers that cannot be written as fractions of whole numbers. For example, we will see why $\sqrt{3}$ is irrational. In Class Activity 12S you will show that $\sqrt{2}$ is irrational. Note, however, that square roots of some numbers are rational. For example, $\sqrt{4}$ is rational because

$$\sqrt{4} = 2 = \frac{2}{1}$$

so $\sqrt{4}$ can be written as a fraction of whole numbers.

If you use a calculator to find $\sqrt{3}$, your calculator's display might read

$$1.73205080757$$

Of course, a calculator can only display a certain number of digits, so the calculator is not telling you that $\sqrt{3}$ is exactly equal to 1.73205080757. Rather, it is telling you the first few digits in the decimal representation of $\sqrt{3}$ (although the last digit may be rounded). When we look at the calculator's display, it appears that the decimal representation of $\sqrt{3}$ does not repeat or terminate, but we can *never tell for sure* by looking at a finite portion of a decimal representation whether or not the decimal repeats or terminates. The decimal might not start to repeat until after 100 places, or the decimal might appear to repeat at first, but in fact not repeat. So to determine whether a number such as $\sqrt{3}$ is rational or irrational we must do something other than look at the decimal representation.

We will now show that $\sqrt{3}$ is irrational by supposing that we could write $\sqrt{3}$ as a fraction of counting numbers and by deducing that this could not be the case. So suppose that

$$\sqrt{3} = \frac{A}{B}$$

where A and B are counting numbers. Then

$$3 = \left(\frac{A}{B}\right)^2$$

so

$$3 = \frac{A^2}{B^2}$$

By multiplying both sides of this equation by B^2, we have

$$3 \cdot B^2 = A^2$$

Now imagine factoring A and B into products of prime numbers. Then A^2 has an even number of prime factors because is has twice as many prime factors as A does. For example, if A were 42, then

$$A = 2 \cdot 3 \cdot 7$$

and

$$A^2 = 2 \cdot 3 \cdot 7 \cdot 2 \cdot 3 \cdot 7$$

which has 6 prime factors, twice as many as A. Similarly $3 \cdot B^2$ has an odd number of prime factors because it has one more than twice as many prime factors as B. For example, if B were 35, then

$$B = 5 \cdot 7$$

and

$$3 \cdot B^2 = 3 \cdot 5 \cdot 7 \cdot 5 \cdot 7$$

which has 5 prime factors, one more than twice as many as A.

But if $3 \cdot B^2$ has an odd number of prime factors and A^2 has an even number of prime factors then it cannot be the case that

$$3 \cdot B^2 = A^2$$

Therefore, it cannot be the case that

$$\sqrt{3} = \frac{A}{B}$$

where A and B are counting numbers. Thus, $\sqrt{3}$ is irrational.

CLASS ACTIVITY   NOW TURN TO CLASS ACTIVITIES MANUAL

12S The Square Root of 2 p. 414

12T Pattern Tiles and the Irrationality of the Square Root of 3 p. 417

PRACTICE PROBLEMS FOR SECTION 12.6

1. Write $0.00\overline{577}$, $1.24\overline{3}$, and $13.2\overline{83}$ as fractions of whole numbers, explaining your reasoning.

2. What is the 103rd digit to the right of the decimal point in the decimal representation of $\frac{13}{101}$? Explain your answer.

3. What is another way to write $3.45\overline{9}$ as a decimal? Explain your answer.

4. We have seen that $0.\overline{9} = 1$. Does this mean that there are other decimal representations for the numbers $0.\overline{8}$, $0.\overline{7}$, and $0.\overline{6}$?

5. Find at least two fractions of whole numbers whose decimal representation begins 0.349205986.

6. Are the following numbers rational or irrational?

   a. .252252225252252225252252225... where the pattern of a 2 followed by a 5, two 2s followed by a 5, and three 2s followed by a 5 continues to repeat.

   b. .252252225222252222252222225... where the pattern of placing more and more 2s between 5s continues.

7. Suppose you have two decimals and each one is either repeating or terminating. Is it possible that the product of these two decimals could be neither repeating nor terminating?

8. Use your calculator to find the decimal representation of $\frac{1}{29}$. When you look at the calculator's display, does the number look as if its decimal representation either repeats or terminates? Does the number in fact have a repeating or terminating decimal representation?

9. a. Find the square root of 5 on a calculator. Make a guess: does it look like it is rational or does it look like it is irrational? Why?

   b. Can you tell for sure whether or not $\sqrt{5}$ is rational just by looking at your calculator's display? Explain your answer.

10. Is $\sqrt{1.96}$ rational or irrational?

11. Is the square root of $1.\overline{7}$ rational or irrational?

ANSWERS TO PRACTICE PROBLEMS FOR SECTION 12.6

1. a. Since

$$\frac{1}{999} = 0.\overline{001}$$

it follows that

$$0.\overline{577} = \frac{577}{999}$$

Dividing by 100, we have

$$0.00\overline{577} = \frac{577}{99,900}$$

b. Since

$$0.\overline{3} = \frac{3}{9} = \frac{1}{3}$$

dividing by 100, we have

$$0.00\overline{3} = \frac{1}{300}$$

Since

$$1.27 = \frac{127}{100}$$

it follows that

$$1.27\overline{3} = 1.27 + 0.00\overline{3} =$$

$$\frac{127}{100} + \frac{1}{300} = \frac{382}{300} = \frac{191}{150}$$

c. Since

$$0.\overline{83} = \frac{83}{99}$$

dividing by 10, we have

$$0.0\overline{83} = \frac{83}{990}$$

Since

$$13.2 = \frac{132}{10}$$

it follows that

$$13.2\overline{83} = 13.2 + 0.0\overline{83} = \frac{132}{10} + \frac{83}{990} = \frac{13,151}{990}$$

2. Since

$$\frac{13}{101} = .\overline{1287} = .128712871287\ldots$$

has a string of 4 digits that repeats, every 4th digit is a 7. So the 4th, the 8th, the 12th, …digit is a 7. Since 100 is divisible by 4, the 100th digit is a 7. The 101st digit is then a 1, the 102nd is a 2, and the 103rd digit is an 8.

3. We know that $0.\overline{9} = 1$. Dividing by 100, we see that $0.00\overline{9} = 0.01$. Therefore,

$$3.45\overline{9} = 3.45 + 0.00\overline{9}$$
$$= 3.45 + 0.01$$
$$= 3.46$$

4. No, it turns out that there is no other way to write decimal representations for these numbers. We can, however, write these numbers as the fractions $\frac{8}{9}$, $\frac{7}{9}$, and $\frac{6}{9} = \frac{2}{3}$.

5. There are many examples. The fraction

$$\frac{349,205,986}{1,000,000,000}$$

is one example. So is

$$\frac{349,205,986}{999,999,999}$$

(This one is repeating.) So is

$$\frac{3,492,059,861,544}{10,000,000,000,000}$$

(This one is terminating.)

6. a.

$$0.252252225252252225252252225\ldots = $$
$$0.\overline{252252225}$$

is a repeating decimal, so it is a rational number.

b.

$$0.252252225222252222252222225\ldots$$

is irrational because it is neither repeating nor terminating.

7. No. To see why not, notice that your two decimals can both be written as fractions of whole numbers because they are repeating or terminating. When we multiply fractions of whole numbers, the result is again a fraction of whole numbers. Therefore, the product of the two decimals must have either a repeating or a terminating decimal representation.

8. On a calculator, the decimal representation looks like it is neither terminating nor repeating. But this is just because a calculator can only display so many digits. In fact, since $\frac{1}{29}$ is a fraction of whole numbers, its decimal representation must be either terminating or repeating. (In fact, it is repeating.)

9. a. The calculator display of the decimal representation of the square root of 5 does not appear to be either terminating or repeating, so it looks like the square root of 5 is irrational. But this is not a proof.

   b. No, you definitely can't tell for sure whether the square root of 5 is irrational just from looking at the calculator's display. Conceivably, its decimal representation might end after 200 digits. Or it might start to repeat after 63 digits. You can't tell for sure what's going to happen just by looking at the calculator. (It turns out that the square root of 5 is irrational. The proof is the same as for the square root of 3 and the square root of 2.)

10.

$$\sqrt{1.96} = \sqrt{\frac{196}{100}} = \frac{14}{10}$$

So the square root of 1.96 is rational.

11.

$$1.\overline{7} = 1 + \frac{7}{9} = \frac{16}{9}$$

So the square root of $1.\overline{7}$ is $\frac{4}{3}$, which is rational.

PROBLEMS FOR SECTION 12.6

1. a. Use long division to determine the decimal representation of $\frac{1}{37}$. Determine whether the decimal representation repeats or terminates; if it repeats, describe the string of repeating digits.

   b. Use long division to determine the decimal representation of $\frac{1}{74}$. Determine whether the decimal representation repeats or terminates; if it repeats, describe the string of repeating digits.

2. a. Use long division to determine the decimal representation of $\frac{1}{101}$. Determine whether the decimal representation repeats or terminates; if it repeats, describe the string of repeating digits.

   b. Use long division to determine the decimal representation of $\frac{1}{505}$. Determine whether the decimal representation repeats or terminates; if it repeats, describe the string of repeating digits.

3. Write the following decimal numbers as fractions of whole numbers, explaining your reasoning (you need not write the fractions in simplest form):

   a. $0.\overline{56}$, $0.000\overline{56}$, $1.111\overline{56}$

   b. $0.\overline{987}$, $0.00\overline{987}$, $0.40\overline{987}$

   c. $1.234\overline{567}$

4. What is the 100th digit to the right of the decimal point in the decimal representation of $\frac{2}{37}$? Explain your reasoning.

5. What is another way to write 1.824 as a decimal? Explain your reasoning.

6. Without actually determining the decimal representation of $\frac{1}{47}$, explain why its decimal representation must either terminate or repeat after at most 46 decimal places to the right of the decimal point.

7. Give an example of an irrational number and explain in detail why your number is irrational.

8. In your own words, prove that the square root of 5 is irrational using the same ideas as we did to prove that the square root of three is irrational.

9. Show how to find the exact decimal representation of

$$0.\overline{027} \times 0.\overline{18} = 0.027027027\ldots \times 0.18181818\ldots$$

without using a calculator. Is the product a repeating decimal? If so, what are the repeating digits? Hint: Write the numbers in a different form.

10. Carl's calculator only displays ten digits. Carl punches some numbers and some operations on his calculator and shows you his calculator's display: 0.034482758. Carl has to figure out whether or not the number whose first nine digits behind the decimal point are displayed on his calculator is rational or irrational.

    a. If you had to make a guess, what would you guess about Carl's number: do you think it is rational or irrational? Why?

    b. Find a fraction of whole numbers whose first nine digits behind the decimal point in its decimal representation agree with Carl's calculator's display. (You do not have to put the fraction in simplest form.)

    c. What if Carl's number was actually

0.03448375800344837580003448375800003448375800003448375800003448375 8 \ldots,

    where the pattern of putting in more and more zeros before a block of 34483758 continues forever. In that case, is Carl's number rational or irrational? Why?

    d. Without any knowledge of how Carl got his number to display on his calculator, is it possible to say for sure whether or not it is rational? Explain.

11. Fran has a calculator that shows at most ten digits. Fran punches some numbers and operations on her calculator and shows you the display: 0.232323232

    a. If you had to make a guess, do you think Fran's number is rational or irrational? Why?

b. Find two different rational numbers whose decimal representation begins 0.232323232.

c. Find an irrational number whose decimal representation begins 0.232323232.

12. Tyrone used a calculator to solve a problem. The calculator gave Tyrone the answer 0.217391304. Without any further information, is it possible to tell if the answer to Tyrone's problem can be written as a fraction of whole numbers? Explain.

13. It is a fact that

$$513{,}239 \times 194{,}841 = 99{,}999{,}999{,}999.$$

Use the multiplication fact above to find the exact decimal representations of

$$\frac{1}{513{,}239}$$

and

$$\frac{1}{194{,}841}$$

(Your answer should not just show the first bunch of digits, as on a calculator, but should make clear what *all* the digits in the decimal representation are.) If the decimal representation is repeating, describe the string of digits that repeat. Explain your reasoning. Hint: Think about rewriting the fractions using the denominator 99,999,999,999.

14. Suppose you have a fraction of the form $\frac{1}{A}$, where A is a whole number. Also, suppose that the decimal representation of this fraction is repeating, that the string of repeating digits starts right after the decimal point, and that this string consists of 5 digits.

a. Explain why $\frac{1}{A}$ is equal to a fraction with denominator 99,999.

b. Factor 99,999 as a product of prime numbers.

c. Use parts (a) and (b) to help you find all the prime numbers A such that $\frac{1}{A}$ has a repeating decimal representation with a 5-digit string of repeating digits.

15. Suppose you have a whole number N that divides evenly into 9,999,999 (in other words 9,999,999 ÷ N is a whole number), but N is not 1, 3 or 9. What can you say about the decimal representation of $\frac{1}{N}$? Explain your reasoning.

16. a. For each of the fractions in the first column of Table 12.7, factor the denominator of the fraction as a product of prime numbers.

b. For each of the fractions in the second column of Table 12.7, factor the denominator of the fraction as a product of prime numbers.

c. Contrast the prime factors that you found in part (a) with the prime factors that you found in part (b). Contrast the decimal representations of the fractions in the first column of Table 12.7 with the decimal representations of the fractions in the second column of Table 12.7. Based on your findings, develop a conjecture (an educated guess) about decimal representations of fractions.

17. a. Use a calculator to calculate

$$\frac{1}{5^{100}}$$

b. Based on your calculator's display, is it possible to tell whether or not $\frac{1}{5^{100}}$ has a repeating decimal representation or a terminating decimal representation? Why or why not?

c. Find a fraction that is equal to $\frac{1}{5^{100}}$ and that has a denominator that is a power of 10.

d. Based on part (c), determine whether $\frac{1}{5^{100}}$ has a repeating or a terminating decimal representation. Explain your reasoning.

18. a. Use a calculator to calculate

$$\frac{1}{7^{100}}$$

b. Based on your calculator's display, is it possible to tell whether or not $\frac{1}{7^{100}}$ has a repeating decimal representation or a terminating decimal representation? Why or why not?

c. Can there be a fraction of whole numbers that is equal to $\frac{1}{7^{100}}$ and that has a denominator that is a power of 10? Why or why not?

d. Based on part (c), determine whether $\frac{1}{7^{100}}$ has a repeating or a terminating decimal representation. Explain your reasoning.

19. a. Suppose that a fraction of counting numbers $\frac{A}{B}$ has a terminating decimal representation and that $\frac{A}{B}$ is in simplest form. Explain why B must divide a power of 10.

b. Suppose that a fraction of counting numbers $\frac{A}{B}$ has a terminating decimal representation and that $\frac{A}{B}$ is in simplest form. Show that when B is factored into a product of prime numbers, the only prime numbers that can appear are 2 and 5.

c. Given a counting number B, suppose that when B is factored into a product of prime numbers, the only prime numbers that appear are 2, or 5, or both 2 and 5. Let A be another counting number. Explain why $\frac{A}{B}$ must have a terminating decimal representation.

d. Does

$$\frac{1}{2^{100}}$$

have a repeating or a terminating decimal representation? Explain your answer.

e. Does

$$\frac{1}{3^{100}}$$

have a repeating or a terminating decimal representation? Explain your answer.

20. a. Find the decimal representations of the fractions that follow. In each case, compare the number of digits in the repeating string and the *denominator minus*

one. (Compare the number of digits in the repeating string of $\frac{1}{7}$ and the number 6, compare the number of digits in the repeating string of $\frac{1}{11}$ and the number 10, etc.) Find a general relationship (other than the fact that the number of digits in the repeating string is less than the denominator minus one). Here are the fractions:

$$\frac{1}{7}, \quad \frac{1}{11}, \quad \frac{1}{13}, \quad \frac{1}{37}, \quad \frac{1}{41},$$

$$\frac{1}{73}, \quad \frac{1}{101}, \quad \frac{1}{137}, \quad \frac{1}{239}, \quad \frac{1}{271}$$

b. The denominators of the fractions in part (a) are prime numbers. Now look at lots of examples of other fractions of the form $\frac{1}{A}$, where A is a whole number that is not a prime number, but where the decimal representation of $\frac{1}{A}$ is repeating. In each case, compare the number of digits in the repeating string and the denominator minus one and see if the relationship you found in part (a) still holds all the time.

TABLE 12.7

Decimal Representations of Various Fractions

$\frac{1}{12}$	=	.0833333333333333	$\frac{1}{4}$	=	.25
$\frac{1}{11}$	=	.0909090909090909	$\frac{2}{5}$	=	.4
$\frac{113}{33}$	=	3.424242424242424	$\frac{37}{8}$	=	4.625
$\frac{491}{550}$	=	.8927272727272727	$\frac{17}{50}$	=	.34
$\frac{14}{37}$	=	.3783783783783783	$\frac{1}{125}$	=	.008
$\frac{35}{101}$	=	.3465346534653465	$\frac{9}{20}$	=	.45
$\frac{1}{41}$	=	.0243902439024390	$\frac{19}{32}$	=	.59375
$\frac{5}{7}$	=	.7142857142857142	$\frac{1}{3200}$	=	.0003125
$\frac{1}{14}$	=	.0714285714285714	$\frac{1}{64,000}$	=	.000015625
$\frac{1}{21}$	=	.0476190476190476	$\frac{1}{625}$	=	.0016

Chapter 13

Functions and Algebra

 lgebra is the study of the structure of arithmetic and of ways to use the structure of arithmetic to describe situations and to solve problems. Using algebra we can study the structure of growing or repeating patterns and we can solve many problems that can be formulated with an equation. To solve equations we must rely on the structure of arithmetic. Although children usually do not study algebra formally until the end of middle school or in high school, the foundations for learning algebra must be laid in elementary school. We have already studied the structure of arithmetic in Chapters 4 and 5 when we learned about the commutative, associative, and distributive properties and their importance to standard computational procedures and to mental calculation methods. In this chapter, we continue our study of algebra by studying patterns, functions, and some elementary ways to solve equations. Algebra is a deep subject that continues far beyond what we discuss in this book. Our focus here is on aspects of algebra that will be most useful to elementary school teaching.

Concerning the learning of algebra, the National Council of Teachers of Mathematics recommends the following (see [44]):

Instructional programs from prekindergarten through grade 12 should enable all students to—

NCTM Standards

- *understand patterns, relations, and functions;*

- *represent and analyze mathematical situations and structures using algebraic symbols;*

- *use mathematical models to represent and understand quantitative relationships;*

- *analyze change in various contexts.*

For NCTM's specific algebra recommendations for grades Pre-K–2, grades 3–5, grades 6–8, and grades 9–12, see

`www.aw-bc.com/beckmann`

13.1 Mathematical Expressions, Formulas, and Equations

Essential skills in algebra include creating and evaluating mathematical expressions, applying formulas, and manipulating and solving equations. In this section, we define some of the basic terms we use in algebra and we study some of the elementary ways that we use these fundamental concepts.

VARIABLES, EXPRESSIONS, AND FORMULAS

variable   A **variable** is a letter or other symbol that stands for any number within a specified (or understood) set of numbers. For example, if we say "a circle of radius r has area πr^2," then r is a variable that stands for any positive real number. In this context, it wouldn't make sense for r to be negative, so it's simply understood that r is restricted to positive numbers.

mathematical expression
expression   A **mathematical expression**, or simply, **expression**, is a meaningful string of numbers, operations (such as $+$, $-$, $\times$, $\div$, or raising to a power, but not $=$), and possibly variables. For example,

$$3 + 5 \times 2 - 4^3$$

and

$$2x + 3y + 7$$

are expressions. A convention in writing expressions with variables is to drop multiplication symbols between numbers and variables or variables and variables. For example,

$$2x \text{ means } 2 \cdot x$$

and

$$3xy \text{ means } 3 \cdot x \cdot y$$

Also, we usually write numbers before variables when numbers and variables are multiplied in an expression. So instead of writing $x \cdot 2$ or $x2$, we prefer to write $2x$. Since multiplication is commutative, we can always write expressions so that the numbers precede the variables.

Even very young children in elementary school learn about simple mathematical expressions, such as

$$2 + 3$$

$$3 \times 5$$

$$35 \div 7$$

Older elementary school children learn to work with more complex expressions, such as

$$5 \cdot 12 + 7$$

$$\frac{3 \cdot 10}{5 \cdot 9}$$

formula   A **formula** is an expression involving variables. A commonly used expression that involves variables is usually called a formula rather than an expression. For example, we talk about the formula for the area of a rectangle of length L and width W,

$$L \times W$$

or the formula for the circumference of a circle of radius r,

$$2\pi r$$

In order to apply algebra to solve story problems we must be able to write expressions or formulas for given situations. For example, suppose that Corinna has two collections of postcards on her wall. One collection is arranged in 5 rows with 6 postcards in each row. The other collection is arranged in 10 rows with 2 postcards in each row. Then the following expression stands for the total number of postcards that Corinna has on her wall:

$$5 \cdot 6 + 10 \cdot 2$$

The expression shows the $5 \cdot 6$ postcards from the 5 groups of 6 added to the $10 \cdot 2$ from the 10 groups of 2 postcards. Notice that we do not need to put parentheses around $5 \cdot 6$ and $10 \cdot 2$ because of the conventions of order of operations. (See Section 5.7 on page 188.)

EQUATIONS

When we have an expression or a formula, we usually want to say that it is equal to some number or some other expression. An **equation** is a statement that an expression or number is equation   equal to another expression or number. Unlike expressions, equations involve an equals ($=$) sign. For example,

$$4x = 8$$

$$3x + 5 = 2x + 6$$

$$y = 4x + 1$$

$$3 + 2 = 5$$

$$3 + 2 = 4 + 1$$

are equations.

The most familiar way that equations arise is to show the result of a calculation, such as

$$2 + 5 = 7$$

or

$$5 \cdot 6 + 2 = 32$$

But equations arise whenever we have two different ways to express the same quantity; they are not used only to signify the result of a calculation. For example, the equation

$$99 + 17 = 100 + 16$$

states that $99 + 17$ and $100 + 16$ are equal, but it does not say what number these two expressions are equal to. Of course, we might use such an equation in the process of evaluating $99 + 17$ because the second expression, $100 + 16$, is easy to evaluate mentally. Similarly, the equation

$$38 \cdot 47 = 47 \cdot 38$$

states that the two expressions, $38 \cdot 47$ and $47 \cdot 38$ are equal, but it does not say what number they are equal to.

Equations often arise from the situation in a story problem. For example, suppose there are D dollars in a bank account initially. When $\frac{3}{4}$ of the money in the account is removed and another \$150 is added, there is \$400 in the account. This situation gives rise to the equation

$$\frac{1}{4}D + 150 = 400$$

Why? Focus first on the left-hand side of the equation (to the left of the equals sign). When $\frac{3}{4}$ of the money in the account is removed, $\frac{1}{4}$ remains, so $\frac{1}{4}D$ remains. When \$150 is added, there is then

$$\frac{1}{4}D + 150$$

dollars in the account. On the other hand, we are told that there is \$400 in the account at this point. Therefore, the two amounts must be equal, so $\frac{1}{4}D + 150 = 400$.

Some situations or story problems give rise to equations that involve more than one variable. For example, suppose that there are 3 times as many books in Brittany's backpack as in Toby's. If B represents the number of books that Brittany has and if T represents the number of books that Toby has, then we have the equation

$$B = 3T$$

Why is it $B = 3T$ and not $3B = T$? To see why, it may help you to draw a simple picture for the situation, such as the one in Figure 13.1.

In problem situations we usually find an equation for the purpose of solving the problem. In this case, we must solve the equation we have found. In Section 13.2 we will study some techniques for solving the most elementary kinds of equations.

FIGURE 13.1

Brittany Has Three Times as Many Books as Toby

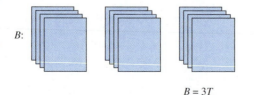

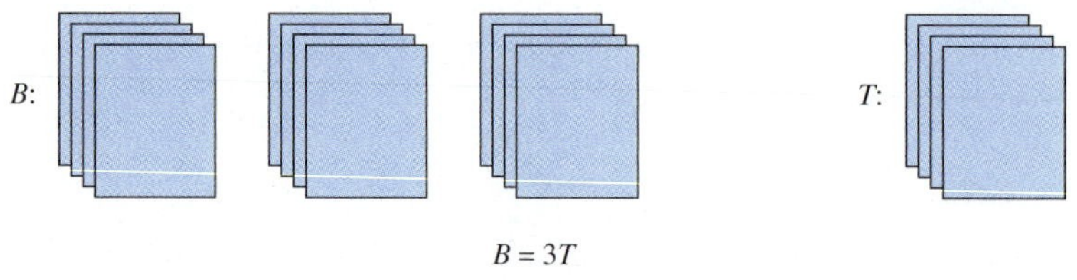

$B = 3T$

EVALUATING EXPRESSIONS AND FORMULAS

evalutate an
expressions

When we have an expression that does not involve variables, we usually want to evaluate it. To **evaluate** an expression that does not involve variables means to determine the number to which the expression is equal. We usually summarize the result in an equation. For example, to evaluate the expression

$$5 \cdot 6 + 10 \cdot 2$$

for the total number of Corinna's postcards, we carry out the indicated operations, following the conventional order of operations. (See Section 5.7 on page 188.) We find that

$$5 \cdot 6 + 10 \cdot 2 = 50$$

Sometimes we may prefer to string together several equations and evaluate the expression in several steps:

$$5 \cdot 6 + 10 \cdot 2 = 30 + 20$$
$$= 50$$

evaluate a
formula

To **evaluate** a formula or an expression that involves variables means to replace the variables with numbers and to determine the number to which the resulting expression is equal. For example, to evaluate

$$3x + 4y$$

at $x = 6$ and $y = 7$, we replace x with 6 and y with 7. We find that

$$3 \cdot 6 + 4 \cdot 7 = 18 + 28$$
$$= 46$$

In order to be able to use algebra effectively, children must become proficient at evaluating expressions. To evaluate expressions correctly and efficiently is just a matter of using the correct order of operations, using the properties of arithmetic, and working with fractions and mixed numbers according to the rules we have studied in previous chapters.

When an expression involves fractions we can sometimes find ways to make evaluating the expression easy. For example, to evaluate

$$\frac{8}{57} \cdot \frac{57}{61}$$

we could write

$$\frac{8}{57} \cdot \frac{57}{61} = \frac{8 \cdot 57}{57 \cdot 61} = \frac{456}{3477}$$

We would then put $\frac{456}{3477}$ in its simplest form, $\frac{8}{61}$. However, a more efficient way to evaluate $\frac{8}{57} \cdot \frac{57}{61}$ is to "cancel" the 57s, namely, to write

$$\frac{8}{57} \cdot \frac{57}{61} = \frac{8}{\cancel{57}} \cdot \frac{\cancel{57}}{61} = \frac{8}{61}$$

Why is this canceling valid? Canceling is just shorthand for several steps, such as the following, that use fraction multiplication, the commutative property of multiplication, and equivalent fractions:

$$\frac{8}{57} \cdot \frac{57}{61} = \frac{8 \cdot 57}{57 \cdot 61} = \frac{8 \cdot 57}{61 \cdot 57} = \frac{8}{61} \cdot \frac{57}{57} = \frac{8}{61}$$

Rather than write all these steps we simply cancel the 57s. In effect, the 57s are just "pulled out" to make $\frac{57}{57}$, which is 1.

How can we make

$$\frac{21}{50} \cdot \frac{1}{35}$$

easy to evaluate? In this case, a factor 7 in the 35 and a factor 7 in the 21 cancel:

$$\frac{\overset{3}{\cancel{21}}}{50} \cdot \frac{1}{\underset{5}{\cancel{35}}} = \frac{3 \cdot 1}{50 \cdot 5} = \frac{3}{250}$$

Again, we can write equations to explain why this canceling is valid:

$$\frac{21}{50} \cdot \frac{1}{35} = \frac{21 \cdot 1}{50 \cdot 35} = \frac{3 \cdot 7}{50 \cdot 5 \cdot 7} = \frac{3}{50 \cdot 5} \cdot \frac{7}{7} = \frac{3}{250}$$

In effect, the 7s are just "pulled out" to make $\frac{7}{7}$, which is 1.

CLASS ACTIVITY   NOW TURN TO CLASS ACTIVITIES MANUAL

13A Writing Expressions and a Formula for a Flower Pattern p. 423

13B Expressions and Formulas for Story Problems p. 426

13C Deriving the Formula for Temperature in Degrees Fahrenheit in Terms of Degrees Celsius p. 428

13D Writing Equations for Square Patterns p. 430

13E Writing Equations for Story Situations p. 431

13F Evaluating Expressions with Fractions Efficiently and Correctly p. 433

PRACTICE PROBLEMS FOR SECTION 13.1

1. Write an expression that uses addition and multiplication to describe the total number of dots in Figure 13.2.

2. Goo-Young has lollipops to distribute among goodie bags. When Goo-Young tries to put L lollipops in each of 6 goodie bags, she is 1 lollipop short.

a. Draw pictures showing how Goo-Young distributes her lollipops if $L = 3$ and if $L = 4$.

b. Write a formula for the total number of lollipops Goo-Young has in terms of L. Relate your formula to your pictures in part (a).

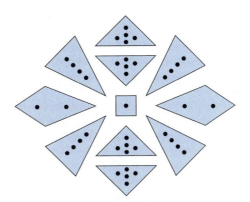

FIGURE 13.2

Write an Expression for the
Total Number of Dots

3. a. If $\frac{1}{3}$ of the gasoline in the can is poured out and then another $\frac{1}{2}$ liter of gasoline is poured into the can, then how much gasoline is in the can? Write a formula in terms of L. Evaluate your formula when $L = 3$ and when $L = 3\frac{1}{2}$.

    b. If $\frac{1}{2}$ liter of gasoline is poured into the can and then $\frac{1}{3}$ of the gasoline in the can is poured out, then how much gasoline is in the can? Write a formula in terms of L. Is your formula the same as in part (a)? Evaluate your formula when $L = 3$ and when $L = 4$.

4. There were P pounds of apples for sale in the produce area of a store. After $\frac{2}{5}$ of the apples were sold, a clerk brought out another 30 pounds of apples. Then $\frac{1}{3}$ of all the apples in the produce area were sold. Write a formula in terms of P for the number of pounds of apples that are in the produce area now.

5. Write two different expressions for the total number of small squares in Figure 13.3. Each expression should use either multiplication or addition, or both. Write an equation by setting these two expressions equal to each other.

6. Let P be the initial population of a town. After 10% of the town leaves, another 1400 people move in to the town. The new population is 5000. Write an equation, involving P, that corresponds to this situation.

7. There are $2\frac{1}{2}$ times as many people in Popperville as in Boppertown. Write an equation that corresponds to this situation.

8. Show how to make the expression

$$\frac{37}{80} \cdot \frac{40}{37} + \frac{17}{30} \cdot \frac{15}{34}$$

    easy to evaluate without a calculator.

9. Write equations that use rules about operating with fractions, as well as properties of arithmetic, to show why it

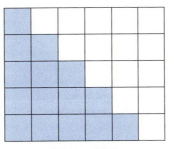

FIGURE 13.3

Write an Equation for
This Picture

is valid to cancel to evaluate the expression

$$\frac{35 + 21}{63}$$

as follows:

$$\frac{\overset{5}{\cancel{35}} + \overset{3}{\cancel{21}}}{\underset{9}{\cancel{63}}} = \frac{5 + 3}{9} = \frac{8}{9}$$

10. Explain why it is *not valid* to evaluate

$$\frac{35 \cdot 21}{63}$$

by canceling as follows:

warning, incorrect : $\dfrac{\overset{5}{\cancel{35}} \cdot \overset{3}{\cancel{21}}}{\underset{9}{\cancel{63}}} = \dfrac{5 \cdot 3}{9} = \dfrac{15}{9} = 1\dfrac{2}{3}$

ANSWERS TO PRACTICE PROBLEMS FOR SECTION 13.1

1. There are 4 triangles (arranged vertically in the center) that each contain 5 dots. These 4 triangles contribute $4 \cdot 5$ dots. There are 4 triangles (the elongated ones at the corners) that each contain 4 dots. These 4 triangles contribute $4 \cdot 4$ dots. There are 2 quadrilaterals (at the left and right) that each contain 2 dots. These quadrilaterals contribute $2 \cdot 2$ dots. The square in the center contributes 1 more dot. All together, the number of dots in the figure is

$$4 \cdot 5 + 4 \cdot 4 + 2 \cdot 2 + 1$$

2. a. See Figure 13.4.

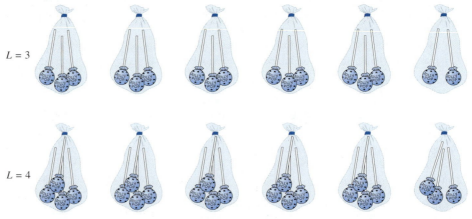

$L = 3$

$L = 4$

FIGURE 13.4

Lollipops Distributed Among 6 Bags with L Lollipops in Each Bag Except the Last

b. If Goo-Young puts L lollipops in each of 6 bags, except that the last bag is 1 short, then there is a total of

$$6L - 1$$

lollipops in the bags. We see this formula arising from Figure 13.4 because there would be $6L$ lollipops in the 6 bags if there were L lollipops in each bag, but the last bag has 1 less, so there are $6L - 1$ lollipops. You could also write the formula

$$5L + (L - 1)$$

for the number of lollipops. This formula is valid because there are L lollipops in each of 5 bags, but the 6th bag contains $L - 1$ lollipops.

3. a. When $\frac{1}{3}$ of the gasoline in the can is poured out, $\frac{2}{3}$ of the gasoline in the can remains. Therefore, after pouring out the gasoline, there are $\frac{2}{3}L$ liters remaining in the can. When another $\frac{1}{2}$ liter is poured in, there are

$$\frac{2}{3}L + \frac{1}{2}$$

liters of gasoline in the can. When $L = 3$, the number of liters of gasoline remaining in the can is

$$\frac{2}{3} \cdot 3 + \frac{1}{2} = 2\frac{1}{2}$$

When $L = 3\frac{1}{2}$, the number of liters of gasoline remaining in the can is

$$\frac{2}{3} \cdot 3\frac{1}{2} + \frac{1}{2} = \frac{2}{3} \cdot \frac{7}{2} + \frac{1}{2} = \frac{17}{6} = 2\frac{5}{6}$$

b. When $\frac{1}{2}$ liter of gasoline is poured into the can there are $L + \frac{1}{2}$ liters in the can. When $\frac{1}{3}$ of this amount is poured out, $\frac{2}{3}$ of the amount remains in the can. Therefore, after pouring out $\frac{1}{3}$ of the gasoline in the can, there are

$$\frac{2}{3}\left(L + \frac{1}{2}\right)$$

liters left in the can. We can use the distributive property to rewrite this formula for the number of liters of gasoline in the can as

$$\frac{2}{3}L + \frac{2}{3} \cdot \frac{1}{2} = \frac{2}{3}L + \frac{1}{3}$$

Notice that this formula, $\frac{2}{3}L + \frac{1}{3}$ is different from the formula $\frac{2}{3}L + \frac{1}{2}$ in part (a). When $L = 3$, the number of liters of gasoline remaining in the can is

$$\frac{2}{3} \cdot 3 + \frac{1}{3} = 2\frac{1}{3}$$

When $L = 3\frac{1}{2}$, the number of liters of gasoline remaining in the can is

$$\frac{2}{3} \cdot 3\frac{1}{2} + \frac{1}{3} = \frac{2}{3} \cdot \frac{7}{2} + \frac{1}{3} = \frac{7}{3} + \frac{1}{3} = 2\frac{2}{3}$$

4. When $\frac{2}{5}$ of the P pounds of apples were sold, $\frac{3}{5}$ remain. So the number of pounds of apples remaining at this point is $\frac{3}{5}P$. When 30 more pounds are brought in, there are $\frac{3}{5}P + 30$ pounds of apples. When $\frac{1}{3}$ of these

apples are sold, $\frac{2}{3}$ remain. Notice that this is $\frac{2}{3}$ of the $\frac{3}{5}P + 30$ pounds of apples. Therefore, there are now

$$\frac{2}{3} \cdot \left(\frac{3}{5}P + 30\right)$$

pounds of apples set out. Using the distributive property, we can also write this formula as

$$\frac{2}{5}P + 20$$

5. On the one hand, since Figure 13.3 consists of 5 rows of small squares with 6 squares in each row, the total number of small squares in the figure is $5 \cdot 6$. On the other hand, the shaded portion of the figure consists of

$$1 + 2 + 3 + 4 + 5$$

small squares, which we see by adding the number of squares in the first, second, third, fourth, and fifth rows. The unshaded portion of the figure is a rotated copy of the shaded portion. Therefore, the unshaded portion has the same number of small squares, so the total number of small squares in the figure is

$$2 \cdot (1 + 2 + 3 + 4 + 5)$$

Thus, the figure gives us the equation

$$2 \cdot (1 + 2 + 3 + 4 + 5) = 5 \cdot 6$$

6. After 10% of the town leaves, 90% of the town remains. So, after the 10% leave, there are $0.9P$ people in the town. When another 1400 people move in, the population rises to $0.9P + 1400$. We are given that this new population is 5000. Therefore,

$$0.9P + 1400 = 5000$$

7. If we let P stand for the population of Popperville and B stand for the population of Boppertown, then

$$P = 2\frac{1}{2} \cdot B$$

The picture in Figure 13.5 on p. 592 may help you see why the correct equation is $P = 2\frac{1}{2} \cdot B$ and not $B = 2\frac{1}{2} \cdot P$.

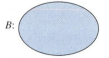

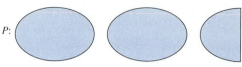

FIGURE 13.5

There are $2\frac{1}{2}$ Times as Many People in Popperville as in Boppertown

8. The 37s cancel each other. The 40 in the numerator cancels with the 40 in the denominator, leaving 2 in the denominator. The 17 in the numerator cancels with a 17 in the denominator, leaving 2 in the denominator. The 15 in the numerator cancels with 30 in the denominator leaving 2 in the denominator. We can thus write

$$\frac{\overset{1}{\cancel{37}}}{\underset{2}{\cancel{80}}} \cdot \frac{\overset{1}{\cancel{40}}}{\underset{1}{\cancel{37}}} + \frac{\overset{1}{\cancel{17}}}{\underset{2}{\cancel{30}}} \cdot \frac{\overset{1}{\cancel{15}}}{\underset{2}{\cancel{34}}} = \frac{1}{2} + \frac{1}{4} = \frac{3}{4}$$

9. We can use the distributive property to "pull out" a 7 in the numerator. We can also "pull out" a 7 in the denominator to make a $\frac{7}{7}$, which is 1. All together, we have

$$\frac{35 + 21}{63} = \frac{(5+3) \cdot 7}{9 \cdot 7} = \frac{5+3}{9} \cdot \frac{7}{7} = \frac{8}{9} \cdot 1 = \frac{8}{9}$$

10. In the expression

$$\frac{35 \cdot 21}{63} = \frac{(5 \cdot 7) \cdot (3 \cdot 7)}{9 \cdot 7}$$

the 7 in the denominator can only cancel one of the 7s in the numerator, not both of them. So it is correct to write

$$\frac{35 \cdot 21}{63} = \frac{35 \cdot (3 \cdot 7)}{9 \cdot 7} = \frac{35 \cdot 3}{9} \cdot \frac{7}{7} = \frac{35}{3} = 11\frac{2}{3}$$

PROBLEMS FOR SECTION 13.1

1. Write an expression that uses addition and multiplication to describe the total number of stars in Figure 13.6.

FIGURE 13.6

Write an Expression for This Picture

2. Tyrone has a marble collection that he wants to divide among some bags. When Tyrone puts M marbles into each of 8 bags, he has 3 marbles left over. Write a formula for the total number of marbles Tyrone has in terms of M. Evaluate your formula when $M = 3$ and when $M = 4$.

3. Kenny has stickers to distribute among some goodie bags. When Kenny tries to put 5 stickers into each of G goodie bags he is 2 stickers short of filling the last bag.

   a. Draw pictures showing how Kenny distributes his stickers if $G = 3$ and if $G = 4$.

   b. Write a formula for the total number of stickers Kenny has in terms of G. Relate your formula to your pictures in part (a).

4. Ted is saving for a toy. Ted has no money now, but he figures that if he saves $\$D$ per week for the next 4 weeks he will have $1 more than he needs for the toy.

   a. Draw a picture to show how much Ted will save and how much he will need for the toy if $D = 3$.

   b. Write a formula for the cost of the toy Ted wants to buy in terms of D. Relate your formula to your picture in part (a).

5. If you pay $7 for a night light that costs $0.50 per year in electricity to operate, then how much will you have

spent on the night light if you operate it continuously for Y years? Write a formula in terms of Y. Evaluate your formula for the cost of the night light when $Y = 4$ and when $Y = 5\frac{1}{2}$.

6. There were Q quarts of liquid in a container. First, $\frac{3}{4}$ of the liquid in the container was removed. Then another $\frac{1}{2}$ quart was poured into the container. Write an expression

in terms of Q for the number of quarts of liquid in the container at the end.

7. Martin had M dollars initially. First, Martin spent $\frac{1}{3}$ of his money. Then Martin spent $\frac{3}{4}$ of what was left. Finally Martin spent another \$20. Write an expression in terms of M for the final amount of money that Martin has.

---

8. Write two different expressions for the total number of small squares in pattern (a) of Figure 13.7. Each expression should use either multiplication or addition, or both. Write an equation by setting these two expressions equal to each other.

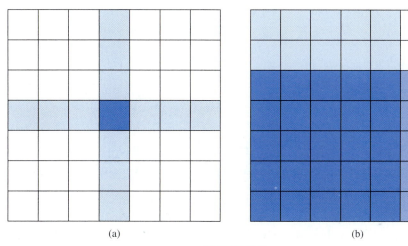

(a)   (b)

FIGURE 13.7

Write Equations for These Square Patterns

---

9. Write two different expressions for the total number of small squares in pattern (b) of Figure 13.7. Each expression should use either multiplication or addition, or both. Write an equation by setting these two expressions equal to each other.

10. Bonnie had B stickers. First, Bonnie gave away $\frac{1}{6}$ of her stickers. Then Bonnie gets 12 more stickers and she has 47 stickers. Write an equation, involving B, that corresponds to this situation. Explain your work.

11. There are F cups of flour in a bag. First, $\frac{3}{8}$ of the flour in the bag was used. Then $\frac{1}{5}$ of the remaining flour was used. At that point, there were 10 cups of flour in the bag. Write an equation, involving F, that corresponds to this situation. Explain your work.

12. There are $3\frac{1}{4}$ times as many children at Abbott School as at Barrow School. Write an equation that corresponds to this situation. Explain why your equation is correct.

13. a. There are 50% as many people in Morgantown as in Carterville. Write an equation that corresponds to this situation. Explain why your equation is correct.

    b. There are 50% more people in Morgantown than in Carterville. Write an equation that corresponds to this situation. Explain why your equation is correct.

14. Show how to make the expression

$$\frac{21}{13} \cdot \frac{13}{56} + \frac{17}{36} \cdot \frac{18}{34}$$

easy to evaluate without using a calculator.

15. Show how to make the expression

$$\left(3 - \frac{1}{11}\right) \div \left(\frac{1}{5} + \frac{1}{11}\right)$$

easy to evaluate without using a calculator.

16. Write equations that use rules about operating with fractions and possibly also properties of arithmetic to show why it is valid to cancel to evaluate the expression

$$\frac{22 \cdot 3}{77}$$

as follows:

$$\frac{\overset{2}{\cancel{22}} \cdot 3}{\underset{7}{\cancel{77}}} = \frac{2 \cdot 3}{7} = \frac{6}{7}$$

17. Explain why it is *not valid* to evaluate

$$\frac{22 + 3}{77}$$

using the canceling shown in the following step:

warning, incorrect : $\dfrac{\overset{2}{\cancel{22}} + 3}{\underset{7}{\cancel{77}}} = \dfrac{2 + 3}{7} = \dfrac{5}{7}$

Show how to evaluate $\frac{22+3}{77}$ correctly. (Is canceling possible or not?)

18. Write equations that use rules about operating with fractions as well as properties of arithmetic to show why it is valid to cancel to evaluate the expression

$$\frac{15}{25 + 10}$$

as follows:

$$\frac{\overset{3}{\cancel{15}}}{\underset{5}{\cancel{25}} + \underset{2}{\cancel{10}}} = \frac{3}{5 + 2} = \frac{3}{7}$$

13.2 Solving Equations with Pictures and with Algebra

A major part of algebra is solving equations. In this section, we will study the fundamental ideas we need to solve equations and we will apply these ideas to solve some of the simplest kinds of equations. We will also see how drawing diagrams can make some algebra story problems easy to solve without using variables. We will then relate the diagram-drawing method of solving algebra story problems to standard algebraic methods. The diagram-drawing method of solving problems we will study is shown in the textbooks used in grades 3 – 6 in Singapore. Of the 38 nations studied in the Third International Mathematics and Science Study, children in Singapore scored highest in math. (See [20].)

SOLUTIONS OF EQUATIONS

We must first say what it means to solve an equation. Some equations are true; for example, the equations

$$3 + 2 = 5$$
$$6 \times 3 = 3 \times 6$$
$$10 = 2 \times 5$$

are true. Some equations are false; for example the equations

$$1 = 0$$
$$3 + 2 = 6$$

are false. Most equations with variables are true for some values of the variable and false for other values of the variable. For example, the equation

$$3 + x = 5$$

is true when $x = 2$, but is false for all other values of x.

solve an equation   To **solve an equation** involving variables means to determine those values for the variables which make the equation true. The values for the variables which make the equation true are **solution**   called the **solutions** to the equation. To solve the equation

$$3x = 12$$

means to find all those values for x for which $3x$ is equal to 12. Only $3 \cdot 4$ is equal to 12; 3 times any number other than 4 is not equal to 12. So, there is only one solution to $3x = 12$, namely, $x = 4$.

Even young children in elementary school learn to solve simple equations. For example, first or second graders could be asked to fill in the box to make the following equation true:

$$5 + \boxed{} = 7$$

One source of difficulty in solving equations is understanding that the equals sign does not mean "calculate the answer." For example, when children are asked to fill in the box to make the equation

$$5 + 3 = \boxed{} + 2$$

true, many will fill in the number 8 because $5 + 3 = 8$. In order to understand how to solve equations, we must understand that we want to make the expressions to the left and right of the equal sign equal to each other. One helpful piece of imagery for understanding equations is a pan balance, as shown in Figure 13.8. If we view the equation $5 + 3 = \boxed{} + 2$ in terms of a pan balance, each side of the equation corresponds to a side of the pan balance. To solve the equation means to make the pans balance rather than tilt to one side or the other.

FIGURE 13.8

Viewing the
Equation
$5 + 3 = ? + 2$
in Terms of a
Pan Balance

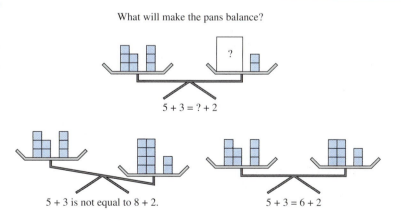

What will make the pans balance?

$5 + 3 = ? + 2$

$5 + 3$ is not equal to $8 + 2$. $5 + 3 = 6 + 2$

Solving Equations Using Algebra

How do we solve equations? The equation

$$2x = 6$$

is easy to solve because it asks

2 times what number is 6?

which we can solve by division:

$$x = 6 \div 2 = 3$$

The equation

$$x + 3 = 7$$

is also easy to solve because it asks

what number plus 3 is 7?

which we can solve by subtraction:

$$x = 7 - 3 = 4$$

But how do we solve a more complex equation like

$$5x + 2 = 3x + 8?$$

The strategy for solving complex equations is to change the equation into a new equation that has the same solutions and is easy to solve. We can change an equation into a new equation that has the same solutions by doing any of the following:

1. Use properties of arithmetic or valid ways of operating with numbers (including fractions) to change an expression on either side of the equals sign into an equal expression. For example, we can change the equation

    $$3x + 4x + 2 = 1$$

    to the equation

    $$7x + 2 = 1$$

    because, according to the distributive property,

    $$3x + 4x = (3 + 4)x = 7x$$

2. Add the same quantity to both sides of the equation or subtract the same quantity from both sides of the equation. For example, we can change the equation

    $$6x - 2 = 3$$

    to the equation

    $$6x - 2 + 2 = 3 + 2$$

which becomes

$$6x = 5$$

by using item (1).

3. Multiply both sides of the equation by the same nonzero number or divide both sides of the equation by the same nonzero number. For example, we can change the equation

$$5x = 3$$

to the equation

$$x = \frac{3}{5}$$

by dividing both sides of the equation by 5. Remember that $5x$ stands for $5 \cdot x$, so when we divide this by 5, we get x.

Why do items (1), (2), and (3) change an equation into a new equation that has the same solutions as the original one? When we use item (1), we don't really change the equation at all, we just write the expression on one side of the equals sign in a different way. Therefore, when we use item (1), the new equation must have the same solutions as the original equation.

To see why item (2) should produce a new equation that has the same solutions as the original one, think in terms of a pan balance. If the two pans are balanced and you add the same amount to both sides or take the same amount away from both sides, then the pans will remain balanced. The same is true with equations.

To see why item (3) should produce a new equation that has the same solutions as the original one, think again in terms of a pan balance. If the two pans are balanced and you put twice as much or three times as much, and so on, on each pan, then the pans will remain balanced. Similarly, if the two pans are balanced and you take $\frac{1}{2}$ or $\frac{1}{3}$, and so on, of each pan, then the pans will remain balanced.

Thus, the strategy for solving an equation such as

$$5x + 2 = 3x + 8$$

is to use items (1), (2), and (3) to change the equation into an equation of the form

$$x = a$$

where a is a number. This equation has the same solutions as the original one. And we can see that this equation has solution a. We can find the equation $x = a$ by getting all the terms with x on one side of the equation and getting all the numbers on the other side. Figure 13.9 on page 598 shows how to solve $5x + 2 = 3x + 8$ this way. Each step is also illustrated in terms of a pan balance. Each small square on the pan balance represents 1. Each larger square with a question mark inside represents an x (because this x is the unknown amount we are looking for). To solve the equation means to determine the value of x that will make the equation true. In terms of the pan balance, we want to determine how many small squares should fill a box with a question mark to make the pans balance. We see from Figure 13.9 that the solution is $x = 3$. In other words, 3 small squares should fill a square with a question mark.

FIGURE 13.9

Solving $5x + 2 = 3x + 8$ with Equations and with a Pan Balance

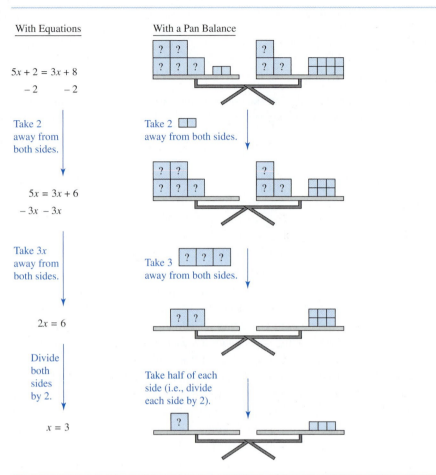

CLASS ACTIVITY NOW TURN TO CLASS ACTIVITIES MANUAL

13G Solving Equations Algebraically and with a Pan Balance p. 435

SOLVING ALGEBRA STORY PROBLEMS WITH SINGAPORE-STYLE STRIP DIAGRAMS

The pan balance view of equations presented previously provides an excellent concrete way to solve equations. However, many algebra story problems, especially those involving fractions or percentages, are not readily formulated in terms of a pan balance. We will now study a way to use diagrams to solve some algebra and other story problems.

In the elementary school mathematics textbooks used in Singapore (see [13]), narrow strips, such as the ones shown in Figure 13.10, are often used to represent quantities. We will solve three different problems from the Singapore textbooks with the aid of such strip diagrams.

FIGURE 13.10

A Strip Diagram to Determine How Many Girls and How Many Children Were at a Concert when There Were 120 Boys and 15 More Girls Than Boys

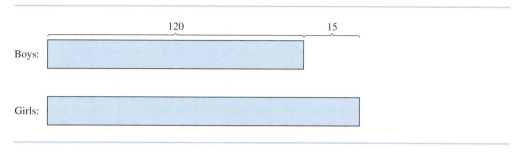

For the second and third problem, we will also see how to formulate the problem algebraically and how the algebraic solution method mirrors the strip diagram solution method.

PROBLEM 1 FROM A THIRD-GRADE WORKBOOK

There were 120 boys at a concert. There were 15 more girls than boys. How many girls were there? How many children were there altogether? (from [14], Volume 3A, part 1, page 20)[1]

Accompanying the problem is a diagram similar to Figure 13.10. This strip diagram makes it clear that we should add $120 + 15 = 135$ to find the number of girls at the concert and then add $120 + 135 = 255$ to find the total number of children at the concert. Notice also that we could calculate the total number of children by adding $120 + 120 + 15$, as we see from the diagram.

PROBLEM 2 FROM A THIRD-GRADE WORKBOOK

Rani had \$47. After paying for 3 kg of prawns, she had \$20 left. Find the cost of 1 kg of prawns. (from [14], Volume 3A, part 1, page 55)

This problem is accompanied by a strip diagram like that shown in Figure 13.11. We can see from the diagram that 3 kg of prawns must cost $47 - 20 = 27$. Therefore, 1 kg of prawns must cost $27 \div 3 = 9$.

We can also formulate this problem algebraically with an equation. Let x be the cost of 1 kg of prawns. Then 3 kg of prawns cost $3x$ dollars. So after paying for the 3 kg of prawns,

FIGURE 13.11

A Strip Diagram to Determine How Much 1 kg of Prawns Cost if Rani Started with \$47 and Had \$20 Left after Paying for 3 kg of Prawns

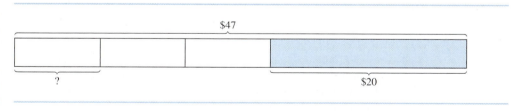

Material throughout this activity is adapted with permission from *Primary Mathematics*, Vol. 4A, pp. 40, 41; Vol. 4B, pp. 73, 100; Vol. 5A, pp. 60, 79, 90; Vol. 6A, pp. 38, 67, 71; Vol. 6B, pp. 34, 58, 63, 64; and from *Primary Mathematics Workbook*, Vol. 3A part 1, pp. 20, 55; Vol. 4A part 1, p. 62; Vol. 4B part 1, p. 38 by Ministry of Education and Times Media Private Limited.

Rani has

$$47 - 3x$$

dollars left, which the problem tells us is $20. Therefore,

$$47 - 3x = 20$$

If we add $3x$ to both sides, we obtain the equation

$$47 = 20 + 3x$$

which fits nicely with the strip diagram in Figure 13.11, where one small white strip represents x. If we subtract 20 from both sides of the equation, we obtain the equation

$$47 - 20 = 3x \quad \text{after simplifying:} \quad 27 = 3x$$

Dividing both sides by 3, we have

$$\frac{27}{3} = x \quad \text{after simplifying:} \quad 9 = x$$

So $x = 9$ is the solution and each kilogram of prawns costs $9. Notice that the calculations we performed when solving the problem algebraically are the same as those we performed when using the strip diagram: in both cases we subtracted 20 from 47 to obtain 27 and then we divided 27 by 3 to obtain 9.

PROBLEM 3 FROM THE FOURTH-GRADE TEXTBOOK

Although the previous two problems can be considered arithmetic problems, most math teachers would probably view the following fourth-grade problem as an algebra problem rather than an arithmetic problem:

> 300 children are divided into two groups. There are 50 more children in the first group than in the second group. How many children are there in the second group? (from [13], Volume 4A, page 40)

The Singapore textbook has not introduced algebra with variables at this point, so the children are expected to solve this problem with the aid of a diagram. They might draw a diagram like the one in Figure 13.12. From the diagram we see that if we subtract 50 from 300 and then divide the resulting amount by 2, we will have the number of children represented by each of the long strips. Since $300 - 50 = 250$ and $250 \div 2 = 125$, the second group has 125 children and the first group has $125 + 50 = 175$ children.

We can also formulate this problem algebraically with equations. If we let A be the number of children in the first group and B be the number of children in the second group, then, because there are 300 children in all, we have

$$A + B = 300$$

We are given that A is 50 more than B, so

$$A = B + 50$$

FIGURE 13.12

A Strip Diagram to Determine the Number of Children in Each Group if 300 Children are Divided into 2 Groups with 50 More Children in the First Group Than the Second Group

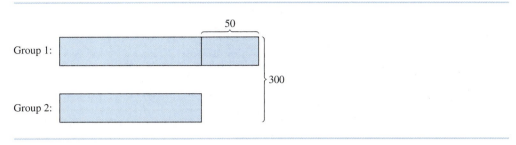

Notice that we can also get these equations from Figure 13.12, where the bottom strip represents B and the top, combined strip represents A. Since the second equation tells us that A is equal to $B + 50$, we can replace A in the first equation with $B + 50$ to obtain the equation

$$B + 50 + B = 300$$

Combining the Bs, we have

$$2B + 50 = 300$$

We can see this equation in Figure 13.12 because the 2 long strips representing B and the strip representing 50 combine to make 300. Subtracting 50 from both sides, we obtain

$$2B = 300 - 50 \quad \text{after simplifying:} \quad 2B = 250$$

Dividing both sides by 2, we have

$$B = \frac{250}{2} \quad \text{after simplifying:} \quad B = 125$$

So there are 125 children in the second group.

We can also solve this problem algebraically by using only one variable instead of two variables. Let C be the number of children in the smaller group, which is the second group. Then, since there are 50 more children in the first group, the number of children in the first group is

$$C + 50$$

Since the total number of children is 300, we have

$$C + 50 + C = 300$$

Combining the Cs gives

$$2C + 50 = 300$$

We solve this equation in the same way that we solved $2B + 50 = 300$, using the letter C instead of B.

Notice that the calculations we performed when solving the problem algebraically are the same as those we performed when using the strip diagram: in both cases we subtracted 50 from 300 to obtain 250 and then divided 250 by 2 to obtain 125.

CLASS ACTIVITY   NOW TURN TO CLASS ACTIVITIES MANUAL

13H   Solving Story Problems with Singapore-Style Strip Diagrams and with Algebra p. 436

13I   Solving Story Problems p. 440

PRACTICE PROBLEMS FOR SECTION 13.2

1. Dante has twice as many jelly beans as Carlos. Natalia has 4 more jelly beans than Carlos and Dante together. All together, Dante, Carlos, and Natalia have 58 jelly beans. How many jelly beans does Carlos have?

   a. Solve this problem with the aid of a diagram. Explain your solution.

   b. Now solve the problem algebraically, letting x stand for the number of jelly beans that Carlos has.

      i. Write an expression in terms of x for the number of jelly beans that Dante has. Relate your expression to your diagram in part (a).

      ii. Write an expression in terms of x for the number of jelly beans that Natalia has. Relate your expression to your diagram in part (a).

      iii. Write an equation involving x which states that the total number of jelly beans is 58. Relate your equation to your diagram in part (a).

      iv. Solve your equation in part (iii). Relate your work to the work you did with your diagram in part (a).

2. A sixth-grade problem (from [13] Volume 6A, p. 71): A tank is $\frac{2}{5}$ filled with water. When another 26 liters of water are poured in, the tank becomes $\frac{5}{6}$ full. Find the capacity of the tank. Solve this problem in two ways: with the aid of a diagram and with equations. Discuss how the two solution methods are related.

3. A sixth-grade problem (from [13] Volume 6B, p. 34): Samy had 130 stickers and Devi had 50 stickers. After Samy gave Devi some stickers, Samy had twice as many stickers as Devi. How many stickers did Samy give Devi? Solve this problem in two ways: with the aid of a diagram and with equations. Discuss how the two solution methods are related.

ANSWERS TO PRACTICE PROBLEMS FOR SECTION 13.2

1. a. We see from Figure 13.13 that the total number of jelly beans consists of 6 equal groups of jelly beans and 4 more jelly beans. If we take the 4 away from 58, we have 54 jelly beans that must be distributed equally among the 6 groups. Therefore, each group has $54 \div 6 = 9$ jelly beans. Since Carlos has one group of jelly beans, he has 9 jelly beans. (Dante has $2 \cdot 9 = 18$ jelly beans, and Natalia has $9 + 18 + 4 = 31$ jelly beans.)

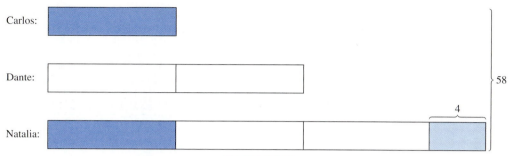

FIGURE 13.13

A Strip Diagram for Determining How Many Jelly Beans Carlos Has

b. i. Since Dante has twice as many jelly beans as Carlos, Dante has $2x$ jelly beans. In Figure 13.13, Carlos's strip represents x jelly beans and since Dante has two such strips, he has $2x$ jelly beans.

  ii. Since Natalia has 4 more than Carlos and Dante combined, Natalia has $x + 2x + 4$ jelly beans, which we can also write as $3x + 4$ jelly beans. We see in Figure 13.13 that Natalia's portion consists of a strip representing x jelly beans, two more copies of this strip representing $2x$ jelly beans, and another strip representing the additional 4 jelly beans. So, all together, Natalia's portion is $3x + 4$ jelly beans.

  iii. We are given that the total number of jelly beans is 58. On the other hand, the total number of jelly beans is Carlos's, Dante's, and Natalia's combined, which is

$$x + 2x + 3x + 4$$

By combining the xs, we can rewrite this expression as

$$6x + 4$$

Since the two ways of expressing the total number of jelly beans must be equal, we have the equation

$$6x + 4 = 58$$

We can also deduce this equation from Figure 13.13 because there are 6 strips which each represent x jelly beans and another small strip representing 4 more jelly beans; the combined amount must be 58, so $6x + 4 = 58$.

  iv. To solve $6x + 4 = 58$, first subtract 4 from both sides to obtain

$$6x = 58 - 4 \quad \text{after simplifying:} \quad 6x = 54$$

Then divide both sides by 6 to obtain

$$x = \frac{54}{6} \quad \text{after simplifying:} \quad x = 9$$

Therefore, Carlos has 9 jelly beans. Notice that the arithmetic we performed in order to solve the equation $6x + 4 = 58$ is the same as the arithmetic we performed in part (a): we first subtracted 4 from 58 and then we divided the resulting amount, 54, by 6.

2. As we see in Figure 13.14 on p. 604, 26 liters fill the difference between $\frac{2}{5}$ of the tank and $\frac{5}{6}$ of the tank. Since

$$\frac{5}{6} - \frac{2}{5} = \frac{5 \cdot 5}{6 \cdot 5} - \frac{2 \cdot 6}{5 \cdot 6} = \frac{25}{30} - \frac{12}{30} = \frac{13}{30}$$

there are 13 parts (each of which is $\frac{1}{30}$ of the tank) that the 26 liters fill. So each of these 13 parts holds 2 liters. The full tank consists of 30 parts, so the full tank holds $30 \cdot 2 = 60$ liters.

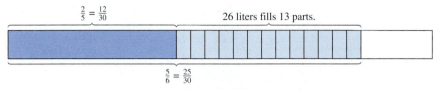

FIGURE 13.14

A Strip Diagram for Determining the Capacity of a Tank if 26 Liters Added to a $\frac{2}{5}$ Full Tank Makes the Tank $\frac{5}{6}$ Full

To solve the problem with equations, let x be the number of liters that the tank holds. Then $\frac{2}{5}x$ is the number of liters in $\frac{2}{5}$ of the tank and $\frac{5}{6}x$ is the number of liters in $\frac{5}{6}$ of the tank. We are given that

$$\frac{2}{5}x + 26 = \frac{5}{6}x$$

Subtracting $\frac{2}{5}x$ from both sides, we have

$$26 = \frac{5}{6}x - \frac{2}{5}x$$

so

$$26 = \left(\frac{5}{6} - \frac{2}{5} \right) x \quad \text{after simplifying:} \quad 26 = \frac{13}{30}x$$

After multiplying both sides by $\frac{30}{13}$ (or dividing both sides by $\frac{13}{30}$), we have

$$\frac{30}{13} \cdot 26 = \frac{30}{13} \cdot \frac{13}{30}x \quad \text{after simplifying:} \quad 60 = x$$

Therefore, $x = 60$ and the full capacity of the tank is 60 liters.

We perform the same arithmetic using both solution methods. With each method, we calculated $\frac{5}{6} - \frac{2}{5}$. When we solved the problem using a diagram, we divided $26 \div 13$ to find the number of liters in each $\frac{1}{30}$ part of the tank. Then we multiplied that result, 2, by the total number of parts in the tank, 30. Similarly, using the algebraic solution method, we had to calculate

$$\frac{30}{13} \cdot 26$$

which we can do by canceling 13s:

$$\frac{30}{\cancel{13}} \cdot \cancel{26}^{2} = 30 \cdot 2 = 60$$

This way, we again divide 26 by 13 and multiply the result, 2, by 30.

3. In Figure 13.15, focus on the amount Samy has after he gives some stickers to Devi. In Samy's strip, we see that if we take away the amount in 2 shaded strips from the 130 stickers, then the remaining amount will be 3 times the number of stickers that Samy gave Devi. The 2 shaded strips make 100 stickers, so taking these 100 stickers away from the 130 stickers leaves 30. These 30 stickers must be 3 times the number of stickers that Samy gave Devi. Therefore, Samy gave Devi 10 stickers.

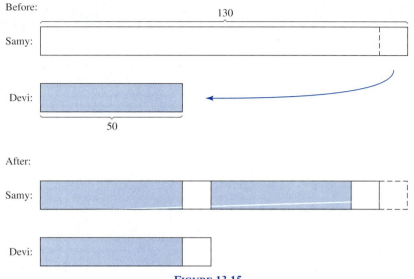

FIGURE 13.15

A Strip Diagram for Determining How Many Stickers Samy Gave Devi

To solve the problem with algebra, let S be the number of stickers that Samy gave Devi. Then after Samy gives away these stickers, he has $130 - S$ stickers and Devi has $50 + S$ stickers. We are told that Samy has 2 times as many as Devi, so

$$130 - S = 2 \cdot (50 + S)$$

Therefore,

$$130 - S = 100 + 2S$$

Adding S to both sides, we have

$$130 = 100 + 3S$$

Subtracting 100 from both sides gives

$$30 = 3S$$

Dividing both sides by 3, we have

$$10 = S$$

Therefore, $S = 10$ and Samy gave Devi 10 stickers.

We perform the same arithmetic using both solution methods. In both cases we multiplied 50 by 2 and then subtracted the resulting amount, 100, from 130. We then divided the result, 30, by 3. We can see the amount, S in the diagram as the small box. We can also see the equation $130 - S = 2 \cdot (50 + S)$ in the strip representing the amount Samy has after giving stickers to Devi.

PROBLEMS FOR SECTION 13.2

1. Solve

$$4x + 5 = x + 8$$

in two ways: with equations and with pictures of a pan balance. Relate the two methods.

2. Solve

$$3x + 2 = x + 8$$

in two ways: with equations and with pictures of a pan balance. Relate the two methods.

3. A fourth-grade problem (from [14] Volume 4B part 1, p. 38): 1650 pupils took part in a parade. There were twice as many boys as girls. How many boys were in the parade?

   a. Solve this problem with the aid of a diagram. Explain your solution.

   b. Now solve the problem algebraically, letting x stand for the number of girls in the parade:

      i. Write an expression in terms of x for the number of boys in the parade. Relate your expression to your diagram in part (a).

      ii. Write an expression in terms of x for the total number of children in the parade. Relate your expression to your diagram in part (a).

      iii. Write an equation involving x which states that the total number of children in the parade is 1650. Relate your equation to your diagram in part (a).

      iv. Solve your equation in part (iii) and use the solution to determine the number of boys in the parade. Relate your work to the work you did with your diagram in part (a).

4. A fifth-grade problem (from [13] Volume 5A, p. 60): After spending $\frac{2}{5}$ of his money on a toy car, Samy had $42 left. How much money did he have at first?

   a. Solve this problem with the aid of a diagram. Explain your solution.

   b. Now solve the problem algebraically, letting x stand for the amount of money Samy had to start with:

      i. Write an expression in terms of x for the amount of money Samy had left after buying the toy car. Relate your expression to your diagram in part (a).

   ii. Write an equation involving x which states that Samy has $42 left after buying his toy. Relate your equation to your diagram in part (a).

   iii. Solve your equation in part (ii). Relate your work to the work you did with your diagram in part (a).

5. A fifth-grade problem (from [13] Volume 5A, p. 60): Mrs. Chen bought some eggs. She used $\frac{1}{2}$ of them to make tarts and $\frac{1}{4}$ of the remainder to make a cake. She had 9 eggs left. How many eggs did she buy?

   a. Solve this problem with the aid of a diagram. Explain your solution.

   b. Now solve the problem algebraically, letting x stand for the number of eggs that Mrs. Chen bought:

      i. Write an expression in terms of x for the number of eggs Mrs. Chen had left after making tarts. Then write an expression in terms of x for the number of eggs Mrs. Chen used to make a cake. Relate your expressions to your diagram in part (a).

      ii. Write an expression in terms of x for the total number of eggs Mrs. Chen used. Write another expression in terms of x for the total number of eggs Mrs. Chen has left. Relate your expressions to your diagram in part (a).

      iii. Write an equation involving x which states that Mrs. Chen has 9 eggs left. Relate your equation to your diagram in part (a).

      iv. Solve your equation in part (iii). Relate your work to the work you did with your diagram in part (a).

6. A fourth-grade problem (from [13] Volume 4A, p. 41): 3000 exercise books are arranged into 3 piles. The first pile has 10 more books than the second pile. The number of books in the second pile is twice the number of books in the third pile. How many books are there in the third pile? Solve this problem in two ways: with the aid of a diagram and with equations. Discuss how the two solution methods are related.

7. A fourth-grade problem (from [14] Volume 4A part 1, p. 62): 2500 people took part in a cross-country race. The number of adults were 4 times the number of children. If there were 1200 men, how many women were

there? Solve this problem in two ways: with the aid of a diagram and with equations. Discuss how the two solution methods are related.

8. A fourth-grade problem (from [13] Volume 4B, p. 73): A piece of rope 3 m 66 cm long was cut into 2 pieces. The longer piece was twice as long as the shorter piece. What was the length of the longer piece? Solve this problem in two ways: with the aid of a diagram and with equations. Discuss how the two solution methods are related.

9. A fourth-grade problem (from [13] Volume 4B, p. 100): $\frac{2}{3}$ of a sum of money is $18. Find the sum of money. Solve this problem in two ways: with the aid of a diagram and with equations. Discuss how the two solution methods are related.

10. A fifth-grade problem (from [13] Volume 5A, p. 60): A hawker sold $\frac{2}{3}$ of his curry puffs in the morning and $\frac{1}{6}$ in the afternoon. He sold 200 curry puffs altogether. How many curry puffs had he left? Solve this problem in two ways: with the aid of a diagram and with equations. Discuss how the two solution methods are related.

11. A fifth-grade problem (from [13] Volume 5A, p. 60): Minghua bought a bag of marbles. $\frac{1}{4}$ of the marbles were blue, $\frac{1}{8}$ were green, and $\frac{1}{5}$ of the remainder were yellow. If there were 24 yellow marbles, how many marbles did he buy? Solve this problem in two ways: with the aid of a diagram and with equations. Discuss how the two solution methods are related.

12. A fifth-grade problem (from [13] Volume 5A, p. 90): Ali saved twice as much as Ramat. Maria saved $60 more than Ramat. If they saved $600 altogether, how much did Maria save? Solve this problem in two ways: with the aid of a diagram and with equations. Discuss how the two solution methods are related.

13. A sixth-grade problem (from [13] Volume 6B, p. 34): $\frac{1}{3}$ of the beads in a box are red, $\frac{2}{3}$ of the remainder are blue and the rest are yellow. If there are 24 red beads, how many yellow beads are there? Solve this problem in two ways: with the aid of a diagram and with equations. Discuss how the two solution methods are related.

14. (See Problem 10 in Section 13.1.) Bonnie had some stickers. First, Bonnie gave away $\frac{1}{6}$ of her stickers. When Bonnie gets 12 more stickers she has 47 stickers. Determine how many stickers Bonnie had at the start. Explain your solution.

15. (See Problem 11 in Secion 13.1.) There was some flour in a bag. First, $\frac{3}{8}$ of the flour in the bag was used. Then $\frac{1}{5}$ of the remaining flour was used. At that point, there were 10 cups of flour in the bag. Determine how much flour was in the bag at the start. Explain your solution.

16. Not every algebra story problem is naturally modeled with either a pan balance or with a strip diagram. Solve the following problem in any way that makes sense and explain your solution.

    Kenny has stickers to distribute among some goodie bags. When Kenny tries to put 5 stickers in each goodie bag he is 2 stickers short. But when Kenny puts 4 stickers in each bag he has 3 stickers left over. How many goodie bags does Kenny have? How many stickers does Kenny have?

17. Not every algebra story problem is naturally modeled with either a pan balance or with a strip diagram. Solve the following problem in any way that makes sense and explain your solution.

    Frannie has no money now, but she plans to save $D every week so that she can buy a toy she would like to have. Frannie figures that if she saves her money for 3 weeks she will have $1 less than she needs to buy the toy, but if she saves her money for 4 weeks she will have $3 more than she needs to buy the toy. How much is Frannie saving every week and how much does the toy Frannie plans to buy cost?

18. A fifth-grade problem (from [13] Volume 5A, p. 79): David cuts a rope 60 m long into two pieces in the ratio 2 : 3. What is the length of the shorter piece of rope? Solve this problem and explain your solution.

19. A fifth-grade problem (from [13] Volume 5A, p. 79): The ratio of Samy's weight to John's weight is 6 : 5. If Samy weighs 48 kg, find John's weight. Solve this problem and explain your solution.

20. A fifth-grade problem (from [13] Volume 5A, p. 79): The ratio of the number of boys to the number of girls is 2 : 5. If there are 100 boys, how many children are there altogether? Solve this problem and explain your solution.

21. A sixth-grade problem (From [13] Volume 6A, p. 38): Susan and Sally had an equal amount of money each. After Sally spent $15 and Susan spent $24, the ratio of Sally's money to Susan's money was 4 : 3. How much money did each girl have at first? Solve this problem and explain your solution.

22. A sixth-grade problem (from [13] Volume 6A, p. 67): There are 10% more boys than girls in a choir. If there are 4 more boys than girls, how many children are there altogether? Solve this problem and explain your solution.

23. A sixth-grade problem (from [13] Volume 6A, p. 67): 60% of the books in a library are for adults, 5% are for young people and the rest are for children. If there are 280 books for children, how many books are there altogether? Solve this problem and explain your solution.

24. A sixth-grade problem (from [13] Volume 6A, p. 67): At a sale, Mrs. Li bought a fan for $140. This was 70% of its usual price. What was the usual price of the fan? Solve this problem and explain your solution.

25. A sixth-grade problem (from [13] Volume 6B, p. 64): Rahim has 30% more books than Gopal. If Rahim has 65 books, how many books does Gopal have? Solve this problem and explain your solution.

26. A sixth-grade problem (from [13] Volume 6B, p. 58): John and Mary had $350 altogether. After John spent $\frac{1}{2}$ of his money and Mary spent $\frac{1}{3}$ of her money, they each had an equal amount of money left. How much did they spend altogether? Solve this problem and explain your solution.

27. A sixth-grade problem (from [13] Volume 6B, p. 63): If $\frac{2}{3}$ of a number is 12, what is the value of $\frac{1}{2}$ of the number? Solve this problem and explain your solution.

28. A sixth-grade problem (from [13] Volume 6B, p. 63): 10 glasses of water can fill $\frac{5}{8}$ of a bottle. How many *more* glasses of water are needed to fill up the bottle? Solve this problem and explain your solution.

29. Jane had a bottle filled with juice. At first, Jane drank $\frac{1}{5}$ of the juice in the bottle. After 1 hour, Jane drank $\frac{1}{4}$ of the remaining juice in the bottle. After another 2 hours, Jane drank $\frac{1}{3}$ of the remaining juice in the bottle. At that point, Jane checked how much juice was left in the bottle: there was $\frac{2}{3}$ of a cup left. No other juice was added to or removed from the bottle. How much juice was in the bottle originally? Solve this problem and explain your solution.

30. A flock of geese on a pond were being observed continuously. At 1:00 P.M., $\frac{1}{5}$ of the geese flew away. At 2:00 P.M., $\frac{1}{8}$ of the geese that remained flew away. At 3:00 P.M., 3 times as many geese as had flown away at 1:00 P.M. flew away, leaving 28 geese on the pond. At no other time did any geese arrive or fly away. How many geese were in the original flock? Solve this problem and explain your solution.

13.3   Sequences

Most people, including young children, enjoy discovering and experimenting with growing and repeating patterns. But patterns aren't just fun. By thinking about how to describe them generally we begin to develop algebraic reasoning. In this section we will study growing and repeating patterns from a mathematical point of view by studying *sequences*.

sequence   A **sequence** is a list of items occuring in a specified order. Sequences of physical objects or figures can give rise to sequences of numbers, which we can study mathematically.

The following example shows a simple way that sequences can arise. Austin is making "trains" out of snap cubes (snap-together cubes). Each train has an engine, which is made of 4 white snap cubes, and one or more train cars. Each train car is made of 3 yellow and 2 red snap cubes. Austin makes a sequence of trains: the first train consists of an engine and 1 train car; the second train consists of an engine and 2 train cars; the third train consists of an engine and 3 train cars, and so on, as indicated in Figure 13.16.

Austin's sequence of trains gives rise to several sequences of numbers by considering the red snap cubes, the yellow snap cubes, and the total number of snap cubes in the trains. The numbers of red snaps cubes in the trains are

$$2, \quad 4, \quad 6, \quad 8, \quad 10, \dots$$

FIGURE 13.16

A Sequence of Snap Cube "Trains"

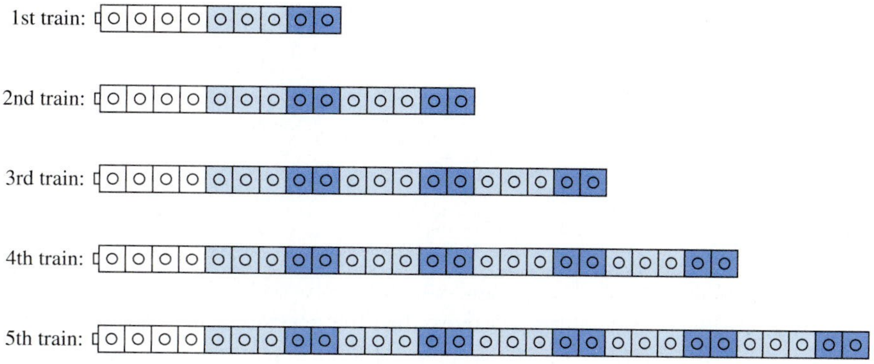

1st train:

2nd train:

3rd train:

4th train:

5th train:

the numbers of yellow snap cubes in the trains are

$$3, \quad 6, \quad 9, \quad 12, \quad 15, \ldots$$

and the total numbers of snap cubes in the trains are

$$9, \quad 14, \quad 19, \quad 24, \quad 29 \ldots$$

Of course, Austin will eventually run out of snap cubes and will only be able to make so many trains. But if we imagine an infinite supply of snap cubes, we can imagine each of these sequences going on forever.

ARITHMETIC SEQUENCES

arithmetic sequence

The sequences of the numbers of red and yellow snap cubes in Austin's trains and the total number of snap cubes in Austin's trains, are examples of *arithmetic sequences*. We create an **arithmetic sequence** by starting with any number as the first entry. Each subsequent entry is obtained from the previous entry by adding the same fixed number, as shown in Figure 13.17.

Every arithmetic sequence is described by a simple formula. This formula is related to the sequence in a nice way. Let's consider the example of the arithmetic sequence

$$9, \quad 14, \quad 19, \quad 24, \quad 29 \ldots$$

FIGURE 13.17

Arithmetic Sequences Are Produced by Repeatedly Adding the Same Amount

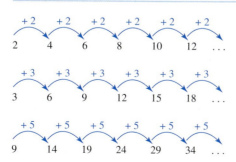

FIGURE 13.18

We Obtain the
Formula
$5N + 4$ by
Starting with the
0th Entry and
Repeatedly
Adding 5

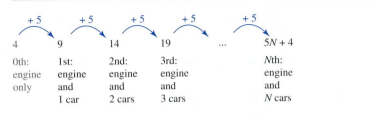

of the total number of snap cubes in Austin's trains. This sequence starts with 9 and increases by 5. That is, each entry is 5 more than the previous entry. Imagine this sequence going on forever, so that there is a 100th entry, a 1000th entry, and so on; for each counting number, N, there is an Nth entry in the sequence. Is there a formula in terms of N for this Nth entry? Yes, there is, namely,

$$5N + 4$$

Although you might be able to guess this formula, there is a systematic way to find it, to explain why it makes sense, and to relate it to the growth of the sequence.

To explain why the formula $5N + 4$ for the Nth entry in the sequence above makes sense, think about how to obtain each entry. Notice that the first entry, 9, was obtained by adding one 5 to 4 because a single train car made of 5 snap cubes was added to an engine made of 4 snap cubes. The second entry, 14, was obtained by adding two 5s to 4 because 2 train cars were added to an engine. The third entry, 19, was obtained by adding three 5s to 4. In general, the Nth entry is obtained by adding N 5s to 4. Therefore, the Nth entry is $5N + 4$. So, it's like starting with the 0th entry (just the engine, no train cars), which is 4, and adding N 5s to get the Nth entry, as indicated in Figure 13.18.

The reasoning to find the formula for the Nth entry of any other arithmetic sequence is similar. Consider a "0th entry" that you obtain by subtracting the increase amount from the initial entry. Then the Nth entry will be obtained by adding the increase amount N times to the 0th entry. Therefore, the Nth entry is

$$(\text{increase amount}) \cdot N \; + \; (\text{0th entry})$$

CLASS ACTIVITY   NOW TURN TO CLASS ACTIVITIES MANUAL

GEOMETRIC SEQUENCES

geometric sequence

Previously, we saw that arithmetic sequences are obtained by repeatedly adding the same amount. Geometric sequences are similar, but are obtained by repeatedly *multiplying* by the same amount. We create a **geometric sequence** by starting with any number as the first entry. Each subsequent entry is obtained from the previous entry by multiplying by the same fixed number, as shown in Figure 13.19.

Geometric sequences have important practical applications. For example, suppose that $1000 is deposited in a bank account that pays 5% interest annually. If no money other than the interest is added or removed, then at the end of each year, the amount of money in the account will be 5% more than the amount at the end of the previous year. Therefore, the amount each year is obtained from the amount the previous year by multiplying by 1.05, as shown in Figure 13.20. (The 1.05 is $1 + .05$, which is 100% and an additional 5%; see Section 4.5.) So the amounts of money in the account at the end of each year form a geometric sequence.

As with arithmetic sequences, geometric sequences can also be described by formulas. Consider again the sequence

$$1050, \ 1102.50, \ 1157.63, \ 1215.51 \ldots$$

which is the dollar amount in the bank account described above after 1 year, 2 years, 3 years, and so on. What is a formula for the number of dollars in the account after N years? To find this formula, think about how to obtain each entry in the sequence by starting with the initial $1000 in the account. After 1 year, there is

$$1.05 \times 1000$$

FIGURE 13.19

Geometric Sequences Are Produced by Repeatedly Multiplying by the Same Amount

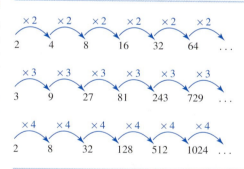

FIGURE 13.20

The Geometric Sequence of the Amounts of Money in a Bank Account after N Years When the Initial Amount Is $1000 and the Interest Rate Is 5%

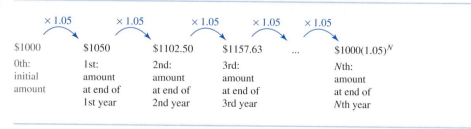

dollars in the account. After another year, this amount is multiplied by 1.05, so after 2 years there is

$$1.05 \times 1.05 \times 1000 = (1.05)^2 \cdot 1000$$

dollars in the account. After another year this amount is multiplied by 1.05 again, so after 3 years there is

$$1.05 \times 1.05 \times 1.05 \times 1000 = (1.05)^3 \cdot 1000$$

dollars in the account. Continuing in this way, after N years there is

$$\underbrace{1.05 \times 1.05 \times \cdots \times 1.05}_{N} \times 1000 = (1.05)^N \cdot 1000$$

dollars in the account.

The reasoning to find the Nth entry of any other geometric sequence is similar. Let's call the factor by which each entry in a geometric sequence is multiplied in order to obtain the next entry the "ratio" of the geometric sequence. Consider a "0th entry" that you obtain from the first entry by dividing by the ratio of the geometric sequence. Then the Nth entry will be obtained by multiplying the 0th entry by the ratio N times. Therefore, the Nth entry is

$$\underbrace{(\text{ratio}) \cdot (\text{ratio}) \cdot \ \cdots \ \cdot (\text{ratio})}_{N} \cdot (\text{0th entry})$$

or

$$(\text{ratio})^N \cdot (\text{0th entry})$$

CLASS ACTIVITY   NOW TURN TO CLASS ACTIVITIES MANUAL

13L   Geometric Sequences p. 447

OTHER SEQUENCES

Some of the sequences that young children encounter first in math generally consist of shapes or colors that are repeated in a pattern, such as the sequence of shapes in Figure 13.21. With such a sequence, we can ask which shape will be in the 100th entry, for example. If we assume that the sequence continues with repetitions of a square followed by a circle, followed by a triangle, then, since there are 3 shapes in the pattern, we can answer this question by dividing 100 by 3. Since

$$100 \div 3 = 33, \text{ remainder } 1$$

FIGURE 13.21

A Sequence of Shapes

FIGURE 13.22

The Sequence Repeats in a Pattern of 3

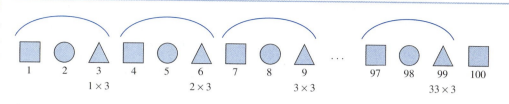

There are 33 repetitions of a square followed by a circle followed by a triangle in the entries 1 through $33 \times 3 = 99$, as indicated in Figure 13.22. The 100th entry is then a square.

Some numerical sequences are neither arithmetic nor geometric. For example, the sequence

$$1, \ 4, \ 9, \ 16, \ 25, \ 36, \ldots$$

is neither arithmetic nor geometric. For each of the 6 entries shown, the Nth entry is N^2.

Fibonacci sequence

One fascinating sequence is the **Fibonacci sequence**, which is the sequence

$$1, \ 1, \ 2, \ 3, \ 5, \ 8, \ 13, \ 21, \ 34, \ldots$$

Each entry in the sequence is obtained by adding the previous two entries. The Fibonacci sequence has surprising connections to nature and art. For example, the number of petals on many flowers is a Fibonacci number (i.e., an entry in the Fibonnaci sequence). If you look carefully at pine cones and pineapples, you will see "swirls" going in two directions. The number of swirls in the two directions are usually consecutive Fibonacci numbers such as 5 and 8. Rectangles whose width and length are in the ratio of consecutive Fibonacci numbers, such as 3 to 5 or 5 to 8, are pleasing to the eye and often used in art and architecture. Links to websites about the Fibonacci sequence are at

www.aw-bc.com/beckmann

CLASS ACTIVITY   NOW TURN TO CLASS ACTIVITIES MANUAL

13M Repeating Patterns p. 449

13N The Fibonacci Sequence in Nature and Art p. 451

PRACTICE PROBLEMS FOR SECTION 13.3

1. Figure 13.23 on page 614 shows a sequence of figures made of small squares. Assume that the sequence continues by adding 2 darkly shaded squares to the top of a figure in order to get the next figure in the sequence.

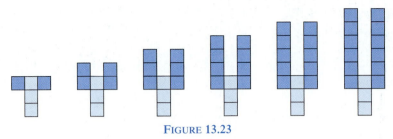

FIGURE 13.23

A Sequence of Figures

   a. Write the first 6 entries in the sequence whose entries are the number of small squares making up the figures.

   b. Find a formula for the number of small squares making up the Nth figure in the sequence. Explain why your formula makes sense by relating it to the structure of the figures.

   c. Will there be a figure in the sequence that is made of 100 small squares? If yes, which one? If no, why not? Determine the answer to these questions in two ways: with algebra and in a way that a third, fourth, or fifth grader might be able to.

   d. Will there be a figure in the sequence that is made of 199 small squares? If yes, which one, if no, why not? Determine the answer to these questions in two ways: with algebra and in a way that a third, fourth, or fifth grader might be able to.

2. Consider the arithmetic sequence whose first few entries are

$$1, 4, 7, 10, 13, 16, \ldots$$

Find a formula for the Nth entry in this sequence and explain in detail why your formula is valid.

3. Draw the first 5 figures in a sequence of figures whose Nth entry is made of $3N + 2$ small circles. Describe

how subsequent figures in your sequence would be formed.

4. Consider the geometric sequence whose first few entries are

$$6, 12, 24, 48, 96, 192, \ldots$$

Find a formula for the Nth entry in this sequence and explain in detail why your formula is valid.

ANSWERS TO PRACTICE PROBLEMS FOR SECTION 13.3

1. a. The number of squares in the figures are

$$5, 7, 9, 11, 13, 15, \ldots$$

   b. The Nth pattern in the sequence is made of $2N + 3$ small squares. We can explain why this formula is valid by relating it to the structure of the figures. Each figure is made of 3 small, lightly shaded squares and 2 "prongs" of darkly shaded squares. In the Nth figure, each prong is made of N squares. So in all, there are $2N + 3$ small squares in the Nth figure.

   c. A third, fourth, or fifth grader might realize that the number of squares making up the figures is always

odd. Since 100 is even, it could not be the number of small squares that a figure in the sequence is made of. A third or fourth grader might also try to find how many squares are in each prong if there were a figure made of 100 squares. There would have to be $100 - 3 = 97$ squares in the two prongs, so each prong would have to have $97 \div 2$ squares. Since 97 is not evenly divisible by 2, there can't be such a figure.

To answer the questions with algebra, notice that we want to know if there is a counting number, N, such that

$$2N + 3 = 100$$

Subtracting 3 from both sides, we have the new equation

$$2N = 97$$

which has the same solution as the original equation. Dividing both sides by 2, we have

$$N = \frac{97}{2}$$

But $\frac{97}{2}$ is not a counting number, so there is no such figure.

d. A third, fourth, or fifth grader might try to find the number of squares in each "prong." There must be $199 - 3 = 196$ squares in the two prongs combined. Therefore, each prong must have $196 \div 2 =$

98 squares. So the 98th figure in the sequence will be made of 199 small squares.

To answer the questions with algebra, notice that we want to solve the equation

$$2N + 3 = 199$$

Subtracting 3 from both sides, we have the new equation

$$2N = 196$$

which has the same solutions as the original equation. Dividing both sides by 2, we have

$$N = 98$$

So the 98th figure is made of 199 small squares.

---

2. The entries in the sequence increase by 3 each time. If we imagine a 0th entry preceding the first entry, this 0th entry would have to be 3 less than the first entry, 1; therefore, the 0th entry would be -2. Starting at the $0th$ entry, the 1st entry is obtained by adding one 3, the second entry is obtained by adding two 3s, the third entry is obtained by adding three 3s, and so on, as indicated in Figure 13.24. The Nth entry is obtained by adding N 3s to -2, so the Nth entry is

$$3N + (-2)$$

which we can also write as

$$3N - 2$$

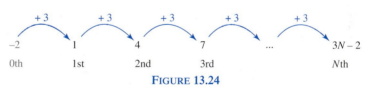

FIGURE 13.24

We Obtain the Formula $3N - 2$ by Starting with the
0th Entry and Repeatedly Adding 3

3. See Figure 13.25. Of course many other sequences of figures are also possible. Each figure is formed by adding one circle to each of the three "prongs" of the previous figure.

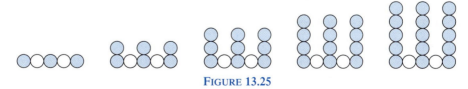

FIGURE 13.25

A Sequence of Figures Whose Nth Entry Is Made of $3N + 2$ Small Circles

---

4. We multiply each entry in the sequence by 2 to obtain the next entry. If we imagine a 0th entry preceding the 1st entry, this 0th entry would have to be 6 divided by 2, namely, 3. Starting at the 0th entry, the 1st entry is obtained by multiplying by 2 once, the second entry is obtained by multiplying by 2 twice, the third entry is

obtained by multiplying by 2 three times, and so on, as indicated in Figure 13.26. The Nth entry is obtained by multiplying N 2s with 3, so the Nth entry is

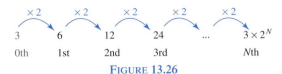

FIGURE 13.26

$$3 \cdot 2^N$$

We Obtain the Formula $3 \cdot 2^N$ by Starting With the 0th Entry and Repeatedly Multiplying by 2

PROBLEMS FOR SECTION 13.3

1. Figure 13.27 shows a sequence of figures made of small circles. Assume that the sequence continues by adding one circle to each of the 5 "arms" of a figure in order to get the next figure in the sequence.

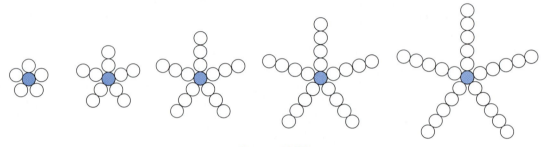

FIGURE 13.27

A Sequence of Figures

a. Find a formula for the number of circles making up the Nth figure in the sequence. Explain why your formula makes sense by relating it to the structure of the figures.

b. Will there be a figure in the sequence that is made of 100 circles? If yes, which one? If no, why not? Determine the answer to these questions in two ways: with algebra and in a way that a third, fourth, or fifth grader might be able to.

c. Will there be a figure in the sequence that is made of 206 circles? If yes, which one? If no, why not? Determine the answer to these questions in two ways: with algebra and in a way that a third, fourth, or fifth grader might be able to.

2. Consider the arithmetic sequence whose first few entries are

$$6, \ 11, \ 16, \ 21, \ 26, \ 31, \ldots$$

a. Determine the 100th entry in the sequence and explain why your answer is correct.

b. Find a formula for the Nth entry in this sequence and explain in detail why your formula is valid.

c. Is 1000 an entry in the sequence or not? If yes, which entry? If no, why not? Determine the answer to these questions in two ways: with algebra and in a way that a third, fourth, or fifth grader might be able to.

d. Is 201 an entry in the sequence or not? If yes, which entry? If no, why not? Determine the answer to these questions in two ways: with algebra and in a way that a third, fourth, or fifth grader might be able to.

3. Consider the arithmetic sequence whose first few entries are

$$3, \ 7, \ 11, \ 15, \ 19, \ 23, \ldots$$

a. Determine the 50th entry in the sequence and explain why your answer is correct.

b. Find a formula for the Nth entry in this sequence and explain in detail why your formula is valid.

c. Is 207 an entry in the sequence or not? If yes, which entry? If no, why not? Determine the answer to these questions in two ways: with algebra and in a way that a third, fourth, or fifth grader might be able to.

d. Is 100 an entry in the sequence or not? If yes, which entry? If no, why not? Determine the answer to these questions in two ways: with algebra and in a way that a third, fourth, or fifth grader might be able to.

4. a. Draw the first 5 figures in a sequence of figures whose Nth entry is made of $4N + 1$ small circles. Describe how subsequent figures in your sequence would be formed.

b. Is there a figure in your sequence in part (a) which is made of 889 small circles or not? If yes, which one? If no, why not? Determine the answer to these questions in two ways: with algebra and in a way that a third, fourth, or fifth grader might be able to.

c. Is there a figure in your sequence in part (a) which is made of 150 small circles or not? If yes, which one? If no, why not? Determine the answer to these questions in two ways: with algebra and in a way that a third, fourth, or fifth grader might be able to.

5. a. Draw the first 5 figures in a sequence of figures whose Nth entry is made of $6N + 3$ small circles. Describe how subsequent figures in your sequence would be formed.

b. Is there a figure in your sequence in part (a) which is made of 102 small circles or not? If yes, which one? If no, why not? Determine the answer to these questions in two ways: with algebra and in a way that a third, fourth, or fifth grader might be able to.

c. Is there a figure in your sequence in part (a) which is made of 333 small circles or not? If yes, which one? If no, why not? Determine the answer to these questions in two ways: with algebra and in a way that a third, fourth, or fifth grader might be able to.

6. Consider an arithmetic sequence whose third entry is 10 and whose fifth entry is 16. Use the most elementary reasoning you can to find the first and second entries of the sequence. Explain your reasoning.

7. Consider an arithmetic sequence whose fourth entry is 2 and whose seventh entry is 4. Use the most elementary reasoning you can to find the first and second entries of the sequence. Explain your reasoning.

8. Consider an arithmetic sequence whose fifth entry is 1 and whose tenth entry is 7. Use the most elementary reasoning you can to find the first and second entries of the sequence. Explain your reasoning.

9. The Widget Company sells boxes of widgets by mail order. The company charges a fixed amount for shipping, no matter how many boxes of widgets are ordered. All boxes of widgets are the same size and cost the same amount. The company is out of state, so there are no taxes charged. You find out that the total cost (including shipping) for 6 boxes of widgets is \$27 and the total cost (including shipping) for 10 boxes of widgets is \$41. Find the shipping cost and the price of one box of widgets. Explain your reasoning.

10. Consider the geometric sequence whose first few entries are

$$2, \ 10, \ 50, \ 250, \ 1250, \ 6250, \ldots$$

Find a formula for the Nth entry in this sequence and explain in detail why your formula is valid.

11. Consider the geometric sequence whose first few entries are

$$\frac{1}{2}, \ \frac{1}{4}, \ \frac{1}{8}, \ \frac{1}{16}, \ \frac{1}{32}, \ \frac{1}{64}, \ldots$$

Find a formula for the Nth entry in this sequence and explain in detail why your formula is valid.

12. Consider the geometric sequence whose first few entries are

$$\frac{1}{6}, \ \frac{1}{12}, \ \frac{1}{24}, \ \frac{1}{48}, \ \frac{1}{96}, \ \frac{1}{192}, \ldots$$

Find a formula for the Nth entry in this sequence and explain in detail why your formula is valid.

13. Consider the geometric sequence whose first few entries are

$$\frac{2}{3}, \ \frac{2}{9}, \ \frac{2}{27}, \ \frac{2}{81}, \ \frac{2}{243}, \ \frac{2}{729}, \ldots$$

Find a formula for the Nth entry in this sequence and explain in detail why your formula is valid.

14. Suppose you owe $500 on a credit card that charges you 1.6% interest on the amount you owe at the end of every month. Assume that you do not pay off any of your debt and that you do not add any more debt other than the interest you are charged.

    a.   Explain why you will owe

$$(1.016)^N \cdot 500$$

dollars at the end of N months.

b.   Use a calculator to determine how much you will owe after 2 years.

15. Suppose you put $1000 in an account whose value will increase by 6% every year. Assume that you do not take any money out of this account and that you do not put any money into this account other than the interest it earns. Find a formula for the amount of money that will be in the account after N years. Explain in detail why your formula is valid.

---

16. Parts (a) and (b) of this problem are similar to some of the problems in *Navigating through Algebra in Prekindergarten – Grade 2* by the National Council of Teachers of Mathematics [46]. Assume that the pattern of a square followed by 2 circles and a triangle continues to repeat in the sequence of shapes in Figure 13.28 and that the numbers below the shapes indicate the position of each shape in the sequence.

FIGURE 13.28

A Repeating Pattern of Shapes

    a.   What shape will be above the number 150? Explain how you can tell.

    b.   How many circles will there be above the numbers 1 through 160? Explain your answer.

    c.   Consider the numbers that are below the squares. What kind of sequence do these numbers form: an arithmetic sequence, a geometric sequence, or neither? Why? If it is either an arithmetic sequence or a geometric sequence, give a formula for the Nth entry.

    d.   Consider the numbers that are below the triangles. What kind of sequence do these numbers form: an arithmetic sequence, a geometric sequence, or neither? Why? If it is either an arithmetic sequence or a geometric sequence, give a formula for the Nth entry.

    e.   Consider the numbers that are below the circles. What kind of sequence do these numbers form: an arithmetic sequence, a geometric sequence, or neither? Why? If it is either an arithmetic sequence or a geometric sequence, give a formula for the Nth entry.

17. Parts (a) and (b) for this problem are similar to some of the problems in *Navigating through Algebra in Prekindergarten – Grade 2* by the National Council of Teachers of Mathematics [46]. Assume that the pattern of a circle followed by 3 squares and a triangle continues to repeat in the sequence of shapes in Figure 13.29 and that the numbers below the shapes indicate the position of each shape in the sequence.

FIGURE 13.29

A Repeating Pattern of Shapes

    a. What shape will be above the number 999? Explain how you can tell.

    b. How many squares will there be above the numbers 1 through 200? Explain your answer.

    c. Consider the numbers that are below the triangles. What kind of sequence do these numbers form: an arithmetic sequence, a geometric sequence, or neither? Why? If it is either an arithmetic sequence or a geometric sequence, give a formula for the Nth entry.

    d. Consider the numbers that are below the circles. What kind of sequence do these numbers form: an arithmetic sequence, a geometric sequence, or neither? Why? If it is either an arithmetic sequence or a geometric sequence, give a formula for the Nth entry.

    e. Consider the numbers that are below the squares. What kind of sequence do these numbers form: an arithmetic sequence, a geometric sequence, or neither? Why? If it is either an arithmetic sequence or a geometric sequence, give a formula for the Nth entry.

18. What day of the week will it be 1000 days from today? Use math to solve this problem. Explain your answer.

19. Assume that the first day of school is a Wednesday and that school runs Monday through Friday every week with no days off. What day of the week will the 100th school day be? Use math to solve this problem. Explain your answer.

20. Suppose a scientist puts a colony of bacteria weighing 1 gram in a large container. Assume that these bacteria reproduce in such a way that their number doubles every 20 minutes.

    a. How much will the colony of bacteria weigh after 1 hour? after 8 hours? after 1 day? after 1 week? In each case, explain your reasoning.

    b. Is it plausible that bacteria in a container could double every 20 minutes for more than a few hours? Why or why not?

13.4 Series

series

Have you ever wondered how banks calculate loan payments? Loan payments are calculated using formulas for mathematical series. A **series** is a sum of the numbers in a sequence. Just as there are arithmetic and geometric sequences, there are arithmetic and geometric series. There are also series that are neither arithmetic nor geometric. We will study some of the basics of series in this section.

ARITHMETIC SERIES

arithmetic series

An **arithmetic series** is a sum of entries in an arithmetic sequence. For example, if we add the first 100 entries in the arithmetic sequence

$$1, \ 2, \ 3, \ 4, \ 5, \ 6, \ldots$$

we have the arithmetic series

$$1 + 2 + 3 + 4 + 5 + \cdots + 97 + 98 + 99 + 100$$

If you solved Practice Problem 4 on page 9 (see also its solution on page 10) you may recall finding a quick way to calculate such a sum. Class Activity 13O will help you find a formula for sums like this one.

GEOMETRIC SERIES

Geometric series are of great practical importance since they are used in calculating loan payments, for example. Although we will not discuss loan payments, we will discuss formulas for adding geometric series, which are the basis for various financial formulas and are widely used in mathematics.

geometric series   A **geometric series** is a sum of some or all of the entries in a geometric sequence. For example,

$$1 + 2 + 4 + 8 + 16 + 32 + 64$$

and

$$\frac{1}{10} + \frac{1}{100} + \frac{1}{1000} + \frac{1}{10,000} + \cdots$$

are geometric series. Notice that the second series is an infinite sum. It may seem surprising that we can consider an infinite sum of numbers, but observe that

$$\frac{1}{10} + \frac{1}{100} + \frac{1}{1000} + \frac{1}{10,000} + \cdots$$

is really just another way to write the decimal

$$0.1111111111\ldots = 0.\overline{1}$$

Consider a finite geometric series such as

$$1 + 3 + 9 + 27 + 81 + 243 + 729 + 2187$$

Although we could add the numbers one by one to determine the sum, there is a quicker way, which we will now see. Let S stand for the sum of the series, so that

$$S = 1 + 3 + 9 + 27 + 81 + 243 + 729 + 2187$$

If we multiply the series by 3, many of the terms will be the same as in the original series (because each term in the series is 3 times the previous term):

$$3S = 3 \cdot (1 + 3 + 9 + 27 + 81 + 243 + 729 + 2187)$$
$$= 3 + 9 + 27 + 81 + 243 + 729 + 2187 + 6561$$

We get the last equation by the distributive property. If we subtract S from $3S$ we have the following two different ways we can write the result, as an expression in terms of S and as a series:

$$
\begin{array}{rl}
\text{in terms of } S: & \text{as a series :} \\
3S & \quad 3 + 9 + 27 + 81 + 243 + 729 + 2187 + 6561 \\
\underline{-\ S} & \quad \underline{-\ 1 - 3 - 9 - 27 - 81 - 243 - 729 - 2187} \\
2S & \quad -\ 1 \qquad\qquad\qquad\qquad\qquad\qquad\qquad 6561
\end{array}
$$

So

$$2S = 6560$$

and

$$S = 6560 \div 2 = 3280$$

So the sum

$$1 + 3 + 9 + 27 + 81 + 243 + 729 + 2187 = 3280$$

Notice how we got this answer: we took the next term that would go in the series, namely, $3 \times 2187 = 6561$, we subtracted the initial term, which is 1, and we divided the result by $3 - 1 = 2$. The reasoning that we used previously will show that for any finite geometric series,

$$\text{sum of finite geometric series} = \frac{\text{next term} - \text{first term}}{\text{ratio} - 1}$$

where "ratio" stands for the number that each term is multiplied by to get the next term in the series.

We can use the same method to determine sums of infinite geometric series. For example, consider the infinite series

$$\frac{1}{10} + \frac{1}{100} + \frac{1}{1000} + \frac{1}{10,000} + \cdots$$

Let S stand for its sum. The ratio of this series is $\frac{1}{10}$ (i.e., we multiply a term in the series by $\frac{1}{10}$ to get the next term in the series). So if we multiply S by $\frac{1}{10}$, many of the terms in S and $\frac{1}{10}S$ will match:

$$
\begin{aligned}
\frac{1}{10}S &= \frac{1}{10} \cdot \left(\frac{1}{10} + \frac{1}{100} + \frac{1}{1000} + \cdots \right) \\
&= \frac{1}{100} + \frac{1}{1000} + \frac{1}{10,000} + \cdots
\end{aligned}
$$

assuming that there is an "infinite distributive property." When we subtract $\frac{1}{10}S$ from S many of the terms will cancel. If we subtract $\frac{1}{10}S$ from S we can write the result in the following two ways, in terms of S and in terms of the series:

in terms of S:

$$
\begin{array}{r}
S \\
-\frac{1}{10}S \\
\hline
\frac{9}{10}S
\end{array}
$$

as a series :

$$
\begin{array}{r}
\frac{1}{10} + \frac{1}{100} + \frac{1}{1000} + \frac{1}{10,000} + \cdots \\
- \frac{1}{100} - \frac{1}{1000} - \frac{1}{10,000} + \cdots \\
\hline
\frac{1}{10}
\end{array}
$$

Therefore,

$$\frac{9}{10}S = \frac{1}{10}$$

so

$$S = \frac{1}{10} \div \frac{9}{10} = \frac{1}{9}$$

and

$$\frac{1}{10} + \frac{1}{100} + \frac{1}{1000} + \frac{1}{10,000} + \cdots = \frac{1}{9}$$

Notice how we got this answer: we took the first term in the series and divided it by 1 minus the ratio. The reasoning that we used previously will show that for any geometric series,

$$\text{sum of infinite geometric series} = \frac{\text{first term}}{1 - \text{ratio}}$$

Notice also that the infinite sum $\frac{1}{10} + \frac{1}{100} + \frac{1}{1000} + \cdots$ is just another way to write

$$0.11111111111\ldots = 0.\overline{1}$$

We showed previously that

$$0.\overline{1} = \frac{1}{9}$$

The reasoning we used earlier is very similar to the reasoning we used on pages 574–575 in Section 12.6 to show how to write repeating decimals as fractions.

CLASS ACTIVITY   NOW TURN TO CLASS ACTIVITIES MANUAL

13R   Sums of Powers of Two p. 460

13S   An Infinite Geometric Series p. 463

13T   Making Payments into an Account p. 465

PRACTICE PROBLEMS FOR SECTION 13.4

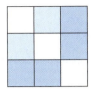 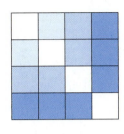 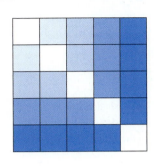 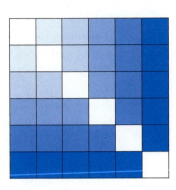

FIGURE 13.30

Sums of Even Numbers

1. a. Using the four square patterns in Figure 13.30, find a way to express the following sums in terms of multiplication and subtraction:

$$2 + 4$$

$$2 + 4 + 6$$

$$2 + 4 + 6 + 8$$

$$2 + 4 + 6 + 8 + 10$$

Write the associated equations.

   b. Based on your results in part (a), predict the sum of the arithmetic series

$$2 + 4 + 6 + 8 + \cdots + 98$$

   c. Find a formula for the sum of the arithmetic series

$$2 + 4 + 6 + 8 + \cdots + 2N$$

Explain why your formula is valid.

2. Suppose that at the beginning of every month, you make a payment of $100 into an account that earns 0.5% interest per month. (That is, the value of the account at the end of the month is 0.5% higher than it was at the beginning of the month.)

   a. Find a series that describes the amount of money in the account at the beginning of the Nth month, right after the $100 payment for that month has been added.

   b. Use the techniques described in this section to find a formula for the amount of money in the account at the beginning of the Nth month, right after the Nth payment into the account.

   c. Determine how much will be in the account at the beginning of the 24th month, right after the 24th payment into the account.

ANSWERS TO PRACTICE PROBLEMS FOR SECTION 13.4

1. a. There are $2 + 4$ small, shaded squares in the first square pattern. The shaded squares fill up all but 3 of the $3 \cdot 3$ squares in the pattern. Therefore,

$$2 + 4 = 3^2 - 3$$

Similarly, there are $2 + 4 + 6$ small, shaded squares in the second square pattern. The shaded squares fill up all but 4 of the $4 \cdot 4$ squares in the pattern. Therefore,

$$2 + 4 + 6 = 4^2 - 4$$

Using the same reasoning for the 3rd and 4th square patterns, we see that

$$2 + 4 + 6 + 8 = 5^2 - 5$$
$$2 + 4 + 6 + 8 + 10 = 6^2 - 6$$

b. Assuming that the pattern we found in part (a) continues, it should be the case that

$$2 + 4 + 6 + 8 + \cdots + 98 = 50^2 - 50 = 2450$$

Why $50^2 - 50$? We can find the expressions on the right-hand side of the equations in part (a) by taking half of the last term in the series (on the left-hand side of the equation) and adding 1. Half of 98 is 49; adding 1 we get 50.

c. If we imagine a square made of $N + 1$ rows with $N + 1$ small squares in each row, and if we imagine shading the small squares as in Figure 13.30, then counting the squares of each color, there will be

$$2 + 4 + 6 + \cdots + 2N$$

small shaded squares. These small shaded squares fill up all but $N + 1$ squares of the $N + 1$ by $N + 1$ square pattern. Therefore, there are

$$(N + 1)^2 - (N + 1)$$

small shaded squares. Since it's the same number of small, shaded squares either way you count them,

$$2 + 4 + 6 + \cdots + 2N = (N + 1)^2 - (N + 1)$$

Notice that since

$$(N + 1)^2 - (N + 1) = N^2 + 2N + 1 - N - 1$$
$$= N^2 + N$$

we can also say that

$$2 + 4 + 6 + \cdots + 2N = N^2 + N$$

2. a. At the beginning of the first month there is $100. At the end of the first month there is therefore

$$(1.005)100$$

dollars in the account, because the amount is increased by 0.5%. (See Section 4.5.) At the beginning of the second month another $100 is added to this amount, so then there is

$$100 + (1.005)100$$

dollars in the account. At the end of the second month there is 1.005 times as much as this in the account; therefore, there is

$$(1.005) \cdot (100 + (1.005)100) =$$
$$(1.005)100 + (1.005)^2100$$

dollars in the account, by the distributive property. At the beginning of the third month another $100 is added to this amount, so then there is

$$100 + (1.005)100 + (1.005)^2100$$

dollars in the account. Continuing in this way, we see that at the beginning of the Nth month, after the Nth payment is added, there is

$$100 + (1.005)100 + (1.005)^2100 +$$
$$(1.005)^3100 + \cdots + (1.005)^{(N-1)}100$$

dollars in the account.

b. Let S be the sum of the geometric series in part (a). So

$$S = 100 + (1.005)100 + (1.005)^2100 +$$
$$(1.005)^3100 + \cdots + (1.005)^{(N-1)}100$$

Then, by the distributive property,

$$(1.005)S = (1.005)100 + (1.005)^2100 +$$
$$(1.005)^3100 + \cdots + (1.005)^N100$$

When we subtract $(1.005)S - S$ we can write the answer in two ways: in terms of S and as a series. In terms of S,

$$(1.005)S - S = (0.005)S$$

In terms of a series, most of the terms in $(1.005)S - S$ cancel:

$$\begin{array}{l} (1.005)100 + (1.005)^2 100 + \ldots + (1.005)^{(N-1)}100 + (1.005)^N 100 \\ -100 - (1.005)100 - (1.005)^2 100 - \ldots - (1.005)^{(N-1)}100 \\ \hline -100 \hspace{6cm} + (1.005)^N 100 \end{array}$$

Therefore,

$$(0.005)S = (1.005)^N 100 - 100$$

and it follows that

$$S = \frac{(1.005)^N 100 - 100}{0.005}$$

c. According to the formula in part (b), at the beginning of 24 months, there will be

$$\frac{(1.005)^{24} 100 - 100}{0.005} = \$2543.20$$

in the account.

PROBLEMS FOR SECTION 13.4

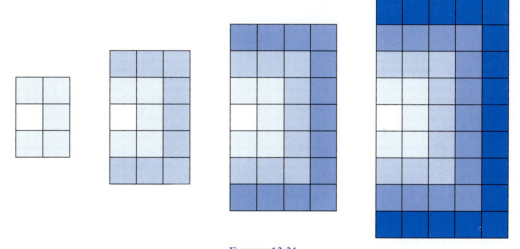

FIGURE 13.31

Patterns for Some Arithmetic Series

1. a. For each of the 4 rectangular patterns in Figure 13.31, find two different ways to express the total number of small squares that the pattern is made of. In each case, write an associated equation.

   b. Find a formula for the sum of the arithmetic series

   $$1 + 5 + 9 + 13 + \cdots + (4N + 1)$$

   Explain why your formula is valid.

---

2. Use the method described in this section to determine the sum of the geometric series

   $$1 + 4 + 16 + 64 + 256 + 1024 + 4096$$

   without adding the terms. Explain your work.

3. Use the method described in this section to determine the sum of the geometric series

   $$1 + 4 + 4^2 + 4^3 + \cdots + 4^N$$

   Explain your work.

4. Use the method described in this section to determine the sum of the geometric series

$$1 + 5 + 25 + 125 + 625 + 3125 + 15{,}625$$

without adding the terms. Explain your work.

5. Use the method described in this section to determine the sum of the geometric series

$$1 + 5 + 5^2 + 5^3 + \cdots + 5^N$$

Explain your work.

6. Use the method described in this section to determine the sum of the infinite geometric series

$$1 + \frac{1}{3} + \left(\frac{1}{3}\right)^2 + \left(\frac{1}{3}\right)^3 + \left(\frac{1}{3}\right)^4 + \cdots$$

Explain your work.

7. Use the method described in this section to determine the sum of the infinite geometric series

$$1 + \frac{1}{4} + \left(\frac{1}{4}\right)^2 + \left(\frac{1}{4}\right)^3 + \left(\frac{1}{4}\right)^4 + \cdots$$

Explain your work.

8. a. Use the method described in this section to determine the sum of the infinite geometric series

$$7\left(\frac{1}{10}\right) + 7\left(\frac{1}{10}\right)^2 + 7\left(\frac{1}{10}\right)^3 + 7\left(\frac{1}{10}\right)^4 + \cdots$$

Explain your work.

b. Rephrase your result in part (a) as a result about a repeating decimal.

9. a. Use the method described in this section to determine the sum of the infinite geometric series

$$12\left(\frac{1}{100}\right) + 12\left(\frac{1}{100}\right)^2 + 12\left(\frac{1}{100}\right)^3 + 12\left(\frac{1}{100}\right)^4 + \cdots$$

Explain your work.

b. Rephrase your result in part (a) as a result about a repeating decimal.

10. Suppose that at the beginning of every month, you make a payment of $100 into an account that earns 0.75% interest per month. (That is, the value of the account at the end of the month is 0.75% higher than it was at the beginning of the month.)

a. Find a series that describes the amount of money in the account at the beginning of the Nth month, right after the $100 payment for that month has been added.

b. Use the techniques described in this section to find a formula for the amount of money in the account at the beginning of the Nth month, right after the Nth payment into the account.

c. Determine how much will be in the account at the beginning of the 36th month, right after the 36th payment into the account.

13.5 Functions

In this section we will extend our study of sequences in Section 13.3 to the study of mathematical functions. We will define the concept of function and we will study the three main ways that functions can be represented: by formulas, by tables, and by graphs. Functions are one of the cornerstones of mathematics and science. By studying and using functions, mathematicians and scientists have been able to describe and predict a variety of phenomena, such as motions of planets, the distance a rocket will travel, or the dose of medicine that a patient needs. The study of functions goes far beyond what we discuss in this section. But even at the most elementary level, functions provide a way to organize, represent, and study information.

function   A **function** is a rule that assigns one output to each allowable input. Most commonly, these "inputs" and "outputs" are numbers. The set of "allowable inputs" is called the **domain** of the function. The set of "allowable outputs" is called the **range** of the function.

Texts for children often portray functions as machines, as in Figure 13.32. With this imagery, it's easy to imagine putting a number into a machine; the machine somehow transforms

FIGURE 13.32

Viewing
Functions as
Machines

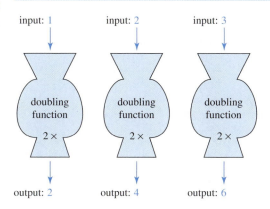

FIGURE 13.33

A Sequence of
Shapes

the input number into a new output number. For example, consider the *doubling function*: to each input number, the assigned output is twice the input. To the input 1, the doubling function assigns output 2; to the input 2, the doubling function assigns the output 4; to the input 3, the doubling function assigns the output 6; and so on.

Every sequence can be considered a function by assigning to each counting number, N, the output which is the Nth entry in the sequence. For example, the sequence

$$10, \ 12, \ 14, \ 16, \ 18, \ldots$$

corresponds to the function which takes the input 1 to the output 10, the input 2 to the output 12, the input 3 to the output 14, and so on. The sequence in Figure 13.33 corresponds to the function which takes the input 1 to a square, the input 2 to a circle, the input 3 to a triangle, the input 4 to a square, and so on.

It can be cumbersome to describe functions in words by repeatedly referring to inputs and outputs. Therefore, there are several standard ways of displaying functions: with tables, with graphs, and with formulas. These methods can help make functions comprehensible. We will now look at these ways of representing functions.

REPRESENTING FUNCTIONS WITH TABLES

A table provides an easily understood, organized way to display a function. To display a function in a table, make two columns: the column on the left consisting of various input values, the column on the right consisting of the corresponding output values. Some people like to call such a table a "t-chart" because you can make such a table by first drawing a large T.

For example, consider the function that assigns to each one of five girls, her height. This function is fully displayed in the table on the left in Table 13.1 on page 628.

TABLE 13.1

Tables for Two
Functions: A
Height Function
and a Doubling
Function

HEIGHT FUNCTION		DOUBLING FUNCTION	
INPUT (GIRL)	OUTPUT (HEIGHT)	INPUT	OUTPUT
		1	2
Kaitlyn	53 in	2	4
Lameisha	56 in	3	6
Sarah	49 in	4	8
Manuela	50 in	5	10
Kelli	54 in	⋮	⋮

Of course, we can also consider the height function that assigns to *each person in the world*, his or her height. In this case it would be neither informative nor realistically possible to make a table for this function that would show all inputs and their outputs.

The table on the right in Table 13.1 represents the doubling function. This table does not show all possible inputs and outputs and therefore does not give complete information about the function. But an incomplete table can still be useful for summarizing a function.

In the case of the doubling function we can make different choices for the inputs that we want to allow. We can decide only to allow counting numbers as inputs for the doubling function. In this case, the doubling function corresponds exactly to the sequence

$$2, 4, 6, 8, 10, \ldots$$

because we assign to the Nth place in the sequence the number $2N$.

If we decide that we want to allow all real numbers as inputs for the doubling function, then the doubling function encompasses more than just the sequence

$$2, 4, 6, 8, 10, \ldots$$

For example, the doubling function assigns to the input 3.25, the output 6.5, but 6.5 is not in the sequence

$$2, 4, 6, 8, 10, \ldots$$

Tables are easy to read and to understand, but they rarely display complete information about a function. Consider the function that we will call *S & P 500*, which assigns to any time since 1957, the value of the Standard and Poor's 500 Index at that time. (The Standard and Poor's 500 Index is a commonly used standard for measuring stock market performance. For further information on the S & P 500, see the website www.aw-bc.com/beckmann.) Table 13.2 displays some inputs and outputs of the S & P 500 function in table form.

Table 13.2 only shows the value of the S & P 500 on January 1st of each year over an 11-year period, but it doesn't show any information about the S & P 500's value in the intervening

Table 13.2
A Table for the Standard and Poor's 500 Function

S & P 500 Function	
Input (time)	**Output (value of S & P 500)**
Jan 1, 2003, 5 P.M.	880
Jan 1, 2002, 5 P.M.	1148
Jan 1, 2001, 5 P.M.	1320
Jan 1, 2000, 5 P.M.	1469
Jan 1, 1999, 5 P.M.	1229
Jan 1, 1998, 5 P.M.	970
Jan 1, 1997, 5 P.M.	740
Jan 1, 1996, 5 P.M.	615
Jan 1, 1995, 5 P.M.	459
Jan 1, 1994, 5 P.M.	466
Jan 1, 1993, 5 P.M.	435

months. If we wanted to display more information, we could make a longer table, but that might become difficult to read and comprehend.

Another good way to display a function is to use a *graph*.

Coordinate Planes and Displaying Functions with Graphs

The graph of a function displays the function in a picture. By graphing a function we can often display more inputs and outputs than we could display in a table, and we can often discern trends and patterns in the function that are less obvious in a table.

Coordinate Planes

coordinate plane

We graph a function that has numerical inputs and outputs in a *coordinate plane*. A **coordinate plane** is a plane, together with two perpendicular number lines in the plane that (usually) meet at the location of 0 on each number line. The two number lines can have different scales. Traditionally, one number line is displayed horizontally and the other is displayed vertically. The two number lines are called the **axes** of the coordinate plane (singular: axis). The horizontal axis is often called the ***x*-axis**, and the vertixal axis is often called the ***y*-axis**.

axis

The main feature of a coordinate plane is that *the location of every point in the plane can be specified by referring to the two axes.* This works in the following way: A pair of numbers, such as

$$(4.5, 3)$$

FIGURE 13.34

Locating
(4.5, 3) in a
Coordinate
Plane

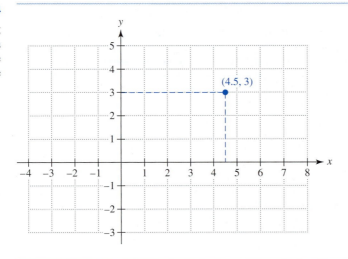

corresponds to the point in the plane that is located where a vertical line through 4.5 on the horizontal axis meets a horizontal line through 3 on the vertical axis, as shown in Figure 13.34. This point is designated (4.5, 3); or we say that 4.5 and 3 are the **coordinates** of the point. Specifically, 4.5 is the **first coordinate**, or **x-coordinate**, of the point and 3 is the **second coordinate**, or **y-coordinate**, of the point.

coordinates

You can use your fingers to locate the point (4.5, 3) by putting your right index finger on 4.5 on the horizontal axis and your left index finger on 3 on the vertical axis, and sliding the right index finger vertically upward and the left index finger horizontally to the right until your two fingers meet. Some people like to use two sheets of paper to locate points in a coordinate plane by holding the edge of one sheet parallel to the x-axis and the edge of the other sheet parallel to the y-axis.

Figure 13.35 shows the coordinates of several different points in a coordinate plane. Notice that the coordinates of a point can include negative numbers.

FIGURE 13.35

Points in a
Coordinate
Plane

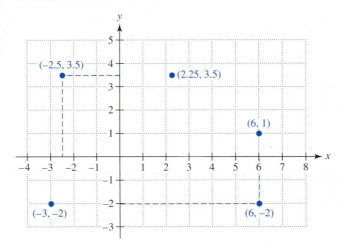

THE GRAPH OF A FUNCTION

graph of a
function

A function whose inputs and outputs are *numbers* has a *graph*. The **graph of a function** consists of all those points in a coordinate plane whose second coordinate is the output of first coordinate. For example, Figure 13.36 shows the graph of the doubling function when only the counting numbers are allowed to be inputs. In this case, the graph of the doubling function consists of the points

$$(1, 2), \ (2, 4), \ (3, 6), \ (4, 8), \ (5, 10), \ldots$$

For each point on the graph, the second coordinate is twice the first coordinate.

Figure 13.37 on page 632 shows the graph of the doubling function when we allow all real numbers as inputs. In this case, the graph consists of all points of the form

$$(N, 2N)$$

where N can be any real number. For example, in this case

$$(3.25, 6.5)$$

is on the graph of the doubling function, and so is

$$(-1.3, -2.6)$$

The graph in Figure 13.36 consists of isolated points, but the graph in Figure 13.37 consists of so many points that the points fill in a solid line.

FIGURE 13.36

Graph of the
Doubling
Function with
Only Counting
Numbers as
Inputs

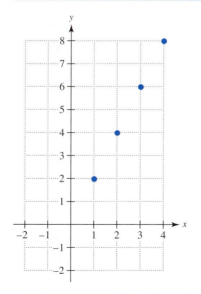

FIGURE 13.37

Graph of the
Doubling
Function When
All Real
Numbers Are
Allowed as
Inputs

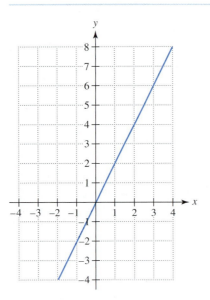

DRAWING THE GRAPH OF A FUNCTION

In order to draw the graph of a function, it often helps to display the function in a table first. For example, suppose that a seed is planted and the height of the plant growing from the seed is recorded every day after planting. This situation gives rise to a *height* function, such as the hypothetical one shown in the table in Table 13.3. Each input and corresponding output produce a point in a coordinate plane. The collection of all such points in a coordinate plane form the graph of the function.

The points obtained from the table for the function can then be plotted in a coordinate plane, as in Figure 13.38. Notice that in this graph, the points obtained from the table have been connected by a curve. Why does it make sense to connect the points? In between the times that the plant's height was measured, the plant continued to grow. For example, between days 7 and 8, the plant grew from 3 cm to 5 cm. So, $7\frac{1}{2}$ days after planting, the plant was probably about 4 cm tall. Therefore, the point (7.5, 4), or a point that is very close to this point, is also on the graph of the plant height function. For every time between 7 days after planting and 8 days after planting, the plant had some height between 3 cm and 5 cm, and this yields points on the graph that lie between the points (7, 3) and (8, 5), and that connect these two points with a curve.

Note the following two features of the graph in Figure 13.38 that help make the graph easy to see and to interpret:

- The scales on the two axes are different: the distances between numbers on the horizontal axis are smaller than the distances between numbers on the vertical axis. This is acceptable because the horizontal and vertical axes represent different kinds of quantities: the horizontal axis represents time, and the vertical axis represents height. When graphing a function, choose scales that make the function easy to see and interpret.

- The axes are labeled. The horizontal axis is labeled "Number of days since planting" and the vertical axis is labeled "Plant height (in cm)" Labeling the axes helps a reader understand and interpret a graph when the function arises from a real or realistic situation.

TABLE 13.3

A Table for a Plant Height Function Yields Points That Can Be Plotted in a Coordinate Plane to Make a Graph of the Function

PLANT HEIGHT FUNCTION		
INPUT # OF DAYS SINCE PLANTING	**OUTPUT PLANT HEIGHT (IN CM)**	
1	0	yields the point (1, 0)
2	0	yields the point (2, 0)
3	0	yields the point (3, 0)
4	0	yields the point (4, 0)
5	.5	yields the point (5, .5)
6	1.5	yields the point (6, 1.5)
7	3	yields the point (7, 3)
8	5	yields the point (8, 5)
9	5.5	yields the point (9, 5.5)
10	6	yields the point (10, 6)
11	6	yields the point (11, 6)
12	6	yields the point (12, 6)

FIGURE 13.38

Graph of the Plant Height Function

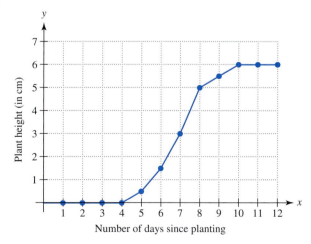

INTERPRETING GRAPHS OF FUNCTIONS

Once you learn how to interpret graphs of functions, you can quickly and easily see qualitative aspects of functions. Look at the graph in Figure 13.38: "read" it from left to right. Moving from left to right along the graph represents time passing: how far a point on the graph is to the right corresponds to how many days have passed since the seed was planted. The height of a point on the graph above the horizontal axis represents the height of the plant at that time (i.e., at the time represented by the horizontal location of the point). Thus, following the graph from left to right we see that the plant doesn't sprout until after 4 days. Between the 6th and 8th days the plant grows more quickly than at other times: we see this because the graph rises steeply during those times, meaning that over a short period of time, a lot of growth has occured. Between days 8 and 10 the the the plant grows less rapidly: the height doesn't increase as much over those two days as it did over the previous two days. After 10 days, the plant has stopped growing: its height no longer increases.

FORMULAS FOR FUNCTIONS

Some functions can be described by formulas; others cannot. To give a formula for a function means to give a formula for the output of the function in terms of the input of the function. For example, consider the doubling function. The output is always 2 times the input. Therefore, if the input is x, the output is $2x$, and so the doubling function is described by the formula

$$2x$$

If we give the doubling function the name "D," then a standard way to describe the formula for the doubling function is to write

$$D(x) = 2x$$

This is shorthand for saying that when x is put into the function D, the output is $2x$.

Not all functions have formulas. For example, consider the plant height function we studied previously. There is no formula to describe the height of the plant in terms of the number of days since the seed was planted.

If you have a formula for a function, then you can make a table and a graph for the function by using the formula. For example, consider the function, f, which is given by the formula

$$f(x) = x^2 - x - 2$$

To make a table for the function, remember what it means to say that $f(x) = x^2 - x - 2$. It means that when x is put into the function, the output is $x^2 - x - 2$. So, if 1 is put into the function, the output is

$$1^2 - 1 - 2 = 1 - 1 - 2 = -2$$

If 2 is put into the function, the output is

$$2^2 - 2 - 2 = 4 - 2 - 2 = 0$$

If 3 is put into the function, the output is

$$3^2 - 3 - 2 = 9 - 3 - 2 = 4$$

And so on. Therefore, the function f has the table given in Table 13.4.

TABLE 13.4

A Table for the
Function $f(x) =$
$x^2 - x - 2$

f	
INPUT	**OUTPUT**
−3	10
−2	4
−1	0
0	−2
.5	−2.25
1	−2
2	0
3	4
4	10

We can use the table in Table 13.4 to graph the function f, as in Figure 13.39. Assuming that all real numbers are allowed as inputs for f, the graph of f will be a curve that connects the points from the table. If you were to plot many more points, you would see that this curve is nice and smooth, as shown in Figure 13.39.

In more advanced mathematics, a central topic of study is the relationship between the nature of a formula for a function and the nature of the graph of the function.

FIGURE 13.39

The Graph of
$f(x) =$
$x^2 - x - 2$

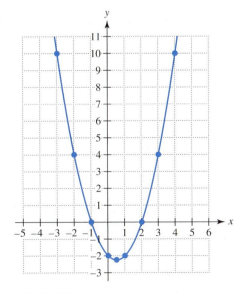

CLASS ACTIVITY NOW TURN TO CLASS ACTIVITIES MANUAL

13U Interpreting Graphs of Functions p. 466

PRACTICE PROBLEMS FOR SECTION 13.5

1. Plot the following points in a coordinate plane:

   $(4, -2.5)$, $(-4, 2.5)$, $(3, 2.75)$, $(-3, -2.75)$

2. For each of the following descriptions, draw the graph of an associated pollen count function, for which the input is time elapsed since the beginning of the week, and the output is the pollen count at that time. In each case, explain why you draw your graph in the shape that you do.

   a. At the beginning of the week, the pollen count rose sharply. Later in the week, the pollen count continued to rise, but more slowly.

   b. The pollen count fell steadily during the week.

   c. At the beginning of the week, the pollen count fell slowly. Later in the week, the pollen count fell more rapidly.

ANSWERS TO PRACTICE PROBLEMS FOR SECTION 13.5

1. See Figure 13.40.

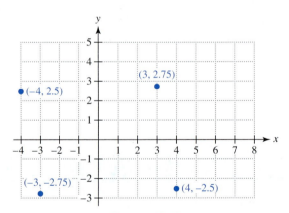

FIGURE 13.40

Points in a Coordinate Plane

2. See Figure 13.41. Moving from left to right along the graph corresponds to time passing during the week.

   a. Graph *a* goes up steeply at first (reading it from left to right) because for each of the first few days of the week, the pollen count is significantly higher than it was the day before. Thus, the graph slopes steeply upward at first. But later in the week, toward the right of the graph, the graph must go up less steeply because for each day later in the week, the pollen count is only a little higher than it was the day before.

   b. Graph *b* slopes downward (reading from left to right) and shows that for each day that goes by, the pollen count drops by the same amount.

c. Graph *c* goes down slowly at first (reading from left to right) because for each of the first few days of the week, the pollen count is only a little less than it was the day before. Thus, the graph slopes gently downward at first. But later in the week, toward the right of the graph, the graph must go down more steeply because for each day later in the week, the pollen count is significantly lower than it was the day before. Thus, towards the right, the graph slopes steeply downward.

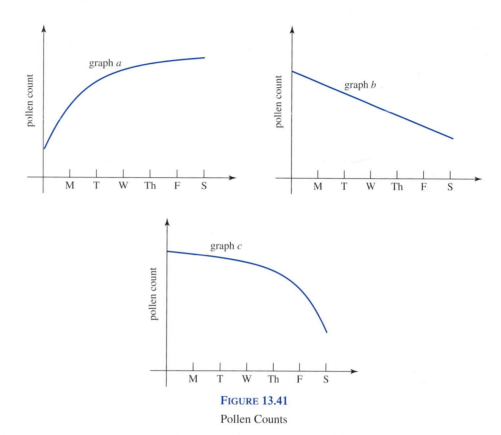

FIGURE 13.41

Pollen Counts

PROBLEMS FOR SECTION 13.5

1. Consider the *doubling plus one* function: the output is two times the input, plus one. For example, if the input is 3, the output is 7. Determine which of the following points are on the graph of the *doubling plus one* function:

$$(2, 5), \quad (2, 8), \quad (-1, -1), \quad (3, 4),$$
$$(-2, -3), \quad (1.5, 4), \quad (.5, 1.5)$$

Explain your reasoning.

2. Consider the *squaring* function: the output is the input times itself. For example, if the input is 3, the output is 9. Determine which of the following points are on the graph of the *squaring* function:

$$(6, 36), \quad (4, 8), \quad (-2, 4), \quad (-3, -6),$$
$$(5, 25), \quad (6, 12), \quad (-6, 36)$$

Explain your reasoning.

3. Consider the function given by $f(x) = x^2 - 2x$.

    a. Make a table in which you show the outputs for the following inputs:

$$-2, \ -1, \ 0, \ 1, \ 2, \ 3, \ 4$$

b. Plot the points that are associated with the inputs and outputs you found in part (a).

c. Based on the points you found in part (b), draw the graph of the function $f(x) = x^2 - 2x$ as best you can.

4. For each of the following descriptions, draw the graph of an associated temperature function, for which the input is time elapsed since the beginning of the day, and the output is the temperature at that time. In each case, explain why you draw your graph in the shape that you do.

a. At the beginning of the day, the temperature dropped sharply. Later on, the temperature continued to drop, but more gradually.

b. The temperature rose steadily at the beginning of the day, remained stable in the middle of the day, and then fell steadily later in the day.

c. The temperature rose quickly at the beginning of the day until it reached its peak. Then the temperature fell gradually throughout the rest of the day.

5. Sally invests $5000 in stock. This situation gives rise to a stock value function, for which the input is the time elapsed since investing, and the output is the value of Sally's stock. The graph of this function is shown in Figure 13.42.

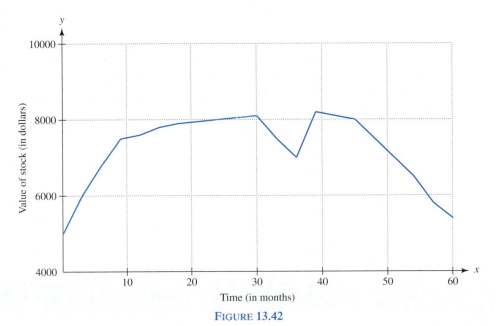

FIGURE 13.42

The Value of Sally's Stock Over Time

a. Compare the rate of growth of the value of Sally's stock in the first 10 months, versus in the next 20 months. During which time period does the value of the stock grow more rapidly? Explain how the graph shows you this.

b. What happens to Sally's stock between 30 months and 40 months? Explain your answer.

c. What happens to Sally's stock between 45 months and 60 months? Explain your answer.

6. Consider the following scenario: Over a period of 5 years, the level of water in a lake drops slowly at first, then much more rapidly. During the next 5 years, the water level in the lake remains stable. Over the next 3 years, the water level rises slowly, and after another 2 years of more rapid increase in water level, the lake is back to its original level from 15 years before.

What function does this scenario give rise to? Sketch a graph that could be the graph of this function. Explain how features of your graph correspond to events described in the previous scenario.

7. Draw *two different* graphs of two different functions that have the following three properties:

- When the input is 1, the output is 3.
- When the input is 2, the output is 6.
- When the input is 3, the output is 9.

8. Find *two different* formulas for two different functions that have the following two properties:

- When the input is 0, the output is 1.
- When the input is 1, the output is 2.

9. During a 6-hour rainstorm, rain fell at a rate of 1 inch every hour for the first 3 hours. During the next hour, $\frac{1}{2}$ inch of rain fell. During the last 2 hours, rain fell at a rate of $\frac{1}{4}$ inch every hour.

This scenario gives rise to a rainfall function in which the input is the time elapsed since the rainstorm began and the output is the amount of rain that fell during the storm up to that time.

a. Make a table for the rainfall function.

b. Draw a graph of the rainfall function.

c. Is there a single formula for the rainfall function?

10. In the country of Taxo, each household pays taxes on their annual household income as determined by the following:

- A household with income between $0 and $20,000 pays no tax.
- For each $1 of income beyond $20,000, and up to $50,000, the household must pay $.20 in tax.
- For each $1 of income beyond $50,000, and up to $100,000, the household must pay $.40 in tax.
- For each $1 of income beyond $100,000 the household must pay $.60 in tax.

This situation gives rise to a tax function, in which the input is annual household income and the output is the taxes paid.

a. Make a table for the tax function.

b. Draw a graph of the tax function.

c. Is there a single formula for the tax function?

13.6 Linear Functions

In this section we will introduce and study linear functions, which will allow us to tie together several topics we have studied: arithmetic sequences, ratios, and similar triangles.

linear function   A **linear function** is a function whose graph is a line in a coordinate plane. We will see that there are several other ways to characterize linear functions, but first we will see how linear functions are related to arithmetic sequences.

Suppose the Tropical Orchid Company sells orchids by mail order. The company charges $7.00 for shipping plus $3.00 for each orchid ordered. This situation gives rise to an arithmetic sequence in which the first entry is the shipping cost of 1 orchid, the second entry is the shipping cost of 2 orchids, the third entry is the shipping cost of 3 orchids, and so on, as shown in Figure 13.43. Each entry in the sequence is $3 more than the previous entry because each additional orchid costs an extra $3 in shipping. Based on our work on arithmetic sequences,

FIGURE 13.43

An Arithmetic Sequence of Shipping Costs of Orchids if There Is a $7 Fee Plus $3 for Each Orchid

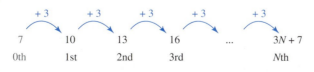

FIGURE 13.44

A Table and a
Graph of the
Orchid Shipping
Sequence

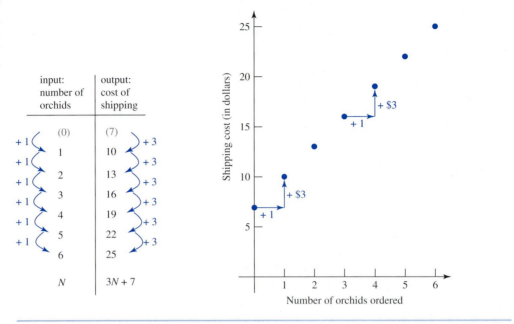

the Nth entry is

$$3N + 7$$

dollars. In other words, the shipping cost for N orchids is $3N + 7$ dollars.

Now let's view the orchid shipping sequence as a function whose input is the number of orchids to be shipped and whose output is the shipping cost of that many orchids. We can make a table for this function and we can graph this function, as in Figure 13.44. Notice that the arithmetic nature of the orchid shipping sequence is reflected in its table in the following way: whenever the input goes up by 1, the output goes up by a fixed amount, namely 3. In other words, the ratio of the change in input to the corresponding change in output is always the same. Notice also that the graph of the orchid shipping sequence appears to lie on a straight line.

In fact, the following are true for all arithmetic sequences:

1. Viewing the sequence as a function, the ratio of the change in input to the corresponding change in output is always the same.

2. There are numbers m and b so that the Nth entry in the sequence is $mN + b$.

3. The graph of the sequence (viewed as a function) lies on a line.

Viewed as functions, arithmetic sequences are linear functions whose input set (domain) is restricted to counting numbers rather than allowing all real numbers as inputs. We will now see that linear functions also have properties (1) and (2) (once property (2) is suitably rephrased).

Consider a linear function. By definition, its graph is a line. We can use this line to form many different right triangles that have a horizontal and a vertical side, as shown in Figure 13.45. Because the horizontal lines are all parallel, the angles they form with the graph of the function are all the same. Since they are right triangles, all three angles of each of these

FIGURE 13.45

For Linear
Functions the
Ratio of the
Change in Input
to the
Corresponding
Change in
Output Is
Always the
Same Due to
Similar
Triangles

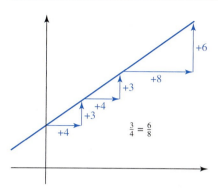

triangles must be equal, and so all such triangles are similar. (See Section 9.4.) Therefore, the ratios of the lengths of the horizontal sides to the lengths of the vertical sides are equal for all these triangles. The lengths of horizontal sides of these triangles represent "changes in inputs" because the horizontal axis represents inputs; the lengths of vertical sides of these triangles represent the corresponding "changes in outputs" because the vertical axis represents outputs. Therefore, for a linear function, the ratio of the change in input to the corresponding change in output is always the same.

We have seen that for linear functions, the ratio of the change in input to the change in output is always the same. We will now use this fact to deduce that every linear function has a formula of a certain type. Let b be the location on the y-axis where the graph of the function crosses the y-axis. Suppose that when the input is increased by 1, the output increases by m. Because the ratio of the change in input to the corresponding change in output is always the same, if the input increases by x, the output increases by mx. Therefore, when the input is x, the output is

$$mx + b$$

as indicated in Figure 13.46. So every linear function has a formula of the form

$$f(x) = mx + b$$

slope

y-intercept

for some numbers m and b. The number m is called the **slope** of the line; the number b is called the **y-intercept** of the line.

FIGURE 13.46

For a Linear
Function, if the
Output
Increases by m
When the Input
Increases by 1,
Then the Output
Increases by mx
When the Input
Increases by x

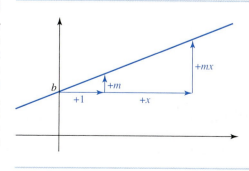

To summarize, for any linear function, the following are true:

1. The ratio of the change in input to the corresponding change in output is always the same.

2. There are numbers m and b so that the linear function is described by the formula $f(x) = mx + b$.

3. The number b in (2) is the output for the input 0. The number m in (2) is the increase in output when the input increases by 1.

CLASS ACTIVITY   NOW TURN TO CLASS ACTIVITIES MANUAL

13V Comparing the Way Functions Change p. 471

13W Functions for Which the Ratio of Change in Input to Change in Output Is Always the Same p. 472

13X Story Problems for Linear Functions p. 474

PRACTICE PROBLEMS FOR SECTION 13.6

1. A nightlight costs $7.00 to buy and uses $.75 of electricity per year to operate. What function does this situation give rise to? Make a table for this function, sketch a graph of this function, and find a formula for this function.

2. For each of the tables that follow, determine whether it could be the table for a linear function. If so, find a formula for the linear function. If not, explain why not.

f	
INPUT	OUTPUT
0	2
1	7
2	12
3	17
4	22

g	
INPUT	OUTPUT
2	5
4	10
6	15
8	20
10	25

h	
INPUT	OUTPUT
1	2
2	4
3	8
4	16
5	32

3. Assuming that the following tables are for linear functions, fill in the blank outputs and find formulas for these functions:

INPUT	OUTPUT
0	
1	
2	
3	4
4	
5	8
6	

INPUT	OUTPUT
0	
1	
2	3
3	
4	
5	
6	6

INPUT	OUTPUT
−3	7
−2	
−1	
0	
1	−1
2	
3	

ANSWERS TO PRACTICE PROBLEMS FOR SECTION 13.6

1. The situation gives rise to a "nightlight function" for which the input is the number of years since purchase and the output is the cost of operating the nightlight for that amount of time (where this cost includes the cost of buying the nightlight). The following is a table for the nightlight function:

NIGHTLIGHT FUNCTION	
INPUT	**OUTPUT**
YEARS SINCE PURCHASE	TOTAL COST OF OPERATING
1	$7.75
2	$8.50
3	$9.25
4	$10.00
5	$10.75

Because the nightlight costs $7.00 to buy and costs $.75 for each year of operation, it costs a total of

$$7 + .75y$$

dollars to operate the nightlight for y years. A sketch of the graph of the nightlight function is in Figure 13.47.

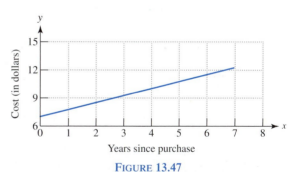

FIGURE 13.47

The Cost of Using a Nightlight

2. The first table is a table of the linear function with formula

$$f(x) = 5x + 2$$

This formula fits with the table because whenever the input increases by 1, the output increases by 5. For a linear function with this data, when the input increases by x, the output must increase by $5x$. When the input is 0, the output is 2. So when the input is x, the output must be $5x$ more than 2, which is $5x + 2$.

The second table is a table of the linear function

$$g(x) = \frac{5}{2}x$$

This formula fits with the table because whenever the input increases by 2, the output increases by 5. For a linear function with this data, the ratio of increase in input to increase in output is 2 to 5. This is the same as the ratio 1 to $\frac{5}{2}$. So, when the input increases by 1, the output increases by $\frac{5}{2}$. Therefore, if the input increases by x, the output must increase by $\frac{5}{2}x$. When the input is 0, the output is 0, so when the input is x, the output is $\frac{5}{2}x$ more than 0, which is $\frac{5}{2}x$.

The third table is not a table for a linear function because when the input increases by 1, the output increases by different amounts. For example, in going from the input 1 to the input 2, the output increases by 2 from 2 to 4. But in going from the input 2 to the input 3, the output increases by 4 from 4 to 8.

3. Formulas for each function are shown at the bottom of each table.

INPUT	OUTPUT	INPUT	OUTPUT	INPUT	OUTPUT
0	−2	0	1.5	−3	7
1	0	1	2.25	−2	5
2	2	2	3	−1	3
3	4	3	3.75	0	1
4	6	4	4.5	1	−1
5	8	5	5.25	2	−3
6	10	6	6	3	−5
x	$2x - 2$	x	$0.75x + 1.5$	x	$-2x + 1$

PROBLEMS FOR SECTION 13.6

1. A long-distance phone company charges $3 for a phone call plus $0.75 for each minute of the call. What function does this situation give rise to? Make a table for this function, sketch a graph of this function, and find a formula for this function.

2. For each of the tables that follow, determine whether it could be the table for a linear function. If so, find a formula for the linear function. If not, explain why not.

f	
INPUT	OUTPUT
0	3
1	6
2	9
3	12
4	15

g	
INPUT	OUTPUT
0	3
2	6
4	9
6	12
8	15

h	
INPUT	OUTPUT
0	1
1	2
2	5
3	10
4	17

3. For each of the tables that follow, determine whether it could be the table for a linear function. If so, find a formula for the linear function. If not, explain why not.

f	
INPUT	OUTPUT
0	3
1	7
2	11
3	15
4	19

g	
INPUT	OUTPUT
2	4
4	7
6	10
8	13
10	16

h	
INPUT	OUTPUT
0	−1
1	0
2	3
3	8
4	15

4. Assuming that the following tables are for linear functions, fill in the blank outputs and find formulas for these functions:

INPUT	OUTPUT
0	
1	
2	7
3	
4	10
5	
6	

INPUT	OUTPUT
0	
1	−2
2	
3	
4	
5	
6	2

INPUT	OUTPUT
−3	13
−2	
−1	
0	
1	
2	−2
3	

5. Consider a function that has the following properties:

   • When the input is 0, the output is 2.

   • Whenever the input increases by 1, the output increases by 4.

   a. Make a table and draw the graph of a function that has the given properties.

   b. Explain how the properties in the two bulleted items are reflected in the graph of the function.

   c. Find a formula for a function that has the given properties.

6. Consider a function that has the following properties:

   • When the input is 2, the output is 4.

   • Whenever the input increases by 3, the output increases by 5.

   a. Make a table and draw the graph of a function that has the given properties.

   b. Explain how the properties in the two bulleted items are reflected in the graph of the function.

   c. Find a formula for a function that has the given properties.

7. Starting from milepost 100, John drove at a steady 60 miles per hour down the straight road. Consider the distance function whose input is the time elapsed since John was at milepost 100, and whose output is the distance that John is from milepost 100 at that time.

   a. Make a table for the distance function.

   b. Draw the graph of the distance function.

   c. Find a formula for the distance function and explain why your formula is valid.

8. Scientists determined that for each inch of rain that falls in a certain area, 50,000 gallons of water will flow into the area's storm sewer system. Consider the water function whose input is the number of inches of rainfall and whose output is the number of gallons of water which will flow into the storm sewer system with that amount of rainfall.

   a. Make a table for the water function.

   b. Draw the graph of the water function.

   c. Find a formula for the water function and explain why your formula is valid.

9. For each dollar spent at Athens stores, Athens receives $.01 in sales tax. Consider the tax function whose input is the number of dollars that might be spent at stores in Athens and whose output is the amount of tax dollars Athens would receive in that case.

   a. Make a table for the tax function.

   b. Draw the graph of the tax function.

   c. Find a formula for the tax function and explain why your formula is valid.

10. At one stage in a factory's candy production, cocoa is mixed with sugar. For each pound of cocoa, 0.65 pounds of sugar is needed. Consider the chocolate function whose input is the number of pounds of cocoa and whose output is the number of pounds of sugar needed for that many pounds of cocoa.

   a. Make a table for the chocolate function.

   b. Draw the graph of the chocolate function.

   c. Find a formula for the chocolate function and explain why your formula is valid.

11. Write a story problem which gives rise to a function, f, that has the formula $f(x) = 0.35x + 2.50$. Explain why your function has that formula.

12. Write a story problem that gives rise to a function, g, having the formula $g(x) = \frac{2}{3}x$.

Chapter 14

Statistics

T he field of statistics provides tools for studying questions that can be answered with data. Increasingly, a variety of data are available for populations, health, financial and business activities, and the environment. It is virtually impossible to think of a social activity or a physical phenomenon for which we cannot collect data. Statistical concepts help us to interpret these data and to recognize trends.

Concerning the learning of statistics and probability, the National Council of Teachers of Mathematics recommends the following (see [44]):

Instructional programs from prekindergarten through grade 12 should enable all students to—

NCTM STANDARDS

- formulate questions that can be addressed with data and collect, organize, and display relevant data to answer them;

- select and use appropriate statistical methods to analyze data;

- develop and evaluate inferences and predictions that are based on data;

- understand and apply basic concepts of probability.

For NCTM's specific statistics and probability recommendations for grades Pre-K–2, grades 3–5, grades 6–8, and grades 9–12, see

www.aw-bc.com/beckmann

14.1　Designing Investigations and Gathering Data

We humans are naturally curious about each other and about the world. We want to know what others around us think about the political questions of the day, about the latest movies, TV shows, and music, and similar topics of interest. Local, state, and national governments want to know facts about population, employment, income, and education in order to design appropriate programs. Scientists want to know how to cure and prevent diseases, as well as to know about the world around us. Therefore, news organizations, government bodies, and scientists regularly design investigations and gather data to help answer specific questions.

Some investigations use surveys to gather data. In a survey, people respond to questions or provide information either by phone, in person, or in writing. Some investigations use observations to gather data, such as by counting the number of turtle eggs in a certain area, or by measuring the height of a plant every day as it grows. Some investigations use controlled experiments to generate data. In this case, there are two or more groups, some given a treatment and some given no treatment, and the outcomes of the groups are compared.

POPULATIONS AND SAMPLES

Regardless of the nature of the investigation, a statistical study focuses on a certain population.
population　The **population** of a statistical study is the full set of people or things that the study is designed to investigate. For example, before a presidential election, a study might investigate how registered voters plan to vote. In this case, the population of the study is all registered voters in the country. Or a study might investigate properties of ears of corn in a cornfield. In this case, the population of the study is all ears of corn in the cornfield. In most cases, it is not possible to study every single member of a population; instead, a sample is chosen for study. A **sample**
sample　of a population of a study consists of some collection of members of the full population. For example, a pollster will select a sample of voters to survey about how they will vote in an upcoming election instead of trying to ask every voter. A scientist studying ears of corn in a cornfield will select a sample of ears of corn to study rather than trying to study every single ear of corn in the cornfield.

In order for a sample of a population under study to be useful, the characteristics of the sample must reflect the characteristics of the full population. In other words, the sample should
representative　be **representative** of the full population. But unless you choose a sample carefully, the sample's
sample　characteristics might not be representative of the full population's characteristics. For example, if a group studying child nutrition in a county took their sample of children from a school in an affluent neighborhood, their findings could differ significantly from the actual status of child nutrition in the county. The best way to choose a representative sample of a population is to
random sample　pick a random sample. To choose a **random sample** means to choose the members of the sample in such a way that every member of the population has an equal chance of becoming a member of the sample. Random samples are often chosen with the aid of a list of random numbers, which can be generated by computers and calculators.

If a random sample of a population is chosen, then the characteristics of the sample are likely to be very close to the characteristics of the full population if the sample is large enough. For example, if 54% of a random sample of voters said they were likely to vote for candidate X, then it is likely that close to 54% of the full population would also vote for candidate X.

We will now look at some of the different types of statistical studies.

SURVEYS

CLASS ACTIVITY   NOW TURN TO CLASS ACTIVITIES MANUAL

14A Challenges in Conducting Good Surveys p. 475

It may seem like a simple matter to determine people's opinions or facts about people by surveying. But if you did Class Activity 14A, then you probably began to realize that it is not.

In addition to the challenge of choosing a representative sample of a population to survey, another challenge in surveying is posing good questions. Different people may interpret the same question in different ways. For example, even a relatively simple question such as "how many children are in your family?" can be interpreted in different ways. Does it mean how many people under the age of 18 are in your family? Does it mean how many siblings you have, plus yourself? Should stepchildren be included? A good survey question must be clear.

EXPERIMENTS

In controlled experiments, such as an experiment to determine if a new drug is safe and effective, a group of volunteers will be divided into two (or more) groups. One group receives the new drug, the other group receives a placebo—an inert substance that is made to look like a real drug. Patients are observed to see if their symptoms improve and if there are any side effects in the group receiving the new drug.

But such an experiment is useful only if it accurately predicts the outcome of the use of the drug in the population at large. Therefore, it is crucial that the two groups, the one receiving the drug and the one receiving the placebo, are as alike as possible in all ways except in whether or not they receive the drug. The two groups should also be as much like the general population who would take the drug as possible. Therefore, the two groups are usually chosen randomly (such as by flipping a coin: heads goes to group 1, tails to group 2). In addition, such experiments are usually double-blind; that is, neither the patient nor the treating physician knows who is in which group.

In controlled drug experiments, the randomness of the assignment of patients to the two groups and the double-blindness ensure that the study will accurately predict the outcome of the use of new drug in the whole population. If the patients were not assigned randomly to the two groups then the person assigning the patients to a group might subconsciously pick patients with certain attributes for one or the other group, thereby making the group that is treated by the new drug slightly different from the group receiving the placebo. This difference could result in inaccurate predictions about the safety and effectiveness of the drug in the population at large. Similarly, if the patient knows whether or not he has received the new drug, his resulting mental state may affect his recovery. If the physician knows whether or not the new drug has been administered, his treatment of the patient may be different and the patient may again be affected differently.

CLASS ACTIVITY NOW TURN TO CLASS ACTIVITIES MANUAL

14B Which Experiment Is Better? p. 477

14C Using Random Samples p. 478

14D Using Random Samples to Estimate Population Size by Marking (Capture-Recapture) p. 479

PRACTICE PROBLEMS FOR SECTION 14.1

1. Kaitlyn, a fifth grader, asked 5 of her friends in class which book is their favorite. All five said *Harry Potter and the Sorcerer's Stone*. Can Kaitlyn conclude that most of the children at her school would say *Harry Potter and the Sorcerer's Stone* is their favorite book? Why or why not?

2. There is a large bin filled with 200 table tennis balls. Some of the balls are white and some are orange. Tyler reaches into the bin and randomly pulls out 10 table tennis balls. Three of the balls are orange and 7 are white. Based on Tyler's sample, what is the best estimate we can give for the number of orange table tennis balls in the bin? Explain your reasoning.

3. There is a large bin filled with table tennis balls, but we don't know how many. There are 40 orange table tennis balls in the bin; the rest are white. Amalia reaches into the bin and randomly picks out 20 table tennis balls. Out of the 20 Amalia picked, 6 are orange. Based on Amalia's sample, what is the best estimate we can give for the number of table tennis balls in the bin? Explain your reasoning.

ANSWERS TO PRACTICE PROBLEMS FOR SECTION 14.1

1. No, Kaitlyn can't conclude that most of the children at her school would say *Harry Potter and the Sorcerer's Stone* is their favorite book. Kaitlyn's friends may all have a similar taste in books and their taste in books may not be representative of the tastes of the whole school. Also, younger children at lower reading levels might prefer different books because *Harry Potter and the Sorcerer's Stone* is too advanced for them to read.

2. Since Tyler picked a random sample, the characteristics of his sample should reflect the characteristics of the full population of the balls in the bin. Three out of the 10 balls that Tyler picked are orange, so 30% of the balls in Tyler's sample are orange. The full population of balls in the bin should also be about 30% orange. Since 30% of 200 is 60, the best estimate is that 60 balls are orange and 140 are white.

3. Since Amalia picked a random sample, the characteristics of her sample should reflect the characteristics of the full population of the balls in the bin. Since 6 out of 20 balls that Amalia picked were orange, 30% of the balls Amalia picked were orange. Therefore, approximately 30% of the balls in the bin should be orange. Since there

are 40 orange table tennis balls in the bin and since these 40 balls should be about 30% of all the balls in the bin, we have

$$30\% \cdot (\text{balls in the bin}) = 40$$

So

$$\text{balls in the bin} = 40 \div 0.30 = 133$$

Therefore, there should be approximately 133 balls in the bin.

PROBLEMS FOR SECTION 14.1

1. Neil, a third grader, asked 10 of his classmates whether they prefer to wear shoes with laces or without laces. Most of the classmates Neil asked prefer to wear shoes without laces. Discuss whether or not it would be reasonable to assume that most of Neil's class or most of the students at Neil's school would also prefer to wear shoes without laces.

2. An announcer of a TV program invited viewers to vote in an internet poll, indicating whether or not they are better off economically this year than last year. Most of the people who participated in the poll indicated they are worse off this year than last year. Based on this information, can we conclude that most people are worse off this year than last year? If so, explain why. If not, explain why not.

3. At a factory that produces doorknobs, a batch of 1500 doorknobs has just been produced. To check the quality of the doorknobs, a random sample of 100 doorknobs is selected to test for defects. Out of these 100 doorknobs, 2 were found to be defective. Based on these results, what is the best estimate you can give for the number of defective doorknobs in the batch of 1500? Explain your reasoning.

4. At a factory that produces switches, a batch of 3000 switches has just been produced. To check the quality of the switches, a random sample of 75 switches are selected to test for defects. Out of these 75 switches, 2 were found to be defective. Based on these results, what is the best estimate you can give for the number of defective switches in the batch of 3000? Explain your reasoning.

5. At a light bulb factory there are 1728 light bulbs that are ready to be packaged. A worker randomly pulls 60 light bulbs out and tests each one. Two light bulbs are found to be defective.

   a. Based on the information given, what is your best estimate for the number of defective light bulbs in the full batch of 1728 light bulbs? Explain your reasoning.

   b. If the light bulbs are put in boxes with 144 light bulbs in each box (in 12 packages of 12), then approximately how many defective light bulbs would you expect to find in a box? Explain your reasoning.

6. Carter has a large marble collection. Carter knows that he has exactly 20 blue marbles in his collection, but does not know how many marbles he has in all. Carter mixes up his marbles and randomly picks out 30 marbles. Of the 30 marbles he picked, 4 were blue. Based on these results, what is the best estimate you can give for the total number of marbles in Carter's collection? Explain your reasoning.

7. The following problem is an example of the *capture-recapture* method of population estimation. A researcher wants to estimate the number of rabbits in a region. The researcher sets some traps and catches 30 rabbits. After tagging the rabbits, they are released unharmed. A few days later the researcher sets some traps again and this time catches 35 rabbits. Of the 35 rabbits trapped, 4 were tagged, indicating that they had been trapped a few days earlier. Based on these results, what is the best estimate you can give for the number of rabbits in the region? Explain your reasoning.

8. A group studying violence wants to determine the attitudes toward violence of all the fifth graders in a county. The group plans to conduct a survey, but the group does not have the time or resources to survey every fifth grader. The group can afford to survey 120 fifth graders. For each of the methods in (a) through (c) of choosing 120 fifth graders to survey, discuss advantages and disadvantages of that method. Based on your answer, which of the three methods will be best for determining the attitudes toward violence of all the fifth graders in the county? (There are 6 elementary schools in the county and each school has 4 fifth grades.)

a. Have each elementary school principal in the county select 20 fifth graders.

b. Have each fifth grade teacher in the county select 5 fifth graders.

c. Obtain a list of all fifth graders in the county. Obtain a list of 120 random numbers from 1 to the number of fifth graders in the county. Use the list of random numbers to select 120 random fifth graders in the county.

14.2   Displaying Data and Interpreting Data Displays

Every day we see the results of surveys in newspapers or on television newscasts. Sometimes results are reported with numbers (such as a number of people or as a percentage), but other times, results are shown in a table, or in a chart or graph. Charts and graphs help us get a sense of what data mean because a chart or graph can show us the "big picture" about data in a way that we can understand quickly and easily. In this section we will study the most common ways to display data in charts and graphs. We will use data displays to help us interpret data and to help us communicate our interpretation to others. We will also look at some common errors in working with data displays.

COMMON DATA DISPLAYS

In this section we discuss several types of data displays: real graphs and pictographs, bar graphs, line plots, line graphs, and pie graphs.

REAL GRAPHS AND PICTOGRAPHS

The most basic kinds of graphs are real graphs and pictographs. Both of these are suitable for use with young children and can act as a springboard for understanding other kinds of graphs. A **real graph** or *object graph* displays actual objects in a graph form; a **pictograph** is like a real graph except that it uses pictures of objects instead of actual objects. In some pictographs, a single picture may represent more than one object. We can use real graphs and pictographs to show how a collection of related objects is sorted into different groups. The real graph or pictograph allows us to see at a glance which groups have more, and which groups have fewer objects in them.

real graph

pictograph

Suppose we have a tub filled with small beads that are alike except that they have different colors: yellow, pink, purple, and green. We can mix up the beads and scoop some out. How many beads of each color do we have? Which color beads do we have the most of? Which do we have the least of? To answer these questions at a glance, we can line the beads up, as indicated in Figure 14.1. This creates a real graph.

A real graph can be turned into a pictograph by replacing the objects with pictures of the objects, as in Figure 14.2.

When we want to represent a large number of objects in a pictograph it usually makes sense to allow each picture to stand for more than one object. Each picture could stand for 2, 5, 10, 100, 1000, or some other convenient number of objects. If each picture stands for more than 1 object, then a key near the graph should show clearly how many objects the picture stands for.

Suppose a fifth grade class of students in Wyoming has penpals in Idaho, Oregon, Washington state, and Georgia. The class can look up the populations of these states on the Internet

FIGURE 14.1

A Real Graph:
Beads of
Different Colors

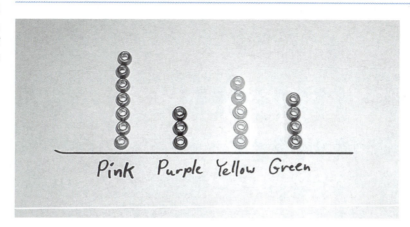

FIGURE 14.2

A Pictograph:
Beads of
Different Colors

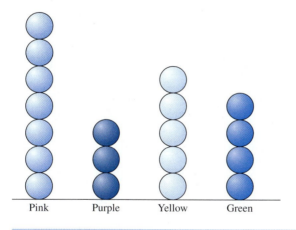

by going to the U.S. Census Bureau homepage (or see www.aw-bc.com/beckmann) and get a list as follows:

Wyoming	494,423
Idaho	1,321,006
Oregon	3,472,867
Washington	5,987,973
Georgia	8,383,915

Although this list shows the (estimated) populations quite precisely, to get a good feel for the relative populations of the states, the class can first round the populations to the nearest half-million, and then draw a pictograph, as in Figure 14.3 on page 654. In this pictograph, the picture of a person represents 1 million people, so half a million people can be represented by half a person. In the pictograph we can see at a glance how much bigger in population some of these states are than others.

Pictographs can be horizontal, as in Figure 14.3, or vertical, as in Figure 14.4 (p. 654).

FIGURE 14.3

Pictograph:
Current
Populations of
Some States

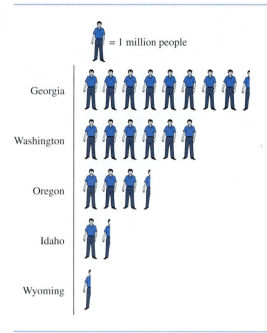

FIGURE 14.4

Vertical
Pictograph:
Current
Populations of
Some States

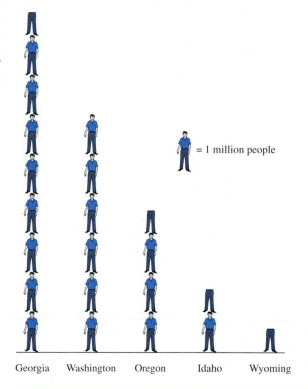

BAR GRAPHS

bar graph

A **bar graph**, or *bar chart*, is essentially just a streamlined version of a pictograph. A bar graph shows at a glance the relative sizes of different categories. The pictographs of Figures 14.3 and 14.4 have been turned into horizontal and vertical bar graphs in Figure 14.5.

double bar graph

A **double bar graph** is a bar graph in which each category has been subdivided into two subcategories. For example, if the bar graph is about people, then the two subcategories might be "male" and "female." Figure 14.6 shows mean annual earnings in 1989 of full-time, year-round workers, 25–29 years old. The different colored bars for men and women show the different earnings of these two groups in each of the different educational levels. The data used in constructing the double bar graph were gathered by the U.S. Census Bureau and are available on the Internet. (See www.aw-bc.com/beckmann.)

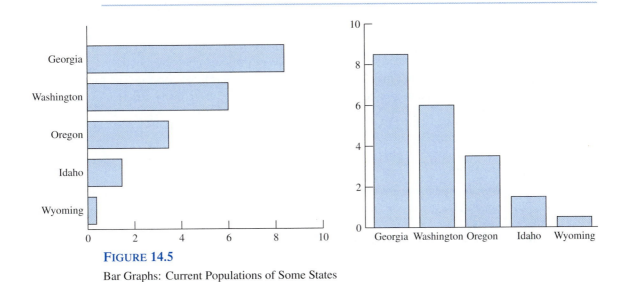

FIGURE 14.5

Bar Graphs: Current Populations of Some States

FIGURE 14.6

Double Bar Graph: Mean Annual Earnings by Education and Sex

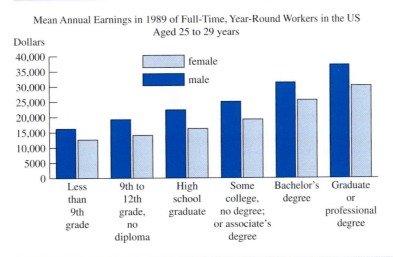

FIGURE 14.7

Line Plot: The
Number of
Black-Eyed
Peas in a
Teaspoon

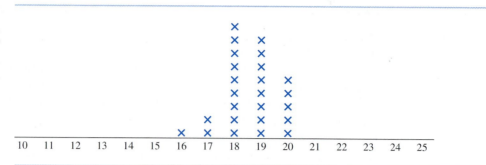

LINE PLOTS

line plot   A **line plot** is a kind of pictograph in which the categories are numbers and the pictures are Xs. The Xs function as tally marks.

Let's say we have a bag of black-eyed peas and a measuring teaspoon. How many peas are in a teaspoon? We can scoop out a teaspoon of peas and count the number of peas. If we do it again and again, we may get different numbers of peas in a teaspoon. The following numbers could be the number of peas in a teaspoon that we get from many different trials:

18,   19,   19,   17,   19,   18,   20,   18,   19,   19,   20,   20,   19

20,   18,   19,   18,   17,   18,   20,   18,   18,   16,   18,   19

We can organize these data in a line plot, as shown in Figure 14.7. Each X in the line plot represents a number in the previous list. We can see right away from the line plot that a teaspoon of black-eyed peas usually consists of 18 or 19 peas.

Line plots are easy to draw, and you can even draw a line plot as you collect data—you don't necessarily have to write your data down first before you draw a line plot. For this reason, line plots can be a quick and handy way to organize numerical data.

LINE GRAPHS

line graph   A **line graph** is a graph in which adjacent data points are connected by a line. Often, a line graph is just a graph of a function, such as the graph of a population of a region over a period of time, or the graph of the temperature of something over a period of time.

Line graphs are appropriate for displaying "continuously varying" data, such as data that varies over a period of time. Figure 14.8 shows that the average math scores of 9-year-old children have been slowly increasing, as measured by NAEP, the National Assessment of Educational Progress. Notice that the points that are plotted in Figure 14.8 indicate that the NAEP math test was given in 1978, 1982, 1986, 1990, 1992, 1994, 1996, and 1999. It makes sense to connect these points because if the test were given at times in between, and if the children's average math scores were plotted as points, these points would probably be close to the lines drawn in the graph of Figure 14.8.

Line graphs are not appropriate for displaying data in categories that don't vary continuously. For example, it would not be appropriate to use a line graph to display the data on state populations that is displayed in the pictographs and bar graphs of Figures 14.4 and 14.5. Figure 14.9 shows an inappropriate line graph of this sort. It doesn't make sense to use a line

FIGURE 14.8

Line Graph:
Math NAEP
Scores of
9-Year-Old
Children

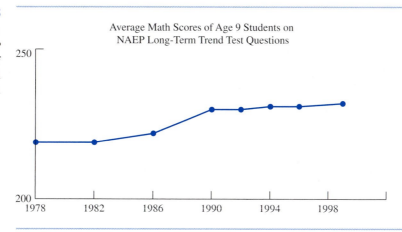

FIGURE 14.9

An
Inappropriate
Use of a Line
Graph:
Connecting
State
Populations

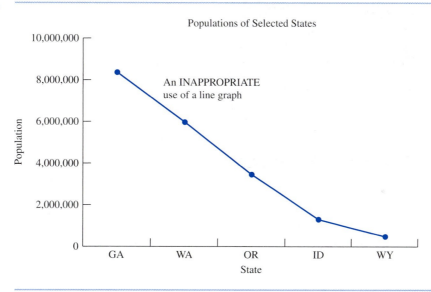

graph in this case because there is no logical reason to connect the state populations—there is no continuous variation between the population in Georgia and the population in Washington, for example.

PIE GRAPHS

pie graph   A **pie graph** or *pie chart* uses a subdivided circle to show how data partitions into categories. Figure 14.10 on page 658 shows in a pie graph how U.S. businesses in 1999 were divided according to how many employees they had. We can see at a glance that the vast majority of businesses had fewer than 20 employees. The data for this pie graph were taken from the U.S. Census Bureau. (See www.aw-bc.com/beckmann.)

By their nature, pie graphs involve percentages or fractions. In fact, simple activities with pie graphs can be used to introduce fractions and percents to children. For example, let's say we have a small bag filled with 6 black snap cubes, 8 purple snap cubes, 6 green snap cubes,

FIGURE **14.10**

Pie Graph: Percent of U.S. Businesses with Various Numbers of Employees, 1999

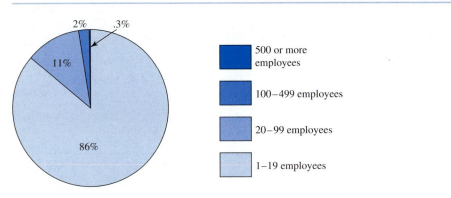

FIGURE **14.11**

A Pie Chart in a Paper Plate

and 4 white snap cubes. We can arrange these snap cubes evenly around a paper plate, keeping like colors together, as in Figure 14.11. We can then make "pie wedges" showing $\frac{1}{2}$, $\frac{1}{3}$, $\frac{1}{4}$, and $\frac{1}{6}$ of the plate. By filling the plate with pie wedges, we can see that $\frac{1}{3}$ of the snap cubes are purple, $\frac{1}{4}$ are black, $\frac{1}{4}$ are green, and $\frac{1}{6}$ are white.

CREATING AND INTERPRETING DATA DISPLAYS

Data displays are created to communicate and interpret data. Therefore, in creating a data display, choose a display that will communicate the point you want to make most clearly. Give your display a title so that the reader will know what it is about. Label your display clearly.

Read data displays carefully before interpreting them. For example, the vertical axis of the line graph in Figure 14.8 starts at 200, not at 0, which can make changes appear greater than they are. For example, you might initially think that the math scores in 1999 were almost 50% higher than they were in 1978. In fact, the math scores in 1999 were only about 6% higher than they were in 1978.

Be careful when drawing conclusions about cause and effect from a data display. For example, in Figure 14.6, we see that women earn less money than men in all categories of education. We might be tempted to conclude that this is due to discrimination against women on the basis of sex. While this could be the case, there could be other reasons for the difference in earnings. For example, many of the women might have taken time off from work in order to raise children. This break in employment could cause many women to have less work experience and therefore earn less. In any case, further study would be needed to determine the cause of the difference in earnings.

Although we can read specific numerical information from data displays, their main value is in showing qualitative information about the data. So, when you read a data display, you should look for qualitative information that it conveys and think about the implications of this information. For example, the pie chart in Figure 14.10 shows that only a tiny fraction of U.S. businesses have 100 or more employees. Therefore, a law that would only apply to businesses with 100 or more employees would not affect the vast majority of businesses. On the other hand, from the display in Figure 14.10, we do not know what percent of all *employees* that work in businesses with 100 or more employees. Further data would need to be gathered to determine this.

"READING" GRAPHS AT DIFFERENT LEVELS

In the book *Developing Data-Graph Comprehension in Grades K–8*, [12], author Frances Curcio describes three different levels of graph comprehension: *reading the data*, *reading between the data*, and *reading beyond the data*. The following descriptions of these three levels are parts of the descriptions given in this resource [12, p. 7]:

Reading the data. This level of comprehension requires a literal reading of the graph. The reader simply "lifts" the facts explicitly stated in the graph, or the information found in the graph title and axes labels, directly from the graph. There is no interpretation at this level.

Reading between the data. This level of comprehension includes the interpretation and integration of the data in the graph. It requires the ability to compare quantities (e.g., greater than, tallest, smallest) and the use of other mathematical concepts and skills (e.g., addition, subtraction, multiplication, division) that allow the reader to combine and integrate data and identify the mathematical relationships expressed in the graph.

Reading beyond the data. This level of comprehension requires the reader to predict or infer from the data by tapping existing schemata (i.e., background knowledge, knowledge in memory) for information that is neither explicitly nor implicitly stated in the graph.

The book gives a number of graphing activities for grades K–8. Each graphing activity includes a list of questions for discussion. Each question is labeled according to the level of graph reading comprehension the question is designed to foster in the children.

For example, one activity is to create a real graph and a pictograph showing the different ways that the children in a class get to school in the morning. (The real graph uses toy cars and buses, and it uses toy shoes to represent walking to school; the pictograph uses pictures of these objects.) The following is an excerpt of the questions for discussion ([12, p. 37]):

- What is a good title for this graph? [Read between the data]
- How many children ride the bus to school? [Read the data]
- Which is the most popular way for children to travel to school? How do you know? How can you be sure? [Read between the data]
- How do you think this graph would change if it were raining or snowing? [Read beyond the data]
- About how many children live close to school? How do we know this? [Read beyond the data]

CLASS ACTIVITY   NOW TURN TO CLASS ACTIVITIES MANUAL

14E   Three Levels of Questions about Graphs p. 481

14F   What Do You Learn from the Display? p. 483

14G   What Is Wrong with These Displays? p. 486

14H   Display These Data p. 489

14I   How Accurately Do Random Samples Reflect the Full Population? p. 491

PRACTICE PROBLEMS FOR SECTION 14.2

1. A class plays a fishing game in which there is a large tub filled with plastic fish that are identical, except that some are red and some are white. A child is blindfolded and pulls 10 fish out of the tub. The child removes the blindfold, writes down how many of each color fish she got, and then puts the fish back in the tub. Each child takes a turn. The results are shown in Table 14.1. Show several different ways to graph these data.

TABLE 14.1

Numbers of Red and White Fish Picked out of a Tub by Children

NAME	FISH	NAME	FISH	NAME	FISH
Michelle	9 red, 1 white	Peter	6 red, 4 white	Sarah	8 red, 2 white
Tyler	7 red, 3 white	Brandon	9 red, 1 white	Adam	7 red, 3 white
Antrice	7 red, 3 white	Brittany	6 red, 4 white	Lauren	6 red, 4 white
Yoon-He	6 red, 4 white	Orlando	4 red, 6 white	Letitia	9 red, 1 white
Anne	6 red, 4 white	Chelsey	7 red, 3 white	Jarvis	7 red, 3 white

2. When is it appropriate to use a line graph?

3. If you have data consisting of percentages, is it always possible to display these data in a single pie graph?

ANSWERS TO PRACTICE PROBLEMS FOR SECTION 14.2

1. One simple way to display the data is with a pictograph, as in Figure 14.12.

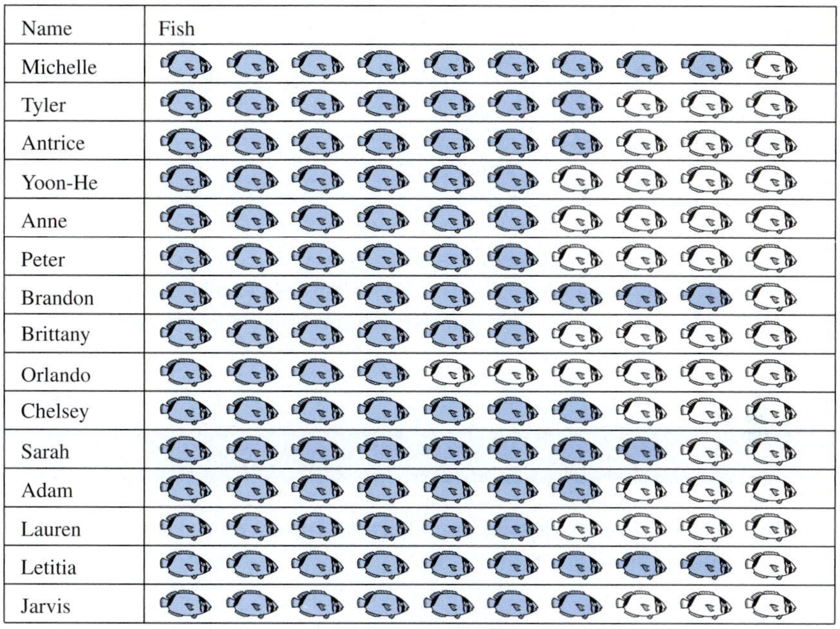

Name	Fish
Michelle	🐟🐟🐟🐟🐟🐟🐟🐟🐟🐟
Tyler	🐟🐟🐟🐟🐟🐟🐟🐟🐟🐟
Antrice	🐟🐟🐟🐟🐟🐟🐟🐟🐟🐟
Yoon-He	🐟🐟🐟🐟🐟🐟🐟🐟🐟🐟
Anne	🐟🐟🐟🐟🐟🐟🐟🐟🐟🐟
Peter	🐟🐟🐟🐟🐟🐟🐟🐟🐟🐟
Brandon	🐟🐟🐟🐟🐟🐟🐟🐟🐟🐟
Brittany	🐟🐟🐟🐟🐟🐟🐟🐟🐟🐟
Orlando	🐟🐟🐟🐟🐟🐟🐟🐟🐟🐟
Chelsey	🐟🐟🐟🐟🐟🐟🐟🐟🐟🐟
Sarah	🐟🐟🐟🐟🐟🐟🐟🐟🐟🐟
Adam	🐟🐟🐟🐟🐟🐟🐟🐟🐟🐟
Lauren	🐟🐟🐟🐟🐟🐟🐟🐟🐟🐟
Letitia	🐟🐟🐟🐟🐟🐟🐟🐟🐟🐟
Jarvis	🐟🐟🐟🐟🐟🐟🐟🐟🐟🐟

FIGURE 14.12

A Pictograph

This pictograph can be turned into a bar graph, as in the text. (In this case, each bar could be part red, part white.)

We can also create a different kind of display: one which shows how many children had each of the possible arrangements of 10 fish, as does the line plot in Figure 14.13. In this line plot, we can see at a glance that most of the children picked 6 or 7 red fish (and 4 or 3 white fish).

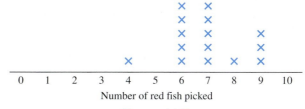

Number of red fish picked

FIGURE 14.13

A Line Plot of the Number of Children Picking
the Given Numbers of Fish

2. See text. Line graphs are only appropriate for displaying data that vary continuously.

3. It is not always possible to display data consisting of percentages in a single pie graph. If the percentages aren't *separate parts of the same whole*, then you can't display them in a pie graph. For example, the following table on children's eating is based on a table found on the Internet. See www.aw-bc.com/beckmann.

FOOD	PERCENT OF 4 TO 6 YEAR OLDS MEETING THE DIETARY RECOMMENDATION FOR THE FOOD
Grains	27%
Vegetables	16%
Fruits	29%
Saturated fat	28%

Even though the percentages add to 100%, they probably do not represent *separate* groups of children. For example, many of the 16% of the children who meet the recommendations on vegetables probably also meet the recommendations on fruits. Therefore, it would not be appropriate to display these data in a pie graph.

PROBLEMS FOR SECTION 14.2

1. The line graph in Figure 14.14 was taken from a website

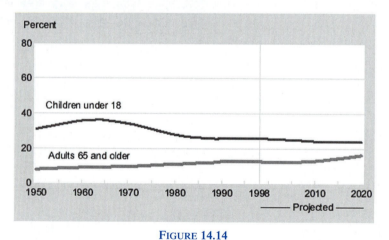

FIGURE 14.14

Percent Population of Children Under 18 and Adults Over 65

maintained by the Federal Interagency Forum on Child and Family Statistics.   (See www.aw-bc.com/beckmann.) Write at least four questions about this graph, including at least one at each of the three levels ("read the data," "read between the data," and "read beyond the data"). Label each question with its level. Answer your questions (to the extent that it is possible), explaining your answers.

2. The double bar graph in Figure 14.15 was taken from a website maintained by the Federal Interagency Forum on Child and Family Statistics. (See www.aw-bc.com/beckmann.) Write at least four questions about this graph, including at least one at each of the three levels ("read the data," "read between the data," and "read beyond the data"). Label each question with its level. Answer your questions (to the extent that it is possible), explaining your answers.

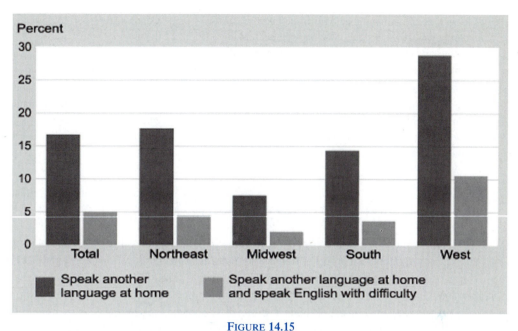

FIGURE 14.15

Percentage of Children Ages 5 to 17 Who Speak a Language Other Than English at Home and Who Have Difficulty Speaking English by Region, 1999

3. The National Assessment of Education Progress (NAEP), often called "the Nation's Report Card," is a national assessment of what children know and can do in various subjects. Information about NAEP is available on the Internet. Using the Internet, find the achievement level results from the 1992–2003 NAEP assessments in reading and mathematics at grades 4 and 8. See www.aw-bc.com/beckmann. Browse through these data. Select some data that interest you, write a paragraph about the data, and make a display of the data to help convey your point.

4. The National Assessment of Education Progress (NAEP), often called "the Nation's Report Card," is a national assessment of what children know and can do in various subjects. Information about NAEP is available on the Internet. Using the internet, find the NAEP 1999 Long-Term Trend Summary Data Tables. See www.aw-bc.com/beckmann. By clicking in the boxes for "scale scores and performance data," you can view long-term trend data in mathematics, reading, and science. Browse through these data. Notice that they include a variety of information, such as data on watching television, time spent on homework, reading practices in and out of school, and the use of scientific instruments. Select some data that interest you, write a paragraph about the data, and make a display of the data to help convey your point.

5. The National Assessment of Education Progress (NAEP), often called "the Nation's Report Card," is a national assessment of what children know and can do in various subjects. Information about NAEP is available on the Internet. Using the Internet, find NAEP data about states. See www.aw-bc.com/beckmann. Gather some data about one or several states that you find interesting, write a paragraph about the data, and make a display of the data to help convey your point.

6. a. Describe *in detail* an activity suitable for use with elementary school children in which the children gather data and create an appropriate display for the data.

   b. Write at least four questions that you could ask the children about the proposed graph in part (a), once the graph is completed. Include at least one question at each of the three levels ("read the data," "read between the data," and "read beyond the data"). Label each question with its level.

7. At a math center in a class, there is a bag filled with 40 red blocks and 10 blue blocks. Each child in the class

of 25 will do the following activity at the math center: randomly pick 10 blocks out of the bag without looking and write on one index card the number of red blocks and the number of blue blocks picked.

a. Describe a good way to display the whole class's data using one of the types of graphical displays discussed in this section. Your proposed display should be a realistic and practical way to show every child's piece of data (in a class of 25).

b. Sketch a graph that could be the graph you proposed in part (a).

c. Write at least four questions about your hypothetical graph in part (b) that the teacher could ask the children. Include at least one question at each of the three levels ("read the data," "read between the data," and "read beyond the data"). Label each question with its level. Answer your questions.

14.3 The Center of Data: Mean, Median, and Mode

When we have a collection of numerical data, such as the collection of scores on a test, or the heights of a group of 6-year-old children, our first questions about the data are usually, "What is typical or representative of these data?" and "What is the center of these data?" When a data set is large, it is especially helpful to know the answers to these questions. For example, if the data consist of the test scores of all students taking the SAT in a given year, no one person would want to look at the entire data set (even if he could); it would be overwhelming and impossible to comprehend. Instead, we want ways of understanding the nature of the data.

To say what is representative, in the center, or typical of a list of numbers, we commonly use the *mean*, the *median*, or the *mode*.

mean

average

To calculate the **mean**, also called the *arithmetic mean* or the **average**, of a list of numbers, add all the numbers and divide this sum by the number of numbers in the list. For example, the average of

$$7, \quad 10, \quad 11, \quad 8, \quad 10$$

is

$$(7 + 10 + 11 + 8 + 10) \div 5 = 46 \div 5 = 9.2$$

We divide by 5 because there are 5 numbers in the list 7, 10, 11, 8, 10.

We can think of the average of a list of numbers as obtained by "leveling out" the numbers. For example, consider the list

$$4, \quad 3, \quad 6, \quad 5, \quad 2$$

as represented by towers of blocks: one 4 blocks tall, one 3 blocks tall, one 6 blocks tall, one 5 blocks tall, and one 2 blocks tall, as shown in Figure 14.16. If we rearrange the blocks so that all 5 towers become the same height, the common height is the average. Why? When we calculate the average numerically, we first add the numbers

$$4 + 3 + 6 + 5 + 2$$

In terms of the block towers, adding the numbers tells us the total number of blocks in all 5 towers. When we then divide the sum of the numbers by 5,

$$(4 + 3 + 6 + 5 + 2) \div 5$$

FIGURE 14.16

The Average as "Leveling Out"

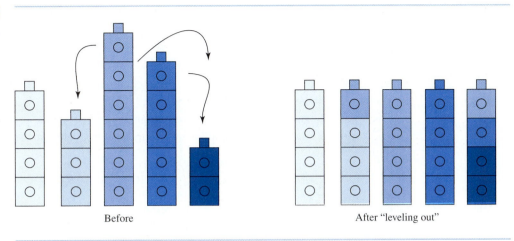

Before   After "leveling out"

we are determining the number of blocks in each tower when the full collection of blocks is divided equally among 5 towers. In other words, we are determining the number of blocks in each tower when the 5 towers are made even.

Another way to think about the average is as a "balance point." With this point of view, we think of placing items on a see-saw at locations representing the numbers in the list to be averaged. The average of the list of numbers is the balance point of the see-saw. For example, we can represent the list

$$4, \quad 4, \quad 4, \quad 12$$

on a see-saw by placing 3 chips at the location labeled 4 and 1 chip at the location labeled 12 as in Figure 14.17. The see-saw balances at 6, which is also the average of the list of numbers. We will not explain why this is a valid way to determine the average. These two ways of thinking about averages are often helpful in solving problems.

median   To calculate the **median** of a list of numbers, organize the list from smallest to largest (or vice versa). The number in the middle of the list is the median. If there is no number in the middle, then the median is halfway between the two middle numbers. For example, to find the median of

$$7, \quad 10, \quad 11, \quad 8, \quad 10$$

rearrange the numbers from smallest to largest as follows:

$$7, \quad 8, \quad \boxed{10}, \quad 10, \quad 11$$

FIGURE 14.17

The Average as "Balance Point"

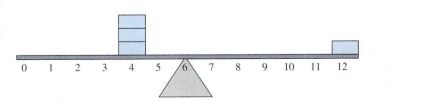

The number 10 is exactly in the middle, so it is the median. There are 6 numbers in the list, namely,

$$3, \quad 3, \quad \boxed{4, \quad 5}, \quad 5, \quad 6$$

so there is no single number in the middle. The median is halfway between the two middle numbers 4 and 5; therefore, the median is 4.5.

mode   To calculate the **mode** of a list of numbers, reorganize the list so as to group equal numbers in the list together. The mode is the number that occurs most frequently in the list. For example, to find the mode of

$$11, \quad 9, \quad 10, \quad 8, \quad 11, \quad 10, \quad 11, \quad 9, \quad 12$$

first reorganize the list as follows:

$$8, \quad 9, \quad 9, \quad 10, \quad 10, \quad 11, \quad 11, \quad 11, \quad 12$$

The number 11 occurs 3 times, which is more than any other number on the list. Therefore, 11 is the mode of this list of numbers. Statisticians usually do not use the mode because relatively small changes in data can produce large changes in the mode. Consequently, we will not work with the mode as a measure of center in the Activities, Practice Problems, or Problems.

CLASS ACTIVITY   NOW TURN TO CLASS ACTIVITIES MANUAL

14J The Average as "Making Even" or "Leveling Out" p. 493

14K The Average as "Balance Point" p. 496

14L Same Median, Different Average p. 498

14M Can More Than Half Be above Average? p. 500

PRACTICE PROBLEMS FOR SECTION 14.3

1. Explain how to use physical objects to describe the average of a data set as "leveling out" the data set.

2. We can determine the average of a list of whole numbers such as 4, 3, 6, 5, and 2 by building 5 block towers made of 4 blocks, 3 blocks, 6 blocks, 5 blocks, and 2 blocks and by redistributing the blocks among the 5 towers until they are all the same height; this common height is the average. We can also determine the average of a list of numbers by adding the numbers and then dividing the sum by the number of numbers in the list. Interpret the numerical procedure for calculating an average in terms of block towers. Explain why the average you determine by physically "leveling out" block towers and the average you determine by calculating numerically must always come out the same.

3. Explain how to show the average of a data set as the "balance point" of the data set.

4. Explain why the average of a list of numbers must always be in between the smallest and largest numbers in the list.

5. Explain why the average of two numbers is exactly halfway between the two numbers.

6. Juanita read an average of 3 books a day for 4 days. How many books will Juanita need to read on the fifth day so that she will have read an average of 5 books a day over 5 days? Solve this problem in several ways and explain your solutions.

7. George's average score on his math tests in the first quarter is 60. George's average score on his math tests in the second quarter is 80. George's semester score in math is the average of all the tests George took in the first and

second quarters. Can George necessarily calculate his semester score by averaging 60 and 80? If so, explain why. If not, explain why not. Explain what other information you would need to calculate George's semester average and show how to calculate this average.

8. Suppose that all fourth graders in a state take a writing competency test that is scored on a 5-point scale. Is it possible (in theory) for 80% of the fourth graders to score below average? If so, show how that could occur. If not, explain why not.

ANSWERS TO PRACTICE PROBLEMS FOR SECTION 14.3

1. See text.

2. See text.

3. See text.

4. If we think of the average of a list of numbers as its "balance point," then clearly the average must be in between the largest and smallest numbers on the list because the balance point must be in between those numbers.

    Similarly, if we think of the average as "leveling out" the numbers in a list, then with this point of view it's also clear that the average must be in between the smallest and largest numbers in the list because when you level out the block towers, the common "leveled out" height must be in between the height of the tallest tower and the height of the shortest tower.

5. If we think of the average of two numbers as the balance point between those numbers, then it's clear that the average is exactly halfway between those numbers because the balance point is exactly halfway between the numbers.

6. Method 1: Imagine putting the books that Juanita has read into 4 stacks, one for each day. Because she has read an average of 3 books per day, the books can be redistributed so that there are 3 books in each of the 4 stacks. This uses the "average as leveling out" point of view. Now picture a fifth stack with an unknown number of books in it—the number of books Juanita must read to make the average 5 books per day. If the books in this fifth stack are distributed among all 5 stacks so that all 5 stacks have the same number of books, then this common number of books is the average number

of books Juanita will have read over 5 days. Therefore, this fifth stack must have 5 books plus another $4 \times 2 = 8$ books, in order to distribute 2 books to each of the other 4 stacks. So Juanita must read $5 + 8 = 13$ books on the fifth day.

    Method 2: To read an average of 5 books a day over 5 days, Juanita must read a total of 25 books. This is because

$$(\text{total books}) \div 5 = \text{average}$$

Therefore,

$$(\text{total books}) \div 5 = 5$$

so

$$\text{total books} = 5 \times 5 = 25$$

Similarly, because Juanita has already read an average of 3 books per day for 4 days, she has read a total of $4 \times 3 = 12$ books. Therefore, Juanita must read $25 - 12 = 13$ books on the fifth day.

7. George cannot necessarily calculate his semester score by averaging 60 and 80. We don't know if George took the same number of tests in the first and second quarters. To calculate George's average score over the whole semester, we need to know how many tests he took in the first and second quarters. Suppose George took 3 tests in the first quarter and 5 tests in the second quarter. Then the sum of his first three test scores is $3 \times 60 = 180$, and the sum of his next 5 test scores is $5 \times 80 = 400$, so the sum of all 8 of his test scores is $180 + 400 = 580$. Therefore, George's average score

on all 8 tests is $580 \div 8 = 72.5$. Notice that this is not the same as the average of 60 and 80, which is only 70.

8. It is theoretically possible for 80% of the fourth graders to score below average. For example, suppose there are 200,000 fourth graders and that

>10,000 score 1 point,
>
>150,000 score 2 points,
>
>10,000 score 3 points,
>
>25,000 score 4 points, and
>
>5000 score 5 points,

as shown in the bar graph of Figure 14.18. Then you can see from the bar graph that the "balance point" is between 2 and 3, and 160,000 children, namely, 80% of the 200,000 children, score below this average.

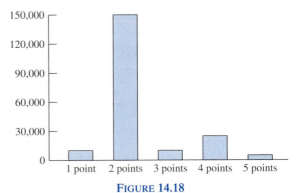

FIGURE 14.18

Scores on a Writing Test

PROBLEMS FOR SECTION 14.3

1. Shante caught 17 ladybugs every day for 4 days. How many ladybugs does Shante need to catch on the fifth day so that she will have caught an average of 20 ladybugs per day over the 5 days. Solve this problem in two different ways and explain your solutions.

2. John's average annual income over a 4-year period was $25,000. What would John's average annual income have to be for the next 3 years so that his average annual income over the 7-year period would be $50,000? Solve this problem in two different ways and explain your solutions.

3. Tracy's times swimming 200 yards were as follows:

$$2{:}45, \quad 2{:}47, \quad 2{:}44$$

How fast will Tracy have to swim her next 200 yards so that her average time for the 4 trials is 2:45? Explain how you can solve this problem without adding the times.

4. Explain how you can quickly calculate the average of the following list of test scores without adding the numbers:

$$83, \quad 79, \quad 81, \quad 76, \quad 81$$

5. Explain how you can quickly calculate the average of the following list of test scores without adding the numbers:

$$84, \quad 79, \quad 81, \quad 78, \quad 83$$

6. Julia's average on her first 3 math tests was 80. Her average on her next 2 math tests was 95. What is Julia's average on all 5 math tests? Solve this problem in two different ways and explain your solutions.

7. A teacher gives a 10-point test to a class of 10 children.

  a. Is it possible for 9 of the 10 children to score above average on the test? If so, give an example to show how. If not, explain why not.

  b. Is it possible for all 10 of the children to score above average on the test? If so, give an example to show how. If not, explain why not.

8. In Ritzy County, the average annual household income is $100,000. In neighboring Normal County, the average annual household income is $30,000. Does it follow that in the two-county area (consisting of Ritzy County and Normal County), the average annual household income is the average of $100,000 and $30,000? If so, explain why; if not, explain why not. What other information would you need in order to calculate the average annual household income in the two-county area? Show how to calculate this average if you had this information.

9. In county A, the average score on the grade 5 Iowa Test of Basic Skills (ITBS) was 50. In neighboring county B, the average score on the grade 5 ITBS was 71. Can you conclude that the average score on the grade 5 ITBS in the two-county region (consisting of counties A and B) can be calculated by averaging 50 and 71? If so, explain

why; if not, explain why not. What other information would you need to know in order to calculate the average grade 5 ITBS score in the two-county region? Show how to calculate this average if you had this information.

10. A teacher gives a 10-point test to a class of 9 children. All scores are whole numbers. If the median score is 8, then what are the highest and lowest possible average scores? Explain your answers.

11. A teacher gives a 10-point test to a class of 9 children. All scores are whole numbers. If the average score is 8, then what are the highest and lowest possible median scores? Explain your answers.

12.

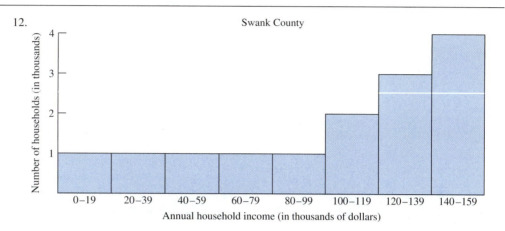

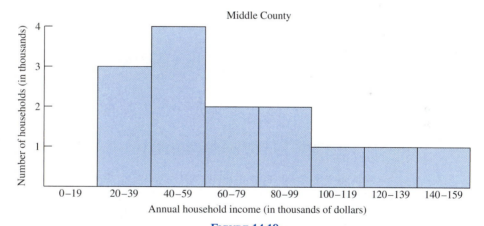

FIGURE 14.19

Annual Household Incomes in Swank and Middle Counties

a. The bar graph at the top of Figure 14.19 shows the distribution of annual household incomes in Swank County. Without calculating the median or average annual household income in Swank County, make an educated guess about which one is greater. Describe your thinking.

b. Determine the approximate values of the median and average annual household income in Swank County from the bar graph at the top of Figure 14.19. Which is greater, the average or the median? Compare to part (a).

c. The bar graph at the bottom of Figure 14.19 shows the distribution of annual household incomes in Middle County. Without calculating the median or average annual household income in Middle County, make an educated guess about which one is greater. Describe your thinking.

d. Determine the approximate values of the median and average annual household income in Middle County from the bar graph at the bottom of Figure 14.19. Which is greater, the average or the median?

13. Ms. Smith needs to figure her students' homework grades. There have been five homework assignments: the first out of 30 points, the second out of 40 points, the third out of 26 points, the fourth out of 50 points, and the fifth out of 25 points. Ms. Smith is debating between two different methods of calculating the homework grades.

METHOD 1:

Add up the points scored on all five of the homework assignments and divide by the total number of points possible. Then multiply by 100 to get the total score out of 100.

METHOD 2:

Find the percentage correct on each homework assignment and find the average of these five percentages.

a. Jodi's homework grades are: $\frac{27}{30}$, $\frac{25}{40}$, $\frac{26}{30}$, $\frac{30}{50}$, $\frac{24}{25}$, where, for example, $\frac{27}{30}$ stands for 27 points out of 30 possible points in a homework assignment. Compare how Jodi does with each of these two methods. Which method gives her a higher score?

b. Make up scores (on the same assignments) for Robert so that Robert scores higher with the other method than the one that gave Jodi the higher score.

14. The *average speed* of a moving object during a period of time is the distance the object traveled, divided by the length of the time period. For example, if you left on a trip at 1:00 P.M., arrived at 3:15 P.M., and drove 85 miles, then the average speed for your trip would be

$$85 \div 2.25 = 37.8$$

miles per hour.

a. Jane drives 100 miles from Philadelphia to New York with an average speed of 60 miles per hour. When she gets to New York, Jane turns around immediately and heads back to Philadelphia along the same route. Using this information, find several different possible average speeds for Jane's *entire trip* from Philadelphia back to Philadelphia.

b. Ignoring practical issues, such as speed limits and how fast cars can go, would it be theoretically possible for Jane to average 100 miles per hour for the whole trip from Philadelphia to New York and back that is described in part (a)? If so, explain how; if not, explain why not.

c. Theoretically, what are the largest and smallest possible average speeds for Jane's entire trip from Philadelphia back to Philadelphia that is described in part (a)? Explain your answers clearly.

14.4 The Spread of Data: Percentiles

The average and the median of numerical data tell us about the center of the data. But data sets that look different can have the same median or average. The line plots in Figure 14.20 represent hypothetical test scores on a 10-point quiz. In all three line plots, the median and the average are both 7. But the distribution of the test scores differs in the three cases. Therefore, it is useful to have ways of summarizing how numerical data are "spread out." In this brief section, we will study one tool for reporting how data is spread out, namely, percentiles. Statisticians often use another concept for studying the spread of data, that of *standard deviation*. We will not study standard deviations here, but almost any book on statistics, such as [21], discusses the topic.

percentile   A **percentile** is a generalization of the median. The 90th percentile of a set of numerical data is the number such that 90% of the data is less than or equal to that number and 10% is greater than or equal to that number. (If there is no single number that satisfies this criterion, then use the midpoint of all numbers that satisfy it, as in computing the median.) Other percentiles are defined similarly. The median of a list of numbers is the 50th percentile, because 50% of the list is at or below the median, and 50% is at or above the median.

FIGURE 14.20

Line Plots of 3
Different
Hypothetical
Test Results
Each with
Median 7 and
Average 7

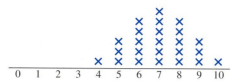

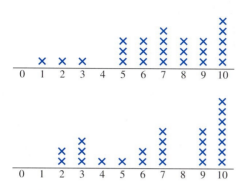

Results of standardized tests are often reported as percentiles. So if a student scores in the 75th percentile on a test, then that student has scored the same as or better than 75% of the students taking the test.

CLASS ACTIVITY   NOW TURN TO CLASS ACTIVITIES MANUAL

14N Determining Percentiles p. 501

14O Percentiles versus Percent Correct p. 504

14P How Percentiles Inform You about Data Spread: The Case of Household Income p. 505

14Q More about Household Income p. 510

PRACTICE PROBLEMS FOR SECTION 14.4

1. Determine the 25th, 50th, and 75th percentiles for the hypothetical test scores shown in the line plots of Figure 14.21 on page 672. Discuss how these percentiles indicate the way the data are spread.

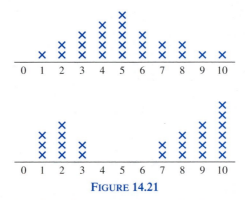

FIGURE 14.21

Line Plots Showing Hypothetical Test Results

2. Figure 14.22 shows bar graphs of hypothetical scores on a 400-point test given at two different schools. Determine approximately the 20th, 40th, 60th, and 80th percentiles for the test at each of the schools. Discuss how the percentiles indicate the way the data are spread.

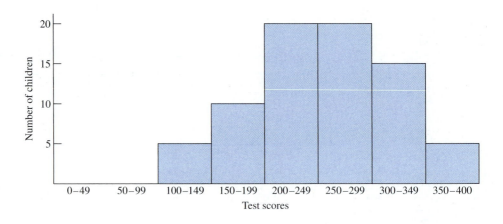

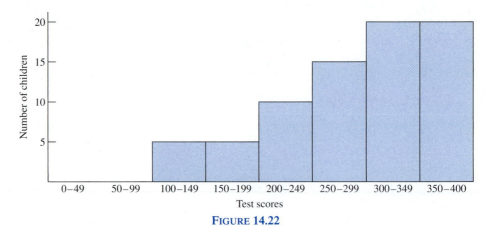

FIGURE 14.22

Bar Graphs Showing Hypothetical Test Results at Two Schools

3. Figure 14.22 shows bar graphs of hypothetical scores on a 400-point test given at two different schools.

   a. At both of the schools, consider a student who scored in the 20th percentile. In each case, approximately what score did such a student get? Did the student get 20% correct, more than 20% correct, or less than 20% correct on the test?

   b. At both of the schools, consider a student who scored in the 80th percentile. In each case, approximately what score did such a student get? Did such a student get 80% correct, more than 80% correct, or less than 80% correct on the test?

ANSWERS TO PRACTICE PROBLEMS FOR SECTION 14.4

1. For both of the hypothetical tests, there are 24 test scores. Since 25% of 24 is 6, the 25th percentile is the score between the score of the 6th lowest and 7th lowest test scores. Therefore, the 25th percentile is 3.5 in the first case and 2 in the second case. The 50th percentile is the score between the 12th and 13th lowest scores, which is 5 in the first case and 8 in the second case. The 75th percentile is between the 18th and 19th lowest scores, which is 6.5 in the first case and 9.5 in the second case. The following table summarizes these results:

	25TH PERCENTILE	50TH PERCENTILE	75TH PERCENTILE
Test 1	3.5	5	6.5
Test 2	2	8	9.5

The large gap between the 25th and 50th percentiles for test 2 corresponds to the large gap between the lower scores and the higher scores. For test 1, the closeness among the 25th, 50th, and 75th percentiles corresponds to the bunching of the test scores around 5. For test 2, the high 50th and 75th percentiles reflect the bunching of the test scores around the higher scores.

2. An easy way to determine the percentiles is to divide the bar graph into squares, as in Figure 14.23 on page 674. Each square represents 5 students. In each bar graph, there are 15 such squares, so 20% of the students is represented by 3 squares. Therefore, for each school, the 20th percentile occurs where 3 squares are to the left and the remaining 12 squares are to the right, the 40th percentile occurs where 6 squares worth are to the left and the remaining 9 squares worth are to the right, and so on for the 60th and 80th percentiles as well.

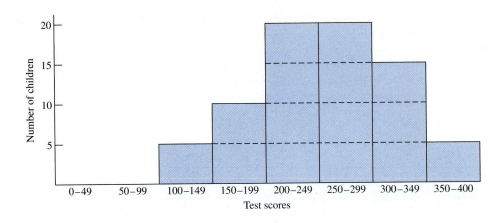

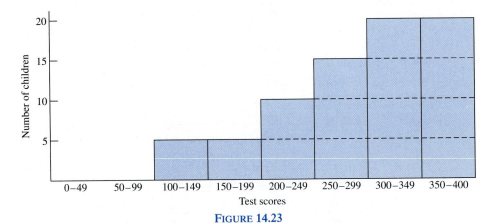

FIGURE 14.23

Hypothetical Test Results at Two Schools

So, in the first school, the 20th percentile is about 200; the 40th percentile should be about $\frac{3}{4}$ of the way between 200 and 250, which is about 240; the 60th percentile should be about halfway between 250 and 300, at about 275; and the 80th percentile should be about one third of the way between 300 and 350, at about 315. The results are similar for the second school. The percentiles are shown in the following table:

	APPROXIMATE VALUES OF PERCENTILES			
	20TH PERCENTILE	40TH PERCENTILE	60TH PERCENTILE	80TH PERCENTILE
School 1	200	240	275	315
School 2	225	285	325	360

The higher values for the 20th, 40th, 60th, and 80th percentiles in the second school correspond to the way the test scores are shifted toward the higher scores.

3. In the first school, a student scoring at the 20th percentile scored about 200 points, which is 50% of the maximum possible score of 400. At the second school, a student scoring at the 20th percentile probably scored about 225 points, or about 56% of the maximum possible score. Thus, we see that scoring in the 20th percentile is not the same as scoring 20% on the test.

In the first school, a student scoring at the 80th percentile probably scored about 315 points, or about 79% on the test. In the second school, a student scoring at the 80th percentile probably scored about 360 points, or about 90%.

PROBLEMS FOR SECTION 14.4

1. What is the difference between scoring in the 90th percentile on a test and scoring 90% correct on a test. Discuss this carefully, giving examples to illustrate.

2. What is the purpose of reporting a child's percentile on a state or national standardized test? How is this purpose different from reporting the child's percent correct on a test?

3. Determine the 25th, 50th, and 75th percentiles for the hypothetical test scores shown in the line plots of Figure 14.24. Discuss how these percentiles indicate the way the data are spread. Each line plot has 28 pieces of data.

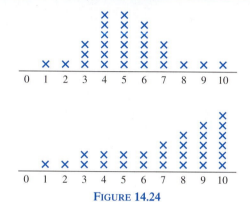

FIGURE 14.24

Line Plots Showing Hypothetical Test Results

4. Figure 14.25 on page 676 shows bar graphs of hypothetical scores on a 400-point test given at two different schools. Determine approximately the 20th, 40th, 60th, and 80th percentiles for the test at each of the schools. Discuss how the percentiles indicate the way the data are spread.

5. Figure 14.22 shows bar graphs of hypothetical scores on a 400-point test given at two different schools.

   a. At both of the schools, consider a student who scored at the 20th percentile. In each case, approximately what score did such a student get? Did such a student get 20% correct, more than 20% correct, or less than 20% correct on the test?

   b. At both of the schools, consider a student who scored at the 80th percentile. In each case, approximately what score did such a student get? Did such a student get 80% correct, more than 80% correct, or less than 80% correct on the test?

6. The bar graphs in Figure 14.26 on page 677 show hypothetical scores on a 100-point test given at two different schools. Each school has 200 children. At each school, consider a student who scored at the 75th percentile. Approximately what score did such a student get? Explain your answer. If a student scores at the 75th percentile, does that mean that the student has gotten 75% correct on the test?

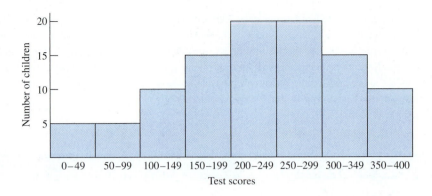

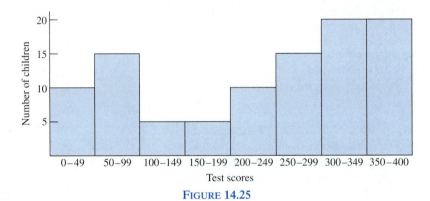

FIGURE 14.25

Bar Graphs Showing Hypothetical Test Results at Two Schools

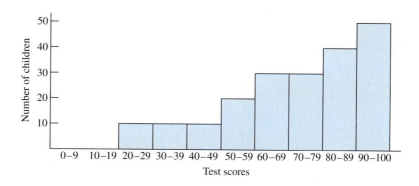

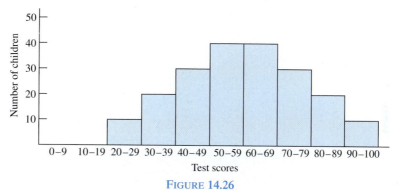

FIGURE 14.26

Bar Graphs Showing Hypothetical Test Results at Two Schools

Chapter 15

Probability

T he study of probability arises naturally in a variety of games. Games in which we flip coins, roll dice, spin spinners, or pick cards all have an element of chance. We don't know what the spinner will land on, and we don't know how the dice will roll. This element of chance adds excitement to the game and makes it easy to design class activities and problems that present probability in a way that can be fun for students to learn.

If probability were only used in analyzing games of chance, it would not be a very important subject. But probability has far wider applications. In business and finance, probability can be used in determining how best to allocate assets. Farmers can use data on the probability of various amounts of rain to determine how best to plant their fields. In medicine, probability can be used to determine how likely it is that a person actually has a certain disease, given the outcomes of test results. Some diagnostic techniques, particularly genetic tests, can be used to determine the likelihood that a patient will contract a disease.

Concerning the learning of statistics and probability, the National Council of Teachers of Mathematics recommends the following (see [44]):

Instructional programs from prekindergarten through grade 12 should enable all students to—

NCTM Standards

- formulate questions that can be addressed with data and collect, organize, and display relevant data to answer them;

- select and use appropriate statistical methods to analyze data;

679

- *develop and evaluate inferences and predictions that are based on data;*

- *understand and apply basic concepts of probability.*

   For NCTM's specific statistics and probability recommendations for grades Pre-K–2, grades 3–5, grades 6–8, and grades 9–12, see

`www.aw-bc.com/beckmann`.

15.1 Basic Principles and Calculation Methods of Probability

In this section we will study some of the basic principles of probability, and we will describe some simple contexts in which to introduce basic ideas of probability. We will also use basic principles of probability to calculate probabilities in relatively simple situations.

In probability, we study experiments or situations where we know which outcomes are possible, but we do not know precisely which outcome will occur at a given time. Examples include flipping a coin and seeing if it lands with a head or a tail facing up, spinning a spinner and seeing which color the arrow lands on, or randomly picking a person out of a group and determining if the person has a certain disease or not. Given an experiment or a situation in which various outcomes are possible, the (theoretical) **probability** of a given outcome is a number quantifying how likely that outcome is. The probability of a given outcome is the fraction or percentage of times that outcome should occur in the ideal. For example, the probability that a spinner lands on green is the fraction of times that the spinner should land on green in the ideal.

probability

Probabilities are always between 0 and 1, or equivalently, when they are given as percentages, between 0% and 100%. A probability of 0% means that outcome will not occur; a probability of 100% means that outcome is certain to occur; a probability of 50% means the outcome is as likely to occur as not to occur. For example, when we flip a coin, the probability of getting *either* a head *or* a tail is 100% because one or the other is certain to occur. The probability of getting a head is 50%, or $\frac{1}{2}$, because heads and tails are equally likely to occur.

If we have a coin in hand, how do we know for sure that a coin is *fair* (i.e., that heads and tails are equally likely to occur when we flip the coin)? In fact, we never know for sure that the coin is fair, but unless the coin has been weighted lopsidedly, there is no reason to believe that one of heads or tails should be more likely to occur than the other. If it were important for the coin to be fair, for example, if the coin were to be used in a state lottery, then the coin would be flipped many times, perhaps thousands of times, to determine if the coin is likely to be fair. There are statistical methods for determining the likelihood that a coin is fair; we will not discuss these methods.

PRINCIPLES FOR DETERMINING PROBABILITY

There are several key principles about probability that we can use to calculate probabilities. We will illustrate and apply these principles in the rest of this section. Here are the principles:

principles of probability

1. If two outcomes of an experiment or situation are equally likely, then their probabilities are equal.

2. If there are several outcomes of an experiment or situation that cannot occur simultaneously, then the probability that one of those outcomes will occur is the sum of the probabilities of each of the individual outcomes.

3. If an experiment is performed many times, then the fraction of times that a given outcome occurs is likely to be close to the probability of that outcome occurring.

EXPERIMENTAL PROBABILITY

Although the theoretical probability that a given outcome of an experiment will occur is the fraction of times that that outcome occurs in the ideal, when you perform an experiment a number of times, the *actual* fraction of times that the given outcome occurs will usually *not* be equal to the probability of that outcome. When you perform an experiment a number of times, the fraction or percentage of times that a given outcome occurs is called the **experimental probability** of that outcome. For example, if you flipped a coin 100 times and you got 54 heads, then the experimental probability of getting heads was 54%.

experimental probability

Principle 3 given earlier says that the experimental probability is likely to be close to the theoretical probability of an outcome of an experiment if the experiment was performed many times. In other words, if you perform an experiment many times, the actual fraction that a given outcome occurs is usually close to the probability of that outcome occurring. This statement is similar to the statement that a random sample of a population is likely to be representative of the full population if the sample is large enough. (See page 648 in Section 14.1.)

Because of principle 3, we can use experimental probabilities to estimate theoretical probabilities. For example, suppose there are 10 blocks in a bag. Some of the blocks are red and some are blue, but we don't know how many of each color block are in the bag. Let's say we reach into the bag 50 times, each time removing one block, recording its color, and putting it back into the bag. If we picked 16 red blocks, then the experimental probability of picking a red block was 32%. This 32% is likely to be close to the theoretical probability of picking a red block, which is

$$\frac{\text{(number of red blocks)}}{10}$$

because in the ideal, out of every 10 blocks we pick, (number of red blocks) of the blocks should be red. If there are 3 red blocks in the bag, the probability of picking red is 30%, which is close to 32%. Our best guess for the number of red blocks in the bag is therefore 3. Even though we based our guess on the available evidence, our guess could turn out to be wrong because of the variability that is inherent in random choices.

PROBABILITIES OF EQUALLY LIKELY OUTCOMES

If an experiment or situation has N distinct possible outcomes, none of which can occur simultaneously, and if all of these outcomes are equally likely, then for each possible outcome, the probability of that outcome is $\frac{1}{N}$. This statement follows from the first two principles of probability given earlier. We will explain why this is so in a simple case.

Suppose you have an ordinary die that is in the shape of a cube and has dots on each of its six faces. Each face has a different number of dots on it, ranging from 1 dot to 6 dots. You can roll the die and count the number of dots on the side that lands face up. Let's assume that the die is not weighted, so that each side is equally likely to land face up. Then by principle 1, the probability of rolling a 1 is equal to the probability of rolling a 2, which is equal to the probability of rolling a 3, and so on, up to 6. In other words,

$$\text{prob. of } 1 = \text{prob. of } 2 = \text{prob. of } 3 = \text{prob. of } 4 = \text{prob. of } 5 = \text{prob. of } 6$$

Now when you roll a die, the probability of rolling either a 1 or 2 or 3 or 4 or 5 or 6 is 1, because *some* face has to land up. And you can't *simultaneously* roll a 2 and a 3 on one die, or any other pair of distinct numbers. Therefore, by principle 2,

$$\text{prob. of } 1 + \text{prob. of } 2 + \text{prob. of } 3 + \text{prob. of } 4 + \text{prob. of } 5 + \text{prob. of } 6 = 1$$

Because all these probabilities are equal and add up to 1, each probability must be $\frac{1}{6}$. That is,

$$\text{prob. of } 1 = \frac{1}{6}$$

$$\text{prob. of } 2 = \frac{1}{6}$$

$$\text{prob. of } 3 = \frac{1}{6}$$

$$\text{prob. of } 4 = \frac{1}{6}$$

$$\text{prob. of } 5 = \frac{1}{6}$$

$$\text{prob. of } 6 = \frac{1}{6}$$

THE PROBABILITY THAT ONE OF SEVERAL POSSIBLE OUTCOMES OCCURS

What is the probability of rolling an odd number with a die? The only way to roll an odd number with a die is to roll a 1, 3, or 5. Since you can't roll any two of those numbers simultaneously on one die, principle 2 applies and tells us that the probability of rolling a 1, 3, or 5 is the sum of the probabilities of rolling a 1, of rolling a 3, and of rolling a 5. Therefore, the probability of rolling an odd number is

$$\frac{1}{6} + \frac{1}{6} + \frac{1}{6} = \frac{3}{6} = \frac{1}{2}$$

INDEPENDENT AND DEPENDENT EVENTS

If you flipped a coin 10 times in a row and all 10 flips came up heads, would you think that your next flip is more likely to be a tail because a tail is "due"? Would you think that you were on a "hot streak" and that you are therefore more likely to get another head on your next coin flip? In fact, on your next flip you are just as likely to get a tail as a head. There is nothing in

independent   the flipping history of the coin that can influence the next flip. We say that the outcome of one coin flip is **independent** of the outcome of any other coin flip. In other words, the outcome of one coin flip has no bearing on the outcome of any other coin flip.

Suppose you have a bag filled with 9 blue marbles and 1 green marble. If you randomly pick a marble out of the bag, record its color, and put it back in the bag, then your next random marble pick is independent of your first marble pick. On the other hand, if you leave the first marble out of the bag after selecting it, then you *do* influence the next marble pick. If the first marble you selected was the green one, then your second marble must be blue. If the first marble you selected was blue, then on your second marble pick, the probability of picking the green marble is $\frac{1}{9}$ instead of $\frac{1}{10}$ because now there are only 9 marbles in the bag. When the
dependent   first marble is left out of the bag, the second marble pick is **dependent** on the first.

USING ORGANIZED LISTS, TREE DIAGRAMS, AND ARRAYS TO CALCULATE PROBABILITIES

Often, we can use techniques we learned for showing multiplicative structure in Section 5.1 to display all possible outcomes of an experiment. If all possible outcomes are equally likely, then we can calculate the probability that one of those outcomes will occur by using principles 1 and 2 on page 681.

For example, suppose you have three coins: a penny, a nickel, and a dime. If you flip all three coins, what is the probability that all three will come up heads? First, let's determine how many different possible outcomes there are when we flip the three coins. For example, three heads is one possible outcome. Heads on the penny and nickel, and a tail on the dime is another possible outcome. We can show all possible outcomes in an organized list, as in Table 15.1, or we can use a tree diagram, as in Figure 15.1.

TABLE 15.1

An Organized List Showing All Possible Outcomes from Flipping 3 Coins

HHH
HHT
HTH
HTT
THH
THT
TTH
TTT

FIGURE 15.1

A Tree Diagram Showing All Possible Outcomes from Flipping 3 Coins

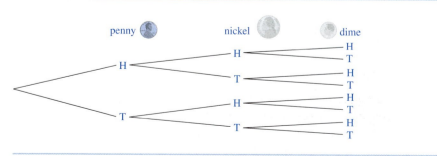

For each of the two outcomes for the penny, the nickel has two possible outcomes (because the outcome of flipping the nickel is independent of the outcome of flipping the penny), so there are

$$2 \times 2$$

possible outcomes for the penny and nickel. For each of the 2×2 possible outcomes for the penny and nickel, there are two possible outcomes for the dime (because the outcome of flipping the dime is independent of the outcome of flipping the penny and nickel). So all together, there are

$$2 \times 2 \times 2 = 8$$

total possible outcomes when we flip the three coins.

Next, since pennies, nickels, and dimes are equally likely to come up heads as tails, all of these eight possible outcomes shown in the tree diagram are equally likely. Therefore, according to principles 1 and 2, the probability of getting three heads when we flip three coins is $\frac{1}{8}$, or 12.5%.

We can use the tree diagram of Figure 15.1 to calculate other probabilities as well. For example, what is the probability of getting two heads and one tails when we toss the three coins? Two heads and one tail occurs in three of the eight possible outcomes, namely HHT, HTH, and THH. According to principle 2, the probability that either one of these three outcomes will occur is

$$\frac{1}{8} + \frac{1}{8} + \frac{1}{8} = \frac{3}{8}$$

or 37.5%.

Some children's games use spinners, such as the one in Figure 15.2, where the spinner is equally likely to land in red, yellow, blue, or green. What is the probability that in 2 spins, the spinner will land first on red and then on blue? Because the outcome of the second spin is independent of the outcome of the first spin, we could make an organized list or a tree diagram to show all possible outcomes, as we did before. We can also make an array to show all possible outcomes, as in Figure 15.3. The array shows 4 rows of 4 possible outcomes, so there are $4 \times 4 = 16$ possible outcomes for the two spins. Each of the 16 spins is equally likely, so the probability of spinning a red and then a blue is $\frac{1}{16} = 6.25\%$.

FIGURE 15.2

A Spinner for a
Children's
Game

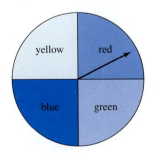

FIGURE 15.3

An Array
Showing All
Possible
Outcomes on
Two Spins of a
Spinner

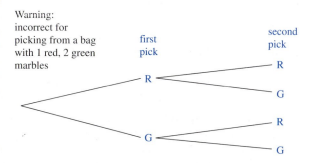

FIGURE 15.4

An Incorrect
Tree Diagram
for Picking Two
Marbles from a
Bag with 1 Red
and 2 Green
Marbles (If the
First Marble Is
Replaced after
Picking)

Warning:
incorrect for
picking from a bag
with 1 red, 2 green
marbles

Be careful about using an organized list, a tree diagram, or an array when working with outcomes that are not equally likely. For example, suppose there are 3 marbles in a bag, 1 red and 2 green. If we randomly pick a marble from the bag, record its color, put it back in the bag, and then randomly pick another marble, can we use the tree diagram in Figure 15.4 to calculate the probability of picking a green marble followed by the red marble? No, we cannot use this tree diagram because even though there are only two possible outcomes when picking a marble: picking red and picking green, these two outcomes are not equally likely. We are more likely to pick the green marble than the red one. If we think of the two green marbles as labeled "green 1" and "green 2," then we can use the tree diagram in Figure 15.5. The 9 outcomes in Figure 15.5 are equally likely. Two of these outcomes are a green marble followed by the red marble. Therefore, the probability of picking green followed by red is $\frac{2}{9} = 22\%$.

FIGURE 15.5

A Correct Tree
Diagram for
Picking Two
Marbles from a
Bag with 1 Red
and 2 Green
Marbles (If the
First Marble Is
Replaced after
Picking)

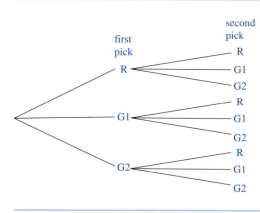

FIGURE 15.6

A Tree Diagram for Picking Two Marbles from a Bag with 1 Red and 2 Green Marbles (If the First Marble Is Not Replaced after Picking)

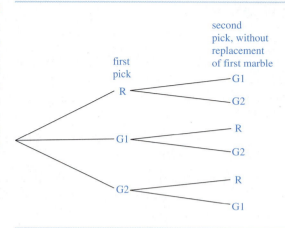

We can use organized lists, tree diagrams, and arrays even when working with dependent outcomes. For example, suppose there are 3 marbles in a bag, 1 red and 2 green. If we randomly pick a marble from the bag, record its color, and *without* returning the marble to the bag, randomly pick another marble, what is the probability of picking a green marble followed by a red marble? There are 6 equally likely outcomes, as shown in Figure 15.6. Two of these outcomes are a green marble followed by the red marble. Therefore, the probability of picking green followed by red is $\frac{2}{6} = \frac{1}{3} = 33\%$.

CLASS ACTIVITY   NOW TURN TO CLASS ACTIVITIES MANUAL

15A Comparing Probabilities p. 511

15B If You Flip 10 Pennies Should Half Come Up Heads? p. 513

15C Experimental versus Theoretical Probability: Picking Cubes from a Bag p. 514

15D Guess the Colors of Cubes in a Bag p. 515

15E Some Probability Misconceptions p. 516

15F Picking Two Marbles From a Bag of 1 Black and 3 Red Marbles p. 517

15G A Dice Rolling Game p. 520

PRACTICE PROBLEMS FOR SECTION 15.1

1. At a math center in a class, there is a bag filled with 30 red blocks and 20 blue blocks. Each child in the class of 25 will complete the following activity at the math center: pick a block out of the bag without looking, record the block's color, and put the block back into the bag. Each child will do this 10 times in a row. Then the child will write the number of blue blocks picked on a sticky note. Describe a good way to display the whole class's data. Show roughly what you expect the display you suggest to look like, and say why.

2. Consider the experiment of rolling a die two times.

    a. Determine the probability of getting a 1 on both rolls of the die. Explain your answer.

    b. Determine the probability of getting a 1 on either the first or the second roll of the die. Explain your answer.

    c. Determine the probability of getting either a 1 or a 2 on both rolls of the die.

3. What is the probability of getting four heads in a row on four tosses of a coin?

4. Suppose you have 3 marbles in a bag: 1 red and 2 green. If you reach into the bag without looking and randomly pick out two marbles at once, what is the probability that both of the marbles you pick will be green?

ANSWERS TO PRACTICE PROBLEMS FOR SECTION 15.1

1. A line plot like the one in Figure 15.7 would be a good way to display these data. The class could make this line plot on the chalkboard, using sticky notes instead of Xs. Since $\frac{2}{5}$ of the blocks in the bag are blue, the probability of picking a blue block is $\frac{2}{5} = 40\%$. So, in the ideal, there would be 4 blue blocks chosen in 10 picks of a block. However, since the choices are random, most children probably won't get exactly 4 blue blocks, but most will probably get somewhere around 4 blocks. A few children may even get 8, 9, or 10 blue blocks in their 10 picks. So the line plot will probably look something like the one in Figure 15.7, where most Xs cluster around 4, but some Xs are not close to 4.

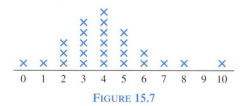

FIGURE 15.7

A Line Plot of the Number of Blue Blocks Picked in 10 Trials Out of a Bag Containing 20 Blue Blocks and 30 Red Blocks

2. a. Since the two rolls of the die are independent, the following array shows all possible outcomes:

(1, 1)	(1, 2)	(1, 3)	(1, 4)	(1, 5)	(1, 6)
(2, 1)	(2, 2)	(2, 3)	(2, 4)	(2, 5)	(2, 6)
(3, 1)	(3, 2)	(3, 3)	(3, 4)	(3, 5)	(3, 6)
(4, 1)	(4, 2)	(4, 3)	(4, 4)	(4, 5)	(4, 6)
(5, 1)	(5, 2)	(5, 3)	(5, 4)	(5, 5)	(5, 6)
(6, 1)	(6, 2)	(6, 3)	(6, 4)	(6, 5)	(6, 6)

Here, (4, 3) stands for a 4 on the first roll and a 3 on the second roll. There are $6 \times 6 = 36$ possible outcomes, all of which are equally likely. So, by principles 1 and 2, the probability of getting a one on both rolls is $\frac{1}{36}$, or 2.8%.

b. Referring to the array in part (a), there are 11 ways to get a 1 on either the first roll or the second roll, or both (the first row together with the first column). Therefore, by principle 2, the probability of getting at least one 1 on two rolls of a die is $\frac{11}{36}$, or 30.6% (so you'd expect to get at least one 1 a little less than one-third of the time).

c. Referring to the array in part (a), there are four ways to get either a 1 or a 2 on both rolls, namely, (1, 1), (1, 2), (2, 1), and (2, 2). Therefore, by principle 2, the probability of getting either a 1 or a 2 each time on two rolls of a die is $\frac{4}{36} = \frac{1}{9}$, or about 11%.

3. The tree diagram in Figure 15.8 shows all possible outcomes of the 4 coin tosses since the coin tosses are independent. There are $2 \times 2 \times 2 \times 2 = 16$ possible outcomes, all of which are equally likely. Only one of those outcomes produces four heads in a row. So, by principles 1 and 2, the probability of getting four heads in a row on four tosses of a coin is $\frac{1}{16}$, or 6.25%.

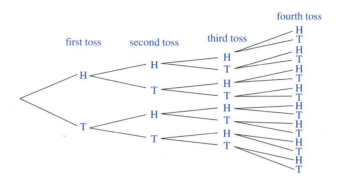

FIGURE 15.8

A Tree Diagram for Tossing a Coin Four Times

4. We can use the tree diagram in Figure 15.6 to solve this problem by thinking of one of the marbles as the first one chosen and the other as the second one chosen (where the first marble is not replaced before the second one is chosen). The tree diagram shows that there are 6 equally likely outcomes. Two of those outcomes correspond to both green marbles being chosen. So the probability of picking the two green marbles is $\frac{2}{6} = \frac{1}{3} = 33\%$.

Another way to solve the problem is to create the following organized list, which shows all possible ways of picking a pair of marbles from the bag:

$$(G1, G2) \quad (G1, R)$$
$$(G2, R)$$

(Notice that $(G2, G1)$ would be the same as $(G1, G2)$. Results are similar for other cases.) These 3 possible ways of picking two marbles are all equally likely. Out of the 3 ways of picking two marbles, there is 1 way to pick two green marbles. Therefore, the probability of picking 2 green marbles is $\frac{1}{3}$.

PROBLEMS FOR SECTION 15.1

1. Some children's games have spinners. When the arrow in a spinner is spun, it can land in any one of several different colored regions. Determine the probability of spinning each of the following on a spinner like the one in Figure 15.9:

a. Red

b. Either red or green

c. Either red or yellow

Explain your answers.

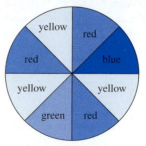

FIGURE 15.9

A Spinner

2. Farmer Brown must plant his crop next week. The ideal day to plant is a day when it doesn't rain, followed by a day when it does rain, followed by a day when it doesn't rain. The probability of rain for each day next week is shown in the table that follows. Based on this weather forecast, what day should farmer Brown plant his crop? Why?

DAY	MON	TUES	WED	THURS	FRI	SAT	SUN	MON
probability of rain	20%	30%	60%	80%	35%	70%	40%	20%

3. Write a paragraph discussing the following:

   a. Michael says that if you flip a coin 100 times it will have to come up heads 50 times out of the 100. Is Michael correct?

   b. If the probability that a certain outcome of an experiment will occur is 30%, does that mean that when you carry out the experiment 100 times, the outcome will occur 30 of those times?

4. A family math night at school features the following game: There are two opaque bags, each containing red blocks and yellow blocks. Bag 1 contains 3 red blocks and 5 yellow blocks. Bag 2 contains 5 red blocks and 15 yellow blocks. To play the game, you pick a bag and then you pick a block out of the bag without looking. You win a prize if you pick a red block. Kate thinks she should pick from bag 2 because it contains more red blocks. Is Kate more likely to pick a red block if she picks from bag 2 rather than bag 1? Explain why or why not.

5. There are 50 small balls in a tub. Some balls are white and some are orange. Without being able to see into the tub, each student in a class of 25 is allowed to pick a ball out of the tub at random. The color of the ball is recorded and the ball is put back into the tub. At the end, 7 orange balls and 18 white balls were picked. What is the best estimate you can give for the number of orange balls and the number of white balls in the tub? Describe how to calculate this best estimate and explain why your method of calculation makes sense in a way that a fifth grader might understand. Is your best estimate necessarily accurate? Why or why not?

6. In a classroom, there are 100 plastic fish in a tub. The tub is hidden from the students' view. Some fish are green, and some are yellow. The students know that there are 100 fish, but they don't know how many of each color there are. The students go fishing, each time picking a random fish from the tub, recording its color, and throwing the fish back in the tub. At the end of the day, 65 fish have been chosen, 12 green, 53 yellow. What is the best estimate you can give for the number of green fish and the number of yellow fish in the tub? Describe

how to calculate this best estimate and explain why your method of calculation makes sense in a way that a fifth grader might understand. Is your best estimate necessarily accurate? Why or why not?

7. There is a bag filled with 4 red blocks and 16 yellow blocks. Each child in a class of 25 will pick a block out of the bag without looking, record the block's color, and put the block back into the bag. Each child will repeat this for a total of 10 times. Then the child will write on the sticky note the number of red blocks picked. Describe a good way to display the whole class's data. Show roughly what you expect the display you suggest to look like, and say why.

8. There are 3 plastic bears in a bag; 2 are red and 1 is blue. The teacher tells Bob that there are 3 bears in the bag, but she doesn't tell Bob what color the bears are. Bob picked a bear out of the bag 3 times and each time he got a red bear. (He puts the bear he picked back in the bag before picking the next bear).

   a. Bob says that the 3 bears in the bag must all be red. Is Bob necessarily correct?

   b. Calculate the probability of picking 3 red bears in 3 picks.

9. A children's game has a spinner that is equally likely to land on any one of four colors: red, blue, yellow, or green. Determine the probability of spinning a red both times on 2 spins. Explain your reasoning.

10. A children's game has a spinner that is equally likely to land on any one of four colors: red, blue, yellow, or green. Determine the probability of spinning a red at least once on 2 spins. Explain your reasoning.

11. A children's game has a spinner that is equally likely to land on any one of four colors: red, blue, yellow, or green. Determine the probability of spinning a red 3 times in a row on 3 spins. Explain your reasoning.

12. A children's game has a spinner that is equally likely to land on any one of four colors: red, blue, yellow, or green. Determine the probability of spinning either a

red followed by a yellow or a yellow followed by a red in 2 spins. Explain your reasoning.

13. A children's game has a spinner that is equally likely to land on any one of four colors: red, blue, yellow, or green. Determine the probability of not spinning a red on either of 2 spins. Explain your reasoning.

14. Determine the probability of spinning a blue followed by a green in 2 spins on the spinner in Figure 15.9. Explain your reasoning.

15. Determine the probability of spinning a blue followed by a red in 2 spins on the spinner in Figure 15.9. Explain your reasoning.

16. Determine the probability of spinning a red followed by a yellow in 2 spins on the spinner in Figure 15.9. Explain your reasoning.

17. Suppose you have a penny, a nickel, a dime, and a quarter in a bag. You shake the bag and dump out the coins. What is the probability that exactly 2 coins will land heads up and 2 coins will land heads down? Is the probability 50% or not? Solve this problem by drawing a tree diagram that shows all the possible outcomes when you dump the coins out.

18. You have a bag containing 2 yellow and 3 blue blocks.

    a. You reach in and randomly pick a block, record its color, and put the block back in the bag. You then reach in again and randomly pick another block. What is the probability of picking two yellow blocks? Explain your answer.

    b. You reach in and randomly pick a block, record its color, and, without replacing the first block, you reach in again and randomly pick another block. What is the probability of picking two yellow blocks? Explain your answer.

19. There are 4 black marbles and 5 red marbles in a bag. If you reach in and randomly select two marbles, what is the probability that both are red? Explain your reasoning clearly.

20. A game at a fundraiser: there are 20 rubber ducks floating in a pool. One of the ducks has a mark on the bottom indicating that the contestant wins a prize. Each contestant pays 25 cents to play. A contestant randomly picks 2 of the ducks. If the contestant picks the duck with the mark on the bottom, the contestant wins a prize that cost $1.

    a. What is the probability that a contestant will win a prize?

    b. If 100 people play the game, about how many people would you expect to win? Why?

    c. Based on your answer to part (b), how much money should the duck game expect to earn for the fundraiser (net) if 100 people play it?

21. You are making up a game for a fundraiser. You take five table tennis balls, number them from 1 to 5, and put all five in a brown bag. Contestants will pick the five balls out of the bag one at a time, without looking, and line the balls up in the order they were picked. If the contestant picks 1, 2, 3, 4, 5, in that order, then the contestant wins a prize of $2. Otherwise, the contestant wins nothing. Each contestant pays $1 to play.

    a. What is the probability that a contestant will win the prize? Explain your reasoning.

    b. If 100 people play your game, then approximately how many people would you expect to win the prize? Why?

    c. Based on your answer to part (b), about how much money would you expect your game to earn (net) for the fundraiser if 100 people play?

22. a. A waitress is serving five people at a table. She has the five dishes they ordered (all five are different), but she can't remember who gets what. How many different possible ways are there for her to give the five dishes to the five people? Explain your answer.

    b. Based on part (a), if the waitress just hands the dishes out randomly, what is the probability that she will hand the dishes out correctly?

23. Social security numbers have 9 digits and are presented in the form

$$xxx - xx - xxxx$$

where each x can be any digit from 0 to 9.

    a. How many different social security numbers can be made this way? Explain.

b. If the population of the U.S. is 275 million people, and if everybody living in the U.S. has a social security number, then what fraction of the possible social security numbers are in current use? Explain your answer. (This fraction is the probability that a random number in the format of a social security number is the social security number of a living person.)

24. Standard dice are shaped like cubes: they have 6 faces, each of which is a square. Because there are 6 faces on a standard die and because each face is equally likely to land face up, the probability that a given face will land up is $\frac{1}{6}$. Could you make other three-dimensional "dice" so that for each face, the probability of that face landing up is $\frac{1}{4}$? $\frac{1}{8}$? $\frac{1}{10}$? $\frac{1}{12}$? Other fractions? Explain your answers. *Hint:* see Section 8.7.

25. Suppose you have 100 light bulbs and one of them is defective. If you pick out two light bulbs at random (either both at the same time, or first one, then another, without replacing the first light bulb), what is the probability that one of your chosen light bulbs is defective? Explain your answer.

15.2 Calculating Probabilities by Considering the Ideal Outcome in a Large Number of Trials

In this section we will see how to calculate the probability of an outcome of an experiment by determining the ideal fraction of times the given outcome should occur if the experiment were performed many times.

USING FRACTION MULTIPLICATION TO CALCULATE PROBABILITIES

We can use basic principles of probability and the meaning of fraction multiplication to calculate certain probabilities. In order to do so, we will rely on the following way of thinking about the outcome of an experiment: Consider what would happen ideally if the experiment were repeated a very large number of times. Calculate directly the ideal fraction of times the outcome you are interested in occurs. This fraction is the probability of that outcome.

For example, if we roll a die two times, what is the probability that we will get two 1s in a row? Think about doing the experiment of rolling a die two times in a row many times over. The numbers 1, 2, 3, 4, 5, and 6 are all equally likely to come up on one roll of a die; therefore, in the ideal, the first roll will be a 1 in $\frac{1}{6}$ of the rolls. Now consider those $\frac{1}{6}$ of the experiments when the first roll is a 1. Of those double rolls in which the first roll is a 1, in the ideal, the second roll will be a one $\frac{1}{6}$ of the time, too, because the outcomes of the first and second rolls are independent. Therefore, in the ideal, in $\frac{1}{6}$ of $\frac{1}{6}$ of all the experiments, both rolls will be a 1. According to the meaning of multiplication,

$\frac{1}{6}$ of $\frac{1}{6}$ of all the experiments

is

$$\frac{1}{6} \cdot \frac{1}{6} = \frac{1}{36}$$

of all the experiments. Therefore, in the ideal, in $\frac{1}{36}$ of all the experiments, you will roll two 1s, so the probability of rolling two 1s in a row on two rolls of a die is $\frac{1}{36}$. (In this case, this probability can also be calculated using an array or a tree diagram.)

THE CASE OF SPOT, THE DRUG-SNIFFING DOG

We will now study a more advanced probability calculation. As in the previous example, we will consider the outcome of a large number of experiments in the ideal. We then divide the number of times the outcome we are interested in occurs by the total number of experiments in order to calculate the probability that the outcome we are interested in occurs.

Spot is a dog who has been trained to sniff luggage to detect drugs. But Spot isn't perfect. His trainers conducted careful experiments and found that in luggage that *doesn't* contain any drugs, Spot will nevertheless bark to indicate the presence of drugs for 1% of this luggage. In luggage that *does* contain drugs, Spot will bark to indicate he thinks drugs are present for about 97% of this luggage. (So he fails to recognize the presence of drugs in about 3% of luggage containing drugs.)

Consider the consequences of putting Spot in a busy airport where police estimate that 1 in 10,000 pieces of luggage (or .01%) passing through that airport contain drugs. Here's a question we should answer before we put Spot in a busy airport:

> If Spot barks to indicate that a piece of luggage contains drugs, what is the probability that the luggage actually does contain drugs? (Remember, Spot is not perfect, sometimes he barks when there are no drugs.)

Before you read on, guess what this probability is. You might be surprised at the actual answer, which we will now calculate. It is not 97%.

Imagine that we made Spot sniff a very large number of pieces of luggage, let's say 1,000,000 pieces of luggage. For these 1,000,000 pieces of luggage, we will first determine in the ideal how many times Spot will bark, indicating that he thinks they contain drugs. Then we will determine how many of these pieces of luggage actually do contain drugs.

According to the police estimates, .01% of the 1,000,000 pieces of luggage at this airport, or about 100 pieces of luggage, will contain drugs. The remaining 999,900 will not contain drugs. Based on the information we have about Spot, of the 100 pieces of luggage that contain drugs, Spot will bark, recognizing the presence of drugs in 97 pieces of luggage, in the ideal. In the ideal, for the 999,900 pieces of luggage that don't contain drugs, Spot will still bark for 1% of these pieces of luggage, which is 9999 pieces of luggage. So, all together, Spot will bark indicating that he thinks drugs are present for

$$9999 + 97 = 10{,}096$$

pieces of luggage, in the ideal. But of those 10,096 pieces of luggage, only 97 actually contain drugs (because the other 9999 of the 10,096 were ones where Spot barked, but they didn't actually contain any drugs). So, of the 10,096 times that Spot barks, indicating he thinks there are drugs, only 97 of those times are drugs *actually present*. Now,

$$\frac{97}{10{,}096} = 0.0096\ldots = 0.96\%$$

so that, of the times when Spot barks, indicating he thinks drugs are present, only 0.96% of the time are drugs actually present. So if Spot barks to indicate he *thinks* drugs are present, the probability that drugs actually *are* present is 0.96%, which is *less than one percent*. Even though Spot's statistics sounded pretty good at the start, we might think twice about putting him on the job, since many innocent people could be delayed at the airport.

CLASS ACTIVITY NOW TURN TO CLASS ACTIVITIES MANUAL

15H Using the Meaning of Fraction Multiplication to Calculate a Probability p. 521

PRACTICE PROBLEMS FOR SECTION 15.2

1. What is the probability of spinning a yellow followed by a blue in 2 spins on the spinner in Figure 15.10? Explain how to solve this problem with fraction multiplication and explain why this method makes sense.

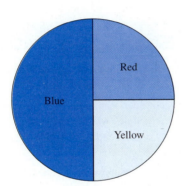

FIGURE 15.10

A Spinner

2. At a doughnut factory, 3 of the 200 boxes of doughnuts are bad. Each bad box contains only 10 doughnuts (instead of the usual 12) and in each bad box, 3 of the 10 doughnuts are tainted. (No doughnuts in any other boxes are tainted.) If a person buys a random box of these doughnuts and eats a random doughnut from the

box, what is the probability that he will eat a tainted doughnut? Explain how to solve this problem with fraction multiplication and explain why the method makes sense.

3. You are on a game show where you can pick one of 3 doors. A prize is placed randomly behind one of the 3 doors; the other doors do not have a prize behind them. You pick a door. The host opens one of the other doors that does not have the prize behind it. The host then offers you the opportunity to switch to the other unopened door before he reveals where the prize is.

    a. What is the probability that you will win the prize if you stick with your door?

    b. What is the probability that you will win the prize if you switch to the other unopened door? Solve this problem by imagining that you could play the game a large number of times; determine what fraction of the time you would win if you switched to the other unopened door each time you played.

    c. If you want to win the prize, what strategy is better: sticking with the first door you chose or switching to the other unopened door? Explain.

ANSWERS TO PRACTICE PROBLEMS FOR SECTION 15.2

1. Imagine spinning the spinner twice in a row many times. In the ideal, in $\frac{1}{4}$ of the times, the first spin will land on yellow. In the ideal, in $\frac{1}{2}$ of those times when the first spin is yellow, the second spin will be blue. Therefore, in the ideal, in $\frac{1}{2}$ of $\frac{1}{4}$ of the times, the spinner will land on yellow followed by blue. Since $\frac{1}{2}$ of $\frac{1}{4}$ is

$$\frac{1}{2} \cdot \frac{1}{4}$$

the probability of spinning yellow followed by blue is

$$\frac{1}{2} \cdot \frac{1}{4} = \frac{1}{8}$$

2. Imagine that the person could buy a random box of doughnuts and eat a random doughnut from the box many times. Then in the ideal, in $\frac{3}{200}$ of the times, the person will get a bad box. Of the $\frac{3}{200}$ of the times the person gets a bad box, the person will pick a tainted

doughnut $\frac{3}{10}$ of the time, in the ideal. Therefore, in $\frac{3}{10}$ of $\frac{3}{200}$ of the times, the person will eat a tainted doughnut from a bad box. Since $\frac{3}{10}$ of $\frac{3}{200}$ is

$$\frac{3}{10} \cdot \frac{3}{200}$$

the probability of eating a tainted doughnut is

$$\frac{3}{10} \cdot \frac{3}{200} = \frac{3 \cdot 3}{10 \cdot 200} = \frac{9}{2000} = 0.45\%$$

3. a. Since the prize is placed randomly behind one of the 3 doors, your probability of winning the prize is $\frac{1}{3}$ if you stick with the door you picked originally.

b. If you played the game many times, then in the ideal, in $\frac{1}{3}$ of those times, the prize would be behind the first door you pick. So in these cases, when you switch, you will not get the prize. But in the $\frac{2}{3}$ of the times when the door you picked does not have the prize behind it, the other unopened door must have the prize behind it. So when you switch doors, you will get the prize. Therefore, when you switch doors, you will win the prize $\frac{2}{3}$ of the times, in the ideal.

c. The switching strategy is better because when you switch, your probability of winning is $\frac{2}{3}$, whereas when you don't switch, your probability of winning is only $\frac{1}{3}$.

PROBLEMS FOR SECTION 15.2

1. A children's game has a spinner that is equally likely to land on any one of four colors: red, blue, yellow, or green. What is the probability of spinning a red followed by a green in 2 spins? Explain how to solve this problem with fraction multiplication and explain why this method makes sense.

2. Suppose you flip a coin and roll a die. What is the probability of getting a head and rolling a 6? Explain how to solve this problem with fraction multiplication and explain why this method makes sense.

3. There are 3 boxes. One of the 3 boxes contains 2 envelopes, the other two boxes contain 3 envelopes each. In the box with the 2 envelopes, one of the 2 envelopes contains a prize. No other envelope contains a prize. If you pick a random box and a random envelope from that box, what is the probability that you will win the prize? Explain how to solve this problem with fraction multiplication and explain why this method makes sense.

4. Due to its high population, China has a stringent policy on having children. In rural China, couples are allowed to have either one or two children according to the following rule: if their first child is a boy, they are not allowed to have any more children. If their first child is a girl, then they are allowed to have a second child. Let's assume that all couples follow this policy and have as many children as the policy allows.

    a. Do you think this policy will result in more boys, more girls, or about the same number of boys as

girls being born? (Answer without performing any calculations—just make a guess.)

b. Now consider a random group of 100,000 rural Chinese couples who will have children. Assume that any time a couple has a child the probability of having a boy is $\frac{1}{2}$.

    i. In the ideal, how many couples will have a boy as their first child, and how many will have a girl as their first child?

    ii. In the ideal, of those couples who have a girl first, how many will have a boy as their second child and how many will have a girl as their second child?

    iii. Therefore, overall, what fraction of the children born in rural China under the given policy should be boys and what fraction should be girls?

5. The Pretty Flower Company starts plants from seed and sells the seedlings to nurseries. They know from experience that about 60% of the calla lily seeds they plant will sprout and become a seedling. Each calla lily seed costs twenty cents, and a pot containing at least one sprouted calla lily seedling can be sold for $2.00. Pots that don't contain a sprouted seedling must be thrown out. The company figures that costs for a pot, potting soil, water, fertilizer, fungicide and labor are $0.30 per pot (whether or not a seed in the pot sprouts). The Pretty Flower Company is debating between planting one or two seeds per

pot. Help them figure out which choice will be more profitable by working through the following problems:

a. Suppose the Pretty Flower Company plants one calla lily seed in each of 100 pots. Using the previous information, approximately how much profit should the Pretty Flower Company expect to make on these 100 pots? Profit is income minus expenses.

b. Now suppose that the Pretty Flower Company plants two calla lily seeds (one on the left, one on the right) in each of 100 pots. Assume that whether or not the left seed sprouts has no influence on whether or not the right seed sprouts. So, the right seed will still sprout in about 60% of the pots in which the left seed does not sprout. Explain why the Pretty Flower Company should expect about 84 of the 100 pots to sprout at least one seed.

c. Using part (b), determine how much profit the Pretty Flower Company should expect to make on 100 pots if two calla lily seeds are planted per pot. Compare your answer to part (a). Which is expected to be more profitable: one seed or two seeds per pot?

d. What if calla lily seeds cost fifty cents each instead of twenty cents each (but everything else stays the same)? Now which is expected to be more profitable: one seed or two seeds per pot?

6. Suppose that in a survey of a large, random group of people, 17% were found to be smokers and 83% were not smokers. Suppose further that .08% of the smokers and .01% of the nonsmokers from the group died of lung cancer. (Be careful: notice the decimal points in these percentages—they are not 8% and 1%.)

a. What percent of the group died of lung cancer?

b. Of the people in the group who died of lung cancer, what percent were smokers? (It may help you to make up a number of people for the large group and to calculate the number of people who died of lung cancer and the number of people who were smokers and also died of lung cancer.)

7. Suppose that 1% of the population has a certain disease. Let's also suppose that there is a test for the disease, but it is not completely accurate: It has a 2% rate of false positives and a 1% rate of false negatives. This means that the test *reports* that 2% of the people who don't have the disease do have it, and the test *reports* that 1% of the people who do have the disease don't have it. This problem is about the following question:

If a person tests positive for the disease, what is the probability that he or she actually has the disease?

Before you start answering the next set of questions, go back and read the beginning of the problem again. Notice that there is a difference between *actually having* the disease and *testing positive* for the disease, and there is a difference between *not having* the disease and *testing negative* for it.

a. What do you think the answer to the previous question is? (Answer without performing any calculations—just make a guess.)

b. For parts (b) through (f), suppose there is a random group of 10,000 people and all of them are tested for the disease.

Of the 10,000 people, in the ideal, how many would you expect to actually have the disease and how many would you expect not to have the disease?

c. Continuing part (b), of the people from the group of 10,000 who do not have the disease, how many would you expect to test positive, in the ideal? (Give a number, not a percentage.)

d. Continuing part (b), of the people from the group of 10,000 who have the disease, how many would you expect to test positive? (Give a number, not a percentage.)

e. Using your work in parts (c) and (d), of the people from the group of 10,000, how many in total would you expect to test positive, in the ideal?

f. Using your previous work, what percent of the people who test positive for the disease should actually have the disease, in the ideal? How does this compare to your guess in part (a)? Are you surprised at the actual answer?

8. Suppose you have two boxes, 50 black pearls and 50 white pearls. You can mix the pearls up any way you like and put all them back into the two boxes. You don't have to put the same number of pearls in each box, but each box must have at least one pearl in it. You will then be blindfolded, and you will get to open a random box and pick a random pearl out of it.

a. Suppose you put 25 black pearls in box one and 25 black pearls and 50 white pearls in box two. Calculate the probability of picking a black pearl by imagining that you were going to pick a random box and a random pearl in the box a large number of times.

In the ideal, what fraction of the time would you pick box 1? What fraction of those times that you picked box 1, would you pick a black pearl from box 1, in the ideal?

On the other hand, in the ideal, what fraction of the time would you pick box 2? What fraction of those times that you picked box 2, would you pick a black pearl from box 2, in the ideal?

Overall, what fraction of the time would you pick a black pearl, in the ideal?

b. Describe at least two other ways to distribute the pearls than what is described in part (a). Find the probability of picking a black pearl in each case.

c. Try to find a way to arrange the pearls so that the probability of picking a black pearl is as large as possible.

Bibliography

[1] Edwin Abbott Abbott. *Flatland*. Princeton University Press, 1991. See `http://eldred.ne.mediaone.net/eaa/FL.HTM`.

[2] W. S. Anglin. *Mathematics: A Concise History and Philosophy*. Springer-Verlag, 1994.

[3] Deborah Loewenberg Ball. Prospective elementary and secondary teachers' understanding of division. *Journal for Research in Mathematics Education*, 21(2):132–144, 1990.

[4] P. Barnes-Svarney, editor. *New York Public Library Science Desk Reference*. Stonesong Press, 1995.

[5] Tom Bassarear. *Mathematics for Elementary School Teachers*. Houghton Mifflin, 1997.

[6] Petr Beckmann. *A History of Pi*. St. Martin's Press, 1971.

[7] E. T. Bell. *The Development of Mathematics*. Dover, 1992. Originally published in 1945.

[8] George W. Bright. Helping elementary- and middle-grades preservice teachers understand and develop mathematical reasoning. In *Developing Mathematical Reasoning in Grades K-12*, pages 256–269. National Council of Teachers of Mathematics, 1999.

[9] California State Board of Education. *Mathematics Framework for California Public Schools*, 1999.

[10] The Carnegie Library of Pittsburgh Science and Technology Department. *Science and Technology Desk Reference*, 1993.

[11] National Research Council. *How People Learn*. National Academy Press, 1999.

[12] Frances Curcio. *Developing Data-Graph Comprehension in Grades K–8*. National Council of Teachers of Mathematics, second edition, 2001.

[13] Singapore Curriculum Planning and Development Division, Ministry of Education. *Primary Mathematics*, volume 1A – 6B. Times Media Private Limited, Singapore, third edition, 2000. Available at http://www.singaporemath.com.

[14] Singapore Curriculum Planning and Development Division, Ministry of Education. *Primary Mathematics Workbook*, volume 1A – 6B. Times Media Private Limited, Singapore, third edition, 2000. Available at http://www.singaporemath.com.

[15] Stanislas Dehaene. *The Number Sense*. Oxford University Press, 1997.

[16] Demi. *One Grain of Rice*. Scholastic, 1997.

[17] Tatiana Ehrenfest-Afanassjewa. *Uebungensammlung zu einer Geometrischen Propaedeuse*. Martinus Nijhoff, 1931.

[18] Euclid. *The Thirteen Books of the Elements, Translated with Commentary by Sir Thomas Heath*. Dover, 1956.

[19] National Center for Education Statistics. Findings from education and the economy: An indicators report. Technical Report NCES 97-939, National Center for Education Statistics, 1997. Available online, see http://nces.ed.gov/pubs97/97939.html.

[20] National Center for Education Statistics. Highlights from the third international mathematics and science study–repeat (timss–r). Technical Report NCES 2001-027, National Center for Education Statistics, 2000. Available online, see the website http://nces.ed.gov/timss/.

[21] David Freedman, Robert Pisani, and Roger Purves. *Statistics*. W. W. Norton and Company, 1978.

[22] K. C. Fuson. Developing Mathematical Power in Whole Number Operations. In J. Kilpatrick, W. G. Martin, and D. Schifter, (Eds.), *A Research Companion to* Principles and Standards for School Mathematics, pages 68–94. National Council of Teachers of Mathematics, 2003.

[23] Georgia Department of Education. *Georgia Quality Core Curriculum*. See the website http://admin.doe.k12.ga.us/gadoe/sla/qcccopy.nsf.

[24] Sir Thomas Heath. *Aristarchus of Samos*. Oxford University Press, 1959.

[25] N. Herscovics and L. Linchevski. A cognitive gap between arithmetic and algebra. *Educational Studies in Mathematics*, 27:59–78, 1994.

[26] Encyclopaedia Britannica Inc. *Encyclopaedia Britannica*. Encyclopaedia Britannica, Inc., 1994. Available at www.britannica.com.

[27] H. G. Jerrard and D. B. McNeill. *Dictionary of Scientific Units*. Chapman and Hall, 1963.

[28] Graham Jones and Carol Thornton. *Data, Chance, and Probability, Grades 1 – 3 Activity Book*. Learning Resources, 1992.

[29] Constance Kamii, Barbara A. Lewis, and Sally Jones Livingston. Primary arithmetic: Children inventing their own procedures. *Arithmetic Teacher*, 41:200–203, 1993.

[30] Felix Klein. *Elementary Mathematics from an Advanced Standpoint*. Dover, 1945. Originally published in 1908.

[31] Morris Kline. *Mathematics and the Physical World*. Dover, 1981.

[32] Liora Linchevski and Drora Livneh. Structure sense: The relationship between algebraic and numerical contexts. *Educational Studies in Mathematics*, 40(2):173–196, 1999.

[33] Edward Manfre, James Moser, Joanne Lobato, and Lorna Morrow. *Heath Mathematics Connections*. D. C. Heath and Company, 1994.

[34] Liping Ma. *Knowing and Teaching Elementary Mathematics*. Lawrence Erlbaum Associates, 1999.

[35] Francis H. Moffitt and John D. Bossler. *Surveying*. Addison-Wesley, 1998.

[36] Joan Moss and Robbie Case. Developing children's understanding of the rational numbers: A new model and an experimental curriculum. *Journal for Research in Mathematics Education*, 30(2):122–147, 1999.

[37] Gary Musser and William F. Burger. *Mathematics for Elementary Teachers*. Prentice Hall, fourth edition, 1997.

[38] Phares G. O'Daffer, Randall Charles, Thomas Cooney, John Dossey, and Jane Schielack. *Mathematics for Elementary School Teachers*. Addison-Wesley, 2004.

[39] Mathematical Association of America (MAA) Committee on the Mathematical Education of Teachers. *A Call For Change: Recommendations For The Mathematical Preparation Of Teachers Of Mathematics*. The Mathematical Association Of America, 1991.

[40] U.S. Department of Education. Mathematics equals opportunity. Technical report, U.S. Department of Education, 1997. Available online, see http://www.ed.gov/pubs/math/index.html.

[41] Bureau of Labor Statistics U. S. Department of Labor. More education: Higher earnings, lower unemployment. *Occupational Outlook Quarterly*, page 40, Fall 1999. Available online, see http://stats.bls.gov/opub/ooq/ooqhome.htm.

[42] Bureau of Labor Statistics U. S. Department of Labor. Core subjects and your career. *Occupational Outlook Quarterly*, pages 26–40, Summer 1999. Available online, see http://stats.bls.gov/opub/ooq/ooqhome.htm.

[43] National Council of Teachers of Mathematics. *Curriculum and Evaluation Standards for School Mathematics*. National Council of Teachers of Mathematics, 1989.

[44] National Council of Teachers of Mathematics. *Principles and Standards for School Mathematics*. National Council of Teachers of Mathematics, 2000. See the website www.nctm.org.

[45] National Council of Teachers of Mathematics. *Navigating through Algebra in Grades 3 – 5*. National Council of Teachers of Mathematics, 2001.

[46] National Council of Teachers of Mathematics. *Navigating through Algebra in Prekindergarten – Grade 2*. National Council of Teachers of Mathematics, 2001.

[47] National Council of Teachers of Mathematics. *Navigating through Geometry in Grades 3 – 5*. National Council of Teachers of Mathematics, 2001.

[48] National Council of Teachers of Mathematics. *Navigating through Geometry in Prekindergarten – Grade 2*. National Council of Teachers of Mathematics, 2001.

[49] Conference Board of the Mathematical Sciences (CBMS). *The Mathematical Education of Teachers*, volume 11 of *CBMS Issues in Mathematics Education*. The American Mathematical Society and the Mathematical Association of America, 2001. See also `http://www.cbmsweb.org/MET_Document/index.htm`.

[50] The Secretary's Commission on Achieving Necessary Skills. *What Work Requires of Schools, A SCANS Report for America 2000*. U. S. Department of Labor, 1991.

[51] A. M. O'Reilley. Understanding teaching/ teaching for understanding. In Deborah Schifter, editor, *What's Happening in Math Class?*, volume 2: Reconstructing Professional Identities, pages 65–73. Teachers College Press, 1996.

[52] Dav Pilkey. *Captain Underpants and the Attack of the Talking Toilets*. Scholastic, 1999.

[53] George Polya. *How To Solve It; A New Aspect of Mathematical Method*. Princeton University Press, 1988. Reissue.

[54] Deborah Schifter. Reasoning about operations, early algebraic thinking in grades K–6. In *Developing Mathematical Reasoning in Grades K–12*, 1999 Yearbook. National Council of Teachers of Mathematics, 1999.

[55] Steven Schwartzman. *The Words of Mathematics*. The Mathematical Association of America, 1994.

[56] John Stillwell. *Mathematics and Its History*. Springer–Verlag, 1989.

[57] Dina Tirosh. Enhancing prospective teachers' knowledge of children's conceptions: The case of division of fractions. *Journal for Research in Mathematics Education*, 31(1):5–25, 2000.

[58] Ron Tzur. An integrated study of children's construction of improper fractions and the teacher's role in promoting learning. *Journal for Research in Mathematics Education*, 30(4):390–416, 1999.

Index

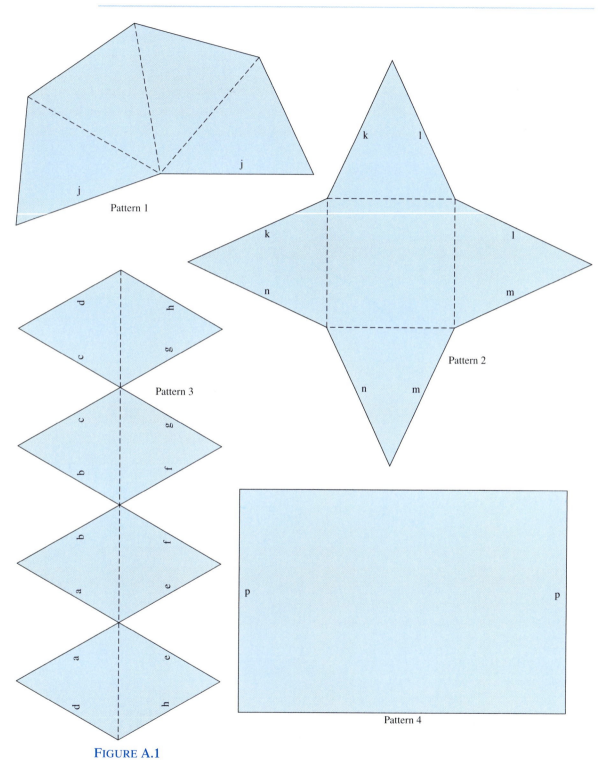

FIGURE A.1

Patterns for Shapes for Exercise 1 on Page 321

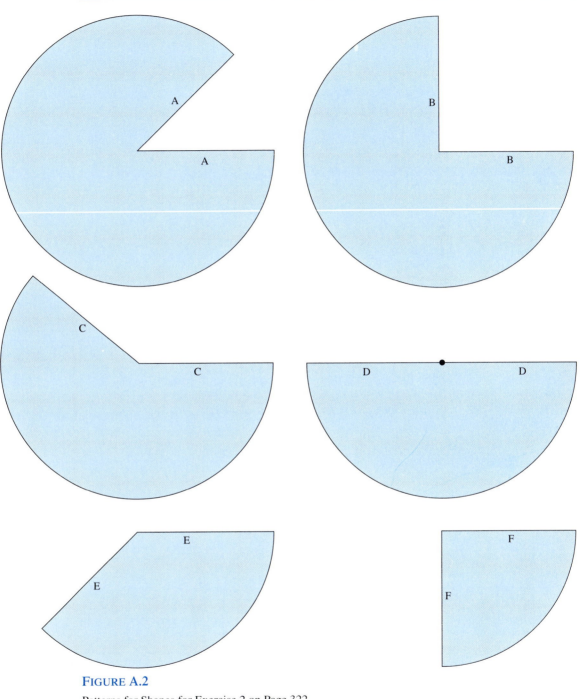

FIGURE A.2

Patterns for Shapes for Exercise 2 on Page 322

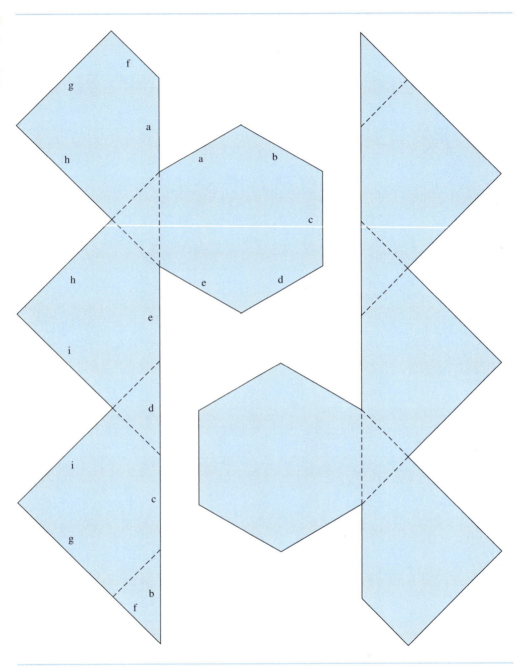

FIGURE A.3

Patterns for a
Sliced Cube for
Exercise 5 on
Page 324

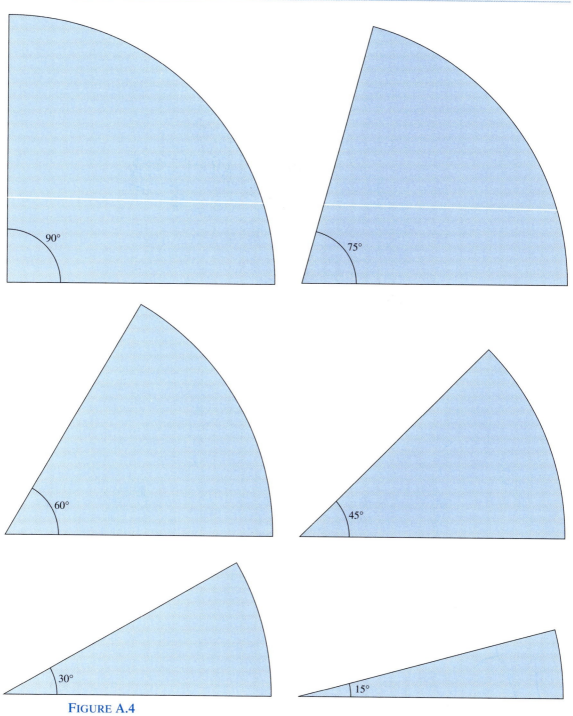

FIGURE A.4

Pie Wedges for Measuring Angles for Problem 1 on Page 339

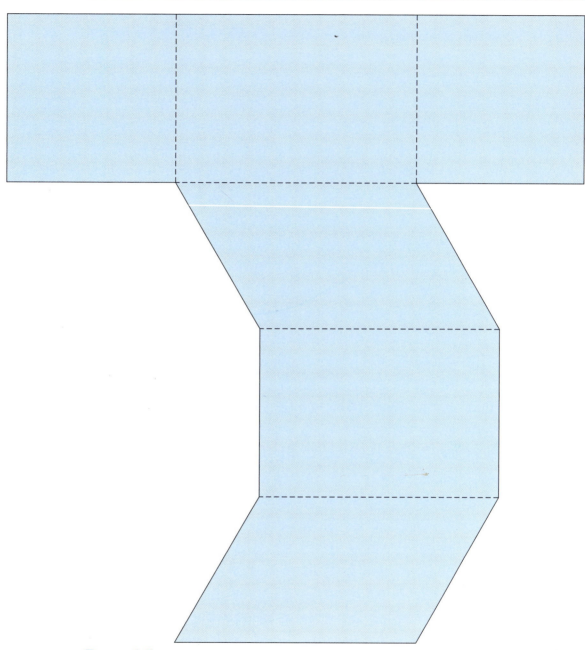

FIGURE A.5

For Exercise 1 on Page 378

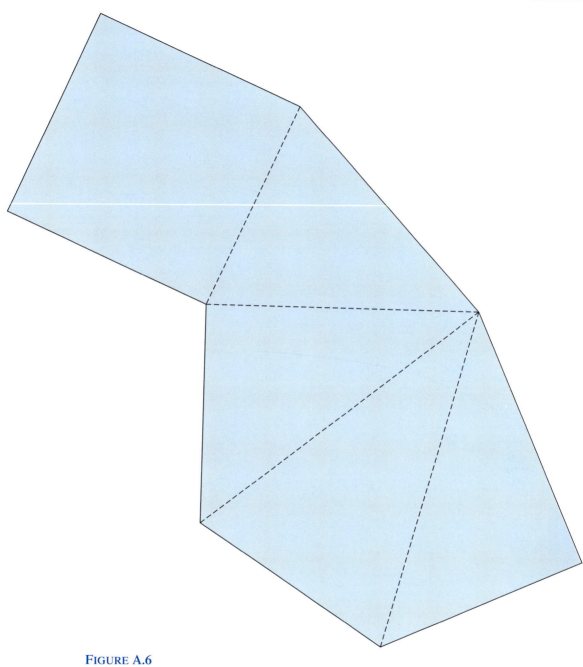

For Exercise 1 on Page 378

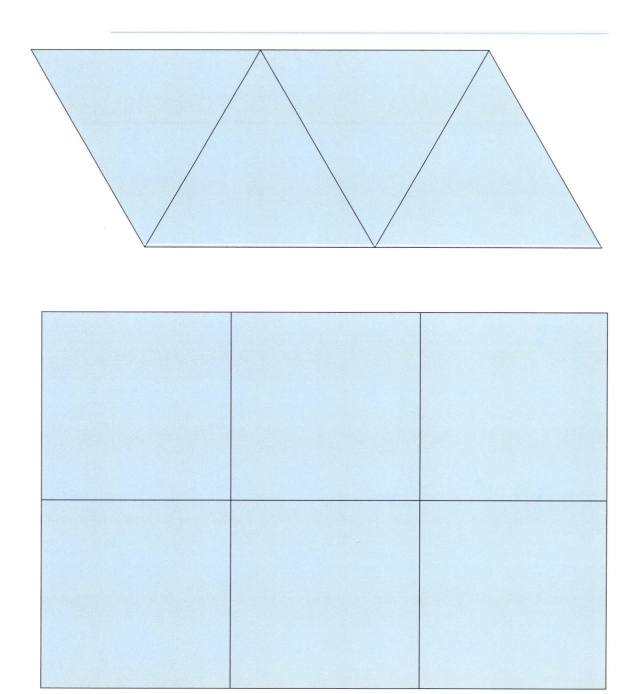

FIGURE A.7

For Exercise 2 on Page 381

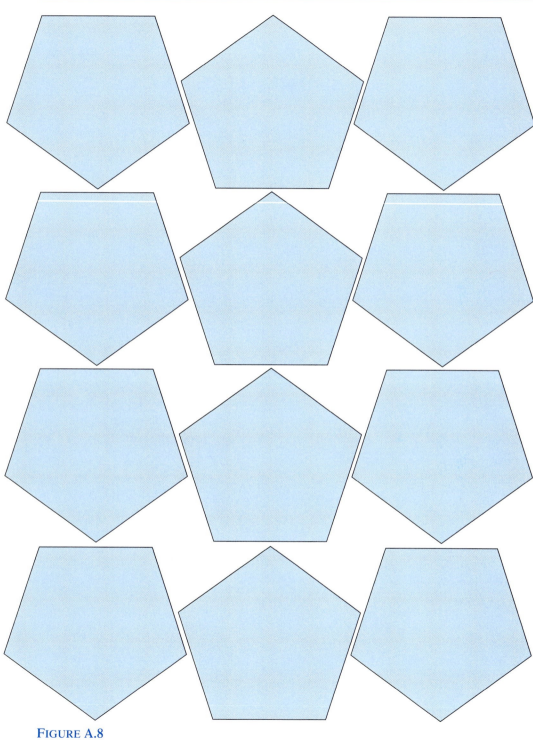

FIGURE A.8

For Exercise 2 on Page 381

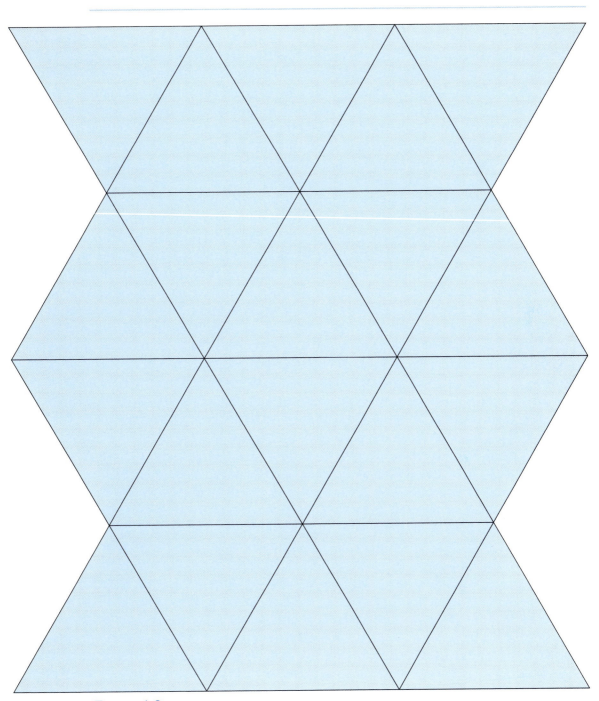

FIGURE A.9

For Exercise 2 on Page 381

FIGURE A.10

Triangles for Problem 13 on Page 382

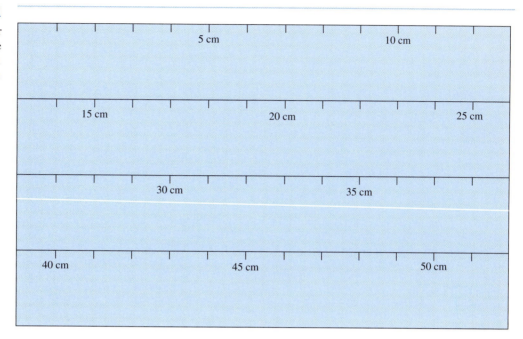

A Centimeter Tape Measure for Problem 11 on Page 541

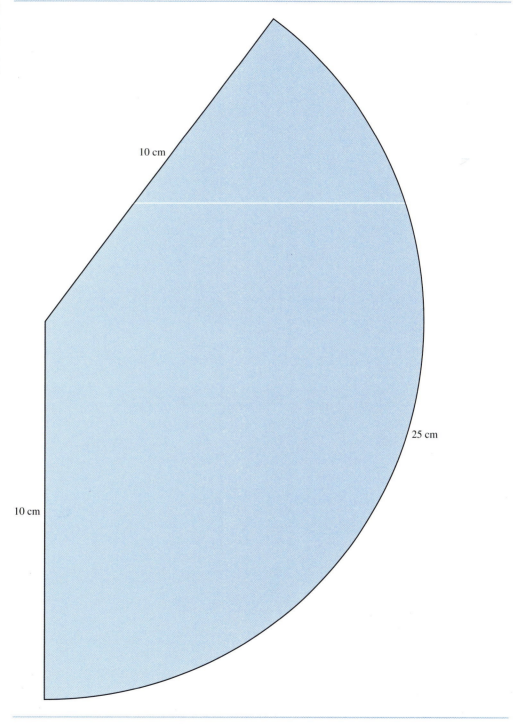

FIGURE A.12

A Pattern for a
Cup for
Problem 12 on
Page 542

10 cm

25 cm

10 cm